# Case Studies in Atomic Collision Physics I

*edited by*

## E. W. McDANIEL
*Georgia Institute of Technology, Atlanta, Georgia, U.S.A.*

## M. R. C. McDOWELL
*University of Durham, Durham, England*

1969

NORTH-HOLLAND PUBLISHING COMPANY – AMSTERDAM · LONDON

1969 © NORTH-HOLLAND PUBLISHING COMPANY – AMSTERDAM

*All rights reserved. No part of this publication may be reproduced, stored in a retrieval system or transmitted, in any form or by any means, electronic, mechanical, photocopying, recording, or otherwise without the prior permission of the copyright owner*

Library of Congress Catalog Card Number 79-91451
S.B.N. 7204 0151 8

*Publishers:*
NORTH-HOLLAND PUBLISHING COMPANY – AMSTERDAM
NORTH-HOLLAND PUBLISHING COMPANY, LTD. – LONDON

*Sole distributors for the Western Hemisphere:*
WILEY INTERSCIENCE DIVISION
JOHN WILEY & SONS, INC. – NEW YORK

PRINTED IN THE NETHERLANDS

# PREFACE

We have long felt that those who have worked on a particular topic in the field of Atomic Physics for a considerable period should have an opportunity from time to time to present a fuller account of the topic than is normally possible in the standard journals. This feeling is the rationale for the present volume. A second volume is in preparation.

The selection of authors and topics was ours. We asked the authors to give a detailed account of the present position in their field, with special emphasis on the work of their own laboratory. They were to feel free to include details of technique or of results taken, which would normally have been omitted from a paper. In reviewing recent work in the field we asked our authors to be selective, and not comprehensive, and to indicate, where possible, what in their opinions were profitable lines of work for the future.

E. W. McDaniel
M. R. C. McDowell

# CONTENTS

## CHAPTER 1. DRIFT TUBE STUDIES OF THE TRANSPORT PROPERTIES AND REACTIONS OF SLOW IONS IN GASES

E. W. McDaniel

| | |
|---|---|
| 1-0. Introduction | 3 |
| 1-1. The Motion of Slow Ions through Gases | 3 |
| 1-2. Classical Mobility Theory | 6 |
|     A. Langevin's theories | 6 |
|     B. The Chapman-Enskog theory | 9 |
|     C. Wannier's theory | 10 |
|     D. Kihara's theory | 12 |
|     E. Mason and Schamp's calculation | 12 |
|     F. Perel's theory | 13 |
| 1-3. Quantum Mechanical Mobility Theory | 13 |
| 1-4. The Mobility of Ions in Binary Gas Mixtures; Blanc's Law | 15 |
| 1-5. Drift Tube Measurements – General Considerations | 17 |
|     A. Drift velocities | 19 |
|     B. Diffusion coefficients | 20 |
|     C. Reaction rate coefficients | 22 |
|     D. Binding energies of molecular ions | 23 |
| 1-6. Experimental Apparatus | 23 |
| 1-7. Mathematical Analysis of the Space-Time Behavior of a Drifting Ion Swarm | 29 |
|     A. The transport equation at low $E/p_0$ | 30 |
|     B. The transport equation at high $E/p_0$ | 34 |
|     C. Solution of the transport equation at low $E/p_0$ | 35 |
| 1-8. Experimental Studies of Nitrogen Ions in Nitrogen | 37 |
|     A. Drift velocities | 38 |
|     B. Longitudinal diffusion coefficients | 40 |
|     C. Transverse diffusion coefficients | 40 |
|     D. Reaction rate coefficients | 42 |
| References | 45 |

## CHAPTER 2. PHOTOIONIZATION AND IONIZATION OF THE ALKALI METALS

M. R. C. McDowell

| | |
|---|---|
| 2-1. Introduction | 49 |
| 2-2. Photoionization | 50 |
|     A. Experimental studies of photoionization | 50 |
|     B. Theoretical studies of photoionization: independent particle models | 55 |
|     C. Correlation effects in photoionization | 63 |
| 2-3. Electron Impact Ionization | 72 |
|     A. Experimental studies | 72 |
|     B. Electron impact ionization: theory | 78 |
| 2-4. Conclusions | 94 |
| Acknowledgement | 95 |
| References | 96 |

## CHAPTER 3. THE IMPULSE APPROXIMATION AND RELATED METHODS IN THE THEORY OF ATOMIC COLLISIONS

J. P. Coleman

| | |
|---|---|
| 3-1. Introduction | 101 |
|     A. Notation | 101 |
|     B. Some results from the formal theory of scattering | 104 |
| 3-2. The Impulse Approximation | 107 |
|     A. A simple model | 107 |
|     B. Formal derivation | 112 |
|     C. Electron capture | 115 |
|         1. Atomic hydrogen target | 115 |
|         2. Helium target | 128 |
|         3. The high-energy limit | 129 |
|     D. Excitation | 131 |
|     E. Ionization | 136 |
|     F. Summary | 137 |
| 3-3. The Vainshtein, Presnyakov and Sobelman Approximation | 138 |
|     A. The post form of the VPS approximation | 139 |
|         1. Derivation of the approximation | 139 |
|         2. The peaking approximation | 143 |
|         3. The inclusion of exchange | 147 |
|         4. The effective charge | 148 |
|     B. The prior form of the VPS approximation | 149 |
|     C. Summary | 152 |
| 3-4. The Extended Impulse Approximation | 154 |
|     A. The post form | 154 |
|     B. The prior form | 159 |
|     C. Summary | 159 |

APPENDIX 1. COULOMB FUNCTIONS . . . . . . . . . . . . . . . 160

APPENDIX 2. NORDSIECK'S INTEGRATION TECHNIQUE . . . . . . . . 161

References . . . . . . . . . . . . . . . . . . . . . . . . . 165

# CHAPTER 4. POSITRON COLLISIONS

### B. H. BRANSDEN

4-1. INTRODUCTION . . . . . . . . . . . . . . . . . . . . . . 171
  A. Positron scattering and annihilation . . . . . . . . . . . 172
    1. Free annihilation . . . . . . . . . . . . . . . . . . . 172
    2. Selection rules . . . . . . . . . . . . . . . . . . . . 176
    3. Positronium . . . . . . . . . . . . . . . . . . . . . . 177
  B. The interaction of positrons with a monatomic gas . . . . . 180
    1. Quenching of positronium . . . . . . . . . . . . . . . . 181
    2. Collision theory and experiment . . . . . . . . . . . . 182

4-2. THE ELASTIC SCATTERING OF POSITRONS BY ATOMS . . . . . . . 184
  A. General formulation . . . . . . . . . . . . . . . . . . . 184
    1. The Green's function . . . . . . . . . . . . . . . . . . 188
    2. Angular momentum decomposition . . . . . . . . . . . . . 189
  B. Approximation schemes . . . . . . . . . . . . . . . . . . 192
    1. The static approximation . . . . . . . . . . . . . . . . 192
    2. The adiabatic polarisation potential . . . . . . . . . . 196
    3. Virtual positronium formation . . . . . . . . . . . . . 200
    4. Variational methods . . . . . . . . . . . . . . . . . . 202
  C. Applications of the variational method . . . . . . . . . . 216
    1. Positron–hydrogen scattering . . . . . . . . . . . . . . 216
    2. Positron–helium scattering . . . . . . . . . . . . . . . 222
  D. Developments of the polarised orbital method . . . . . . . 225
    1. Positron scattering by hydrogen . . . . . . . . . . . . 226
    2. Positron scattering by the rare gases . . . . . . . . . 231

4-3. THE FORMATION AND SCATTERING OF POSITRONIUM . . . . . . . 235
  A. The formation of positronium by positron impact on hydrogen and helium 236
    1. General formalism . . . . . . . . . . . . . . . . . . . 237
    2. Trial functions and applications . . . . . . . . . . . . 239
  B. The scattering of positronium by hydrogen and helium . . . 243

Acknowledgement . . . . . . . . . . . . . . . . . . . . . . . 245
References . . . . . . . . . . . . . . . . . . . . . . . . . 245

# CHAPTER 5. EXPERIMENTS WITH COLLIDING CHARGED-PARTICLE BEAMS

### K. T. DOLDER

5-1. INTRODUCTION . . . . . . . . . . . . . . . . . . . . . . 251

5-2. GENERAL PRINCIPLES OF EXPERIMENTS WITH COLLIDING CHARGED-PARTICLE
BEAMS . . . . . . . . . . . . . . . . . . 251
   A. Typical experimental arrangements . . . . . . . . . . . . 251
   B. The calculation of cross sections from experimental data . . . . . 252
   C. Form factors . . . . . . . . . . . . . . . . . . . 254
   D. Some effects of space charge in electron and ion beams . . . . . . 257
   E. The separation of signals from backgrounds . . . . . . . . . 260
   F. The separation of beams of ions and atoms . . . . . . . . . 266
   G. Initial states of excitation of parent ions . . . . . . . . . 269

5-3. THE IONIZATION OF POSITIVE IONS BY ELECTRON IMPACT . . . . . . . 272
   A. Experiments which have been performed with crossed beams . . . . 272
   B. Four experimental approaches . . . . . . . . . . . . . 272
   C. Results of the experiments . . . . . . . . . . . . . . 283

5-4. THE EXCITATION OF POSITIVE IONS BY ELECTRON IMPACT . . . . . . 292
   A. Introduction . . . . . . . . . . . . . . . . . . . 292
   B. The measurements of the excitation function $He^+$ (1S–2S) by Dance et al. 292
      1. The method. . . . . . . . . . . . . . . . . . . 292
      2. The detector for 304 Å photons . . . . . . . . . . . . 293
      3. Backgrounds . . . . . . . . . . . . . . . . . . 294
      4. Results . . . . . . . . . . . . . . . . . . . . 296
      5. Discussion . . . . . . . . . . . . . . . . . . . 298
   C. Methods which rely upon the direct observation of light emitted by colliding
      electron and ion beams . . . . . . . . . . . . . . . 300
   D. The experiment of Bacon and Hooper . . . . . . . . . . . 301
      1. Backgrounds . . . . . . . . . . . . . . . . . . 301
      2. Other experimental consideration . . . . . . . . . . . 302
   E. The experiment of Lee and Carleton . . . . . . . . . . . 304

5-5. THE DETACHMENT OF ELECTRONS FROM NEGATIVE IONS BY ELECTRON IMPACT 306
   A. Introduction . . . . . . . . . . . . . . . . . . . 306
   B. The experiment of Dance, Harrison and Rundel . . . . . . . . 306
   C. The experiment of Tisone and Branscomb . . . . . . . . . . 308
   D. Results and discussion . . . . . . . . . . . . . . . . 309

5-6. MEASUREMENTS OF CROSS SECTIONS FOR THE PRODUCTION OF PROTONS BY
   COLLISIONS BETWEEN ELECTRONS AND $H_2^+$ IONS . . . . . . . . . 311
   A. Introduction . . . . . . . . . . . . . . . . . . . 311
   B. The experiments of Dunn, Van Zyl and Zare [5, 78, 79] . . . . . . 312
   C. The experiment of Dance et al. [14] . . . . . . . . . . . 314
   D. Results and discussion . . . . . . . . . . . . . . . . 315

5-7. EXPERIMENTS WITH COLLIDING ION BEAMS . . . . . . . . . . . 317

5-8. ALTERNATIVE METHODS USED TO STUDY COLLISIONS BETWEEN ELECTRONS
   AND IONS . . . . . . . . . . . . . . . . . . . . 323
   A. Sequential mass spectrometry . . . . . . . . . . . . . 323
   B. Collisions between electron beams and low-density plasmas . . . . 324
   C. Collisions between electrons and highly-charged ions . . . . . . 325

5-9. SOME EXPERIMENTAL TECHNIQUES . . . . . . . . . . . . . 325
   A. Ion sources . . . . . . . . . . . . . . . . . . . 325
   B. Electron beam sources . . . . . . . . . . . . . . . . 326
   C. Electron and ion optics . . . . . . . . . . . . . . . 327

D. Vacuum techniques . . . . . . . . . . . . . . . 327
E. Particle detectors . . . . . . . . . . . . . . . 327
F. Intensity calibration of UV detectors . . . . . . . . . 330
G. Miscellaneous techniques . . . . . . . . . . . . . 330

Acknowledgements . . . . . . . . . . . . . . . . . . 330
References . . . . . . . . . . . . . . . . . . . . . 330

# Chapter 6. BINARY-ENCOUNTER AND CLASSICAL COLLISION THEORIES

L. Vriens

6-1. Introduction . . . . . . . . . . . . . . . . . . 337

6-2. Historical Survey . . . . . . . . . . . . . . . . 338
  A. Early work . . . . . . . . . . . . . . . . . . 338
    1. Scattering by nuclei of atoms . . . . . . . . . . . 338
    2. Scattering by atomic electrons . . . . . . . . . . 339
  B. Recent developments . . . . . . . . . . . . . . 344

6-3. Two Particle Collisions . . . . . . . . . . . . . . 346
  A. Basic relations . . . . . . . . . . . . . . . . 346
  B. Some special applications . . . . . . . . . . . . 349

6-4. Charged Particle–Atom Collisions . . . . . . . . . . 353
  A. Collision models . . . . . . . . . . . . . . . . 353
  B. Single ionization . . . . . . . . . . . . . . . . 354
  C. Excitation . . . . . . . . . . . . . . . . . . 357
  D. Double ionization . . . . . . . . . . . . . . . 358
  E. Charge transfer . . . . . . . . . . . . . . . . 359
  F. Electronic stopping . . . . . . . . . . . . . . . 361
  G. Velocity distribution of atomic electrons . . . . . . . 362

6-5. Comparison with the Born Approximation . . . . . . . 363
  A. Double differential cross sections . . . . . . . . . . 363
  B. The cross section per unit energy transfer . . . . . . . 366
  C. The total ionization cross section . . . . . . . . . . 368
  D. Stopping power . . . . . . . . . . . . . . . . 369
  E. Sum rules and total cross section . . . . . . . . . . 370
  F. Alternative method of deriving binary-encounter formulae . . . . . 372

6-6. The Classical Theory . . . . . . . . . . . . . . . 376
  A. Differential cross sections . . . . . . . . . . . . 380
  B. Total cross sections . . . . . . . . . . . . . . . 381

6-7. Comparison with Experiment . . . . . . . . . . . . 382
  A. Differential cross sections . . . . . . . . . . . . 382
  B. Total cross sections . . . . . . . . . . . . . . . 384
  C. Similarity of scaled cross sections . . . . . . . . . . 385

6-8. Resonances . . . . . . . . . . . . . . . . . . 390

6-9. Summary and Conclusions . . . . . . . . . . . . . 393

References . . . . . . . . . . . . . . . . . . . . 395

## Chapter 7. COINCIDENCE MEASUREMENTS

Q. C. Kessel

7-1. Introduction . . . . . . . . . . . . . . . . . . . 401

7-2. The Theory of Measurement . . . . . . . . . . . . 404
    A. The inelastic energy loss . . . . . . . . . . . . . 404
    B. Charge state analysis . . . . . . . . . . . . . . 405

7-3. The Experiment . . . . . . . . . . . . . . . . . 407
    A. Apparatus . . . . . . . . . . . . . . . . . . 407
    B. Procedure . . . . . . . . . . . . . . . . . . 410

7-4. Coincidence Scattering Data . . . . . . . . . . . . 412
    A. Argon–argon . . . . . . . . . . . . . . . . . 413
        1. Inelastic energy loss . . . . . . . . . . . . . 413
        2. Ionization states . . . . . . . . . . . . . . 419
        3. Correlation between scattered and recoil charge states . . . . 423
        4. Correlation of the triple-peak region . . . . . . . 426
    B. Neon–neon . . . . . . . . . . . . . . . . . . 430
        1. Inelastic energy loss . . . . . . . . . . . . . 430
        2. Ionization states . . . . . . . . . . . . . . 431
        3. Correlation between scattered and recoil charge states . . . . 434
    C. Krypton–krypton . . . . . . . . . . . . . . . . 436
    D. Neon–argon . . . . . . . . . . . . . . . . . . 438
    E. Oxygen–argon . . . . . . . . . . . . . . . . . 440
    F. Related, non-coincident data . . . . . . . . . . . 440
        1. Anomalies in the total differential cross section . . . . 440
        2. Energies of ionized electrons . . . . . . . . . . 441
    G. Summary . . . . . . . . . . . . . . . . . . 445

7-5. Models Concerning Ionization in Heavy Ion Collisions . . . 445
    A. Collective oscillations . . . . . . . . . . . . . . 446
    B. Energy level crossings of molecular wave functions . . . . 447
        1. The $Ar^+$–Ar collision . . . . . . . . . . . . 450
        2. The $Ne^+$–Ne collision . . . . . . . . . . . . 451
    C. Statistical models for ionization . . . . . . . . . . 451

7-6. Additional Coincidence Measurements . . . . . . . . . 452
    A. Total scattering cross sections . . . . . . . . . . 453
    B. Electron–ion coincidence experiments . . . . . . . . 456

7-7. Appendix . . . . . . . . . . . . . . . . . . . 456
    A. Calculation of $T_1$, $T_2$ and $Q$ . . . . . . . . . . 456
    B. Some experimental details . . . . . . . . . . . . 458
        1. Charge analysis of coincident ions . . . . . . . . 458
        2. Coincidence resolving time . . . . . . . . . . 459

Acknowledgement . . . . . . . . . . . . . . . . . . 460

References . . . . . . . . . . . . . . . . . . . . 460

## Chapter 8. RECOMBINATION OF RARE GAS IONS WITH ELECTRONS

H. J. Oskam

8-1. Introduction . . . . . . . . . . . . . . . . . . . 465

8-2. Mechanisms of Electron-Ion Recombination . . . . . . . . 468
    A. Dissociative recombination . . . . . . . . . . . . . 468
        1. Dependence of the recombination coefficient on electron temperature . 469
        2. The direct dissociative recombination process . . . . . . . . 471
        3. The indirect dissociative recombination process . . . . . . . 472
    B. Collisional-radiative recombination . . . . . . . . . . . 473

8-3. Experimental Methods . . . . . . . . . . . . . . . 476
    A. Microwave cavity method . . . . . . . . . . . . . . 476
        1. Theory of the method . . . . . . . . . . . . . . . 476
        2. Experimental technique . . . . . . . . . . . . . . 484
    B. Light spectrometer techniques . . . . . . . . . . . . 487
        1. Light emission . . . . . . . . . . . . . . . . . 488
        2. Spectral line shape . . . . . . . . . . . . . . . . 489
    C. Mass spectrometer techniques . . . . . . . . . . . . 494

8-4. Recombination Processes Involving Helium Ions . . . . . . . 503

8-5. Recombination Processes Involving Rare Gas Ions other than Helium Ions . . . . . . . . . . . . . . . . . . . 512
    A. Neons ions . . . . . . . . . . . . . . . . . . . 512
        1. Electron density studies . . . . . . . . . . . . . . 512
        2. Light emission studies . . . . . . . . . . . . . . . 514
        3. Spectral line-shape studies . . . . . . . . . . . . . 516
        4. Dependence of $\alpha(Ne_2^+)$ on the electron temperature . . . . . 517
    B. Argon ions . . . . . . . . . . . . . . . . . . . 518
    C. Krypton and xenon ions . . . . . . . . . . . . . . 520

Acknowledgement . . . . . . . . . . . . . . . . . . 521
References . . . . . . . . . . . . . . . . . . . . 521

## Chapter 9. HIGH RESOLUTION ELECTRON BEAMS AND THEIR APPLICATION

L. Kerwin, P. Marmet, J. D. Carette

9-1. Introduction . . . . . . . . . . . . . . . . . . . 527

9-2. Some History . . . . . . . . . . . . . . . . . . 528

9-3. Instruments . . . . . . . . . . . . . . . . . . 529
   A. The RPD method . . . . . . . . . . . . . . . 529
   B. The 127° electrostatic selector . . . . . . . . . . 530
   C. The spherical electrostatic selector . . . . . . . . . 536
   D. The parallel-plane electrostatic selector . . . . . . . . 537
   E. The axial-source cylindrical electrostatic selector . . . . 538
   F. The monokinetron . . . . . . . . . . . . . . . 539
   G. Magnetic selectors . . . . . . . . . . . . . . . 540
   H. Time-of-flight selectors . . . . . . . . . . . . . 541
   I. Miscellany . . . . . . . . . . . . . . . . . 542

9-4. Electron Spectroscopy . . . . . . . . . . . . . . 544
   A. The electron spectroscope . . . . . . . . . . . . 544
   B. Elastic scattering, resonances . . . . . . . . . . . 546
   C. Inelastic scattering . . . . . . . . . . . . . . 555
      1. Second mode . . . . . . . . . . . . . . . 555
      2. Third mode . . . . . . . . . . . . . . . . 561
      3. Rotational levels . . . . . . . . . . . . . . 562

9-5. Ionization . . . . . . . . . . . . . . . . . . . 562
   A. Atoms . . . . . . . . . . . . . . . . . . 563
      1. Hydrogen . . . . . . . . . . . . . . . . 563
      2. Helium . . . . . . . . . . . . . . . . . 566
      3. Ne, Ar, Kr, Xe . . . . . . . . . . . . . . . 568
      4. Double ionization of inert gases . . . . . . . . . 571
      5. Higher degree of ionization of inert gases . . . . . . 572
   B. Diatomic molecules . . . . . . . . . . . . . . 573
      1. Hydrogen . . . . . . . . . . . . . . . . 573
      2. Nitrogen . . . . . . . . . . . . . . . . . 574
      3. Oxygen . . . . . . . . . . . . . . . . . 575
   C. Triatomic molecules . . . . . . . . . . . . . . 575
   D. The negative ion $SF_6^-$ . . . . . . . . . . . . . 576

9-6. Conclusion . . . . . . . . . . . . . . . . . . 576
References . . . . . . . . . . . . . . . . . . . . 577

Author Index . . . . . . . . . . . . . . . . . . . 582
Subject Index . . . . . . . . . . . . . . . . . . . 591

# CHAPTER 1

# DRIFT TUBE STUDIES OF THE TRANSPORT PROPERTIES AND REACTIONS OF SLOW IONS IN GASES

BY

EARL W. McDANIEL

*Georgia Institute of Technology,*
*Atlanta, Georgia,*
*U.S.A.*

are normally required for ions to attain a steady-state condition after they are produced in the gas. It is easy to show that the ratio of the electric field intensity to the gas number density, $E/N$, is the parameter that determines the average ionic energy acquired from the field in steady-state drift, above the energy associated with the thermal motion. The electric force on an ion of charge $e$ is $eE$, and the resulting acceleration is $eE/m$, where $m$ is the mass of the ion. We shall make the crude assumption that when an ion undergoes a collision it loses, on the average, all the energy it acquired from the field during the preceding free path. Then, if $\tau$ denotes the collision period, or mean free time, the velocity acquired just before a collision is $eE\tau/m$. Since $\tau \sim 1/N$, the energy obtained between collisions from the field is thus seen to be proportional to $(E/N)^2$. Rigorous calculations also show $E/N$ to be the parameter that determines the "field energy" of the ions.

Although $E/N$ is the more fundamental quantity, until recently most experimentalists have reported the results of their work in terms of $E/p$, where $p$ is the gas pressure, or in terms of $E/p_0$, where $p_0$ is the *reduced pressure*, normalized to 0°C. That is,

$$p_0 = p(273/T) \tag{1-1-1}$$

where $T$ is the absolute temperature at which the measurement was made. This convention has not been wholly satisfactory as $p_0$ has also been used in some instances to represent normalizations to temperatures other than 0°C. If one uses the parameter $E/N$, however, there is no ambiguity in comparing experimental results. The conversion relations are

$$E/N = (1.0354 \times T \times 10^{-2})(E/p) \tag{1-1-2}$$

or

$$E/N = 2.828 \times E/p_0 \tag{1-1-3}$$

where $E/N$ is in units of $10^{-17}$ V cm$^2$, $T$ in degrees Kelvin, and $E/p$ or $E/p_0$ in V/cm Torr. Huxley, Crompton and Elford [3] have suggested that the units of $E/N$ be denoted by the "Townsend", or "Td", where 1 Td $= 10^{-17}$ V cm$^2$, and this designation is attaining widespread usage.

If $E/N$ is small and constant, and if steady-state conditions have been achieved, the motion of an ensemble, or *swarm*, of ions of a given kind consists of a slow uniform drift in the field direction superimposed on the much faster random motion which produces diffusion*. Under such con-

---

* The present discussion is concerned only with the drift motion of the ions. The combined effects of drift and diffusion are analyzed mathematically in section 1-7, where the effects of ion-molecule reactions are also taken into account.

ditions, the average energy that the ions have acquired from the field is small with respect to their thermal energy, and their mean *drift velocity* $v_d$ in the field direction is proportional to the field strength $E$. Thus

$$v_d = \kappa E \qquad (1\text{-}1\text{-}4)$$

where the constant $\kappa$ is called the *mobility* of the ions and is usually expressed in units of cm$^2$/V sec. Theory predicts the validity of eq. (1-1-4) in the "low-field" region ($E/p_0 \lesssim 2$V/cm Torr), where the ionic energy is close to thermal, and at higher ionic energies provided that the collision frequency does not depend on the energy of the ions. This relationship has been verified experimentally for cases where a nonreacting ionic species is involved in the drift velocity measurement.

The field energy is negligible compared to the thermal energy if

$$\left(\frac{M}{m} + \frac{m}{M}\right) eE\lambda \ll kT \qquad (1\text{-}1\text{-}5)$$

where $M$ and $m$ are the molecular and ionic masses, respectively, and $eE\lambda$ is the energy acquired by an ion in moving a mean free path $\lambda$ in the field direction. The factor involving the masses accounts for the ability of the ions to store the acquired energy over many collisions if the masses are significantly different. Using the relationships $NkT = p$ and $\lambda = 1/Nq$, where $N$ is the molecular number density and $q$ is the ion-molecule collision cross section, we may express the foregoing inequality as $(M/m + m/M) eE \ll pq$. Taking a singly charged ion moving through the parent gas and making the reasonable assumption that $q = 50 \times 10^{-16}$ cm$^2$, we find that the field energy is much less than the thermal energy if $E/p \ll 5 \times 10^{-6}$ (statvolt/cm) per (dyne/cm$^2$) $\approx 2$ V/cm Torr. The electric field is said to be "low" when the criterion (1-1-5) is satisfied and "high" when the inequality is reversed. It should be noted that a given field in a gas of given density may change from "low" to "high" if the gas temperature is lowered sufficiently. The velocity distribution of the ions is approximately Maxwellian provided $E/p$ is small. By contrast, when "high-field" conditions prevail, the ionic velocity distribution is non-Maxwellian and cannot be accurately calculated at present.

Experimental mobility data are useful in several respects. First, numerical values of mobilities, and particularly their dependence on the temperature, can provide information about ion-molecule interaction potentials at greater separation distances than are accessible in scattering experiments. Second, mobilities are required for the calculation of ion-ion recombination coefficients, dispersion in a gas due to mutual repulsion, and the characteristics of electrical discharges. Third, the most accurate experimental diffusion coefficients for thermal energy ions are obtained from measured mobilities by use of the Einstein relationship between the two quantities (see section 1-2-B).

The mobility of a given ionic species in a given gas is inversely proportional to the number density of the molecules but is relatively insensitive to small changes (a few degrees Kelvin) in the gas temperature if the number density is held constant. To facilitate the comparison and use of data, a measured mobility $\kappa$ is usually converted to a *reduced mobility* $\kappa_0$ defined by the equation

$$\kappa_0 = \kappa \frac{p}{760} \frac{273}{T} = \kappa \frac{p_0}{760} \qquad (1\text{-}1\text{-}6)$$

where $p$ is the gas pressure in Torr and $T$ is the gas temperature in degrees Kelvin at which the mobility $\kappa$ was obtained. Under the standard conditions of pressure and temperature (760 Torr and 273°K), the gas number density is $2.69 \times 10^{19}/\text{cm}^3$. It must be emphasized that the use of eq. (1-1-6) merely provides a standardization or normalization with respect to the molecular number density; the temperature to which the reduced mobility actually refers is the temperature of the gas during the measurement. For ions of atmospheric interest, the reduced mobility is of the order of several cm²/Vsec. In the modern literature, when a single value is quoted as "the mobility" of an ion in a gas, the value cited is the reduced mobility extrapolated to zero field strength.

## § 1-2. Classical mobility theory

Here we shall describe several classical treatments of the mobility problem. The quantum mechanical theory of mobilities is discussed in section 1-3.

### A. LANGEVIN'S THEORIES

The first theoretical treatment of mobilities was published by Langevin in 1903 [4, 5]; it was based on the kinetic theory of gases, which was just beginning to be widely accepted. Langevin considered the ions and molecules to be solid elastic spheres, the ions differing from the molecules only by possession of an electric charge. Only repulsive forces acting at the instant of impact were taken into account, and $E/N$ was assumed to be small, so that the field energy would be negligible compared with the thermal energy. The ion density was taken to be low in order that ion-ion interactions could be ignored. The result* obtained by Langevin in 1903 was

$$\kappa = e\lambda/(m\bar{v}) \qquad (1\text{-}2\text{-}1)$$

where $\lambda$ is the common mean free path for the molecules and ions, $m$ is the common molecular and ionic mass, and $\bar{v}$ is the mean thermal velocity.

Equation (1-2-1) proved to be deficient in several important respects when its predictions were compared with experiment. In the first place, the calculated values are always too high by about a factor of 4. In addition, this equation incorrectly predicts that $\kappa$ should vary directly with the ionic charge and mean free path and inversely with $T^{\frac{1}{2}}$. Also, the mobility is not found experimentally to be independent of the dielectric constant of the gas, as (1-2-1) predicts. However, the observed inverse variation of the mobility with density is correctly predicted. Langevin concluded that one source of error was the assumption that the mean free path of the ion is the same as that of a molecule of the same species. Only by the introduction of a mechanism to shorten substantially the ionic mean free path could the discrepancy between theory and experiment be removed. Furthermore, the crude nature of the mean free path calculations was apparent, and it was evident that more rigorous methods should be applied to the problem.

Langevin immediately set about to improve upon his first theory, and in 1905 published a rigorous theory [7] based on Maxwell's momentum transfer method [8]. The theory applied to the low-field region and took into account the elastic scattering of ions by the inverse-fifth-power attractive forces between ions and molecules as well as by rigid sphere repulsion. An ion attracts neutral molecules by polarization forces in the case of nonpolar gases and, in addition, by direct attraction of the permanent dipoles if the gas is polar. The force of attraction $f$ is

$$f = \frac{(K-1)e^2}{2\pi N r^5} \qquad (1\text{-}2\text{-}2)$$

where $K$ is the dielectric constant of the gas, $e$ is the ionic charge, $N$ is the molecular number density, and $r$ is the distance between the centers of the ion and molecule [9]. This equation holds when $r$ is large compared with the charge separation of the dipole. A decrease in the ionic mean free path results from the increased collision rate, and momentum exchanges take place between the ions and molecules even when they do not actually collide.

Langevin's 1905 paper unfortunately lay unnoticed for about twenty

---

* This equation and two slight extensions of it are derived in the text on kinetic theory by Loeb [6].

years, until Hassé published a paper referring to it in 1926 [10]. This classic work is now readily accessible, however, as it has been translated into English and published as an appendix in the book by McDaniel [2].

The 1905 paper by Langevin [11] provides a complicated equation for the mobility which involves the gas pressure and density, the dielectric constant of the gas, the ionic and molecular masses, the ionic charge and the sum of the radii of the ion and gas molecule. The uncertainty which arises in the specification of the last-mentioned quantity makes the use of the full Langevin equation somewhat doubtful. However, if it is assumed that the polarization effects predominate over those of elastic sphere scattering in determining the mobility, the full equation reduces to a much simpler one which does not involve the sum of the radii of the ion and gas molecule. This equation, which gives the *Langevin polarization limit*, is

$$\kappa_p = \frac{0.5105}{\sqrt{\rho(K-1)}} \left(1 + \frac{M}{m}\right)^{\frac{1}{2}} \tag{1-2-3}$$

where $\rho$ is the gas density, $K$ the dielectric constant of the gas, $M$ the mass of a molecule, and $m$ the ionic mass. Fortunately, it appears that in most cases the effects of elastic sphere scattering are negligible compared to those of polarization scattering, so that one is usually entitled to use the polarization limit equation. In fact, this simple equation has usually given better agreement with experiment than the full Langevin equation. The Langevin equation in the polarization limit, (1-2-3), appears to be the most generally satisfactory mobility equation yet derived that does not contain adjustable parameters. It has been more widely used than any other for the calculation of mobilities, and in many cases its predictions agree closely with experimental results.

The charge-independence of the mobility expressed in (1-2-3) may be explained as follows. Although the force on the ion due to the electric field is directly proportional to the charge, the momentum loss of the ion due to impacts produced by the inverse-fifth-power electrostatic forces is also proportional to the charge, with the result that the charge dependence cancels out. The temperature independence is also explained on the basis of two effects canceling. An increase in the temperature tends to decrease the mobility by increasing the thermal velocity of the ions, but the momentum loss is also decreased by a factor sufficient to eliminate the temperature dependence. Note that since $\rho$ is directly proportional to $M$ at constant gas number density, (1-2-3) predicts that $\kappa_0 \sqrt{M_r}$ should be determined only by the dielectric constant of the gas and therefore be independent of the nature of the ion. (The reduced mass of the ion-molecule system is denoted by $M_r$.) This prediction, combined with that of temperature independence, is of considerable use in the evaluation of experimental data in terms of the polarization limit of the theory. An inverse-fourth-power potential is the only potential of the form $V \sim r^{-n}$ for which temperature independence is expected, as we shall see in section 1-2-B.

## B. The Chapman-Enskog Theory

In 1916–1917, Chapman and Enskog developed a rigorous kinetic theory for gases composed of spherically symmetric (monatomic) particles [12]. Their work was motivated by an interest in transport phenomena involving un-ionized gases, but their expression for the mutual diffusion coefficient can be evaluated for ion-atom as well as atom-atom interaction potentials and utilized for the calculation of ionic mobilities. This is true because the mobility of an ion at low $E/p_0$ is related to the ion-atom mutual diffusion coefficient $\mathscr{D}_{12}$ by the *Einstein relation*

$$\kappa = e\mathscr{D}_{12}/(kT). \qquad (1\text{-}2\text{-}4)$$

In the Chapman-Enskog theory, $\mathscr{D}_{12}$ is given to second order by the equation

$$\mathscr{D}_{12} = \frac{3\sqrt{\pi}}{16} \left(\frac{2kT}{M_r}\right)^{\frac{7}{2}} \frac{1+\varepsilon_0}{(N_1+N_2) P_{12}} \qquad (1\text{-}2\text{-}5)$$

where

$$P_{12} = \int_0^\infty v_0^5 \, q_D(v_0) \, e^{-M_r v_0^2/(2kT)} \, dv_0 \qquad (1\text{-}2\text{-}6)$$

and

$$q_D(v_0) = \int (1 - \cos \Theta) \, I_s(\Theta) \, d\Omega_{CM}. \qquad (1\text{-}2\text{-}7)$$

Here $M_r$ is the reduced mass of the ion-atom system, $N_1$, the gas, and $N_2$, the ion number density, and $\varepsilon_0$ is a second-order correction which is usually less than experimental errors ($\varepsilon_0$ is zero for an inverse-fourth-power potential and has a maximum value of 0.136 for a hard-sphere interaction). The ionic number density $N_2$ is generally much smaller than $N_1$ and may be ignored. We shall also ignore the second-order correction; then (1-2-5) becomes identical to the equation for the diffusion coefficient which Langevin derived in 1905.

The collision integral $P_{12}$ is an average of the diffusion cross section $q_D(v_0)$ over a Maxwellian velocity distribution. The diffusion cross section depends on the detailed nature of the ion-atom interaction through the differential cross section for elastic scattering $I_s(\Theta)$ and is a function of $v_0$, the relative velocity of approach at large separation. $\Theta$ is the scattering angle in the center-of-mass system, and $d\Omega_{CM}$ is the element of solid angle in this frame of reference.

Equation (1-2-5) shows that $\mathscr{D}_{12}$ should be proportional to $(T/M_r)^{\frac{1}{2}}$ if $q_D$ is independent of $v_0$. The variation of $\mathscr{D}_{12}$ with $M_r^{-\frac{1}{2}}$ is predicted by rigorous classical theory for all interactions. Dimensional considerations [13] show that, for a potential of the form $V(r) \sim r^{-n}$, $q_D$ varies as $v_0^{-4/n}$ and $\mathscr{D}_{12}$ as $T^{2/n}T^{\frac{1}{2}}$ if the gas density is held constant. Since $\kappa \sim \mathscr{D}_{12}/T$, it follows that

$$\kappa \sim T^{2/n}\,T^{-\frac{1}{2}}. \tag{1-2-8}$$

Hence observations on the temperature variation of mobilities can lead to considerable information on ion-atom interactions. Note that the mobility should be temperature-independent for a pure $r^{-4}$ potential. The scattering of an ion at low temperatures is determined mainly by the long-range attractive polarization potential, which varies as $r^{-4}$. Thus we should expect all mobilities to become essentially independent of the temperature, provided sufficiently low temperatures can be reached before quantum effects set in. At high temperatures, on the other hand, the scattering is determined principally by the short-range forces. If we represent these forces by a repulsive $r^{-12}$ potential, then we would predict the mobility to vary approximately as $T^{-\frac{1}{2}}$ at high temperatures. At intermediate temperatures there will be some cancellation of the short- and long-range forces, and the mobility should pass through a maximum as the temperature is varied.

The interaction potential between an ion and molecule is usually expressed in modern work as

$$V(r) = A\,e^{-\alpha r} + \sum_{n=4} C_n r^{-n}. \tag{1-2-9}$$

The exponential term represents the short-range repulsion due to electron cloud interpenetration and other quantum mechanical effects; the series represents the long-range interactions.

## C. Wannier's Theory

The theory thus far discussed is valid only for the region of low $E/N$, in which the ionic energy distribution is essentially Maxwellian and the average energy of the ions is equal to the average energy of the molecules. An important advance was made in 1951 when Wannier [14] developed a theory of mobility, based on the Boltzmann transport equation, which is applicable to the high-field as well as the low-field region. Wannier considered three types of ion-molecule interaction: rigid sphere repulsion, symmetry forces, and polarization attraction.

The symmetry effects are purely quantum mechanical. They arise when the cores of the ions and the gas molecules are identical, and include resonance attraction and repulsion and resonant charge transfer [2]. Wannier used the elastic sphere model to describe the symmetry effects. This model is characterized by isotropic scattering in the center-of-mass system and by a mean free path and collision cross section which are constant, i.e. independent of the relative velocity of the collision partners. On the other hand, the cross section for polarization attraction varies inversely with the relative velocity and thus is characterized by a constant mean free time between collisions rather than a constant mean free path.

At high $E/N$, the polarization cross section is small, and consequently we have a situation which can be described approximately by a constant mean free path. In this case, Wannier's analysis predicts that

$$v_d \sim \sqrt{E/N} \quad \text{(high } E/N, \text{ constant } \lambda\text{)}. \tag{1-2-10}$$

If, on the other hand, a constant mean free time situation is assumed at high $E/N$, the drift velocity is predicted to vary linearly with $E/N$:

$$v_d \sim E/N \quad \text{(high } E/N, \text{ constant } \tau\text{)}. \tag{1-2-11}$$

In the low $E/N$ region, Wannier's analysis shows that the drift velocity should vary directly with $E/N$ regardless of the interaction assumed:

$$v_d \sim E/N \quad \text{(low } E/N\text{)}. \tag{1-2-12}$$

Wannier also shows that the polarization limit of the Langevin equation is exact, not only at low $E/N$, but at high $E/N$ as well.

Perhaps the most useful feature of Wannier's theory is that it yields an expression for the total energy of an ion at high $E/N$ if the motion is assumed to be characterized by a constant mean free time:

$$\tfrac{1}{2}m\overline{v_i^2} = \tfrac{1}{2}mv_d^2 + \tfrac{1}{2}Mv_d^2 + \tfrac{3}{2}kT \tag{1-2-13}$$

where $m$ and $M$ are the ionic and molecular masses, respectively, $\overline{v_i^2}$ is the mean square of the total ionic velocity, and $v_d$ is the drift velocity. The first term on the right side is the field energy associated with the drift motion of the ion, whereas the second term is the random part of the field energy. The last term represents the thermal energy. Equation (1-2-13) illustrates the capacity that light ions in a heavy gas have for storing energy in the form of random motion. For ions traveling in the parent gas the ordered and random field energies are equal. For heavy ions in a light gas the random field energy is negligible. It has already been pointed out that drift tubes are now being widely used to determine ion-molecule reaction rates, as well as drift velocities. These quantities are experimentally determined as functions of $E/N$, but what is really desired in the case of ion-molecule reaction rates is the rate coefficient as a function of the average energy of the ions. Equation (1-2-13) permits this functional dependence to be deduced from the measured variation of $v_d$ with $E/N$. It must be realized, however, that eq. (1-2-13) is based on a somewhat unrealistic model; unfortunately no such simple relationship has been derived for a constant mean free path model.

At high $E/N$, the ions do not have a Maxwellian distribution, and, in fact, the distribution has not been calculated except for the constant mean free time model. At intermediate values of $E/N$ the polarization forces become more important, and neither the mean free path nor the mean free time is constant. When the low $E/N$ region is reached, the ions approach a Maxwellian distribution, and the drift velocity depends on the temperature as well as on $E/N$. At low $E/N$ the polarization forces may not actually predominate in a given gas at room temperature, but they are bound to prevail at a sufficiently low temperature.

### D. Kihara's Theory

Kihara [15] has extended the methods developed by Chapman and Enskog for the solution of the Boltzmann equation and derived a theory in which the dependence of the low-field mobility on the temperature and field strength was considered. Only elastic scattering collisions were taken into account. He showed the mobility to be independent of the field strength as well as the temperature for a pure $r^{-4}$ interaction but to vary with $E$ for other types of interaction. Kihara calculated the mobility in higher approximation than did Langevin but showed the higher order corrections to be small.

### E. Mason and Schamp's Calculation

Mason and Schamp [16] have used Kihara's extension of the Chapman-Enskog theory to obtain the second- and third-order approximations to the mobility in a weak electric field as a function of the temperature and field strength. Charge transfer and quantum mechanical effects were ignored.

The mobility was expressed as a series in ascending powers of the square of the field strength with coefficients which are complicated functions of the temperature, the ratio of the ionic and molecular masses, and the force law assumed for the ion-molecule interaction. The force law they used takes into account the point charge-induced dipole, the point charge-induced quadrupole, the London dispersion forces, and an inverse-twelfth-power repulsive potential. Three parameters in the potential energy function specify the depth and position of the minimum and the relative contributions of the various terms. Only two of these parameters are disposable, since the polarization force is known if the dielectric constant is. Moreover, of these two, one is disposable only within fairly narrow limits, since the charge-induced quadrupole and the dispersion force can be calculated approximately. These detailed calculations verify the type of temperature dependence for the mobility suggested in section 1-2-B, namely, constancy at very low temperatures, a $T^{-\frac{1}{2}}$ variation at high temperatures, and a maximum at an intermediate temperature. A new effect also appears when other long-range forces are included (i.e., the dispersion and charge-induced quadrupole terms) – a minimum can now appear at low temperatures. The importance of this minimum is that low-temperature mobility measurements may be extrapolated to the wrong zero-temperature limit if this effect is ignored, hence an apparent disagreement with the polarization limit of the theory may result.

## F. Perel's theory

Perel [17] has developed a mobility theory for positive ions in their parent gas in which he considers resonant charge transfer to be the dominant effect. The charge transfer cross section is taken to be independent of the velocity. The mobility was calculated for the atomic noble gas ions, and the results are consistent with experimental data over a wide range of $E/p$.

## § 1-3. Quantum mechanical mobility theory

The discussion in section 1-2-B indicates that the classical calculation of the mobility by the Chapman-Enskog method involves two steps. The first is the derivation of the mutual diffusion coefficient $\mathscr{D}_{12}$ in terms of the diffusion cross section $q_D$. This is a problem in kinetic theory, whose solution is a general result applicable to all ion-gas combinations which may be described by spherically symmetric interaction potentials. The second part of the calculation is the evaluation of $q_D$ for the particular ion-gas system under consideration. This is a problem in collision theory.

If we now wish to calculate the mobility quantum mechanically, the first step indicated above is still satisfactory, i.e. equations (1-2-5) and (1-2-6) may again be used. The question then arises about the proper method of calculating $q_D$. This question has been discussed in considerable detail by Dalgarno, McDowell, and Williams in two papers [18], the first of which applies to ions in unlike gases, the second to ions in their parent gas. They point out that quantal methods must be employed to obtain $q_D$ if the gas temperature is very low because then the classical concept of a well-defined trajectory is meaningless [19]. Furthermore, $q_D$ must be calculated quantum mechanically whatever the temperature if the ion and gas atom have identical cores, for then the interaction potential may arise from states either symmetric or antisymmetric in the nuclei. Dalgarno et al. display the expressions for the diffusion cross section in terms of the quantum mechanical phase shifts both for ions in unlike gases and ions in the parent gas. They then show that in the former case, the classical and quantum theories give identical results, except at very low temperatures. The success of the classical theory in this case is the result of the presence in the expression for $q_D$ of the factor $(1-\cos\Theta)$, which suppresses the contribution of small-angle scattering. It is precisely this scattering at angles near $\Theta=0$ which is usually invalid in the classical treatment.

The effect of charge transfer is important for an ion $X^+$ diffusing in its parent atomic gas. Dalgarno [18] showed that in this case the diffusion cross section is approximately equal to twice the cross section for the resonance charge transfer process $X^+ + X \rightarrow X + X^+$ except at very low temperatures, at which the contribution of long-range attractive forces cannot be ignored. Since at low impact energies, resonance charge transfer cross sections decrease slowly with increasing energy, the mobility should decrease monotonically with increasing gas temperature. This behavior is to be contrasted with that of ions in unlike gases and is helpful in identifying the ion whose mobility is being measured.

The role of charge transfer in determining the mobility of positive atomic ions in their parent gas has also been discussed by Holstein [20]. He calculated quantum mechanically the two possible energies of interaction between the ion and atom and showed how the results can be used in the computation of low-field mobilities. He showed that the mobility calculation itself can be made on a classical basis once the two energy curves and the charge transfer probability have been given a suitable quantum mechanical treatment. Holstein applied his general method to the specific cases of neon and argon.

The charge transfer and mobility of atomic ions in their parent gas have also been recently studied in great detail by Heiche [21].

Most discussions of the mobility of ions in molecular gases have been based on the assumption that it is permissible to average the interaction over all orientations before computing the appropriate elastic collision cross sections. Arthurs and Dalgarno [22] have developed a theory of scattering which makes it possible to eliminate this assumption, and they have derived a formula for the low-field mobility of an ion in a diatomic molecular gas and obtained quantitative results for mobilities in the limit of vanishing temperature. As an example of the detailed application of the theory, they calculated the ion mobilities in molecular hydrogen and deuterium for which it is necessary to take account of the rotational distributions. It was shown that in contrast to atomic gases, for which the low-temperature mobilities are independent of the temperature, the low-temperature mobilities in molecular gases decrease as the temperature decreases, ultimately passing through a minimum at some very low temperature, because of the interaction between the charge of the ion and the permanent quadrupole of the molecule.

The analysis by Arthurs and Dalgarno [22] discussed above has been extended by Dalgarno and Henry [23], who considered the effects of rotational transitions on the transport properties of molecular gases. They point out that these properties are usually interpreted assuming single channel (elastic) scattering, and they show that this is justifiable as a first approximation if the elastic cross sections are replaced by total cross sections. They also state that the distinction between total and elastic cross sections is especially significant for ion-molecule interactions at thermal velocities, since it implies a form for the long range interaction different from that which has been adopted. Dalgarno and Henry then describe some consequences of this difference in long range behavior.

A number of quantum mechanical mobility calculations have been made taking account of the specific interactions between the ion and the particles composing the gas. The reader is referred to the book by McDaniel [11] for a discussion of these calculations.

## § 1-4. The mobility of ions in binary gas mixtures; Blanc's law

The question now arises concerning the mobility of an ion in a binary mixture of gases. A number of measurements have been made to investigate this question [11], and it has been frequently observed that the reciprocal of the mobility is a linear function of the fractional concentration of either constituent of the mixture. This relationship is known as *Blanc's Law* and may be derived as follows. Let us assume that (a) low field conditions prevail; (b) the ionic and molecular number densities are low enough that ion-ion interactions and three-body collisions may be neglected; and (c) the ionic identity does not change with the composition of the gas.

We first consider a single pure gas. Since it is known that the mobility is inversely proportional to the density of the gas, it follows that $\kappa$ may be expressed as $1/\kappa = GN$, where $\kappa$ is the mobility at temperature $T$ and pressure $p$, $G$ is a constant, and $N$ is the number of molecules per unit volume. Similarly, if $\kappa_0$ is the mobility at standard conditions, then $1/\kappa_0 = GN_L$, where $N_L$ is the Loschmidt number.

Now consider a binary mixture composed of $N_A$ molecules per unit volume of a gas A and $N_B$ molecules of a gas B, with constants $G_A$ and $G_B$. Since $1/\kappa_{OA} = G_A N_L$ and $1/\kappa_{OB} = G_B N_L$, $G_A = 1/\kappa_{OA} N_L$ and $G_B = 1/\kappa_{OB} N_L$. For an ion of given drift velocity the rate of transfer of momentum to one component of the gas is considered here to be independent of the presence of the other component. Thus, since it is assumed that the nature of the ions does not change when the gases are mixed, the mobility of the ions in the mixture $\kappa_{AB}$ is given by

$$\frac{1}{\kappa_{AB}} = G_A N_A + G_B N_B = \frac{N_A}{\kappa_{OA} N_L} + \frac{N_B}{\kappa_{OB} N_L}.$$

If we now assume a total pressure of 1 atm, so that $N_A + N_B = N_L$, then $\kappa_{AB}$ becomes $\kappa_{OAB}$, and the mobility of the ions in the mixture at standard conditions is given by the equation

$$\frac{1}{\kappa_{OAB}} = \frac{f_A}{\kappa_{OA}} + \frac{f_B}{\kappa_{OB}}$$

where $f_A$ and $f_B$ are the fractional concentrations of the A and B molecules, respectively. Since $f_A + f_B = 1$, this equation may be expressed as

$$\frac{1}{\kappa_{OAB}} = \frac{f_A \kappa_{OB} + (1-f_A) \kappa_{OA}}{\kappa_{OA} \kappa_{OB}} \tag{1-4-1}$$

which is Blanc's law.

Physically, Blanc's law implies the additivity of cross sections in the binary-collision limit, and hence follows from first-order mobility theory.

Sufficient confidence has often been placed in Blanc's law that deviations from it have been interpreted as evidence that the ion changes its identity by clustering [11] or chemical ion-molecule reactions [1] as the composition of the mixture is varied. However, deviations from Blanc's law are to be expected in any more accurate theory, even if the ion retains its identity. Such deviations correspond theoretically to the weak composition dependence of the ordinary binary gas diffusion coefficient, which appears only in the second order of the Chapman-Enskog kinetic theory. Sandler and Mason [24] have derived explicit expressions for the deviations from Blanc's law in terms of the composition of the mixture and properties of the molecules of the gases. They show that in certain cases, for example for $r^{-4}$ potentials, Blanc's law is exact, but that in general small deviations are to be expected. An earlier variational calculation by Holstein [25] proved that the deviations from Blanc's law are always positive, and of the same magnitude as the second-order corrections to the pure-component mobilities. The results of Sandler and Mason are in harmony with this general theorem, but they now permit explicit numerical calculations to be made.

## § 1-5. Drift tube measurements – general considerations

The apparatus normally used for drift tube measurements consists of an enclosure containing gas at a "high" pressure, an ion source at one end of the enclosure, a set of electrodes which establish a uniform axial electrostatic field along which the ions drift, and a detector at the end of the ionic drift path. Usually, the ion source is operated in a pulsed mode, and the spectrum of arrival times of the ions at the detector is measured electronically. Such measurements are intrinsically capable of providing much more information than are d.c. measurements, although d.c. techniques are ideally suited for certain types of studies (e.g., see the discussion of the determination of transverse diffusion coefficients in section 1-5-B).

The pressures at which ionic drift tube experiments have been performed have ranged from as low as $2 \times 10^{-2}$ Torr to as high as $10^3$ Torr. On the assumption of a cross section of $50 \times 10^{-16}$ cm$^2$ for collisions of the ions with gas molecules, the range of ionic mean free path corresponding to this pressure range is 0.2 cm to $4 \times 10^{-6}$ cm. The drift field $E$ causes the ions of any given molecular composition to flow through the gas with a drift velocity and diffusion rate which are characteristic of the particular ion-gas combination, the field strength, and the gas temperature and pressure. The product of the gas pressure $p$ and the drift distance $d$ should be large enough resolving power may be unable to separate peaks in the arrival-time spectrum produced by the various ionic species. Furthermore, a given peak may be produced by ions which spent a significant fraction of their drift time in another molecular form, so that what is measured is some unknown kind of average of the properties of the various species which are involved.

Since the early days of mobility research, the importance of excluding impurities from drift tubes has been realized because it has long been known that the primary ions produced in the gas intended for study can be efficiently converted to ions of other types in collisions with impurity molecules. Special attention was paid to polar or highly polarizable impurities (which can produce clustering) and to impurities of low ionization potential (which can become ionized by charge transfer at the expense of the primary ions). Strangely enough, however, only within the last few years have the complicating effects of reactions with molecules of the *parent* gas been fully realized. In the case of nitrogen, for example, two primary ions, $N^+$ and $N_2^+$, are usually produced in the ion source, and both of these ions can be converted into secondary species by collisions with $N_2$ molecules. The conversion proceeds mainly by the three-body reactions

$$N^+ + 2N_2 \to N_3^+ + N_2$$

and

$$N_2^+ + 2N_2 \to N_4^+ + N_2$$

and reconversion to the original ions $N^+$ and $N_2^+$ is possible by the reverse reactions when $E/p$ is high enough to permit the collisional dissociation of the secondary ions. The simultaneous presence of four kinds of nitrogen ions even in the pure gas made it impossible to disentangle the behavior of the individual species by conventional techniques. Difficulties of this type led to the development of a new kind of apparatus which can provide a positive mass spectrometric identification of the ions being studied *simultaneous* with the measurement of their drift velocity. An example of these so-called "drift tube mass spectrometers" will be described in section 1-6.

### B. Diffusion Coefficients

The mobility $\kappa$ and diffusion coefficient $\mathscr{D}$ of a given kind of ion in a given gas are closely related, since both quantities are a measure of the ease with which the ions can flow through the gas. As pointed out in section 1-2-B, if the ions are essentially in thermal equilibrium with the gas molecules, $\kappa$ and $\mathscr{D}$ are related by the Einstein relation [27]*

$$\mathscr{D}/\kappa = kT/e \tag{1-5-1}$$

where $k$ is Boltzmann's constant and $e$ is the charge on each ion. At low $E/p$, the ionic motion is largely the random thermal motion produced by the heat energy of the gas, with a small drift component superimposed in the direction of the applied field. Hence, under low-field conditions, a single diffusion coefficient suffices to characterize the diffusion of the ions both in the direction of the applied drift field and in directions at right angles to it. At high $E/p$, on the other hand, the thermal motion is negligible but there are two large components of motion produced by the drift field: a directed component along the field lines and a random component representing energy acquired from the drift field but converted into random form by collisions with molecules. Consequently, under these conditions, diffusion proceeds at a different rate in the longitudinal direction (i.e., along $E$) than in the transverse directions (i.e., perpendicular to $E$). Both a longitudinal diffusion coefficient, $\mathscr{D}_L$, and a transverse diffusion coefficient, $\mathscr{D}_T$, are required to specify the rate of dispersion of the ions.

Recently drift tubes have been utilized to obtain both kinds of diffusion coefficients. The technique for measuring $\mathscr{D}_L$ is to map the spectrum of arrival times of the ions of a given type at the detector located at the end of the drift space. If these ions do not undergo a change of molecular composition because of reactions as they drift through the gas, their arrival time spectrum should consist of a single, smooth peak whose location on the time axis gives the drift velocity and whose width depends on the longitudinal diffusion coefficient. Consequently, an analysis of the peak shape can yield an accurate value of $\mathscr{D}_L$ if the experiment is carefully done and if the other factors contributing to a finite width of the peak are considered in the analysis. The main experimental requirement is to use a sufficiently small ion current that mutual repulsion among the ions is negligible, because the width and shape of a peak in the ionic arrival time spectrum are seriously affected by space charge. Considerably more care must be used in measuring $\mathscr{D}_L$ than in measuring $\kappa$, as the diffusion coefficient is much more sensitive to space charge than is the mobility. It is interesting to note that as the ion current is increased from an initially low value, the peaks in the arrival time spectrum are appreciably broadened by space charge repulsion long before the

---

* Sandler and Mason [24] have recently shown that contrary to previous conjectures [11, 28], the Einstein relation is not limited to the Chapman-Enskog first approximation but is exact in higher kinetic theory approximations.

mobility has begun to be seriously affected. A good way to determine whether the ion current is high enough to produce space charge effects is to measure independently both the mobility and the diffusion coefficient at low $E/p$ and see whether the two results are related by the Einstein equation. If the ratio $\mathscr{D}/\kappa$ is greater than predicted by this equation, then space charge effects are in evidence, and the ion current should be reduced.

As stated above, a drift tube can also be used to measure $\mathscr{D}_T$. One technique is to collect the ions at the end of the drift path on a movable, closely spaced set of electrodes mounted perpendicular to the drift tube axis, and to measure separately the current to each of these electrodes as a function of the lateral position of the set. These current measurements give information about the spatial distribution of the ions in a plane perpendicular to the drift tube axis at a known drift distance and hence permit the diffusion coefficient to be computed. Such measurements have been made by Skullerud [29] and Dutton et al. [30]; a circular disc split along a diameter into two semi-circular electrodes was used in both experiments to collect the ions. Since no time measurement was required, the ion source in each experiment was operated in the d.c. mode. The ions were not subjected to mass analysis in these experiments. A technique which has been used with a drift tube mass spectrometer to obtain $\mathscr{D}_T$ for mass-identified ions will be described in section 1-8.

## C. REACTION RATE COEFFICIENTS

The complicating effect of ion-molecule reactions on the measurement of drift velocities has already been mentioned in section 1-5-A. Such reactions must be regarded as the rule rather than as the exception in drift tube measurements, and in only a small fraction of the experiments performed without mass analysis has it been possible to demonstrate that the measured values of $v_d$ pertain to ions of an unvarying, known composition. There is a desirable feature of this situation, however, which compensates for the difficulties created in the measurement of drift velocities, namely that the drift tube provides an excellent means of *determining* the rates of ion-molecule reactions. The technique basically is to compare the currents of ions of the various types, or alternatively the shapes of the ionic arrival-time spectra, with what would be expected if the ions suffered only non-reactive collisions on the way down the drift tube and deduce reaction rates from the observed differences. A fairly involved mathematical analysis of the combined drift-diffusion-reaction behavior of the ions is required, but such analyses have been carried out, and accurate rate coefficients have been obtained by this method. The details of a typical analysis are presented in section 1-7.

Drift tube measurements of reaction rates are especially valuable at low $E/p$, because in this regime, the energy distribution of the ions is known to be Maxwellian. In certain of the alternative methods of determining rate coefficients [1], the ionic energy distribution is not known, and even the average energy of the ions is uncertain in many cases. At the time of this writing, the rates of 15 reactions have been determined by drift tube techniques [1], and the drift tube is rapidly becoming one of the most important means of obtaining accurate information on reaction rates.

### D. Binding Energies of Molecular Ions

It may happen that two species of ions are coupled by a reversible reaction and are present in equilibrium concentrations in a drift tube. In this case, a measurement of the temperature dependence of the equilibrium constant can be used to estimate the energy required to dissociate the more complex ion into the less complex one. Data have been obtained in this manner by Munson and Hoselitz [31] for alkali-noble gas complexes, by Varney [32] for $N_4^+$, by Pack and Phelps [33] for $CO_4^-$, and by Voshall, Pack and Phelps [34] for $O_4^-$. It is to be expected that data on many other ions will be obtained by this method in the future.

## § 1-6. Experimental apparatus

Dozens of drift tubes, differing greatly in details of design and construction, have been used for studies of slow ions in gases. Descriptions of many of these instruments appear in books on atomic collisions [1, 4, 11] and will not be repeated here. Instead, we shall give a fairly detailed description of one of the newest instruments, a drift tube mass spectrometer recently built at the Georgia Institute of Technology. This apparatus incorporates most of the design features which experience has shown to be important and illustrates the trend which drift tube research has taken in recent years. The reader is referred to the original literature for descriptions of other drift tube mass spectrometers which have been built [35–43].

The experimental facility [44] now in use at the Georgia Institute of Technology consists of a large ultra-high-vacuum enclosure containing a drift tube and ion-sampling apparatus, plus associated circuitry (see Figs. 1-6-1 and 1-6-2). The gas to be studied is admitted to the drift tube through a servo-controlled leak, and the sample gas continuously flows from the tube

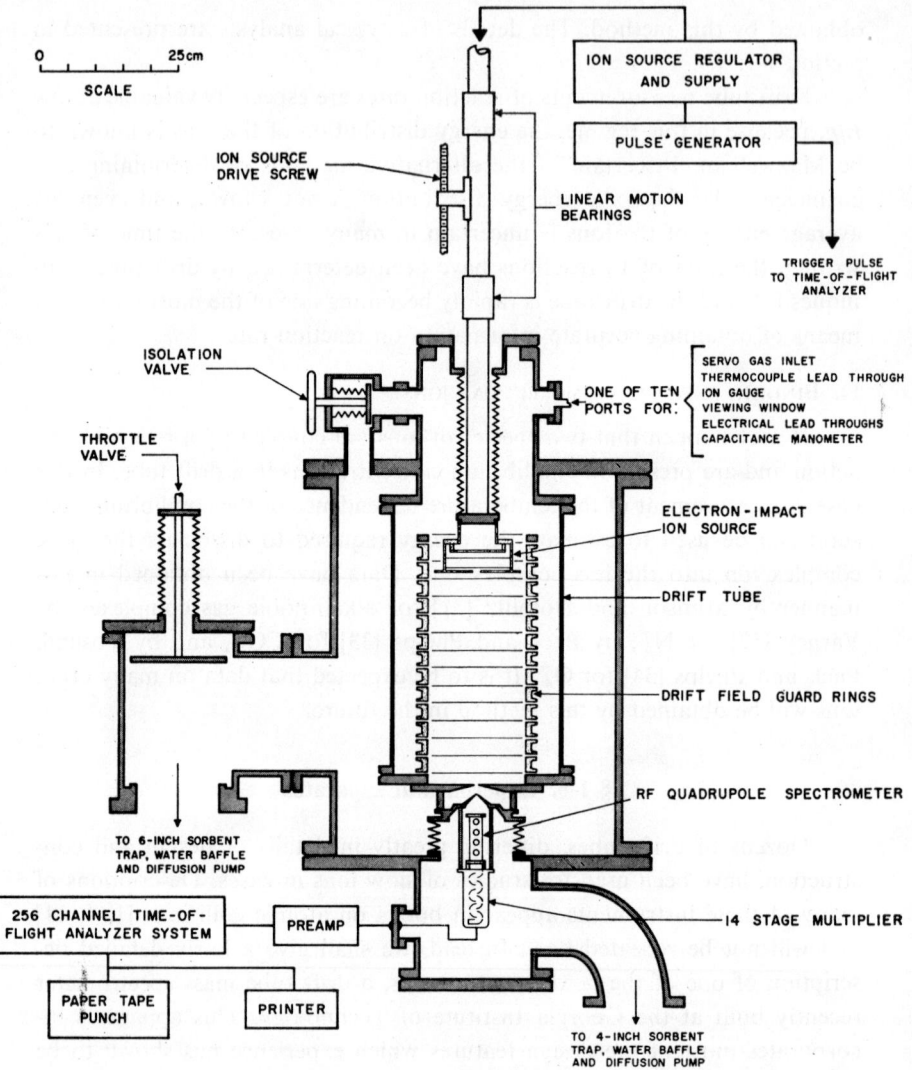

Fig. 1-6-1. Overall schematic view of the drift tube mass spectrometer in use at the Georgia Institute of Technology.

through an exit aperture on the axis at the bottom of the tube. The pressure in the drift tube is held constant during operation at some desired value in the range 0.02 to 1.0 Torr. A pulsed electron-impact ion source is used to create repetitive, short bursts of primary ions at a selected source position on the drift tube axis. Each burst of ions moves downward out of the source

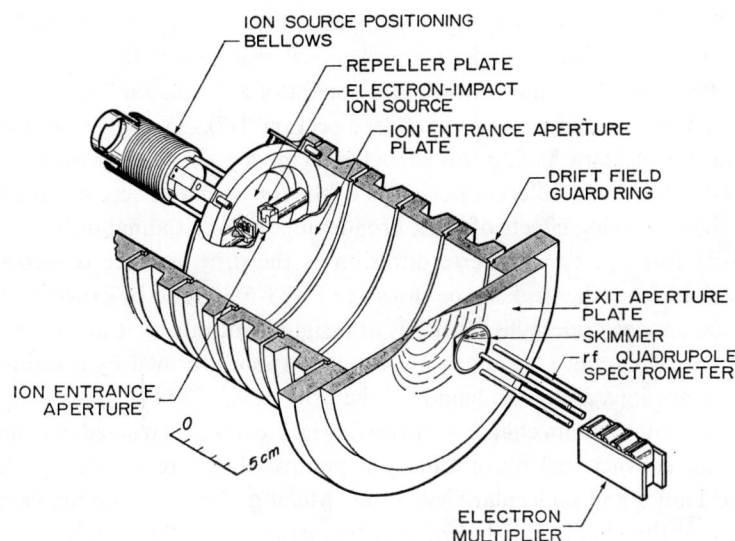

Fig. 1-6-2. Isometric view of the drift tube, ion source, and sampling system shown in Fig. 1-6-1.

and migrates along the axis of the drift tube under the influence of a weak uniform electric field produced by electrodes (the drift field guard rings) inside the tube. When the ions reach the bottom of the drift tube, those close to the axis are swept out through the exit aperture, and the core of the emerging jet of ions and gas molecules is cut out by a conical skimmer and allowed to pass into an r.f. quadrupole mass spectrometer. Ions of a selected charge-to-mass ratio traverse the length of the spectrometer, all other ions being rejected in the mass selection process. The selected ions are then detected individually by a nude electron multiplier operated as a pulse counter, and the resulting pulses are electronically sorted as to their arrival time by a 256-channel time-of-flight analyzer. Because of diffusion and geometrical losses, only a vanishingly small fraction of the number of ions originally present in each burst reach the detector. However, a histogram of arrival times can be built up by superimposing the data from $10^5$–$10^6$ ion bursts for a given source position. This procedure is repeated for various other positions of the source along the axis of the drift tube. Following this, the mass spectrometer is tuned successively to other ionic masses, and arrival time spectra are acquired for each type of ion present in the drift tube. Finally, the sequence of measurements is repeated for other values of gas pressure and drift field intensity.

If a given type of primary ion travels all the way from the ion source to the detector without undergoing chemical reactions with the gas molecules, its arrival time spectrum should consist of a "Gaussian" peak slightly skewed toward later arrival times (see section 1-7). Typical experimental arrival time spectra having this expected shape are shown superimposed in Fig. 1-6-3 for seven different positions of the ion source. These spectra illustrate the increasing effects of peak broadening by longitudinal diffusion and intensity loss due to transverse diffusion as the drift distance is increased. Deviations from the peak shape shown in Fig. 1-6-3 are to be expected if the detected species undergoes reactions at a significant rate as it drifts through the gas or if the detected ion is a secondary species formed by reactions of the primary ions along the length of the drift tube.

The main vacuum chamber shown in Fig. 1-6-1 is constructed of stainless steel and is evacuated by oil diffusion pumps which are baffled by water-cooled baffles and molecular sieve traps. Metal gaskets provide the vacuum seals, and the chamber is baked at a temperature of 200°C during pumpdown prior to measurements. The stainless steel drift tube inside this chamber is heated to 300°C during pumpdown, and base vacua below $10^{-9}$ Torr are achieved within it. The background drift tube pressure after the isolation

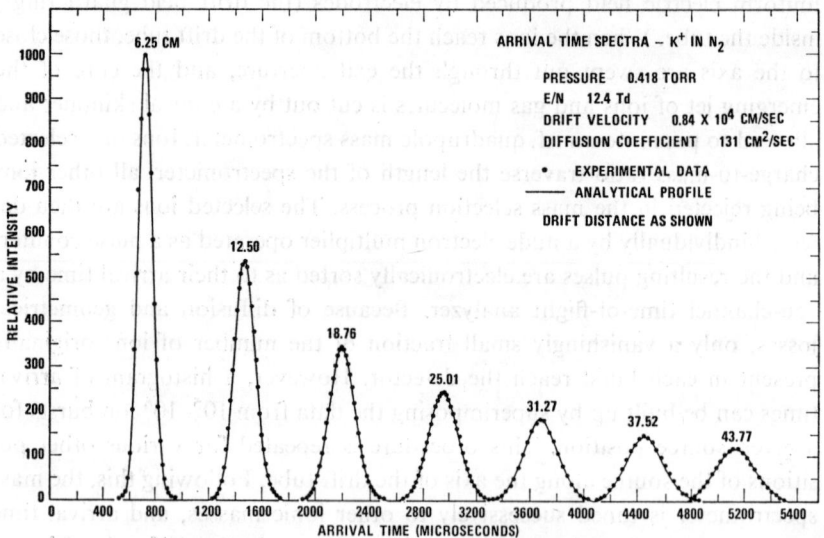

Fig. 1-6-3. Arrival time spectra of non-reacting $K^+$ ions in nitrogen recorded at seven ion source positions (dots) compared with the spectra calculated by the analysis of section 1-7 (solid curves).

valve is closed never exceeds $2 \times 10^{-8}$ Torr. Important to the achievement of low base pressures are the facts that (a) the use of organic materials is scrupulously avoided, (b) all insulators are made either of alumina ($Al_2O_3$) or steatite ($MgO \cdot SiO_2$), and (c) all welds are heliarc welds made from the inside. After the drift tube has been evacuated and filled with the gas to be studied, the pressure in the drift tube is measured with a capacitance manometer which has been calibrated by a trapped McLeod gauge. Thermocouples attached to the exterior of three of the guard rings indicate the temperature inside the drift tube.

Details of the ion source construction are shown in Fig. 1-6-2. The source contains two nonmagnetic stainless steel boxes, one mounted on each side of a ring magnet which produces a field of about 100 gauss in the magnet gap. Electrons are evaporated from a thoriated-iridium filament in the box on the left side, and are periodically admitted into the ionization region through a slit in a control plate which is used to gate the passage of the electrons. Because of the magnetic field, the electrons are constrained to move in tight helices. The electron beam has the shape of a narrow ribbon perpendicular to the drift tube axis, and thus the primary ionization is restricted to a narrow, well defined region in the gas. After traversing the ionization gap, the electrons are collected in the box on the right side of the source. The source frame is maintained at the local equipotential in the drift field, and a suitable potential is applied to the repeller plate at the top of the source to cause the ions which are formed to move toward the ion entrance aperture plate. A Tyndall grid, not shown in the figure, is mounted in the $\frac{3}{4}$-inch hole in this plate. This grid consists of two closely spaced wire meshes mounted perpendicular to the drift tube axis. Normally, a potential is applied between these meshes to prevent the ions from passing through the hole. Periodically, however, this potential is removed so that ions may flow through the grid and enter the drift space. The repetition rate of the pulses applied to the control plate and the Tyndall grid is $10^2$–$10^4$/sec, and the width of the pulses is usually less than 1 microsecond, a time which is negligible compared to typical drift times. Short pulse widths and an electron beam current of less than 1 $\mu$A during each pulse are utilized in order to restrict the number of ions in each burst to a value which does not produce appreciable space charge effects in the drift space.

The electron beam in the ion source has a fairly small energy spread (several eV), and its energy can be set at any desired value within wide limits. This feature provides closer control on the production of excited and multiply-charged ions than is possible with many other kinds of ion sources.

Both positive and negative ions can be produced by the ion source. Positive ions are created by the ejection of bound electrons into the continuum. In an electronegative gas, negative ions can be produced directly by beam electrons in various processes or as the result of capture by gas molecules of electrons ejected from other gas molecules by the ion source beam.

The ion source is mounted at the bottom of a long stainless steel bellows whose length can be varied from outside the apparatus during operation. By changing the length of the bellows, the ion source can be placed within a few thousandths of an inch at any of 16 positions along the drift tube axis to yield drift distances which vary from 1 to 44 cm. A variable drift distance is very useful because end effects produced both in the ion source and in the mass selection and detection system can be eliminated by comparing results obtained with different distances.

The drift space which the ions enter after leaving the ion source is bounded by a set of 14 guard rings with 17.5 cm i.d. The rings are similar to those used by Crompton, Elford and Gascoigne [45] and maintain an axial electric field which is free of distortion to a fraction of a per cent in the region traversed by the ion swarm. (Calculations of the field established by these electrodes may be found in the report by Albritton et al. [44].) Concealed alumina spacers and dowel pins electrically separate the guard rings and provide alignments to a few thousandths of an inch. All surfaces exposed to the ions here and in other parts of the apparatus are gold plated to reduce surface potentials.

Ions leave the drift space through a knife-edged hole in the exit aperture plate at the bottom of the drift tube. This aperture is similar to a simple molecular beam effusion orifice, and its design minimizes mass discrimination effects in the ionic mass sampling. Strong differential pumping is applied between the drift tube exit aperture and the skimmer, and between the skimmer and the mass spectrometer so that very few collisions occur after the ions have passed out of the drift tube. The apparatus may be operated with no guiding or focusing field applied between the drift tube exit aperture plate and the skimmer in order that the ions not be accelerated until they have passed through the skimmer and entered the analysis region where the pressure is below $10^{-6}$ Torr. This procedure guards against the possibility that weakly bound ions may be dissociated in energetic collisions with gas molecules. However, when tests on a given ion in a given gas indicate that it is safe to do so, a drawout potential of several volts is usually applied to the skimmer in order to increase the detected ion current and reduce the spread of transit times between the drift tube and the mass spectrometer.

The holes in the exit aperture plate and in the tip of the skimmer are both quite large (0.079 cm diameter), so that practically no ions that have grazed a surface are detected in the mass spectrometer.

After the ions have passed through the skimmer, they are brought to an energy of about 4 eV for analysis in a Varian radio-frequency quadrupole mass filter. This device is 7 inches long and has a maximum resolving power of about 50.

An EMI venetian blind electron multiplier with 17 stages is used as a pulse counter to detect the ions transmitted through the mass spectrometer. Pulse counting techniques permit the collection of data even though the signal may be extremely weak. This capability allows the apparatus to be operated at very low $E/p_0$ and with sufficiently small ion currents that space charge effects are negligible. A few counts per second is the practical lower limit on the detector sensitivity. The maximum counting rate which may be used is about $10^5$ per second.

The time interval between the creation of an ion swarm and the detection of one of its members is measured and stored by a 256-channel time-of-flight analyzer (channel widths: 0.25 to 64 $\mu$sec). The sweep of the analyzer is triggered by the pulse applied to the Tyndall grid at the upper boundary of the drift space.

The drift tube may be operated over a wide pressure range (about 0.02 to 1 Torr). The lower limit is imposed by the requirement that the product of gas pressure and drift distance be sufficiently large that steady state drift conditions obtain over most of the drift distance. The upper limit arises from the inability of the diffusion pumps to handle the gas load imposed on them at higher pressures. It is important to be able to operate over a wide range of pressure so that the pressure dependence of ion-molecule reactions occurring in the drift tube may be accurately determined. Furthermore, the reaction pattern in a given gas might make it mandatory to operate at either a high or a low gas pressure in order to obtain meaningful results for a given ionic species in that gas.

## § 1-7. Mathematical analysis of the space-time behavior of a drifting ion swarm

We have already discussed in section 1-5 the general types of measurements which may be made with drift tubes in order to obtain information on the transport properties and reactions of slow ions in gases. It now remains to show how such information may be extracted from the experimental data. To do this, we shall first analyze the space-time behavior of a swarm of ions which is diffusing and reacting with molecules as it drifts through the gas filling the drift tube. Then, in section 1-8, we shall demonstrate how this analysis may be applied to the evaluation of drift velocities, diffusion coefficients and reaction rates, using ions in nitrogen as an example. In the analysis presented here, a simple case is chosen; by this choice, we can illustrate the salient features which must characterize the solution of any drift-diffusion-reaction problem in a drift tube, yet we avoid the mathematical intricacies which arise in the choice of a more complex example. This analysis is abstracted from a calculation by J. T. Moseley and I. R. Gatland which is presented in full in a report by Moseley et al. [46]. It applies to the geometry of the drift tube mass spectrometer described in section 1-6.

## A. The Transport Equation at Low $E/p_0$

Consider an ion population created at one end of a cylindrically symmetric drift space in which there exists an externally applied uniform electric field $E$ along the axis. Let the drift space be filled with a gas of uniform pressure $p$, and for the moment suppose that the electric field is weak enough that the ion energy due to the field is small compared to the thermal energy (low $E/p_0$). Further assume that the number density, $\rho$, of the ion swarm is low enough that the space charge field is negligible. Then the ions will move with a constant average velocity $v_d$ down the drift space due to the field, will spread by diffusion through the gas, and, possibly, will react with the gas molecules with a resulting loss of the original ion species and the creation of new ions. Under the assumption of low $E/p_0$, the ionic current density $\boldsymbol{J}$ is the sum of the diffusion contribution and the drift contribution:

$$\boldsymbol{J}(t, z, r, \theta) = -\mathscr{D}\nabla\rho(t, z, r, \theta) + \boldsymbol{v}_d\rho(t, z, r, \theta) \qquad (1\text{-}7\text{-}1)$$

where $\mathscr{D}$ is the scalar diffusion coefficient of the ions, $t$ is the time, and $(z, r, \theta)$ are the space coordinates. If we do not allow for the possibility of gain of the species under consideration during the movement down the drift space, but do consider loss of these ions through ion-molecule reactions at the frequency $\alpha$, then the ion swarm is subject to a continuity equation of the form

$$\frac{\partial \rho}{\partial t} + \nabla \cdot \boldsymbol{J} + \alpha\rho = 0 \qquad (1\text{-}7\text{-}2)$$

or

$$\frac{\partial \rho}{\partial t} - \mathscr{D}\nabla^2\rho + \boldsymbol{v}_d \cdot \nabla\rho + \alpha\rho = 0 \qquad (1\text{-}7\text{-}3)$$

where $\alpha$ is usually expressed in units of [1/sec].

If we further add a source term, $\beta(t, z, r, \theta)$ to represent an input of ions at the initial end of the drift space and, possibly, an input due to ion-molecule reactions as the swarm drifts, the continuity equation becomes

$$\frac{\partial \rho}{\partial t} = \mathscr{D}\nabla^2\rho - \mathbf{v}_\mathrm{d}\cdot\nabla\rho - \alpha\rho + \beta. \tag{1-7-4}$$

Assume that the ions enter the drift space through a circular aperture lying in a plane normal to the drift tube axis and centered on the axis. If the coordinate system is assumed to have its origin at the center of this aperture and the electric field is along $z$ in such a direction as to cause the ions to drift in the positive $z$ direction, then the solution to eq. (1-7-4) in unbounded space* is

$$\rho(x, y, z, t) = \int_{-\infty}^{t} dt' \int_{-\infty}^{\infty} dx' \int_{-\infty}^{\infty} dy' \int_{-\infty}^{\infty} dz' \frac{\beta(x', y', z', t')}{[4\pi\mathscr{D}(t - t')]^{\frac{3}{2}}}$$
$$\times \exp\left(-\alpha(t - t') - \frac{(x - x')^2 + (y - y')^2 + [z - z' - v_\mathrm{d}(t - t')]^2}{4\mathscr{D}(t - t')}\right). \tag{1-7-5}$$

That (1-7-5) is indeed a solution to (1-7-4) is shown in section 1-7-C. To express (1-7-5) in cylindrical coordinates, let

$$x' = r'\cos\theta'; \quad y' = r'\sin\theta'; \quad x = r\cos\theta; \quad y = r\sin\theta. \tag{1-7-6}$$

Then

$$\rho(r, \theta, z, t) = \int_0^\infty r'\, dr' \int_0^{2\pi} d\theta' \int_{-\infty}^\infty dz' \int_{-\infty}^t dt' \frac{\beta(r', \theta', z', t')}{[4\pi\mathscr{D}(t - t')]^{\frac{3}{2}}}$$
$$\times \exp\left(-\alpha(t - t') - \frac{r^2 + r'^2 - 2rr'\cos(\theta - \theta') + [z - z' - v_\mathrm{d}(t - t')]^2}{4\mathscr{D}(t - t')}\right). \tag{1-7-7}$$

Clearly if the input $\beta$ is cylindrically symmetric, then the ion number density $\rho$ will be cylindrically symmetric. Suppose

$$\beta(r', z', t') = \frac{b}{\pi r_0^2} S(r_0 - r')\,\delta(z')\,\delta(t') \tag{1-7-8}$$

---

* If the radius of the drift field electrodes is small enough that ions reach these electrodes by diffusion, radial boundary conditions must be applied to take this fact into account. Such is not the case for the apparatus described in section 1-6. For discussion of this point, see J. H. Whealton and S. B. Woo, Phys. Rev. Letters **20** (1968) 1137.

where $S(\varphi)=0$ if $\varphi<0$, $S(\varphi)=1$ otherwise. This function describes an axially thin disk source of $b$ ions with uniform surface density and radius $r_0$, being created instantaneously at $t'=0$ in the plane $z'=0$. Let $b/\pi r_0^2 = C$, the planar source density. Then

$$\rho(r,z,t) = \frac{C e^{-\alpha t} \exp[-(z-v_d t)^2/(4\mathcal{D}t)]}{(4\pi\mathcal{D}t)^{\frac{3}{2}}}$$

$$\times \int_0^{r_0} r'\, dr' \int_0^{2\pi} d\theta' \exp\left[-\frac{r^2 + r'^2 - 2rr'\cos\theta'}{4\mathcal{D}t}\right]. \qquad (1\text{-}7\text{-}9)$$

Now [50]

$$I_0(x) = \frac{1}{\pi}\int_0^\pi d\theta\, e^{\pm x\cos\theta} = \sum_{m=0}^\infty \frac{(\tfrac{1}{2}x)^{2m}}{(m!)^2}. \qquad (1\text{-}7\text{-}10)$$

Hence

$$\rho(r,z,t) = \frac{2\pi C e^{-\alpha t}\exp[-(z-v_d t)^2/(4\mathcal{D}t)]}{(4\pi\mathcal{D}t)^{\frac{3}{2}}}$$

$$\times \int_0^{r_0} r'\, dr' \exp\left(-\frac{r^2+r'^2}{4\mathcal{D}t}\right)\sum_{m=0}^\infty \frac{1}{(m!)^2}\left(\frac{rr'}{4\mathcal{D}t}\right)^{2m}. \qquad (1\text{-}7\text{-}11)$$

If we let $x=r'^2/(4\mathcal{D}t)$, then

$$\rho(r,z,t) = \frac{C}{(4\pi\mathcal{D}t)^{\frac{1}{2}}}\exp\left[-\alpha t - \frac{(z-v_d t)^2 + r^2}{4\mathcal{D}t}\right]$$

$$\times \sum_{m=0}^\infty \frac{[r^2/(4\mathcal{D}t)]^m}{(m!)^2}\int_0^{r_0^2/(4\mathcal{D}t)} x^m e^{-x}\, dx. \qquad (1\text{-}7\text{-}12)$$

The remaining integral can be done by integrating by parts $m$ times. The result is

$$\int_0^a x^m e^{-x}\, dx = m!\left[1 - e^{-a}\sum_{n=0}^m \frac{a^n}{n!}\right]. \qquad (1\text{-}7\text{-}13)$$

Substituting in (1-7-12) we obtain

$$\rho(r,z,t) = \frac{C}{(4\pi\mathcal{D}t)^{\frac{1}{2}}}\left\{\exp\left[-\alpha t - \frac{(z-v_d t)^2 + r^2}{4\mathcal{D}t}\right]\right\}$$

$$\times \sum_{m=0}^\infty \frac{[r^2/(4\mathcal{D}t)]^m}{m!}\left[1 - e^{-r_0^2/(4\mathcal{D}t)}\sum_{n=0}^m \frac{[r_0^2/(4\mathcal{D}t)]^n}{n!}\right] \qquad (1\text{-}7\text{-}14)$$

or

$$\rho(r, z, t) = \frac{C}{(4\pi \mathscr{D} t)^{\frac{1}{2}}} \exp\left[-\alpha t - \frac{(z - v_d t)^2}{4\mathscr{D} t}\right]$$
$$\times \left[1 - \sum_{m=0}^{\infty} \sum_{n=0}^{m} \frac{1}{m!n!} \left(\frac{r^2}{4\mathscr{D} t}\right)^m \left(\frac{r_0^2}{4\mathscr{D} t}\right)^n \exp\left(-\frac{r_0^2 + r^2}{4\mathscr{D} t}\right)\right]. \quad (1\text{-}7\text{-}15)$$

Equation (1-7-15) is an expression for the ion number density at any given time at any point in space for an ion swarm which (1) was instantaneously created with uniform density across an axially thin disk, (2) drifts in unbounded space under "low $E/p_0$" conditions and (3) possibly undergoes a depleting reaction with the neutral gas molecules.

Since the ions detected in the apparatus described in section 1-6 are those which exit the drift tube on the axis, the result of interest here is the axial ionic number density

$$\rho(0, z, t) = \frac{C e^{-\alpha t}}{(4\pi \mathscr{D} t)^{\frac{1}{2}}} \{1 - \exp[-r_0^2/(4\mathscr{D} t)]\} \exp[-(z - v_d t)^2/(4\mathscr{D} t)] \quad (1\text{-}7\text{-}16)$$

where $C$ is the initial ion surface density of the delta-function input of ions and $r_0$ is the radius of the ion entrance aperture. The total number of ions entering the drift tube during each pulse is $C\pi r_0^2$. The quantity experimentally measured is the flux of ions leaving the drift tube at a fixed drift distance $z$:

$$\Phi(0, z, t) = AJ(0, z, t) \quad (1\text{-}7\text{-}17)$$

where $A$ is the area of the exit aperture, and $J(0, z, t)$ is the $z$-component of the ionic current density in the drift tube, on the axis, at the end of the drift distance $z$.

The ionic current density is related to the ionic number density by eq. (1-7-1), so that

$$J(0, z, t) = -\mathscr{D}(\partial \rho/\partial z) + v_d \rho \quad (1\text{-}7\text{-}18)$$

where $\rho$ here is given by eq. (1-7-16). Equation (1-7-17) then becomes

$$\Phi(0, z, t) = -A\mathscr{D}(\partial \rho/\partial z) + Av_d \rho. \quad (1\text{-}7\text{-}19)$$

Differentiation of (1-7-16) with respect to $z$ gives

$$\partial \rho/\partial z = -\rho(z - v_d t)/(2\mathscr{D} t). \quad (1\text{-}7\text{-}20)$$

Substitution of (1-7-20) into (1-7-19) provides the final result,

$$\Phi(0, z, t) = \tfrac{1}{2} A (v_d + z/t) \rho(0, z, t) \qquad (1\text{-}7\text{-}21)$$

or in full

$$\Phi(0, z, t) = \frac{A C \, e^{-\alpha t}}{4(\pi \mathscr{D} t)^{\frac{1}{2}}} (v_d + z/t) \{ 1 - \exp[-r_0^2/(4\mathscr{D} t)] \}$$
$$\times \exp[-(z - v_d t)^2/(4\mathscr{D} t)]. \qquad (1\text{-}7\text{-}22)$$

Equation (1-7-22) is an expression for the flux of primary ions reaching the detector as a function of the drift distance and the time elapsed since creation of these ions. Low-field conditions are assumed. Other analyses which provide expressions for the low-field arrival time spectrum of the secondary ions formed by reactions are described in refs. [47] and [48], and the use of these expressions to obtain reaction rates is outlined there. Ref. [49] contains an analysis somewhat similar to the one presented here.

In most of the measurements made with the apparatus described in section 1-6, source bursts of less than 1 $\mu$sec duration are used, while the drift times are of the order of hundreds or thousands of microseconds. Consequently, the description of the initial ion burst by a delta function in time causes little error. However, one need not make this approximation if extreme accuracy is desired, since the analysis above is easily extended to the case of a step function input burst of finite duration [46].

Figure 1-7-1 illustrates the degree of agreement between eq. (1-7-22) and a typical experimental arrival time histogram for a case where reactions occur at a negligibly slow rate. The smooth curve was obtained by plotting (1-7-22) with $\alpha = 0$, with $v_d$ set equal to the experimentally determined value, and with $\mathscr{D}$ determined from this value of $v_d$ by use of the Einstein relation (1-5-1). The curve was normalized to the histogram at its peak.

## B. THE TRANSPORT EQUATION AT HIGH $E/p_0$

When low-field conditions are not satisfied, the equations in the analysis presented above must be modified to take account of the tensor character of the diffusion coefficient (see section 1-5-B). For the geometry which has been chosen in the analysis, the continuity equation (1-7-4) must be replaced by the equation

$$\frac{\partial \rho}{\partial t} = \mathscr{D}_T \left[ \frac{\partial^2 \rho}{\partial r^2} + \frac{\partial \rho/\partial r}{r} \right] + \mathscr{D}_L \frac{\partial^2 \rho}{\partial z^2} - v_d \frac{\partial \rho}{\partial z} - \alpha \rho + \beta \qquad (1\text{-}7\text{-}23)$$

where $\mathscr{D}_T$ and $\mathscr{D}_L$ are the transverse and longitudinal diffusion coefficients,

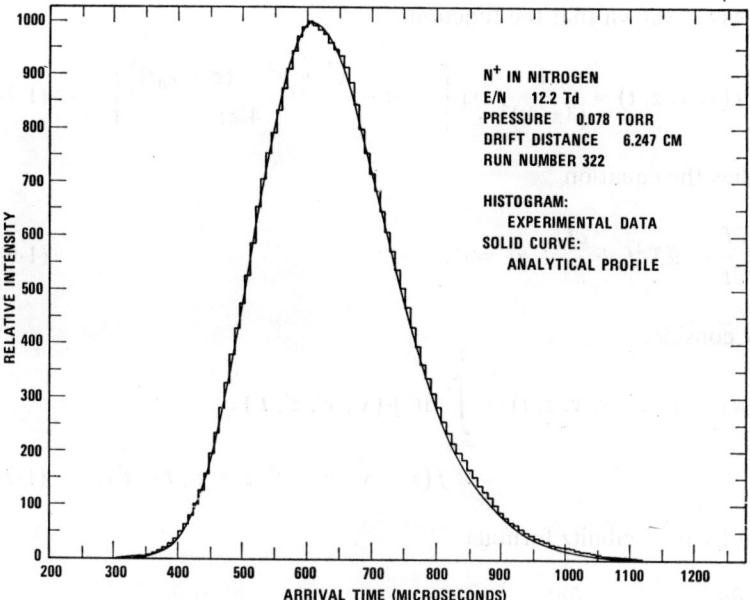

Fig. 1-7-1. Comparison of prediction of analysis with a typical experimental arrival time spectrum.

respectively [14]. If we make the same assumptions as were made in section 1-6-A above, the analysis proceeds as before, and the flux through the drift tube exit aperture is seen to be [46]

$$\Phi(0, z, t) = \frac{AC\, e^{-\alpha t}}{4(\pi \mathscr{D}_L t)^{\frac{1}{2}}} (v_d + z/t) \left[ 1 - \exp\left(-\frac{r_0^2}{4\mathscr{D}_T t}\right) \right] \\ \times \exp\left[-\frac{(z - v_d t)^2}{4\mathscr{D}_L t}\right]. \qquad (1\text{-}7\text{-}24)$$

### C. Solution of the Transport Equation at Low $E/p_0$

The transport equation assumed for the ions is (1-7-4). Under the assumption that $v_d$ is directed along the positive $z$ direction, this equation may be written

$$\frac{\partial \rho(x, y, z, t)}{\partial t} = \mathscr{D} \nabla^2 \rho - \frac{\partial \rho}{\partial z} - \alpha \rho + \beta(x, y, z, t). \qquad (1\text{-}7\text{-}25)$$

It is easily shown that the function

$$f(x, y, z, t) = \frac{1}{(4\pi\mathscr{D}t)^{\frac{3}{2}}} \exp\left[-\alpha t - \frac{x^2 + y^2 + (z - v_d t)^2}{4\mathscr{D}t}\right] \quad (1\text{-}7\text{-}26)$$

satisfies the equation

$$\frac{\partial f}{\partial t} - \mathscr{D}\nabla^2 f + \frac{\partial f}{\partial z} + \alpha f = 0. \quad (1\text{-}7\text{-}27)$$

Now consider

$$g(x', y', z', x, y, z, t) = \int_{-\infty}^{t} dt' \, \beta(x', y', z', t')$$
$$\times f(x - x', y - y', z - z', t - t'). \quad (1\text{-}7\text{-}28)$$

Then by the Leibnitz formula

$$\frac{\partial g}{\partial t} - \mathscr{D}\nabla^2 g + \frac{\partial g}{\partial z} + \alpha g$$

$$= \int_{-\infty}^{t} dt' \left\{ \beta(x', y', z', t') \left[ \frac{\partial f}{\partial t} - \mathscr{D}\nabla^2 f + \frac{\partial f}{\partial z} + \alpha f \right] \right\}$$

$$+ \lim_{t-t' \to 0} \beta(x', y', z', t') f(x - x', y - y', z - z', t - t'). \quad (1\text{-}7\text{-}29)$$

Let $t_0 = t - t'$.
The first term on the right vanishes by virtue of (1-7-27). Hence

$$\frac{\partial g}{\partial t} - \mathscr{D}\nabla^2 g + \frac{\partial g}{\partial t} + \alpha g$$

$$= \beta(x', y', z', t) \lim_{t-t' \to 0} \left\{ e^{-\alpha t_0} \frac{\exp[-(x - x')^2/(4\mathscr{D}t_0)]}{\sqrt{4\pi\mathscr{D}t_0}} \right.$$

$$\left. \times \frac{\exp[-(y - y')^2/(4\mathscr{D}t_0)]}{\sqrt{4\pi\mathscr{D}t_0}} \frac{\exp[-(z - z' - v_d t_0)^2/(4\mathscr{D}t_0)]}{\sqrt{4\pi\mathscr{D}t_0}} \right\}. \quad (1\text{-}7\text{-}30)$$

But

$$\lim_{\sigma \to 0} \frac{e^{-\mu^2/\sigma}}{\sqrt{\pi\sigma}} = \delta(\mu). \quad (1\text{-}7\text{-}31)$$

Therefore

$$\frac{\partial g}{\partial t} - \mathscr{D}\nabla^2 g + \frac{\partial g}{\partial t} + \alpha g = \beta(x', y', z', t)\,\delta(x-x')\,\delta(y-y')\,\delta(z-z').$$
(1-7-32)

Now consider

$$\rho(x,y,z,t) = \int_{-\infty}^{\infty} dx' \int_{-\infty}^{\infty} dy' \int_{-\infty}^{\infty} dz' \int_{-\infty}^{t} dt'\, \beta(x',y',z',t')$$

$$\times f(x-x', y-y', z-z', t-t') = \int_{-\infty}^{\infty} dx' \int_{-\infty}^{\infty} dy' \int_{-\infty}^{\infty} dz'\,[g]$$
(1-7-33)

$$\frac{\partial \rho}{\partial y} - \mathscr{D}\nabla^2 \rho + v_\mathrm{d} \frac{\partial \rho}{\partial z} + \alpha \rho$$

$$= \int_{-\infty}^{\infty} dx' \int_{-\infty}^{\infty} dy' \int_{-\infty}^{\infty} dz' \left[ \frac{\partial g}{\partial t} - \mathscr{D}\nabla^2 g + v_\mathrm{d} \frac{\partial g}{\partial t} + \alpha g \right]$$

$$= \int_{-\infty}^{\infty} dx' \int_{-\infty}^{\infty} dy' \int_{-\infty}^{\infty} dz' \,[\beta(x',y',z',t)\,\delta(x-x')\,\delta(y-y')\,\delta(z-z')]$$

$$= \beta(x,y,z,t).$$
(1-7-34)

Thus

$$\rho(x,y,z,t) = \int_{-\infty}^{\infty} dx' \int_{-\infty}^{\infty} dy' \int_{-\infty}^{\infty} dz' \int_{-\infty}^{t} dt'\, \beta(x',y',z',t')$$

$$\times f(x-x', y-y', z-z', t-t') \quad (1\text{-}7\text{-}35)$$

satisfies (1-7-25) with the boundary conditions $\rho \to 0$ as $x \to \pm\infty$, $y \to \pm\infty$, $z \to \pm\infty$ or $t \to -\infty$.

## § 1-8. Experimental studies of nitrogen ions in nitrogen

Space limitations do not permit a general review of the experimental work which has been done on drift velocities, diffusion coefficients, and reaction rates with drift tubes. The reader is referred instead to refs. [1, 2, 4, 44 and 46] for such reviews. The purpose of this section is much more restricted. What we shall attempt to do is to illustrate the kinds of measurements that can be made by discussing a single recent set of experiments performed on nitrogen ions in nitrogen by Moseley et al. [46], with the drift tube mass spectrometer described in section 1-6. These experiments were chosen for discussion because they illustrate the use of certain modern techniques and because they represent as nearly complete a set of measurements as has been made to date on a single gas.

A. DRIFT VELOCITIES

A very large number of measurements have been made to determine the drift velocities of nitrogen ions in nitrogen [46]. Because of the complicating effects of reactions between the ions and gas molecules discussed in section 1-5-A, these measurements yielded results which for the most part were very discordant and difficult to interpret. Only the data to be presented here and the recent data of McKnight et al. [36], are in close agreement and can be clearly interpreted. This agreement and the unambiguous nature of the results are due to the fact that in both experiments mass analysis of the ions was performed and careful attention was paid to the effects of ion-molecule reactions.

The drift velocity data of Moseley et al. [46], are shown in terms of ionic mobilities in Figs. 1-8-1 and 1-8-2. Four species of ions were observed: $N^+$, $N_2^+$, $N_3^+$ and $N_4^+$. The drift velocities were determined from the average arrival time of the ions by the method discussed in section 1-5-A. In the case

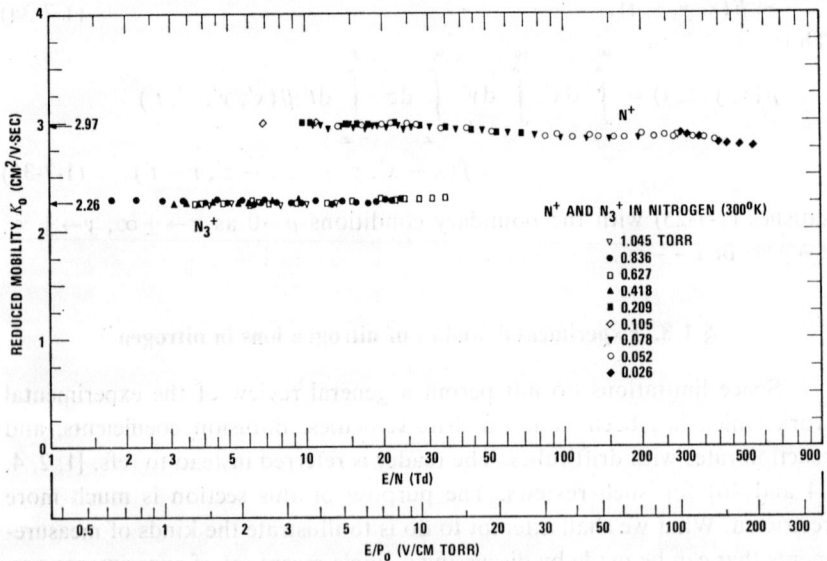

Fig. 1-8-1. Reduced mobilities of $N^+$ and $N_3^+$ ions in nitrogen.

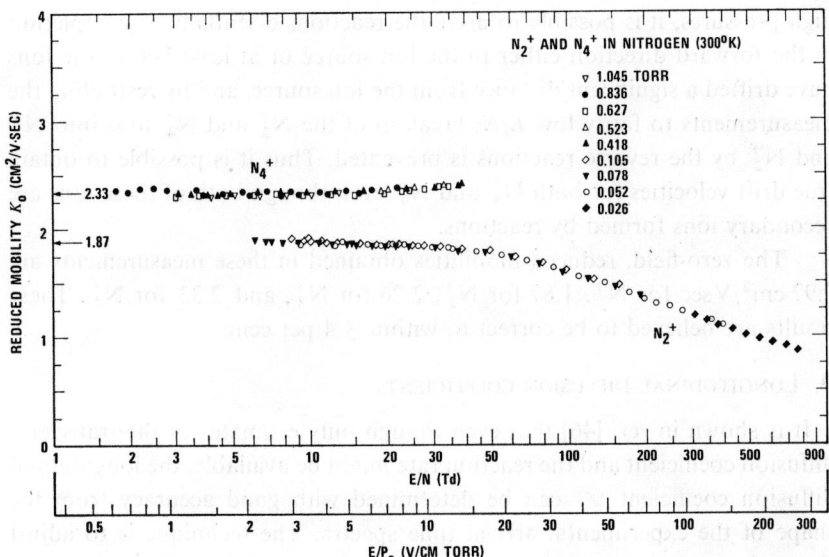

Fig. 1-8-2. Reduced mobilities of $N_2^+$ and $N_4^+$ ions in nitrogen.

of the primary ions $N^+$ and $N_2^+$ generated by electron impact in the ion source*, the data were taken at short drift distances and very low pressures (below about 0.2 Torr for $N^+$ and 0.1 Torr for $N_2^+$) in order that the correction to the mean arrival time necessary because of reactions converting these ions to the secondary species would be negligible. The fact that reactions did not affect the measurements was verified by comparing the experimental arrival time spectra with the predictions of the analysis of section 1-7. Figure 1-7-1 shows such a comparison for the case of the $N^+$ ion.

$N_3^+$ and $N_4^+$ ions are formed from the primary $N^+$ and $N_2^+$ ions by the three-body reactions

$$N^+ + 2N_2 \to N_3^+ + N_2 \tag{1-8-1}$$

and

$$N_2^+ + 2N_2 \to N_4^+ + N_2. \tag{1-8-2}$$

Each of these reactions is reversible, the forward reactions being favored by high pressures and low $E/N$ and the reverse reactions favored by high $E/N$. By making measurements on the $N_3^+$ and $N_4^+$ ions at long drift distances and

---

* Some $N_2^+$ ions were also formed in the drift region by dissociation of $N_4^+$ above an $E/N$ of 40 Td. The use of short drift distances and low pressures kept the amount of dissociation small, and an approximate correction was made for it [46].

high pressures, it is possible to drive the reactions essentially to completion in the forward direction either in the ion source or at least before the ions have drifted a significant distance from the ion source, and by restricting the measurements to fairly low $E/N$, breakup of the $N_3^+$ and $N_4^+$ ions into $N^+$ and $N_2^+$ by the reverse reactions is prevented. Thus it is possible to obtain true drift velocities for both $N_3^+$ and $N_4^+$ even though both of these ions are secondary ions formed by reactions.

The zero-field, reduced mobilities obtained in these measurements are 2.97 cm$^2$/V sec for $N^+$, 1.87 for $N_2^+$, 2.26 for $N_3^+$, and 2.33 for $N_4^+$. These results are believed to be correct to within $\pm 4$ per cent.

### B. Longitudinal Diffusion Coefficients

It is shown in ref. [46] that even though only estimates of the transverse diffusion coefficient and the reaction rate might be available, the longitudinal diffusion coefficient $\mathcal{D}_L$ can be determined with good accuracy from the shape of the experimental arrival time spectra. The technique is to adjust $\mathcal{D}_L$ in the analytical expression for the flux, eq. (1-7-24), until the analytical profile and the experimental data best correspond when normalized to agree at their points of maximum intensity. A computer program was written to perform a least squares fit of $\Phi$ to an experimental spectrum by varying $\mathcal{D}_L$ in eq. (1-7-24). In the expression for $\Phi$, $v_d$ is known from the work described in section 1-8-A, $\alpha$ is assumed to have its thermal value $\alpha_t$, the determination which will be discussed in section 1-8-D, and $\mathcal{D}_T$ is assumed to have its low field value, i.e., to be equal to the value of $\mathcal{D}_L$ at low $E/N$.

Note that each determination of $\mathcal{D}_L$ depends on the use of data taken with a single ion source position; the differencing technique used in obtaining $v_d$ is not applicable here. This fact limits the determination of $\mathcal{D}_L$ to ions which are formed entirely in the ion source. In the present experiment, these ions are $N^+$ and $N_2^+$. Data for these ions are shown in Fig. 1-8-3. Since $\mathcal{D}_L \sim 1/N$, the results are displayed in the form $N\mathcal{D}_L$ versus $E/N$. For each ion, $N\mathcal{D}_L$ approaches a limiting value at low $E/N$ which is in good agreement with the value predicted from the zero-field mobilities by the Einstein relation: $2.05 \times 10^{18}$/cm sec for $N^+$ and $1.29 \times 10^{18}$/cm sec for $N_2^+$. The results presented here appear to be the first values of $\mathcal{D}_L$ determined for any ions by direct measurement, and hence no other data are available for comparison.

### C. Transverse Diffusion Coefficients

The total intensity of a given species of primary ion which reaches the detector of the apparatus described in section 1-6 will depend on the po-

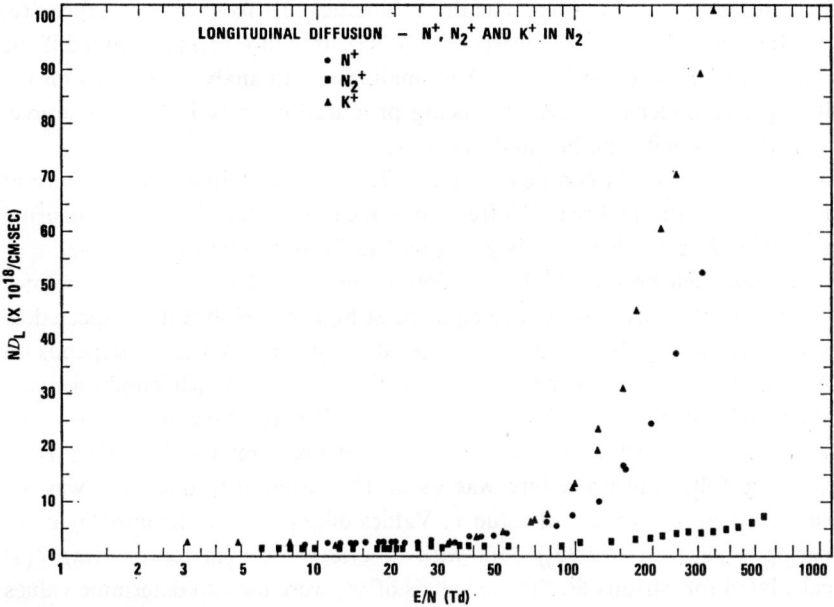

Fig. 1-8-3. Longitudinal diffusion coefficients for N⁺ and N₂⁺ ions in nitrogen. Data for K⁺ ions are shown for comparison.

sition of the ion source. As the drift distance is increased, the increased effect of transverse diffusion will cause a smaller fraction of the ions to leave the drift tube through the exit aperture. Furthermore, depleting reactions with the gas molecules may produce an attenuation of the primary ion population as it drifts through the gas. The method used to determine $\mathscr{D}_T$ consists of comparing the total number of ions composing the experimental spectra at various drift distances $I'(z_j)$ with the integrated intensity on the axis

$$I(z) = \int_0^\infty \Phi(0, z, t) \, dt \qquad (1\text{-}8\text{-}3)$$

calculated for corresponding $E/N$ and pressure conditions. Since the absolute intensity of the entering ion swarm is not known, $I'(z_j)$ and $I(z)$ are normalized to agree at the shortest drift distance used ($z = 6.25$ cm). It is shown in ref. [46] that only a reasonable estimate of $\mathscr{D}_L$ is necessary in order to determine $\mathscr{D}_T$, but in view of the work described in section 1-8-B, $\mathscr{D}_L$ can be considered to be known. The drift velocities are also known. Hence $I(z)$ can be calculated and $\mathscr{D}_T$ varied until a best fit with $I'(z_j)$ is obtained if the reaction frequency $\alpha$ is known or is small. Such an analysis depends on the ion species under consideration being produced entirely in the ion source, as is the case with the $N^+$ and $N_2^+$ ions.

At low $E/N$, $\mathscr{D}_T$ can be calculated from the Einstein relation. Hence at low $E/N$, the thermal reaction frequency $\alpha_t$ can be determined by comparing $I(z)$ with $I'(z_j)$, where $\mathscr{D}_T$ is given by the Einstein relation. This value of $\alpha = \alpha_t$ can then be used in $I(z)$ to determine $\mathscr{D}_T$ at higher $E/N$. Of course, $\alpha$ is not necessarily expected to equal $\alpha_t$ at higher $E/N$, but it is expected to vary fairly slowly. In addition, since $\mathscr{D}_T$ depends on $1/N$ while $\alpha$ depends on $N^2$, data taken at low pressures (less than 0.1 Torr) will emphasize the effects of diffusion and minimize the effects of the reaction. All of the transverse diffusion coefficient data were taken at pressures less than 0.08 Torr.

The following procedure was used. The reaction frequency $\alpha$ was assumed to equal its thermal value $\alpha_t$. Values of $\mathscr{D}_T$ were determined by comparing experimental $I'(z_j)$ data with a series of curves representing $I(z)$ calculated for various $\mathscr{D}_T$. These values of $\mathscr{D}_T$ were used to determine values of $\alpha$ at higher $E/N$ (see section 1-8-D). Then these higher-$E/N$ values of $\alpha$ were used to recalculate the $I(z)$ curves, and a second comparison with the $I'(z_j)$ data was made. However, the use of more accurate values for $\alpha$ caused only a negligible change in the various $I(z)$ curves at the low pressures used to determine $\mathscr{D}_T$, and no change in $\mathscr{D}_T$ from the values determined by assuming $\alpha = \alpha_t$ was required.

The results obtained for $N^+$ and $N_2^+$ ions in nitrogen are shown in Fig. 1-8-4. In both cases, the data approach the $N\mathscr{D}_T$ values predicted by the Einstein relation at low $E/N$. Again, no other data are available for comparison with the results presented here.

D. REACTION RATE COEFFICIENTS

Rate coefficients were determined for the reactions (1-8-1) and (1-8-2), which lead to the production of $N_3^+$ from $N^+$ and $N_4^+$ from $N_2^+$, respectively. The intermediate pressure range, 0.1 to 0.4 Torr was chosen for operation. At lower pressures the reaction frequencies are small and the effect of the reactions is masked by transverse diffusion. At higher pressures, the reactions go essentially to completion before the required observations can be made.

The basic method used is the same as that used for the determination of $\mathscr{D}_T$ (see section 1-8-C). A comparison is made between the total number of

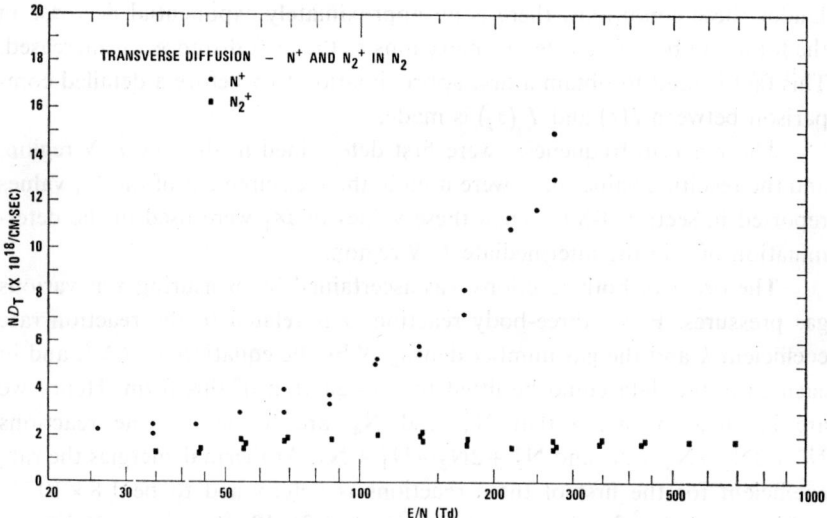

Fig. 1-8-4. Transverse diffusion coefficients for $N^+$ and $N_2^+$ ions in nitrogen.

$N^+$ or $N_2^+$ ions composing the experimental spectra at various drift distances, $I'(z_j)$, and the integrated intensity on the axis, $I(z)$. If this comparison is to yield values of the reaction frequency $\alpha$, then the transverse diffusion coefficient must be known. $\mathscr{D}_L$ is known from the results of section 1-8-B, and $v_d$ is known from section 1-8-A.

The requirement that $\mathscr{D}_T$ be known divides the determination of $\alpha$ into three $E/N$ regions. In the low $E/N$ region (below 30 Td), $\mathscr{D}_T$ can be calculated from the Einstein relation and can thus be considered to be known to within the accuracy that the zero-field mobility is known. In the intermediate $E/N$ region (30 to 100 Td), $\mathscr{D}_T$ can be considered known on the basis of the determination discussed in section 1-8-C, but is not known as accurately as in the low $E/N$ region. In the high $E/N$ region (above 100 Td), the reaction rate was not determined because to obtain data in this region, it would have been necessary to operate at such low pressures that the effects of the reaction would have been overshadowed by diffusion.

We may note from eq. (1-7-24) that in the limit of small $\mathscr{D}_L$ and $\mathscr{D}_T$ (i.e. high gas pressures) and/or large $\alpha$, the reaction will entirely dominate the spectra intensities and $I(z)$ may be written

$$I(z) \approx I_0 \, e^{-\alpha z/v_d}. \tag{1-8-4}$$

Under these conditions, there is an approximately exponential decrease in the total number of detected primary ions as the drift distance $z$ is increased. This fact is used to obtain a first approximation to $\alpha$ before a detailed comparison between $I(z)$ and $I'(z_j)$ is made.

The reaction frequencies were first determined in the low $E/N$ region, and the resulting values of $\alpha_t$ were used in the measurement of the $\mathscr{D}_T$ values reported in section 1-8-C. Then these values of $\mathscr{D}_T$ were used in the determination of $\alpha$ in the intermediate $E/N$ region.

The order of both reactions was ascertained by measuring $\alpha$ at various gas pressures. For a three-body reaction, $\alpha$ is related to the reaction rate coefficient $k$ and the gas number density $N$ by the equation $\alpha = kN^2$, and in each case, the data could be fitted by an equation of this form. Hence we are justified in saying that $N_3^+$ and $N_4^+$ are formed in the reactions $N^+ + 2N_2 \rightarrow N_3^+ + N_2$ and $N_2^+ + 2N_2 \rightarrow N_4^+ + N_2$. At thermal energies the rate coefficient for the first of these reactions is determined to be $1.8 \times 10^{-29}$ cm$^6$/sec, and that for the second reaction is $5.0 \times 10^{-29}$ cm$^6$/sec. Values of $k$ for the complete range of $E/N$ covered are presented in Fig. 1-8-5. The results obtained here are in general agreement with those obtained by other investigators [46].

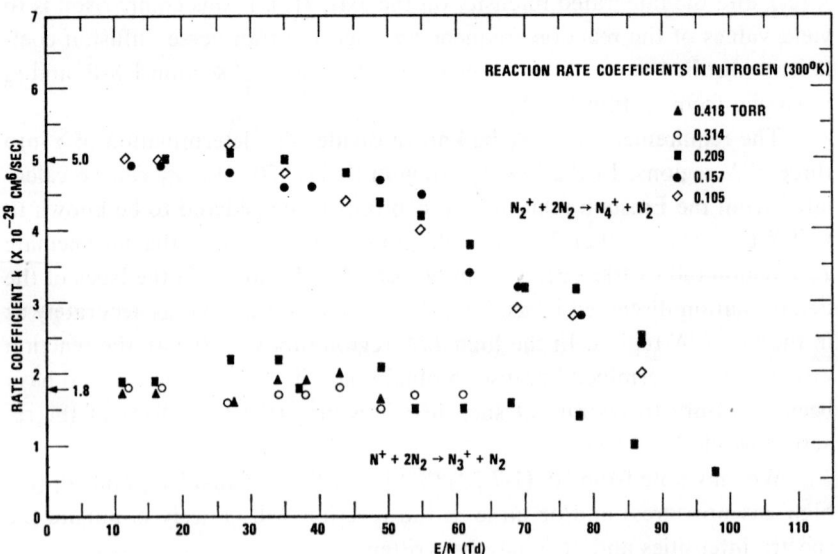

Fig. 1-8-5. Rate coefficients for the reactions $N^+ + 2N_2 \rightarrow N_3^+ + N_2$ and $N_2^+ + 2N_2 \rightarrow N_4^+ + N_2$.

## References

1. E. W. McDaniel, V. Čermák, A. Dalgarno, E. E. Ferguson and L. Friedman, Ion-molecule Reactions (Wiley, New York, 1969).
2. E. W. McDaniel, Collision Phenomena in Ionized Gases (Wiley, New York, 1964).
3. L. G. H. Huxley, R. W. Crompton and M. T. Elford, Bull. Inst. Physics and Physical Society **17** (1966) 251.
4. L. B. Loeb, Basic Processes of Gaseous Electronics, 2nd ed. (University of California Press, Berkeley, 1960) Ch. 1.
5. P. Langevin, Ann. Chim. Phys. **28** (1903) 289.
6. L. B. Loeb, Kinetic Theory of Gases, 3rd ed. (Dover, New York, 1961) pp. 547–552.
7. P. Langevin, Ann. Chim. Phys. **5** (1905) 245.
8. R. D. Present, Kinetic Theory of Gases (McGraw-Hill, New York, 1958).
9. E. W. McDaniel, Collision Phenomena in Ionized Gases (Wiley, New York, 1964) Section 1-8.
10. H. R. Hassé, Phil. Mag. **1** (1926) 139.
11. E. W. McDaniel, Collision Phenomena in Ionized Gases (Wiley, New York, 1964) Ch. 9.
12. S. Chapman and T. G. Cowling, The Mathematical Theory of Non-uniform Gases, 2nd ed. (Cambridge University Press, London, 1952).
13. E. W. McDaniel, Collision Phenomena in Ionized Gases (Wiley, New York, 1964) Section 3-9.
14. G. H. Wannier, Bell System Technical Journal **32** (1953) 170–254. A short version of parts of this paper appeared in Phys. Rev. **83** (1951) 281; **87** (1952) 795.
15. T. Kihara, Rev. Mod. Phys. **24** (1952) 45; **25** (1953) 844.
16. E. A. Mason and H. W. Schamp, Ann. Phys. (New York) **4** (1958) 233.
17. V. I. Perel, Soviet Phys.-JETP **5** (1957) 440.
18. A. Dalgarno, M. R. C. McDowell and A. Williams, Phil. Trans. Roy. Soc. (London) **A-250** (1958) 411;
    A. Dalgarno, Phil. Trans. Roy. Soc. (London) **A-250** (1958) 426. A summary of these papers appears in ref. [11].
19. E. W. McDaniel, Collision Phenomena in Ionized Gases (Wiley, New York, 1964) Section 3-10.
20. T. Holstein, J. Phys. Chem. **56** (1952) 832.
21. G. Heiche, Ph.D. Thesis, University of Maryland, 1967.
22. A. M. Arthurs and A. Dalgarno, Proc. Roy. Soc. **A-256** (1960) 540, 552.
23. A. Dalgarno and R. J. W. Henry, in: Atomic Collisions Processes, ed. M. R. C. McDowell (North-Holland Publishing Co., Amsterdam, 1964) p. 914.
24. S. I. Sandler and E. A. Mason, J. Chem. Phys. **48** (1968) 2873.
25. T. Holstein, Phys. Rev. **100** (1955) 1230 (A).
26. E. C. Beaty, J. C. Browne and A. Dalgarno, Phys. Rev. Letters **16** (1966) 723.
27. E. W. McDaniel, Collision Phenomena in Ionized Gases (Wiley, New York, 1964) Chs. 10 and 12.
28. A. Dalgarno, in: Atomic and Molecular Processes, ed. D. R. Bates (Academic Press, New York, 1962) p. 655.
29. H. R. Skullerud, Proceedings of the Seventh International Conference on Phenomena in Ionized Gases, Belgrade, 1965, Vol. 1 (Gradevinska Knjiga Publishing House, Belgrade, 1966) p. 50.
30. J. Dutton, F. Llewellyn-Jones, W. D. Rees and E. M. Williams, Phil. Trans. Roy. Soc. (London) **A-259** (1966) 339.
31. R. J. Munson and K. Hoselitz, Proc. Roy. Soc. London **A-172** (1939) 43.
32. R. N. Varney, J. Chem. Phys. **31** (1959) 1314; **33** (1960) 1709;

R. N. VARNEY, Phys. Rev. **174**, No. 1 (1968) 165.
33. J. L. PACK and A. V. PHELPS, Bull. Amer. Phys. Soc. **7** (1962) 636.
34. R. E. VOSHALL, J. L. PACK and A. V. PHELPS, J. Chem. Phys. **43** (1965) 1990; see also D. C. CONWAY and L. E. NESBITT, J. Chem. Phys. **48** (1968) 509.
35. E. W. MCDANIEL, D. W. MARTIN and W. S. BARNES, Rev. Sci. Instr. **33** (1962) 2; D. W. MARTIN, W. S. BARNES, G. E. KELLER, D. S. HARMER and E. W. MCDANIEL, Proceedings of the Sixth International Conference on Ionization Phenomena in Gases, Paris, 1963, Vol. 1 (SERMA, Paris, 1963) p. 295;
    G. E. KELLER, D. W. MARTIN and E. W. MCDANIEL, Phys. Rev. **140** (1965) 1535 A.
36. K. B. MCAFEE and D. EDELSON, Proceedings of the Sixth International Conference on Ionization Phenomena in Gases, Paris, 1963, Vol. 1 (SERMA, Paris, 1963) p. 299; D. EDELSON and K. B. MCAFEE, Rev. Sci. Instr. **35** (1964) 187;
    K. B. MCAFEE, D. SIPLER and D. EDELSON, Phys. Rev. **160**, No. 1 (1967) 130;
    L. G. MCKNIGHT, K. B. MCAFEE and D. P. SIPLER, Phys. Rev. **164**, No. 1 (1967) 62.
37. M. SAPOROSCHENKO, Phys. Rev. **139** (1965) A-349, A-352.
38. J. M. MADSON, H. J. OSKAM and L. M. CHANIN, Phys. Rev. Letters **15** (1965) 1018; G. F. SAUTER, R. A. GERBER and H. J. OSKAM, Rev. Sci. Instr. **37** (1966) 572.
39. J. L. MORUZZI and A. V. PHELPS, J. Chem. Phys. **45** (1966) 4617.
40. Y. KANEKO, L. R. MEGILL and J. B. HASTED, J. Chem. Phys. **45** (1966) 3741; D. K. BOHME, M. M. NAKSHBANDI, P. P. ONG and J. B. HASTED, Eighth International Conference on Phenomena in Ionized Gases, Vienna, 1967;
    D. K. BOHME, P. P. ONG, J. B. HASTED and L. R. MEGILL, Planet. Space Sci. **15** (1967) 1777.
41. D. E. GOLDEN, G. SINNOTT and R. N. VARNEY, Phys. Rev. Letters **20** (1968) 239; G. SINNOTT, D. E. GOLDEN and R. N. VARNEY, Phys. Rev. **170**, No. 1 (1968) 272;
    R. N. VARNEY, Phys. Rev. **174**, No. 1 (1968) 165.
42. M. A. BIONDI, to be published.
43. M. T. ELFORD and R. W. CROMPTON, to be published.
44. D. L. ALBRITTON, D. W. MARTIN, E. W. MCDANIEL, T. M. MILLER and J. T. MOSELEY, Georgia Institute of Technology Technical Report, May 10, 1967;
    T. M. MILLER, D. W. MARTIN, E. W. MCDANIEL, J. T. MOSELEY and R. M. SNUGGS, Georgia Institute of Technology Technical Report (February, 1968);
    D. L. ALBRITTON, T. M. MILLER, D. W. MARTIN and E. W. MCDANIEL, Phys. Rev. **171**, No. 1 (1968) 94;
    T. M. MILLER, J. T. MOSELEY, D. W. MARTIN and E. W. MCDANIEL, Phys. Rev. **173**, No. 1 (1968) 115.
45. R. W. CROMPTON, M. T. ELFORD and J. GASCOIGNE, Australian J. Phys. **18** (1965) 409.
46. J. T. MOSELEY, D. W. MARTIN, E. W. MCDANIEL, R. M. SNUGGS and T. M. MILLER, Georgia Institute of Technology Technical Report (August, 1968);
    J. T. MOSELEY, R. M. SNUGGS, D. W. MARTIN and E. W. MCDANIEL, Phys. Rev. Letters **21** (1968) 873;
    J. T. MOSELEY, I. R. GATLAND, D. W. MARTIN and E.W. MCDANIEL, Phys. Rev. **178**, No. 1 (1969) 234;
    J. T. MOSELEY, R. M. SNUGGS, D. W. MARTIN and E. W. MCDANIEL, Phys. Rev. **178**, No. 1 (1969) 240.
47. E. C. BEATY and P. L. PATTERSON, Phys. Rev. **137** (1965) A346.
48. D. EDELSON, J. A. MORRISON, L. G. MCKNIGHT and D. P. SIPLER, Phys. Rev. **164**, No. 1 (1967) 71.
49. W. S. BARNES, Phys. of Fluids **10** (1967) 1941.
50. A. GRAY, G. B. MATHEWS and T. M. MACROBERT, Bessel Functions, 2nd ed. (MacMillan and Company, Ltd., London, 1952) pp. 20, 46.

CHAPTER 2

# PHOTOIONIZATION AND IONIZATION OF THE ALKALI METALS

BY

## M. R. C. McDOWELL[*]

*NASA-Goddard Space Flight Center,
Greenbelt, Maryland*

---

[*] NRC-NASA Senior Resident Research Associate, on leave from the University of Durham, England 1967–68. Now at Royal Holloway College (University of London), Englefield Green, Surrey, England.

## Contents

| | Page |
|---|---|
| 2-1. Introduction | 49 |
| 2-2. Photoionization | 50 |
|     A. Experimental studies of photoionization | 50 |
|     B. Theoretical studies of photoionization: independent particle models | 55 |
|     C. Correlation effects in photoionization | 63 |
| 2-3. Electron impact ionization | 72 |
|     A. Experimental studies | 72 |
|     B. Electron impact ionization: theory | 78 |
| 2-4. Conclusions | 94 |
| Acknowledgement | 95 |
| References | 96 |

## § 2-1. Introduction

The purpose of this article is to review the available experimental and theoretical information on ionization of the alkali metals (Li, Na, K, Rb, Cs) by photons and by atomic projectiles, with emphasis on attempting to resolve the present discrepancies.

The experimental aspect of photoionization of the alkalis has been briefly reviewed in a recent article by Samson [1]. The principal experiments up to 1965 were those of Ditchburn and his colleagues [2–5] on Li and Na. These have now been superseded by the work of Hudson and Carter [6, 7]. Some work on the more difficult cases of K and Cs has been performed [8, 9]. Early experiments depended on the use of unreliable data on the vapour pressures of the alkalis, and we shall take the Hudson and Carter results as establishing the experimental cross sections, within the quoted errors to be discussed further below, for Li, Na, K. There are no reliable, published, measurements for the higher members of the sequence.*

There has been much less experimental work on ionization by projectile impact. The most important set of measurements are the electron impact data of Brink and McFarland [10–14]. Some measurements of near threshold ionization have been reported by a Russian group [15, 67], and there have been two additional experiments on Cs [16, 17].

Theoretical studies of photoionization of the alkalis have again been mainly restricted to Li and Na, with some work on K. Theoretical results for Li [18] and Na [19] are in good agreement with experiment at ejected electron energies from zero to 6 eV. Work up to 1965 has been reviewed by Stewart [62].

Calculations on projectile ionization (e.g. electrons and protons) have been carried out in the first Born approximation (with, in the case of electron impact, some allowance for exchange) both for Li [20–24] and Na [22, 25, 26] and are extended in this paper. We shall argue that the first Born approximation should give reliable results at energies in excess of forty times threshold (approx. 200 eV for the alkalis). The theoretical results (in models which yield accurate photoionization cross sections) lie approximately a factor of two below the reported experimental results, both for Li and Na, in the range of validity of the Born approximation.

In section 2-2 we discuss the experiments and theoretical calculations on photoionization, while the experimental data and published theoretical

---

* See, however, a recent paper in Proc. Roy. Soc. A **304** (1968) 233 by G. V. Marr and D. M. Creek, on photoionization of K, Rb and Cs.

material on electron impact ionization is reviewed in §2-3, and present our conclusions in §2-4.

## § 2-2. Photoionization

### A. Experimental Studies of Photoionization

The problems involved in an accurate measurement of the continuous absorption of light by a gas or a metal vapour are discussed in detail elsewhere [1, 6], and may be illustrated by reference to the work of Hudson and Carter [6, 7]. Their apparatus is shown schematically in Fig. 2-2-1. A grating monochromator with an inverse dispersion of 7.5 Å/mm was employed to prepare a photon beam with a bandwidth of 0.75 Å. The beam was split by a half-coated mirror, one part acting as an intensity monitor, and the other passing through the absorbing gas, both parts then passing through windows into photomultipliers. Both windows and splitting mirror were kept at a constant temperature, and hence constant reflectivity and transmissivity, by water cooling. The matched phototubes were insensitive to infra-red and visible background radiation ($\lambda > 3200$ Å). Then if $I_0$ is the intensity of the photon beam when no vapour is present, its intensity when vapour is admitted is given by

$$I_v = I_0 \, e^{-a_v n Z} \tag{2-2-1}$$

where $a_v$ is the required photoabsorption cross section at frequency $v$, $n$ the number density of Li atoms in the chamber and $Z$ the path length in cm.

The photon beam is split, one part being monitored and the other passing through the furnace, which is held at a temperature $T$, in the range

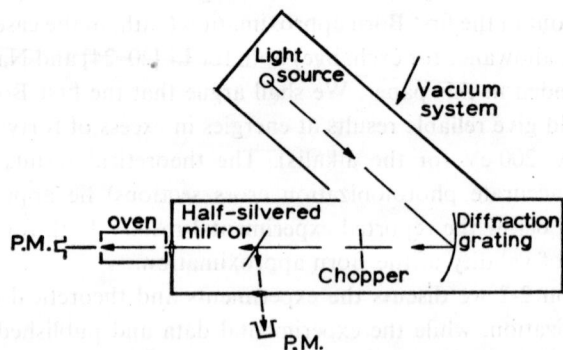

Fig. 2-2-1. Schematic representation of Hudson and Carter's photoabsorption apparatus. (P.M. indicates a photomultiplier.)

$450 \leqslant T \leqslant 1250\,°K$. The intensity $I_0$ is taken to be that when the furnace is at $450\,°K$, the vapour pressure of Li then being less than $10^{-5}$ mm Hg. Data were taken at temperatures above $630\,°K$ where the vapour pressure of Li exceeds $10^{-3}$ mm Hg (Table 2-2-1). In this temperature range the partial pressure of $Li_2$ is less than $5 \times 10^{-5}$ mm Hg (Table 2-2-2) and since both vary with temperature runs at six different temperatures at each wave length allowed a determination of both atomic and molecular absorption cross sections, $a_v$ and $\sigma_v$. Then

$$\ln(I_0/I_v) = Z(n_a a_v + n_m \sigma_v) \qquad (2\text{-}2\text{-}2)$$

where $n_a$, $n_m$ are the atomic and molecular concentrations respectively, which are obtained from the adopted vapour pressure data by use of the perfect gas law*.

The molecular absorption cross section of $Li_2$ was $4.8 \pm 0.7$ Mb in the region from 2300 Å to 2400 Å below the atomic threshold, but could not be reliably determined at shorter wavelengths, because while both $a_v$ and $\sigma_v$ were of the same order, $n_m$ was less than 1% of $n_a$.

The measured atomic absorption cross sections, which in the energy range considered is solely due to photoionization in the case of Li (though the results for Na may be affected by autoionization lines to the short wave length side of 800 Å) are given in Table 2-2-3.

TABLE 2-2-1
Vapour pressure of Li in mm Hg

| $T(°K)$ | 800 | 900 | 1000 | 1100 |
|---|---|---|---|---|
| JANAF [31] | 6.77, −3 | 8.78, −2 | 6.74, −1 | 3.547 |
| Nesmeyanov [32] | 7.39, −3 | 9.52, −2 | 7.17, −1 | 3.858 |
| Hicks [31] | 7.2, −3 | 8.97, −2 | 7.22, −1 | 3.88 |

TABLE 2-2-2
Vapour pressure of $Li_2$ in mm Hg

| $T(°K)$ | 800 | 900 | 1000 | 1100 |
|---|---|---|---|---|
| JANAF [31] | 5.115, −5 | 1.306, −3 | 1.702, −2 | 1.357, −1 |
| Nesmeyanov [32] | 6.056, −5 | 1.516, −3 | 2.043, −2 | 1.623, −1 |

* In the presence of a buffer gas, whose partial pressure is large compared with that of the alkali vapour considered.

### TABLE 2-2-3

Measured atomic absorption cross sections for Na and Li [7] (in $Mb = 10^{-18}$ cm$^2$) at wave length $\lambda$, $E_{h\nu}$ being the incident photon energy in eV. The Na threshold value is 0.130 Mb at $\lambda = 2412$ Å

| $\lambda$(Å) | $E_{h\nu}$(eV) | $a_\nu$(Na) | $a_\nu$(Li) |
|---|---|---|---|
| 2400 | 5.167 | 0.126 | – |
| 2350 | 5.277 | 0.110 | – |
| 2300 | 5.391 | 0.092 | 1.54 |
| 2200 | 5.636 | 0.045 | 1.62 |
| 2100 | 5.905 | 0.008 | 1.69 |
| 2000 | 6.200 | 0.000 | 1.75 |
| 1800 | 6.888 | 0.006 | 1.84 |
| 1600 | 7.749 | 0.072 | 1.84 |
| 1400 | 8.816 | 0.122 | 1.69 |
| 1200 | 10.331 | 0.166 | 1.51 |
| 1000 | 12.398 | 0.192 | 1.33 |
| 800 | 15.497 | 0.207 | 1.14 |
| 600 | 20.663 | 0.230 | 0.84 |

The principal sources of error are in the vapour pressure data for the alkalis, photoabsorption due to impurities and photon detector efficiencies. The vapour pressures are uncertain by $\pm 5\%$ (Tables 2-2-1, 2-2-2): errors in temperature calibration and path length measurement introduce additional systematic error of $\pm 5\%$*. Modulation of the photon beam at 200 cps and monitoring the unabsorbed portion should reduce signal intensity errors to $\pm 1\%$. At the high temperature of the furnace the metal vapour reacts with outgassed water vapour to produce $H_2$ with a high absorption coefficient below 800 Å. Previous outgassing and cold-trapping of the furnace tube should, the authors claim [7], eliminate this problem. Nevertheless the total quoted systematic error of $\pm 10\%$ should be treated with caution to the short wave length side of 800 Å. The random error associated with the measured data points was $\pm 10\%$, giving an overall error determination of $\pm 20\%$ ($\lambda > 800$ Å).

The situation of shorter wave lengths is more uncertain. Contributions from $H_2$ absorption may, in spite of experimental precautions, be significant, and the molecular absorption cross section may be as much as an order of magnitude larger than the atomic in this region. (In particular the quoted

---

* Dr. R. A. McFarland has pointed out that convection within the oven (height 2″) may produce temperature gradients within it of more than 25 °K, and consequently additional uncertainties of perhaps 10% in the actual vapour pressure present. To some extent this is already averaged out by taking averages of runs at six different temperatures for each $\nu$.

atomic absorption cross section for Na is anomalously large to the short wave length side of 800 Å.) It would be rash to suppose the quoted results to have overall uncertainties of less than ±30% in this range.

The earlier experiments of Tunsted [4] and Marr [5] are less satisfactory. Marr's work is essentially an extension and replacement of Tunsted's, extending it to 1550 Å. Marr used vapour pressure data of Honig [33], which is 20% higher than the values adopted here, hence his quoted cross section should be increased by 20% to a threshold value (Li) of $2.0 \pm 0.4$ Mb, with a maximum of $2.3 \pm 0.5$ Mb at $\lambda = 1900$ Å. The two experimental curves are shown (from threshold to 1500 Å) in Fig. 2-2-2 with error bars. They are clearly compatible, and indicate that (giving a weighting of 3:2 to Hudson and Carter's work compared to Marr's) an acceptable curve rising from a threshold value of 1.72 Mb to a maximum of 2.02 Mb at 1850 Å and then falling sharply to shorter wave lengths may be postulated.

Hudson and Carter [6] have also measured the atomic and molecular absorption by potassium. The molecular absorption falls steeply from a

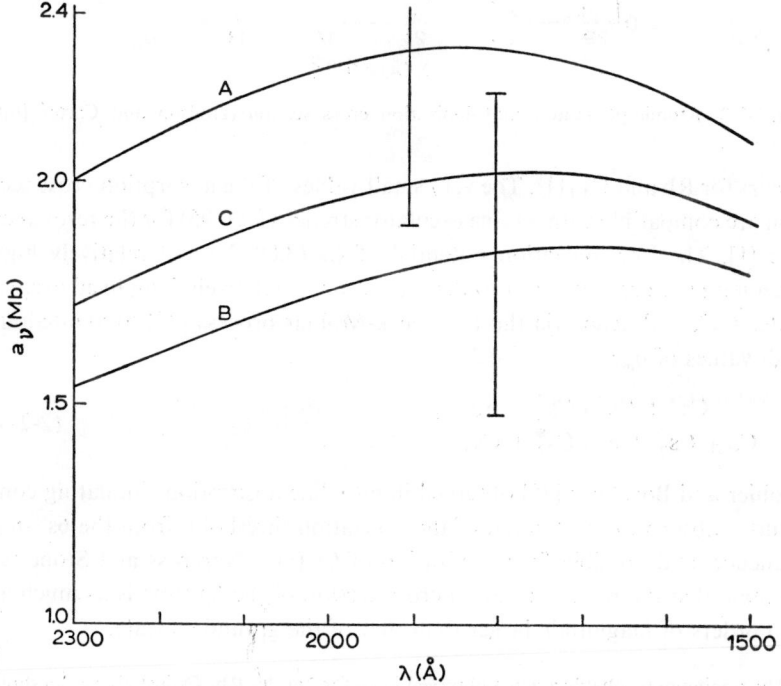

Fig. 2-2-2. Experimental values for photoionization of Li from threshold to 1500 Å in Mb. (A) Marr [5]; (B) Hudson and Carter [7]; (C) Adopted mean values (see text).

threshold value of $1.5 \pm 0.2$ Mb at 3000 Å to a negligible value at 2300 Å, but was not determined at shorter wave lengths. The values for atomic potassium are shown in Fig. 2-2-3. The threshold value is less than 0.01 Mb and the cross section decreases to a very small value with a minimum at 2650 Å, becoming quite large to shorter wave lengths. A similar behaviour

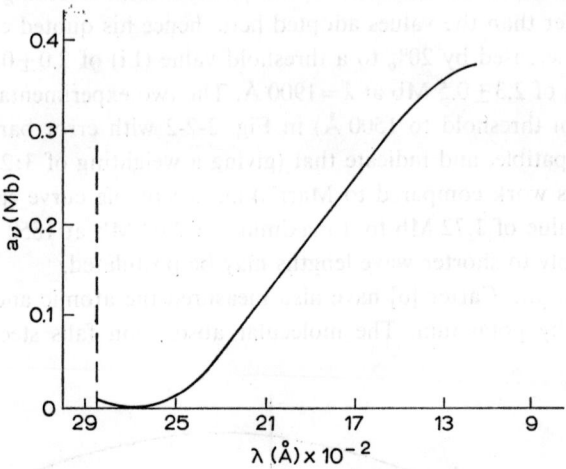

Fig. 2-2-3. Atomic potassium photoionization cross section (Hudson and Carter [6]).

occurs for Rb and Cs [1]*. The very small values of the absorption cross section are compatible with a large oscillator strength ($\approx 0.95$) for the resonance line [1]. The low ionization potential of Cs (3.89 eV) and relatively high oven temperatures may lead to the formation of relatively long lived excited states Cs*, and hence via the Hornbeck-Molnar process [34] to anomalous high values of $n_m$,

$$\begin{aligned} \text{Cs}^* + \text{Cs} &\to \text{Cs}_2^+ + e, \\ \text{Cs} + \text{Cs}_2^+ + e &\to \text{Cs}_2^* + \text{Cs}. \end{aligned} \qquad (2\text{-}2\text{-}3)$$

Mohler and Boeckner [35] observed intense line absorption simulating continuous absorption to the red of the ionization threshold, from the $6s \to mp$ sequence, and possibly from ionization of Cs (6p). Norcross and Stone [9] estimate that the photoionization cross section of the 6p state is as much as two orders of magnitude larger than that of the ground 6s state.

* The minimum (probably a zero) observed in $a_v$ for Na, K, Rb, Cs just above threshold for photoionization, occurs for Li in the oscillator strength distribution below threshold, between $n=2$ and $n=3$ (cf. Cooper and Fano [36] section 4.5).

## B. Theoretical Studies of Photoionization: Independent Particle Models

The general theoretical formulation of atomic photoionization cross sections has been reviewed recently by Cooper and Fano [36]. Detailed studies of Li and Na have been undertaken by Chang and McDowell [18] and by Manson and Cooper [19] respectively. Some preliminary calculations have been carried out for potassium [27, 38] and rubidium [27, 38, 39].

For an $N$-electron atom, the photoionization cross section at frequency $v$, is given by [40]

$$a_v = \frac{2\pi e^2 h^2}{m^2 cv} \int d\Omega_v \left| \sum_j \int d\mathbf{r}_N \Psi_k^*(N) e^{i\mathbf{k}_v \cdot \mathbf{r}_j} \mathbf{p}_j \Psi_0(N) \right|^2 \qquad (2\text{-}2\text{-}4)$$

where $e$, $h$, $m$ and $c$ have their usual meanings, $\Psi_0(N)$ and $\Psi_k(N)$ are the initial and final states of the atomic system respectively, $\mathbf{k}_v$ is the photon momentum, $\mathbf{p}_j$ and $\mathbf{r}_j$ the momentum and position (with respect to the target nucleus) of the $j^{\text{th}}$ electron, and $d\Omega_v$ indicates an integration over all directions of ejection. If more than one final state is possible, a suitably weighted sum over such states must be introduced. In the visible region of the electromagnetic spectrum $\mathbf{k}_v \cdot \mathbf{r}$ is of order $10^{-3}$ atomic units, and the exponential may be replaced by unity. Taking the values of the various atomic constants from Bethe and Salpeter [40] (2-2-4) may be expressed as

$$a_v = 8.56 \times 10^{-19} \frac{1}{\omega(I+k^2)} S_{if} \text{ cm}^2 \qquad (2\text{-}2\text{-}5)$$

where, expressing energies in Rydbergs (1 Ry = 13.6 eV)

$$hv = I + k^2$$

and $I$ is the ionization potential of the target, $k^2$ the energy of the ejected electron. Here $\omega$ is the statistical weight of the initial atomic state, and the crucial quantity of interest is

$$S_{if} = 4 \sum_f |\Psi_k^*(N) \nabla_N \Psi_0(N) d\mathbf{r}_N|^2 \qquad (2\text{-}2\text{-}6)$$

the sum being over final states, and

$$\nabla_N = \sum_{j=1}^{N} \nabla_j \qquad (2\text{-}2\text{-}7)$$

$$\mathbf{p}_j = i\nabla_j.$$

For the present we may assume the final state to be unique; and note that for alkali atoms in their ground $^2S_{\frac{1}{2}}$ state, $\omega = 1$,

$$S_{if} = 4|\sigma_V|^2 \qquad (2\text{-}2\text{-}8)$$

where

$$\sigma_V = \int \Psi_k^*(N)\,\nabla_N\Psi_0(N)\,d\mathbf{r}_N \qquad (2\text{-}2\text{-}9)$$

is the $N$-electron "velocity" form of the photoionization matrix element. An alternative expression may be developed using the commutation relation [37]

$$[H, \mathbf{r}_N] = -\nabla_N, \quad \mathbf{r}_N = \sum_{j=1}^{N} \mathbf{r}_j \qquad (2\text{-}2\text{-}10)$$

and the exact Schrödinger equation of the atom, to obtain

$$S_{if} = (I + k^2)^2\,|\sigma_L|^2 \qquad (2\text{-}2\text{-}11)$$

with

$$\sigma_L = \int \Psi_k^*(N)\,\mathbf{r}_N\Psi_0(N)\,d\mathbf{r}_N \qquad (2\text{-}2\text{-}12)$$

being the "length" form of the matrix element. If the many-body atomic wave functions $\Psi_0(N)$, $\Psi_k(N)$ were known exactly, then the evaluation of (2-2-8) and (2-2-11) would yield identical results.

The remaining theoretical problem is to find sufficiently good approximations to the many-body wave functions. Zero order calculations [38] assume that the electrons are separable (i.e. correlation plays no role) and represent $\Psi_0$, $\Psi_k$ by a linear combination of Slater determinants composed of one electron spin-orbitals. For the alkalis (provided we consider the core as inert) a single such determinant suffices for the ground state,

$$\Psi_0(N) = \frac{1}{\sqrt{N!}} |[\phi_{1s}(\alpha)\,\phi_{1s}(\beta)\ldots]\,\phi_{\text{valence}}(\alpha)| \qquad (2\text{-}2\text{-}13)$$

where only the diagonal of the determinant is written, and the square bracket $[\phi_{1s}(\alpha)\,\phi_{1s}(\beta)\ldots]$ indicates the core electron spin-orbitals, $\alpha$ and $\beta$ denoting the possible spin directions of each electron, and $\phi_{\text{valence}}(\alpha)$ is the spin orbital for the outer valence electron. For Li this is

$$\phi_{2s}(\alpha) = \frac{1}{r} P_{2s}(r)\,Y_{00}(\hat{r})\,\alpha \qquad (2\text{-}2\text{-}14)$$

and for Na

$$\phi_{3s}(\alpha) = \frac{1}{r} P_{3s}(r) Y_{00}(\hat{r}) \alpha \qquad (2\text{-}2\text{-}15)$$

where $Y_{00}(\hat{r}) = 1/\sqrt{4\pi}$ and $P_{ns}(r)$ is a radial wave function, normalized such that

$$\int_0^\infty P_{ns}^2(r)\, dr = 1.$$

If we consider only photoionization of the valence electron from its ground $ns$ orbital, then the only optically allowed transition is to the $kp$ state of the continuum

$$\Psi_k(N) = \frac{1}{\sqrt{N!}} \left| [\phi_{1s}(\alpha)\, \phi_{1s}(\beta) \ldots ]\, \phi_{kp}(\alpha) \right| \qquad (2\text{-}2\text{-}16)$$

and the normalization is now such that

$$\phi_{kp}(\alpha) = \frac{1}{r} P_{kp}(r) Y_{1m}(\hat{r}) \alpha$$

$$P_{kp}(0) = 0, \quad P_{kp}(r) \sim k^{-\frac{1}{2}} \sin[\chi(r) + \delta_1(k)] \qquad (2\text{-}2\text{-}17)$$

$$\int_0^\infty P_{kp}(r) P_{k'p}(r)\, dr = \delta(k^2 - k'^2)$$

where $\chi(r)$ is slowly varying and $\delta_1(k)$ is the p-wave phase shift for scattering of an electron of angular momentum $l=1$ and energy $k^2$ Ry by the positive ion core.

Elementary treatments [38, 41] assume that $P_{ns}(r)$ may be parametrized in some simple way, for example by hydrogen-like wave functions with Slater screening parameters [42]. The continuum function $P_{kp}(r)$ is usually replaced by an undistorted regular Coulomb wave function, equivalent to assuming the core to be a point charge. The values obtained, particularly for the alkalis, are sensitive to the choice of screening parameters. A careful calculation by Omidvar [41] for Li($2s\ ^2S_{\frac{1}{2}}$) is compared with the experimental result of Hudson and Carter [7] in Fig. 2-2-4. The elementary theoretical prediction is almost a factor of four low at threshold and shows much too large a maximum cross section, but provides a reasonable order of magnitude estimate.

A best-possible result in the separable-electron fixed core model may

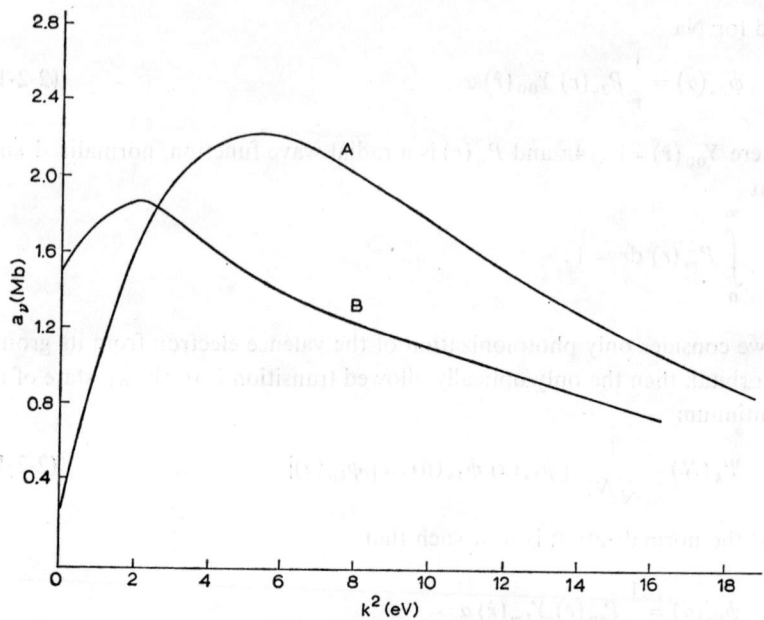

Fig. 2-2-4. Theoretical and experimental values of $a_v$ for Li. (A) Omidvar [42], screened hydrogenic functions; (B) Experiment, Hudson and Carter [7].

be obtained by using Hartree-Fock wave functions for the electrons. For Li and Na the overlap between the ion and atomic core orbitals is essentially unity, and $S_{if}$ may be reduced to the forms

$$S_{if}^{(0)} = (I + k^2)^2 |\sigma_{0,L}|^2 \tag{2-2-18a}$$

or

$$S_{if}^{(0)} = 4 |\sigma_{0,V}|^2 \tag{2-2-18b}$$

where

$$\sigma_{0,L} = \int_0^\infty P_{kp}(r) \, r P_{ns}(r) \, dr \tag{2-2-19}$$

and

$$\sigma_{0,V} = \int_0^\infty P_{kp}(r) \left[ \frac{1}{r} - \frac{d}{dr} \right] P_{ns}(r) \, dr \tag{2-2-20}$$

are the length and velocity forms of the one-electron matrix element.

The Hartree-Fock equation for the valence electron, assuming an undistorted core, may be written

$$[h(i) - \varepsilon_{nl}] \phi_{nlm_l}(\mathbf{r}_i) = 0 \tag{2-2-21}$$

where

$$h(i) = -\nabla_i^2 - \frac{2Z}{r_i} + 2\sum_q \langle \phi_q^{(c)}(j)| \frac{1}{r_{ij}} (2 - \mathscr{P}_{ij}) |\phi_q^{(c)}(j)\rangle \tag{2-2-22}$$

$$\mathscr{P}_{ij} f(\mathbf{r}_i, \mathbf{r}_j) = f(\mathbf{r}_j, \mathbf{r}_i) \tag{2-2-23}$$

$$\phi_{nlm_l}(\mathbf{r}_i) = \frac{1}{r} P_{nl}(r) Y_{lm_l}(\hat{r}) \tag{2-2-24}$$

and $\phi_q^{(c)}(j)$ is a core-orbital.

Calculations for Li, in this approximation were first carried out by Stewart [30], who approximated $P_{2s}(r)$ by the tabulated function of Fock and Petrachen [43] for $r < 5a_0$ and a fit to a Bates-Damgaard Coulomb function for $r > 5a_0$. Such an approximation should be less severe for the velocity form (2-2-20) of the matrix element. Sewell [28] repeated the calculations and found substantial disagreement with Stewart, but his length result was in very close accord with experiment. Chang and McDowell [18] again repeating the calculation found their velocity result was identical to Stewart's, but their length result was somewhat larger, due to the rather slower fall off of their 2s wave function with increasing $r$ than Stewart's approximation to it. They proved that $\sigma_{0,L}$ and $\sigma_{0,V}$ were not independent quantities, but that their difference was given by certain two-electron integrals. Chang and McDowell evaluated these and showed that their results satisfied the theorem to 1 part in $10^4$. Sewell's result is probably in error, though his continuum radial function (private communication; 1968) is almost identical to that of Chang and McDowell and both used Roothan analytic Hartree-Fock functions for the ground state. However Sewell's values of $\sigma_{0,L}; \sigma_{0,V}$ violate the Chang-McDowell theorem.

Chang and McDowell obtain a second theorem giving an acceleration form of the matrix element in terms of the velocity form. They show that the differences are now large and that

$$|\sigma_{0,A}| - |\sigma_{0,V}|^2 \approx Z^2 |\sigma_{0,L}|^2$$

so that the acceleration form is always poor (within the context of a Hartree-Fock calculation) for $Z > 1$. These theorems emphasize the fact that agreement between the results obtained in calculations using alternative forms of the matrix element is a necessary, but not a sufficient, condition that such a result is close to the exact result.

The various authors Hartree-Fock results for Li are shown in Fig. 2-2-5. The most reliable is the Stewart-Chang-McDowell velocity result (the corresponding length result being a consequence of the Chang-McDowell theorem), and while in contrast to the elementary result [41] it has the general shape of the experimental curve, it lies about 25%~30% too low.

A similar Hartree-Fock calculation for photoionization of Na was per-

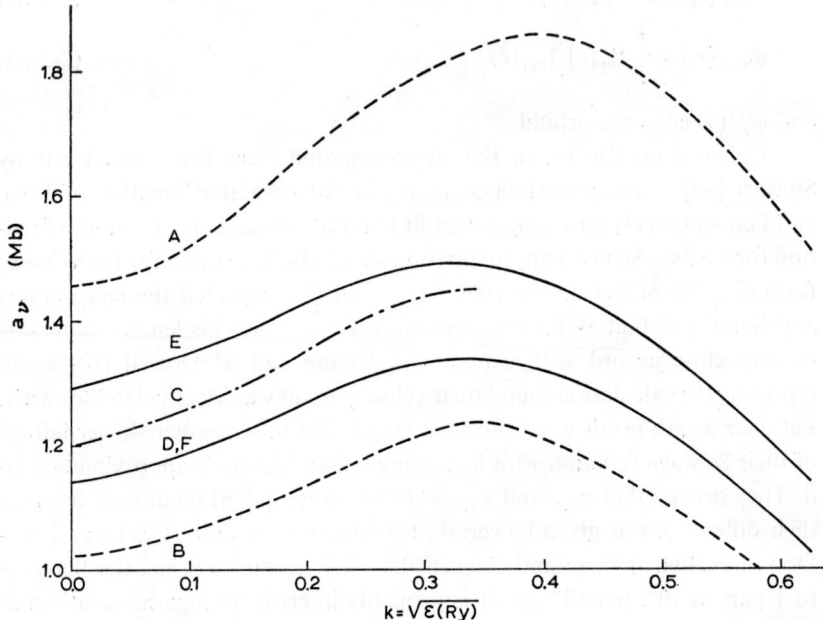

Fig. 2-2-5. Hartree-Fock calculations of $a_\nu$ for Li. (A), (B) Sewell [28], A = length, B = velocity; (C), (D) Stewart [30], C = length, D = velocity; (E), (F) Chang and McDowell [18], E = length, F = velocity = D.

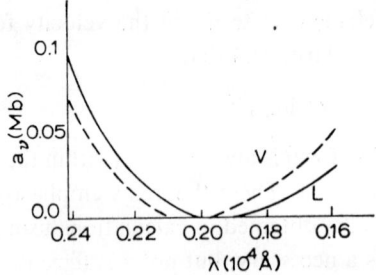

Fig. 2-2-6. Hartree-Fock calculation of $a_\nu$ for Na (Seaton [45]).

formed by Seaton [39] on a hand-calculating machine (Fig. 2-2-6). Cancellation was a severe problem. If $D$ is the ratio of the value of the radial integral to the absolute value of the larger of its positive or negative parts, then at threshold Seaton finds $D_L = 0.082$ and $D_v = 0.111$ (whereas for Li, $D_v = 0.375$): the corresponding values to the blue of the spectral head are even smaller. His calculated results again showed the same general behavior as the experimental curve (see Fig. 2-2-8) from the threshold to just beyond the minimum, but fail badly to the short wave length side of 1600 Å.

Several less general one-electron calculations exist in the literature. Cooper and Manson [44] and Boyd [26] simplify the Hartree-Fock equations, replacing the exchange terms by an effective local potential. Boyd's results for Na are in somewhat better agreement with experiment than the more detailed Hartree-Fock calculations of Seaton [39], but since she gives different values for the length and velocity formulations, and these must be identical in any central field calculation, some doubt attaches to the reliability of her numerical work. We adopt her length values, and they are shown in Fig. 2-2-8 below, as the curve marked (B).

More detailed notice must be taken of the very general quantum defect method (Q.D.M.) developed by Seaton [45]. In principle this uses a Bates-Damgaard Coulomb function for the bound electron, and replaces the continuum radial function by a linear combination of regular and irregular Coulomb functions

$$P_{kl}(r) \propto \{F_{kl}(r) + (-1)^l \tan \eta_l(k) G_{kl}(r)\} \tag{2-2-25}$$

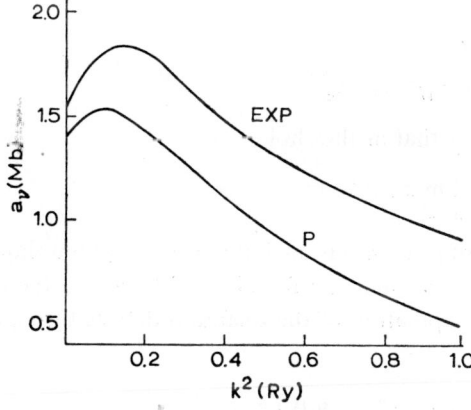

Fig. 2-2-7. (P) Quantum defect calculation of $a_v$ for Li (Peach [46]). The experimental curve shown (EXP) is that of Hudson and Carter [7].

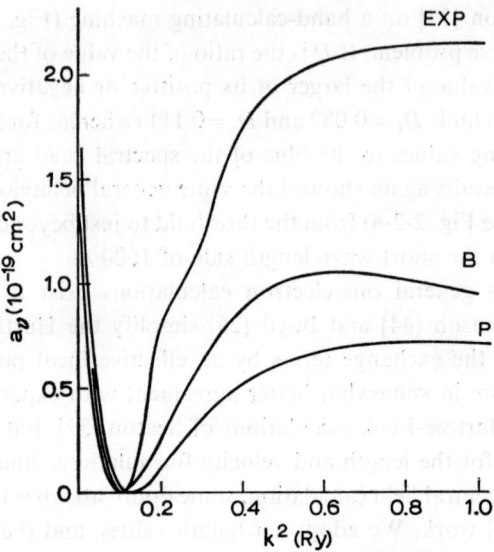

Fig. 2-2-8. Photoionization of atomic sodium. (B) Boyd [26] central field approximation; (P) Peach [46] quantum defect method. The upper curve is the experiment of Hudson and Carter [7].

where $\eta_l$ is the $l^{th}$ phase shift. This function has exactly the correct asymptotic form, and the Coulomb functions are modified to obtain a reasonable behaviour for small $r$.

If the bound state energies $E_{nl}$ of the target are known accurately (from spectroscopic data) for the relevant values of $l$ and large $n$, and are expressed as

$$E_{nl} = -\frac{Z^2}{(n-\mu_{nl})^2} = -\frac{Z^2}{v_{nl}^2}, \tag{2-2-26}$$

then Seaton shows that at threshold

$$\lim_{k \to 0} \eta_l(k^2) = \lim_{n \to \infty} \pi\mu_{nl} \tag{2-2-27}$$

and that therefore $\mu_{nl}(-k^2)$ is analytic in the neighbourhood of $k^2 = 0$. Further since $\eta_l$ is an analytic function of $k^2$ above the elastic scattering threshold, an extrapolation of the quantum defects through threshold may be made

$$\eta_l(k^2) = \eta_l(0) + bk^2, \quad \eta_l(0) = a \tag{2-2-28}$$

where $a$, $b$ are obtained from a fit to the $\mu_{nl}$.

Peach [46] has recently used this method to calculate photoionization cross sections for many systems. Her results for Li are shown (and compared with experiment) in Fig. 2-2-8. The general shape is again good, the results being slightly above the Hartree-Fock length values, but lying appreciably below experiment. These calculations essentially belong in the separable-electron class since although apparently many body effects are taken into account by using the experimental energy levels, such effects can be shown to produce only a small energy shift in the alkalis, and to enter the photoionization matrix element in a more sophisticated way (cf. section C, below). Peach's results for Na show good agreement with experiment near threshold (but about 50% too large a threshold value), a slight displacement of the minimum, and are about a factor of four lower than Hudson and Carter by ejection energies of 0.8 Ry.

A method similar to the Q.D.M. has been introduced by McGuire [27]. He solves a central field equation

$$\left\{\frac{d^2}{dr^2} + k^2 - \frac{2}{r^2} + V(r)\right\} P_{k^2 p}(r) = 0 \qquad (2\text{-}2\text{-}29)$$

with

$$V(r) = \frac{2Z_e}{r} - \Delta, \quad r \leqslant r_1$$
$$= \frac{2}{r}, \quad r > r_1 \qquad (2\text{-}2\text{-}30)$$

and $Z_e$, $r_1$ and $\Delta$ chosen to fit the $np$ bound energy level series. His result for Li lies between the Hartree-Fock length and Hartree-Fock velocity values, while for Na he can fit the shape of the experimental curve near the minimum quite well. The high energy maximum of 0.14 Mb obtained lies about 30% below experiment ($k^2 = 1.0$). McGuire extends his calculations to K, Rb and Cs, and for suitable choices of parameters can obtain values consistent with experiment at low ejection energies.

## C. Correlation effects in photoionization

The calculations discussed in §2-2-B assume the electrons are uncorrelated and the core is inert. Chang and McDowell [18] have shown that intrashell correlation is important in Li, and that when properly included, removes the discrepancy between theory and experiment. The dominant effect is a polarization of the core by the valence electron, though when the valence electron is close to the core an additional type of short range correlation enters, and is particularly important in a velocity formulation.

Chang and McDowell's approach is via Brueckner-Goldstone perturbation theory [47], in which the expansion is made in terms of a complete set of Hartree-Fock orbitals of the target. It is convenient to introduce a diagrammatic technique to describe their work. They denote the Hartree-Fock matrix element $\langle 2s| O |kp \rangle$ by the diagram of Fig. 2-2-9, where the photon operator $O$ may be either $\nabla_N$ or $\mathbf{r}_N$.

First order correlation effects enter the initial state wave function $\Psi_0(N)$ via the correlation operator $v = 1/r_{12}$ between a valence electron (1) and a core electron (2), denoted by a dashed line in Fig. 2-2-10 in which the initial $|1s\,1s\,2s\rangle$ state contains through $v$ a component $|1s\,k'p\,kp\rangle$ or $|1s\,np\,kp\rangle$, i.e. a virtual doubly excited state which is deexcited to the final $|1s\,1s\,kp\rangle$ state by the photon. Similar effects arise in the final $|1s\,1s\,kp\rangle$ state in which the photon first excites the core, leaving the system in the virtual $|1s\,k'p\,2s\rangle$ or $|1s\,np\,2s\rangle$ states, which are then deexcited by the correlation $v$.

Mathematically, the contribution of the diagram of Fig. 2-2-10 to the

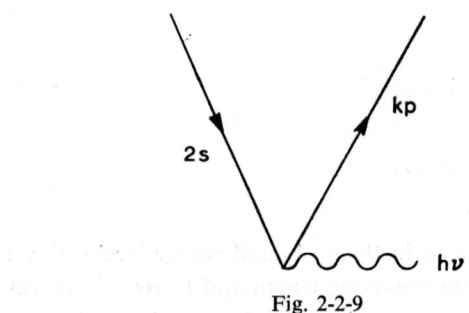

Fig. 2-2-9

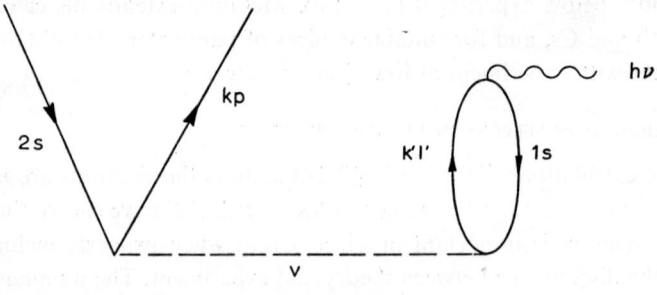

Fig. 2-2-10

photoionization matrix element may be represented as $S_1$,

$$S_1 = 2 \times \frac{4}{\pi} \int_0^\infty k_s \, dk_s \langle 1s| O |k_s p\rangle \frac{\langle k_s p, kp| v |1s\, 2s\rangle}{\varepsilon_{1s} + \varepsilon_{2s} - k_s^2 - k_p^2} \quad (2\text{-}2\text{-}31)$$

where the integration is over a continuum $|k_s p\rangle$ and includes a sum over bound intermediate states $|np\rangle$ (of which $n=2,\ldots,8$ were included explicitly, and an $n^{-3}$ rule used for the remainder). The continuum integration may be represented as $(2/\pi)\int_0^\infty k_s\, dk_s$, and additional factors of two arise from (1) expressing the energies occurring in the denominator in Rydbergs rather than a.u., (2) summing over the two 1s electrons. (The corresponding exchange contribution does not have this last factor, as only parallel spins contribute, and takes a negative sign.) Rules for obtaining the diagram energy denominator follow from standard perturbation theory, for if

$$E_0 = 2\varepsilon_{1s} + \varepsilon_{2s}$$

and

$$E_{k_s} = \varepsilon_{1s} + k_s^2 + k_p^2$$

then

$$E_0 - E_{k_s} = \varepsilon_{1s} + \varepsilon_{2s} - k_s^2 - k_p^2 \quad (2\text{-}2\text{-}32)$$

and this is dominated by $\varepsilon_{1s}(\approx 5.2 \text{ Ry})$.

The effect of such correlation contributions arises almost entirely from values of $k_s$ greater than 1.5, and hence corresponds to small values of $r_{12}$. The effect is almost twice as large in the final state as in the initial state, increasing the contribution to the photoionization amplitude from regions close to the nucleus by almost 10% in the velocity formulation, but is fairly unimportant in the more distant regions which give the main contribution to the length matrix element. Consequently the length results are scarcely affected by the inclusion of this type of correlation, but the velocity results are everywhere increased, and are now in good agreement with the length values. However these values remain inconsistent with experiment.

Long range correlation effects arise in second order and correspond to the virtual excitation of configurations $|1s\, 1s\, k',(l\pm1)\rangle_{l=1}$ in the initial state, and $|1s\, k'p\, 1s\rangle$ in the final state. A typical second order diagram is shown in Fig. 2-2-11.

Reading time upwards from the vertex, this diagram shows polarization of the initial $|1s\, 1s\, 2s\rangle$ state, which goes through the intermediate configuration $|1s\, k''\, l''\, k'\, l'\rangle$ to a configuration $|1s\, 1s'\, k',(l\pm1)\rangle_{l=1}$ also marked by an arrow in the figure, which is then photoionized with the ejection of a $kp$

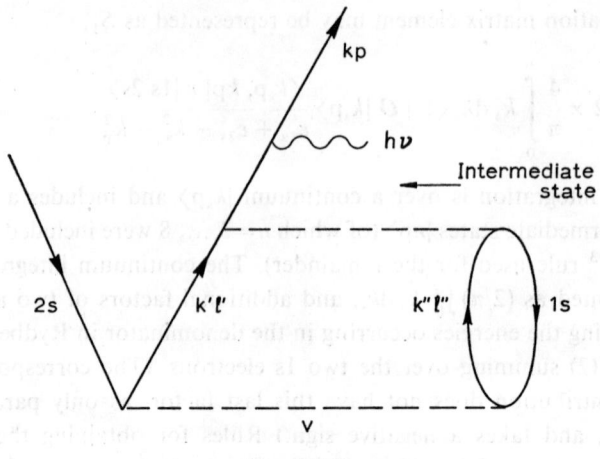

Fig. 2-2-11

electron. The sub-diagram shown in Fig. 2-2-12 represents a polarization of the $1s^2$ core by the $kp$ electron, and contributes 90% of the dipole polarizability. The effects are small in the initial state, but appreciable in the final state.

The contribution of the diagram of Fig. 2-2-11 to the matrix element is, say

$$A = 4\left(\frac{2}{\pi}\right)^2 \int k_s \, dk_s \langle kp| \, O \, |k_s s\rangle \int k_1 \, dk_1 \int k_2 \, dk_2$$
$$\times \frac{\langle 1s, k_s s| \, v \, |k_1 l_1, k_2 l_2\rangle \langle k_1 l_1, k_2 l_2| \, v \, |1s \, 2s\rangle}{(\varepsilon_{1s} + \varepsilon_{2s} - k_1^2 - k_2^2)(\varepsilon_{2s} - k_s^2)} \quad (2\text{-}2\text{-}33)$$

which may be reduced by use of the closure theorem to

$$A = 4\left(\frac{2}{\pi}\right)^2 \int k_s \, dk_s \langle kp| \, O \, |k_s s\rangle \int k_t \, dk_t$$
$$\times \frac{\langle k_s s| \mathscr{I}_t^2 \, |2s\rangle}{(\varepsilon_{1s} + \varepsilon_{2s} - k_t^2 - k_s^2)(\varepsilon_{2s} - k_s^2)} \quad (2\text{-}2\text{-}34)$$

with

$$\mathscr{I}_t(1) = \langle 1s(2)| \frac{1}{r_{12}} |k_t(2)\rangle. \quad (2\text{-}2\text{-}35)$$

Consider the Schrödinger equation

$$[h(i) + \mathscr{V}(r) + \bar{\varepsilon}_{2s}] \chi_{2s}(r) = 0 \quad (2\text{-}2\text{-}36)$$

where

$$\mathscr{V}(r) = \frac{4}{\pi} \int k_t \, dk_t \, \frac{\mathscr{I}_t^2}{(\varepsilon_{1s} + \varepsilon_{2s} - k_t^2 - k_s^2)}. \tag{2-2-37}$$

To first order in $\mathscr{V}$ the solution of (2-2-36) is

$$\chi_{2s}(r_1) = \phi_{2s}(r_1) + \int dr_2 \, G(r_1, r_2, \bar{\varepsilon}_{2s}) \, \mathscr{V}(r_2) \, \phi_{2s}(r_2) \tag{2-2-38}$$

where the Green's function is

$$G(r_1, r_2, \bar{\varepsilon}_{2s}) = 2 \sum_t{}' \frac{|\phi_t\rangle \langle \phi_t|}{\bar{\varepsilon}_{2s} - \varepsilon_t}, \tag{2-2-39}$$

hence the contribution to the matrix element is

$$\langle kp| \, O \, |2s\rangle - \langle kp| \, O \, |\chi_{2s}\rangle$$
$$= \langle kp| \, O \, | \int dr_2 G(r_1, r_2, \bar{\varepsilon}_{2s}) \, \mathscr{V}(r_2) \, \phi_{2s}(r_2)\rangle$$
$$= \frac{4}{\pi} \int k_s \, dk_s \, \frac{\langle kp| \, O \, |k_s s\rangle}{\bar{\varepsilon}_{2s} - \varepsilon_{ks}} \langle k_s| \, \mathscr{V} \, |2s\rangle \tag{2-2-40}$$

which is (A). Hence to first order in $\mathscr{V}(r)$, "A" may be evaluated by solving (2-2-36) for $\chi_{2s}(r)$. The potential $\mathscr{V}(r)$ is velocity dependent but may be approximated by the adiabatic potential

$$V_{ad}(r) = \frac{4}{\pi} \int k_t \, dk_t \, \frac{\mathscr{I}_t^2}{\varepsilon_{1s} - k_t^2} \tag{2-2-41}$$

since in the range of $k_s$ which contribute, the denominator is slowly varying with $k_s$. The potential (2-2-41) is numerically almost exactly the two electron Bethe potential corresponding to the diagram polarizability of $0.174 a_0^3$.

All multipoles of the inverse power expansion of the potential are included. Detailed calculations show that only the monopole, dipole and quadrupole contributions are significant, and that of these the monopole and dipole are nearly cancelled by the non-adiabatic exchange polarization potential. It is sufficient to retain only the dipole term of $V_{ad}(r)$, and this is accurately represented by the Bethe form

$$V_{pol}^{(1)}(r) = \frac{9}{\kappa^4} \{1 - \tfrac{1}{3}[1 + 2\kappa + \kappa^2 + 6\tfrac{2}{3}\kappa^3 + \tfrac{4}{3}\kappa^4] \, e^{-2\kappa} - e^{-4\kappa}(1 + \kappa)^4\} \tag{2-2-42}$$

$$\kappa = Z_1 r, \quad Z_1^4 = 9/\alpha_0, \quad \alpha_0 = 0.174 a_0^3. \tag{2-2-43}$$

At large $r$,

$$V_{\text{pol}}^{(1)}(r) \xrightarrow[r \to \infty]{} \alpha_0/r^4 \tag{2-2-44}$$

whereas at small $r$

$$V_{\text{pol}}^{(1)}(r) \xrightarrow[r \to 0]{} \beta r$$

where $\beta$ is a constant.

We see therefore that the effect of the second order diagram Fig. 2-2-11 simulates a polarization potential, but introduces this in a systematic way. Note that asymptotically this potential is not proportional to the core polarizability but only to that portion of it which arises from this class of second order diagrams. Other contributions to the polarizability arise e.g. from fourth order diagrams in which the sub diagram in Fig. 2-2-12 occurs twice: such additional terms contribute about 10% of the polarizability.

The correction obtained to the initial state from polarizability diagrams is in general small, and decreases the matrix element in both length and velocity formulations, but by less than 2%. However the final state is more strongly affected, particularly in the calculation of the length matrix element, which depends on the details of the wave function at moderately large values of $r$. The p-wave phase shift (which is a slowly varying function of ejected electron momentum) changes by almost 50%, from a threshold value of

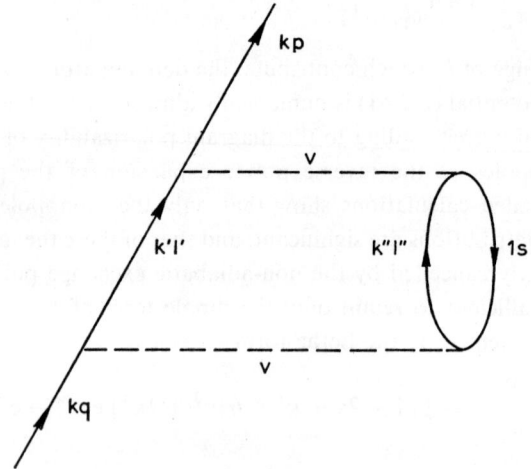

Fig. 2-2-12

0.110 in the Hartree-Fock approximation to 0.170 when the core polarization term is included. From (2-2-17), $P_{kp}(r)$ is asymptotically, very approximately, of the form $\sin(kr+\delta_1(k))$ and the change in the position of the zeros of the wave function is of order

$$\Delta r = -\frac{\Delta\delta_1(k)}{k} \approx -\frac{0.06}{k}.$$

Since the 2s-radial function is essentially unchanged, the change in the length form of the matrix element is large near threshold, but rapidly decreases with increasing $k$. It is increased by almost 15% at threshold, but by only 1% by $k^2 = 0.4$ Ry. The Hartree-Fock dipole length matrix element, and the initial and final state polarization corrections to it are shown in Table 2-2-4.

TABLE 2-2-4

Second order polarization corrections $\Delta\sigma_0(i)$, $\Delta\sigma_0(f)$ in initial (i) and final (f) states to the Hartree-Fock length matrix element $\sigma_0(L)$, as a function of $k = \sqrt{E_{ej}}$ (Ry)

| $k$ | $\sigma_0(L)$ | $\Delta\sigma_0(i)$ | $\Delta\sigma_0(f)$ |
|---|---|---|---|
| 0.05 | 1.963 | −0.041 | 0.279 |
| 0.25 | 1.931 | −0.030 | 0.211 |
| 0.55 | 1.464 | −0.008 | 0.093 |

The calculated results for the photoionization cross section $a_v$ in both length and velocity formulations are compared with experiment in Fig. 2-2-13. They include all first order correlation effects and the dominant, second order "polarization" type correlations, evaluated in the adiabatic-dipole-exchange approximation. The main effects neglected are (i) second-order short range correlation contributions, which are estimated to be small compared with the first order contributions of this type, (ii) higher order polarization corrections and (iii) non-adiabatic effects. The overall accuracy is thought to be $\pm 10\%$, while length and velocity formulations give results in agreement to $\pm 5\%$.

The agreement with the experimental values of Hudson and Carter [7] is everywhere within the limits of experimental error. If a mean experimental curve (giving allowance to Marr's work) is defined as in Fig. 2-2-2, the calculated curve (allowing $\pm 10\%$ uncertainty) remains compatible with experiment near threshold, but is perhaps somewhat too low at shorter wavelengths.

An alternative method of including short range correlation was intro-

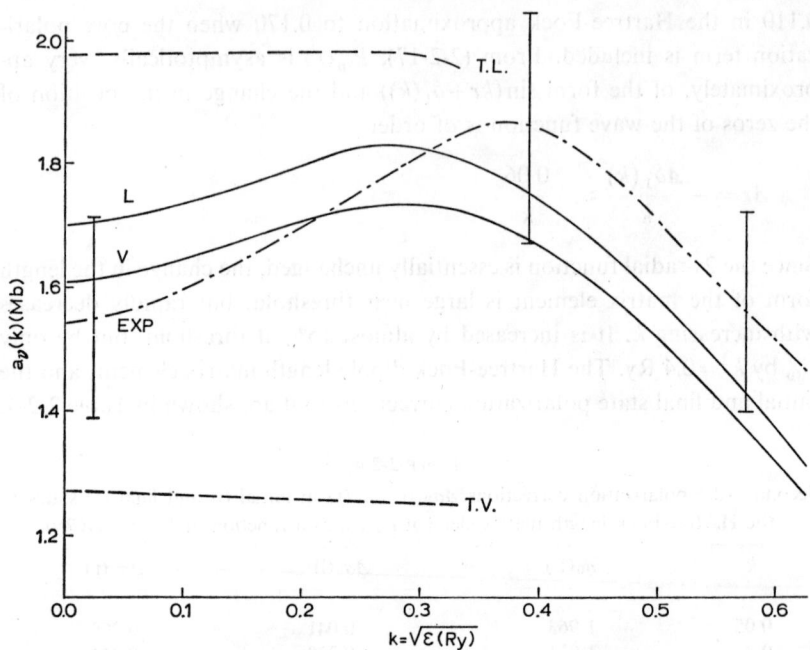

Fig. 2-2-13. Brueckner-Goldstone second order calculations of $a_v$ for Li (Chang and McDowell [18]). L = length, V = velocity; T.L. and T.V. are the corresponding calculations of Tait [29]. The curve marked EXP is the result of Hudson and Carter [7] with ±20% error bars.

duced by Tait, who used a Hylleras-type wave function for the ground state. This includes terms of the form

$$r_{ij}^p r_i^l r_j^m r_k^n \exp\{-\alpha_i r_i - \beta_j r_j - \gamma_k r_k\}$$

and is fully antisymmetrized. (Here $\alpha$, $\beta$, $\gamma$ are fixed constants, and $p$, $l$, $m$, $n$ are integers in the set (0, 1, 2).) Tait however did not include short range correlation in the final state, nor did he attempt to include the more important long range correlation. Indeed for this final state he chose a simple one-parameter function for the $1s^2$ core, and Stewart's numerical tabulation of the $P_{kp}(r)$ Hartree-Fock function for the ejected electron. He then expresses his velocity formulation many-electron matrix element in a form which involves derivatives of his rather crude $Li^+$ core function. Tait's results are also shown in Fig. 2-2-13. The large differences between his length and velocity results are probably explained by the need to take derivatives of his core function in the latter. His length result appears too large, and does not reproduce the shape of the experimental curves correctly.

An apparently different approach to including correlation has been used in a study of Na by Manson and Cooper [19]. With various choices of unperturbed basis set, they use second order perturbation theory to include the effects of mixing the direct $2p^6 3s \to 2p^6 \varepsilon p$ process with the virtual $2p^5 3s \,\varepsilon s$ and $2p^5 3s \,\varepsilon d$ final states. This is equivalent to including some of the final state long range correlations. Their results expressed in terms of an effective matrix element $T$ are given in Table 2-2-5, where

$$a_v = 2.553(I + k^2)|T|^2 \text{ Mb}$$

and for Na, $I \approx 0.380$.

TABLE 2-2-5

Matrix elements $T(L), T(V)$ for $3s \to kp$ in Na (Manson and Cooper [19])

| $k^2$ | (1) | $\Delta L_1$ | $\Delta V_1$ | (2) | (3) | $\Delta L_2$ | $\Delta V_3$ | EXP. [7] |
|---|---|---|---|---|---|---|---|---|
| 0   | −0.295 | 0.045 | 0.638 | −0.324 | −0.311 | 0.044 | 0.629 | −0.366 |
| 0.2 | +0.184 | 0.038 | 0.364 | +0.171 | +0.181 | 0.038 | 0.357 | +0.222 |
| 0.4 | 0.234 | 0.034 | 0.241 | 0.201 | 0.233 | 0.033 | 0.236 | 0.285 |
| 0.6 | 0.221 | 0.031 | 0.175 | 0.184 | 0.206 | 0.030 | 0.171 | 0.270 |
| 0.8 | 0.196 | 0.028 | 0.135 | 0.164 | 0.181 | 0.027 | 0.131 | 0.259 |
| 1.0 | 0.173 | 0.026 | 0.108 | 0.145 | 0.158 | 0.025 | 0.104 | 0.251 |

(1) Herman-Skillman central field. (2) Hartree-Fock "length". (3) Hartree-Fock "velocity". (In (1), $T(L) \equiv T(V)$.)

The first three columns refer to a calculation in which the basis set uses the Herman-Skillman potential, and the next four to the Hartree-Fock basis. They have also carried out other variants of these procedures. The correction to either model in the velocity formulation at low ejection energies is astonishingly large, and gives the wrong sign to the matrix element. In a more complete calculation one would expect this correction to the Herman-Skillman or Hartree-Fock matrix element to be largely cancelled by short range correlation effects.

Manson and Cooper's final Hartree-Fock corrected results are shown in Fig. 2-2-14. It is clear that the length result is in better agreement with experiment than any previous calculation, but the discrepancy for $k^2 > 0.6$ Ry lies outside reasonable error bounds on the experiment. The short range part of the wave function used is clearly very poor and gives quite a wrong form to the velocity result. Further work is in progress.

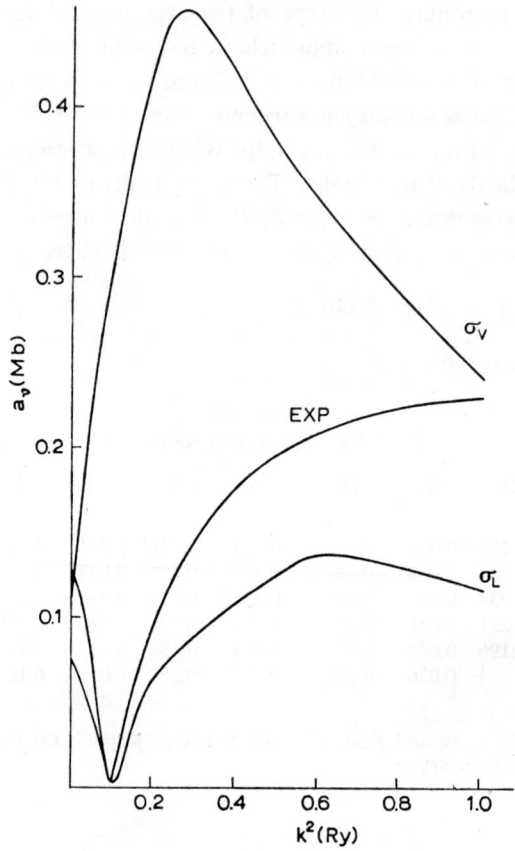

Fig. 2-2-14. Corrected Hartree-Fock results for photoionization of Na (Manson and Cooper [19]).

## § 2-3. Electron impact ionization

### A. Experimental studies

The most detailed experimental studies of electron impact ionization of the alkalis has been carried out by the Livermore group [11–14]. Relative measurements by Brink [10, 11] have been extended, and absolute total ionization cross sections obtained by McFarland and Kinney [12, 13] and McFarland [14]. Single ionization cross sections were deduced from the total cross section data using the multiple ionization factors measured by Tate and Smith [48]. A measurement of the electron impact ionization cross section of Cs has been reported by Heil and Scott [16], while results on lithium** close to threshold were given by Aleksakhin and collaborators [15, 67]. Very recently, Nygaard has measured the ionization cross section of Cs at electron energies up to 100 eV [17].

A block diagram of the apparatus used by McFarland and Kinney [12] is shown in Fig. 2-3-1. A beam of lithium (or other alkali) vapour from an

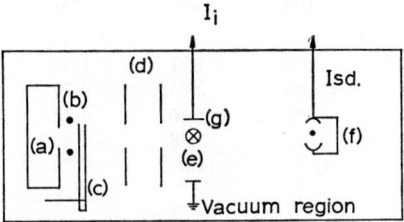

Fig. 2-3-1. Schematic diagram of the apparatus of McFarland and Kinney [12] for measurement of the total absolute ionization cross section of the alkali by electron impact (electronics not shown). (a) oven, (b) ion deflectors, (c) chopper, (d) defining slits, (e) electron beam, (f) surface ionization detector, (g) ion collector plates.

oven at approximately 1000 °K, exits with a number density of under $10^9$ particles cm$^{-3}$, passes a 100 cps chopper and enters a high vacuum region (pressure $<5 \times 10^{-8}$ Torr), where it is crossed at right angles by a d.c. electron beam. The electron beam was run at energies of 50, 100 (100) 500 eV, but no indication is given of energy spread.* The alkali metal beam was detected on a hot wire, the ionization efficiency of the surface detector was low, decreasing for Li from about 20% at 1925 °K to approximately 2% at 2375 °K. Uncertainties in the detection efficiency represent the largest uncertainty in the experiment and may be as large as ±25%.

The experiment does not distinguish between atomic and molecular forms of the vapour. For Li at an oven temperature of 1000 °K the ratio $n(\mathrm{Li}_2)/n(\mathrm{Li})$ may be as large as 5%. The ionization cross section of the molecule is unknown, but it could be large, particularly if formed in a distribution of rotational and vibrational states at 1000 °K. The effect would however be to decrease the observed cross section if dimers were present, as they dissociate on the hot wire detector.

The total ionization cross sections measured at the given energies are listed in Table 2-3-1. Single ionization cross sections may be obtained by using the ratios of single to multiple ionization measured by Tate and Smith,

---

* McFarland (1968, private communication) has informed us that it was 0.25 eV at 19.3 eV.
** This work has now been extended to the other alkalis [67].

as tabulated by Keiffer and Dunn [56] and are listed in Table 2-3-2 (in units of $\pi a_0^2$).

TABLE 2-3-1

Total measured ionization cross sections of the alkalis (McFarland and Kinney [12]) in units of $\pi a_0^2$

| $E$(eV) | 50 | 100 | 200 | 300 | 400 | 500 |
|---|---|---|---|---|---|---|
| Li | 3.52 | 2.50 | 1.71 | 1.36 | 1.13 | 0.94 |
| Na | 4.68 | 3.60 | 2.85 | 2.29 | 1.95 | 1.71 |
| K | 7.30 | 6.82 | 5.46 | 4.22 | 3.52 | 3.11 |
| Rb | 9.57 | 9.32 | 7.58 | 6.11 | 5.13 | 4.62 |
| Cs | 12.9 | 12.3 | 11.0 | 9.78 | 8.18 | 7.10 |

TABLE 2-3-2

Single ionization cross sections of the alkalis (in $\pi a_0^2$) obtained from Table 2-3-1 using (A) Tate and Smith's data as presented by Keiffer and Dunn, (B) Tate and Smith's data reduced by McFarland (private communication, 1968)

| $E$(eV) | 50 | 100 | 200 | 300 | 400 | 500 | |
|---|---|---|---|---|---|---|---|
| Li | 3.5 | 2.5 | 1.7 | 1.4 | 1.1 | 0.94 | |
| Na | 4.7 | 3.5 | 2.6 | 2.1 | 1.7 | 1.45 | |
| K | 6.7 | 5.4$_5$ | 4.4 | 3.4 | 2.9 | 2.5$_5$ | (A) |
| Rb | 8.2 | 6.1$_5$ | 4.8 | 3.9 | 3.2 | 2.7 | |
| Cs | 11.3$_5$ | 10.0 | 8.5 | 7.1 | 6.0 | 5.0 | |
| Li | 3.5 | 2.5 | 1.7 | 1.4 | 1.1 | 0.94 | |
| Na | 4.7 | 3.3 | 2.4 | 1.9 | 1.65 | 1.5 | |
| K | 6.7 | 5.1 | 4.0 | 3.1 | 2.6 | 2.4 | (B) |
| Rb | 8.3 | 6.0 | 4.9 | 4.0 | 3.2 | 2.8 | |
| Cs | 11.5 | 9.9 | 8.4 | 7.1 | 5.8$_5$ | 5.0 | |

The differences between sets (A) and (B) do not exceed 10%, and set (A) is adopted throughout this paper.

The total cross section measured by McFarland and Brink is derived from the equation

$$Q = C \frac{\bar{v} E I_i}{I_e I_d} \tag{2-3-1}$$

where $\bar{v}$ is the mean velocity of atoms in the beam, $I_i$ the ion current produced by the electron beam, $I_e$ the electron current, $I_d$ the surface detector ion current and $E$ that detector's efficiency.

The currents can be accurately measured ($\pm 1\%$) (since they enter only as a ratio, $O(10^{-4})$) and while there are somewhat larger uncertainties in $\bar{v}$ and the geometric factor $C$ the main source of error clearly is in the determination of the efficiency $E$ of the surface ionization detector, which is estimated at $\pm 25\%$. In determining $E$, the ratio $N_+/N_i$ of the number of atoms ionized by the hot wire to the number incident, it was assumed that $r_i$, the reflection coefficient of the incident beam was zero, and that the ratio $(1-r_a)/(1-r_+)$ (which involves the reflection coefficients of the ion and the atom in temperature equilibrium on the hot wire), was unity. McFarland and Kinney [13] therefore undertook an independent measurement of their mean beam density by loading the oven with a known quantity of the alkali metal and monitoring the surface detector current until it fell to zero. A correction for oven material deposited on slits, chopper wheel and beam definition slits is necessary.

The directly measured values, corrected values, those used in the earlier experiment and these obtained with the assumptions outlined above for $r_i$, $r_a$ and $r_+$ are shown in Table 2-3-3. The values all assume that the molecular weight of the species in the beam was that of the appropriate atom. If the mean molecular weight was higher, then the directly measured value of $E$ would be too low. If on the other hand the actual detector efficiency $E_a$ was higher than that assumed in the earlier work [12] the cross sections tabulated in Tables 2-3-1 and 2-3-2 would be underestimates as would be the case if $Li_2^+$ ions dissociate on the wire. The uncertainty in the case of Li does not appear to exceed 15%, and 30% for Na. For K and the higher alkalis the relatively high vapour pressures cause the correction for deposits on slits and shields to become large, and the method was not useful. It is difficult to see how systematic error of more than $\pm 25\%$ in the case of Li and perhaps $\pm 30\%$ in other cases can occur in this experiment. The random error at 100 eV was $\pm 8\%$. We can for the moment allocate an error $\pm 33\%$ to the Li results, $\pm 40\%$ to the Na values, and a similar uncertainty to the total ionization measurements for the other alkalis.

TABLE 2-3-3

Alkali beam surface ionization detector efficiency $E$, ref. [13]

|    | Direct | Corrected | Old experiment [12] | Theory |
|----|--------|-----------|---------------------|--------|
| Li | 2.1    | 1.6       | 1.4                 | 1.7    |
| Na | 7.5    | 6.0       | 4.1                 | 5.7    |

Heil and Scott [16] have measured the total electron impact ionization cross section for Cs at energies up to 26 eV by firing an electron beam through the metal vapour, whose density is again measured by a surface ionization detector whose efficiency is assumed to be 100%. Their derived single ionization cross section is shown in Fig. 2-3-2 and reaches a maximum value of $7\pi a_0^2$ at energies above 15 eV. Brink's relative measurements normalized to McFarland and Kinney's absolute measurement at 500 eV, and reduced to a single ionization cross section by use of Tate and Smith's data give a value of approximately $12\ \pi a_0^2$ near 25 eV. This is in close agreement with a value of $13\ \pi a_0^2$ obtained by Witting [49] using an ionization gauge and normalized to the well-established value for $N_2$. McFarland [14] has criticized Heil and Scott's estimate of their detector efficiency, and has pointed out that independent attempts [50] at measuring Cs vapour density with a hot wire detector immersed in the vapour did not give satisfactory results.

Nygaard [17] using a modern version of Tate and Smith's apparatus has recently remeasured the single ionization cross section of Cs from threshold to 100 eV. The shape of the cross section curve obtained is close to the

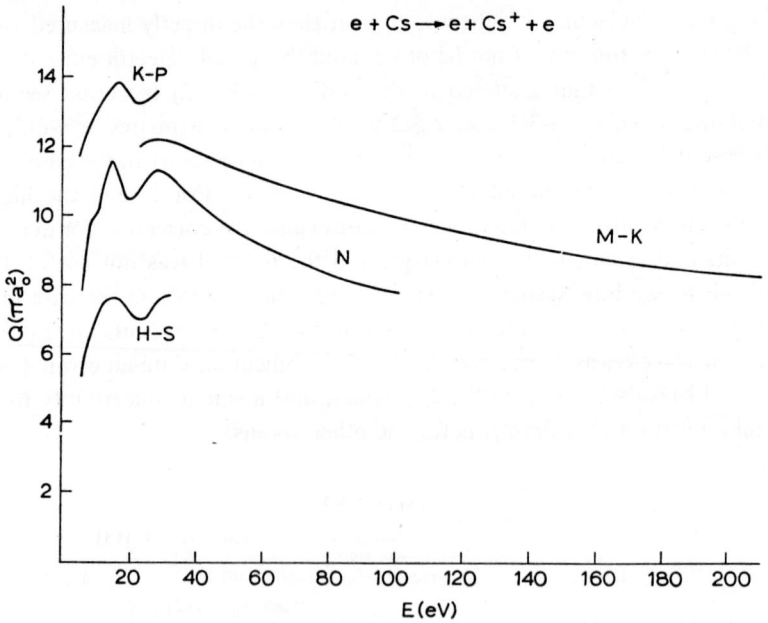

Fig. 2-3-2. Electron impact ionization of Cs. The curves represent the following sets of experimental results: N – Nygaard [17], M-K – McFarland and Kinney [12], H-S – Heil and Scott [16], K-P – Korchevoi and Przonski [57].

original result of Tate and Smith, when that latter is normalized to Nygaard's result at 28 eV. Some structure, possibly from low lying autoionization levels, is present about 15 eV above threshold, and has also been observed independently by Korchevoi and Przonski [57].

Nygaard's results are also shown in Fig. 2-3-2, and lie some 20% below the values we attribute to McFarland and Kinney,* but the maximum measured cross section of $11.5\pi a_0^2$ is in much better agreement with McFarland and Kinney than with Heil and Scott. If McFarland and Kinney's data on total ionization cross section are reduced by supposing Tate and Smith's data refer to relative ionization probabilities, lower values are obtained, which are in close agreement with Nygaard's results.

Aleksakhin et al. [15] have briefly reported on a crossed beam measurement of ionization of Li at electron energies up to 14 eV.[†] They give no detailed discussion of their beam density or current measurements, and report a cross section rising linearly to a maximum of $4.8\pi a_0^2$ at 12.5 eV, with a quoted error of $\pm 20\%$. McFarland and Kinney's maximum value is approximately $5.2\pi a_0^2$ ($\pm 20\%$) at an energy below 25 eV. These values seem in reasonable accord. We may therefore tentatively conclude that the single ionization values of Table 2-3-2 are reliable to within $\pm 40\%$, and perhaps to within $\pm 25\%$ at energies above 200 eV.

An additional result strengthening this conclusion comes from the recent measurement by Martin et al. [51] of the ionization cross section for the process

$$Mg^+ + e \rightarrow Mg^{++} + e + e \qquad (2\text{-}3\text{-}2)$$

which is iso-electronic with

$$Na + e \rightarrow Na^+ + e + e \qquad (2\text{-}3\text{-}3)$$

and does not require a density measurement. If a scaling law derived from classical dynamics

$$Q_{red} = \left(\frac{I_x}{I_H}\right)^2 Q_{obs} \qquad (2\text{-}3\text{-}4)$$

is used to obtain a reduced cross section, where $I_x$ is the ionization potential of the species studied, then the values compared at equal reduced energies $E_r = E/I_x$ the results for $Mg^+$ are within a few percent of McFarland and Kinney's values for Na for $E_r > 30$ (Fig. 2-3-3). This is not conclusive, since

---

* Obtained, following Keiffer and Dunn, by interpreting Tate and Smith's ordinates as ion currents proportional to $NQ(Cs^{n+})$.
[†] See also ref. [67].

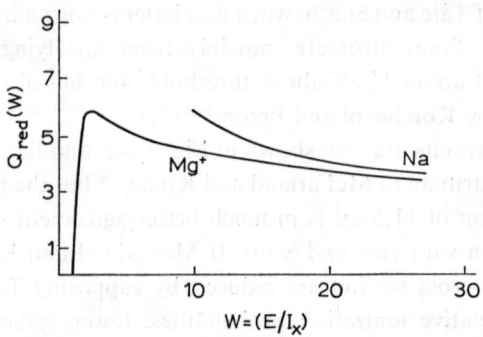

Fig. 2-3-3. Electron impact ionization of Mg⁺ (Martin et al. [51]). The upper curve shows the classically scaled result of McFarland and Kinney [12] for Na.

the validity of the classical scaling law (2-3-4) has not been demonstrated for the processes concerned.*

## B. ELECTRON IMPACT IONIZATION: THEORY

All quantal theoretical studies of electron impact ionization of the alkalis have been restricted to the first Born approximation (including variants in which some allowance is made for electron exchange). Preliminary studies [20] by Peach and McDowell indicated that unlike ionization of H and of closed shell systems (He, Ne, etc.) the process does not proceed primarily via an allowed s→p transition, but that at energies below 250 eV the continuum d (and higher $l$) partial wave contributions dominate, producing an energy dependence of the form $E^{-\alpha}(\alpha \approx 0.7)$ in the range of laboratory measurements, rather than the expected $E^{-1} \log E$ behaviour associated with an allowed transition.

The first detailed study was of ionization of Li [21]. These calculations have been extended to Na by Peach [24] who also considers the effect of exchange. Further work is reported below. Bates et al. have carried out a similar calculation for Na [25]. A less detailed analysis by Omidvar and Sullivan is available for comparison [41]. Classical Gryzinski-type calculations have been carried out by many authors. Among the most reliable are those of Catlow and McDowell [22]: values for higher impact energies are reported below.

We now briefly review the theoretical analysis. The cross section for electron impact ionization of an atom by an incident electron of energy

---

* See further, the review by Dolder in this book.

$k_i^2$ Ry is given by (the wavefunctions being normalized as in §2-2-B),

$$Q(k_i^2) = \frac{1}{4\pi^2 k_i} \int \widehat{dk_f} \int \widehat{dk_e} k_e k_f \int_0^{\varepsilon_{max}} d\varepsilon \, |T_{if}(k_f, k_e)|^2 \qquad (2\text{-}3\text{-}5)$$

in units of $\pi a_0^2$, where $k_e$ is the momentum of the ejected electron ($\varepsilon = k_e^2$ Ry being its energy), $k_f$ that of the scattered electron, and the integrations are over all directions of ejection and of scattering, and all ejected electron energies $\varepsilon$ consistent with conservation of energy and momentum.

The transition matrix element $T_{if}(k_f, k_e)$ is given by

$$T_{if} = \langle \chi_f, V_f \Psi_i^+ \rangle \qquad (2\text{-}3\text{-}6)$$

where $\chi_f$ is the final unperturbed wave function, $V_f$ the interaction potential in the final channel, and $\Psi_i^+$ the total scattering function with incoming wave boundary conditions. Let $\Phi_i$, $\Phi_f$ be the initial and final unperturbed target wave functions. Then, neglecting exchange, and considering Li in particular

$$\chi_f = \Phi_f(1, 2, 3) \, e^{i k_f \cdot r_4} \qquad (2\text{-}3\text{-}7)$$

(labelling the bound electrons by $i = 1, 2, 3$, the incident electron by $i = 4$). If $H$ is the total Hamiltonian of the four-electron system, $H_0$ that of the target

$$\Psi_i^+ = (1 + G V_f) \chi_i \qquad (2\text{-}3\text{-}8)$$

and

$$G = \frac{1}{E - H + i\eta} \qquad (2\text{-}3\text{-}9)$$

$$\chi_i = \Phi_i(1, 2, 3) \, e^{i k_i \cdot r_4}. \qquad (2\text{-}3\text{-}10)$$

The first Born approximation is obtained on approximating $\Psi_i^+$ by $\chi_i$, so that

$$T_{if}^{(B)} = \langle \chi_f, V_f \chi_i \rangle = 2\pi f(k_f, k_e) \qquad (2\text{-}3\text{-}11)$$

where on integration over the co-ordinates of the incident electron

$$f(k_f, k_e) = -\frac{2}{K^2} \int \Psi_i^*(1, 2, 3) \sum_{j=1}^{3} e^{i K \cdot r_j} \Psi_f(1, 2, 3) \, dr_N \qquad (2\text{-}3\text{-}12)$$

and

$$K = k_i - k_f \qquad (2\text{-}3\text{-}13)$$

is the momentum transfer.

McDowell et al. [21] represent $\Psi_i$, $\Psi_f$ by single Slater determinants, in which the radial parts of the single electron spin-orbitals are solutions of the Hartree-Fock equation (2-2-21).

Integrating over spins, and all co-ordinates except those of the ejected electron, $f(k_f, k_e)$ becomes*

$$f(k_f, k_e) = -\frac{2}{K^2} \int \phi_{2s}^* e^{i\mathbf{K}\cdot\mathbf{r}} \phi_\varepsilon(r) \, d\mathbf{r} \qquad (2\text{-}3\text{-}14)$$

where

$$\phi_{2s}(r) = \frac{1}{r} P_{2s}(r) Y_{00}(\hat{r}) \qquad (2\text{-}3\text{-}15)$$

and

$$\phi_\varepsilon(r) = \frac{1}{(2\pi)^{3/2} k_e^{1/2}} \sum_{l=0}^{\infty} i^l (2l+1) \frac{P_{k_e l}(r)}{r} P_l(\hat{k}_e \cdot \hat{r}) \qquad (2\text{-}3\text{-}16)$$

in the notation of §2-2. Choosing the positive Z-axis along $\hat{\mathbf{K}}$, the integral in (2-3-14) becomes

$$\mathscr{I}(\mathbf{K}, k_e) = \frac{1}{\pi\sqrt{2k_e}} \sum_{l=0}^{\infty} \sum_{m=-l}^{l} I_{l,m}(\varepsilon, K) Y_{lm}(\hat{k}_i \cdot \hat{\mathbf{K}}) \qquad (2\text{-}3\text{-}17)$$

with

$$I_{l,m}(\varepsilon, K) = \iint P_{2s}^*(r) P_{k_e l}(r) Y_{00}^*(\Omega) Y_{lm}^*(\Omega) e^{i\mathbf{K}\cdot\mathbf{r}} \, d\mathbf{r} \, d\Omega$$

$$= (4\pi)^{1/2} \sum_{\lambda\mu} i^\lambda (2\lambda + 1) J_{l\varepsilon}(K) \int Y_{00}^* Y_{lm} Y_{\lambda\mu} \, d\Omega \qquad (2\text{-}3\text{-}18)$$

$$= i^l (2l+1)^{1/2} (4\pi)^{1/2} J_{l\varepsilon}(K) \qquad (2\text{-}3\text{-}19)$$

where

$$J_{l\varepsilon}(K) = \int_0^\infty P_{2s}^*(r) j_l(Kr) P_{k_e l}(r) \, dr \qquad (2\text{-}3\text{-}20)$$

and $j_l(x)$ is a spherical Bessel function of order $l$. Introducing (2-3-14) to (2-3-20) in (2-3-5) and integrating over all angles, noting that

$$\int d\hat{k}_f = \int d(\cos\theta) \, d\phi, \qquad \theta = \cos^{-1}(\hat{k}_f \cdot \hat{k}_i)$$

and that

$$K^2 = k_i^2 + k_f^2 - 2k_i k_f \cos\theta$$

---

* Neglecting, for simplicity the small term due to ejection of a core electron, followed by relaxation of the ion to the $1s^2$ ground state. This term is in the numerical work.

we have

$$Q^{(B)}(k_i^2) = \frac{8}{\pi k_i^2} \sum_{l=0}^{\infty} (2l+1) \int_0^{\varepsilon_{max}} d\varepsilon \int_{K_{min}}^{K_{max}} |J_{l\varepsilon}(K)|^2 K^{-3} dK. \qquad (2\text{-}3\text{-}21)$$

The limits on momentum transfer are

$$K_{min} = k_i - k_f, \qquad K_{max} = k_i + k_f \qquad (2\text{-}3\text{-}22)$$

where by conservation of energy

$$k_i^2 = k_f^2 + I_x + \varepsilon \qquad (2\text{-}3\text{-}23)$$

and

$$\varepsilon_{max} = k_i^2 - I_x. \qquad (2\text{-}3\text{-}24)$$

Peterkop [53] has noted that in the no-exchange case in calculating a cross section for comparison with the experimental cross section, in which the faster of the outgoing electrons is considered to be the scattered electron, it is more consistent to modify the last result to

$$\varepsilon_{max} = \tfrac{1}{2}(k_i^2 - I_x). \qquad (2\text{-}3\text{-}25)$$

We shall use this throughout: it has little effect at impact energies greater than five times the threshold, but reduces the maximum calculated cross section by approximately 30%. This approximation will be refered to as the modified Born approximation.

The calculation of the ionization cross section to this approximation is thus reduced to the evaluation of the matrix elements $J_{l\varepsilon}(k)$. We write

$$Q^{(B)}(k_i^2) = \sum_{l=0}^{\infty} Q_l \qquad (2\text{-}3\text{-}26)$$

where the $l^{th}$ partial cross section is given by*

$$Q_l = \frac{8(2l+1)}{\pi k_i^2} \int_0^{\varepsilon_{max}} d\varepsilon \int_{K_{min}}^{K_{max}} |J_{l\varepsilon}(K)|^2 K^{-3} dK. \qquad (2\text{-}3\text{-}27)$$

---

* Exchange effects have been examined by Peach [24] in the context of the first Born approximation. They lead to a substantial reduction in the calculated cross sections at impact energies below 100 eV, but in view of the very doubtful validity of first order approximations of Born type at energies below 20 times threshold, they are not discussed in detail here.

The results for Li($1s^2$ $2s$ $^2S_{\frac{1}{2}}$) when Hartree-Fock orbitals are used for both the 2s and ejected electrons are presented in Table 2-3-4. The s-wave of the continuum function is not orthogonal to the 1s and 2s orbitals, and it is necessary to impose this restriction by replacing $P_{\varepsilon 0}(r)$ by

$$\bar{P}_{\varepsilon 0}(r) = P_{\varepsilon 0}(r) - \int_0^\infty P_{1s}(r') P_{\varepsilon 0}(r') \, dr' \, P_{1s}(r) - \int_0^\infty P_{2s}(r') P_{\varepsilon 0}(r') \, dr' \, P_{2s}(r). \tag{2-3-28}$$

However the contribution from the s-wave is small, and this can scarcely introduce substantial error. The largest contributions come from the s→d transition, and partial waves up to $l=7$ are important. As in Peach's calculations contributions from $1s \to kl$ are included, but are not important in this energy range. All partial cross sections contributing more than $10^{-3}\pi a_0^2$ have been included. Our results for the $l=0, 1, 2$ partial waves are slightly larger than those given by Peach, who used a Coulomb wave for the ejected electron. The Coulomb approximation is valid for $l > 2$.

At $E=3$ Ry, our calculated result of $3.36\pi a_0^2$ is almost 40% higher than Peach's value [24] in the unmodified Born approximation and at $E=5$ Ry our value of $2.36\pi a_0^2$ is still 40% above her value. The reason for this discrepancy is not clear. The numerical procedures were totally different, Peach tabulating $J_{\varepsilon l}(K)$ at 61 values of $K$ and 17 values of $\varepsilon$ and interpolating for repeated Gaussian quadrature. We divided the range of $K$ (for each $k_1$) into 100 steps, and calculated $J_{l\varepsilon}(K)$ for 11 values of $\varepsilon$ in the range $(0, 0.25\varepsilon_{max})$ and 11 values of $\varepsilon$ in the range $(0.25\varepsilon_{max}, \varepsilon_{max})$, the integrations being carried out by a repeated Simpson's rule. The calculation was particularly sensitive to the limiting value $J_{0l}(K)$ as $\varepsilon \to 0$. We describe an alternative formulation, which provides a check on our numerical accuracy, below.

Similar calculations for Na have been reported by Bates et al. [25], who use Coulomb wave functions for the ejected electron partial waves for $l \geq 2$, but approximate Hartree-Fock functions for $l=0, 1$, as in Boyd's calculation of the Na photoionization cross section [26]. Peach [24] uses Coulomb functions for all partial waves, but allows for non-orthogonality of the s-wave. Their results are compared in Fig. 2-3-4, where the upper curve A is the unmodified Born approximation ($l \leq 5$), the curve marked B is the result of Bates et al. [25] and C is Peach's modified Born calculation. It is clear that the discrepancies of some 20% between curves B and C at low impact energies are partially due to the non-Coulomb effects in the $l=0, 1$ partial waves. Again, the effect here is to increase the cross section from that ob-

TABLE 2-3-4
Partial and total Born cross sections for electron impact ionization of Li ($\pi a_0^2$)

| Ry | eV | l=0 | 1 | 2 | 3 | 4 | 5 | 6 | 7 | 8 | 9 | 10 | Total |
|---|---|---|---|---|---|---|---|---|---|---|---|---|---|
| 1 | 13.6 | 0.38 | 0.584 | 2.41 | 1.17 | 0.260 | 0.035 | 0.005 | 0.002 | – | – | – | 4.85 |
| 2 | 27.2 | 0.36 | 0.548 | 1.72 | 0.962 | 0.333 | 0.105 | 0.038 | 0.016 | 0.004 | 0.001 | – | 4.09 |
| 3 | 40.8 | 0.345 | 0.510 | 1.28 | 0.740 | 0.298 | 0.108 | 0.052 | 0.018 | 0.009 | 0.004 | 0.002 | 3.36 |
| 4 | 54.2 | 0.32 | 0.472 | 1.01 | 0.587 | 0.250 | 0.103 | 0.051 | 0.021 | 0.010 | 0.005 | 0.002 | 2.82 |
| 5 | 68.0 | 0.265 | 0.436 | 0.790 | 0.478 | 0.209 | 0.096 | 0.046 | 0.022 | 0.010 | 0.005 | 0.002 | 2.36 |
| 10 | 136.0 | 0.14 | 0.325 | 0.404 | 0.238 | 0.104 | 0.052 | 0.027 | 0.013 | 0.006 | 0.003 | 0.001 | 1.31 |
| 15 | 204.0 | 0.0975 | 0.274 | 0.276 | 0.163 | 0.069 | 0.031 | 0.015 | 0.007 | 0.003 | 0.001 | – | 0.94 |
| 20 | 272.0 | 0.050 | 0.237 | 0.212 | 0.117 | 0.042 | 0.020 | 0.010 | 0.004 | 0.001 | – | – | 0.69 |

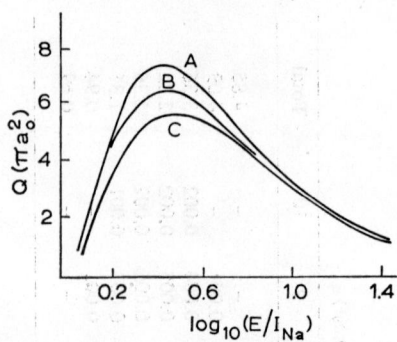

Fig. 2-3-4. Electron impact ionization of Na, Born's approximation. (A) Peach [24], (B) Bates et al. [25], (C) Peach, half-range Born [24].

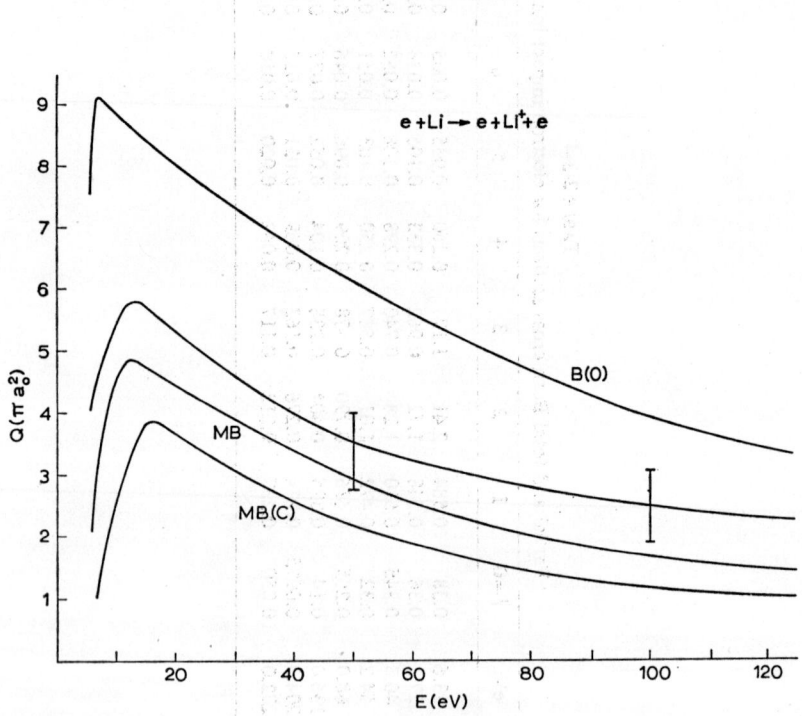

Fig. 2-3-5. Electron impact of Li, threshold to 120 eV. The curve with error bars is the experimental result of McFarland and Kinney [12]. The Born calculations shown are the simple wave function calculations of Omidvar and Sullivan [23], B(O); the modified (half-range) Born of Peach [24] with Hartree-Fock ground state and Coulomb continuum, MB (C); and the full Hartree-Fock modified Born of this paper, MB.

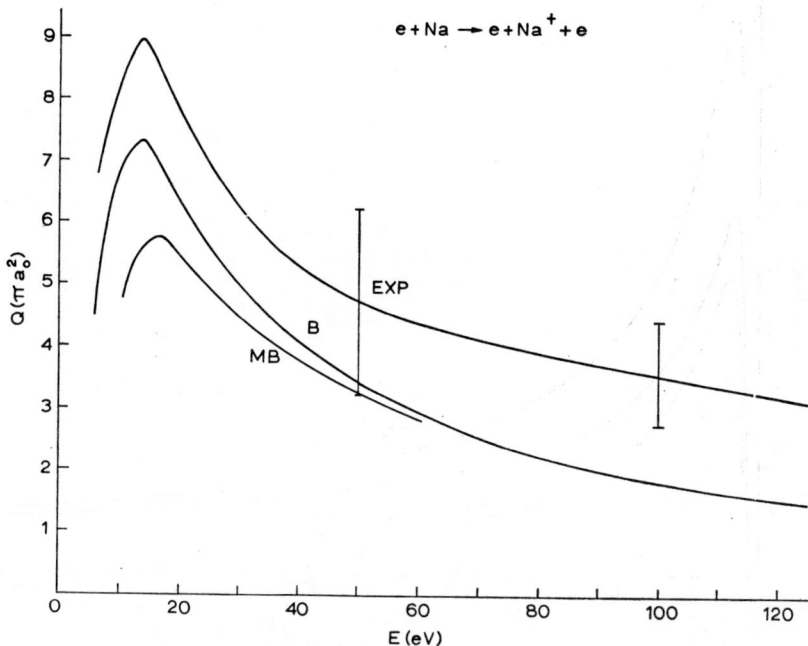

Fig. 2-3-6. Electron impact ionization of Na. The lower curves are the Born calculation, of Peach [24], B; and her modified Born calculation, MB. The experimental curve is that of McFarland and Kinney [12]. The divergence between theory and experiment is already noticeable at 100 eV.

tained with undistorted Coulomb waves. Bates et al. use radial functions obtained from a central field equation with the potential of Biermann and Lübeck [54] which are not necessarily an improvement on pure Coulomb functions when used to evaluate ionization matrix elements, as they may have quite the wrong phase shift. We noted earlier (§2-2) that there are some discrepancies in Boyd's photoionization calculation using these wave functions. The results of McDowell et al. [21] for the $l=1$ partial cross section for Li using a Hartree-Fock radial are 20% higher than those obtained by them (and by Peach [24]) using undistorted Coulomb functions, and agree well with the results reported in this paper.

The theoretical and experimental results for Li and Na are compared in Figs. 2-3-5 and 2-3-6 respectively, at energies up to 120 eV. Although the calculations for Li reported in this paper are in satisfactory agreement with the experiments of Brink and of McFarland and Kinney at energies up to 80 eV (16 times threshold) theory and experiment gradually diverge (Fig.

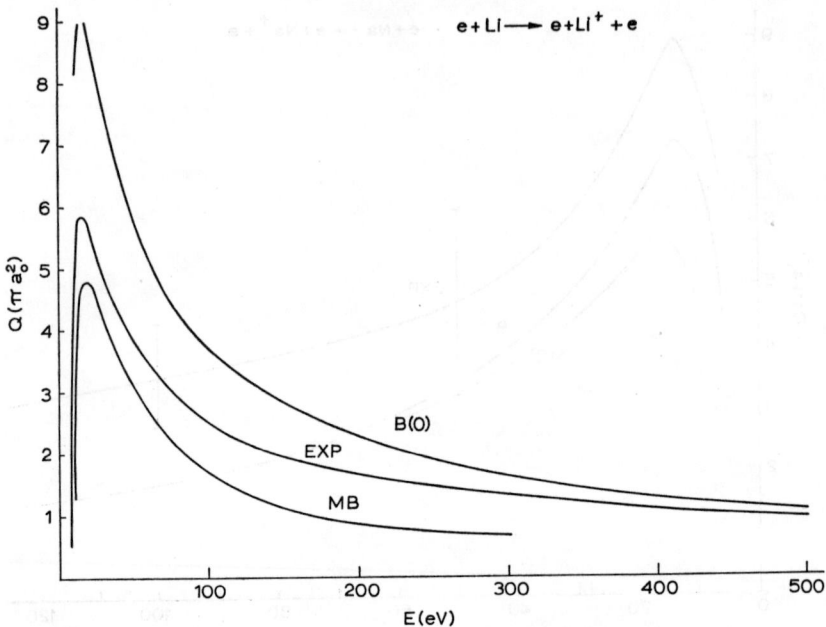

Fig. 2-3-7. Electron impact ionization of Li, threshold to 500 eV. As Fig. 2-3-5.

2-3-7) until at 300 eV (60 times threshold) where the first Born approximation should be reliable, the theoretical value is now a factor of two below experiment.

Omidvar and Sullivan [23] have used a model in which screened hydrogenic wave functions were used for the target ground state, and undistorted but orthogonalized Coulomb functions for the ejected electron to calculate the first Born approximation to the ionization cross section for Li. Their results (shown as the curve labelled B(O) in Figs. 2-3-5 and 2-3-7) lie considerably above the results of this paper, and of Peach, at all energies, but approach the experimental values by 500 eV (100 times threshold). As we have seen in §2-2 the photoionization cross section obtained in their model is a factor of four low at threshold and does not reproduce the shape of the experimental curve, in contrast to that obtained with the Hartree-Fock model. However the total area under their photoionization curve is in fair accord with that under the experimental curve, and this area is a measure of the oscillator strength available for the process. We suggest the apparent agreement of their calculation with experiment shown in Fig. 3-2-4 should be treated with reserve.

## 2-3 ELECTRON IMPACT IONIZATION

Inner shell ionization, followed by core relaxation,

$$e + Li(1s^2\, 2s\, {}^2S_{\frac{1}{2}}) \rightarrow e + Li^+(1s^2\, {}^1S_0) + e \tag{2-3-29}$$

has been included in our calculation, and an additional effect

$$e + Li(1s^2\, 2s\, {}^2S_{\frac{1}{2}}) \rightarrow e + Li^+(1s\, 2s) + e \tag{2-3-29a}$$

is not expected to be significant in Li. Its effects may be estimated using a classical model. Catlow and McDowell [22] show that in a classical binary encounter model the ionization cross section at impact energy $s^2/u^2$ is given by (where $u^2$ is the ionization potential in Ry)

$$Q(s^2) = N \int_0^\infty f(t)\, Q(s, t)\, t^{\frac{1}{2}}\, dt \quad \pi a_0^2 \tag{2-3-30}$$

in which $f(t)$ is the momentum distribution of the target electrons, $Q(s, t)$ the cross section for ionization when the bound electron has energy $t^2/u^2$ Ry, and $N$ the number of electrons in the shell considered. They take $f(t)$ to be the momentum distribution conjugate to the Roothaan Hartree-Fock electron distribution, and their results for ionization of the 2s-electron of Li are compared with the quantal calculations of this paper in Fig. 2-3-8. The

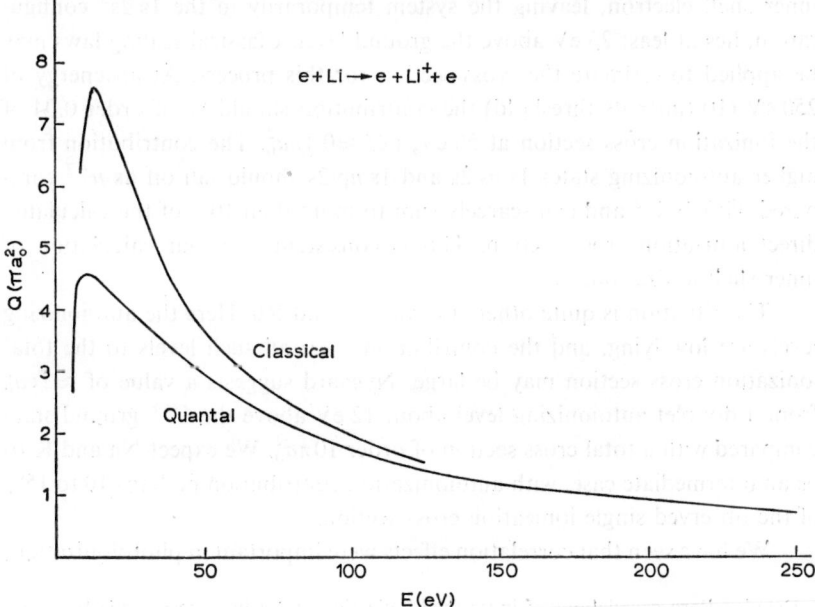

Fig. 2-3-8. Quantal (modified Born) and classical (Catlow and McDowell [22]) calculations of ionization of Li by electron impact, the classical using the same target electron momentum distribution as the quantal.

TABLE 2-3-5

Cross sections for inner shell ionization in the classical (velocity-averaged) binary encounter model, in units of $\pi a_0^2$.

| $E$(eV)   | 101   | 203   | 268   | 324   | 446   |
|-----------|-------|-------|-------|-------|-------|
| Li(2s)    | 0.060 | –     | 0.104 | –     | 0.079 |
| Na(2s)    | 0.028 | 0.076 | 0.071 | 0.063 | 0.050 |
| Na(2p)    | 0.355 | 0.577 | 0.572 | 0.543 | 0.473 |
| Na(total) | 0.383 | 0.653 | 0.643 | 0.606 | 0.523 |

agreement at energies above 50 eV is excellent. The work has therefore been extended to estimate the inner shell contributions for Li and for Na. The results are given in Table 2-3-5.

The effect is negligible for Li, but quite substantial in the case of Na. It may be sufficient to bring theory and experiment into agreement at the higher energies (>400 eV) for Na, but the total calculated cross sections would remain substantially below the experimental values of Table 2-3-2 at lower energies.*

The lowest autoionizing level of Li, corresponding to excitation of an inner shell electron, leaving the system temporarily in the 1s 2s² configuration, lies at least 25 eV above the ground level. Classical scaling laws may be applied to estimate the cross section for this process. At an energy of 250 eV (10 times its threshold) the contribution should be of order 0.04 of the ionization cross section at 50 eV, i.e. $\approx 0.1\pi a_0^2$. The contribution from higher autoionizing states 1s $ns$ 2s and 1s $np$ 2s should fall off as $n^{-3}$ compared with 1s 2s² and can scarcely sum to more than 10% of the calculated direct ionization cross section. This is consistent with our calculation of inner shell ionization.

The situation is quite otherwise with Cs and Rb. Here the autoionizing levels are low lying, and the contribution from all such levels to the total ionization cross section may be large. Nygaard suggests a value of $\approx 2\pi a_0^2$ from a doublet autoionizing level about 12 eV above the Cs⁺ ground state compared with a total cross section of order $10\pi a_0^2$. We expect Na and K to be an intermediate case, with autoionization contribution perhaps 10 to 15% of the observed single ionization cross section.

We have seen that correlation effects were important in photoionization,

---

* Detailed Born calculations of inner shell ionization in Na have been made by Peach [1969, private communication]. Her results, to be presented at the VI[th] Intern. Conf. on the Physics of Electronic and Atomic Collisions, are smaller than our classical estimates, and go about half way towards removing the discrepancy between theory and experiment for Na.

and that while they did not change the overall magnitude of the calculated photoionization cross section by more than 25%, they significantly affected the shape of the curve. The dominant correlation effect in the length formulation which corresponds to the formulation of Born's approximation above, was a long range polarization effect in the final state. Its effect in electron impact ionization cross sections may be estimated by using the solutions of (2-2-36) rather than Hartree-Fock functions to compute the matrix elements. The corresponding phase shifts for $l=0, 1$ are appreciably different from the Hartree-Fock values, as may be seen from Table 2-3-6 where they are com-

TABLE 2-3-6

s and p wave phase shifts for $e + Li^+$ scattering.
(a) Hartree-Fock [61], (b) eq. (2-2-6), (c) quantum defect (Doughty et al. [60])

|  | $l=0$ | | | $l=1$ | | |
| --- | --- | --- | --- | --- | --- | --- |
|  | (a) | (b) | (c) | (a) | (b) | (c) |
| $k^2 = 0$ | 1.234 | 1.261 | 1.254 | 0.110 | 0.167 | 0.148 |
| $k^2 = 1.0$ | 1.106 | 1.178 | 1.169 | 0.161 | 0.224 | 0.220 |

pared with these and with quantum defect values. They are in close agreement with the latter values. Corresponding values of the p-wave ionization matrix element $J_{1\varepsilon}(k)$ integrated over momentum transfer $K$,

$$F_1(\varepsilon) = \int_{K_{min}}^{K_{max}} |J_{1\varepsilon}(K)|^2 \, K^{-3} \, dK \qquad (2\text{-}3\text{-}31)$$

with (upper curve) and without (lower curve) polarization are shown in Fig. 2-3-9 at an energy of $W = 15$ Ry, as a function of ejected electron energy in units of 0.219 Ry. The effect is, as expected, largest for very low energies of ejection, $F_1(0)$ increasing from 1.57 to 1.83. The effect rapidly decreases with $W$. The partial wave cross sections $Q_0$ and $Q_1$ with and without polarization are shown below, in Table 2-3-7.

TABLE 2-3-7

Partial cross sections for ionization of Li (modified Born approximation) with radial functions obtained with (pol.) and without (no pol.) a polarization potential $(\pi a_0^2)$

| $W$(Ry) | 2 | 5 | 10 | 15 | 20 |
| --- | --- | --- | --- | --- | --- |
| $Q_0$ (no pol.) | 0.36 | 0.27 | 0.14 | 0.097 | 0.050 |
| $Q_0$ (pol.) | 0.29 | – | 0.13 | 0.102 | 0.056 |
| $Q_1$ (no pol.) | 0.548 | 0.436 | 0.325 | 0.274 | 0.237 |
| $Q_1$ (pol.) | 0.513 | 0.407 | 0.325 | 0.294 | 0.262 |

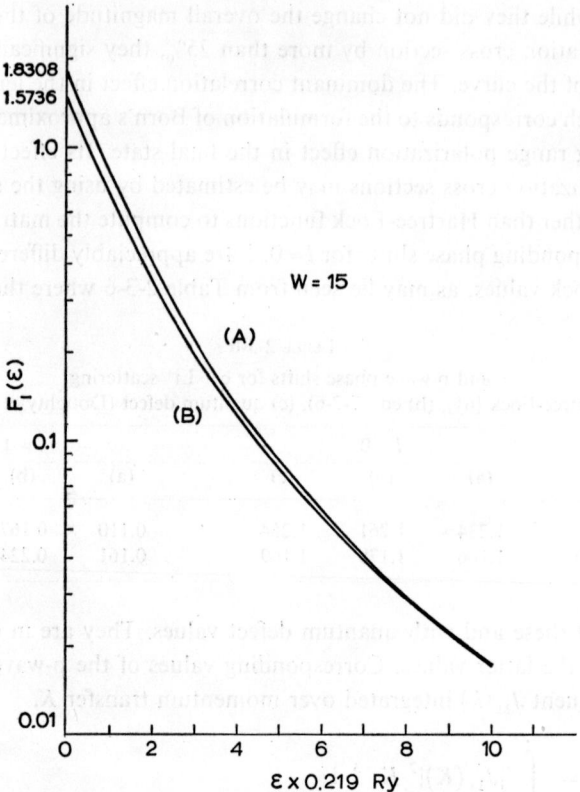

Fig. 2-3-9. The dominant $l=2$ squared and integrated matrix element $F_1(\varepsilon)$ for electron ionization of Li, (A) with polarization potential, (B) without, as a function of ejected electron energy in units of 0.219 Ry at $W=15$ Ry.

Polarization decreases these cross section at low impact energies, but increases them for $W>10$ Ry, the effect being of order 10% by $E=20$ Ry. However the higher partial cross sections $Q_l(l>2)$ are unaltered, since for example though polarization effects change the d-wave phase shift from $3\times10^{-3}$ to $6\times10^{-3}$ at threshold, a displacement of the nodes of the continuum radial function by $3\times10^{-3}$ radians $(k=1.0)$ cannot significantly affect the corresponding matrix element. The total calculated ionization cross section is increased by 2% at $W=15$ Ry and by 4% at $W=20$ Ry, which is insignificant by comparison with the discrepancy of a factor of two between our calculated results and experiment at these energies.

Assuming for the moment that given exact atomic wave functions Born's approximation would be valid at say 40 times threshold (say 200 eV), then the fact that the present calculations lie almost a factor of two below experiment at this energy can be attributed to

a) Errors of order $\pm 50\%$ in the determination of the absolute experimental values,

b) Inadequacy of the Hartree-Fock and polarized Hartree-Fock wave functions,

c) Some combination of (a) and (b), or

d) Some physical process neglected by either the theoreticians or the experimentalists.

We now show that hypothesis (b) is not satisfactory. We can reformulate Born's approximation by assuming the exactness of the Hartree-Fock wave functions and using the commutator

$$\left[H, \sum_{j=1}^{3} e^{i\mathbf{K}\cdot\mathbf{r}_j}\right]$$

to obtain an equivalent velocity form.

Let

$$I_{0n} = \int \Psi_i^*(1,2,3) \sum_{j=1}^{3} e^{i\mathbf{K}\cdot\mathbf{r}_j} \Psi_f(1,2,3) \, d\mathbf{r}_N \qquad (2\text{-}3\text{-}32)$$

$$= -\tfrac{1}{2}K^2 f(k_f, k_e)$$

and noting that

$$\left[\sum_{j=1}^{3} e^{i\mathbf{K}\cdot\mathbf{r}_j}, H\right] = \left(\sum_{j=1}^{3} e^{i\mathbf{K}\cdot\mathbf{r}_j}\right)\sum_{t}\nabla_t^2 - \left(\sum_{t}\nabla_t^2\right)\left(\sum_{j=1}^{3} e^{i\mathbf{K}\cdot\mathbf{r}_j}\right) \qquad (2\text{-}3\text{-}33)$$

we have

$$I_{0n} = \frac{1}{(E_0 - E_f)}\left[-K^2 I_{0n} + 2iK \int \Psi_f^* \sum_{j=1}^{3} e^{i\mathbf{K}\cdot\mathbf{r}_j} \frac{\partial \Psi_0}{\partial Z_j} \, d\mathbf{r}_N\right] \qquad (2\text{-}3\text{-}34)$$

for Li. Since all the orbitals in $\Psi_0$ have angular momentum zero,

$$\frac{\partial}{\partial Z_j}\phi_{ns}(r_j) = \frac{\cos\theta}{r_j}\left[\frac{d}{dr_j} - \frac{1}{r_j}\right]P_{ns}(r_j)Y_{00}(\hat{r}_j) = \bar{X}_{ns} \qquad (2\text{-}3\text{-}35)$$

so that it replaces an s-orbital by a modified p-orbital, $\bar{X}_{ns}$.

In particular for $l=1, 2$, on carrying out the angular integrations we find

$$J_{1\varepsilon}(K) = \frac{1}{(I+\varepsilon)}\left[K^2 J_{1\varepsilon}(K) + 2K\left\{\tfrac{2}{3}\bar{J}_{2\varepsilon}(K) - \tfrac{1}{3}\bar{J}_{0\varepsilon}(K)\right\}\right] \qquad (2\text{-}3\text{-}36)$$

and

$$J_{2\varepsilon}(K) = \frac{1}{(I+\varepsilon)}[K^2 J_{2\varepsilon}(K) + 2K\{\tfrac{3}{5}\bar{J}_{3\varepsilon}(K) - \tfrac{2}{3}\bar{J}_{1\varepsilon}(K)\}] \qquad (2\text{-}3\text{-}37)$$

where $J_{l\varepsilon}(K)$ is defined by (2-3-16) and $\bar{J}_{l\varepsilon}(K)$ is the same matrix element with $P_{2s}(r)$ replaced by the radial part of $X_{ns}$. These results may be readily checked by noting that as $K \to 0$ that for $l=1$ reduces to

$$\sigma_L = \frac{2}{I+\varepsilon}\sigma_V \qquad (2\text{-}3\text{-}38)$$

and similarly for $l=2$.

We denote the partial cross section evaluated by replacing $J_{l\varepsilon}(K)$, ($l=1, 2$) by the right hand side of (2-3-36) or (2-3-37) by $\bar{Q}_l(k^2)$, the "velocity" form. If the atomic wave functions were exact then

$$Q_l(k^2) = \bar{Q}_l(k^2), \quad \text{all } l.$$

Agreement between them to a given accuracy is of course merely a necessary, but not a sufficient condition on the accuracy of the wave functions. It would however, if that accuracy were better than 10%, make it difficult to advance hypothesis (b) alone.

TABLE 2-3-8
Radial matrix element $J_{2\varepsilon}(K)$, $\varepsilon=0$; $W=20$ Ry.

| $N$ (see text) | Length | Velocity |
|---|---|---|
| 1 | 6.55, $-3$ | 6.46, $-3$ |
| 2 | 5.82, $-2$ | 5.73, $-2$ |
| 3 | 1.60, $-1$ | 1.58, $-1$ |
| 4 | 3.01, $-1$ | 2.96, $-1$ |
| 5 | 4.49, $-1$ | 4.42, $-1$ |
| 10 | 4.76, $-1$ | 4.69, $-1$ |
| 15 | 1.19, $-1$ | 1.20, $-1$ |

In Table 2-3-8 we compare the two sides of (2-3-37) above for $W=20$ Ry, $\varepsilon=0$ as a function of $K=N(K_{\max}-K_{\min})/100$. The differences are small, the velocity result being slightly smaller. The velocity matrix elements are always smaller than the corresponding length matrix elements, and since the integral over ejected electron energy is dominated by small values of $\varepsilon$, the overall reduction in $\bar{Q}_2(k^2)$ as compared with $\bar{Q}_1(k^2)$ is about 5% for $W>5$, whereas at low energies ($W<5$), $\bar{Q}_2(k^2)$ is about 3% greater than $Q_2(k^2)$. Similar results are obtained for $l=1$. Values of the momentum transfer integrated squared matrix element $F_2(\varepsilon)$, which is proportional to the ejected electron energy-differential of the cross section, in both length and velocity forms are tabulated in Table 2-3-9 for incident energies of 2, 10 and 20 Ry. The percentage difference drops from 7% at threshold ($\varepsilon=0$) to approximately 3% at high values of $\varepsilon$. This is quite satisfactory agreement in the present context, and indicates that the source of the discrepancy between the Born approximation results, calculated with Hartree-Fock wave functions, and experiment if unlikely to be the inadequacy of the target wave functions.

There remains the possibility that the experimental and theoretical results are both correct (to within say $\pm 30\%$) but that the Born series does not converge to its first term at high energies [55]. The much more accurate experimental results for He [56] ($\pm 10\%$) are in good agreement with comparable Born calculations [24], but it may be that the convergence of the Born series follows a different pattern for collisions with open shell systems.*

TABLE 2-3-9A

$F_2(\varepsilon)$, $W=2$ Ry

| $\varepsilon$ | Length $F_2(\varepsilon)$ | Velocity $\bar{F}_2(\varepsilon)$ |
|---|---|---|
| 0 | 1.4482 | 1.4964 |
| 0.024 | 1.2256 | 1.2636 |
| 0.048 | 1.0441 | 1.0742 |
| 0.072 | 0.8956 | 0.9197 |
| 0.096 | 0.7732 | 0.7926 |
| 0.120 | 0.6716 | 0.6873 |
| 0.144 | 0.5866 | 0.5995 |
| 0.168 | 0.5152 | 0.5257 |
| 0.192 | 0.4547 | 0.4633 |
| 0.216 | 0.4031 | 0.4103 |
| 0.240 | 0.3590 | 0.3649 |
| 0.32 | 0.2640 | – |
| 0.40 | 0.1909 | – |
| 0.48 | 0.1424 | – |
| 0.56 | 0.1089 | – |
| 0.64 | 0.0852 | – |
| 0.72 | 0.0677 | – |
| 0.80 | 0.0548 | – |

* See also for He, Sloan [63], for H⁻, Bely and Schwartz [64], for Li⁺, Peart and Dolder [65] and for a general discussion Mc Dowell and Economides [66].

TABLE 2-3-9B

| | Ejected electron energy $\varepsilon$(Ry) | Length $F_2(\varepsilon)$ | Velocity $F_2(\varepsilon)$ | % Difference |
|---|---|---|---|---|
| | 0.000 | 1.6432 | 1.5375 | 6.8 |
| | 0.144 | 0.6452 | 0.6105 | 5.6 |
| | 0.288 | 0.3096 | 0.2952 | 4.8 |
| $W = 10$ Ry | 0.432 | 0.1704 | 0.1634 | 4.3 |
| | 0.576 | 0.1034 | 9.957, $-2$ | 3.8 |
| | 0.720 | 6.739, $-2$ | 6.510, $-2$ | 3.5 |
| | 0.864 | 4.638, $-2$ | 4.491, $-2$ | 3.3 |
| | 1.008 | 3.333, $-2$ | 3.232, $-2$ | 3.1 |
| | 0.000 | 1.6490 | 1.5411 | 7.0 |
| | 0.294 | 0.3035 | 0.2877 | 5.5 |
| | 0.588 | 0.1004 | 9.587, $-2$ | 4.7 |
| $W = 20$ Ry | 0.882 | 4.484, $-2$ | 4.303, $-2$ | 4.2 |
| | 1.176 | 2.391, $-2$ | 2.303, $-2$ | 3.8 |
| | 1.470 | 1.431, $-2$ | 1.382, $-2$ | 3.5 |
| | 1.764 | 9.280, $-3$ | 8.980, $-3$ | 3.3 |
| | 2.058 | 6.385, $-3$ | 6.187, $-3$ | 3.2 |

## § 2-4. Conclusions

We have considered the available measurements and theoretical predictions of photoionization and electron impact ionization of the alkalis. There is good accord between theories which take account of correlation effects and experiment for photoionization of Li and Na from threshold to about 900 Å, where the experimental results appear to high. There are no reliable theoretical calculations for the other alkalis.

The situation is not so clear in electron impact ionization. The experimental results, which are consistent among all but one of the various experiments lie in general a factor of two above the theoretical predictions (which, however refer only to Li and to a lesser extent, Na) in the region where the Born approximation might be expected to be valid, when the calculations are carried out with wave functions of sufficient complexity to yield accurate values for the photoionization cross section*. However theoretical predictions made with crude wave functions which give poor values for $a_v$, are in close accord with experiment for electron impact ionization of Li.

The matrix element for the dominant s→d ($l=2$) transition in the ion-

---

* I am indebted to Dr. A. Temkin for pointing out that theoretical close coupling calculations on total elastic scattering of slow electrons by Li [58] also give results lying a factor of two below experiment [59].

ization process seems very sensitive to the details of the continuum wave function at small distances from the nucleus. Matrix elements computed with a continuum Hartree-Fock wave function and a Roothaan Hartree-Fock ground state, are 10% larger than those calculated with a simpler ground state wave function [21]; and 20% larger than those compared with a pure Coulomb continuum function, even though the Hartree-Fock functions phase shift $(l=2)$ at threshold is only $3.8 \times 10^{-3}$. However at small $r(<2a_0)$ the Hartree-Fock radial function differs appreciably from a pure Coulomb function, and the radial bound 2s function is fairly compact, so that it is precisely such small values of $r$ which affect the matrix element. We believe that the agreement between length and velocity forms obtained here, indicates that most of these uncertainties have been removed, and that within the validity of Born's approximation, the theoretical values of the outer shell, single ionization cross section are reliable to better than $\pm 10\%$ for Li and Na.

We suggest that 80% higher values obtained in the experiments at high energies ($>200$ eV) are caused by uncertainties in the determination of the efficiency of an unoxidized hot tungsten wire detector, when this efficiency is low, and to the presence (especially in the case of Li) of diatoms of the target species in the beam, together with contributions from autoionization and inner shell ionization, which may be substantial (except in the case of Li).

Secondary electrons produced at collimating slit edges may, at impact energies in excess of 200 eV, be an important source of ionization and would lead to excessively large cross sections at higher energies. In the range of the measurements the cross section has not approached its asymptotic $E^{-1} \log E$ behaviour, so the slope cannot be compared with calculations of the oscillator strength. Measurements at electron energies up to 2 keV would allow such a check. These difficult experiments have been carried out with considerable care, but further work is now desirable. Detailed theoretical calculations of autoionization (induced by electron impact) in Li and Na would be of considerable value.

## Acknowledgement

I am indebted to Drs. R. A. McFarland, J. W. Cooper, K. Omidvar and R. Hudson for their helpful criticism of an initial draft, and for permission to use some unpublished results. York University, Toronto, Canada provided hospitality and secretarial assistance during the later part of the work. Mr. G. Catlow carried out the calculations on inner shell ionization.

## References

1. J. A. R. SAMSON, Advances in Atomic and Molecular Physics **2**, eds. D. R. Bates and I. Estermann (Academic Press, N.Y., 1966) pp 177–261 (especially pp. 244–253).
2. R. W. DITCHBURN and P. J. JUTSUM, Nature **165** (1950) 723.
3. R. W. DITCHBURN and G. V. MARR, Proc. Roy. Soc. (London) **A 219** (1953) 89.
4. J. TUNSTED, Proc. Phys. Soc. (London) **A 66** (1953) 304.
5. G. V. MARR, Proc. Phys. Soc. (London) **81** (1963) 9.
6. R. D. HUDSON and V. L. CARTER, Phys. Rev. **137** (1965a) A 1648.
7. R. D. HUDSON and V. L. CARTER, J. Opt. Soc. Am. **57** (1967) 651.
8. R. D. HUDSON, Phys. Rev. **135** (1964) A 1212;
   A. R. WILLIAMS, J. Chem. Phys. **47** (1967) 4281.
9. K. J. NYGAARD, Sperry Rand Corp., Tech. Memo. 457-26 (1966).
10. G. O. BRINK, Phys. Rev. **127** (1962) 1204.
11. G. O. BRINK, Phys. Rev. **134** (1964) A 345.
12. R. A. MCFARLAND and J. D. KINNEY, Phys. Rev. **137** (1965) A 1058.
13. R. A. MCFARLAND and J. D. KINNEY, Proc. 7th Intern. Conf. on Ioniz. Gas, Belgrade, Vol. 1 (1960) p. 254.
14. R. A. MCFARLAND, Phys. Rev. **159** (1967) 20.
15. I. S. ALEKSAKHIN, I. P. ZAPESOCHNYI and O. B. SHPENIK, Abstracts 5th Intern. Conf. on the Physics of Electronic and Atomic Collisions (Nauk, Leningrad, 1967) p. 499.
16. H. HEIL and B. SCOTT, Phys. Rev. **145** (1966) 279.
17. K. J. NYGAARD, Sperry Rand Report, SRRC-CR-68-4 (1968), and J. Chem. Phys. (1968).
18. E. S. CHANG and M. R. C. MCDOWELL, Phys. Rev. **176** (1968) 126.
19. S. MANSON and J. COOPER, private communication (1968).
20. G. PEACH and M. R. C. MCDOWELL, Phys. Rev. **121** (1961) 1383.
21. M. R. C. MCDOWELL, V. P. MYERSCOUGH and G. PEACH, Proc. Phys. Soc. (London) **85** (1965) 703.
22. G. CATLOW and M. R. C. MCDOWELL, Proc. Phys. Soc. (London) **92** (1967) 875. (See also E. BAUER and C. D. BARTHY, J. Chem. Phys. **43** (1965) 2466.)
23. K. OMIDVAR and E. SULLIVAN, to be submitted to Phys. Rev., see a brief version in Abstracts 5th Intern. Conf. on the Physics of Electronic and Atomic Collisions (Nauk, Leningrad, 1967) p. 446.
24. G. PEACH, Proc. Phys. Soc. **87** (1966) 375 and 381.
25. D. R. BATES, A. H. BOYD and S. S. PRASAD, Proc. Phys. Soc. (London) **85** (1965) 1121.
26. A. H. BOYD, Planetary Space Sci. **12** (1964) 729.
27. E. J. MCGUIRE, Phys. Rev. **161** (1967) 51.
28. K. G. SEWELL, J. Chem. Phys. **57** (1967) 1058.
29. J. H. TAIT, in: Atomic Collision Processes, ed. M. R. C. McDowell (North-Holland Publishing Co., Amsterdam, 1964) p. 586.
30. A. L. STEWART, Proc. Phys. Soc. (London) **A67** (1954) 64.
31. JANEF Thermochemical Tables (Dow Chemical Co., Midland, Mich., 1962);
    W. T. HICKS, J. Chem. Phys. **38** (1962) 1873.
32. A. N. NESMEYANOV, Vapour Pressure of the Chemical Elements (Elsevier, N.Y., 1963).
33. R. E. HORNIG, RCA Rev. **18** (1957) 195.
34. J. A. HORNBECK and J. P. MOLNAR, Phys. Rev. **84** (1951) 621.
35. F. L. MOHLER and C. BOECKNER, Bur. Standard J. Res. **5** (1930) 51.
36. J. COOPER and U. FANO, Rev. Mod. Phys. **40** (1968) 441.
37. L. D. LANDAU and E. M. LIFSHITZ, Quantum Mechanics (Addison-Wesley, N.Y.) p. 43.
38. D. R. BATES, Monthly Notices Roy. Astron. Soc. **106** (1946) 423 and 432.

39. M. J. SEATON, Proc. Roy. Soc. A **208** (1951) 418.
40. H. A. BETHE and E. E. SALPETER, Quantum Mechanics of One and Two Electron Systems (Academic Press, N.Y., 1957).
41. K. OMIDVAR, private communication (1968).
42. H. EYING, J. WALTER and G. E. KIMBALL, Quantum Chemistry (Wiley, N.Y., 1944) p. 162.
43. V. FOCK and M. J. PETRASHEN, Phys. Zeit. f. Sowjet. **8** (1935) 547.
44. J. COOPER and S. MANSON, Phys. Rev. (1967).
45. M. J. SEATON, Monthly Notices Roy. Astron. Soc. **118** (1958) 504.
46. G. PEACH, Mem. Roy. Astron. Soc. **71** (1967) 1 and 12.
47. J. S. BELL and E. SQUIRES, Adv. in Phys. **10** (1961) 211, is expository and gives full references to earlier work.
48. J. T. TATE and P. T. SMITH, Phys. Rev. **46** (1934) 773.
49. H. L. WITTING, reported in [14].
50. L. L. MORINO, A. C. H. SMITH and E. CAPLINGER, Phys. Rev. **128** (1962) 2243.
51. S. O. MARTIN, B. PEART and K. T. DOLDER, J. Phys. B. (Proc. Phys. Soc.), (1968).
52. G. PEACH and M. R. C. MCDOWELL, in: Atomic Collision Processes, ed. M. R. C. McDowell (North-Holland Publishing Co., Amsterdam, 1964) p. 277.
53. R. PETERKOP, Proc. Phys. Soc. **77** (1961) 1220.
54. L. BIERMAN and K. LUBECK, Z. Astrophys. **25** (1948) 325.
55. K. DETTMANN and G. LEIBFRIED, Z. Physik **210** (1968) 43.
56. L. J. KEIFFER and G. H. DUNN, Rev. Mod. Phys. **38** (1966) 1.
57. YU. P. KORCHEVOI and A. M. PRZONSKI, Sov. Phys. JETP **24** (1967) 1089.
58. R. MARRIOT and M. ROTENBERG, Abstracts 5th Intern. Conf. on the Physics of Electronic and Atomic Collisions (Nauk, Leningrad, 1967) p. 379.
59. J. PEREL, P. ENGLANDER and B. BEDERSON, Phys. Rev. **128** (1962) 1148.
60. N. A. DOUGHTY, M. J. SEATON and V. SHEORY, J. Phys. B (Proc. Phys. Soc.) **1** (1969) 802.
61. M. R. C. MCDOWELL, Phys. Rev. **175** (1968) 189.
62. A. L. STEWART, Advances in Atomic and Molecular Physics **3** (1967) 1.
63. I. H. SLOAN, Proc. Phys. Soc. **85** (1966) 435.
64. O. BELY and S. B. SCHWARTZ, J. Phys. B (Atom. and Mol. Phys.), Ser. 2, **2** (1969) 159.
65. B. PEART and K. DOLDER, submitted to J. Phys. B (Atom. and Mol. Phys.).
66. M. R. C. MCDOWELL and D. G. ECONOMIDES, submitted to J. Phys. B (Atom. and Mol. Phys.).
67. I. P. ZAPESOCHNYI and I. S. ALEKSAKHIN, J.E.P.T. **28** (1969) 41.

CHAPTER 3

# THE IMPULSE APPROXIMATION AND RELATED METHODS IN THE THEORY OF ATOMIC COLLISIONS

BY

J. P. COLEMAN

*Department of Mathematics,*
*University of Durham, England*

## Contents

|  | Page |
|---|---|
| 3-1. Introduction | 101 |
|    A. Notation | 101 |
|    B. Some results from the formal theory of scattering | 104 |
| 3-2. The impulse approximation | 107 |
|    A. A simple model | 107 |
|    B. Formal derivation | 112 |
|    C. Electron capture | 115 |
|       1. Atomic hydrogen target | 115 |
|       2. Helium target | 128 |
|       3. The high-energy limit | 129 |
|    D. Excitation | 131 |
|    E. Ionization | 136 |
|    F. Summary | 137 |
| 3-3. The Vainshtein, Presnyakov and Sobelman approximation | 138 |
|    A. The post form of the VPS approximation | 139 |
|       1. Derivation of the approximation | 139 |
|       2. The peaking approximation | 143 |
|       3. The inclusion of exchange | 147 |
|       4. The effective charge | 148 |
|    B. The prior form of the VPS approximation | 149 |
|    C. Summary | 152 |
| 3-4. The extended impulse approximation | 154 |
|    A. The post form | 154 |
|    B. The prior form | 159 |
|    C. Summary | 159 |
| Appendix 1. Coulomb functions | 160 |
| Appendix 2. Nordsieck's integration technique | 161 |
| References | 165 |

## § 3-1. Introduction

Two approximations which have been used extensively in atomic collision calculations are examined in this review. The quantum mechanical impulse approximation is derived and discussed in §3-2, and §3-3 deals with an approximation proposed by Vainshtein et al. [1]. The evolution of both methods is described and particular attention is given to the errors and over-simplifications which have been detected and rectified, and the shortcomings which still remain. Both approximations have achieved a moderate degree of agreement with experiments. Here, however, greater emphasis is placed on what still remains to be done and an attempt is made to suggest directions which future research may take. It is shown in §3-4 that the methods discussed in §3-2 and §3-3 may be regarded as approximate versions of an improved impulse approximation and the consequences of this are explored.

The exclusion of all other approximations from this review does not imply that the methods discussed here are in any sense superior to all others. At low energies eigenfunction expansion methods have achieved notable successes for electron atom scattering and similar methods should be satisfactory for heavy particle collisions. However, as the projectile energy is increased the number of states which must be retained in the expansion becomes prohibitively large (see e.g. Burke et al. [2, 3]). At very high energies it is generally believed that the Born approximation correctly predicts the cross section for any direct process, but there is still a need for methods which fully or partially bridge the gap between the energies at which the Born approximation can be assumed reliable and the low energies where close coupling methods can be applied without undue effort. For rearrangement collisions the situation is less satisfactory and, although the correct high-energy limit has not been rigorously established for any rearrangement process, the present consensus of opinion is that it is not given by the first Born approximation. The methods discussed here may be regarded as high-energy approximations which have been proposed either to provide excitation cross sections when the Born approximation may be unreliable, or in an attempt to elucidate the high-energy behaviour of cross sections for rearrangement processes.

A. NOTATION

Throughout this article we shall be concerned with collisions between a structureless particle of mass $M_1$ and charge $Z_1$ and an atom consisting of a nucleus of mass $M_2$ and one or more electrons. Since the target most frequently considered is atomic hydrogen we shall only describe the notation used in that case. Atomic units will be used for all except two quantities, cross sections which will be expressed in units of $\pi a_0^2$ ($=8.8 \times 10^{-17}$ cm$^2$) and energies for which a number of different units will be used depending on the problem under consideration. The symbol $M$ without a subscript will be used to denote the proton mass in units of the electron mass.

The position vectors of the atomic electron, labelled particle 3, with respect to the projectile, 1, and the target nucleus, 2, are denoted by $x$ and $r$ respectively and $R$ is the position vector of the projectile with respect to the target nucleus (see Fig. 3-1-1). The position vector of particle 1 with respect to the centre of mass of 2 and 3 is $\sigma$, and $\rho$ is that of the centre of mass of 1 and 3 with respect to particle 2. The following relationships are readily deduced from Fig. 3-1-1:

$$R = r - x, \qquad \sigma = br - x, \qquad \rho = r - ax \qquad (3\text{-}1\text{-}1)$$

where

$$a = \frac{M_1}{M_1 + 1}, \qquad b = \frac{M_2}{M_2 + 1}. \qquad (3\text{-}1\text{-}2)$$

In the centre of mass frame of reference the Hamiltonian of the system is

$$H = H_0 + V_{12} + V_{13} + V_{23} \qquad (3\text{-}1\text{-}3)$$

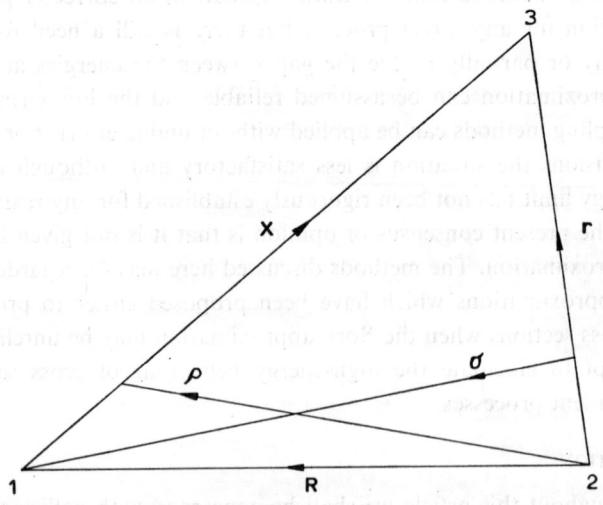

Fig. 3-1-1. Coordinate system for a collision of a structureless particle with a hydrogen atom (not to scale).

where $V_{ij}$ is the interaction potential between the particles labelled $i$ and $j$, and $H_0$ is the kinetic energy operator. $H_0$ takes its simplest form when expressed in terms of the relative coordinates of any two particles and the coordinates of the third particle with respect to the centre of mass of the other two. Thus $r$ and $\sigma$, and $x$ and $\rho$ are suitable sets of independent coordinates and

$$H_0 = -\frac{1}{2b}\nabla_r^2 - \frac{1}{2\mu}\nabla_\sigma^2 = -\frac{1}{2a}\nabla_x^2 - \frac{1}{2\mu'}\nabla_\rho^2 \tag{3-1-4}$$

where

$$\mu = \frac{M_1(M_2+1)}{M_1+M_2+1}, \quad \mu' = \frac{M_2(M_1+1)}{M_1+M_2+1}. \tag{3-1-5}$$

When the collision considered does not involve rearrangement $\mu$ is the reduced mass of the system before and after the collision, and in the case of electron capture $\mu$ and $\mu'$ are respectively the initial and final reduced masses.

It is occasionally necessary to express functions of $r$ and $\sigma$ in terms of $x$ and $\rho$ or vice versa. The required relationships are readily deduced from (3-1-1) and since

$$(1-ab) = \frac{a}{\mu} = \frac{b}{\mu'} \tag{3-1-6}$$

they may be written in the form

$$x = br - \sigma, \quad \rho = a\left(\frac{1}{\mu}r + \sigma\right) \tag{3-1-7a}$$

$$r = ax + \rho, \quad \sigma = b\left(\rho - \frac{1}{\mu'}x\right). \tag{3-1-7b}$$

The interaction between the projectile and the target atom tends to zero more rapidly than $1/\sigma$ as $\sigma \to \infty$. Thus, at infinite separation of the target and projectile before the collision the motion of the system is governed by the "unperturbed" Hamiltonian

$$H_i = H_0 + V_{23} \tag{3-1-8}$$

and the corresponding three-particle wave function satisfies the Schrödinger equation

$$(H_i - E)\psi_i = 0 \tag{3-1-9}$$

and takes the form

$$\psi_i = \exp(i\mathbf{k}_i \cdot \boldsymbol{\sigma})\varphi_i(r). \tag{3-1-10}$$

Here $\phi_i(r)$ is a target eigenfunction, $k_i$ is the initial relative momentum of the projectile and target, and $E$ is the total energy of the system in atomic units, i.e.

$$E = \frac{1}{2\mu} k_i^2 - \mathscr{E}_i \qquad (3\text{-}1\text{-}11)$$

where $\mathscr{E}_i$ is the (positive) binding energy of the initial bound state of the target. The normalization has been chosen so that $\psi_i$ represents a projectile beam of unit density impinging on the target. The form of the final unperturbed wave function $\psi_f$ depends on the type of transition considered.

## B. Some Results from the Formal Theory of Scattering

The formal theory of scattering is treated in many books, e.g. by Goldberger and Watson [4], Roman [5], Newton [6] and McDowell and Coleman [7]. For convenient reference and to establish notation some results which will be required later are briefly discussed in this section. For a more detailed treatment the reader is referred to any of the above-mentioned works.

The Schrödinger equation

$$(H - E)\Psi = 0 \qquad (3\text{-}1\text{-}12)$$

may be written as

$$(E - H_i)\Psi = V_i\Psi \qquad (3\text{-}1\text{-}13)$$

where $H_i$ is defined by (3-1-8) and

$$V_i = V_{12} + V_{13}. \qquad (3\text{-}1\text{-}14)$$

The eigenvalues of the Hamiltonian $H_i$ are all real and consequently, when $\mathrm{Re}(\varepsilon) \neq 0$, the operators $(E - H_i \pm i\varepsilon)$ have well defined inverses which we shall write as

$$G_i^{\pm} = \frac{1}{E - H_i \pm i\varepsilon}. \qquad (3\text{-}1\text{-}15)$$

Without loss of generality $\varepsilon$ may be assumed to be real and positive. Eq. (3-1-13) is therefore equivalent to the integral equation

$$\Psi^{\pm} = \psi_0 + \lim_{\varepsilon \to 0+} G_i^{\pm} V_i \Psi^{\pm} \qquad (3\text{-}1\text{-}16)$$

where $\psi_0$ is any solution of (3-1-9) and may be regarded as an "integration constant" to be determined from the boundary conditions imposed on the relevant solution of (3-1-13). The superscripts "$\pm$" serve to distinguish be-

## 3-1 INTRODUCTION

tween the two distinct classes of solutions of (3-1-13) obtained from (3-1-16) in the limit as $\varepsilon \to 0$; both $\Psi^+$ and $\Psi^-$ contain the incident wave $\psi_0$ but $\Psi^+$ contains *outgoing* scattered waves and $\Psi^-$ contains *ingoing* scattered waves. Thus the wave function which evolves from the state $\psi_i$ and which describes the colliding systems in the presence of all their interactions is

$$\Psi_i^+ = \psi_i + G_i^+ V_i \Psi_i^+ . \tag{3-1-17}$$

As in this equation, we shall frequently omit the symbols denoting limits when dealing with equations involving Green's function operators such as $G_i^\pm$, but in all cases it is understood that the limit as $\varepsilon \to 0$ is to be taken when the relevant integrations have been carried out.

If $P$ and $Q$ are operators for which inverse operators exist it is obvious that

$$P^{-1} = Q^{-1} + P^{-1}(Q - P) Q^{-1} \tag{3-1-18a}$$
$$= Q^{-1} + Q^{-1}(Q - P) P^{-1} . \tag{3-1-18b}$$

Therefore the Green's function operators

$$G^\pm = \frac{1}{E - H \pm i\varepsilon} \tag{3-1-19}$$

satisfy the integral equations

$$G^\pm = G_i^\pm + G^\pm V_i G_i^\pm \tag{3-1-20a}$$
$$= G_i^\pm + G_i^\pm V_i G^\pm . \tag{3-1-20b}$$

Let us now operate with $G^+ V_i$ on both sides of (3-1-17) and use (3-1-20a) to obtain

$$G^+ V_i \Psi_i^+ = G^+ V_i \psi_i + G^+ V_i G_i^+ V_i \Psi_i^+$$
$$= G^+ V_i \psi_i + (G^+ - G_i^+) V_i \Psi_i^+ ; \tag{3-1-21}$$

it follows that

$$G^+ V_i \psi_i = G_i^+ V_i \Psi_i^+ \tag{3-1-22}$$

and therefore a formal solution of (3-1-17) is

$$\Psi_i^+ = (1 + G^+ V_i) \psi_i . \tag{3-1-23}$$

The cross section for any collision process in which all the particles are distinguishable takes the form

$$Q = \frac{\mu \mu_f}{4\pi^3} \left(\frac{k_f}{k_i}\right) \int |T_{if}|^2 \, d\Omega \, (\pi a_0^2), \tag{3-1-24}$$

where $d\Omega$ is an element of solid angle about the direction of the vector $\mathbf{k}_f$, the relative momentum after the collision, and $\mu_f$ is the final reduced mass. The transition matrix element in (3-1-24) is

$$T_{if} = \langle \psi_f | V_f | \Psi_i^+ \rangle \tag{3-1-25a}$$
$$= \langle \Psi_f^- | V_i | \psi_i \rangle \tag{3-1-25b}$$

where $V_f$ is the perturbation in the final state and

$$\Psi_f^- = (1 + G^- V_f) \psi_f. \tag{3-1-26}$$

Energy conservation demands that the wave functions $\psi_i$, $\psi_f$, $\Psi_i^+$ and $\Psi_f^-$ all correspond to the same total energy $E$.

It is frequently convenient to replace the angular integration in (3-1-24) by integration with respect to the magnitude of a momentum transfer vector which arises in the analysis. For direct transitions (i.e. excitation or ionization of the target) the appropriate momentum transfer vector is

$$\mathbf{q} = \mathbf{k}_i - \mathbf{k}_f \tag{3-1-27}$$

and since in this case $\mu_f = \mu$, (3-1-24) becomes

$$Q = \frac{1}{2\pi^2 v^2} \int_{q_{min}}^{q_{max}} |T_{if}|^2 \, q \, dq \tag{3-1-28}$$

where $v = k_i/\mu$. The integration limits in (3-1-28) are

$$q_{min} = |k_i - k_f|, \quad q_{max} = k_i + k_f \tag{3-1-29}$$

and $k_f$ is determined by the equation

$$k_i^2 - k_f^2 = \mu \, \Delta E \tag{3-1-30}$$

where $\Delta E$ is the difference, in Rydbergs, between the binding energies of the initial and final bound states. When one is dealing with electron capture it is more convenient to express $Q$ as an integral with respect to the magnitude of

$$\mathbf{p} = a\mathbf{k}_f - \mathbf{k}_i. \tag{3-1-31}$$

In this case $\mu_f = \mu'$ and

$$Q = \frac{1}{2\pi^2 v^2 a} \left(\frac{\mu'}{\mu}\right) \int_{p_{min}}^{p_{max}} |T_{if}|^2 \, p \, dp \tag{3-1-32}$$

with
$$p_{\min} = |ak_f - k_i|, \quad p_{\max} = ak_f + k_i \qquad (3\text{-}1\text{-}33)$$
and
$$\frac{k_i^2}{\mu} - \frac{k_f^2}{\mu'} = \Delta E. \qquad (3\text{-}1\text{-}34)$$

## § 3-2. The impulse approximation

In classical mechanics an impulsive force is a force of very large magnitude which acts for an infinitesimal length of time. More precisely, if $F(t)$ is an impulsive force which acts from time $t_0$ to $t$, $\lim_{t \to t_0} \int_{t_0}^{t} F(t) \, dt$ is finite and represents the instantaneous change in the linear momentum of the system on which the force acts. Clearly when considering the effects of an impulsive force all finite forces can be neglected. The analogous situation in the scattering of a particle by a bound system occurs when the duration of the collision is short compared with some time interval characteristic of the target, for example the orbital period of a bound particle. In that case it may be assumed that, apart from determining the momentum distributions of the bound states, the target binding forces do not play an important role in the collision. This assumption, which will be termed the *impulse hypothesis*, forms the basis of the quantum mechanical impulse approximation, but in general the impulse approximation also involves some subsidiary assumptions which enable one to reduce the cross section formulae to tractable form.

If impact parameters greater than $\rho_0$ provide a negligible contribution to the cross section for the process considered, a convenient measure of the collision time is the time taken by the projectile to travel the distance $\rho_0$. In particular, if $\rho_0$ is of the order of one atomic unit, a collision may be regarded as impulsive if $v \gg 1$, in other words if the projectile energy greatly exceeds $M_1$ Rydbergs. This crude criterion which is certainly sufficient, but may not be necessary to justify the impulse hypothesis, at least shows that the impulse approximation may be expected to become increasingly accurate as the projectile energy is increased.

### A. A SIMPLE MODEL

The impulse approximation was first introduced by Chew [8] who applied it to high-energy neutron-deuteron scattering. To simplify the problem he considered the scattering of a particle (particle 1 of Fig. 3-1-1) by a target consisting of a single particle, 3, bound to a fixed centre of force, 2, and assumed that there was no interaction between the projectile and the centre of force. This model provides a useful introduction to the impulse approximation because, by avoiding complications which arise in more complex problems, and which necessitate the introduction of additional assumptions, it demonstrates very clearly the main features of the method.

The Hamiltonian for the system under consideration is simply

$$H = H_0 + V_{23} + V_{13} \tag{3-2-1}$$

and the exact $T$ matrix element for a transition of the particle 3 from a state i to a state f within the potential well is

$$T_{if} = \langle \psi_f | V_{13} | \Psi_i^+ \rangle. \tag{3-2-2}$$

In the Born approximation the corresponding matrix element is

$$T_{if}^B = \langle \psi_f | V_{13} | \psi_i \rangle$$
$$= \int d\mathbf{R} \int d\mathbf{r} \exp(i\mathbf{q} \cdot \mathbf{R}) \, \varphi_f^*(\mathbf{r}) \, V_{13}(x) \, \varphi_i(\mathbf{r}) \tag{3-2-3}$$

in the notation of §3-1. If the equation

$$g_n(\mathbf{k}) = \int \exp(-i\mathbf{k} \cdot \mathbf{r}) \, \varphi_n(\mathbf{r}) \, d\mathbf{r}, \tag{3-2-4}$$

which defines the Fourier transform of the wave function $\varphi_n$, is inverted we obtain

$$\varphi_n(\mathbf{r}) = (2\pi)^{-3} \int \exp(i\mathbf{k} \cdot \mathbf{r}) \, g_n(\mathbf{k}) \, d\mathbf{k}. \tag{3-2-5}$$

Consequently

$$T_{if}^B = (2\pi)^{-6} \int d\mathbf{k}_1 \int d\mathbf{k}_2 \, g_f^*(\mathbf{k}_2) \, g_i(\mathbf{k}_1) \int d\mathbf{R} \int d\mathbf{r}$$
$$\times \exp\{-i(\mathbf{k}_f \cdot \mathbf{R} + \mathbf{k}_2 \cdot \mathbf{r})\} \, V_{13}(x) \exp\{i(\mathbf{k}_i \cdot \mathbf{R} + \mathbf{k}_1 \cdot \mathbf{r})\}. \tag{3-2-6}$$

The integral over coordinate space in (3-2-6) is simply the matrix element which describes, in the Born approximation, the scattering of two free particles, with initial momenta $\mathbf{k}_1$ and $\mathbf{k}_i$, and final momenta $\mathbf{k}_2$ and $\mathbf{k}_f$, which interact via the potential $V_{13}(x)$. It is clear from (3-2-6) that $T_{if}^B$ is obtained by averaging this two-particle matrix element over the momentum distributions of the initial and final bound states. Consequently the Born approximation for this problem involves the following two assumptions:

(i) The particle 3 is regarded as free for the duration of the collision, but its momentum distribution is taken to be that of the appropriate bound state. This is the impulse hypothesis.

(ii) The interaction between the two particles is assumed to be so weak that the two-particle scattering is adequately described by the Born approximation.

When the coordinate integral in (3-2-6) is expressed in terms of $x$ and $\rho$, the variables appropriate for the description of scattering of the free particles 1 and 3, that equation becomes

$$T_{if}^B = (2\pi)^{-3} \int dk_1 \int dk_2 \, g_f^*(k_2) \, g_i(k_1) \, \delta(q + k_1 - k_2) \, T_{kK}^B \qquad (3\text{-}2\text{-}7)$$

where

$$k = ak_2 - (1-a) k_f, \quad K = ak_1 - (1-a) k_i. \qquad (3\text{-}2\text{-}8)$$

The two-particle Born approximation matrix element

$$T_{kK}^B = \int \exp\{i x \cdot (K - k)\} \, V_{13}(x) \, dx \qquad (3\text{-}2\text{-}9)$$

describes the scattering in the centre of mass frame of the two particles. However, in (3-2-7) the values of the momenta $k$ and $K$ are restricted by the momentum-conserving delta function, $\delta(q + k_1 - k_2)$ and are in general unequal, whereas if (3-2-9) described real two-particle scattering the magnitudes $k$ and $K$ would be equal because of energy conservation. Matrix elements such as $T_{kK}^B$, in which $k \neq K$, are said to be "off the energy shell".

The impulse approximation retains the assumption that particle 3 can be regarded as free but the restriction (ii) is removed by replacing the Born two-particle matrix element by the exact matrix element for the same process. Thus

$$T_{if}^{IMP} = (2\pi)^{-3} \int dk_1 \int dk_2 \, g_f^*(k_2) \, g_i(k_1) \, \delta(q + k_1 - k_2) \, T_{kK} \qquad (3\text{-}2\text{-}10)$$

where

$$T_{kK} = \int \exp(-i k \cdot x) \, V_{13}(x) \, \psi_K^+(x) \, dx \qquad (3\text{-}2\text{-}11)$$

and $\psi_K^+(x)$ is the solution of the equation

$$\left\{ -\frac{1}{2a} \nabla_x^2 + V_{13}(x) - \frac{1}{2a} K^2 \right\} \psi_K^+(x) = 0 \qquad (3\text{-}2\text{-}12)$$

which satisfies outgoing-wave boundary conditions. In particular, if

$$V_{13}(x) = -Z_1/x,$$

$$\psi_K^+(x) = \exp\left(\frac{aZ_1\pi}{2K}\right)\Gamma\left(1 - \frac{iaZ_1}{K}\right)$$

$$\times \exp(i\mathbf{K}\cdot\mathbf{x})\,_1F_1\left[\frac{iaZ_1}{K}, 1, i(Kx - \mathbf{K}\cdot\mathbf{x})\right]. \quad (3\text{-}2\text{-}13)$$

Whenever the Born approximation is valid so is the impulse approximation, but the latter is also valid when the two-particle interaction cannot be assumed to be weak, provided the target binding forces can be neglected.

We began this discussion by writing down the Born approximation matrix element as an integral in coordinate space and then transformed this to momentum space to clarify the nature of the assumptions made. Let us now complete the cycle by transforming (3-2-10) back to coordinate space, in order to determine the form of the impulse approximation wave function. The $k_2$ integral is trivial because of the delta function and, if $g_f^*(k_2)$ and $T_{kK}$ are written as coordinate integrals as in (3-2-4) and (3-2-11), (3-2-10) becomes

$$T_{if}^{\text{IMP}} = \int d\mathbf{x} \int d\mathbf{r}\, \varphi_f^*(\mathbf{r}) \exp(-i\mathbf{k}_f\cdot\mathbf{R})\, V_{13}(x)\, \Psi_i^{\text{IMP}} \quad (3\text{-}2\text{-}14)$$

where

$$\Psi_i^{\text{IMP}} = (2\pi)^{-3} \exp\{i\mathbf{k}_f\cdot\mathbf{R}\} \int d\mathbf{k}_1\, g_i(\mathbf{k}_1)$$

$$\times \exp\{i\mathbf{r}\cdot(\mathbf{q}+\mathbf{k}_1) - i\mathbf{x}\cdot(\mathbf{K}+\mathbf{q})\}\,\psi_K^+(x)$$

$$= (2\pi a)^{-3} \int d\mathbf{K}\, g_i\left(\frac{1}{a}\mathbf{K}+\mathbf{v}\right) \exp\left\{i\boldsymbol{\rho}\cdot\frac{1}{a}(\mathbf{K}+\mathbf{k}_i)\right\}\psi_K^+(x).$$

$$(3\text{-}2\text{-}15)$$

Thus the impulse approximation for this problem involves the replacement of the exact wave function $\Psi_i^+$ in (3-2-2) by the approximate wave function $\Psi_i^{\text{IMP}}$ defined by (3-2-15).

In his pioneering work on the impulse approximation Chew [8] was primarily interested in deducing the free neutron–neutron scattering cross section from measurements of n–d inelastic cross sections. For this reason he introduced a further approximation, usually called a peaking approximation, which greatly simplifies the dependence of $T_{if}^{\text{IMP}}$ on the two-particle scattering amplitude. The peaking approximation is based on the fact that, if a function $F_1(\mathbf{K})$ has a sharp peak at $\mathbf{K}=\mathbf{K}_0$ and is negligible for values of $K$ far removed from the peak, and $F_2(\mathbf{K})$ varies slowly in the vicinity of

$K_0$, one can write

$$\int F_1(K) F_2(K) \, dK \approx F_2(K_0) \int F_1(K) \, dK. \qquad (3\text{-}2\text{-}16)$$

Clearly (3-2-16) becomes an identity if $F_1(K)$ is a Dirac delta function or $F_2$ is independent of $K$. If in the present problem it is assumed that $g_f^*(k_1 + q) \times g_i(k_1)$ is sharply peaked when $k_1 = -q$ and that $T_{kK}$ varies slowly near the peak, the above reasoning applied to (3-2-10) yields

$$T_{if}^{IMP} \approx (2\pi)^{-3} T_{k'_0 K'_0} \int dk_1 \, g_f^*(k_1 + q) \, g_i(k_1) \qquad (3\text{-}2\text{-}17)$$
$$= T_{k'_0 K'_0} \Phi(q)$$

where

$$k'_0 = -\frac{a}{\mu} k_f, \qquad K'_0 = -a(q + v) \qquad (3\text{-}2\text{-}18)$$

and

$$\Phi(q) = \int \varphi_f^*(r) \, \varphi_i(r) \exp(iq \cdot r) \, dr. \qquad (3\text{-}2\text{-}19)$$

By transforming to coordinate space it is seen that the wave function which replaces $\Psi_i^+$ in this case is

$$\Psi_i^{Peak} = \exp(i\rho \cdot k_i + iaq \cdot x) \, \psi_{K'_0}^+(x) \, \varphi_i(r). \qquad (3\text{-}2\text{-}20)$$

It is perhaps worthy of note that, because of the relationship $k - K = q$, which is enforced by the delta function in (3-2-10), (3-2-17) is exact if $T_{kK}$ depends only on the difference between $k$ and $K$.

The reason for the assumption that the major contribution to the integral in (3-2-10) comes from the vicinity of $k_1 = -q$ will be discussed later. At this stage it will suffice to point out that although Chew's peaking approximation has enjoyed considerable popularity among nuclear physicists its applicability to atomic collision problems is severely limited. Recent calculations [9] have shown that for the process

$$H^+ + H(1s) \to H^+ + H(2s) \qquad (3\text{-}2\text{-}21)$$

the peaking approximation yields cross section values which are seriously at variance with the correct impulse approximation results. In fact, in that case $T_{if}^B$ is a much better approximation to $T_{if}^{IMP}$ than is the peaking approximation matrix element $T_{k'_0 K'_0} \Phi(q)$. It is unfortunate that some recent reviews [10, 11] appear to suggest that the peaking approximation is an integral part of the impulse approximation, rather than a subsidiary approximation of very doubtful validity.

## B. Formal derivation

The extension of Chew's ideas to the scattering of a particle by a target containing an arbitrary number of particles was carried out by Chew and Goldberger [12]. Their formulation can be applied to *direct* atomic collisions but is not directly applicable to rearrangement collisions because it involves the assumption that the initial and final perturbations are identical. A slightly different version of the impulse approximation was used by Pradhan [13] in his work on electron capture, and the formulation which has been used in all subsequent applications to atomic collision problems is due to McDowell [14]. For the collision depicted in Fig. 3-1-1 the exact three-particle wave function which obeys outgoing-wave boundary conditions is

$$\Psi_i^+ = \Omega^+ \psi_i \qquad (3\text{-}2\text{-}22)$$

where

$$\Omega^+ = 1 + G^+ V_i. \qquad (3\text{-}2\text{-}23)$$

The impulse approximation will be obtained by truncating an expansion of $\Omega^+$ in terms of the simpler operators $\omega_{ij}^+$ defined below.

Let $\chi_m$ be a member of the complete set of free-particle wave functions which satisfy the Schrödinger equation

$$(H_0 - E_m)\chi_m = 0. \qquad (3\text{-}2\text{-}24)$$

If the operators $\omega_{ij}^+(m)$ are defined by the equation

$$\omega_{ij}^+(m)\chi_m = \left(1 + \frac{1}{E_m - H_0 - V_{ij} + i\varepsilon} V_{ij}\right)\chi_m, \qquad (3\text{-}2\text{-}25)$$

where $V_{ij}$ is the interaction potential between the particles labelled $i$ and $j$, it is clear that the function $\psi_m^+(ij) = \omega_{ij}^+(m)\chi_m$ satisfies the differential equation

$$(H_0 + V_{ij} - E_m)\psi_m^+(ij) = 0 \qquad (3\text{-}2\text{-}26)$$

provided that

$$\lim_{\varepsilon \to 0+} \varepsilon \psi_m^+(ij) = 0. \qquad (3\text{-}2\text{-}27)$$

Mapleton [15] has shown that the condition (3-2-27) is not satisfied when

## 3-2 THE IMPULSE APPROXIMATION

$V_{ij}$ is a Coulomb potential and in that case (3-2-26) will be taken as the equation which defines $\psi_m^+(ij)$.

Use of the operator identity (3-1-18a) gives

$$G^+ = \frac{1}{E_m - H_0 - V_{ij} + i\varepsilon}$$
$$+ G^+ \{(E_m - E) + V_{12} + V_{13} + V_{23} - V_{ij}\} \frac{1}{E_m - H_0 - V_{ij} + i\varepsilon}$$
(3-2-28)

so

$$G^+ V_{ij} = b_{ij}^+(m) + G^+ \{(E_m - E) + V_{12} + V_{13} + V_{23} - V_{ij}\} b_{ij}^+(m)$$
(3-2-29)

where

$$b_{ij}^+(m) = \omega_{ij}^+(m) - 1 \tag{3-2-30}$$

and the plane wave basis $\chi_m$ is understood. Operating on $\psi_i$ and using the fact that

$$(E_m - E)\langle \chi_m | \psi_i \rangle = \langle E_m \chi_m | \psi_i \rangle - \langle \chi_m | E \psi_i \rangle = -\langle \chi_m | V_{23} | \psi_i \rangle$$
(3-2-31)

we obtain

$$G^+ V_{ij} \psi_i = \sum_m G^+ V_{ij} \chi_m \langle \chi_m | \psi_i \rangle$$
$$= \{b_{ij}^+ + G^+ [V_{23}, b_{ij}^+] + G^+ (V_{12} + V_{13} - V_{ij}) b_{ij}^+\} \psi_i$$
(3-2-32)

where $[a, b]$ denotes the commutator of the operators $a$ and $b$ and

$$b_{ij}^+ \equiv \sum_m b_{ij}^+(m) |\chi_m\rangle \langle \chi_m|. \tag{3-2-33}$$

Consequently

$$\Omega^+ = (\omega_{13}^+ + \omega_{12}^+ - 1) + G^+ [V_{23}, (b_{13}^+ + b_{12}^+)] + G^+ (V_{13} b_{12}^+ + V_{12} b_{13}^+)$$
(3-2-34)

with

$$\omega_{ij}^+ = b_{ij}^+ + 1. \tag{3-2-35}$$

Up to this point no approximations have been made and substitution of

(3-2-34) in (3-1-25a) yields the following exact expression for $T_{\text{if}}$:

$$\begin{aligned}T_{\text{if}} &= \langle\psi_{\text{f}}|\,V_{\text{f}}\,|(\omega_{13}^+ + \omega_{12}^+ - 1)\,\psi_{\text{i}}\rangle \\ &\quad + \langle\psi_{\text{f}}|\,V_{\text{f}}\,|G^+\,[V_{23},(b_{13}^+ + b_{12}^+)]\,\psi_{\text{i}}\rangle \\ &\quad + \langle\psi_{\text{f}}|\,V_{\text{f}}\,|G^+\,(V_{13}b_{12}^+ + V_{12}b_{13}^+)\,\psi_{\text{i}}\rangle.\end{aligned} \qquad (3\text{-}2\text{-}36)$$

The impulse hypothesis leads to the neglect of the commutator involving $V_{23}$. This does not imply any restriction on the strength of the binding potential, but simply requires that $V_{23}$ should be a sufficiently slowly varying function of the coordinates of the bound particle in the region of space which provides the major contribution to the matrix element; in particular, if $V_{23}$ were constant the second term in (3-2-36) would vanish identically. If the impulse hypothesis is invoked and the third term in (3-2-36) is neglected, leaving aside for the moment the reason for this neglect, (3-2-36) reduces to

$$T_{\text{if}}^{\text{IMP}} = \langle\psi_{\text{f}}|\,V_{\text{f}}\,|(\omega_{13}^+ + \omega_{12}^+ - 1)\,\psi_{\text{i}}\rangle \qquad (3\text{-}2\text{-}37)$$

which is the impulse approximation to $T_{\text{if}}$. Thus the impulse approximation is obtained by replacing the exact wave function $\Psi_{\text{i}}^+$ in (3-1-25a) by

$$\Psi_{\text{i}}^{\text{IMP}} = (\omega_{13}^+ + \omega_{12}^+ - 1)\,\psi_{\text{i}}. \qquad (3\text{-}2\text{-}38)$$

The third term in (3-2-36) is a contribution from multiple scattering and it would vanish if $V_{12}$ were zero. In the case of heavy-particle impact there is reason to believe that the neglect of this term should not in general introduce a serious error. Briefly, the argument is that if the masses of the projectile and the target nucleus were infinite their interaction potential $V_{12}$ would be a known function of time and could therefore be removed from the Schrödinger equation by a canonical transformation. Thus, when the projectile is a heavy particle, the contribution to $T_{\text{if}}$ from the potential $V_{12}$ is expected to be of order $1/M$ compared with the contributions from other potentials, and therefore negligible in general. For more detailed discussions see Bransden [16] or McDowell and Coleman [7]. If the argument that the effects of $V_{12}$ are negligible is used to eliminate the multiple scattering terms one should also neglect $V_{12}$ in $V_{\text{f}}$ and replace $\omega_{12}^+$ by unity. Workers in this field have willingly complied with these conditions which reduce the complexity of the problem, but they have chosen to ignore the fact that if $V_{12}$ were zero in an atomic scattering problem the projectile would experience an unscreened Coulomb interaction at infinity, and that for consistency the unperturbed wave functions $\psi_{\text{i}}$ and $\psi_{\text{f}}$ should be modified accordingly. The consequences of this omission are difficult to assess but it seems reasonable to expect that any error introduced in this way decreases as the impact energy is increased.

The version of the impulse approximation given by (3-2-37) is called the "post" form of the approximation, and the "prior" form is obtained from (3-1-25b) by expanding $\Psi_f^-$ in terms of the operators

$$\bar{\omega}_{ij} \equiv \sum_m \bar{\omega}_{ij}(m) |\chi_m\rangle \langle \chi_m| \tag{3-2-39}$$

where

$$\bar{\omega}_{ij}(m) \chi_m = \left(1 + \frac{1}{E_m - H_0 - V_{ij} - i\varepsilon} V_{ij}\right) \chi_m. \tag{3-2-40}$$

The terms "post" and "prior" here refer to the potential, $V_f$ or $V_i$, which appears in the corresponding $T$ matrix element. The expression for the $T$ matrix element in the prior form of the approximation depends on the type of transition considered. In general the cross sections obtained from the post and prior forms are unequal and their difference is called the post-prior discrepancy.

## C. ELECTRON CAPTURE

### 1. *Atomic hydrogen target*

The impulse approximation $T$ matrix element for the electron capture reaction

$$X^+ + H(i) \rightarrow X(f) + H^+ \tag{3-2-41}$$

is

$$T_{if}^{IMP} = \langle \psi_f | V_{12} + V_{23} | (\omega_{12}^+ + \omega_{13}^+ - 1) \psi_i \rangle \tag{3-2-42}$$

with

$$\psi_i = \exp(i\mathbf{k}_i \cdot \boldsymbol{\sigma}) \varphi_i(\mathbf{r}), \quad \psi_f = \exp(i\mathbf{k}_f \cdot \boldsymbol{\rho}) \varphi_f(\mathbf{x}) \tag{3-2-43}$$

where $\varphi_i$ and $\varphi_f$ are hydrogenic wave functions. If the projectile is a heavy particle, for example a proton or an alpha particle, the arguments mentioned in the last section indicate that the effects of the potential $V_{12}$ may be neglected, and therefore

$$T_{if}^{IMP} \approx I_{if}(23, 13) \tag{3-2-44}$$

where

$$I_{if}(jk, lm) \equiv \langle \psi_f | V_{jk} | \omega_{lm}^+ \psi_i \rangle. \tag{3-2-44a}$$

We shall first discuss this case in detail and later consider the possibility of including some of the other terms of (3-2-42).

In accordance with the definition of $\omega^+_{13}$, the impulse approximation wave function in (3-2-44) is

$$\omega^+_{13}\psi_i = \sum_m \psi^+_m(13) \langle \chi_m | \psi_i \rangle \tag{3-2-45}$$

and a convenient set of plane wave solutions of (3-2-24) is

$$\chi_m = (2\pi)^{-3} \exp\{i(\boldsymbol{K}\cdot\boldsymbol{x} + \boldsymbol{k}\cdot\boldsymbol{\rho})\}; \tag{3-2-46}$$

then

$$E_m = \frac{1}{2}\left(\frac{K^2}{a} + \frac{k^2}{\mu'}\right) \tag{3-2-47}$$

and the summation over the index $m$ in (3-2-45) implies integration over all values of $\boldsymbol{K}$ and $\boldsymbol{k}$. When written out explicitly, with $i=1$ and $j=3$, (3-2-26) becomes

$$\left\{\frac{1}{2\mu'}\nabla^2_\rho + \frac{1}{2a}\nabla^2_x + \frac{Z_1}{x} + E_m\right\}\psi^+_m(13) = 0 \tag{3-2-48}$$

and consequently

$$\psi^+_m(13) = N(v)\,\chi_m\,{}_1F_1[iv, 1, i(Kx - \boldsymbol{K}\cdot\boldsymbol{x})] \tag{3-2-49}$$

with $v = aZ_1/K$ and

$$N(v) = \exp(\tfrac{1}{2}\pi v)\,\Gamma(1 - iv) \tag{3-2-50}$$

(see Appendix 1). Furthermore, with the help of (3-1-1) it is seen that the scalar product in (3-2-45) is

$$\langle \chi_m | \psi_i \rangle = \delta(\boldsymbol{k}_i + \boldsymbol{K} - a\boldsymbol{k})\,g_i(\boldsymbol{k} - b\boldsymbol{k}_i) \tag{3-2-51}$$

and in view of (3-1-6) the argument of the Fourier transform $g_i$ may be written simply as

$$\boldsymbol{t}_1 = \frac{1}{a}\boldsymbol{K} + \boldsymbol{v}. \tag{3-2-52}$$

Since $\delta(at) = a^{-3}\,\delta(t)$ substitution of (3-2-51) and (3-2-49) in (3-2-45) yields

$$\begin{aligned}\omega^+_{13}\psi_i &= (2\pi a)^{-3} \int d\boldsymbol{K}\,N(v)\,g_i(\boldsymbol{t}_1) \\ &\quad \times \exp\left\{i\boldsymbol{K}\cdot\boldsymbol{x} + \frac{i}{a}(\boldsymbol{k}_i + \boldsymbol{K})\cdot\boldsymbol{\rho}\right\}{}_1F_1[iv, 1, i(Kx - \boldsymbol{K}\cdot\boldsymbol{x})] \\ &= (2\pi)^{-3} \exp(i\boldsymbol{k}_i\cdot\boldsymbol{\sigma}) \int d\boldsymbol{t}_1\,N(v)\,g_i(\boldsymbol{t}_1) \\ &\quad \times \exp(i\boldsymbol{t}_1\cdot\boldsymbol{r})\,{}_1F_1[iv, 1, i(Kx - \boldsymbol{K}\cdot\boldsymbol{x})]. \end{aligned} \tag{3-2-53}$$

It is interesting to note that the right hand side of (3-2-53) is exactly the same as the result obtained by substituting (3-2-13) in (3-2-15); consequently, when $V_{12}=0$, the physical arguments of §3-2-A and the formal theory of §3-2-B yield the same approximation for the wave function $\Psi_i^+$.

To facilitate the reduction of $T_{if}^{IMP}$ the order of the momentum and coordinate integrals is reversed and we obtain

$$I_{if}(23, 13) = -(2\pi a)^{-3} \int d\mathbf{K}\, N(v)\, g_i(t_1) \int d\mathbf{x} \int d\mathbf{r}\, \varphi_f^*(\mathbf{x}) \frac{1}{r}$$
$$\times \exp(i\mathbf{x}\cdot\mathbf{p} + i\mathbf{r}\cdot\mathbf{t}) \,_1F_1[iv, 1, i(Kx - \mathbf{K}\cdot\mathbf{x})] \qquad (3\text{-}2\text{-}54)$$

with

$$\mathbf{t} = \frac{1}{a}(\mathbf{K} - \mathbf{p}). \qquad (3\text{-}2\text{-}55)$$

The $\mathbf{r}$ integral in (3-2-54) is Bethe's integral,

$$\int \frac{\exp(i\mathbf{t}\cdot\mathbf{r})}{r}\, d\mathbf{r} = \frac{4\pi}{t^2} \qquad (3\text{-}2\text{-}56)$$

and therefore

$$I_{if}(23, 12) = -(2\pi^2 a^3)^{-1} \int d\mathbf{K}\, N(v)\, g_i(t_1)\, \mathscr{F}(f, \mathbf{K}, \mathbf{p})\, t^{-2} \qquad (3\text{-}2\text{-}57)$$

where

$$\mathscr{F}(f, \mathbf{K}, \mathbf{p}) = \int \varphi_f^*(\mathbf{x})\, e^{i\mathbf{p}\cdot\mathbf{x}} \,_1F_1[iv, 1, i(Kx - \mathbf{K}\cdot\mathbf{x})]\, d\mathbf{x}. \qquad (3\text{-}2\text{-}58)$$

Integrals of the form (3-2-58) are normally evaluated by using an integral representation of the confluent hypergeometric function and reversing the order of integration. In particular, if the final state is an s-state an expression for $\mathscr{F}(f, \mathbf{K}, \mathbf{p})$ may be deduced from Appendix 2. For example, since

$$\varphi_{1s}(\mathbf{x}) = \frac{(aZ_1)^{\frac{3}{2}}}{\pi^{\frac{1}{2}}} \exp(-aZ_1 x) \qquad (3\text{-}2\text{-}59)$$

it follows from eq. (A2-19) of Appendix 2 that

$$\mathscr{F}(1s, \mathbf{K}, \mathbf{p}) = -\frac{(aZ_1)^{\frac{3}{2}}}{\pi^{\frac{1}{2}}} \frac{\partial}{\partial \beta} \mathscr{I}(v, 0, \beta, -\mathbf{K}, \mathbf{p})\bigg|_{\beta=aZ_1}$$
$$= 8\pi^{\frac{1}{2}}(aZ_1)^{\frac{3}{2}} \left\{\frac{(1-iv)\beta}{T^2} + \frac{iv(\beta - iK)}{T(T-2\delta)}\right\} \left(\frac{T}{T-2\delta}\right)^{iv}\bigg|_{\beta=aZ_1}$$
$$\qquad (3\text{-}2\text{-}60)$$

where
$$T = \beta^2 + p^2, \quad \delta = \boldsymbol{p}\cdot\boldsymbol{K} + i\beta K; \tag{3-2-61}$$
similarly
$$\mathscr{F}(2s, \boldsymbol{K}, \boldsymbol{p}) = -\frac{\beta^{\frac{3}{2}}}{\sqrt{\pi}}\left(\frac{\partial}{\partial\beta} + \beta\frac{\partial^2}{\partial\beta^2}\right)\mathscr{I}(v, 0, \beta, -\boldsymbol{K}, \boldsymbol{p})\bigg|_{\beta = \frac{1}{2}aZ_1}. \tag{3-2-62}$$

Pradhan [13] attempted to apply the impulse approximation to the symmetric resonant process
$$H^+ + H(1s) \to H(1s) + H^+ \tag{3-2-63}$$
but to simplify the calculation he replaced $\langle\psi_f|\,V_{23}\,|\omega_{13}^+\psi_i\rangle$ by $\langle\psi_f|\,V_{13}\,|\omega_{13}^+\psi_i\rangle$ which can be evaluated in closed form. Despite the fact that, as was pointed out by Bassel and Gerjuoy [17], these matrix elements describe completely different physical processes, Pradhan argued that they are equal "to fairly reasonable approximation" at high energies. This claim was repeated by Pradhan and Tripathy [18] who corrected an error in Pradhan's numerical work; that it is invalid is most readily seen by comparing the two matrix elements in the high-energy limit (see §3-2-C-3).

McDowell [14] also carried out calculations for the process (3-2-63). He used the correct form of the impulse approximation matrix element, but to reduce the complexity of the calculation he argued that since $(1/a-b) = O(1/M)$ the term $(1/a-b)\,\boldsymbol{k}_i$ could be neglected in the argument of $g_i$. This amounts to replacing $g_i(\boldsymbol{t}_1)$ by $g_i(\boldsymbol{K})$, and the advantage is that when the initial state is an s-state it removes the angular dependence of $g_i$; the azimuthal integration in (3-2-57) then becomes trivial. However, as it is evident from (3-1-6) that $(1/a-b)\,\boldsymbol{k}_i = \boldsymbol{v}$, McDowell's simplification is invalid, and the resulting error increases with increasing energy.

Impulse approximation cross sections for (3-2-63) were finally evaluated by Cheshire [19] without any further approximations other than those inherent in the numerical methods used. Cheshire's results are in close agreement with the results of more recent calculations by Coleman and McDowell [20], which are compared with other theoretical predictions in Fig. 3-2-1. For electron capture there are two versions of the first Born approximation. The simplest of these, which we shall call the OBK approximation [21, 22] neglects the interaction potential $V_{12}$, and the other version, in which $V_{12}$ is retained, will be termed the Born approximation; thus
$$T_{if}^{OBK} = \langle\psi_f|\,V_{23}\,|\psi_i\rangle, \quad T_{if}^B = \langle\psi_f|\,V_{12} + V_{23}\,|\psi_i\rangle. \tag{3-2-64}$$

Also shown in Fig. 3-2-1 are the results of a two-state calculation by McCarroll [25]. Because of the resonant nature of the process (3-2-63) the direct

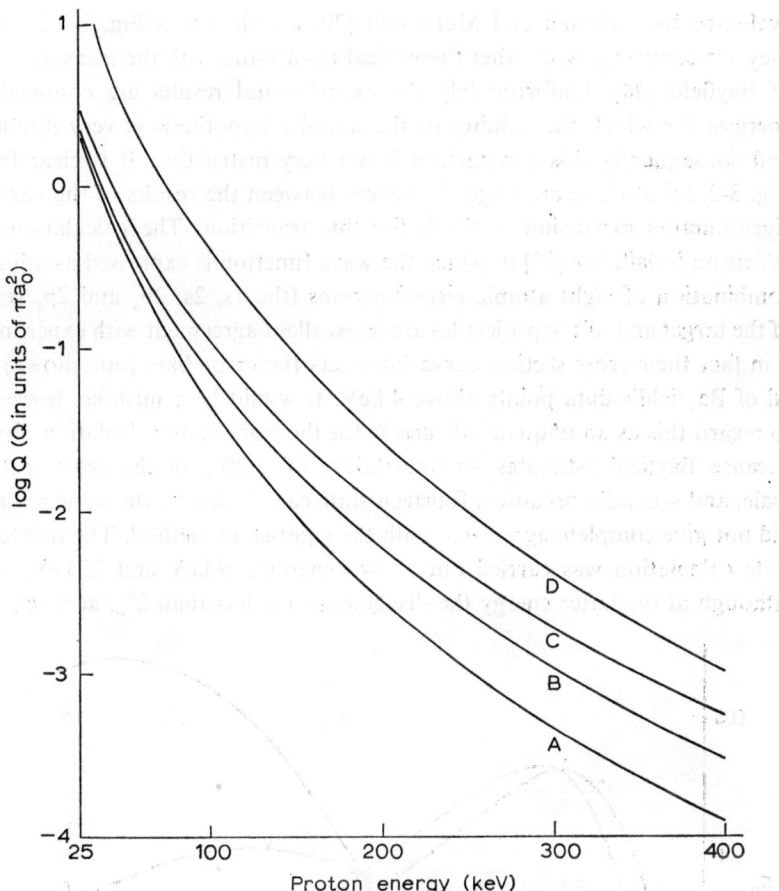

Fig. 3-2-1. Cross sections for $H^+ + H(1s) \to H(1s) + H^+$.
Curve A. Impulse approximation [20]. Curve B. Born approximation [23, 24]. Curve C. Two-state approximation [25]. Curve D. OBK approximation.

reaction path dominates and one would expect the coupling to other bound intermediate states to be relatively unimportant at low energies; for this reason the results of more elaborate calculations [26, 27] based on atomic eigenfunction expansions do not differ significantly from the two-state results in the energy range of the calculations ($\leqslant 100$ keV). There are no experimental results for this process but it provides the major contribution to the total capture cross section which is discussed below.

Impulse approximation cross sections for

$$H^+ + H(1s) \to H(2s) + H^+ \qquad (3\text{-}2\text{-}65)$$

evaluated by Coleman and McDowell [20] are shown in Fig. 3-2-2 where they are compared with other theoretical results and with the measurements of Bayfield [28]. Unfortunately the experimental results are confined to energies for which the validity of the impulse hypothesis is very doubtful and consequently this comparison is not very instructive. It is clear from Fig. 3-2-2 that there are large differences between the results of the various eigenfunction expansion methods for this transition. The calculations of Wilets and Gallaher [27] in which the wave function is expressed as a linear combination of eight atomic eigenfunctions (the 1s, 2s, $2p_x$ and $2p_z$ states of the target and of the projectile) are in excellent agreement with experiment – in fact their cross section curve intersects the error bars (not shown) on all of Bayfield's data points above 4 keV. It would be a mistake, however, to regard this as an unqualified success for the eight-state calculation, firstly because Bayfield estimates an uncertainty of ±30% in the cross section scale, and secondly because a fourteen-state calculation by the same authors did not give complete agreement with the eight-state method. The fourteen-state calculation was carried out at two energies, 9 keV and 25 keV, and, although at the latter energy the disagreement is less than 2%, at 9 keV the

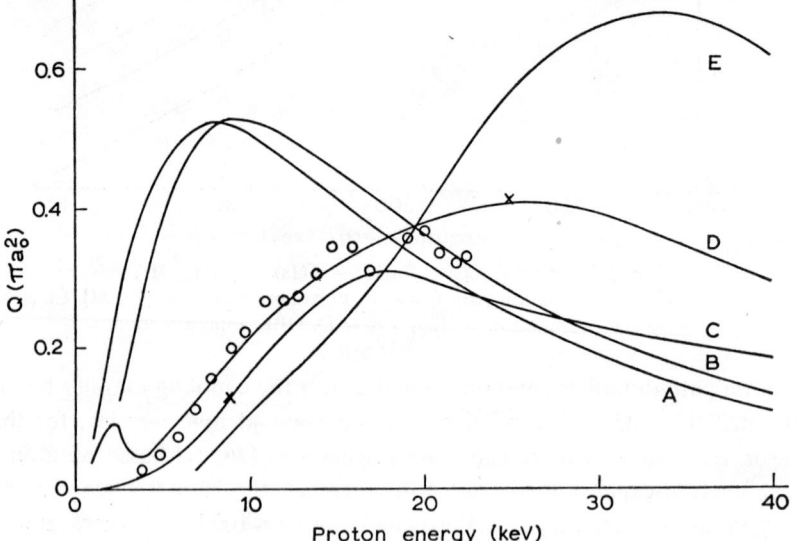

Fig. 3-2-2. Cross sections for $H^+ + H(1s) \rightarrow H(2s) + H^+$.
Curve A. Impulse approximation [20]. Curve B. Born approximation estimate [30]. Curve C. Two-state approximation, hydrogenic [26]. Curve D. Eight-state approximation, hydrogenic [27]. Curve E. Eight-state approximation, Sturmian [29]. ×. Fourteen-state approximation, hydrogenic [27]. ○. Experiment [28].

result of the more elaborate calculation lies below the eight-state result by about 30% and is in fact in close agreement with the two-state calculations of Lovell and McElroy [26]. Cross section measurements in the energy range 40–200 keV for the reaction (3-2-65) have been reported by Ryding et al. [31] but as these are not absolute measurements they do not provide any evidence on the validity of any particular theory.*

If either of the bound states involved in a collision problem is not an s-state it is necessary to make a choice of axis of quantization for one or both of the atoms. When one is interested in transitions between states of definite magnetic quantum number the atomic wave functions must be referred to a fixed axis of quantization which is usually taken in the direction of the incident beam. However, it is frequently easier to evaluate differential cross sections using a frame of reference with polar axis in the direction of the momentum transfer vector $p$, and other rotating frames of reference may be more convenient at earlier stages of the calculation. A knowledge of the transformation properties of the spherical harmonics $Y_{lm}(\theta, \varphi)$ enables one to relate quantities calculated in one frame of reference to the corresponding quantities in another frame. In particular let us consider the process

$$X^+ + H(1s) \to X(nlm) + H^+ \qquad (3\text{-}2\text{-}66)$$

where the axis of quantization for the final state is chosen in the direction of $k_i$. We shall show first of all that the cross section for this process can be deduced from differential cross sections calculated with the axis of quantization in the direction of $p$, and then discuss the use of a different frame of reference in evaluating $\mathscr{F}(nlm, K, p)$.

Let $Oxyz$ be a frame of reference with $z$ axis along $k_i$ having the plane of $p$ and $k_i$ as $xz$ plane, and let $OXYZ$ be a second frame with $Z$ axis along $p$ and the plane of $p$ and $k_i$ as $XZ$ plane. The axes $Oy$ and $OY$ are coincident and the two frames can be brought into coincidence by rotating $OXYZ$ about this common axis through an angle

$$\beta = \cos^{-1}\frac{p \cdot k_i}{pk_i} = \cos^{-1}\frac{\Delta E + p^2/\mu}{2vp}. \qquad (3\text{-}2\text{-}67)$$

In the frame $Oxyz$ the wave function of the final bound state is

$$\varphi_f(x) = \varphi_{nlm}(x, \theta, \varphi) = F_{nl}(x) Y_{lm}(\theta, \varphi) \qquad (3\text{-}2\text{-}68)$$

---

* Gaily [78] has recently proposed a means of normalizing the date of Ryding et al. The resulting cross section values are in close agreement with the results obtained by Wilets and Gallagher [27] and greatly exceed the Born and impulse approximation cross sections.

and spherical polar coordinates in $OXYZ$ will be denoted by $(x, \Theta, \Phi)$. The relationship between the spherical harmonics in the two reference frames is

$$Y_{lm}(\theta, \varphi) = \sum_{\mu=-l}^{l} Y_{l\mu}(\Theta, \Phi) R^l_{\mu m}(\alpha, \beta, \gamma) \qquad (3\text{-}2\text{-}69)$$

(ref. 32, p. 1068) where the functions $R^l_{\mu m}(\alpha, \beta, \gamma)$ are the elements of the rotation matrix and $\alpha, \beta$ and $\gamma$ are the Euler angles of the rotation which brings $OXYZ$ into coincidence with $Oxyz$. In the present case $\alpha = 0 = \gamma$ and $\beta$ is given by (3-2-67).

To avoid an excessively cumbersome notation we shall write $I_{if}$ (23, 13) for the reaction (3-2-66) as $I(1s-nlm)$ and the vector taken as axis of quantization will be indicated by a superscript on $I$ and $\mathscr{F}$. Application of (3-2-69) to (3-2-58) yields

$$\mathscr{F}^{k_1}(nlm, K, p) = \int F_{nl}(x) Y^*_{lm}(\theta, \varphi) \exp(i p \cdot x) \, {}_1F_1[iv, 1, i(Kx - K \cdot x)] \, dx$$

$$= \sum_{\mu=-l}^{l} R^l_{\mu m}(0, \beta, 0) \mathscr{F}^p(nl\mu, K, p) \qquad (3\text{-}2\text{-}70)$$

and, since $\beta$ is independent of $K$, it follows from (3-2-57) that

$$I^{k_1}(1s-nlm) = \sum_{\mu=-l}^{l} R^l_{\mu m}(0, \beta, 0) I^p(1s-nl\mu). \qquad (3\text{-}2\text{-}71)$$

The rotation matrix is unitary and therefore an immediate consequence of (3-2-71) is

$$\sum_{m=-l}^{l} |I^{k_1}(1s-nlm)|^2 = \sum_{\mu=-l}^{l} |I^p(1s-nl\mu)|^2; \qquad (3\text{-}2\text{-}72)$$

this is a statement of the well-known fact that if the quantity of interest is the cross section for a transition to a state $nl$, irrespective of the value of the magnetic quantum number, the calculation can be carried out as if the axis of quantization were in the direction of the rotating vector $p$.

Nordsieck's integration technique may be used to evaluate $\mathscr{F}(nlm, K, p)$ for arbitrary values of $nlm$. However, the feature which makes Nordsieck's analysis so neat in the case studied in Appendix 2 is the fact that each contour integral can be expressed in terms of the residue at a single pole, and this feature is lost when $l \neq 0$ because of additional non-zero contributions from the circle at infinity. Nordsieck's method then offers no obvious advantage over the method described by Coleman and McDowell [20] for $n=2, l=1$, which is based on a real integral representation of the confluent hypergeometric function. When evaluating $\mathscr{F}(nlm, \mathbf{K}, \mathbf{p})$ it is convenient to use a frame of reference $OX'Y'Z'$ with $OZ'$ in the direction of $\mathbf{t}$ and the plane of $\mathbf{t}$ and $\mathbf{p}$ as the $X'Z'$ plane. The integral is greatly simplified by taking the axis of quantization to be $OZ'$ and use of (3-2-69) shows that

$$\mathscr{F}^p(nlm, \mathbf{K}, \mathbf{p}) = \sum_{\mu=-l}^{l} R_{\mu m}^{l*}(\alpha', \beta', \gamma') \mathscr{F}^t(nl\mu, \mathbf{K}, \mathbf{p}) \qquad (3\text{-}2\text{-}73)$$

where

$$\alpha' = 0, \qquad \beta' = \cos^{-1}(\hat{\mathbf{p}} \cdot \hat{\mathbf{t}}) \qquad (3\text{-}2\text{-}74)$$

and $\gamma'$ is the angle which the plane of $\mathbf{p}$ and $\mathbf{t}$ makes with that of $\mathbf{p}$ and $\mathbf{k}_1$. The matrix elements which arise in (3-2-73) for $l=1$ are shown in Table 3-2-1. The angles $\beta'$ and $\gamma'$ depend on $\mathbf{K}$ so the rotation matrix elements cannot be taken outside the $\mathbf{K}$ integral; thus these matrix elements, unlike $R_{\mu m}^{l}(0, \beta, 0)$ in (3-2-70), enter the calculation even if one is not interested in distinguishing between the possible values of $m$.

In the first application of the impulse approximation to transitions of the form (3-2-66), Coleman and McDowell [20] chose the axis of quantization in the direction of $\mathbf{t}$ but failed to realize the nontrivial nature of the resulting modification of the $\mathbf{K}$ integral. This error was corrected by Coleman and Trelease [33] and their results for

$$H^+ + H(1s) \rightarrow H(2p) + H^+ \qquad (3\text{-}2\text{-}75)$$

are shown in Fig. 3-2-3 in the energy range for which experimental results are available. The close agreement obtained with the measurements of Stebbings et al. [34] at energies as low as 5 keV is surprising since one would not expect the impulse hypothesis to be a realistic assumption at such low energies. Somewhat better agreement with experiment is obtained by Gallaher and Wilets [29] using an 8-state expansion in Sturmian functions. However this agreement is marred by the severe disagreement of the 8-state

TABLE 3-2-1

The rotation matrix elements $R_{\mu m}^1(0, \beta', \gamma')$

| $\mu$ \ $m$ | −1 | 0 | 1 |
|---|---|---|---|
| −1 | $\tfrac{1}{2}(1+\cos\beta')\exp(i\gamma')$ | $\sin\beta'/\sqrt{2}$ | $\tfrac{1}{2}(1-\cos\beta')\exp(-i\gamma')$ |
| 0 | $-\sin\beta'\exp(i\gamma')/\sqrt{2}$ | $\cos\beta'$ | $\sin\beta'\exp(-i\gamma')/\sqrt{2}$ |
| 1 | $\tfrac{1}{2}(1-\cos\beta')\exp(i\gamma')$ | $-\sin\beta'/\sqrt{2}$ | $\tfrac{1}{2}(1+\cos\beta')\exp(-i\gamma')$ |

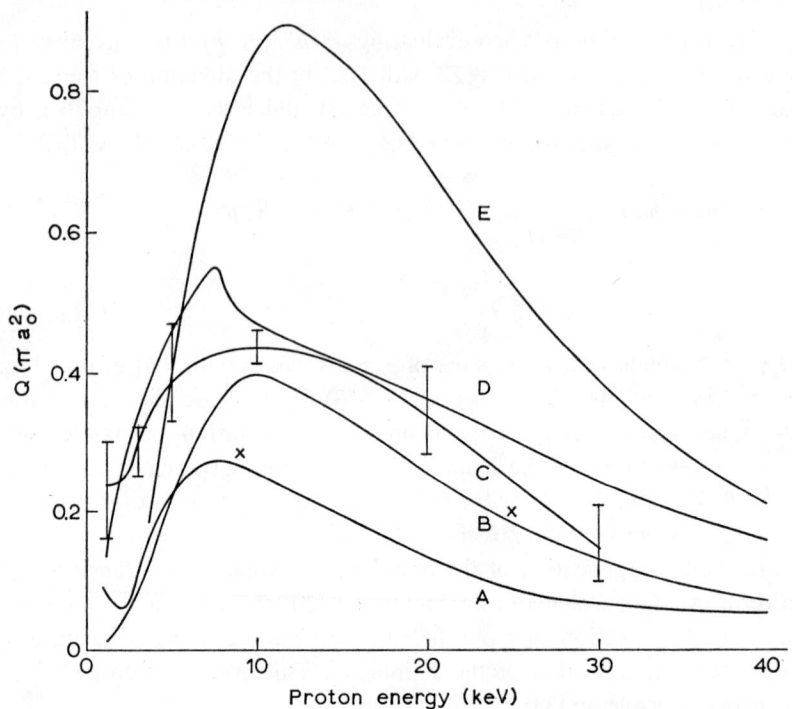

Fig. 3-2-3. Cross sections for $H^+ + H(1s) \rightarrow H(2p) + H^+$.
Curve A. Eight-state approximation, hydrogenic [27]. Curve B. Impulse approximation [33]. Curve C. Experiment [34]. Curve D. Eight-state approximation, Sturmian [29]. Curve E. Born approximation estimate [30]. ×. 16-state approximation, Sturmian [29].

results with those of a 16-state Sturmian calculation by the same authors, which casts considerable doubt on the convergence of the Sturmian expansion. The situation is further complicated by the fact that the accuracy of the experimental results is in doubt. Gaily [74] has found that photoabsorption data for $O_2$ used by Stebbings et al. in deducing cross section values were incorrect, and use of more reliable data increases the cross section at 5 keV by approximately one sixth. The effect at higher energies is not yet known. Furthermore, preliminary results of an independent measurement by Gaily and Geballe [75] at energies up to 6 keV suggest that the cross section for this process may be considerably smaller than that deduced by Stebbings et al.

Jackson and Schiff [24] and Mapleton [35] observed that, although the magnitudes of the Born and OBK cross sections differ considerably, the ratios of cross sections for different transitions are very similar in the two approximations, particularly at high energies. This is illustrated in Fig. 3-2-4 which shows that there is also close agreement between the ratios predicted by the OBK and impulse approximations. If it is assumed that this behaviour is maintained for capture into more highly excited states, we have a means of estimating the contribution to the total capture cross section from these states, and can therefore obtain theoretical data for comparison with measurements of the cross section $Q(1s-\Sigma)$ for the process

$$H^+ + H(1s) \rightarrow H + H^+. \qquad (3\text{-}2\text{-}76)$$

The OBK approximation predicts that at high energies the cross section $Q(1s-n)$ for capture into a state of principal quantum number $n$ is proportional to $n^{-3}$ (see ref. 36) and this has frequently been used to estimate $Q(1s-\Sigma)$ in other approximations. However, it is clear from Fig. 3-2-4 that the approach to the $n^{-3}$ behaviour, which would make $R_2 = 0.125$ and $R_3 = 0.037$, is very slow; consequently it is probably better to use the actual OBK ratios rather than the $n^{-3}$ rule. Thus for example from calculated cross

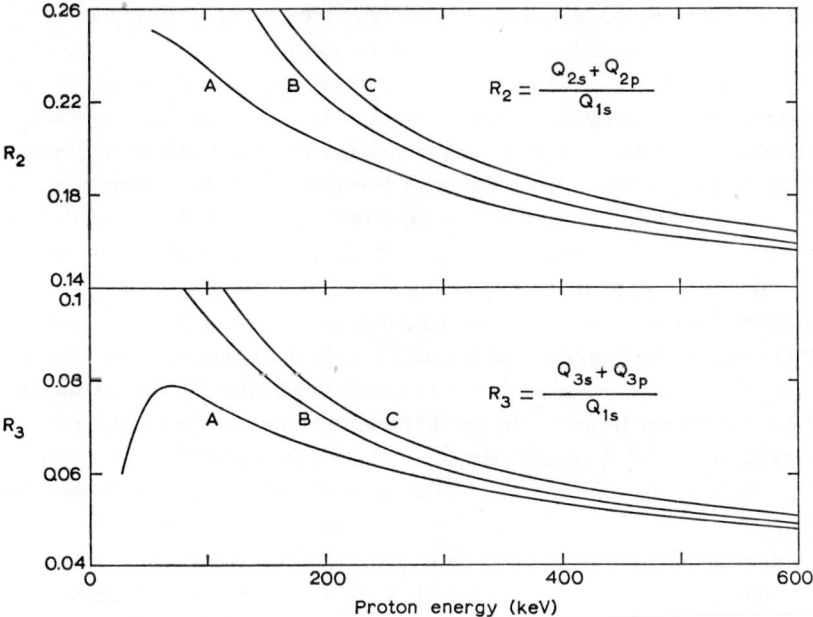

Fig. 3-2-4. Cross section ratios for electron capture by protons from atomic hydrogen (from ref. 33, fig. 3). Curve A. Impulse approximation [33]. Curve B. Born approximation [35]. Curve C. OBK approximation [35].

sections for capture into the $n=2, 3$ states one obtains, as an estimate for the impulse approximation cross section for the process (3-2-76),

$$Q_{IMP}(1s-\Sigma) \approx Q_{IMP}(1s-1s) + Q_{IMP}(1s-2)$$
$$+ \frac{Q_{IMP}(1s-3)}{Q_{OBK}(1s-3)} \sum_{n=3}^{\infty} Q_{OBK}(1s-n) \qquad (3\text{-}2\text{-}77)$$

and a sum rule derived by May [36] permits the evaluation of the sum of the OBK cross sections to any desired accuracy.

Impulse approximation cross sections for the reactions

$$H^+ + H(1s) \to H(3s) + H^+, \qquad (3\text{-}2\text{-}78a)$$
$$H^+ + H(1s) \to H(3p) + H^+, \qquad (3\text{-}2\text{-}78b)$$

were evaluated by Coleman and Trelease [33]. Capture into the 3d state was not considered since estimates based on the OBK approximation indicate that at energies above 120 keV it contributes less than 3% of the cross section for capture into the $n=3$ level. These results, together with the results for the processes (3-2-63), (3-2-65) and (3-2-75), were used in the formula (3-2-77) to obtain curve A of Fig. 3-2-5. The Born approximation results shown in the figure were obtained in a similar manner, and the OBK cross section was evaluated exactly from May's sum rule. The extent of the disagreement between the impulse approximation and the measurements of Wittkower et al. [37] – approximately a factor of two throughout the energy range of the experiments – is very puzzling since one would expect the impulse approximation to become more accurate as the projectile energy is increased.

Another surprising feature of Fig. 3-2-5 is the close agreement between the experimental results and the predictions of the first Born approximation. It has already been argued that for heavy particle collisions the contribution to $T_{if}$ from the potential $V_{12}$ must vanish in the limit as $1/M \to 0$. However, the contribution from $V_{12}$ in the Born approximation does not have this behaviour; in fact it cancels part of $T_{if}^{OBK}$ with the result that $|T_{if}^{B}| < |T_{if}^{OBK}|$. One fault of the conventional OBK approximation is that, although the potential $V_{12}$ has been eliminated, no account is taken of the resulting Coulomb phase in $\psi_i$ and $\psi_f$. This has been rectified by Cheshire [39] whose calculations for the reaction (3-2-63) show that the modified OBK cross section, $Q_{MOBK}$, is less than $Q_{OBK}$ by a factor of 4.41 at 25 keV, 2.8 at 50 keV, 1.66 at 400 keV and 1.16 at 1 MeV, but $Q_{MOBK}$ also exceeds the Born cross section by a considerable amount. The arguments which support the neglect

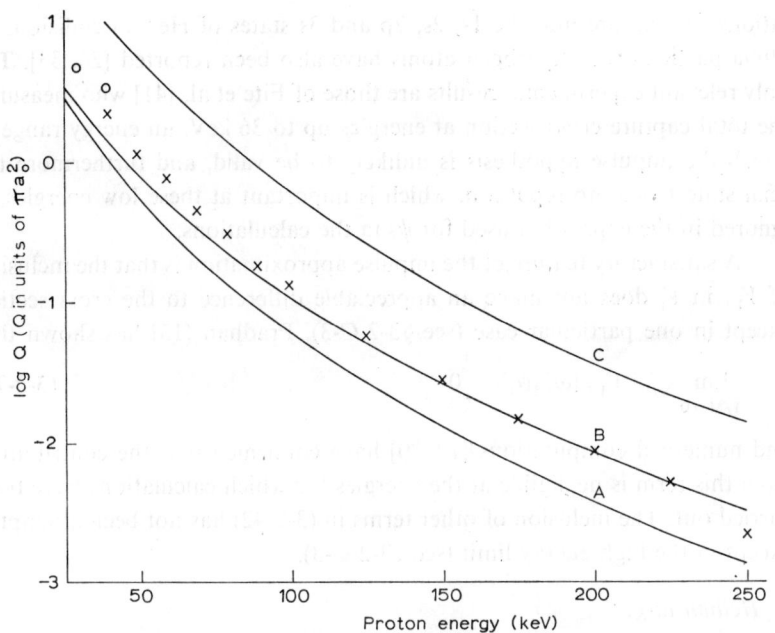

Fig. 3-2-5. Cross sections for $H^+ + H(1s) \rightarrow H + H^+$.
Curve A. Impulse approximation [33]. Curve B. Born approximation [35]. Curve C. OBK approximation. ×. Experiment [37]. ○. Experiment [38].

of $V_{12}$ show that by means of a canonical transformation one can introduce into the Schrödinger equation an arbitrary function $W(R)$ of the internuclear separation alone and, provided that proper account is taken of any Coulomb phase introduced by the transformation, this cannot affect the results of an *exact* calculation. This lack of dependence on the choice of $W(R)$ does not hold when $\Psi_i^+$ is replaced by $\psi_i$ in (3-1-25a); in particular, the choices $W(R)=0$ and $W(R)=1/R$ yield respectively the OBK and Born approximations. Thus different choices of $W(R)$ lead to quite different cross section values in this case, and there appears to be no obvious reason for adopting one choice rather than another. For this reason any agreement between the Born approximation and experiment for rearrangement collisions can only be regarded as fortuitous.

Impulse approximation cross sections for positronium formation in positron-hydrogen collisions have been evaluated by Cheshire [40]. However, since in this case the projectile is not a heavy particle, the neglect of multiple scattering is difficult to justify (see Chapter 4 of this book). Calculations for capture into the 1s, 2s, 2p and 3s states of He$^+$ in collisions of alpha particles with hydrogen atoms have also been reported [20, 33]. The only relevant experimental results are those of Fite et al. [41] who measured the total capture cross section at energies up to 36 keV, an energy range in which the impulse hypothesis is unlikely to be valid, and furthermore the final state Coulomb repulsion, which is important at these low energies, is ignored in the expression used for $\psi_f$ in the calculations.

A satisfactory feature of the impulse approximation is that the inclusion of $V_{12}$ in $V_f$ does not make an appreciable difference to the cross section except in one particular case (see §3-2-C-3). Pradhan [13] has shown that

$$\lim_{1/M \to 0} \langle \psi_f | V_{12} | \omega_{13}^+ \psi_i \rangle = 0 \qquad (3\text{-}2\text{-}79)$$

and numerical computations [19, 20] have confirmed that the contribution from this term is negligible at the energies for which calculations have been carried out. The inclusion of other terms in (3-2-42) has not been attempted except in the high-energy limit (see §3-2-C-3).

## 2. *Helium target*

The only application of the impulse approximation to a problem involving a target other than atomic hydrogen is the work of Bransden and Cheshire [42] on the reaction

$$H^+ + He(1s^2) \to H(1s) + He^+(1s). \qquad (3\text{-}2\text{-}80)$$

If a second electron (particle 4) is added to the configuration depicted in Fig. 3-1-1 an obvious extension of the analysis of §3-2-B shows that the impulse approximation matrix element for the capture of electron 3 is

$$T_{if}^{IMP} = \langle \psi_f | V_{12} + V_{23} + V_{14} + V_{34} |(\omega_{12}^+ + \omega_{13}^+ + \omega_{14}^+ - 2) \psi_i \rangle. \qquad (3\text{-}2\text{-}81)$$

If it is assumed that the distortion due to the potential $V_{12}$ is negligible, $\omega_{12}^+$ may be replaced by unity, and if the effect of the interaction $V_{14}$ between the projectile and the passive electron is treated in the same way (3-2-81) becomes

$$T_{if}^{IMP} \approx \langle \psi_f | V_{12} + V_{23} + V_{14} + V_{34} | \omega_{13}^+ \psi_i \rangle \qquad (3\text{-}2\text{-}82)$$

which is the matrix element used by Bransden and Cheshire.

In calculations for reactions such as (3-2-80) there is a source of uncertainty which is not present when the target is hydrogenic – the exact bound state wave functions of the target are not known. Bransden and Cheshire [42] chose, as the ground state wave function of helium, the simple one-parameter variational wave function

$$\varphi_i(r_3, r_4) = \frac{\lambda^3}{\pi} \exp\{-\lambda(r_3 + r_4)\}, \quad \lambda = 1.6875. \tag{3-2-83}$$

They evaluated the cross section for (3-2-80) in the energy range 25 keV–1 MeV and estimated the cross section for

$$H^+ + He(1s^2) \to H + He^+ \tag{3-2-84}$$

by multiplying their results by the ratio of the Born approximation cross sections for (3-2-84) and (3-2-80) obtained from the work of Mapleton [43]. The results obtained in this way lie below the experimental results of Barnett and Reynolds [44], Welsh et al. [45] and Toburen et al. [46] whereas the average of the Born post and prior cross sections calculated by Mapleton [43, 47] is in close agreement with the measurements in the energy range 200 keV–10 MeV. It has been suggested [16] that the use of a more accurate wave function for the helium ground state may increase the impulse approximation cross section, but in the absence of calculations the amount of the increase is not known.

3. *The high-energy limit*

The high-energy behaviour of electron capture cross sections has aroused considerable interest in recent years. In the limit as $v \to \infty$ cross section calculations based on non-relativistic quantum mechanics do not have any physical relevance. However, the non-relativistic Schrödinger equation still presents a well-defined mathematical problem and from its solution, if it were known, one could determine the exact non-relativistic cross section. A reasonable requirement on any high-energy approximation is that its prediction should agree in the high-energy limit with this exact cross section. A number of methods which have some claim to validity for electron capture at high energies yield very different cross sections in the high-energy limit, but even for the simplest rearrangement collision the correct high-energy limit is not yet known with any certainty. In the present discussion we shall concentrate on the reaction

$$H^+ + H(1s) \to H(1s) + H^+. \tag{3-2-85}$$

The OBK cross section for the process (3-2-85) is

$$Q_{\text{OBK}} = \frac{2^{18}}{5v^2(v^2 + 4)^5} \underset{v \to \infty}{\sim} \frac{2^{18}}{5v^{12}} (\pi a_0^2) \tag{3-2-86}$$

whereas, if terms of order $1/M$ are neglected compared with unity, the asymptotic form of the cross section in the Born approximation is [24]

$$Q_B \underset{v\to\infty}{\sim} 0.661 Q_{OBK}. \tag{3-2-87}$$

Drisko [48] using the second Born approximation obtained

$$Q_{B2} \underset{v\to\infty}{\sim} \left(0.2946 + \frac{5\pi v}{2^{12}}\right) Q_{OBK} \tag{3-2-88}$$

and therefore $Q_{B2}$, unlike $Q_{OBK}$ and $Q_B$, varies ultimately as $v^{-11}$.

In the case of capture from the ground state the Fourier transform in (3-2-57) is

$$g_{1s}(t_1) = \frac{8\pi^{\frac{1}{2}} b^{\frac{5}{2}}}{(b^2 + t_1^2)^2} \tag{3-2-89}$$

which is sharply peaked when $t_1 = 0$. If it is assumed that the remainder of the integrand in (3-2-57) varies slowly in the vicinity of this peak, we may write

$$I_{if}(23, 13) \approx -\frac{N(v_0)}{2\pi^2} \mathscr{F}(\mathbf{f}, \mathbf{K}_0, \mathbf{p}) \int dt\, g_{1s}\left(t + v + \frac{1}{a}p\right) t^{-2}$$

$$= -4\pi^{-\frac{1}{2}} a^{\frac{3}{2}} N(v_0) \mathscr{F}(\mathbf{f}, \mathbf{K}_0, \mathbf{p}) \left\{a^2 + \left(v + \frac{1}{a}p\right)^2\right\}^{-1} \tag{3-2-90}$$

where

$$v_0 = aZ_1/K_0, \quad \mathbf{K}_0 = -a\mathbf{v}. \tag{3-2-91}$$

This result may also be obtained by applying a peaking approximation to (3-2-53) which gives

$$\omega_{13}^+ \psi_i \approx \Psi_i^{\text{Peak}} = (2\pi)^{-3} \exp(i\mathbf{k}_i \cdot \boldsymbol{\sigma}) N(v_0) \,_1F_1[iv_0, 1, i(K_0 x - \mathbf{K}_0 \cdot \mathbf{x})]$$

$$\times \int dt_1\, g_{1s}(t_1) \exp(it_1 \cdot \mathbf{r})$$

$$= \exp(i\mathbf{k}_i \cdot \boldsymbol{\sigma}) N(v_0) \,_1F_1[iv_0, 1, i(K_0 x - \mathbf{K}_0 \cdot \mathbf{x})] \varphi_{1s}(r). \tag{3-2-92}$$

Then it is readily seen that

$$\langle \psi_f | V_{23} | \Psi_i^{\text{Peak}} \rangle = -4\pi^{-\frac{1}{2}} a^{\frac{3}{2}} N(v_0) \mathscr{F}(\mathbf{f}, \mathbf{K}_0, \mathbf{p}) \left\{a^2 + \left(v + \frac{1}{a}p\right)^2\right\}^{-1} \tag{3-2-93}$$

in agreement with (3-2-90).

Bransden and Cheshire [42] found that the cross section deduced from

(3-2-90) has the asymptotic form

$$Q_{\text{IMP}} \underset{v \to \infty}{\sim} \left(0.2946 + \frac{5\pi v}{2^{11}}\right) Q_{\text{OBK}} \qquad (3\text{-}2\text{-}94)$$

which differs from (3-2-88) by a factor of two in the coefficient of $v^{-11}$. The origin of this discrepancy has not been explained but it may be due to the fact that in this version of the impulse approximation the wave function $\Psi_i^{\text{Peak}}$ given by (3-2-92) has a Coulomb phase at infinity, as is consistent with the neglect of $V_{12}$, but $\psi_f$ is still assumed to be given by (3-2-43). The cross section obtained by Pradhan [13], in contrast to $Q_{\text{IMP}}$, tends to $Q_{\text{OBK}}$ as $v \to \infty$.

Mapleton [49] has shown that if terms of order $1/M$ are not neglected, and the protons are regarded as distinguishable, the contribution to the cross section for the reaction (3-2-85) from the potential $V_{12}$ eventually dominates and

$$Q_B \underset{v \to \infty}{\sim} \frac{16}{3M^2 v^6}. \qquad (3\text{-}2\text{-}95)$$

Exactly the same result is obtained in the impulse approximation [50, 51] whether or not $\omega_{12}^+$ is replaced by unity in (3-2-42).

The normal procedure of regarding elastic scattering and the reaction (3-2-85) as independent processes stems from the fact that the angular distribution of scattered protons displays non-overlapping peaks at scattering angles near $0°$ and $90°$ in the laboratory frame. The peak in the forward direction is associated with elastic scattering and that at $90°$ is attributed to electron capture; this is equivalent to assuming that the major contribution to the integral in (3-1-32) comes from the vicinity of $p_{\text{min}}$, which is certainly correct in the limit as $1/M \to 0$. However, the term which varies as $v^{-6}$ in the Born and impulse approximations comes from the forward direction and therefore interferes with elastic scattering. As a result the electron capture cross section cannot be defined in this case but, since the non-relativistic treatment breaks down before the terms of order $1/M$ become important, these considerations are of little practical importance.

D. EXCITATION

When a hydrogen atom is excited from state i to state f the appropriate $T$ matrix element in the impulse approximation is

$$T_{if}^{\text{IMP}} = \langle \psi_f | V_{12} + V_{13} | (\omega_{12}^+ + \omega_{13}^+ - 1) \psi_i \rangle \qquad (3\text{-}2\text{-}96)$$

with

$$\psi_i = \exp(i\mathbf{k}_i \cdot \boldsymbol{\sigma}) \, \varphi_i(\mathbf{r}), \quad \psi_f = \exp(i\mathbf{k}_f \cdot \boldsymbol{\sigma}) \, \varphi_f(\mathbf{r}). \tag{3-2-97}$$

Here it is assumed that electron exchange does not arise or can be neglected. In particular, if $V_{12}$ is neglected, use of (3-2-53) and (3-2-97) in (3-2-96) gives

$$\begin{aligned}
T_{if}^{\text{IMP}} &= \langle \psi_f | V_{13} | \omega_{13}^+ \psi_i \rangle \\
&= -(2\pi a)^{-3} Z_1 \int d\mathbf{x} \int d\mathbf{r} \exp(i\mathbf{q} \cdot \boldsymbol{\sigma}) \, \varphi_f^*(\mathbf{r}) \frac{1}{x} \int d\mathbf{K} \, N(v) \\
&\quad \times g_i(\mathbf{t}_1) \exp(i\mathbf{t}_1 \cdot \mathbf{r}) \, {}_1F_1[iv, 1, i(Kx - \mathbf{K} \cdot \mathbf{x})] \\
&= -(2\pi a)^{-3} Z_1 \int d\mathbf{K} \, N(v) \, g_i(\mathbf{t}_1) \, g_f^*(\mathbf{t}_2) \, \mathscr{I}(v, 0, 0, -\mathbf{K}, -\mathbf{q})
\end{aligned} \tag{3-2-98}$$

where

$$\mathbf{t}_2 = \mathbf{t}_1 + b\mathbf{q} \tag{3-2-99}$$

and the notation of Appendix 2 has been used for the $x$ integral. From eq. (A2-19) it is seen that

$$\mathscr{I}(v, 0, 0, -\mathbf{K}, -\mathbf{q}) = \lim_{\beta \to 0} \frac{4\pi}{\beta^2 + q^2} \left\{ \frac{\beta^2 + q^2}{\beta^2 + q^2 + 2\mathbf{q} \cdot \mathbf{K} - 2i\beta K} \right\}^{iv}. \tag{3-2-100}$$

Furthermore, since $\beta$ and $K$ are non-negative,

$$\arg(\beta^2 + q^2 + 2\mathbf{q} \cdot \mathbf{K} - 2i\beta K)$$
$$= \begin{cases} O(\beta) & \text{if } q^2 + 2\mathbf{q} \cdot \mathbf{K} > 0 & (3\text{-}2\text{-}101a) \\ -\pi + O(\beta) & \text{if } q^2 + 2\mathbf{q} \cdot \mathbf{K} < 0 & (3\text{-}2\text{-}101b) \end{cases}$$

and consequently

$$\mathscr{I}(v, 0, 0, -\mathbf{K}, -\mathbf{q}) = \frac{4\pi}{q^2} \left| 1 + \frac{2K}{q} \cos\theta \right|^{-iv} A(\cos\theta) \tag{3-2-102}$$

where $\cos\theta = \hat{\mathbf{K}} \cdot \hat{\mathbf{q}}$ and

$$A(\cos\theta) = \begin{cases} 1 & \text{if } \cos\theta > -q/(2K) & (3\text{-}2\text{-}103a) \\ \exp(-\pi v) & \text{if } \cos\theta < -q/(2K). & (3\text{-}2\text{-}103b) \end{cases}$$

The first attempt to apply the impulse approximation to direct collision processes was made by Akerib and Borowitz [52] who calculated cross sections for ionization and for excitation of the 2s and 2p levels of atomic hydrogen by electron impact. In view of the fact that $g_i(\mathbf{t}_1)$ is sharply peaked when $\mathbf{t}_1 = 0$ they suggested that $N(v)\mathscr{I}(v, 0, 0, -\mathbf{K}, -\mathbf{q})$ could be regarded as constant throughout the region of $K$ space which provides the major contribution to the integral in (3-2-98). Since

$$(2\pi a)^{-3} \int dK \, g_i(t_1) \, g_f^*(t_2) = \int d\mathbf{r} \exp(i b \mathbf{q} \cdot \mathbf{r}) \, \varphi_i(\mathbf{r}) \, \varphi_f^*(\mathbf{r}) \equiv \Phi(b\mathbf{q}) \tag{3-2-104}$$

(see (3-2-19)) the resulting approximation to $T_{if}^{IMP}$ is

$$T_{if}^0 = -\frac{4\pi Z_1}{q^2} \left| 1 + \frac{2K_0}{q} \cos\theta_0 \right|^{-iv_0} A(\cos\theta_0) \, \Phi(b\mathbf{q}) \, N(v_0) \tag{3-2-105}$$

which is equivalent to replacing $\Psi_i^+$ by the function $\Psi_i^{\text{Peak}}$ defined by (3-2-92). Here $K_0 = -a v$, $v_0 = a Z_1 / K_0 = Z_1/v$, and from (3-1-27) and (3-1-30)

$$\cos\theta_0 = \hat{K}_0 \cdot \hat{q} = -\frac{(q^2/\mu + \Delta E)}{2 v q}. \tag{3-2-106}$$

Therefore

$$A(\cos\theta_0) = \begin{cases} 1 & \text{if } q^2 > \Delta E/b \\ \exp(-\pi v_0) & \text{if } q^2 < \Delta E/b \end{cases} \tag{3-2-107a}$$
$$\tag{3-2-107b}$$

and since

$$|N(v_0)|^2 = \frac{\pi v_0 \exp(\pi v_0)}{\sinh \pi v_0} \tag{3-2-108}$$

it follows that

$$|T_{if}^0|^2 = \frac{B(q) \, \pi v_0}{\sinh \pi v_0} |T_{if}^B|^2 \tag{3-2-109}$$

where

$$B(q) = \begin{cases} \exp(\pi v_0) & \text{if } q^2 > \Delta E/b \\ \exp(-\pi v_0) & \text{if } q^2 < \Delta E/b \end{cases} \tag{3-2-110a}$$
$$\tag{3-2-110b}$$

and

$$T_{if}^B = -\frac{4\pi Z_1}{q^2} \Phi(b\mathbf{q}) \tag{3-2-111}$$

is the result obtained in the Born approximation when the potential $V_{12}$ is neglected.

The Fourier transform $g_f(t_2)$ has a peak when $t_2 = 0$ and the assumption that the major contribution to the integral in (3-2-98) comes from this peak leads to the peaking approximation introduced by Chew [8]. However, at high energies the major contribution to the cross section comes from small values of $q$ so it follows from (3-2-99) that for the relevant values of $q$

the peaks of $g_i(t_1)$ and $g_f(t_2)$ are practically coincident. Consequently there should not be a significant difference between the cross section values obtained from the two peaking approximations. This conclusion is confirmed by the calculations of Coleman and McDowell [53].

The work of Akerib and Borowitz [52] is in error because, instead of using $\psi_m^+(13)$ as given by (3-2-49), they adopted a normalization for the Coulomb function which has the effect of replacing $N(v)$ by $N(v)/K$. That this is incorrect is most easily seen by noting that in the limit as $Z_1 \to 0$, $\omega_{13}^+ \psi_i$ must reduce to $\psi_i$. In this limit $v \to 0$ and (3-2-53) becomes

$$\omega_{13}^+ \psi_i = (2\pi)^{-3} \exp(i\mathbf{k}_i \cdot \boldsymbol{\sigma}) \int d\mathbf{t}_1 \, g_i(\mathbf{t}_1) \exp(i\mathbf{t}_1 \cdot \mathbf{r}) = \psi_i \qquad (3\text{-}2\text{-}112)$$

as required, but if the normalization used by Akerib and Borowitz is accepted an extra factor of $K^{-1}$ appears in the integral in (3-2-112), and the correct limit is not obtained. It is also evident from the work of Mapleton [15] that the correct normalization is that adopted in (3-2-49).

The cross section values reported by Akerib and Borowitz [52] for 1s–2p excitation of atomic hydrogen by electron impact were in reasonable agreement with experiment, and their method seemed superior to the Born approximation at low energies, although the results for 1s–2s excitation were not quite as impressive. Coleman and McDowell [53] repeated the calculations for these processes using the correct normalization for $\psi_m^+(13)$ and obtained results which bear no resemblance to those of Akerib and Borowitz; in fact, as can be seen from Table 3-2-2, they greatly exceed the Born approximation cross sections which are, in turn, somewhat greater than the measured values for both transitions. Electron exchange effects are readily incorporated in the theory [52, 54], but calculations show that this only introduces minor corrections which do not appreciably reduce the discrepancy between theory and experiment. Any agreement between experiment and the theory of Akerib and Borowitz [52] must therefore be regarded as accidental.

The neglect of multiple scattering which is inherent in the impulse approximation, and the neglect of $V_{12}$ in (3-2-42) are difficult to justify for electron impact, and consequently one might conclude that Table 3-2-2 simply demonstrates the inadequacy of the impulse approximation for these transitions. However, Coleman and McDowell [53] also used the peaking approximation for the processes

$$H^+ + H(1s) \to H^+ + H(2s) \qquad (3\text{-}2\text{-}113a)$$
$$H^+ + H(1s) \to H^+ + H(2p) \qquad (3\text{-}2\text{-}113b)$$

TABLE 3-2-2

Cross sections in units of $\pi a_0^2$ for $e + H(1s) \rightarrow e + H(2s \text{ or } 2p)$ in the Born approximation, $Q_B$, and the impulse approximation with a peaking approximation, $Q_P$. ($v^2$ is the electron energy in Rydbergs.)

| $v^2$ | $Q_B(2s)$ | $Q_P(2s)$ | $Q_B(2p)$ | $Q_P(2p)$ |
|---|---|---|---|---|
| 1.0 | 2.48, −1* | 9.9, −1 | 1.04 | 4.0 |
| 2.0 | 1.83, −1 | 5.9, −1 | 1.31 | 4.7 |
| 5.0 | 8.28, −2 | 1.9, −1 | 9.31, −1 | 2.6 |
| 10.0 | 4.29, −2 | 7.8, −2 | 6.15, −1 | 1.4 |
| 20.0 | 2.18, −2 | 3.3, −2 | 3.75, −1 | 6.9, −1 |
| 40.0 | 1.10, −2 | 1.5, −2 | 2.14, −1 | 3.4, −1 |
| 100.0 | 4.42, −3 | 5.3, −3 | 9.44, −2 | 1.3, −1 |

* The integer following the comma indicates the power of 10 by which the number is to be multiplied.

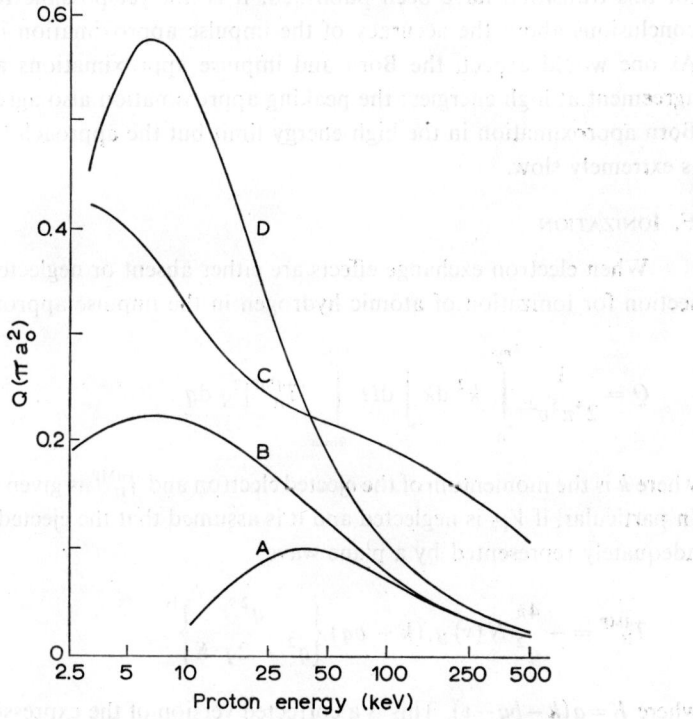

Fig. 3-2-6. Cross sections for $H^+ + H(1s) \rightarrow H^+ + H(2s)$.
Curve A. Prior form of VPS approximation [63]. Curve B. Impulse approximation without peaking approximation [9]. Curve C. Impulse approximation with peaking approximation [53]. Cross section multiplied by ten. Curve D. Born approximation.

for which the impulse approximation should be satisfactory, and once again the results were quite unrealistic, but for these processes they were much less than the Born cross sections (see Fig. 3-2-6).

The failure of the method for proton impact excitation strongly suggested that the fault lay in the peaking approximation rather than in the basic impulse approximation. To examine this possibility the present author evaluated the integral in (3-2-98) for the process (3-2-113a) without making any further approximations. Considerable care was taken in the numerical work, which is described in detail elsewhere [9], to ensure that any discrepancy could be confidently attributed to the failure of the peaking approximation rather than to computational errors. The results are plotted in Fig. 3-2-6 which clearly shows the inadequacy of the peaking approximation. The correct impulse approximation results are certainly more realistic than the peaking approximation would suggest but, since no experimental results for this transition have been published, it is not yet possible to draw any conclusions about the accuracy of the impulse approximation in this case. As one would expect, the Born and impulse approximations are in close agreement at high energies; the peaking approximation also agrees with the Born approximation in the high-energy limit but the approach to that limit is extremely slow.

### E. IONIZATION

When electron exchange effects are either absent or neglected the cross section for ionization of atomic hydrogen in the impulse approximation is

$$Q = \frac{1}{2^4 \pi^5 v^2} \int_0^{k_{max}} k^2 \, dk \int d\Omega \int_{q_{min}}^{q_{max}} |T_{if}^{IMP}|^2 q \, dq \qquad (3\text{-}2\text{-}112)$$

where $k$ is the momentum of the ejected electron and $T_{if}^{IMP}$ is given by (3-2-96). In particular, if $V_{12}$ is neglected and it is assumed that the ejected electron is adequately represented by a plane wave

$$T_{if}^{IMP} = -\frac{4\pi}{q^2} N(v) \, g_i(\mathbf{k} - b\mathbf{q}) \left\{ \frac{q^2}{q^2 - 2\mathbf{q} \cdot \mathbf{K}} \right\}^{iv} \qquad (3\text{-}2\text{-}113)$$

where $\mathbf{K} = a(\mathbf{k} - b\mathbf{q} - \mathbf{v})$. This is a corrected version of the expression used by Akerib and Borowitz [52] in their work on electron impact ionization.

An undesirable consequence of the use of a plane wave for the ejected electron is that the initial and final unperturbed wave functions are not orthogonal. In the Bethe approximation (ref. 10, p. 495), which is the high-energy limit of the Born approximation, the dipole term provides the major contribution to the ionization cross section which for high incident energies takes the form

$$Q \underset{v\to\infty}{\sim} \frac{A}{v^2} \log Bv$$

where $A$ and $B$ are constants. However, if non-orthogonal wave functions are used the monopole term does not vanish and the resulting cross section tends to a finite (non-zero) value as $v \to \infty$. Thus the cross section obtained by using (3-2-113) in (3-2-112) does not vanish as $v \to \infty$. It must therefore be concluded [79] that the agreement with experiment obtained by Akerib and Borowitz [52] is entirely fortuituous as their results are inconsistent with the model which they used.

Attempts to overcome the difficulties due to non-orthogonality by modifications of the plane wave approximation have not been successful [79] and it would appear that there is little hope of obtaining reliable ionization cross sections from (3-2-112) unless a more accurate wave function is used for the final state. However, the computational effort required to evaluate the integrals in (3-2-112), when the ejected electron is described by a Coulomb wave, is formidable and, in view of the additional uncertainty associated with the choice of a final state wave function for ionization [80], is hardly justified until the validity of the impulse approximation is better understood.

## F. SUMMARY

The present author takes the view that the impulse approximation described here is basically a heavy-particle approximation. The impulse hypothesis is no less valid for electron impact than for heavy-particle impact, but other simplifications, such as the neglect of multiple scattering, which are readily justified when the projectile is a heavy particle, are not obviously valid for electron or positron impact.

It is not possible to make a definitive statement on the accuracy of the impulse approximation for heavy-particle impact because of the lack of experimental evidence for hydrogenic targets at high and medium energies. For excitation there are no experimental results to compare with the only calculation which has been carried out, and measurements for individual electron capture processes are so far restricted to energies below 40 keV.*

---

* See footnote on p. 121.

Admittedly a considerable amount of experimental information is available for collisions involving helium and other gaseous targets, but calculations for such collisions inevitably involve the use of approximate bound state wave functions.

If the measurements of Wittkower et al. [37] are accurate, the impulse approximation is not particularly successful in predicting total cross sections for electron capture by protons from atomic hydrogen at energies up to 250 keV. There is some uncertainty in the method used to allow for capture into highly excited states but it is difficult to see how this could account for the large discrepancy between the theory and the measurements. The results of Bransden and Cheshire [42] for proton-helium collisions also underestimate the measured cross sections but in this case it is not known how much of the discrepancy can be attributed to the use of a rather crude wave function for the ground state of helium.

From a theoretical point of view the impulse approximation is much more satisfactory than the Born approximation for high-energy electron capture. At high energies capture cross sections are very small compared with the cross sections for ionization and excitation of the target. Therefore coupling to the excitation and ionization channels must play an important role in electron capture processes. All such couplings are ignored in the first Born approximation but the impulse approximation takes some account of them as it is readily seen that an expansion of $\omega_{13}^{+}\psi_i$ in terms of target eigenfunctions contains contributions from all bound and continuum states of the target. It would therefore appear that the Born approximation cannot be correct in the high-energy limit; this is further discussed in §3-4. Another advantage of the impulse approximation, which it shares with the method of Bates [55], is the absence of ambiguity concerning the inclusion of the potential $V_{12}$ in the $T$ matrix element.

## § 3-3. The Vainshtein, Presnyakov and Sobelman approximation

Vainshtein et al. [1] proposed a method for the calculation of electron impact excitation cross sections, which bears some relationship to the impulse approximation. This approach is frequently called the Vainshtein approximation, but in the present article it will be termed the VPS approximation. Since its introduction this approximation has been applied, in a number of different forms, to electron impact excitation and ionization of a variety of atomic targets, and also to excitation and ionization of atomic hydrogen by proton impact.

The basic idea behind the VPS approximation is to improve on the Born approximation by taking account of the distortion due to the interaction between the incident and bound electrons, but at the same time ensuring that the approximate wave functions used satisfy the correct boundary conditions. The different versions of this approximation which have been used are all variants of two basic forms which will be referred to as the prior and post forms (see §3-2-B). In the post form an approximate expression is obtained for the outgoing-wave solution $\Psi_i^+$ of the Schrödinger equation for the system, and an approximation to $T_{if}$ is obtained by inserting this expression in (3-1-25a). In the prior form, on the other hand, $T_{if}$ is approximated by substituting an approximate expression for $\Psi_f^-$ in (3-1-25b). The terminology used here differs from that used by e.g. Crothers [56] and Presnyakov et al. [57] who associate the terms "prior" and "post" with the wave functions approximated rather than with the potentials.

## A. THE POST FORM OF THE VPS APPROXIMATION

### 1. *Derivation of the approximation*

The starting point of the discussion is the Schrödinger equation for the projectile and target which, in the notation of §3-1, takes the form

$$\left( H_0 + \frac{Z_1}{R} - \frac{Z_1}{x} - \frac{1}{r} - E \right) \Psi_i^+ = 0. \tag{3-3-1}$$

The wave function $\varphi_i(r)$ of the initial state of the target satisfies the equation

$$\left( \frac{1}{2b} \nabla_r^2 + \frac{1}{r} - \mathscr{E}_i \right) \varphi_i(r) = 0 \tag{3-3-2}$$

and consequently if we write

$$\Psi_i^+ = \varphi_i(r) f(x, \rho) \tag{3-3-3}$$

eq. (3-3-1) becomes

$$\varphi_i(r) \left( H_0 + \frac{Z_1}{R} - \frac{Z_1}{x} - E - \mathscr{E}_i \right) f = \frac{1}{b} \nabla_r \varphi_i \cdot \nabla_r f. \tag{3-3-4}$$

A slight rearrangement of this equation and insertion of $Zf/\rho$ on both sides yields

$$\left( \frac{1}{2a} \nabla_x^2 + \frac{1}{2\mu'} \nabla_\rho^2 + \frac{Z_1}{x} - \frac{Z}{\rho} + \frac{k_i^2}{2\mu} \right) f = \mathscr{L} f \tag{3-3-5}$$

where

$$\mathscr{L} \equiv \frac{Z_1}{R} - \frac{Z}{\rho} - \frac{1}{b}\nabla_r \ln \varphi_i \cdot \nabla_r \ln f \tag{3-3-6}$$

and, for the present, $Z$ is regarded as arbitrary.

If the right hand side of (3-3-5) is now neglected, a solution of the resulting equation is

$$f_0(x,\rho) = A f_1(x) f_2(\rho) \tag{3-3-7}$$

where $A$ is a normalization constant and

$$\left(\frac{1}{2a}\nabla_x^2 + \frac{Z_1}{x} + \frac{1}{2a}k^2\right)f_1(x) = 0 \tag{3-3-8a}$$

$$\left(\frac{1}{2\mu'}\nabla_\rho^2 - \frac{Z}{\rho} + \frac{1}{2\mu'}K^2\right)f_2(\rho) = 0 \tag{3-3-8b}$$

with

$$\frac{k^2}{a} + \frac{K^2}{\mu'} = \frac{k_i^2}{\mu}. \tag{3-3-9}$$

Thus $f_1$ and $f_2$ are Coulomb functions and, since $\Psi_i^+$ must obey outgoing-wave boundary conditions, the appropriate solutions of (3-3-8a) and (3-3-8b) are

$$f_1(x) = \exp\left(\frac{aZ_1}{2k}\pi\right)\Gamma\left(1 - \frac{iaZ_1}{k}\right)$$

$$\times \exp(i\mathbf{k}\cdot\mathbf{x}) \, _1F_1\left[\frac{iaZ_1}{k}, 1, i(kx - \mathbf{k}\cdot\mathbf{x})\right] \tag{3-3-10a}$$

$$f_2(\rho) = \exp\left(-\frac{\mu'Z}{2K}\pi\right)\Gamma\left(1 + \frac{i\mu'Z}{K}\right)$$

$$\times \exp(i\mathbf{K}\cdot\boldsymbol{\rho}) \, _1F_1\left[-\frac{i\mu'Z}{K}, 1, i(K\rho - \mathbf{K}\cdot\boldsymbol{\rho})\right] \tag{3-3-10b}$$

(see Appendix 1). The values of $Z$, $k$ and $K$ are now determined by stipulating that $f_0(x,\rho)$ should have the same asymptotic form as $f(x,\rho)$.

The asymptotic form of $f_0(x,\rho)$ as $x$ and $\rho$ tend to infinity simultaneously, which is obtained from (A1-3), is

$$f_0(x,\rho) \sim A \exp(i\mathbf{k}\cdot\mathbf{x} + i\mathbf{K}\cdot\boldsymbol{\rho}) \exp\left\{-\frac{iaZ_1}{k}\ln(kx - \mathbf{k}\cdot\mathbf{x})\right.$$

$$\left. + \frac{i\mu'Z}{K}\ln(K\rho - \mathbf{K}\cdot\boldsymbol{\rho})\right\} + \text{outgoing scattered waves}. \tag{3-3-11}$$

However, in the problem under consideration

$$f(x, \rho) \underset{\sigma \to \infty}{\sim} \exp(i k_i \cdot \sigma) + \text{outgoing scattered waves}. \tag{3-3-12}$$

Consequently

$$k_i \cdot \sigma = k \cdot x + K \cdot \rho \tag{3-3-13}$$

and with the help of (3-1-6) and (3-1-7a) this becomes

$$k_i \cdot \sigma = br \cdot \left(k + \frac{1}{\mu'} K\right) + \sigma \cdot (aK - k). \tag{3-3-14}$$

Since $r$ and $\sigma$ may be varied independently it follows that

$$K = -\mu' k = bk_i \tag{3-3-15}$$

and it is readily seen that these values satisfy (3-3-9). In order to bring (3-3-11) into agreement with (3-3-12) it is also necessary to eliminate the Coulomb phases in (3-3-11). This may be done by taking

$$Z = \frac{aK}{\mu' k} Z_1 = a Z_1 \tag{3-3-16}$$

because since

$$\frac{a Z_1}{k} = \frac{a\mu' Z_1}{b k_i} = \frac{Z_1}{v} \tag{3-3-17}$$

the phase factor in question then becomes

$$\exp\left\{\frac{i Z_1}{v} \ln\left(\frac{K\rho - K\cdot\rho}{kx - k\cdot x}\right)\right\} \underset{\sigma \to \infty}{\sim} \exp\left\{\frac{i Z_1}{v} \ln\left(\frac{Ka}{k}\right)\right\} = (a\mu')^{iv_0} \tag{3-3-18}$$

where

$$v_0 = Z_1/v. \tag{3-3-19}$$

Finally if

$$A = (a\mu')^{-iv_0} \tag{3-3-20}$$

the leading terms in the asymptotic forms (3-3-11) and (3-3-12) are in complete agreement and the required approximation to $f(x, \rho)$ is

$$f_0(x, \rho) = N_1(v_0) \exp(i k_i \cdot \sigma) \, {}_1F_1\left[i v_0, 1, i a (vx + v \cdot x)\right] \\ \times {}_1F_1\left[-i v_0, 1, i b (k_i \rho - k_i \cdot \rho)\right] \tag{3-3-21}$$

where

$$N_1(v_0) = (a\mu')^{-iv_0} |\Gamma(1 + iv_0)|^2. \qquad (3\text{-}3\text{-}22)$$

Comparison with (3-2-92) shows that when the peaking approximation based on the peak of $g_i$ is used in the impulse approximation the result is equivalent to the VPS approximation with $Z=0$. The relationship between the VPS approximation and an improved impulse approximation is discussed in §3-4.

If it is now assumed that $f(x, \rho)$ can be replaced by $f_0(x, \rho)$ and that the interaction potential $V_{12}$ can be neglected in (3-1-25a), one obtains as an approximation to $T_{if}$,

$$T_{if}^{VPS} = - Z_1 N_1(v_0) \int d\mathbf{x} \int d\boldsymbol{\rho} \exp(i\mathbf{q}\cdot\boldsymbol{\sigma}) \, \varphi_f^*(\mathbf{r}) \, \varphi_i(\mathbf{r}) \frac{1}{x}$$
$$\times {}_1F_1[iv_0, 1, ia(vx + \mathbf{v}\cdot\mathbf{x})] \, {}_1F_1[-iv_0, 1, ib(k_i\rho - \mathbf{k}_i\cdot\boldsymbol{\rho})]. \qquad (3\text{-}3\text{-}23)$$

It is convenient at this point to express the product of the bound state wave functions in terms of the Fourier transform

$$\Phi(\mathbf{s}) = \int \exp(i\mathbf{s}\cdot\mathbf{r}) \, \varphi_f^*(\mathbf{r}) \, \varphi_i(\mathbf{r}) \, d\mathbf{r} \qquad (3\text{-}3\text{-}24)$$

so that (3-3-23) becomes

$$T_{if}^{VPS} = - \frac{Z_1 N_1(v_0)}{(2\pi)^3} \int d\mathbf{s} \, \Phi(\mathbf{s}) \int d\mathbf{x} \, \frac{1}{x} \exp\{-i\mathbf{x}\cdot(a\mathbf{s} + a\mathbf{q}/\mu)\}$$
$$\times {}_1F_1[iv_0, 1, ia(vx + \mathbf{v}\cdot\mathbf{x})] \int d\boldsymbol{\rho} \exp\{-i\boldsymbol{\rho}\cdot(\mathbf{s} - b\mathbf{q})\}$$
$$\times {}_1F_1[-iv_0, 1, ib(k_i\rho - \mathbf{k}_i\cdot\boldsymbol{\rho})] \qquad (3\text{-}3\text{-}25a)$$

$$= \frac{Z_1 N_1(v_0)}{(2\pi)^3} \int d\mathbf{s} \, \Phi(\mathbf{s}) \mathscr{I}(v_0, 0, 0, av, \mathbf{P}_1)$$
$$\times \frac{\partial}{\partial \beta} \mathscr{I}(-v_0, 0, \beta, -bk_i, \mathbf{P}_2)|_{\beta=0} \qquad (3\text{-}3\text{-}25b)$$

in the notation of Appendix 2, with

$$\mathbf{P}_1 = -a(\mathbf{s} + \mathbf{q}/\mu), \qquad \mathbf{P}_2 = b\mathbf{q} - \mathbf{s}. \qquad (3\text{-}3\text{-}26)$$

From (A2-19) it is readily seen that (3-3-25) reduces to

$$T_{if}^{VPS} = \frac{4bk_i Z_1 v_0 N_1(v_0)}{\pi} \int \frac{ds\, \Phi(s)}{P_1^2 P_2^2 (P_2^2 - 2bP_2 \cdot k_i)} \left\{ \frac{P_1^2 (P_2^2 - 2bP_2 \cdot k_i)}{P_2^2 (P_1^2 + 2aP_1 \cdot v)} \right\}^{iv_0}. \tag{3-3-27}$$

## 2. The peaking approximation

Rather than evaluate the integral in (3-3-27), Vainshtein et al. [1] argued that the major contribution to this integral comes from small values of $P_2$, that is from the region $s \approx bq$. On the assumption that $\Phi(s)$ is a slowly varying function of $s$ in this region they replaced $\Phi(s)$ by $\Phi(bq)$. With this replacement the $s$ integral in (3-3-25a) becomes

$$\int ds\, \exp\{-is \cdot (ax + \rho)\} = (2\pi)^3\, \delta(ax + \rho) \tag{3-3-28}$$

and we obtain as an approximation to $T_{if}^{VPS}$,

$$T_{if}^{(1)} = -Z_1 N_1(v_0)\, \Phi(bq) \int \frac{dx}{x} \exp(-iq \cdot x)$$
$$\times {}_1F_1[iv_0, 1, ia(vx + v \cdot x)]\, {}_1F_1[-iv_0, 1, iab\mu(vx + v \cdot x)]$$
$$= -Z_1 N_1(v_0)\, \Phi(bq)\, \mathscr{I}(v_0, b\mu, 0, av, -q). \tag{3-3-29}$$

It is shown in Appendix 2 that the functional form of this result depends on the value of the quantity $z$ defined by (A2-15). In the present case

$$\delta = aq \cdot v = \frac{a}{2\mu}(q^2 + \mu\, \Delta E), \quad T = q^2 \tag{3-3-30}$$

and

$$z = z_1 \equiv \frac{a^2 b (q^2 + \mu\, \Delta E)^2}{\mu \{q^2 (1 - a/\mu) - a\, \Delta E\}\{q^2(1 - ab) - ab\mu\, \Delta E\}}$$
$$= \frac{(q^2 + \mu\, \Delta E)^2}{(q^2 - \Delta E/b)(q^2 - b\mu^2\, \Delta E)}. \tag{3-3-31}$$

When $\Delta E > 0$, $|z_1|$ will be greater than unity for some values of $b$, $\mu$ and $q$, and less than unity for others. Use of (A2-18) and (A2-32) in turn yields

$$T_{if}^{(1)} = \frac{-4\pi Z_1 N_1(v_0)}{q^2}\left\{\frac{q^2 - b\mu^2\, \Delta E}{\mu(bq^2 - \Delta E)}\right\}^{iv_0} \Phi(bq)\, {}_2F_1[-iv_0, iv_0, 1, z_1],$$
$$|z_1| < 1 \tag{3-3-32a}$$

$$T_{if}^{(1)} = \frac{-4\pi^{\frac{1}{2}} Z_1 N_1(v_0)}{q^2}\left\{\frac{q^2 - b\mu^2\, \Delta E}{\mu(bq^2 - \Delta E)}\right\}^{iv_0} \Phi(bq)$$
$$\times \{U \cosh \pi v_0 \pm iV \sinh \pi v_0\}, |z_1| > 1 \tag{3-3-32b}$$

where

$$U + iV = (4z_1)^{iv_0} \frac{\Gamma(\tfrac{1}{2} + iv_0)}{\Gamma(1 + iv_0)} \,_2F_1\left[-iv_0, -iv_0, 1 - 2iv_0, z_1^{-1}\right]. \tag{3-3-33}$$

In particular for electron impact $|z_1| > 1$ for all relevant values of $q$.

Vainshtein et al. [1], ostensibly to reduce the error incurred in the region $s \approx 0$ by replacing $\Phi(s)$ by $\Phi(bq)$, replaced $q$ by $-q$ in (3-3-29) and consequently replaced $T_{if}^{(1)}$ by

$$T_{if}^{(2)} = -Z_1 N_1(v_0)\, \Phi(bq)\, \mathscr{I}(v_0, b\mu, 0, a\nu, q). \tag{3-3-34}$$

In this case $\delta = -a(q^2 + \mu\, \Delta E)/(2\mu)$ and when $\delta$ is negative and $\alpha$ and $T$ are positive

$$(T - 2\delta)(T - 2\alpha\delta) > 4\alpha\delta^2. \tag{3-3-35}$$

Therefore

$$z = z_2 = \frac{a^2 b (q^2 + \mu\, \Delta E)^2}{\mu\{q^2(1 + a/\mu) + a\, \Delta E\}\{q^2(1 + ab) + ab\mu\, \Delta E\}} < 1 \tag{3-3-36}$$

and

$$T_{if}^{(2)} = -\frac{4\pi}{q^2} Z_1 N(v_0)\, \Phi(bq)\, X^{iv_0}\, _2F_1\left[-iv_0, iv_0, 1, z_2\right] \tag{3-3-37}$$

where

$$X = \frac{q^2(1 + ab) + ab\mu\, \Delta E}{q^2(1 + a/\mu) + a\, \Delta E}. \tag{3-3-38}$$

For electron impact $a = \tfrac{1}{2}$, $b \approx 1 \approx \mu$ and

$$z_2 \approx \left(\frac{q^2 + \Delta E}{3q^2 + \Delta E}\right)^2, \quad X \approx 1, \tag{3-3-39a}$$

whereas for heavy-particle impact $a \approx 1 \approx b$, $\mu \gg 1$ and

$$z_2 \approx \frac{\Delta E}{q^2 + \Delta E}, \quad X \approx \mu z_2. \tag{3-3-39b}$$

Incautious use of (A2-18) for $|z| > 1$ would suggest that $T_{if}^{(1)}$ is singular, but it can be seen from Appendix 2 that this is not so. Vainshtein et al. by replacing $q$ by $-q$ forced $|z|$ to be less than unity and thereby removed this apparent singularity. The idea of using the analytic continuation of $_2F_1[-iv_0, iv_0, 1, z]$ for $|z| > 1$ was first explored by Omidvar [58] but, because he incorrectly assumed that $(-z)^{iv_0} = \{(-z)^{-iv_0}\}^*$ his result is equivalent to (3-3-32) with $V=0$. The correct expression for $T_{if}^{(1)}$ was first obtained by Crothers [56] in the particular case of electron impact.

For excitation and ionization (but not for collisional de-excitation) $\Delta E > 0$, and $X$ as defined by (3-3-38) is positive for all relevant values of $q$. Also

$$|N_1(v_0)| = |\Gamma(1 + iv_0)|^2 = \frac{\pi v_0}{\sinh \pi v_0} \qquad (3\text{-}3\text{-}40)$$

and consequently the cross section obtained by using $T_{if}^{(2)}$ as an approximation to $T_{if}$ in (3-1-28) is

$$Q^{(2)} = \frac{8Z_1^2}{v^2} \int_{q_{min}}^{q_{max}} \frac{dq}{q^3} |\Phi(bq)|^2 |h^{(2)}(z_2, v_0)|^2 \qquad (3\text{-}3\text{-}41)$$

where

$$h^{(2)}(z, v) = \frac{\pi v}{\sinh \pi v} {}_2F_1[-iv, iv, 1, z]. \qquad (3\text{-}3\text{-}42)$$

Alternatively, if $T_{if}$ is replaced by $T_{if}^{(1)}$ the cross section is

$$Q^{(1)} = \frac{8Z_1^2}{v^2} \int_{q_{min}}^{q_{max}} \frac{dq}{q^3} |\Phi(bq)|^2 |h^{(1)}(z_1, v_0)|^2 \qquad (3\text{-}3\text{-}43)$$

where

$$h^{(1)}(z_1, v_0) = \begin{cases} Y^{iv_0} h^{(2)}(z_1, v_0), & |z_1| < 1 \qquad (3\text{-}3\text{-}44a) \\ \pi^{\frac{1}{2}} v_0 Y^{iv_0} (U \coth \pi v_0 \pm iV), & |z_1| > 1 \qquad (3\text{-}3\text{-}44b) \end{cases}$$

and

$$Y = \frac{q^2 - b\mu^2 \Delta E}{\mu(bq^2 - \Delta E)}. \qquad (3\text{-}3\text{-}45)$$

For electron impact $Y \approx 1$ and (3-3-43) reduces to the expression obtained by Crothers [56]. If the projectile is a heavy particle the situation is slightly more complicated since in that case

$$z_1 \approx \frac{\Delta E}{\Delta E - q^2}, \qquad Y \approx \mu z_1. \qquad (3\text{-}3\text{-}46)$$

Thus for $q < (2\Delta E)^{\frac{1}{2}}$ the function $h^{(1)}(z_1, v_0)$ is given by (3-3-44b) and for larger values of $q$ the expression in (3-3-44a) must be used. Also, provided that $q_{min}^2 < \Delta E$, $|Y^{iv_0}|$ is discontinuous in the range of integration. However, the discontinuity is a finite one, similar to that which arises in excitation calculations in the impulse approximation, and would present no difficulty in numerical calculations.

The cross section $Q^{(2)}$ has been evaluated for a large number of electron induced transitions in hydrogen and the alkali metals [1, 59, 60, 76] and also for 1s–2s and 1s–2p excitation of hydrogen by proton impact [77]. If one believes that the peaking approximation is accurate, $Q^{(2)}$ must be regarded as an approximation to $Q^{(1)}$ and the replacement of $q$ by $-q$, which is basic to the derivation of (3-3-41), is justified only if $Q^{(1)}$ and $Q^{(2)}$ are in close agreement. The extent of the disagreement between these cross sections can be seen by comparing curves B and D of Fig. 3-3-1 which show the results of calculations of $Q^{(1)}$ (ref. 56) and $Q^{(2)}$ (ref. 1) with $v_0$ replaced by $\zeta/v$ where $\zeta$ is the effective charge defined in eq. (3-3-58) below. However, recent work [9, 61] has cast considerable doubt on the reliability of peaking approximations in cross section calculations and, while the replacement of $q$ by $-q$ does not seem well founded, one cannot at this stage rule out the possibility

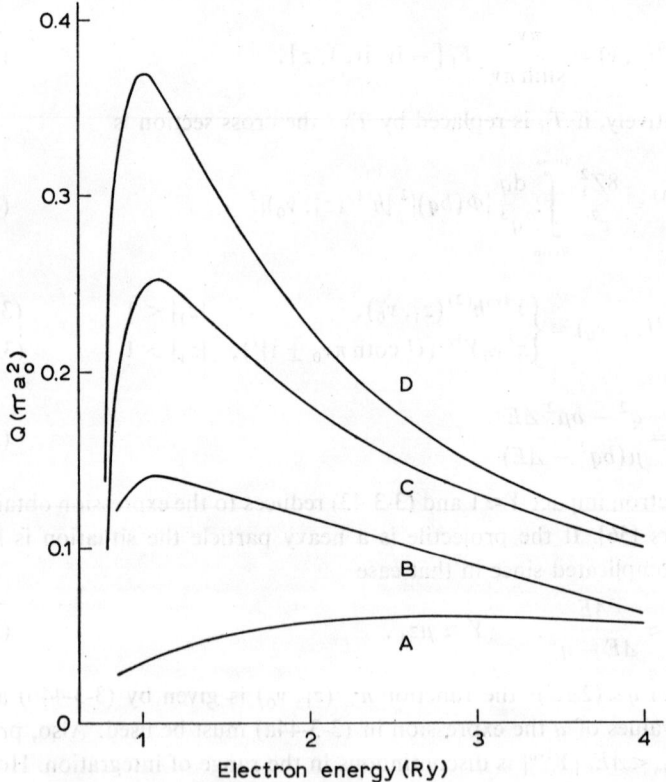

Fig. 3-3-1. Cross sections for $e + H(1s) \rightarrow e + H(2s)$ without exchange.
**Curve A.** $Q^{(2)}$ with $v_0 = -1/v$ [1]. **Curve B.** $Q^{(2)}$ with $v_0 = -1/(v + \sqrt{\mathscr{E}_1})$ [1]. **Curve C.** Born approximation. **Curve D.** $Q^{(1)}$ with $v_0 = -1/(v + \sqrt{\mathscr{E}_1})$ [56].

## 3. The inclusion of exchange

For electron impact $a=\frac{1}{2}$, $b\approx 1$, $k_i \approx v$ and in the limit as $1/M \to 0$ the $T$ matrix element which describes excitation accompanied by exchange of the electrons 2 and 3 is

$$T_{if}^{ex} = \langle \psi_f' | V_f' | \Psi_i^+ \rangle \qquad (3\text{-}3\text{-}47)$$

with

$$V_f' = V_{13} + V_{23}, \qquad \psi_f' = \exp(i\mathbf{k}_f \cdot \mathbf{r}) \varphi_f(\mathbf{R}). \qquad (3\text{-}3\text{-}48)$$

Then the cross section for excitation from state i to state f is

$$Q = \frac{1}{8\pi^2 v^2} \int_{q_{min}}^{q_{max}} \{|T_{if} + T_{if}^{ex}|^2 + 3|T_{if} - T_{if}^{ex}|^2\} q \, dq \qquad (3\text{-}3\text{-}49)$$

(see ref. 10, p. 419) where $T_{if}$ is the direct amplitude. Vainshtein et al. [1] neglected $V_{23}$ in $V_f'$ on the plea that this corresponds to neglecting $V_{12}$ in the direct term. Thus, in the VPS approximation $T_{if}^{ex}$ is replaced by

$$T_{if}^{exVPS} = (2\pi)^{-6} N_1(v_0) \int d\mathbf{s} \int d\mathbf{s}_1 \, g_f^*(\mathbf{s}) g_i(\mathbf{s}_1) \int d\mathbf{x} \frac{1}{x}$$
$$\times \exp\{\tfrac{1}{2} i \mathbf{x} \cdot (-\mathbf{k}_f + \mathbf{s} + \mathbf{s}_1 - \mathbf{k}_i)\} {}_1F_1[iv_0, 1, \tfrac{1}{2} i(vx + \mathbf{v} \cdot \mathbf{x})]$$
$$\times \int d\boldsymbol{\rho} \exp\{i\boldsymbol{\rho} \cdot (\mathbf{q} - \mathbf{s} + \mathbf{s}_1)\} {}_1F_1[-iv_0, 1, i(v\rho - \mathbf{v} \cdot \boldsymbol{\rho})]$$

$$(3\text{-}3\text{-}50)$$

where $g_i$ and $g_f$ are defined by (3-2-4).

To simplify this expression Vainshtein et al. argued that $g_i(\mathbf{s}_1)$ could be replaced by $g_i(\mathbf{s}-\mathbf{q})$ since the $\rho$ integral becomes unbounded when $\mathbf{s}_1 - \mathbf{s} + \mathbf{q} = 0$. The $\mathbf{s}_1$ integral then becomes

$$\int d\mathbf{s}_1 \exp\{i\mathbf{s}_1 \cdot (\tfrac{1}{2}\mathbf{x} + \boldsymbol{\rho})\} = (2\pi)^3 \delta(\tfrac{1}{2}\mathbf{x} + \boldsymbol{\rho}) = (4\pi)^3 \delta(\mathbf{x} + 2\boldsymbol{\rho}) \qquad (3\text{-}3\text{-}51)$$

so $T_{if}^{exVPS}$ is replaced by

$$T_{if}^{ex(1)} = \frac{N_1(v_0)}{2\pi^3} \int d\mathbf{s} \, g_f^*(\mathbf{s}) g_i(\mathbf{q} - \mathbf{s}) \int d\boldsymbol{\rho} \frac{1}{\rho} \exp\{2i\boldsymbol{\rho} \cdot (\mathbf{v} - \mathbf{s})\}$$
$$\times {}_1F_1[-iv_0, 1, i(v\rho - \mathbf{v} \cdot \boldsymbol{\rho})] {}_1F_1[iv_0, 1, i(v\rho - \mathbf{v} \cdot \boldsymbol{\rho})]. \qquad (3\text{-}3\text{-}52)$$

Since the major contribution to the $s$ integral comes from small values of $s$ it was also decided to neglect the expression $\exp(-2i\rho\cdot s)$. In this case the two integrals in (3-3-52) are uncoupled and this equation reduces to

$$T_{if}^{ex(1)} = 4N_1(v_0)\,\Phi(\boldsymbol{q})\,\mathscr{I}(v_0, 1, 0, -v, 2v) \tag{3-3-53}$$

and it follows from eq. (A2-32) that

$$|T_{if}^{ex(1)}| = \frac{16\pi^{\frac{1}{2}}}{q^2}|N_1(v_0)\,\Phi(\boldsymbol{q})\{U\cosh\pi v_0 \pm iV\sinh\pi v_0\}| \tag{3-3-54}$$

where

$$U + iV = \frac{\Gamma(\tfrac{1}{2} + iv_0)}{\Gamma(1 + iv_0)}\lim_{z\to\infty}(4z)^{iv_0}. \tag{3-3-55}$$

The indeterminate phase factor in (3-3-55) arises because $b$ and $\mu$ were replaced by unity and if the exact values of $b$ and $\mu$ are retained $U$ and $V$ may be determined unambiguously. Vainshtein et al. [1] obtained a result which differs from (3-5-54) because of an error which has the effect of replacing $\mathscr{I}(v_0, 1, 0, -v, 2v)$ by $\mathscr{I}(v_0, 1, 0, -v, -2v)$. This error was detected by Omidvar [58] but the only expression for $T_{if}^{exVPS}$ which has been used in calculations is the incorrect one given by Vainshtein et al. [1].

## 4. The effective charge

A possible means of improving the VPS approximation is the introduction of an effective charge $\zeta$ by rewriting (3-3-5) as

$$\left(\frac{1}{2a}\nabla_x^2 + \frac{1}{2\mu'}\nabla_\rho^2 + \frac{\zeta}{x} - \frac{a\zeta}{\rho} + \frac{k_i^2}{2\mu}\right)f = \mathscr{L}_1 f \tag{3-3-56}$$

where

$$\mathscr{L}_1 = \frac{Z_1}{R} - \frac{a\zeta}{\rho} + \frac{\zeta - Z_1}{x} - \frac{1}{b}\nabla_r \ln\varphi_i \cdot \nabla_r \ln f. \tag{3-3-57}$$

The solution of the equation obtained by neglecting the right hand side of (3-3-56) is given by (3-3-21) with $v_0$ replaced by $\zeta/v$. One may then choose $\zeta$ so as to reduce the error incurred in the neglect of $\mathscr{L}_1 f$. As a result of examining the asymptotic form of $\mathscr{L}_1$ as $x$ and $\rho$ tend to infinity Vainshtein et al. [1] choose, in the case of electron impact,

$$\zeta = -\frac{v}{v + \sqrt{\mathscr{E}_i}} \tag{3-3-58}$$

which serves the dual purpose of eliminating a singularity at $v=0$ in this asymptotic form and ensuring that $\mathscr{L}_1$ contains no terms of order $\sigma^{-1}$ asymptotically. Of course this choice is by no means unique (cf. refs. 56, 57, 62). The calculations of Vainshtein et al. [1] show that the cross section in this approximation is rather sensitive to the choice of effective charge (see Fig. 3-3-1).

## B. THE PRIOR FORM OF THE VPS APPROXIMATION

If the ingoing-wave solution of the Schrödinger equation for the system is written as

$$\Psi_f^- = \varphi_f(r) f'(x, \rho) \tag{3-3-59}$$

the differential equation satisfied by $f'$ is

$$\left( \frac{1}{2a} \nabla_x^2 + \frac{1}{2\mu'} \nabla_\rho^2 + \frac{Z_1}{x} - \frac{Z}{\rho} + \frac{k_f^2}{2\mu} \right) f' = \mathscr{L}' f' \tag{3-3-60}$$

where

$$\mathscr{L}' = \frac{Z_1}{R} - \frac{Z}{\rho} - \frac{1}{b} \nabla_r \ln \varphi_f \cdot \nabla_r \ln f'.$$

The VPS approximation consists of neglecting the right hand side of this equation, and a solution of the resulting equation is

$$f_0'(x, \rho) = B f_1'(x) f_2'(\rho) \tag{3-3-61}$$

where $B$ is a normalization constant and

$$f_1'(x) = \exp\left( \frac{aZ_1\pi}{2k} \right) \Gamma\left( 1 + \frac{iaZ_1}{k} \right)$$

$$\times \exp(i\mathbf{k} \cdot \mathbf{x}) \,_1F_1\left[ -\frac{iaZ_1}{k}, 1, -i(kx + \mathbf{k} \cdot \mathbf{x}) \right] \tag{3-3-62a}$$

$$f_2'(\rho) = \exp\left( -\frac{\mu' Z\pi}{2K} \right) \Gamma\left( 1 - \frac{i\mu' Z}{K} \right)$$

$$\times \exp(i\mathbf{K} \cdot \boldsymbol{\rho}) \,_1F_1\left[ \frac{i\mu' Z}{K}, 1, -i(K\rho + \mathbf{K} \cdot \boldsymbol{\rho}) \right] \tag{3-3-62b}$$

with

$$\frac{k^2}{a} + \frac{K^2}{\mu'} = \frac{k_f^2}{\mu}. \tag{3-3-63}$$

The appropriate values of $k$, $K$ and $Z$ are determined by stipulating that $f'_0$ should have the same asymptotic form as $f'$, i.e.

$$f'(x, \rho) \underset{\sigma \to \infty}{\sim} \exp(i k_f \cdot \sigma) + \text{ingoing scattered waves}, \qquad (3\text{-}3\text{-}64)$$

and in this way we obtain

$$f'_0(x, \rho) = N_2(v') \exp(i k_f \cdot \sigma) \, {}_1F_1[-iv', 1, -ia(v_f x - v_f \cdot x)]$$
$$\times {}_1F_1[iv', 1, -ib(k_f \rho + k_f \cdot \rho)] \qquad (3\text{-}3\text{-}65)$$

with $v' = Z_1/v_f$, $v_f = k_f/\mu$ and

$$N_2(v') = (a\mu')^{iv'} |\Gamma(1 + iv')|^2. \qquad (3\text{-}3\text{-}66)$$

As in the post form of the approximation it is also assumed that the potential $V_{12}$ can be neglected and therefore the exact $T$ matrix element is replaced by

$$\mathcal{T}_{if}^{VPS} = -(2\pi)^{-3} Z_1 N_2^*(v') \int d\mathbf{x} \int d\boldsymbol{\rho}$$
$$\times \exp\{i\mathbf{q} \cdot (b\boldsymbol{\rho} - a\mathbf{x}/\mu)\} \frac{1}{x} {}_1F_1[iv', 1, ia(v_f x - v_f \cdot x)]$$
$$\times {}_1F_1[-iv', 1, ib(k_f \rho + k_f \cdot \rho)] \int d\mathbf{s} \exp\{-i\mathbf{s} \cdot (\boldsymbol{\rho} + a\mathbf{x})\} \Phi(\mathbf{s})$$
$$= (2\pi)^{-3} Z_1 N_2^*(v') \int d\mathbf{s} \, \Phi(\mathbf{s}) \mathcal{I}(v', 0, 0, -a v_f, -a\mathbf{s}, -a\mathbf{q}/\mu)$$
$$\times \frac{\partial}{\partial \beta} \mathcal{I}(-v', 0, \beta, b k_f, b\mathbf{q} - \mathbf{s})|_{\beta = 0}. \qquad (3\text{-}3\text{-}67)$$

No calculations based on this expression have yet been reported but some results have been obtained by using a peaking approximation to simplify the computation. If it is assumed that the major contribution to the integral in (3-3-67) comes from $\mathbf{s} = b\mathbf{q}$, $\mathcal{T}_{if}^{VPS}$ may be approximated by

$$\mathcal{T}_{if}^{(1)} = -Z_1 N_2^*(v') \Phi(b\mathbf{q}) \mathcal{I}(v', b\mu, 0, -a v_f, -\mathbf{q}). \qquad (3\text{-}3\text{-}68)$$

In this case (see Appendix 2)

$$\delta = -a v_f \cdot \mathbf{q} = -\tfrac{1}{2} a(\Delta E - q^2/\mu), \quad T = q^2 \qquad (3\text{-}3\text{-}69)$$

and

$$z = \frac{(q^2 - \mu \Delta E)^2}{(q^2 + \Delta E/b)(q^2 + b\mu^2 \Delta E)} < 1, \quad \Delta E > 0. \qquad (3\text{-}3\text{-}70)$$

It follows from (A2-18) that

$$\mathcal{T}_{if}^{(1)} = -\frac{4\pi Z_1 N_2^*(v')}{q^2} \Phi(b\mathbf{q}) \left\{\frac{q^2 + b\mu^2 \Delta E}{\mu(bq^2 + \Delta E)}\right\}^{iv'} {}_2F_1[-iv', iv', 1, z]. \qquad (3\text{-}3\text{-}71)$$

In particular, for electron impact, when terms of order $1/M$ are neglected compared with unity,

$$z = \left(\frac{q^2 - \Delta E}{q^2 + \Delta E}\right)^2 \tag{3-3-72}$$

and

$$\mathcal{T}_{if}^{(1)} \approx \frac{4\pi N_2^*(v')}{q^2} \Phi(q) \,_2F_1[-iv', iv', 1, z] \tag{3-3-73}$$

as obtained by Crothers and McCarroll [62], whereas for proton impact

$$z = \frac{\Delta E}{q^2 + \Delta E} \tag{3-3-74}$$

and

$$\mathcal{T}_{if}^{(1)} = -\frac{4\pi N_2^*(v')}{q^2} \Phi(q) (\mu z)^{iv'} \,_2F_1[-iv', iv', 1, z] \tag{3-3-75}$$

which agrees with McCarroll and Salin's expression [63] apart from a phase factor. From (3-1-30), (3-3-19) and the definition of $v'$ it is evident that

$$v' = v_0\left(1 + \frac{\Delta E}{2\mu v^2} + \cdots\right). \tag{3-3-76}$$

Consequently, for proton impact $\mathcal{T}_{if}^{(1)}$ and the post matrix element $T_{if}^{(2)}$ yield the same cross section values except at very low energies. However, in the absence of detailed calculations it is not known if $\mathcal{T}_{if}^{(1)}$ and $T_{if}^{(1)}$ will give similar results.

The first calculations using the prior form of the approximation were those of Crothers and McCarroll [62] for 1s–2s and 1s–2p excitation of atomic hydrogen and 3s–3p excitation of sodium by electron impact. They included exchange effects by means of approximations very similar to those used by Vainshtein et al. [1] and also attempted, by introducing an effective charge, to reduce the error caused by neglecting the right hand side of (3-3-60). The asymptotic form of $\mathcal{L}'$ is singular at the excitation threshold (i.e. when $k_f=0$) and to remove this singularity Crothers and McCarroll used the complex effective charge

$$\zeta' = -\frac{k_f}{k_f - i\sqrt{\mathcal{E}_f}}. \tag{3-3-77}$$

From Fig. 3-3-2, which shows the results for 1s–2p excitation, it is evident that the introduction of the effective charge has a far more significant effect on the cross section than has the inclusion of electron exchange. In fact electron exchange makes very little difference except below 20 eV where, for reasons discussed by Crothers [56], the effective charge given by (3-3-77) is unsatisfactory. There is, of course, no guarantee that electron exchange would be equally unimportant if the peaking approximation were not used.

McCarroll and Salin [63] also used the prior form in their calculations for 1s–2s and 1s–2p excitation of atomic hydrogen by protons. Their results for the 1s–2s transition are shown as curve A of Fig. 3-2-6. No attempt was made in this case to improve the approximation by introducing an effective charge.

## C. Summary

The underlying concept of the VPS approximation is undoubtedly appealing. It seems reasonable to suppose that the interaction of major importance in inducing an electronic transition is that between the projectile and the bound electron, which is taken into account exactly in this method. The projectile–nucleus interaction is approximated but in such a way as to ensure

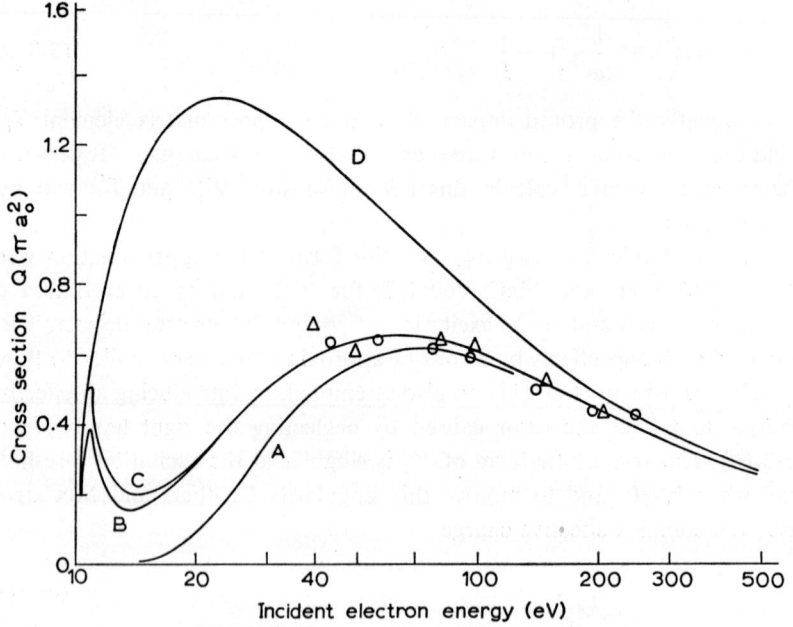

Fig. 3-3-2. Cross sections for $e + H(1s) \to e + H(2p)$ [from ref. 62, fig. 1]. Curve A. $v' = -1/v_t$, no exchange, prior form of VPS approximation. Curve B. $v' = -1/(v_t - i\sqrt{\mathscr{E}_t})$, no exchange, prior form of VPS approximation. Curve C. $v' = -1/(v_t - i\sqrt{\mathscr{E}_t})$, with exchange, prior form of VPS approximation. Curve D. Born approximation. ○, △. Experiments [68, 69].

that the physical boundary conditions are satisfied. Furthermore, the use of an effective charge provides a means of reducing the importance of the neglected terms in (3-3-5) or (3-3-60).

However, all applications of this method published to date involve additional approximations, the validity of which has not been established. The peaking approximation used in the evaluation of the direct term is a source of considerable uncertainty particularly at low and moderate projectile energies. A very similar peaking approximation used in a Coulomb-Born approximation for electron-hydrogen scattering has been examined by Kyle and McDowell [61]; they found that it leads to an error of approximately a factor of two in the 1s–2s excitation cross section near the excitation threshold, but results for energies greater than 1.5 Rydbergs have not been published.

There appears to be no a priori reason for the neglect of the projectile-nucleus interaction $V_{12}$ in the case of electron impact, although it can be justified for heavy-particle impact. In the Born approximation the contribution from $V_{12}$ vanishes, irrespective of the magnitude of the projectile mass, if the target nucleus is assumed to be infinitely massive. However, this cannot be regarded as a justification for the neglect of the corresponding contribution in a more elaborate approximation, as the corrections to the contributions from $V_{12}$ and from $V_{13}$ may well be of the same order of magnitude.

The evaluation of the exchange contribution in the case of electron impact is also unsatisfactory. The neglect of the contribution from $V_{23}$ has not been justified and the peaking approximation used to evaluate the $s_1$ integral in (3-3-50) is similar to that used for the direct term. Furthermore, although the assumption that $\exp(-2i\rho \cdot s)$ may be neglected in (3-3-52) is certainly valid at very high energies, where exchange effects are negligibly small for transitions not involving spin flip, it is not obviously valid in the energy range where exchange is likely to be important.

Despite the satisfactory agreement between the measured cross sections and those calculated by Vainshtein et al. [1], Presnyakov [60], Crothers and McCarroll [62] and Presnyakov et al. [57] it is clear that much remains to be done before any definite conclusion can be reached on the usefulness of the VPS approximation. The obvious next step is a careful investigation of the validity of the peaking approximation used in the direct term. In view of the additional uncertainties involved in electron impact calculations it would seem more satisfactory to carry out this investigation for proton impact, and having done that one would be in a position to decide on an optimum choice of effective charge. If the VPS approximation proved satisfactory for proton impact one would then be encouraged to consider more carefully the effect of the potential $V_{12}$ in the case of electron impact (see however §3-4). Electron exchange could be incorporated either by using justifiable methods to evaluate the integrals in (3-3-50), or more simply by using an Ochkur-Rudge estimate, although this is a somewhat arbitrary procedure [64].

## § 3-4. The extended impulse approximation

### A. The post form

The impulse approximation discussed in §3-2 was obtained by expanding the operator $\Omega^+$ defined by (3-2-23) in terms of the operators $\omega_{ij}^+$. A drawback of this method, particularly for electron and positron impact, is that one must neglect multiple scattering terms in order to obtain tractable expressions for the relevant $T$ matrix elements. A more satisfactory approach, from this point of view, is to expand in terms of the operators

$$\omega_i^+ = \sum_m \omega_i^+(m) |\chi_m\rangle \langle \chi_m| \qquad (3\text{-}4\text{-}1)$$

where

$$\omega_i^+(m)\chi_m = \left(1 + \frac{1}{E_m - H_0 - V_i + i\varepsilon} V_i\right)\chi_m \equiv \psi_m^+(i) \qquad (3\text{-}4\text{-}2)$$

and $\chi_m$ is defined by (3-2-46). Then by invoking the impulse hypothesis, but making no assumptions about multiple scattering effects, an approximation is obtained which we shall call the extended impulse approximation. From this approximation one can derive both the ordinary impulse approximation and the post form of the VPS approximation by making additional assumptions.

From the operator identity (3-1-18a) it is evident that

$$G^+ V_i = b_i^+(m) + G^+(E_m - E + V_{23}) b_i^+(m) \qquad (3\text{-}4\text{-}3)$$

where

$$b_i^+(m) = \omega_i^+(m) - 1 \qquad (3\text{-}4\text{-}4)$$

and, by arguments similar to those which led to (3-2-32),

$$G^+ V_i \psi_i = b_i^+ \psi_i + G^+ [V_{23}, b_i^+] \psi_i \qquad (3\text{-}4\text{-}5)$$

with

$$b_i^+ = \omega_i^+ - 1. \qquad (3\text{-}4\text{-}6)$$

It follows that

$$T_{if} = \langle \psi_f | V_f | \omega_i^+ \psi_i \rangle + \langle \psi_f | V_f | G^+ [V_{23}, b_i^+] \psi_i \rangle. \qquad (3\text{-}4\text{-}7)$$

The impulse hypothesis now leads to the neglect of the commutator involving $V_{23}$ and (3-4-7) reduces to

$$T_{if}^{EIMP} = \langle \psi_f | V_f | \omega_i^+ \psi_i \rangle \qquad (3\text{-}4\text{-}8)$$

which is the required approximation to $T_{if}$.

The approximation derived above is useful only if one can obtain a sufficiently simple expression for $\omega_i^+ \psi_i$. This is not always possible but for some important problems calculations based on this approach may be feasible. Let us consider first the case where particles 1 and 2 in Fig. 3-1-1 are heavy and particle 3 is, as usual, an electron. Then $a \approx 1$ and $1/R$ can be replaced by $1/\rho$ with the result that the differential equation satisfied by $\psi_m^+(i)$ becomes

$$\left( -\frac{1}{2a} \nabla_x^2 - \frac{1}{2\mu'} \nabla_\rho^2 + \frac{Z_1}{\rho} - \frac{Z_1}{x} - E_m \right) \psi_m^+(i) = 0. \qquad (3\text{-}4\text{-}9)$$

Consequently

$$\psi_m^+(i) = N(v) N(-v_1) \chi_m {}_1F_1[-iv_1, 1, i(k\rho - \mathbf{k}\cdot\boldsymbol{\rho})]$$
$$\times {}_1F_1[iv, 1, i(Kx - \mathbf{K}\cdot\mathbf{x})] \qquad (3\text{-}4\text{-}10)$$

where

$$v = aZ_1/K, \quad v_1 = \mu' Z_1/k \qquad (3\text{-}4\text{-}11)$$

and

$$N(v) = \exp(\tfrac{1}{2}\pi v) \Gamma(1 - iv). \qquad (3\text{-}4\text{-}12)$$

Use of (3-2-51) and the definition (3-2-52) then yields

$$\omega_i^+ \psi_i = (2\pi)^{-3} \exp(i\mathbf{k}_i \cdot \boldsymbol{\sigma}) \int dt_1 \, g_i(t_1) \exp(i\mathbf{t}_1 \cdot \mathbf{r}) N(v) N(-v_1)$$
$$\times {}_1F_1[-iv_1, 1, i(k\rho - \mathbf{k}\cdot\boldsymbol{\rho})] {}_1F_1[iv, 1, i(Kx - \mathbf{K}\cdot\mathbf{x})] \qquad (3\text{-}4\text{-}13)$$

with

$$\mathbf{k} = \frac{1}{a}(\mathbf{K} + \mathbf{k}_i). \qquad (3\text{-}4\text{-}14)$$

At this point one might argue that having replaced $1/R$ by $1/\rho$ we should also replace $a$ by unity and $\mu'$ by $M_1 M_2/(M_1 + M_2)$. This is indeed correct but we have neglected to do so in the present case in order to facilitate comparison with other approximations.

When $\omega_i^+ \psi_i$ is inserted in (3-1-25a) as an approximation for $\Psi_i^+$ and $1/R$ is replaced by $1/\rho$ in $V_f$, the coordinate integrals are readily evaluated for both direct and rearrangement collisions and in each case the resulting approximation to $T_{if}$ involves two integrals over momentum space. For example, for direct excitation, neglecting terms of order $1/M$ compared with unity, we obtain

$$T_{if}^{\text{EIMP}} = \frac{Z_1}{(2\pi)^6} \int dt_1 \int dt_2 \, g_i(t_1) \, g_f^*(t_2) \, N(v) \, N(-v_1)$$
$$\times \int d\mathbf{x} \int d\mathbf{R} \left(\frac{1}{R} - \frac{1}{x}\right) \exp\{i\mathbf{R} \cdot (\mathbf{q} + t_1 - t_2) + i\mathbf{x} \cdot (t_1 - t_2)\}$$
$$\times {}_1F_1[-iv_1, 1, i(kR - \mathbf{k} \cdot \mathbf{R})] \, {}_1F_1[iv, 1, i(Kx - \mathbf{K} \cdot \mathbf{x})]$$
(3-4-15)

and closed expressions for the $x$ and $R$ integrals may be deduced from eq. (A2-19). It will probably be necessary to evaluate the remaining integrals numerically.

If $v_1$ is replaced by zero in (3-4-13), $\omega_i^+ \psi_i$ reduces to the impulse approximation wave function $\omega_{13}^+ \psi_i$ (see (3-2-53)). At high energies the major contribution to the integral in (3-4-13) is expected to come from the peak of $g_i(t_1)$, that is where

$$\mathbf{K} = \mathbf{K}_0 \equiv -a\mathbf{v} \quad \text{and} \quad \mathbf{k} = \frac{1}{a}(-a\mathbf{v} + \mathbf{k}_i) = b\mathbf{k}_i \quad (3\text{-}4\text{-}16)$$

and in this case, when $a$ is replaced by unity,

$$v = v_1 = Z_1/v \equiv v_0. \quad (3\text{-}4\text{-}17)$$

Eq. (3-4-16) indicates that the relevant values of $k$ are very large, but it is clear from (3-4-17) that this does not justify the replacement of $v_1$ by zero. This serves to illustrate an unsatisfactory feature of the impulse approximation which was mentioned previously.

If it is assumed that the major contribution to the integral in (3-4-13) comes from the vicinity of $t_1 = 0$, a peaking approximation reduces (3-4-13) to

$$\omega_i^+ \psi_i \approx (2\pi)^{-3} \exp(i\mathbf{k}_i \cdot \boldsymbol{\sigma}) \, |\Gamma(1 + iv_0)|^2 \, {}_1F_1[-iv_0, 1, ib(k_i\rho - \mathbf{k}_i \cdot \boldsymbol{\rho})]$$
$$\times {}_1F_1[iv_0, 1, ia(vx + \mathbf{v} \cdot \mathbf{x})] \int dt_1 \, g_i(t_1) \exp(it_1 \cdot \mathbf{r}). \quad (3\text{-}4\text{-}18)$$

Since

$$(2\pi)^{-3} \int dt_1 \, g_i(t_1) \exp(it_1 \cdot r) = \varphi_i(r)$$

this is, apart from the factor $(a\mu')^{-iv_0}$, the wave function used in the post form of the VPS approximation (see (3-3-21)). However the peaking approximation used in obtaining (3-4-18) is the same as that considered by Coleman [9] and it is likely to be valid only at extremely high energies. Many approximations in scattering theory may be derived in more than one way, and not all derivations show the approximation in the same favourable light. Nevertheless, this author feels that the above argument casts considerable doubt on the validity of the VPS approximation, except at very high energies where the peaking approximation provides a sufficiently accurate estimate of the integral in (3-4-13).

It is interesting to note that the derivation of (3-4-13) holds only for heavy-particle impact. If the projectile is an electron or positron it is no longer correct to replace $1/R$ by $1/\rho$, and it does not seem possible to establish a simple connection between the extended impulse approximation and the VPS approximation, at least in the post form. Perhaps this explains why the introduction of an effective charge in the VPS approximation makes such a large difference in the case of electron impact, and it also suggests that the use of a similar effective charge would not greatly affect the proton-impact results.

By arguments quite different from those given here McCarroll and Salin [65] obtained the expression in (3-4-13) as an approximation for the exact wave function $\Psi_i^+$. Their approach was based on the work of Dodd and Greider [66] who proposed a means of removing from the kernel of the Lippmann-Schwinger integral equation the so-called disconnected terms which correspond to scattering of two particles while the third particle propagates freely. The absence of these undesirable terms makes it probable that the Neumann series solutions of the integral equations derived by Dodd and Greider will converge at high energies, but the convergence of any of these series has not been proved. The matrix element

$$\langle \psi_f | V_f (1 + G_x^+ V_i) | \psi_i \rangle \qquad (3\text{-}4\text{-}19)$$

with

$$G_x^+ = (E - H + V_x + i\varepsilon)^{-1} \qquad (3\text{-}4\text{-}20)$$

is the first term in one of the resulting expansions for $T_{if}$. The potential $V_x$

must satisfy certain conditions (see e.g. ref. 7, ch. 6) but a possible choice is $V_x = V_i$. The expression (3-4-19) is one of three approximations to $T_{if}$ considered by McCarroll and Salin [65] in an attempt to deduce the exact high-energy dependence of the cross section for the reaction

$$H^+ + H(1s) \rightarrow H(1s) + H^+, \qquad (3\text{-}4\text{-}21)$$

but they introduced a further simplification which has the effect of replacing $(1 + G_x^+ V_i) \psi_i$ by $\omega_i^+ \psi_i$ (see ref. 67). The derivation of (3-4-13) as described here gives a clearer indication of the precise nature of the physical assumption involved in the replacement of $\Psi_i^+$ by $\omega_i^+ \psi_i$. The assumption in question is simply the impulse hypothesis, which one would certainly expect to be valid at high energies, and consequently $T_{if}^{EIMP}$ should yield the correct high-energy behaviour.

Using two peaking approximations to evaluate the integrals which arise, McCarroll and Salin deduced that $T_{if}^{EIMP}$ and $T_{if}^{IMP}$ yield cross sections with the same high-energy behaviour. From a different approximation to $T_{if}$, also based on the analysis of Dodd and Greider, the same authors obtained a slightly different high-energy behaviour in agreement with the second Born approximation. They regarded this as the correct result and attributed the discrepancy to the failure of a peaking approximation used in the evaluation of $T_{if}^{EIMP}$. The situation is still far from clear as it has not been proved that the approximation schemes proposed by Dodd and Greider [66] converge, and, if they do, it is not obvious that each of them converges to its first term; thus there is no guarantee that the three matrix elements considered by McCarroll and Salin [65] will lead to the same result. The extended impulse approximation seems to have much to recommend it at high energies and a careful determination of the resulting cross section in the high-energy limit would be of considerable interest.

The approximate evaluation of cross sections in the high-energy limit involves many pitfalls, and a trivial example will serve to illustrate how approximations which appear compatible can lead to conflicting results. It seems reasonable in view of (3-4-16) to suppose that as $v \rightarrow \infty$ the factor $N(-v_1)\,_1F_1[-iv_1, 1, i(k\rho - \boldsymbol{k}\cdot\boldsymbol{\rho})]$ may be replaced by unity [65] so that $T_{if}^{EIMP}$ reduces to $T_{if}^{IMP}$. However for the most important values of $K$, $v \approx v_1$ and the same reasoning suggests that $N(v)\,_1F_1[iv, 1, i(Kx - \boldsymbol{K}\cdot\boldsymbol{x})]$ may also be replaced by unity. But if this is done $T_{if}^{IMP}$ reduces to $T_{if}^{OBK}$, which contradicts the result obtained by Bransden and Cheshire [42]. Because of the ease with which errors can arise in such calculations it is suggested that the peaking approximations generally employed in the evaluation of high-energy limits must be put on a more rigorous basis before the high-energy limit in any of the more elaborate approximations, such as the impulse or extended impulse approximations, can be regarded as established with absolute certainty.

### B. The Prior Form

A prior form of the extended impulse approximation may also be derived in an obvious way, and the resulting approximation to $T_{if}$ is

$$\mathcal{T}_{if}^{EIMP} = \langle \omega_f^- \psi_f | V_i | \psi_i \rangle$$

where

$$\omega_f^- \psi_f = \sum_m \psi_m^-(f) \langle \chi_m | \psi_f \rangle$$

and $\psi_m^-(f)$ is the solution of the equation

$$(H_0 + V_f - E_m) \psi_m^-(f) = 0$$

which satisfies incoming-wave boundary conditions. For excitation by heavy-particle impact $\psi_m^-(f)$, like $\psi_m^+(i)$, is a product of Coulomb functions in the limit as $1/M \to 0$, and a peaking approximation applied to the resulting expression for $\omega_f^- \psi_f$ yields the prior form of the VPS approximation.

If the collision depicted in Fig. 3-1-1 results in capture of the electron 3 by the projectile, the interaction potential in the final state is $V_f = Z_1/R - 1/r$, and since particle 2 is heavy $1/R$ may be replaced by $1/\sigma$. Thus, irrespective of the magnitude of the projectile mass $\psi_m^-(f)$ may be written as a product of Coulomb functions and consequently $\mathcal{T}_{if}^{EIMP}$ can be reduced to, at worst, a six-dimensional integral in momentum space. In particular this version of the approximation is applicable to positronium formation, a process for which the ordinary impulse approximation is unsatisfactory because of the neglect of multiple scattering effects.

### C. Summary

The extended impulse approximation is an improvement on the conventional impulse approximation as it does not involve the assumption that multiple scattering can be neglected, but this improvement greatly increases the difficulty of cross section calculations. The evaluation of cross sections for heavy-particle impact should however be well within the scope of modern computing facilities, and the same applies to calculations for positronium formation based on the prior form of the approximation. For electron impact neither version of this approximation has yet been reduced to tractable form.

Recognition of the relationship between the extended impulse approximation and the approximations discussed in §3-2 and §3-3 helps to clarify some of the shortcomings of these approximations. In particular the indications are that the VPS approximation may be satisfactory only at very high energies. Also, the unsatisfactory nature of the assumption required to reduce $T_{if}^{EIMP}$ to $T_{if}^{IMP}$ casts some doubt on the validity of the conventional impulse approximation; this criticism would not apply to a consistent treatment of the model problem in which $V_{12}$ is taken to be zero but no such consistent calculation has yet been carried out.

## Appendix 1. Coulomb functions

The Schrödinger equation for a particle of mass $m$ in the potential $Z/r$ is

$$\left(\nabla^2 - \frac{2mZ}{r} + k^2\right)\psi = 0 \tag{A1-1}$$

where $k^2/(2m)$ is the energy of the particle at an infinite distance from the origin. Some properties of the solutions of this equation are given here for convenient reference but for derivations and further details the reader is referred to Messiah [32] or McDowell and Coleman [7].

The solution of eq. (A1-1) which satisfies outgoing-wave boundary conditions is

$$\psi^+ = \exp(-\tfrac{1}{2}v\pi)\,\Gamma(1+iv)\exp(i\mathbf{k}\cdot\mathbf{r})\,_1F_1[-iv, 1, i(kr - \mathbf{k}\cdot\mathbf{r})] \tag{A1-2}$$

$$\underset{r\to\infty}{\sim} \exp\{i\mathbf{k}\cdot\mathbf{r} + iv\ln(kr - \mathbf{k}\cdot\mathbf{r})\} + \frac{1}{r}\exp\{ikr - iv\ln 2kr\}\,f_c \tag{A1-3}$$

where $v = mZ/k$, $f_c$ is the Coulomb scattering amplitude i.e.

$$f_c = \frac{-v}{2k^2 \sin^2\tfrac{1}{2}\theta}\exp\{2i\arg\Gamma(1+iv) - iv\ln(\sin^2\tfrac{1}{2}\theta)\} \tag{A1-4}$$

with $\theta = \cos^{-1}(\hat{\mathbf{k}}\cdot\hat{\mathbf{r}})$, and

$$_1F_1[\alpha, \beta, z] = 1 + \frac{\alpha}{\beta}z + \frac{\alpha(\alpha+1)}{\beta(\beta+1)}\frac{z^2}{2!} + \cdots. \tag{A1-5}$$

The corresponding ingoing-wave solution $\psi^-$ is obtained by taking the complex conjugate of $\psi^+$ and reversing the direction of the vector $\mathbf{k}$. Thus

$$\psi^- = \exp(-\tfrac{1}{2}v\pi)\,\Gamma(1-iv)\exp(i\mathbf{k}\cdot\mathbf{r})\,{}_1F_1[iv, 1, -i(kr+\mathbf{k}\cdot\mathbf{r})] \quad \text{(A1-6)}$$

$$\underset{r\to\infty}{\sim}\ \exp\{i\mathbf{k}\cdot\mathbf{r}-iv\ln(kr+\mathbf{k}\cdot\mathbf{r})\} + \frac{1}{r}\exp\{-ikr+iv\ln 2kr\}\,f_c^*. \quad \text{(A1-7)}$$

## Appendix 2. Nordsieck's integration technique

This appendix is concerned with the evaluation of the class of integrals

$$\mathscr{I}(v,\alpha,\beta,\mathbf{K},\mathbf{P}) = \int d\mathbf{r}\,\exp(-\beta r + i\mathbf{P}\cdot\mathbf{r})\frac{1}{r}\,{}_1F_1[iv,1,i(Kr+\mathbf{K}\cdot\mathbf{r})]$$
$$\times\,{}_1F_1[-iv,1,i\alpha(Kr+\mathbf{K}\cdot\mathbf{r})] \quad \text{(A2-1)}$$

where $v$, $\alpha$, $\beta$ are real constants and $\beta \geqslant 0$. This class includes the integrals involving two Coulomb functions which arise in both prior and post forms of the VPS approximation (see §3-3), and some simpler integrals which frequently occur in atomic scattering problems are obtained by taking $\alpha=0$. Integrals of the form (A2-1) with $\alpha=0$ were evaluated by Wentzel [70] by using a real integral representation for the confluent hypergeometric function and changing the order of integration. An alternative method, also based on a real integral representation, was used by Massey and Mohr [71]. The method described here is a more powerful and more elegant contour integral technique due to Nordsieck [72].

The starting point of Nordsieck's method is the contour integral representation

$$_1F_1[a,1,z] = \frac{1}{2\pi i}\oint_C e^{tz}\,t^{a-1}(t-1)^{-a}\,dt \quad \text{(A2-2)}$$

where C is a closed contour which encircles the points $t=0$ and $t=1$ once in the positive direction, and all powers in the integrand have their principal values (see ref. 73, p. 272). Use of (A2-2) in (A2-1) yields

$$\mathscr{I}(v,\alpha,\beta,\mathbf{K},\mathbf{P}) = -\frac{1}{4\pi^2}\oint_{C_1} dt_1\oint_{C_2} dt_2\left\{\frac{t_1(t_2-1)}{t_2(t_1-1)}\right\}^{iv}\frac{1}{t_1 t_2}\,V(t_1,t_2) \quad \text{(A2-3)}$$

where

$$V(t_1,t_2) = \int d\mathbf{r}\,\frac{1}{r}\exp\{-\beta r + i\mathbf{P}\cdot\mathbf{r} + i(t_1+\alpha t_2)(Kr+\mathbf{K}\cdot\mathbf{r})\}$$
$$= \frac{4\pi}{T - 2\delta(t_1+\alpha t_2)} \quad \text{(A2-4)}$$

with

$$T = \beta^2 + P^2, \quad \delta = i\beta K - \mathbf{P}\cdot\mathbf{K} \tag{A2-5}$$

if the change in the order of integration is valid, that is provided $V(t_1, t_2)$ converges uniformly with respect to $t_1$ and $t_2$. A sufficient condition for uniform convergence is that the inequality

$$\beta + 2K \operatorname{Im}(t_1 + \alpha t_2) > 0 \tag{A2-6}$$

be satisfied for all values of $t_1$ and $t_2$ within and on $C_1$ and $C_2$. It will henceforth be assumed that the contours are chosen so that this condition is satisfied. In particular, when $t_2$ is real

$$\beta + 2K \operatorname{Im} t_1 > 0 \tag{A2-7}$$

at all points within and on $C_1$.

The singularities of the integrand in (A2-3), regarded as a function of $t_1$, are branch points at 0 and 1, both of which lie inside $C_1$, and a simple pole at

$$t_1 = T_1 \equiv \frac{T - 2\alpha\delta t_2}{2\delta}. \tag{A2-8}$$

When $t_2$ is real it is readily seen that

$$\beta + 2K \operatorname{Im} T_1 \leqslant 0 \tag{A2-9}$$

and therefore $t_1 = T_1$ lies outside $C_1$. The integration with respect to $t_1$ is carried out by applying Cauchy's theorem to the region bounded by the contour $C_1$ and a circle S of large radius $R$. As $R \to \infty$ the integral on S tends to zero as $R^{-1}$; therefore

$$\mathscr{I}(v, \alpha, \beta, \mathbf{K}, \mathbf{P})$$
$$= -2i \oint_{C_2} \frac{dt_2}{t_2} \left(\frac{t_2 - 1}{t_2}\right)^{iv} (T - 2\alpha\delta t_2)^{iv-1} (T - 2\delta - 2\alpha\delta t_2)^{-iv}. \tag{A2-10}$$

The integrand in (A2-10) has singularities at 0 and 1 inside $C_2$ and at

$$t_2 = \frac{1}{a} \equiv \frac{T}{2\alpha\delta} \quad \text{and} \quad t_2 = \frac{1}{b} \equiv \frac{T - 2\delta}{2\alpha\delta} \tag{A2-11}$$

outside $C_2$. The transformation $t = t_2^{-1}$ maps $C_2$ onto a contour $C_3$ which encircles the points $a$ and $b$ in the $t$ plane. Thus

$$\mathscr{I}(v, \alpha, \beta, \mathbf{K}, \mathbf{P}) = -\frac{2i}{T}\left(\frac{T}{T - 2\delta}\right)^{iv} \oint_{C_3}^{(a+, b+)} dt(1-t)^{iv}(t-a)^{iv-1}(t-b)^{-iv} \tag{A2-12}$$

and by using a further transformation

$$\tau = \frac{t-a}{b-a} \tag{A2-13}$$

this becomes

$$\mathscr{I}(v, \alpha, \beta, \mathbf{K}, \mathbf{P}) = -\frac{2i}{T}\left(\frac{T-2\alpha\delta}{T-2\delta}\right)^{iv} \oint^{(0+,1+)} d\tau (1-\tau z)^{iv} \tau^{iv-1}(\tau-1)^{-iv} \tag{A2-14}$$

where

$$z = \frac{4\alpha\delta^2}{(T-2\delta)(T-2\alpha\delta)}. \tag{A2-15}$$

The hypergeometric function

$$_2F_1[a, b, 1, z] \equiv \frac{1}{\Gamma(a)\Gamma(b)} \sum_{n=0}^{\infty} \frac{\Gamma(a+n)\Gamma(b+n)}{(n!)^2} z^n \tag{A2-16}$$

converges for $|z|<1$ and has the contour integral representation

$$_2F_1[a, b, 1, z] = -\frac{i}{2\pi} \oint^{(0+,1+)} d\tau\, \tau^{b-1}(\tau-1)^{-b}(1-\tau z)^a \tag{A2-17}$$

(ref. 73, p. 60). Consequently for $|z|<1$

$$\mathscr{I}(v, \alpha, \beta, \mathbf{K}, \mathbf{P}) = \frac{4\pi}{T}\left(\frac{T-2\alpha\delta}{T-2\delta}\right)^{iv} {}_2F_1[-iv, iv, 1, z]. \tag{A2-18}$$

The result is particularly simple when $\alpha=0$; then $z=0$ and consequently

$$\int \frac{1}{r} \exp(-\beta r + i\mathbf{P}\cdot\mathbf{r}) \, _1F_1[iv, 1, i(Kr+\mathbf{K}\cdot\mathbf{r})] \, d\mathbf{r}$$

$$= \mathscr{I}(v, 0, \beta, \mathbf{K}, \mathbf{P}) = \frac{4\pi}{T}\left(\frac{T}{T-2\delta}\right)^{iv}. \tag{A2-19}$$

Eq. (A2-19) can be obtained more easily by means of a modified version of the above analysis (see e.g. ref. 7).

If $|z|>1$ the series in (A2-16) diverges, but the integral in (A2-14) still converges and is the analytic continuation of $_2F_1[-iv, iv, 1, z]$ into the region $|z|>1$. Following Crothers [56] we shall show how this analytic continuation can be expressed in terms of known functions. To this end let us consider the integral

$$I = \oint^{(0+,1+)} (1-\tau z)^{iv} \left(\frac{\tau}{\tau-1}\right)^{iv} \frac{d\tau}{\tau} \tag{A2-20}$$

which by means of the transformations $u = \tau^{-1}$ and $v = \tau z$ may be expressed in the forms

$$I = (-z)^{iv} \int^{(0+,1+)} (1-u)^{-iv}(1-u/z)^{iv} u^{-iv-1} du \tag{A2-21a}$$

$$= (-z)^{-iv} \int^{(0+,1+)} (1-v)^{iv}(1-v/z)^{-iv} v^{iv-1} dv. \tag{A2-21b}$$

From ref. 73, p. 60 it is seen that

$$\int^{(0+,1+)} (1-u)^{c-b-1}(1-uZ)^{-a} u^{b-1} du$$
$$= \frac{2i\Gamma(b)\Gamma(c-b)\sin \pi b}{\Gamma(c)\exp(-i\pi b)} {}_2F_1[a,b,c,Z] \tag{A2-22}$$

so, when $|z| > 1$,

$$I = (-z)^{iv} \frac{2i\Gamma(-iv)\Gamma(1-iv)\sin(-\pi iv)}{\Gamma(1-2iv)\exp(-\pi v)} {}_2F_1[-iv, -iv, 1-2iv, 1/z] \tag{A2-23a}$$

$$= (-z)^{-iv} \frac{2i\Gamma(iv)\Gamma(1+iv)\sin(\pi iv)}{\Gamma(1+2iv)\exp(\pi v)} {}_2F_1[iv, iv, 1+2iv, 1/z]. \tag{A2-23b}$$

It is important to realise that $(-z)^{-iv} \neq [(-z)^{iv}]^*$ unless $z$ is real and either negative or zero. In fact, since $-1 = \exp(\pm i\pi)$,

$$(-z)^{iv} = z^{iv} e^{\mp \pi v}, \quad (-z)^{-iv} = z^{-iv} e^{\pm \pi v}. \tag{A2-24}$$

For reasons which will appear later the ambiguity of sign in (A2-24) is irrelevant in the application which we have in mind.

To simplify the expressions (A2-23) the identity

$$\Gamma(z)\Gamma(1-z) = \pi \operatorname{cosec} \pi z \tag{A2-25}$$

is applied to each of the gamma functions. Then

$$\frac{\Gamma(iv)\Gamma(1+iv)\sin(\pi iv)}{\Gamma(1+2iv)\exp(\pi v)} = \frac{2\pi\Gamma(-2iv)\cosh \pi v}{\Gamma(1-iv)\Gamma(-iv)\exp(\pi v)} \tag{A2-26}$$

and, since

$$\frac{1}{2}\left(\frac{e^{\pi v}}{\cosh \pi v} + \frac{e^{-\pi v}}{\cosh \pi v}\right) = 1 \tag{A2-27}$$

(A2-23a) and (A2-23b) may be combined to give

$$I = 2\pi i \left\{ e^{\mp \pi v} \frac{z^{iv}\Gamma(2iv)}{\Gamma(1+iv)\Gamma(iv)} {}_2F_1[-iv, -iv, 1-2iv, 1/z] \right. $$
$$\left. + e^{\pm \pi v} \frac{z^{-iv}\Gamma(-2iv)}{\Gamma(1-iv)\Gamma(-iv)} {}_2F_1[iv, iv, 1+2iv, 1/z] \right\}. \tag{A2-28}$$

Finally, since

$$\Gamma(2z) = 2^{2z-1}\Gamma(z)\Gamma(z+\tfrac{1}{2})\pi^{-\tfrac{1}{2}} \tag{A2-29}$$

this result may be written as

$$I = i\pi^{\tfrac{1}{2}}[e^{\mp \pi v}(U+iV) + e^{\pm \pi v}(U-iV)]$$
$$= 2i\pi^{\tfrac{1}{2}}[U \cosh \pi v \mp iV \sinh \pi v] \tag{A2-30}$$

where

$$U + iV = (4z)^{iv} \frac{\Gamma(\tfrac{1}{2}+iv)}{\Gamma(1+iv)} {}_2F_1[-iv, -iv, 1-2iv, 1/z] \tag{A2-31}$$

and $U$ and $V$ are real. Substitution of (A2-30) in (A2-14) then yields

$$\mathscr{I}(v, \alpha, \beta, \mathbf{K}, \mathbf{P}) = \frac{4\pi^{\tfrac{1}{2}}}{T}\left(\frac{T-2\alpha\delta}{T-2\delta}\right)^{iv}[U \cosh \pi v \pm iV \sinh \pi v], \quad |z| > 1. \tag{A2-32}$$

In the post form of the VPS approximation, where this result is required, the quantity of interest is $|\mathscr{I}(v, \alpha, \beta, \mathbf{K}, \mathbf{P})|^2$. Therefore the question of which of the two signs in (A2-32) is appropriate does not arise.

## References

1. L. VAINSHTEIN, L. PRESNYAKOV and I. SOBELMAN, Zh. Eksperim. i Teor. Fiz. **45** (1963) 2015 (English transl. Soviet Phys. JETP **18** (1964) 1383).
2. P. G. BURKE, S. ORMONDE, A. J. TAYLOR and W. WHITAKER, in: Abstracts of the Fifth International Conference on the Physics of Electronic and Atomic Collisions (Nauka, Leningrad, 1967) p. 368.
3. P. G. BURKE, S. ORMONDE and W. WHITAKER, Proc. Phys. Soc. **92** (1967) 319.
4. M. L. GOLDBERGER and K. M. WATSON, Collision Theory (Wiley, N.Y., 1964).
5. P. ROMAN, Advanced Quantum Theory (Addison Wesley, Reading, Mass., 1965).
6. R. G. NEWTON, Scattering Theory of Waves and Particles (McGraw-Hill, N.Y., 1966).

7. M. R. C. McDowell and J. P. Coleman, An Introduction to the Theory of Ion-Atom Collisions (North-Holland, Amsterdam, 1969).
8. G. F. Chew, Phys. Rev. **80** (1950) 196.
9. J. P. Coleman, J. Phys. B. (Proc. Phys. Soc.) [2], **1** (1968) 567.
10. N. F. Mott and H. S. W. Massey, The Theory of Atomic Collisions (Oxford University Press, 3rd edition 1965).
11. R. Peterkop and V. Veldre, Advan. At. Mol. Phys. **2** (1966) 263.
12. G. F. Chew and M. L. Goldberger, Phys. Rev. **87** (1952) 778.
13. T. Pradhan, Phys. Rev. **105** (1957) 1250.
14. M. R. C. McDowell, Proc. Roy. Soc. **A264** (1961) 277.
15. R. A. Mapleton, J. Math. Phys. **2** (1961) 482.
16. B. H. Bransden, Advan. At. Mol. Phys. **1** (1965) 85.
17. R. H. Bassel and E. Gerjuoy, Phys. Rev. **117** (1960) 749.
18. T. Pradhan and D. N. Tripathy, Phys. Rev. **130** (1963) 2317.
19. I. M. Cheshire, Proc. Phys. Soc. **82** (1963) 113.
20. J. P. Coleman and M. R. C. McDowell, Proc. Phys. Soc. **85** (1965) 1097.
21. J. R. Oppenheimer, Phys. Rev. **31** (1928) 349.
22. H. C. Brinkman and H. A. Kramers, Proc. Acad. Sci. Amsterdam **33** (1930) 973.
23. D. R. Bates and A. Dalgarno, Proc. Phys. Soc. **A65** (1952) 919.
24. J. D. Jackson and H. Schiff, Phys. Rev. **89** (1953) 359.
25. R. McCarroll, Proc. Roy. Soc. **A264** (1961) 547.
26. S. E. Lovell and M. B. McElroy, Proc. Roy. Soc. **A283** (1965) 100.
27. L. Wilets and D. F. Gallaher, Phys. Rev. **147** (1966) 13.
28. J. E. Bayfield, Phys. Rev. Letters **20** (1968) 1223.
29. D. F. Gallaher and L. Wilets, Phys. Rev. **169** (1968) 139.
30. D. R. Bates and A. Dalgarno, Proc. Phys. Soc. **A66** (1953) 972.
31. G. Ryding, A. B. Wittkower, and H. B. Gilbody, Proc. Phys. Soc. **89** (1966) 547.
32. A. Messiah, Quantum Mechanics (North-Holland, Amsterdam, 1962).
33. J. P. Coleman and S. Trelease, J. Phys. B (Proc. Phys. Soc.) [2], **1** (1968) 172.
34. R. F. Stebbings, R. A. Young, C. L. Oxley and H. Ehrhardt, Phys. Rev. **138** (1965) A1312.
35. R. A. Mapleton, Phys. Rev. **126** (1962) 1477.
36. R. M. May, Phys. Rev. **136** (1964) A669.
37. A. B. Wittkower, G. Ryding and H. B. Gilbody, Proc. Phys. Soc. **89** (1966) 541.
38. W. L. Fite, R. F. Stebbings, D. G. Hummer and R. T. Brackmann, Phys. Rev. **119** (1960) 663.
39. I. M. Cheshire, Proc. Phys. Soc. **84** (1964) 89.
40. I. M. Cheshire, Proc. Phys. Soc. **83** (1964) 227.
41. W. L. Fite, A. C. H. Smith and R. F. Stebbings, Proc. Roy. Soc. **A268** (1962) 527.
42. B. H. Bransden and I. M. Cheshire, Proc. Phys. Soc. **81** (1963) 820.
43. R. A. Mapleton, Phys. Rev. **122** (1961) 528.
44. C. F. Barnett and H. K. Reynolds, Phys. Rev. **109** (1958) 355.
45. L. M. Welsh, K. H. Berkner, S. N. Kaplan and R. V. Pyle, Phys. Rev. **158** (1967) 85.
46. L. H. Toburen, M. Y. Nakai and R. A. Langley, Phys. Rev. **171** (1968) 114.
47. R. A. Mapleton, J. Phys. B. (Proc. Phys. Soc.) [2], **1** (1968) 529.
48. R. M. Drisko, Ph.D. Thesis, Carnegie Institute of Technology, 1955 (unpublished).
49. R. A. Mapleton, Proc. Phys. Soc. **83** (1964) 895.
50. J. P. Coleman and M. R. C. McDowell, Proc. Phys. Soc. **83** (1964) 907.
51. J. P. Coleman and M. R. C. McDowell, Proc. Phys. Soc. **84** (1964) 334.
52. R. Akerib and S. Borowitz, Phys. Rev. **122** (1961) 1177.
53. J. P. Coleman and M. R. C. McDowell, Proc. Phys. Soc. **87** (1966) 879.

54. J. P. COLEMAN, Ph.D. Thesis, University of London, 1965 (unpublished).
55. D. R. BATES, Proc. Roy. Soc. **A247** (1958) 294.
56. D. S. F. CROTHERS, Proc. Phys. Soc. **91** (1967) 855.
57. L. PRESNYAKOV, I. SOBELMAN and L. VAINSHTEIN, Proc. Phys. Soc. **89** (1966) 511.
58. K. OMIDVAR, Phys. Rev. Letters **18** (1967) 153.
59. L. VAINSHTEIN, V. OPYKHTIN and L. PRESNYAKOV, Zh. Eksperim. i Teor. Fiz. **47** (1964) 2306 (English transl. Soviet Phys. JETP **20** (1965) 1542).
60. L. PRESNYAKOV, Zh. Eksperim. i Teor. Fiz. **47** (1964) 1134 (English transl. Soviet Phys. JETP **20** (1965) 760).
61. H. L. KYLE and M. R. C. MCDOWELL, J. Phys. B (Proc. Phys. Soc.) [2], **2** (1969) 15.
62. D. S. F. CROTHERS and R. MCCARROLL, Proc. Phys. Soc. **86** (1965) 753.
63. R. MCCARROLL and A. SALIN, Ann. Phys. (Paris) **1** (1966) 283.
64. T. A. GREEN, Proc. Phys. Soc. **92** (1967) 1144.
65. R. MCCARROLL and A. SALIN, Proc. Roy. Soc. **A300** (1967) 202.
66. L. R. DODD and K. R. GREIDER, Phys. Rev. **146** (1966) 675.
67. J. P. COLEMAN, J. Phys. B. (Proc. Phys. Soc.) [2], **1** (1968) 172.
68. W. L. FITE and R. T. BRACKMANN, Phys. Rev. **112** (1958) 1151.
69. W. L. FITE, R. F. STEBBINGS and R. T. BRACKMANN, Phys. Rev. **116** (1959) 356.
70. G. WENTZEL, Z. Physik **58** (1929) 348.
71. H. S. W. MASSEY and C. B. O. MOHR, Proc. Roy. Soc. **A140** (1933) 613.
72. A. NORDSIECK, Phys. Rev. **93** (1954) 785.
73. A. ERDELYI, W. MAGNUS, F. OBERHETTINGER and F. G. TRICOMI, Higher Transcendental Functions (McGraw-Hill, N.Y. 1953) Vol. I.
74. T. D. GAILY, Thesis, University of Washington, Seattle, 1968 (unpublished).
75. T. D. GAILY and R. GEBALLE, private communication (1968).
76. K. L. KYLE and K. OMIDVAR, Phys. Rev. **176** (1968) 164.
77. B. N. ROY, to be published.
78. T. D. GAILY, Phys. Rev. **178** (1969) 207.
79. J. P. COLEMAN and D. E. ECONOMIDES, unpublished.
80. M. R. H. RUDGE and M. J. SEATON, Proc. Roy. Soc. **A283** (1965) 262.

# CHAPTER 4

# POSITRON COLLISIONS

BY

## B. H. BRANSDEN

*Physics Department, University of Durham, England*

# Contents

| | Page |
|---|---|
| 4-1. Introduction | 171 |
|    A. Positron scattering and annihilation | 172 |
|       1. Free annihilation | 172 |
|       2. Selection rules | 176 |
|       3. Positronium | 177 |
|    B. The interaction of positrons with a mon-atomic gas | 180 |
|       1. Quenching of positronium | 181 |
|       2. Collision theory and experiment | 182 |
| 4-2. The elastic scattering of positrons by atoms | 184 |
|    A. General formulation | 184 |
|       1. The Green's function | 188 |
|       2. Angular momentum decomposition | 189 |
|    B. Approximation schemes | 192 |
|       1. The static approximation | 192 |
|       2. The adiabatic polarisation potential | 196 |
|       3. Virtual positronium formation | 200 |
|       4. Variational methods | 202 |
|    C. Applications of the variational method | 216 |
|       1. Positron–hydrogen scattering | 216 |
|       2. Positron–helium scattering | 222 |
|    D. Developments of the polarised orbital method | 225 |
|       1. Positron scattering by hydrogen | 226 |
|       2. Positron scattering by the rare gases | 231 |
| 4-3. The formation and scattering of positronium | 235 |
|    A. The formation of positronium by positron impact on hydrogen and helium | 236 |
|       1. General formalism | 237 |
|       2. Trial functions and applications | 239 |
|    B. The scattering of positronium by hydrogen and helium | 243 |
| Acknowledgement | 245 |
| References | 245 |

## § 4-1. Introduction

The relativistic quantum mechanics of fermions (particles of spin $\tfrac{1}{2}\hbar$) introduced by Dirac [1] predicts that for each kind of particle with a certain mass and charge there should exist a different kind of particle, termed an 'antiparticle', with the same mass and with a charge of equal magnitude, but opposite sign.*

The subsequent discovery of the anti-particle to the electron, known as the positron, by Anderson [2] in 1932 and Blackett and Occhialine [3] in 1933, is rightly regarded as one of the triumphs of modern physics. From the time of their discovery, positrons have been studied from many points of view. At first attention was concentrated on verifying that the fundamental properties of the particles were in accord with those predicted by theory, but later interest has centred on applications in which the positron is employed as a probe into problems of atomic or solid state physics, and it is on one of these applications that we shall concentrate.

The theory of collisions plays a central role in many branches of physics and has achieved considerable success in interpreting the interactions of electrons with atoms and molecules and the interactions between heavy particles. It is important, in evaluating the various approximations that have been developed, to confront the theory with as many different examples of collision phenomena as possible. Because of its positive charge and because it is not identical with the bound atomic electrons, the scattering of positrons by atoms is sufficiently different from electron scattering by atoms, that it provides an interesting and distinct testing ground for collision theory, and this is the topic to be discussed in this chapter.

The importance of the study of positron interactions for collision theory was perhaps first emphasized by Massey and Mohr [4] in 1954. Since that time, important advances have occurred, both theoretically and experimentally, but the points of contact between the experiments and theory are not as many as could be wished and are somewhat indirect. The reason for this is that in contrast to the position in electron scattering, there are at present no controlled mono-energetic beams of low energy positrons and the experiments have been performed with positrons arising from nuclear beta decay. The positrons produced from beta decay have an energy distribution which is peaked about some rather high energy. In our context, high and

---

* This definition must be extended if it is to include the strongly interacting elementary particles. Not only the charge, but the hypercharge and the third component of the isotopic spin charge sign in going from a particle to the corresponding antiparticle.

low energies are with respect to typical binding energies of electrons in atoms. For example, positrons from the source $^{22}$Na have a most probable energy of 120 keV and a maximum energy of 540 keV, so that the great majority of positrons are of extremely high energy compared with binding energies of electrons in atoms. The quantity that can be studied experimentally is the annihilation radiation produced when the positrons slow down in the material being studied. As we shall see, the annihilation rate depends on the collision cross-section between the positrons and the atoms of the material and allows an estimate of this quantity to be made. It is possible that in the near future, methods of producing low energy mono-energetic beams of positrons will be devised and in this case direct measurements of the differential cross-sections for the processes of interest will be possible, and a rapid advance in understanding of the subject will result.

## A. POSITRON SCATTERING AND ANNIHILATION

At non-relativistic velocities ($v \ll c$) the interaction of a positron with an electron or with a nucleus is represented accurately by a Coulomb potential. Apart from the opposite sign of the interaction, electron–positron scattering differs from electron–electron scattering in two respects. As the particles are different, the wave function need not be antisymmetrical and there are no exchange terms in the scattering amplitude and secondly, the electron-positron pair can annihilate (subject to certain selection rules), as a rule into two photons. In principle, this absorptive process modifies the potential acting between the electron and positron by the addition of a positive imaginary part, but as the annihilation cross-section is much smaller than the cross-sections associated with characteristic atomic scattering, this modification of the Coulomb potential can be ignored.

### 1. *Free annihilation*

A free electron–positron pair cannot decay into a single photon, because energy and momentum cannot be conserved simultaneously in such a reaction. It is however possible for single photon annihilation to occur when the pair is in the Coulomb field of a nucleus as the nucleus can absorb the excess momentum and we shall comment on this process below.

Annihilation of an electron–positron pair into two photons is allowed kinematically. As the annihilation rate is inversely proportional to the relative velocity of the electron and positron most annihilations occur when the particles have a small kinetic energy (in the centre of mass system) so that the total energy available to the photons is approximately equal to the rest energy of the system, $2mc^2$, where $m$ is the mass of an electron or positron. The two photons are emitted in opposite directions and each of them has the same energy, $mc^2$, which is 510 keV.

The two Feynmann diagrams involved in pair annihilation are shown in Fig. 4-1-1. The interaction at each vertex is proportional to $(\sqrt{\alpha})$ where $\alpha$ is the fine structure constant, $\alpha = e^2/\hbar c$ where $e$ is the charge on the electron, so that the total cross-section is proportional to $\alpha^2$. If an additional photon is produced, the corresponding Feynmann diagram must possess an additional vertex, giving rise to an additional factor of $\alpha$ in the cross-section and it follows that annihilation into three or more photons is unimportant compared with two photon annihilation, except in special circumstances when the latter is forbidden by selection rules.

At low velocities the positron–electron system will be in a state of zero orbital angular momentum in the centre of mass system. In these circumstances, annihilation into two photons from the triplet ($S=1$) spin state is forbidden and the cross-section for two photon annihilation from the singlet ($S=0$) spin state is given by the expression*

$$q_2 = \frac{4\hbar^2 \pi \alpha^2}{vcm^2} = \frac{4\pi c r_0^2}{v} \qquad (4\text{-}1\text{-}1)$$

where $v$ is the relative velocity of the electron and positron and $r_0$ is the classical electron radius, $r_0 = e^2/mc^2$. This expression was first obtained by Dirac (1930) and is based on a perturbation theory in which the electron and positron are described by unperturbed plane waves. This is clearly not a good approximation at very low velocities, because the Coulomb attraction between the particles will modify the wave-function considerably. A correction is easily obtained. As the form of the cross-section suggests, the

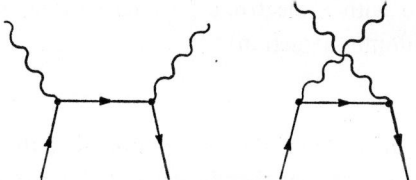

Fig. 4-1-1. Feynmann diagrams for the two photon annihilation of an electron–positron pair.

---

* The calculation of this cross-section, and the selection rules are discussed in books on quantum electrodynamics, see particularly Jauch and Rohrlich [5] and Drell and Bjorken [6].

annihilation takes place when the electron and positron are within distances of the order of $r_0$ from each other. As $r_0$ is small compared with the distances over which the wave-function of the electron–positron pair changes appreciably, the annihilation can be considered as occurring only when the electron is at the same position as the positron. If $\psi(r)$ is the electron–positron wavefunction in the centre of mass system, normalised to a plane wave of unit amplitude at large distances of separation $(r\to\infty)$, the probability of finding the electron and positron at the same position is $|\psi(0)|^2$ and the cross-section becomes

$$\bar{q}_2 = |\psi(0)|^2 \left(\frac{4\pi r_0^2 c}{v}\right). \tag{4-1-2}$$

For annihilation of a free positron–electron pair, $\psi(r)$ must be taken to be the solution of the Schrödinger equation that describes scattering by the Coulomb interaction between the particles, so that*

$$\left(\frac{h^2}{m}\nabla_r^2 + \frac{e^2}{r} + E\right)\psi(r) = 0. \tag{4-1-3}$$

The Coulomb function $\psi(r)$, satisfying this equation is given explicitly in terms of a hypergeometric function (see any text on quantum mechanics) and

$$|\psi(0)|^2 = \left[\frac{2\pi\alpha c/v}{1 - \exp(-2\pi\alpha c/v)}\right]. \tag{4-1-4}$$

It is at once seen that if $2\pi\alpha c/v$ is large then $\bar{q}_2 > q_2$ so that, as one would expect, the Coulomb attraction increases the annihilation rate at small velocities. At large velocities, when $2\pi\alpha c/v < 1$, we see that $\bar{q}_2$ is approximately equal to $q_2$, and the Coulomb attraction can be ignored.

The rate of annihilation into two photons when a beam of positrons traverses a medium with $n$ electrons (treated as free) per unit volume is (neglecting the Coulomb attraction)

$$R_2 = \tfrac{1}{4} q_2 n v \tag{4-1-5}$$

where the factor $\tfrac{1}{4}$ arises from the assumption that the electron spins are distributed at random and that therefore $\tfrac{1}{4}$ of the electrons are in a relative singlet ($S=0$) spin state with the positrons and it is from the singlet state that two photon annihilation occurs. If the material contains $N$ atoms per

---

* Note that the reduced mass of the electron–positron pair which appears in (4-1-3) is equal to $\tfrac{1}{2}m$.

unit volume each with $Z$ electrons, the electron density is $n = NZ$, in which case, if the binding of the electrons can be neglected, the annihilation rate is

$$R_2 = \tfrac{1}{4} q_2 NvZ = 1.84 ZN \times 10^{-5} \text{ sec}^{-1}. \tag{4-1-6}$$

It should be noted that because $q_2$ is inversely proportional to $v$, $R_2$ is independent of $v$ in this approximation. It is also useful to introduce a mean life-time $\tau_2$, by

$$\tau_2 = 1/R_2 = \frac{0.54}{ZN} \times 10^{15} \text{ sec}. \tag{4-1-7}$$

As an example, the mean life time of positrons in air at atmospheric pressure predicted by the formula is about $3 \times 10^{-7}$ sec and this is of the same order of magnitude as the results given by experimental measurements. To obtain quantitative results, the fact that the electrons are not free must be taken into account. To do this a velocity dependent parameter $Z_{\text{eff}}(v)$ is introduced in place of $Z$, that gives the number of electrons per atom effective in annihilation. This parameter can be calculated from the wave-function describing the scattering of the positron by the atom, as will be seen later. In terms of this parameter, the annihilation cross-section per atom $q_2^A$ can be expressed as

$$q_2^A = \tfrac{1}{4} q_2 Z_{\text{eff}}(v) \tag{4-1-8}$$

for an unpolarised positron beam. This cross-section is very small compared with elastic or excitation cross-sections, even when the positrons are moving with thermal velocities and because of this, the scattering of positrons by atoms can be investigated by considering the Schrödinger equation for the system of positrons, electrons and the nucleus, interacting through Coulomb potentials, without introducing an effective absorptive potential to describe annihilation.

The cross-section for the three photon annihilation of a free positron-electron pair has been calculated by Ore and Powell [7]. As we noted earlier, it contains an extra factor of $\alpha$, and it has the form

$$q_3 = \frac{\pi^2 - 9}{\pi} \alpha q_2. \tag{4-1-9}$$

Three photon annihilation takes place from the triplet spin state, so that for an unpolarised beam of positrons, the rate of annihilation is

$$R_3 = \tfrac{3}{4} q_3 nv. \tag{4-1-10}$$

The ratio $R_2/R_3$ is about 370, confirming that three photon annihilation is unimportant in a free positron beam.

Single photon production, by annihilation of a pair in the field of a nucleus takes place through the Feynmann diagrams shown in Fig. 4-1-2. The cross-section per atom is of the order $(\alpha Z)^4 Z\pi r_0^2$, which may be compared with the two photon annihilation rate of $Z\pi r_0^2$. It follows that this process is unimportant except for the heaviest elements, and need not be considered further.

## 2. Selection rules

An electron–positron pair can be in a simultaneous eigenstate of the total spin and its $z$ component with quantum numbers $s$ and $m_s$,

$$S^2 = s(s+1)\hbar^2, \quad S_z = m_s\hbar, \tag{4-1-11}$$

together with the total angular momentum and its $z$ component, with quantum numbers $j$ and $m_j$

$$J^2 = j(j+1)\hbar^2, \quad J = m_j\hbar. \tag{4-1-12}$$

In addition, the system can be in an eigenstate of the parity operator $P$ and the charge conjunction parity operator $C$, and a complete commuting set of variables includes $S^2$, $S_z$, $J^2$, $J_z$, $P$, $C$. The relative internal parity of a particle and the corresponding anti-particle is $-1$ (see Jauch and Rohrlich [5]) so that if the electron and positron are in a state with relative orbital angular momentum $L$ the total parity of the system is

$$P = (-1)^{L+1}. \tag{4-1-13}$$

The charge conjunction parity operator $C$ effects an interchange of the charge of the two particles, and this is equivalent to interchanging the spins of the particles and at the same time inverting the system about the centre of mass ($r \to -r$). Under the spin interchange, the wave-function is even in

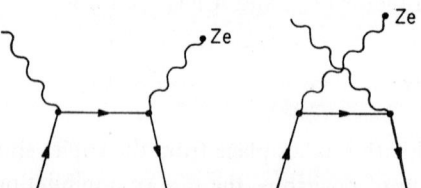

Fig. 4-1-2. Feynmann diagrams for the single photon annihilation of an electron–positron pair in the field of a nucleus.

the triplet spin state and odd in the singlet spin state, while inversion about the centre of mass is achieved by the parity operator and we can write $C$ as

$$C = (-1)^{S+1} P = (-1)^{L+S}. \qquad (4\text{-}1\text{-}14)$$

The charge conjunction parity of a system of $n$ photons is

$$C = (-1)^n \qquad (4\text{-}1\text{-}15)$$

(see Jauch and Rohrlich [5]), and if the pair annihilates into an $n$ photon state, it is required that

$$(-1)^n = (-1)^{s+L}. \qquad (4\text{-}1\text{-}16)$$

Thus singlet states with even orbital angular momentum (s, d, g,... states) and triplet states with odd orbital angular momentum (p, f,... states) must annihilate into an even number of photons, while singlet states with odd orbital angular momentum, and triplet states with even orbital angular momentum, annihilate into an odd number of photons.

3. *Positronium*

The attractive Coulomb interaction between an electron and a positron supports a series of bound states. The bound system of a positron and an electron was named positronium by Ruark [8]. The energy spectrum is (apart from the fine structure) similar to that of the hydrogen atom and the energy levels are given by

$$E_n = -\frac{\mu e^4}{2\hbar^2 n^2} \qquad (4\text{-}1\text{-}17)$$

where $n$ is the principal quantum number ($n = 1, 2, 3,...$) and $\mu$ is the reduced mass. For positronium $\mu = \tfrac{1}{2}m$ while for hydrogen $\mu = m$, so that each level in positronium has half the energy of the corresponding level in hydrogen. No account of the interesting fine structure of positronium will be given here, as it has no bearing on the scattering of positrons by atoms. It is of course quite different from that of the hydrogen atom and the details may be found in articles by Deutsch [9] and De Benedetti and Corben [10].

The positronium atom can be in a triplet $(s=1)$ spin state, in which case it is called orthopositronium, or in a singlet $(s=0)$ spin state, known as parapositronium. As we are neglecting the small spin dependent forces that determine the fine structure, these two states are degenerate in energy. The magnitude of the splitting between the ortho and para states has been calculated by Ferrell [11] to be $8.45 \times 10^{-4}$ eV, so that it is a very good approximation to neglect the splitting in the present context.

Systems composed of two electrons and a positron, two positrons and an electron or two positrons and two electrons are all stable by a fraction of an electron volt (Wheeler [12], Hylleras [13] and Sharma [14]), but while positronium formation in positron–atom collisions can occur readily, the formation of these more complicated systems is quite unlikely and need not be considered. On the other hand, the existence of stable positive ions formed by the attachment of a positron to a neutral atom, would have important implications for the scattering problem. This problem has been solved for some simple systems by Spruch and his collaborators [15, 16] who have shown that neither a bound state of a positron with a hydrogen atom nor a bound state of a positron with a helium atom can exist. It is interesting to note that if the positron mass were increased to a few electron masses these bound states would certainly exist. For example a variational calculation by Frost et al. [17] shows that a bound state of a singlely charged particle with a hydrogen atom certainly exists when the mass of the particle is greater than 2.625 electron masses and there is, for example, a bound state of a $\mu^+$ meson with a hydrogen atom.

The annihilation rates for positronium can be calculated with the help of equations (4-1-2) and (4-1-9). Single photon annihilation is forbidden as before, and because of the selection rules parapositronium in the ground state $(L=0)$ must decay into two photons, while orthopositronium in the ground state decays into three photons. The two photon annihilation rate of parapositronium is given by

$$R_2^{Ps} = |\psi(0)|^2 \, 4\pi r_0^2 c \qquad (4\text{-}1\text{-}18)$$

which is obtained from eq. (4-1-5) by setting $n=1$. In addition the factor $\tfrac{1}{4}$ has been removed, because, in the present case, the system is entirely in the singlet spin state. The wave-functions $\psi_{n,l}$ of the positronium atom, differ from those of the hydrogen atom only in scale. The mass of the electron $m$ has to be replaced by $\tfrac{1}{2}m$, the reduced mass of the system, which is equivalent to replacing $a_0$, the Bohr radius of the hydrogen atom, by $2a_0$, which is the Bohr radius of positronium. For example, the ground state wave-function of hydrogen is

$$\psi_{0,0}^H(r) = \sqrt{\frac{1}{\pi a_0^3}} \, e^{-r/a_0} \qquad (4\text{-}1\text{-}19)$$

while that of positronium is

$$\psi^{Ps}_{0,0}(r) = \sqrt{\frac{8}{\pi a_0^3}}\, e^{-r/2a_0}. \tag{4-1-20}$$

The density $|\psi_{n,l}(0)|^2$ at $r=0$, which occurs in the expression for the annihilation rate eq. (4-1-2) vanishes except for states with $l=0$, and for these states

$$|\psi_{n,0}(0)|^2 = \frac{1}{(2a_0)^3\, \pi n^3}. \tag{4-1-21}$$

The two photon annihilation rate of parapositronium, from s states with principal quantum number $n$, is then

$$R^{Ps}_2 = \frac{r_0^2 c}{2\pi a_0^3 n^3} \tag{4-1-22}$$

and the corresponding life-time is

$$\tau^{Ps}_2 = \frac{1}{R^{Ps}_2} = 1.25 n^3 \times 10^{-10}\ \text{sec}. \tag{4-1-23}$$

This annihilation rate is much faster than the rate for an optical transition from an s state. For example, the 2s state is metastable while the life-time for the 3s→2p optically allowed transition is $3.2 \times 10^{-7}$ sec. On the other hand, the annihilation rate from states with $l \neq 0$ is extremely small since $\psi_{n,l}(0)$ vanishes for $l \neq 0$.

The selection rules do not allow the triplet orthopositronium s states to decay into two photons, so that three photon decay is the most likely process. As before, the annihilation rate from states other than s states is extremely small and using eq. (4-1-10), the three photon annihilation rate of orthopositronium $R^{Ps}_3$ can be expressed in terms of $R^{Ps}_2$ by

$$R^{Ps}_3 = \alpha\, \frac{\pi^2 - q}{\pi} R^{Ps}_2 \approx \frac{1}{1115} R^{Ps}_2 \tag{4-1-24}$$

giving a life-time of

$$\tau^{Ps}_3 = 1.4 n^3 \times 10^{-7}\ \text{sec}. \tag{4-1-25}$$

In general, this is a long life-time compared with those for optical transitions and orthopositronium normally will decay to the ground state before annihilation.

The velocity of positronium with one atomic unit of kinetic energy (27.2 eV) is $2.18 \times 10^8$ cm/sec. Assuming an interaction region of the order of $10a_0$ for a positronium–atom collision, a characteristic collision time of positronium with this energy with an atom will be of the order $\sim 2 \times 10^{-16}$ sec. If the kinetic energy is reduced to say 0.01 eV, the characteristic collision time will be about $4 \times 10^{-14}$ sec. These times are extremely short compared with the life-time of positronium against annihilation and this implies that when calculating the cross-section for a positronium–atom reaction the possibility of annihilation can be ignored and the usual collision theory for stable particles employed.

### B. The Interaction of Positrons with a Mon-atomic Gas

Energetic positrons entering a gas will be slowed rapidly by inelastic collisions (chiefly ionising collisions) with the gas atoms. For gases such as helium with only a few electrons per atom, the time taken for the positron to slow from energies of a few hundred keV to below the ionisation threshold of the atoms is short compared with the free annihilation life-time, but for gases with many electrons per atom, for example argon, a large fraction of the positrons annihilate before reaching energies below the ionisation threshold. Below the ionisation threshold the positrons continue to lose energy, at first mainly by discrete excitations, and at the same time the chance that a positron will capture an atomic electron to form positronium will increase. For many materials the threshold energy for an inelastic collision that leaves the gas atom in its first excited state is above the threshold for positronium formation. In this case, there is always an energy range, known as the 'Ore gap', in which the only kind of inelastic collision possible is positronium formation. Some remaining fraction of the positrons that have neither annihilated or formed positronium will continue to lose energy through elastic collisions and will eventually become thermalised before annihilating.

Of the positronium atoms formed one quarter will be in the singlet spin state (parapositronium) and will annihilate rapidly, but the remaining three quarters will be in the triplet spin state (orthopositronium) and these will decay slowly through three photon annihilation.

Experimentally it is found (Deutsch [18]) that the decay of positrons in a gas envolves three distinct periods. The mean life associated with the first period is very short and is of the order of the time in which a positron slows to energies below the ionisation threshold of the gas atoms. The second is associated with a life-time of the order expected from free annihilation into two photons (eq. (4-1-7)), while the third period is very long and corresponds to a mean life of about $1.0 \times 10^{-7}$ sec. The first short period is attributed to the decay of parapositronium formed in positron–atom collisions and the second period to free annihilations. Referring to eq. (4-1-7), it is seen that the free annihilation rate should be proportional to the pressure of the gas. The correspondingly life-time should be inversely proportional to the pressure and the life-time associated with the second period has exactly this property. The third long period is attributed to the slow three photon decay of orthopositronium and, as expected, it is practically independent of the pressure. The presence of orthopositronium can be confirmed by directly observing all three photons from the decay in a coincidence type of experiment. When the rate of three photon to two photon decay is greater then the $\frac{1}{372}$ ratio predicted for the free annihilation of an unpolarised positron beam, the presence of orthopositronium is signalled.

1. *Quenching of positronium*

The decay rate of the orthopositronium formed when positrons are passed through a gas may be increased over the natural decay rate by several "quenching" mechanism:

(a) *Pick-off quenching*. During a collision with an atom, the positron may annihilate with an atomic electron with which it is in a relative state of zero spin. This process can be studied through its pressure dependence, because the rate of pick-off quenching depends on the number of atoms encountered.

(b) *Ortho-para conversion*. When positronium collides with an atom, it is possible by electron exchange for the spin state to change, converting the ortho-form to the para-form. This process is not always energetically possible, for example in positronium–helium collisions, the reaction is

$$\mathrm{Ps}\,(s=1) + \mathrm{He}\,(s=0) \to \mathrm{Ps}\,(s=0) + \mathrm{He}\,(s=1). \qquad (4\text{-}1\text{-}26)$$

The lowest lying triplet state of helium is some 20 eV above the ground state, so that positronium atoms with kinetic energies less then this amount cannot be so converted. When it can occur, the cross-section for this conversion reaction is large, being of the order of $\pi a_0^2$ and since parapositronium decays rapidly it is an efficient source of quenching. When an atom has total spin $s=\frac{1}{2}$, the reaction can occur at all energies, for example in atomic hydrogen

$$\mathrm{Ps}\,(s=1) + \mathrm{H}\,(s=\tfrac{1}{2}) \to \mathrm{Ps}\,(s=0) + \mathrm{H}\,(s=\tfrac{1}{2}) \qquad (4\text{-}1\text{-}27)$$

no excitation of the hydrogen atom occurs and the process is elastic. For this reason, gases such as nitric oxide or oxygen, with an unpaired electron, quench orthopositronium rapidly. This has been demonstrated experimentally by introducing nitric oxide into other gases in which orthopositronium has been formed. Even small quantities are sufficient to remove the orthopositronium, which may be observed by the disappearance of the three-photon decays.

(c) *Chemical quenching.* If reactions occur in which the positronium atom becomes bound with an atom or molecule, the decay rate will be increased because of the greater chance of the positron finding itself in the proximity of an electron with the correct spin.

## 2. *Collision theory and experiment*

Until mono-energetic positron beams of reasonable intensity are available, the connection between the cross-sections computed from collision theory and experiment is indirect. The experiments are concerned the observation of the two and three photon decay rates when positrons are passed through various gases. These rates can be investigated as a function of pressure, temperature and applied electric fields. As we have seen, the observed rates depend on the cross-sections for a great many processes including the elastic scattering cross-sections, the positronium formation cross-section and the cross-sections for the various quenching mechanisms.

A simple case is that of the decay of positrons that are known to be thermalised. The annihilation cross-section per atom is given by eq. (4-1-8) in terms of $Z_{\text{eff}}(v)$, the number of electrons effective in annihilation per atom. This parameter can be calculated from the total wave-function of the positron–atom system. This wave-function is denoted by $\Psi(r_1, r_2, ..., r_Z, x)$ where the $r_i$ are electron coordinates and $x$ the positron coordinate, measured from the nucleus of the atom (which is assumed to be infinitely heavy compared with the masses of the light particles). At thermal energies only elastic scattering is possible, and the asymptotic form of $\Psi$, for large $x$, is

$$\Psi(r_i, x) \sim \phi(r_i) \mathscr{F}(x) \tag{4-1-28}$$

where $\phi(r_i)$ is the normalised wave-function of the atom and $\mathscr{F}(x)$ has the form of an incident plane wave and an outgoing spherical wave*,

$$\mathscr{F}(x) = \exp(i\mathbf{k} \cdot \mathbf{x}) + x^{-1} \exp(ikx) f(\theta). \tag{4-1-29}$$

---

* Some knowledge of elementary collision theory is necessary to follow the bulk of this chapter, such as may be found in the standard texts on quantum mechanics, for example those by Merzbacher [19] or by Landau and Lifshitz [20]. An advanced and comprehensive account of the theory of atomic collisions has been given by Mott and Massey [21].

With this normalisation, the wave-function represents one positron per unit volume moving freely at large distances $x$, with a velocity $v$ where

$$v = \hbar k/m. \tag{4-1-30}$$

The cross-section for elastic scattering is determined by the absolute square of the scattering amplitude $f(\theta)$ and the calculation of $f(\theta)$ is discussed in detail in § 4-2. The probability density for finding the electron $i$ and the positron at the same positron is $P_i$, where

$$P_i = \int d\mathbf{r}_1 \int d\mathbf{r}_2 \ldots \int d\mathbf{r}_Z \int dx\, \delta(\mathbf{r}_i - x) |\Psi(\mathbf{r}_1, \mathbf{r}_2, \ldots, \mathbf{r}_Z, x)|^2 \tag{4-1-31}$$

and summing over all the electrons we have

$$Z_{\text{eff}}(v) = \sum_{i=1}^{Z} P_i. \tag{4-1-32}$$

At large velocities, the Born approximation is expected to hold, which implies that the wave-function is approximately of the form

$$\Psi(\mathbf{r}_i, x) \approx \phi(\mathbf{r}_i) \exp(i\mathbf{k}\cdot x) \tag{4-1-33}$$

and in this case

$$Z_{\text{eff}}(v) = \sum_{i=1}^{Z} \int d\mathbf{r}_1 \int d\mathbf{r}_2 \ldots \int d\mathbf{r}_Z |\phi(\mathbf{r}_i)|^2 = Z. \tag{4-1-34}$$

Accordingly, it is expected that as $v$ increases the annihilation rate will approach the Dirac rate for the annihilation of positrons by free electrons.

The calculation of $Z_{\text{eff}}$ is the meeting point of collision theory and experiment. In order to estimate $Z_{\text{eff}}$ experimentally it is necessary to know the velocity or velocity distribution of the positrons. In the case of thermalised positrons this is known because the distribution will be the Maxwellian distribution appropriate to the temperature $T$, it is

$$\phi(v) = 4v^2 \sqrt{\frac{m^2}{2\pi K^3 T^3}} \exp(-mv^2/2KT^2) \tag{4-1-35}$$

where $K$ is Boltzman's constant. The observed annihilation rate then determines an average cross-section per atom, $Q$, where

$$Q = \int_0^\infty q_2^A(v)\, \phi(v)\, dv$$
$$= \tfrac{1}{4} \int_0^\infty q_2(v)\, Z_{\text{eff}}(v)\, \phi(v)\, dv. \tag{4-1-36}$$

By applying an electric field, the mean energy of the thermalised positrons can be increased (Falk et al. [22]) and further information about $Z_{\text{eff}}(v)$ obtained.

As our main object is the discussion of the calculation of the various cross-sections the interesting subject of the determination of $Z_{\text{eff}}$ from experiment will not be further discussed. A much fuller account will be found in a recent excellent survey by Fraser [23].

## § 4-2. The elastic scattering of positrons by atoms

At energies below the first excitation threshold of the gas atoms and below the threshold for positronium formation, the annihilation rate of positrons is controlled by the wave-function, introduced in the last paragraph, that describes elastic scattering of the positrons by the gas atoms. In describing the calculation of this wave-function and the scattering cross-section we shall start with the case of positron scattering by atomic hydrogen. This is not a case of practical interest, but because of the simplicity of the positron–hydrogen atom system, it serves to develop theoretical methods that can later be applied to the more complicated systems, for which experimental results are available. In tracing the development of the theory, after a short account of the general formulation of the problem, we shall discuss the various approximation schemes that have been introduced.

### A. GENERAL FORMULATION

The Schödinger equation describing the system of an electron and a positron moving with respect to a proton (which may be taken to be infinitely heavy, and fixed at the origin of the coordinate system) is

$$(H - E)\Psi = 0 \qquad (4\text{-}2\text{-}1)$$

where

$$H = \left[ -\frac{\hbar^2}{2m}\nabla_r^2 - \frac{\hbar^2}{2m}\nabla_x^2 + \frac{e^2}{x} - \frac{e^2}{r} - \frac{e^2}{|r-x|} \right]. \qquad (4\text{-}2\text{-}2)$$

Where, as in the last paragraph, $r$ denotes the position vector of the electron and $x$ that of the positron with respect to the proton. In what follows, it is convenient to remove unnecessary factors by employing atomic units, for which $e = m = \hbar = 1$. In this system, the unit of length is the Bohr radius of the hydrogen atom $a_0 = 0.53 \times 10^{-8}$ cm$^2$, and the unit of energy is 27.2 eV which is twice the binding energy of the ground state of hydrogen. Scattering cross-sections have the dimensions of area and the unit cross-section is $(a_0)^2$ cm$^2$, that is $0.282 \times 10^{-16}$ cm$^2$.

In this system, the Hamiltonian $H$ becomes

$$H = -\tfrac{1}{2}\nabla_r^2 - \tfrac{1}{2}\nabla_x^2 - \frac{1}{r} + V(r, x) \tag{4-2-3}$$

where

$$V(r, x) = \frac{1}{x} - \frac{1}{|r - x|}. \tag{4-2-4}$$

The lowest inelastic threshold is that for positronium formation, which occurs when the kinetic energy of the positron is 0.25 a.u. (6.8 eV). Below this energy, if the target atoms are in the ground state, only elastic scattering is possible, and for large $x$ the wave-function must represent a positron moving freely with respect to the hydrogen atom in the ground state. This implies that $\Psi$ for large $x$ can be written in product form:

$$\Psi(r, x) \underset{x \to \infty}{\sim} \phi_0(r) \mathscr{F}(x) \tag{4-2-5}$$

where $\phi_0(r)$ is the ground state wave-function of hydrogen which satisfies the Schrödinger equation

$$\left[\nabla^2 + \frac{2}{r} + 2\varepsilon_0\right] \phi_0(r) = 0 \tag{4-2-6}$$

where $\varepsilon_0$ is the ground state energy.

The function $\mathscr{F}(x)$ must represent a combination of an incoming plane wave with spherical outgoing scattered waves and must therefore be of the form given in the previous section (§ 4.1) in equation (4-1-19). The amplitude of the scattered wave is a function of the polar angles of $x$, but if the direction of incidence $\hat{k}$ is taken as axis, the system is axially symmetric and $f(\theta, \phi)$ will depend on $\theta$ only. In this case the differential cross-section for elastic scattering, $I(\theta)$, which is defined as the number of particles scattered into a solid angle $d\Omega$ in direction $(\theta, \phi)$ per unit time, when the incident beam contains one particle crossing unit cross-sectional area normal to the beam, per unit time, is found to be [19–21]

$$I(\theta) = |f(\theta)|^2 \tag{4-2-7}$$

and the total cross-section is

$$q = 2\pi \int_{-1}^{+1} I(\theta) \, d(\cos\theta). \tag{4-2-8}$$

It will prove convenient to formulate the boundary conditions* in terms of projection operators $P$ and $Q$ with the properties

$$P + Q = 1$$
$$P^2 = P, \quad Q^2 = Q, \quad QP = PQ = 0. \qquad (4\text{-}2\text{-}9)$$

The complete wave-function can be written as

$$\Psi(r, x) = P\Psi(r, x) + Q\Psi(r, x). \qquad (4\text{-}2\text{-}10)$$

The operators $P$ and $Q$ are now specified (but not uniquely) by requiring that $P\Psi$ should contain the open channel** and satisfy the boundary condition (4-2-5), so that

$$P\Psi(r, x) \sim \phi_0(r)\,\mathscr{F}(x). \qquad (4\text{-}2\text{-}11)$$

The closed channel wave-function $Q\Psi$, must not contribute to the flux of particles at large distances of separation and this requires that $Q\Psi$ decreases for large $x$, faster than $x^{-1}$,

$$xQ\Psi(r, x) \to 0, \quad x \to \infty. \qquad (4\text{-}2\text{-}12)$$

A particular choice for $P$ is

$$P = \phi_0(r) \int dr'\,\phi_0(r') \qquad (4\text{-}2\text{-}13)$$

so that

$$P\Psi(r, x) = \phi_0(r)\,F(x) \qquad (4\text{-}2\text{-}14)$$

where

$$F(x) = \int dr'\,\phi_0(r')\,\Psi(r', x). \qquad (4\text{-}2\text{-}15)$$

The reason for the introduction of $P$ and $Q$ is that the Schrödinger equation

$$(H - E)(P + Q)\Psi = 0 \qquad (4\text{-}2\text{-}16)$$

can be expressed as a pair of coupled equations by operating either with $P$

---

\* The systematic use of projection operators in collision theory is due to Feshbach [24].
\*\* An open channel is a possible final state of the system, while a closed channel is a final state that cannot be reached because the total energy is insufficient. Below 6.83 eV, the e$^+$–H (1s) channel is open, but the e$^+$–H (2s) and Ps(1s)–H$^+$ channels are closed.

or with $Q$ from the left, giving

$$P(H - E) P\Psi = - PHQ\Psi \tag{4-2-17}$$
$$Q(H - E) Q\Psi = - QHP\Psi. \tag{4-2-18}$$

The second of these equations can be solved formally, with the boundary condition (4-2-12), by

$$Q\Psi = - G^Q(E) QHP\Psi \tag{4-2-19}$$

where $G^Q(E)$ is the Green's function or resolvent

$$G^Q(E) \equiv \frac{1}{Q(H - E) Q}.$$

This expression for $Q\Psi$ can be used in (4-2-17) giving an equation for the open channel function $P\Psi$

$$P(H - E) P\Psi = PHQG^Q(E) QHP\Psi. \tag{4-2-20}$$

With the particular choice of $P$ represented by (4-2-13), this equation is an equation for $F(x)$ defined by (4-2-15) and the problem is reduced to obtaining the solution of a one-body Schrödinger equation for scattering by a potential.

Making use of the wave-equation (4-2-6) satisfied by $\phi_0(r)$ and noting that

$$E = \varepsilon_0 + \tfrac{1}{2} k^2 \tag{4-2-21}$$

we find that

$$P(H - E) P\Psi = \phi_0(r) \left[ -\tfrac{1}{2}\nabla_x^2 + \tfrac{1}{2} U(x) - \tfrac{1}{2} k^2 \right] F(x) \tag{4-2-22}$$

where $U(x)$ is defined as

$$U(x) = 2 \int d\mathbf{r} |\phi_0(r)|^2 V(r, x). \tag{4-2-23}$$

This is the average over the hydrogen atom ground state of the interaction between the positron and the hydrogen atom and is called the static interaction.

The term on the right hand side of (4-2-20) can be written as

$$- \tfrac{1}{2}\phi_0(r) \int W(x, x') F(x') dx'$$

where the non-local potential operator $W(x, x')$ is defined as

$$W(x, x') = \int dr \int dr' \, \phi_0(r) \, H(x, r) \, G^Q(E; x, r; x', r') \, H(x', r') \, \phi_0(r'). \tag{4-2-24}$$

The equation satisfied by $F(x)$ is then

$$[\nabla_x^2 + k^2 - U(x)] F(x) = \int W(x, x') F(x') \, dx' \tag{4-2-25}$$

and the problem has been reduced to solving this one-body equation for scattering by the sum of the local potential $U(x)$ and the non-local potential $W(x, x')$. The boundary condition satisfied by $F(x)$ follows from (4-2-11), it is

$$F(x) \underset{x \to \infty}{\sim} \mathscr{F}(x) \tag{4-2-26}$$

where $\mathscr{F}(x)$ is given by (4-1-29).

Of course the difficulties in the problem are now concealed in the very complicated operator $W(x, x')$ and approximations have to be made.

### 1. *The Green's function*

In order to have some idea of the form of $W(x, x')$ we must discuss some of the properties of the Green's function $G^Q(E)$, which is the resolvent of the Hamiltonian $(QHQ)$. A set of functions $\Phi_n$ which are eigenfunctions of $(QHQ)$ with eigenvalues $\varepsilon_n^Q$ can be introduced which satisfy

$$(QHQ - \varepsilon_n^Q) \Phi_n = 0. \tag{4-2-27}$$

These functions form a complete set in that part of the Hilbert space of the complete problem, that is projected out by the operator $Q$ and we may write

$$S_n |\Phi_n\rangle \langle \Phi_n| = Q \tag{4-2-28}$$

where $S_n$ denotes a sum over the discrete and an integration over the continuous spectrum.

Below the energy of the positronium formation threshold at $E = E_{Ps}$ the only part of the wave-function which is non-zero at large $x$ or at large $r$ is by definition $P\Psi$ and it follows that $(QHQ)$ possesses only a discrete spectrum in this energy region with eigenfunctions $\Phi_n$ that vanish at infinity. For $E > E_{Ps}$, $Q\Psi$ must describe the free motion of positronium with respect to the proton and accordingly the $\Phi_n$ for $\varepsilon_n^Q > E_{Ps}$ do not vanish at large $x$ and large $r$, and the eigenvalues $\varepsilon_n^Q$ form a continuous spectrum.

The Green's function $G^Q(E)$ can be expanded in terms of the $\Phi_n$ giving

$$G^Q(E, x, r; x', r') = S_n \left[ \frac{\Phi_n(x, r) \Phi_n^*(x', r')}{(\varepsilon_n^Q - E)} \right]. \quad (4\text{-}2\text{-}29)$$

The analytic structure of $G^Q(E)$ follows immediately from our knowledge of the spectrum $\varepsilon_n^Q$. For $E < E_{\text{Ps}}$, if $(QHQ)$ possesses a discrete spectrum (bound states), $G^Q(E)$ will have simple poles at the points $E = \varepsilon_n^Q$, while for $E > E_{\text{Ps}}$, $G^Q(E)$ has a branch cut along the real energy axis. (When $E > E_{\text{Ps}}$, $G^Q(E)$ requires further definition, and to correspond with outgoing wave boundary conditions, it is necessary to use the values obtained by approaching the cut from above, that is we use $G^Q(E) = \lim_{\varepsilon \to 0^+} G(E + i\varepsilon)$.)

For $E < E_{\text{Ps}}$, the Green's function $G^Q(E)$ is real and if $E < \varepsilon_0^Q$, where $\varepsilon_0^Q$ is lowest discrete eigenvalue of $(QHQ)$, $G^Q(E)$ is seen to be a negative definite operator, because each individual term in the sum (4-2-29) is negative. This is very important, because, refering to equation (4-2-24), we see immediately that $W(x, x')$ must also be a negative definite operator in this energy region, in other words $W$ is an attractive non-local potential. This fact will be of great interest to us when we consider variational approximations in a later section.

## 2. Angular momentum decomposition

The problem can be further simplified by expanding $F(x)$ in a series of Legendre polynomials, which are eigenfunctions of angular momentum,

$$F(x) = \sum_{l=0}^{\infty} x^{-1} f_l(x) P_l(\cos \theta). \quad (4\text{-}2\text{-}30)$$

This series is called a partial wave series. By expanding the kernel $W(x, x')$ in a similar series

$$W(x, x') = \sum_{l=0}^{\infty} (x x')^{-1} w_l(x, x') P_l(\hat{x} \cdot \hat{x}') \tfrac{1}{2}(2l + 1) \quad (4\text{-}2\text{-}31)$$

it is straightforward to verify that the radial functions satisfy the equation

$$\left[ \frac{d^2}{dx^2} - \frac{l(l+1)}{x^2} + k^2 \right] f_l(x) = U(x) f_l(x) + \int_0^{\infty} w_l(x, x') f_l(x') \, dx'. \quad (4\text{-}2\text{-}32)$$

As $U(x)$ and $W_l(x, x')$ vanish for large $x$, $f_l(x)$ is asymptotic to a linear combination of the solutions of the equation

$$\left[\frac{d^2}{dx^2} - \frac{l(l+1)}{x^2} + k^2\right] g(x) = 0. \tag{4-2-33}$$

The regular and irregular solutions of this equation may be denoted by $s_l(kx)$ and $c_l(kx)$, which are related to the spherical Bessel functions $j_l(x)$, $n_l(x)$ (see for example Morse and Feshbach [25]),

$$s_l(x) = xj_l(x), \quad c_l(x) = -xn_l(x). \tag{4-2-34}$$

They have the asymptotic properties

$$s_l(x) \sim \sin(x - \tfrac{1}{2}l\pi), \quad c_l(x) \sim \cos(x - \tfrac{1}{2}l\pi). \tag{4-2-35}$$

The boundary conditions satisfied by $f_l(x)$ may be specified by

$$f_l(0) = 0, \quad f_l(x) \sim s_l(kx) + \tan\eta_l\, c_l(kx). \tag{4-2-36}$$

The parameter $\eta_l$ is known as the phase shift of order $l$.

The scattering amplitude can be expressed in terms of the parameters by [19–21]

$$f(\theta) = \frac{1}{k} \sum_{l=0}^{\infty} (2l+1)\, e^{i\eta_l} \sin\eta_l\, P_l(\cos\theta). \tag{4-2-37}$$

The total cross-section is then given by

$$q = \frac{4\pi}{k^2} \sum_{l=0}^{\infty} (2l+1) \sin^2\eta_l. \tag{4-2-38}$$

The importance of the partial wave expansion is that the functions $f_l(x)$ satisfy an ordinary integro-differential equation which can be integrated without too much trouble on a computer, whereas the functions $F(x)$ satisfied a difficult partial differential equation. Nothing would be gained if the partial wave series for the amplitude $f(\theta)$ converged slowly, but the following argument suggests that convergence is very rapid in the energy region of interest. If a potential is of limited range $a$, particles with impact parameters $b > a$ are not scattered classically. As $b$ is equal to $L/mv$ where $L$ is the orbital angular momentum, we may expect scattering to be small when

$$L > mva.$$

Identifying $L^2$ with $l(l+1)\hbar^2$ and $\hbar mv$ with $k$, we may expect scattering to be small for $l > ka$, that is we expect $\eta_l$ to be small for $l > ka$. This suggests that for small $k$ (compared with $1/a$), only the first few terms in the partial wave series will be important and this turns out to be the case.

The phase shifts have the important property of being monotonic in the potential. Let us illustrate this result, for $s$ wave-scattering by local monotonic potentials. Consider the radial equations for scattering by potentials $(\hbar^2/2m)\,U^{(1)}$ and $(\hbar^2/2m)\,U^{(2)}$:

$$\left(\frac{d^2}{dx^2} - U^{(1)}(x) + k^2\right) f^{(1)}(x) = 0$$

$$\left(\frac{d^2}{dx^2} - U^{(2)}(x) + k^2\right) f^{(2)}(x) = 0$$

(4-2-39)

where $f^{(1)}(x)$ and $f^{(2)}(x)$ have the asymptotic forms

$$f^{(1)}(x) \sim \sin(kx) + \tan\eta^{(1)} \cos(kx)$$
$$f^{(2)}(x) \sim \sin(kx) + \tan\eta^{(2)} \cos(kx).$$

(4-2-40)

Multiplying the first of equations (4-2-39) by $f^{(2)}(x)$ and the second by $f^{(1)}(x)$, subtracting and integrating we find the relation (remember $f^{(1)}(0) = f^{(2)}(0) = 0$),

$$f^{(2)}(x)\frac{d}{dx}f^{(1)}(x) - f^{(1)}(x)\frac{d}{dx}f^{(2)}(x)$$

$$= \int_0^x [U^{(1)}(x) - U^{(2)}(x)]\,f^{(1)}(x)\,f^{(2)}(x)\,dx. \quad (4\text{-}2\text{-}41)$$

Using the asymptotic forms of $f^{(1)}(x)$ and $f^{(2)}(x)$ and taking the limit $x \to \infty$, we find that

$$\tan\eta^{(2)} - \tan\eta^{(1)} = k^{-1} \int_0^\infty f^{(1)}(x)\,f^{(2)}(x)[U^{(1)}(x) - U^{(2)}(x)]\,dx.$$

(4-2-42)

When the difference $[U^{(1)}(x) - U^{(2)}(x)]$ is small, so that $f^{(1)}(x)\,f^{(2)}(x) \approx [f^{(1)}(x)]^2$, we see the sign of the difference $[\tan\eta^{(2)} - \tan\eta^{(1)}]$ is the same as that of $[U^{(1)} - U^{(2)}]$. By constructing a series of comparison potentials between $U^{(1)}$ and $U^{(2)}$, it follows that this result is also valid for finite differences $[U^{(1)} - U^{(2)}]$ and that in general

$$\eta^{(1)} < \eta^{(2)} \quad \text{if} \quad U^{(2)} > U^{(1)}. \tag{4-2-43}$$

This theorem is easily extended to all $l$ (Messiah [26]).

## B. Approximation schemes

### 1. *The static approximation*

If the interaction between positron and the target atom were weak, the atom would remain undistorted during the collision, and wave-function would be given accurately by a product of the form $\phi_0(r) F(x)$. Under these circumstances, the closed channel part of the wave-function $Q\Psi$ would be small and could be neglected, and $F(x)$ would satisfy the equation

$$(\nabla^2 + k^2) F(x) = U(x) F(x) \tag{4-2-44}$$

where $W(x, x')$ has been set equal to zero.

Even in the first calculations of elastic scattering of positrons, by Ore [27] in 1949 and Massey and Moussa [28] in 1958 it was realised that this approximation, the static approximation, was inadequate. This was seen by examining the corresponding situation in electron scattering, which had received much more study.* In that case, it is known that two effects modify the results of the simple static approximation, in the energy range below the first excitation threshold. The first and most important is electron exchange. This has no exact counterpart in positron scattering, but corresponds to some extent to positronium formation, which for $E < E_{\text{Ps}}$ is of course a virtual process. The other effect of importance, is the dipole distortion induced in the hydrogen atom by the electric field of the incident electron. This gives rise to a long range effective potential, which behaves like $-\alpha/x^4$ for large $x$. This distortion, or polarisation as it is often called, of the hydrogen atom also occurs in positron scattering, and we shall see that it is most important, the effect increasing with decreasing energy. Before discussing these distortion effects in detail, it is interesting to see what the static approximation predicts, for comparison.

By inserting the explicit form of the hydrogen ground state wavefunction $\phi_0(r) = \pi^{-\frac{1}{2}} \exp(-r)$ into the integrand in eq. (4-2-23) the integral may be performed with the result

$$U(x) = 2\left(1 + \frac{1}{x}\right) \exp(-2x). \tag{4-2-45}$$

---

* A comprehensive review of electron scattering by hydrogen atoms has been given by Burke and Smith [29], covering work up to 1962. This may be supplemented for more recent work, by an article by Peterkop and Veldre [30].

This interaction is repulsive (positive) and is equal in magnitude, but opposite in sign, to the static interaction between an electron and a hydrogen atom. The phase shifts obtained by numerical integration of the radial Schrödinger equation with this potential are shown in Table 4-2-1 column (a) for the s, p and d partial waves ($l=0$, 1 and 2). In the case of a monotonic potential the phase shifts are of opposite sign to the potential* and are therefore negative in this case.

For a potential that decreases faster than any power of $x$, such as $U(x)$, it can be shown that the phase shifts behave like

$$\eta_l(k) \propto k^{2l+1} \tag{4-2-46}$$

for small values of $k$. Defining the s wave scattering length**

$$A = \lim_{k \to 0} (\eta_0(k)/k) \tag{4-2-47}$$

it follows from eq. (4-2-28), that the total elastic cross-section at zero momentum is given in terms of $A$ by

$$q(k=0) = 4\pi A^2. \tag{4-2-48}$$

The value of the scattering length is given in Table 4-2-2.

*Helium, neon and argon.* The static approximation can also be applied to scattering by helium, neon and argon for which some experimental results are available for comparison. Marder et al. [31] studied the annihilation rate of positronium atoms formed in positron collisions in these gases, as a function of applied electric field. From this data Teutsch and Hughes [32] were able to deduce the positron–atom elastic scattering cross-section[†] for positrons with a mean kinetic energy of about 18 eV ($k=1.2$). The values found were 0.023, 0.12 and 1.5 in units of $\pi a_0^2$ for helium, argon and neon respectively. The accuracy is estimated as $\pm 25\%$. This work is consistent with a later measurement by Falk et al. [22] in argon. No other cross-

---

* This follows from the inequalities (4-2-43) by setting $U^{(2)}$ and $\eta^{(2)}$ equal to zero.
** The scattering length is sometimes defined as $-\lim_{k\to 0}(\eta_0(k)/k)$, but our choice of sign is generally the more convenient one.
† Actually it was the momentum transfer or diffusion cross-section defined as $q_D = 2\pi \int_{-1}^{+1}(1 - \cos\theta) I(\theta) \, d(\cos\theta)$ that was calculated, but this is a close approximation to the elastic scattering cross-section at the energies concerned.

TABLE 4-2-1

Low energy phase shifts for the elastic scattering of positrons by hydrogen atoms in various approximations

| $k$(a.u.) | (a) | (b) | (c) | (d) | (e) | (f) | (g) | (h) | (i) |
|---|---|---|---|---|---|---|---|---|---|
| s waves ($l = 0$) | | | | | | | | | |
| 0.1 | −0.058 | 0.098 | −0.0054 | 0.0072 | — | — | 0.1318 | 0.128 | 0.151 |
| 0.2 | −0.1145 | 0.114 | −0.0426 | −0.0251 | −0.046 | 0.1494 (0.182) | 0.1621 | 0.158 | 0.188 |
| 0.3 | −0.168 | 0.087 | −0.0931 | −0.0748 | — | — | 0.1377 | 0.135 | 0.168 |
| 0.4 | −0.218 | 0.041 | −0.1472 | −0.1295 | −0.134 | 0.0818 (0.119) | 0.0813 | 0.089 | 0.120 |
| 0.5 | −0.264 | −0.010 | −0.1990 | −0.1829 | −0.186 | — | 0.0327 | 0.034 | 0.062 |
| 0.6 | −0.304 | −0.062 | −0.2461 | −0.2317 | −0.238 | 0.0272 (0.009) | −0.0244 | −0.022 | 0.007 |
| 0.7 | — | — | — | — | −0.286 | — | −0.0783 | −0.074 | −0.054 |
| p waves ($l = 1$) | | | | | | | | | |
| 0.2 | −0.0017 | 0.027 | 0.0127 | 0.0162 | — | 0.02893 (0.032) | 0.0306 | 0.0263 | 0.033 |
| 0.3 | −0.0055 | 0.049 | 0.0201 | 0.0257 | — | 0.0547 (0.066) | 0.0576 | 0.0578 | 0.065 |
| 0.4 | −0.0121 | 0.067 | 0.0201 | 0.0300 | — | 0.0801 (0.11) | 0.0821 | 0.0764 | 0.102 |
| 0.5 | −0.020 | 0.079 | 0.0183 | 0.0282 | — | 0.0994 (0.14) | 0.0987 | 0.0952 | 0.132 |
| 0.6 | −0.0322 | 0.084 | 0.0101 | 0.0210 | — | 0.1116 (0.17) | 0.1054 | 0.106 | 0.156 |
| 0.7 | — | — | — | — | — | 0.1189 (0.19) | 0.1046 | 0.109 | — |
| d waves ($l = 2$) | | | | | | | | | |
| 0.4 | −0.005 | — | 0.0112 | 0.0134 | — | — | 0.0222 | — | — |
| 0.5 | −0.0013 | — | 0.0152 | 0.0186 | — | 0.0313 (0.046) | 0.0334 | — | — |
| 0.6 | −0.0028 | — | 0.0183 | 0.0231 | — | 0.057 (0.0601) | 0.0449 | — | — |
| 0.7 | — | — | — | — | — | — | 0.0552 | — | — |

(a) The static approximation*.
(b) The polarised orbital approximation (dipole interaction only (4–3)).
(c) The 1s–2s–2p truncated hydrogenic expansion* [65].
(d) The 1s–2s–2p–3s–3p–3d truncated hydrogenic expansion* [66].
(e) The (1s) hydrogen–(1s) positronium two state approximation* [44].
(f) Variational method of Spruch and the collaborators* [68] (figures in brackets are extrapolated as described in the text in 4-2).
(g) The extended polarised orbital method of Callaway et al.*.
(h) The variational method of Drachman*.
(i) The Kohn variational calculations of Schwartz [63] ($l=0$) and Armstead [64] ($l=1$).

\* These methods give lower bounds.

sections for elastic scattering have been estimated, but there are several measurements of the annihilation rate of thermalised positrons in various substances including helium, neon and oxygen, which, as explained earlier, lead to estimates of the parameter $Z_{\text{eff}}(v)$.

Among the most recent values obtained (for the gas at room temperature) are for helium: $3.92 \pm 0.04$ (Falk et al. [22]), 2.46 (Osmon [33]), $3.43 \pm 0.19$ (Roellig and Kelly [34]), for neon: 3.6 (Osmon [33]) and for argon: $28.8 \pm 1.1$ (Falk et al. [22]), 15.1 (Osmon [33]), $30.7 \pm 1.0$ (Paul [35]). References to earlier measurements may be found in Fraser [23].

The static approximation is generalised to a many electron atom in a straightforward manner. The wave-function is again written as the product $\phi_0(r_1, r_2, ..., r_z) F(x)$, where $\phi_0$ is the ground state wave-function of an

TABLE 4-2-2

Scattering lengths $A(=\lim_{k\to 0} \eta_0(k)/k)$ for positron hydrogen collisions in various approximations

| Approximation | Scattering length $A$ |
|---|---|
| Static | −0.582* |
| Two state with Virtual positronium formation (H(s), Ps(1s)) | −0.1704* |
| Polarised orbital (dipole term only) | 1.267 |
| Polarised orbital based on eq. (4-2-164) | 1.0* |
| Extended polarised orbital | 1.9 |
| Polarised orbital (variational eq. (4-2-174)) | 1.85* |
| Kohn variational method with many terms (Schwartz) | 2.10 |
| Virtual positronium formation plus dipole polarisation | 3.06 |

\* Lower bound

atom containing $Z$ electrons. The static interaction is

$$U(x) = 2 \int d\mathbf{r}_1 \int d\mathbf{r}_2 \ldots \int d\mathbf{r}_Z |\phi_0(\mathbf{r}_1, \mathbf{r}_2, \mathbf{r}_Z)|^2 \left\{ \frac{Z}{x} - \sum_{i=1}^{Z} \frac{1}{|\mathbf{x} - \mathbf{r}_i|} \right\}$$
(4-2-49)

and the radial Schrödinger equations can be solved with this potential to find the phase shifts. For a many electron atom the wave-functions $\phi_0$ are not known exactly. In the work of Massey and Moussa [28], wave-functions in the self-consistent field approximation were used and the cross-sections calculated in this way are shown in Table 4-2-3 for a number of values of the momentum.

TABLE 4-2-3

Total cross-sections (in units of $\pi a_0^2$) in the static approximation for the elastic scattering of positrons by the rare gases

| Atom | $k$(a.u) | | | |
|---|---|---|---|---|
| | 0.2 | 0.5 | 1.0 | 1.5 |
| He | 0.709 | 0.656 | 0.531 | 0.416 |
| Ne | 2.952 | 2.204 | 2.247 | 1.84 |
| A | 7.018 | 6.266 | 5.208 | 4.613 |

In each case, the cross-sections in the static approximation are too large, which shows that the effective potential $W$ cannot be neglected. The potential $U(x)$ is repulsive and the corresponding phase shifts are quite large and negative. We have already seen (at least for energies below the lowest excitation threshold) that the effective potential $W$ is attractive and this will tend to reduce the phase shifts, leading in turn to smaller cross-sections.*

## 2. *The adiabatic polarisation potential*

To discuss the polarisation or distortion of the target atom during the collision, it is necessary to obtain an approximation for the closed channel wave-function $Q\Psi$ that occurs in the coupled equations (4-2-17) and (4-2-18) and which determines the effective potential $W$. As usual, we shall take the case of positron scattering by hydrogen as our example. With $P$ given by (4-2-13), and making use of the Schrödinger equation satisfied by $\phi_0(r)$

---

* The truth of this statement depends on the extension of the monotonicity theorem (4-2-43) to non-local potentials, such as $W$. For a proof see, for example, the paper by Blankenbecler and Sugar [36].

(4-2-6), equation (4-2-18) can be written in the explicit form

$$Q[-\tfrac{1}{2}\nabla_r^2 - \tfrac{1}{2}\nabla_x^2 + V(r, x) - \tfrac{1}{2}k^2 + \varepsilon_0] Q\Psi(r, x)$$
$$= [\tfrac{1}{2}U(x) - V(r, x)] \phi_0(r) F(x) \quad (4\text{-}2\text{-}50)$$

where $V(r, x)$ and $U(x)$ are defined by (4-2-4) and (4-2-23) respectively.

In the adiabatic approximation, it is assumed that the positron is moving so slowly that it is possible to neglect the kinetic energy terms $(-\tfrac{1}{2}\nabla_x^2 - \tfrac{1}{2}k^2)$ on the left hand side of (4-2-50). The resulting equation can be solved by expanding $Q\Psi$ in a complete set of hydrogenic wave-functions $\phi_j(r)$, with eigenenergies $\varepsilon_j$,

$$Q\Psi(r, x) = \sum_{j \neq 0} \phi_j(r) F_j(x) \quad (4\text{-}2\text{-}51)$$

and computing the coefficients $F_j(x)$ to lowest order in perturbation theory. It should be noticed that because of the presence of the operator $Q$, there is no term in $\phi_0$ in the expansions and the functions $\phi_j$, $j \neq 0$, span $Q$ space, with our particular choice of $P$. The result of this calculation is

$$Q\Psi^A(r, x) = \sum_{j \neq 0} (\varepsilon_0 - \varepsilon_j)^{-1} \phi_j(r)$$
$$\times \left[ \int dr' \phi_j(r') \{\tfrac{1}{2}U(x) - V(r', x)\} \phi_0(r') \right] F^A(x) \quad (4\text{-}2\text{-}52)$$

where the superscript A has been added to $\Psi$ and to $F$ to denote that these functions are now approximations. As the functions $\phi_j$ are all orthogonal to $\phi_0$, the term in $U(x)$ makes no contribution and can be omitted. Inserting this expression for $Q\Psi$, into the first of the coupled equations, (4-2-17), the approximate function $F^A(x)$ is found to satisfy

$$[-\tfrac{1}{2}\nabla_x^2 + \tfrac{1}{2}U(x) + \tfrac{1}{2}W_{\text{pol}}(x) - \tfrac{1}{2}k^2] F^A(x) = 0 \quad (4\text{-}2\text{-}53)$$

where

$$W_{\text{pol}}(x) = 2 \sum_{j \neq 0} (\varepsilon_0 - \varepsilon_j)^{-1} \left| \int dr\, \phi_j^*(r) \phi_0(r) V(r, x) \right|^2. \quad (4\text{-}2\text{-}54)$$

The adiabatic polarisation potential $W_{\text{pol}}(x)$ is a (local) approximation to the full effective potential $W(x, x')$. Because $V(r, x)$ enters as a square, the polarisation potential $W_{\text{pol}}(x)$ is the same for electron–atom as for positron–atom scattering, in both cases it is attractive, in accordance with the general property of $W(x, x')$ discussed earlier. The effect of polarisation turns out to be very different in the two systems, being much more marked for positron scattering than for electron scattering. This is due in part to the difference in sign of the static interactions. For positron scattering $U$ and $W_{\text{pol}}$ subtract, the net potential being $+\{|U(x)|-|W_{\text{pol}}(x)|\}$ while for electron scattering the two potentials add, giving a net potential $\{-|U(x)|-|W_{\text{pol}}(x)|\}$. In addition, the strong exchange interaction in electron scattering dominates the inner region of the effective potential and masks the effect of the long range attraction.

The form of $W_{\text{pol}}(x)$ for large $x$ is easily established, by examining the matrix elements $\langle \phi_j | V | \phi_0 \rangle$ where

$$\langle \phi_j | V | \phi_0 \rangle = \int d\mathbf{r}\, \phi_j^*(\mathbf{r}) \phi_0(\mathbf{r}) V(\mathbf{r}, \mathbf{x}). \tag{4-2-55}$$

The expansion of $V(\mathbf{r}, \mathbf{x})$ in Legendre polynomials is (Morse and Feshbach [25])

$$V(\mathbf{r}, \mathbf{x}) = \sum_{l=0}^{\infty} \left[ \delta_{l0} \frac{1}{x} - \gamma_l(r, x) \right] P_l(\hat{\mathbf{r}} \cdot \hat{\mathbf{x}}) \tag{4-2-56}$$

where

$$\gamma_l(r, x) = \begin{cases} \dfrac{1}{r^{l+1}} x^l & r > x \\ \dfrac{1}{x^{l+1}} r^l & x > r. \end{cases}$$

As $\phi_0(r)$ is spherically symmetrical, we have that if $\phi_j = \phi_{n,l}$ where $n$ is the principal and $l$ the orbital angular momentum quantum number, the form of $(\phi_j | V | \phi_0)$ for large $x$ is

$$(\phi_{n,l} | V | \phi_0) \sim -\frac{1}{x^{l+1}} \left[ \int d\mathbf{r}\, \phi_{n,l}^*(\mathbf{r}) \phi_0(\mathbf{r}) r^l P_l(\hat{\mathbf{r}} \cdot \hat{\mathbf{x}}) \right] \tag{4-2-57}$$

except for the case of s states with $l=0$, when $(\phi_{n,0} | V | \phi_0)$ vanishes exponentially. The polarisation potential can be expressed as a series of multipoles, in which the term of order $l$ arises from the hydrogenic state in the expansion (4-2-51) of orbital angular momentum $l$

$$W_{\text{pol}}(x) = \sum_{l=0}^{\infty} w_l(x). \tag{4-2-58}$$

The monopole $w_0(x)$ vanishes exponentially, while the remaining terms (using eq. (4-2-57)) behave like

$$w_l(x) \sim -\alpha_l/x^{2l+2} \tag{4-2-59}$$

for large $x$. The leading term is the dipole term arising from the p states of hydrogen and for large $x$ is

$$w_1(x) \sim \alpha_1/x^4 \qquad (4\text{-}2\text{-}60)$$

where $\alpha_1$ is the polarisability of hydrogen defined by

$$\alpha_1 = \sum_{n \neq 0} 2(\varepsilon_n - \varepsilon_0)^{-1} \left| \int d\mathbf{r}\, \phi_{n,1}^*(\mathbf{r})\, r \cos\theta\, \phi_0(\mathbf{r}) \right|^2. \qquad (4\text{-}2\text{-}61)$$

A remarkable property of the adiabatic polarisation potential is that it can be shown to have the same leading term as the *exact* effective potential (Castellejo et al. [37]). In other words, the exact form of the interaction between a positron (or an electron) and a neutral atom at energies below the first excitation threshold is, for sufficiently large $x$, of the form (4-2-60). The approximate form $W_{\text{pol}}(x)$ is of course in error at small $x$, corrections to the adiabatic approximation to take into account the neglected kinetic energy of the positron introduce terms of the order $1/x^6$ at large $x$, and working to higher order in perturbation theory brings in additional terms of this order*.

These results can be extended to scattering by complex atoms. In positron hydrogen scattering it is seen that the approximate wave function $(P\Psi^A + Q\Psi^A)$ is of the form $\{\phi_0(\mathbf{r}) + \phi_0^{\text{Pol}}(\mathbf{r}, \mathbf{x})\} F^A(\mathbf{x})$ where $\phi_0^{\text{Pol}}(\mathbf{r}, \mathbf{x})$ is the first order perturbation correction to the wave-function of a hydrogen atom in the field of a charge, fixed at the point $\mathbf{x}$. For complex atoms the approximate wave-functions used are generally of the form of a sum of products of single electron wave-functions. A typical term might be $U_1(\mathbf{r}_1) U_2(\mathbf{r}_2) \ldots U_Z(\mathbf{r}_Z)$ and it is then possible to proceed by considering the adiabatic distortion of each orbital $U_i(\mathbf{r}_i)$, so that $U_i(\mathbf{r}_i) \to \{U_i(\mathbf{r}_i) + U_i^{\text{Pol}}(\mathbf{r}_i, \mathbf{x})\}$. Again it is found that the leading term is a dipole distortion giving rise to a polarisation potential behaving like $-\alpha/x^4$ for large $x$.

For positron scattering by helium, neon and argon, Massey and Moussa in their 1958 paper [28] were able to demonstrate the importance of polarisation empirically, by representing the polarisation potential by the parametric form

$$W_{\text{pol}}(x) = -\frac{\alpha}{(x^2 + d^2)^2} \qquad (4\text{-}2\text{-}62)$$

where $\alpha$ was the known polarisability of the atom and $d$ was a variable parameter, which should take values of the order of the radius of the atom.

---

* The calculation of the terms of order $1/x^6$ in the exact potential has been discussed by Kleinman et al. [38].

For reasonable values of $d$ it was shown that the large negative phase shifts predicted by the static approximation could indeed be reduced to values consistent with the data of Marder et al.

The general class of approximations based on expressions similar to (4-2-52) for $Q\Psi$ will be termed polarised orbital methods. The later and important development of these methods stems from the work of Temkin [39] and Temkin and Lamkin [40], who, in particular, showed for electron scattering how both exchange and polarisation effects could be allowed for at the same time. We shall describe the polarised orbital method and its results in some detail at a later stage, but at this point we note the phase shifts obtained by retaining the dipole term $w_1$ alone. The terms $w_0$, $w_1$ and $w_2$ have been calculated explicitly by Reeh [41] and the complete adiabatic potential has been given by Dalgarno and Lynn [42]. The potential $w_1$ has the explicit form, in atomic units,

$$w_1(x) = -\frac{9}{2x^4}\{1 - \tfrac{1}{3}e^{-2x}(1 + 2x + 6x^2 + \tfrac{20}{3}x^3 + \tfrac{4}{3}x^4) - \tfrac{2}{3}(1+x)^4 e^{-4x}\}. \qquad (4\text{-}2\text{-}63)$$

The calculated phase shifts* shown in Table 4-2-1 illustrate the very pronounced effect of polarisation. The effective potential at low energies is converted from a repulsion to an attraction, giving rise to positive rather than negative phase shifts. A long range potential is relatively more effective in states with $l > 0$ so that the absolute magnitude of the P wave phases, for example, is greatly enhanced.

### 3. Virtual positronium formation

The closed channel wave-function for positron scattering by hydrogen atoms has an *exact* expansion like that of (4-2-51)

$$Q\Psi(r, x) = \sum_{j \neq 0}^{\infty} \phi_j(r) F_j(x), \qquad (4\text{-}2\text{-}64)$$

where the coefficients $F_j(x)$ satisfy the boundary conditions, below the first inelastic threshold,

$$xF_j(x) \to 0, \quad x \to \infty. \qquad (4\text{-}2\text{-}65)$$

Each term in the expansion represents a positron moving with respect to an

---

* The phase shifts shown were calculated by Bransden and Jundi [43], employing the potential $w_1(x)$ given in eq. (4-2-63). Similar results were obtained earlier by Cody et al. [44], who employed an approximation to $w_1(x)$ used by Temkin and Lamkin [40].

excited state of hydrogen, but as transitions to these states are not allowed energetically, the states are referred to as virtual. It is, as we have seen, the virtual excitation of the p states of hydrogen that gives rise to the long range dipole polarisation. This is of course not the only possible expansion of $Q\Psi$; we can equally well expand in terms of the complete set of positronium functions $\psi_m(R)$

$$Q\Psi(x, r) = \sum_m \psi_m(R) H_m(y). \qquad (4\text{-}2\text{-}66)$$

The position vectors $R$ and $y$ where $R = r - x$ and $y = \frac{1}{2}(r + x)$ form a set of centre of mass coordinates. The coefficients $H_m(y)$ describe the motion of a positronium atom in the state $m$, moving freely with respect to a proton. This process is virtual below the positronium formation threshold, so that $yH_m(y)$ must vanish for large $y$. The coefficients are restrained by the condition that $Q\Psi$ is orthogonal to $\phi_0(r)$,

$$\int \phi_0(r) \left[ \sum_m \psi_m(r - x) H_m\{\tfrac{1}{2}(r + x)\} \right] dr = 0.$$

Although both the expansions (4-2-64) and the expansion (4-2-66) are complete, certain sub-sets of terms in one expansion more readily describe certain physical aspects of $Q\Psi$, than the corresponding terms in the other expansion. If configurations, in which the system is in the ground state of positronium moving with respect to proton, are very important, then $Q\Psi$ will be well represented by the first term in expansion (4-2-66)

$$Q\Psi \approx Q[\psi_0(R) H_0(y)]. \qquad (4\text{-}2\text{-}67)$$

Such a term leads to an effective non-local interaction of short range that is qualitatively different from the long range polarisation, which is most naturally described by including the p states of the hydrogenic expansion.

One way of describing distortion effects, whether due to long range dipole polarisation or to virtual positronium formation to introduce a parametric form to represent the total wave-function, having the correct asymptotic behavior and containing terms which describe the effect that it is desired to investigate. A simple function of this kind might be [45], for s wave-scattering

$$\Psi(r, x) = x^{-1}\phi_0(r) \left[ \sin kx + (1 - e^{-\delta x}) \tan \eta_0 \cos kx \right]$$
$$+ B \exp(-\tfrac{1}{2}R) \exp(-\lambda y), \qquad (4\text{-}2\text{-}68)$$

where $\delta$ and $B$ are (energy dependent) parameters. The term in square brackets represents an approximation to the open channel part of the wavefunction while the last term is supposed to represent the closed channel function $Q\Psi$ in a configuration in which positronium in the ground state (with wave-function $\psi_0(R) = (8/\pi)^{\frac{1}{2}} \exp(-\frac{1}{2}R)$) moves with respect to the free proton. The question now arises as to the most effective ways of fixing the numerical values of the parameters. This problem is solved by the variational method which will be discussed in the next section. An alternative procedure is to estimate that part of $Q\Psi$ that belongs to the configuration (positronium in the ground state plus proton) by first order perturbation methods. This was done by Bransden [46] who found a large effect, for example at $k = 0.1$, the $l = 0$ phase shift was changed from the static value of $-0.058$ to a value of $0.044$. It turns out that the perturbation theory employed overestimates the effect (although the results may be reasonably accurate for $l \neq 0$ and large $k$), but the results indicate in a qualitative way that virtual positronium formation is important. Of course care must be taken in any method that attempts to describe long range polarisation and virtual positronium formation simultaneously, because the two expressions (4-2-64) and (4-2-66) span the same space and it is difficult, except in the context of the variational method, to find a consistent approximation.

## 4. *Variational methods*

Although the Rayleigh-Ritz variational method for bound state problems was known and used from the early days of quantum mechanics, variational methods for scattering problems were developed much later, in particular by Hulthén [46], Kohn [47] and Schwinger [48]. A short account of some variational methods will be given here and the applications to positron scattering will be discussed in § 4-2-C. For a fuller development of the subject, the interested reader is referred to the articles by Spruch [16, 49], which contain clear accounts of the variational method and the associated problem of obtaining bounds on scattering parameters, and also to the monographs by Demkov [50] and Moiseiwitsch [51].

We shall again study the positron–hydrogen atom system at energies below the first inelastic threshold and we shall concentrate on the calculation of the s-wave phase shift $\eta_0(k)$. The total orbital angular momentum $L$ is a constant of the motion, and as the hydrogen atom ground state is an s state, $L$ coincides with $l$, the orbital angular momentum of the positron. In place of the total wave-function $\Psi(r, x)$, it is convenient to work with $\Psi^L(r, x)$ where

$\Psi^L(r, x)$ is an eigenfunction of the total orbital angular momentum

$$\Psi = \sum_{L=0}^{\infty} \Psi^L(r, x). \tag{4-2-69}$$

If $(\alpha, \beta)$ are the polar angles of $x$, we must have that

$$P\Psi^L(r, x) = \phi_0(r) x^{-1} f_L(x) P_L(\cos \alpha) \tag{4-2-70}$$

where $f_L(x)$ is a radial function which can be taken to satisfy the boundary conditions (4-2-36). For the case $l=0$

$$f_0(0) = 0, \quad f_0(x) \sim \sin(kx) + D \cos(kx) \tag{4-2-71}$$

where $D = \tan \eta_0$. With this normalisation $Q\Psi^L$ and $P\Psi^L$ are both real, and as usual

$$Q\Psi^L \to 0 \quad \text{as} \quad x \to \infty \quad \text{or as} \quad r \to \infty. \tag{4-2-72}$$

A trial function $\Psi_t^0(r, x)$ is introduced such that $\Psi_t^0(r, x)$ obeys the same boundary conditions as $\Psi^0(r, x)$, (4-2-71) and (4-2-72), but with a possibly incorrect value of $D$, denoted by $D_t$. Let $I[\Phi]$ be defined by

$$I[\Phi^L] \equiv \int d\mathbf{x} \int d\mathbf{r} \{\Phi^L(r, x) [H - E] \Phi^L(r, x)\} \tag{4-2-73}$$

where $\Phi^L$ is an eigenfunction of total orbital angular momentum with eigenvalue $L$.

Then we have that

$$I[\Psi_t^0] = \int d\mathbf{x} \int d\mathbf{r} [P\Psi_t^0 \{P(H - E) P\Psi_t^0 + Q\Psi_t^0\}$$
$$+ Q\Psi_t^0 \{Q(H - E) Q\Psi_t^0 + P\Psi_t^0\}]. \tag{4-2-74}$$

If, for a certain trial function $\Psi_t^0$, the error in the wave-function is $\Delta\Psi^0 \equiv \Psi_t^0 - \Psi^0$, then $Q\Delta\Psi^0 \to 0$ as $x \to \infty$ and

$$P \Delta\Psi^0 \sim \phi_0(r) (\Delta D) x^{-1} \cos(kx)$$
$$\Delta D \equiv D_t - D. \tag{4-2-75}$$

We now have that

$$I[\Psi_t^0] = I[\Delta\Psi^0] + \int d\mathbf{x} \int d\mathbf{r} \, \Psi^0 (H - E) \Delta\Psi^0 \tag{4-2-76}$$

where we have used the fact that

$$(H - E) \Psi^0 = 0.$$

Again using this fact, the last term can be re-expressed as

$$\int dx \int dr \, [\Psi H \, \Delta\Psi^0 - \Delta\Psi H \Psi^0]. \tag{4-2-77}$$

Since $\Psi^0$ and $\Delta\Psi^0$ vanish for large $r$, and as $Q\Psi^0$ and $Q\Delta\Psi^0$ vanish for large $x$ as well, this reduces to

$$-\tfrac{1}{2} \int dx \int dr \, [(P\Psi^0) \nabla_x^2 (P \, \Delta\Psi^0) - (P \, \Delta\Psi^0) \nabla_x^2 P\Psi^0]$$

$$= -\frac{4}{2}\pi \int_0^\infty dx \left[ f_0(x) \frac{d^2}{dx^2} \Delta f_0(x) - \Delta f_0(x) \frac{d^2}{dx^2} f_0(x) \right] \tag{4-2-78}$$

$$= -2\pi \lim_{x \to \infty} [f_0(x) \Delta f_0'(x) - \Delta f_0(x) f_0'(x)]_0^x = -2\pi k \, \Delta D,$$

where the asymptotic forms of $f_0(x)$ and $\Delta f_0(x)$ have been used in the last line. This identity, which can be written as

$$D = D_t - \frac{1}{2\pi k} I[\Psi_t^0] + \frac{1}{2\pi k} I[\Delta\Psi^0] \tag{4-2-79}$$

leads to a variational principle, because if $\Delta\Psi^0$ is a small quantity of the first order the last term is of second order and can be dropped. In other words $[2\pi k D + I[\Psi^0]]$ is stationary under the variation $\Psi \to \Psi + \delta\Psi$. One way of using this result is to take a parametric form, such as (4-2-68) as a trial function. Generally if $\Psi_t^0$ is a function of a number of parameters $C_1$, $C_2, \ldots, C_N$ together with $D$ and if it is possible to make $\Psi_t^0$ coincide with $\Psi^0$ by adjusting the $C_i$ and $D$, then we may expand $I(C_i, D)$ by Taylor's theorem and write

$$\delta I(C_i, D) = \sum_{i=1}^{N} \frac{\partial I}{\partial C_i} \delta C_i + \frac{\partial I}{\partial D} \delta D. \tag{4-2-80}$$

As the variations in the $C_i$ and $D$ are independent, we have from the stationary property that

$$\frac{\partial I}{\partial C_i} = 0 \quad i = 1, 2, \ldots N$$

$$\frac{\partial I}{\partial D} = -2\pi k. \tag{4-2-81}$$

Having found values $C_i$ and $D$ by solution of these equations, we can correct $D$, since up to quantities of the second order,

$$D = D_t - \frac{1}{2\pi k} I(C_i, D_t). \qquad (4\text{-}2\text{-}82)$$

This is the method of Kohn. Alteratively, the condition $\delta D = 0$ can be satisfied, if in place of the second of equations (4-2-81) we require that

$$I(C_i, D) = 0. \qquad (4\text{-}2\text{-}83)$$

The solution of this equation again provides $D$ up to terms of the second order and this is the method of Hulthén [46]. Both methods may be trivially extended to the determination of $\eta_L(k)$ with $L \neq 0$.

One disadvantage of the variational approach is that unless the trial function chosen is capable of representing the exact wave-function accurately, completely erroneous results may be obtained. Worse still, there is no test as to which of two trial functions is superior. Because of this, there has been much interest in the development of minimum or maximum principles which enable such a test to be made. An important example of such a principle has been discovered by Spruch and Rosenberg [52], who show that if the Kohn variational method is applied to determine the zero energy scattering length $A$, defined by (4-2-47), then the error term in the identity corresponding to (4-2-79) can be bounded, provided no bound states of the whole system formed by projectile and the target exist,* and $I[\Delta \Psi^0] \geq 0$. Under these circumstances the value of $A$ obtained is a lower bound.

A parametric form of wave-function is not the only kind of trial function that can be envisaged. For example let us take a trial function having the form of a truncated eigenfunction expansion

$$\Psi(r, x) = \sum_{j=0}^{N} \phi_j(r) F_j(x) \qquad (4\text{-}2\text{-}84)$$

where the first term represents $P\Psi$, $F_0(x) \equiv F(x)$ and the remaining terms represent $Q\Psi$. The equations determining the functions $F_j(x)$ can be determined from the condition that $I[\Psi]$ is stationary under the independent variations $F_j(x) \to F_j(x) + \delta F_j(x)$. Just as the condition $\delta I[\Psi^0]$ ensures from that $D$ is stationary, the condition $\delta I[\Psi] = 0$, where we have not made a partial wave expansion, ensures that the scattering amplitude is stationary.

---

* An extension to the case where the composite system formed by the target and the projectile has a finite number of bound states is possible, if the trial function contains components representing these bound states [53].

The change in $I$ under the variation is

$$\delta I = \sum_{j,j'} \int d\mathbf{x} \int d\mathbf{r}\, \phi_j^*(r)\, \delta F_j^*(x)\, (H-E)\, \phi_{j'}(r)\, F_{j'}(x) \qquad (4\text{-}2\text{-}85)$$

and the condition $\delta I = 0$ requires that the coefficient of each of the $\delta F_j^*(x)$ vanishes separately giving rise to the coupled equations:

$$\sum_{j'} \int d\mathbf{r}\, \phi_j^*(r)\, (H-E)\, \phi_{j'}(r)\, F_{j'}(x) = 0, \qquad j = 0, 1, 2, \ldots, N. \qquad (4\text{-}2\text{-}86)$$

These equations are known as the close-coupling equations and have the explicit form

$$[\nabla_x^2 + k^2]\, F_0(x) = U(x)\, F_0(x) + \sum_{j'=1}^{N} V_{0j'}(x)\, F_{j'}(x)$$
$$[\nabla_x^2 - \lambda_j^2]\, F_j(x) = \sum_{j'=0}^{N} V_{jj'}(x)\, F_{j'}(x), \qquad j = 1, 2, 3, \ldots, N \qquad (4\text{-}2\text{-}87)$$

where $\lambda_j^2 = k^2 + 2\varepsilon_0 - 2\varepsilon_j$ and

$$V_{jj'}(x) = 2 \int d\mathbf{r}\, \phi_j^*(r)\, \phi_{j'}(r)\, V(\mathbf{r}, \mathbf{x}). \qquad (4\text{-}2\text{-}88)$$

The boundary conditions placed on the functions $F_j(x)$ follow from our earlier discussion, $F_0(x)$ has the form (4-2-26) and belongs to the $P$, or open channel, subspace, while the functions $F_j(x), j \neq 0$ must vanish at infinity, as these are closed channel, or $Q$ space, functions. The first set of equations represents the first of the coupled equations (4-2-17) and (4-2-18), while the second set of equations with $j$ running from $j=1$ to $j=N$ represents the second of the coupled equations. The coupled equations (4-2-87) and (4-2-88) can be expanded into partial waves and solved numerically provided not too many terms are retained.

As we have already discovered that the long range interaction is very important in our problem, we must ask the question how far such a truncated expansion will be able to represent this effect. In the sum, (4-2-61) from which the polarisability $\alpha_1$ is calculated, the 2p state of hydrogen accounts for 65.8% of the total. If all the discrete p states of hydrogen are included, just over 80% of the polarisability is obtained, leaving nearly 20%, which arises from the continuum p states. This suggests that after the 2p state of hydrogen is included in the expansion, adding a few further discrete p states will not effect a material improvement in the representation of the long range potential. It is not a practical proposition to allow for the continuum p-wave terms in the truncated expansion so that the method can never represent the long range potential very effectively. This can be seen in Table 4-2-1, where the phase-shifts from the approximation in which the 1s, 2p and 2s states of hydrogen are retained, are shown in column (c). Comparing these results with those of column (b), it is seen that the effective potential in this approximation is less attractive than that given by the polarised orbital approximation. Of course at this stage, we do not know whether the results in the polarised orbital method are accurate, and we shall see later that in fact the (1s–2s–2p) approximation is inadequate. The corresponding position in electron scattering is rather different, in that case exchange is the dominant effect, and this is allowed for by symmetrising the truncated eigenvalue expansion

$$\Psi^{\pm}(r, x) = \sum_{i=1}^{N} \{\phi_i(x) F_i^{\pm}(r) \pm \phi_i(r) F_i^{\pm}(x)\} \tag{4-2-89}$$

where the $+$ sign refers to the singlet spin state and the $-$ sign to the triplet. Even the first term in this expansion provides rather good phase shifts at low energies, and adding the 2p and 2s state of hydrogen produces a relatively small correction, except at points just below the $n=2$ threshold, where resonances associated with the discrete spectrum of $(QHQ)$ occur.

As the analogue of exchange in positron scattering is virtual positronium formation, it might be thought that a trial function which includes the positronium channel explicity would be effective. This can be achieved by constructing a trial function of the form

$$\Psi_t(r, x) = \sum_{j=0}^{N} \phi_j(r) F_j(x) + \sum_{i=0}^{M} \psi_i(R) H_i(y) \tag{4-2-90}$$

where the $\psi_i(R)$ are positronium wave functions and the functions of relative motion $H_i(y)$ vanish for large $y$. To obtain equations for the functions $F_j(x)$ and $H_i(y)$ the condition $\delta I[\Psi]=0$ is imposed, where as before $I=\int d\mathbf{x} \int d\mathbf{r}\, \Psi_t(H-E)\,\Psi_t$, under independent variations of the form

$$\begin{aligned} F_j(x) &\to F_j(x) + \delta F_j(x) \\ H_i(y) &\to H_i(y) + \delta H_i(y) \end{aligned} \tag{4-2-91}$$

where $\delta F_j(x) \to 0$, $x \to \infty$ and $\delta H_i(y) \to 0$ as $y \to \infty$. The resulting equations are more complicated than those of (4-2-87), being integro-differential rather than differential equations. These equations have the general form [54]

$$(\nabla_x^2 + k^2 - U(x)) F_0(x) = \sum_{j'=1}^{N} V_{0j'}(x) F_{j'}(x) + \sum_{i=0}^{M} \int dy\, K_{0i}(x, y) H_0(y)$$

$$(\nabla_x^2 - \lambda_j^2) F_j(x) = \sum_{j'=0}^{N} V_{jj'}(x) F_{j'}(x) + \sum_{i=0}^{M} \int dy\, K_{ji}(x, y) H_i(y)$$
$$j = 1, 2, ..., N \qquad (4\text{-}2\text{-}92)$$

$$(\nabla_y^2 - \eta_i^2) H_i(y) = \sum_{i'=0}^{M} \bar{V}_{ii'}(y) H_{i'}(y) + \sum_{j=0}^{N} \int dx\, \bar{K}_{ij}(y, x) F_j(x)$$
$$i = 1, 2, ..., M$$

where $V_{jj'}$ and $\lambda_j^2$ were defined in equations (4-2-88) and $\eta_i^2 = 2k^2 + 4\varepsilon_0 - 4\bar{\varepsilon}_i$, where $\bar{\varepsilon}_i$ is the energy of the $i$th state of positronium. The potential matrix $\bar{V}_{ii'}(y)$ is defined in a similar way to $V_{jj'}(x)$,

$$\bar{V}_{ii'}(y) = 4 \int d\mathbf{R}\, \psi_i^*(\mathbf{R}) \psi_{i'}(\mathbf{R}) \left\{ \frac{1}{|y + \tfrac{1}{2}R|} - \frac{1}{|y - \tfrac{1}{2}R|} \right\}. \qquad (4\text{-}2\text{-}93)$$

Because the centre of mass of the positronium atom coincides with the centre of charge, the diagonal elements $\bar{V}_{ii}(y)$ vanish identically, as is obvious from the antisymmetry of the integrand in (4-2-93). The kernels $K_{ji}$ and $\bar{K}_{ij}$ are defined as

$$K_{ij}(x, y) = 2\phi_j^*(r)(H - E)\psi_i(\mathbf{R})$$
$$\bar{K}_{ji}(y, x) = 32\psi_i^*(\mathbf{R})(H - E)\phi_j(r). \qquad (4\text{-}2\text{-}94)$$

The kernels represent an effective short range interaction, and vanish exponentially for large $x$ or large $y$.

The coupled equations can again be expanded in partial waves, and coupled equations obtained for the radial functions. This reduction, which is comparitively complicated, has been discussed by Smith [54].

Parametric trial functions and trial functions containing unknown functions, for which equations are sought, do not include all possibilities. For example a number of parametric terms can be added to a trial function of the truncated eigenfunction expansion type. Consider a trial function of the type

$$\Psi_t(r, x) = \sum_{j=0}^{N} \phi_j(r) F_j(x) + \Phi(C_i, r, x) \qquad (4\text{-}2\text{-}95)$$

where $F_0(x)$ has the boundary condition (4-2-26) and $F_j(x) \to 0$ as $x \to \infty$, $j \neq 0$, and where $\Phi(C_i, r, x)$ depends on a number of parameters $C_i$ and vanishes both for large $r$ and for large $x$. The condition that $I[\Psi]$ be stationary under the variations $F_j \to F_j + \delta F_j$ and $C_i \to C_i + \delta C_i$ then leads to the mixed differential-algebraic equations

$$\left( \frac{\partial^2}{\partial x^2} + k_0^2 - U(x) \right) F_0(x)$$
$$= \sum_{j'=1}^{N} V_{0j'}(x) F_{j'}(x) + 2 \int dr \, \phi_0^*(r) [H - E] \Phi(C_i, r, x)$$

$$(\nabla_x^2 - \lambda_j^2) F_j(x)$$
$$= \sum_{j'=0}^{N} V_{jj'}(x) F_{j'}(x) + 2 \int dr \, \phi_j^*(r) [H - E] \Phi(C_i, r, x), \qquad (4\text{-}2\text{-}96)$$
$$j = 1, 2, \ldots, N$$

$$\frac{\partial}{\partial C_i} \left[ \int dr \int dx \, \Phi^*(C_i, r, x) [H - E] \right.$$
$$\left. \times \left\{ \Phi(C_i, r, x) + \sum_{j=0}^{N} \phi_j(r) F_j(x) \right\} \right] = 0, \qquad i = 1, 2, \ldots, M.$$

If $\Phi$ depends linearily on the parameters $C_i$, the second set of equations can be simplified. Suppose we look for a real standing wave solution to the equations and take

$$\Phi = \sum_{i=1}^{M} C_i \chi_i(r, x); \qquad (4\text{-}2\text{-}97)$$

further we can choose the functions $\chi_i$ to be orthogonal to the first sum in the trial function (4-2-95)

$$\int \phi_j^*(r) \chi_i(r, x) \, dr = 0, \qquad i = 1, 2, \ldots, M; \quad j = 1, 2, \ldots, N. \qquad (4\text{-}2\text{-}98)$$

Then the last set of equations (4-2-96), using a matrix notation, becomes $\sum_j (H - EN) C_j = L$ where $H$, $N$ are matrices and $L$ is a column vector, with components

$$H_{ij} = \int dr \int dx \, \chi_i(r, x) H \chi_j(r, x)$$

$$N_{ij} = \int dr \int dx \, \chi_i(r, x) \chi_j(r, x) \qquad (4\text{-}2\text{-}99)$$

$$L_i = \int dr \int dx \left[ \chi(r, x) H \left\{ \sum_{j=0}^{N} \phi_j(r) F_j(x) \right\} \right].$$

These simultaneous linear equations for $C_j$ can be solved and the solution substituted into the first set of equations (4-2-96), we have

$$C = (H - EN)^{-1} L \qquad (4\text{-}2\text{-}100)$$

and

$$(\nabla^2 + k^2 - U(x)) F_0(x)$$
$$= \sum_{j'=1}^{N} \left\{ V_{0j'}(x) + \int dx' \, M_{0j'}(x, x') F_{j'}(x) \right\}$$

$$(\nabla^2 - \lambda_j^2) F_j(x) \qquad (4\text{-}2\text{-}101)$$
$$= \sum_{j'=0}^{N} \left\{ V_{jj'}(x) + \int dx' \, M_{jj'}(x, x') F_{j'}(x') \right\}$$

where

$$M_{jj'}(x, x') = \sum_{i,k} 2 \int dr \int dr' \, \phi_j^*(r) H \chi_i(r, x)$$
$$\times \left( \frac{1}{H - EN} \right)_{ik} \chi_k(r', x') H \phi_j(r'). \qquad (4\text{-}2\text{-}102)$$

These equations have been studied by Gailitis [55] and by Burke and Taylor [56]. The potential matrix $M_{jj'}$ has the form of a sum of separable potentials, so that further simplifications are possible.

*Minimum principles.* With the exception of the purely algebraic type of trial function, such as (4-2-68), all the trial functions we have considered can be put in the form

$$\Psi_t(r, x) = P\Psi_t(r, x) + Q\Psi_t(r, x) \qquad (4\text{-}2\text{-}103)$$

where as before $P$ projects out the open channel part of the wave function, and if $P$ is given by (4-2-13) we have

$$\Psi_t = \phi_0(r) F_0(x) + Q\Psi_t(r, x) \qquad (4\text{-}2\text{-}104)$$

where $Q\Psi_t$ can be represented either as a truncated expansion, or algebraically, or as a mixture of the two. We now recall the exact effective potential $W$ given by (4-2-24), which can be expressed in terms of the eigenfunctions $\Phi_n$ of $(QHQ)$ using eqs. (4-2-29). We noted that

$$W(x, x') \qquad (4\text{-}2\text{-}105)$$
$$= S_n(\varepsilon_n^Q - E)^{-1} \left[ \int dr \, \phi_0(r) H \Phi_n(x, r) \right] \left[ \int dr' \, \Phi_n^*(x', r') H \phi_0(r') \right]$$

if $E < \varepsilon_0^Q$, where $\varepsilon_0^Q$ is the lowest eigenvalue of $(QHQ)$ then $W$ was a negative definite operator. Further each term in the expansion (4-2-105) is itself a negative definite operator, so that if $W_N$ is an approximation to $W$ formed by truncating the expansion of $Q\Psi$ to $N$ terms, so that

$$Q\Psi = \sum_{n=1}^{N} a_n \Phi_n \qquad (4\text{-}2\text{-}106)$$

we must have

$$W_{N+1} < W_N. \qquad (4\text{-}2\text{-}107)$$

The monotonicity theorem for the phase-shift, then shows that if $\eta_l^N$ is the phase shift of order $l$ produced by the effective potential $[U(x) + W_N]$, then we must have

$$\delta_l^{N+1} > \delta_l^N. \qquad (4\text{-}2\text{-}108)$$

It follows that $\delta_l^N$ is a lower bound to the exact phase shift of order $l$. A special case is obtained by setting $Q\Psi = 0$ which leads to the static approximation. The phase shifts in this approximation (see column (a) to Table 4-2-1) must be lower bounds to the true phase shifts.

The closed channel subspace, or $Q$ space, is spanned by the functions $\Phi_n$, but equally if any other linearly independent set of functions is chosen to span $Q$ space, it again follows that at each stage of approximation a bound on the phase shifts is obtained. For example, the hydrogenic functions $\phi_j (j \neq 0)$ span $Q$ space, when $P$ is given by (4-2-13). It follows that the close coupling method provides, at each stage of approximation, a bound on the phase shifts. In the same way, the mixed algebraic trial function (4-2-95) and (4-2-97) in which the constants $C_i$ together with the $F_j$ are determined by the Kohn variation method, has the property that as a number of linearly independent terms $\chi_i(\mathbf{r}, \mathbf{x})$ are increased, the portion of $Q$ space spanned by the trial function will be systematically enlarged, and in this case also a bound is obtained on the phase shifts and the best trial function is the one that produces the largest phase shift.

It may now be asked whether a trial function of the form (4-2-90) also provides a bound. In this case the second sum in (4-2-90), that describes the proton and positronium arrangement of the system, is not orthogonal to $\phi_0(r) F_0(x)$ and therefore does not entirely belong to $Q$ space, at least with our choice of $P$. Burke and Taylor [56] have shown that nevertheless a bound is still obtained. This is because the only important property of $P$ is that it should have the asymptotic property (4-2-11) and another $P$ can be found with the same asymptotic property, but such that the positronium terms all belong to $Q$ space. In fact Hahn [57] has shown how to construct projection operators $P$, $Q$ of this kind explicitly. The trick is to consider the Schrödinger equation as a two component matrix equation,

$$\begin{pmatrix} H-E & H-E \\ H-E & H-E \end{pmatrix} \begin{pmatrix} \Psi_1 \\ \Psi_2 \end{pmatrix} = 0 \qquad (4\text{-}2\text{-}109)$$

where $\Psi_1$ is defined so that for large $x$, it has the form of a sum of terms of the type $\phi_j(r) F_j(x)$, while $\Psi_2$ has the asymptotic form, for large $y$, of a sum of products like $\psi_i(R) H_i(y)$.

A matrix projection operator

$$P = \begin{pmatrix} P_1 & 0 \\ 0 & 0 \end{pmatrix} \qquad (4\text{-}2\text{-}110)$$

is introduced where

$$P_1 = \phi_0(r) \int \phi_0^*(r') \, dr' \qquad (4\text{-}2\text{-}111)$$

and $Q$ is defined as

$$Q = 1 - P, \qquad QP = PQ = 0.$$

If we take trial functions

$$\begin{aligned} \Psi_{1_t} &= \sum_j \phi_j(r) F_j(x) \\ \Psi_{2_t} &= \sum_i \psi_i(R) H_i(y), \end{aligned} \qquad (4\text{-}2\text{-}112)$$

then

$$P\Psi_t = \begin{pmatrix} \phi_0(r) F_0(x) \\ 0 \end{pmatrix} \qquad (4\text{-}2\text{-}113)$$

$$Q\Psi_t = \begin{pmatrix} \sum_{j \neq 0} \phi_j(r) F_j(x) \\ \sum_i \psi(R) H_i(y) \end{pmatrix} \qquad (4\text{-}2\text{-}114)$$

and it is clear that $Q\Psi_t$ is orthogonal to $P\Psi_t$. The Kohn variational method applied to the matrix form of the equations requires that

$$I = \int dx \int dr (\Psi^\dagger, (H-E) \Psi) \qquad (4\text{-}2\text{-}115)$$

be stationary and with trial functions of the form (4-2-112) we know that the conditions for a bound are satisfied. The equations for $F$ and $H$ found in this way are identical with equations (4-2-92) obtained from the trial function

(4-2-90), establishing that a bound is obtained, even though the two sums in (4-2-90) are not orthogonal in the usual sense.

It is impossible to exaggerate the importance of bound principles to scattering theory, because they offer at the same time a method of approximation and a method of choice between two trial functions. We know that the function producing the greater phase shift is superior.* A further discussion of minimum principles can be found in the papers of Hahn et al. [58]. Progress has also been made in finding methods that give an upper [59] rather than lower bound to the phase shifts, but as a rule these methods are difficult to apply in such a way as to give useful upper bounds and they have received no application in the theory of positron scattering.

One restriction on the minimum principle, as we have described it, is that it applies for energies below the lowest eigenvalue of $(QHQ)$. This is not as severe a restriction as it appears, because it is a matter of experience that the discrete spectrum of $(QHQ)$, if it exists, does not extend far below the threshold for excitation. If the energy is increased so that $E$ approaches the lowest eigenvalue, $\varepsilon_0^Q$, the potential $W$ is singular at that point. It will be shown below, that if this singularity occurs in the partial wave $l$, and if $\delta_l$ is normalised to lie in the interval $0 \leqslant \delta_l \leqslant \pi$, then $\delta_l$ rises through $\frac{1}{2}\pi$ near $E \approx \varepsilon_0^Q$, giving rise to a resonance in the cross-section. When $E$ is greater than $\varepsilon_0^Q$, but less than $\varepsilon_1^Q$, the phase shift will be in the region $\pi \leqslant \delta_l \leqslant 2\pi$ and will again vary monotonically with the non-resonant part of $W$. In general we can expect the bound properties of the Kohn variational method to hold provided that the trial function is sufficiently flexible to account for the eigenvalues of $(QHQ)$, at least approximately. (See Hahn et al. [58].)

In the case of electron scattering by atoms, a discrete spectrum of the closed channel Hamiltonian exists and is responsible for a series of resonances in electron–atom scattering that occur, not only below the lowest excitation threshold, but beneath higher thresholds as well. Physically, resonant scattering can be pictured as being due to the formation of a metastable, or autoionising, state composed of the projectile and target, which has a life time against decay much greater than the natural collision time. Mittleman [60] has shown that such resonances exist within an electronvolt or so of excitation thresholds in positron–hydrogen scattering, and we shall discuss the possibility of resonances below the positronium formation threshold in a latter section.

To make these remarks more definite, consider a case in which $E$ is close

---

* As presented here, the bound is strictly on $\tan \eta_l$ rather than $\eta_l$, and holds only if $\tan \eta_l$ is a continuous function of the effective potential.

to the lowest eigenvalue of $(QHQ)$, and let us assume that the associated eigenfunction $\Phi_0$ is a state of zero angular momentum. The Schrödinger equation for the zero angular momentum open channel function $P\Psi^0$ can then be written (following O'Malley and Geltman [61]) as

$$(H' - E) P\Psi^0 = -\left[\frac{|PH|\Phi_0\rangle \langle\Phi_0|H|}{(E - \varepsilon_0^Q)}\right] P\Psi^0 \qquad (4\text{-}2\text{-}116)$$

with

$$H' = PHP + \sum_{n \neq 0}\left[\frac{|PH|\Phi_n\rangle \langle\Phi_n| HP|}{E - \varepsilon_n^Q}\right]. \qquad (4\text{-}2\text{-}117)$$

The effective Hamiltonian $H'$ is a slowly varying function of energy near $E \approx \varepsilon_0^Q$, and the term in $W$, containing $\Phi_0$, which is a rapidly varying function of energy, has been displayed explicitly on the right-hand side of (4-2-116). Because the terms on the right-hand side of (4-2-116) are separable, it is possible to obtain a solution in terms of a function $\chi$ which satisfies

$$(H' - E)\chi(r, x) = 0 \qquad (4\text{-}2\text{-}118)$$

with the boundary condition

$$\chi(r, x) \underset{x \to \infty}{\sim} x^{-1}\phi_0(r)\sqrt{\frac{2\pi}{k}} \sin(kx + \delta_0). \qquad (4\text{-}2\text{-}119)$$

From (4-2-116) we have immediately that

$$P\Psi^0 = \chi - \left(\frac{1}{H' - E}\right)\left[\frac{|PH|\Phi_0\rangle \langle\Phi_0| HP|}{E - \varepsilon_0^Q}\right] P\Psi^0. \qquad (4\text{-}2\text{-}120)$$

The Green's function $(H' - E)^{-1}$ has the asymptotic form in configuration space

$$G_1(x, r; x', r') \equiv \langle x, r|(H' - E)^{-1}|x', r'\rangle$$

$$\underset{x \to \infty}{\sim} \sqrt{\frac{2\pi}{k}}\,\phi_0(r)\,x^{-1} \cos(kx + \delta_0)\,\chi(x', r') \qquad (4\text{-}2\text{-}121)$$

which can be verified by expanding $(H' - E)^{-1}$ in terms of the eigenfunctions of $H'$. Taking the scalar product of (4-2-120) from the left with $(\Phi_0|H|$ it is seen that

$$(\Phi_0|H| P\Psi^0) = (\Phi_0|H|\chi) - (E - \varepsilon_0^Q)^{-1} (\Phi_0|HPG_1PH|\Phi_0)(\Phi_0|H| P\Psi^0). \qquad (4\text{-}2\text{-}122)$$

This equation can be solved for $(\Phi_0|H|P\Psi^0)$ and this enables $P\Psi^0$ to be expressed as (from (4-2-120))

$$P\Psi^0 = \chi + (E - \varepsilon_0^Q - \Delta)^{-1}(\Phi_0|H|P\chi)\, G_1 P H \Phi_0 \qquad (4\text{-}2\text{-}123)$$

where $\Delta$ is defined as

$$\Delta = (\Phi_0|HPG_1PH|\Phi_0). \qquad (4\text{-}2\text{-}124)$$

The asymptotic form of $P\Psi^0$ can now be determined from (4-2-119) and (4-2-121)

$$P\Psi^0 \sim x^{-1}\phi_0(r)\sqrt{\frac{2\pi}{k}}\,[\sin(kx+\delta_0) + \cos(kx+\delta_0)\tan\eta_0] \qquad (4\text{-}2\text{-}125)$$

where

$$\tan\eta_0 = -\frac{\tfrac{1}{2}\Gamma}{E - \varepsilon_0^Q - \Delta} \qquad (4\text{-}2\text{-}126)$$

and

$$\Gamma = 2\pi |(\chi|H|\Phi_0)|^2.$$

The phase shift $\eta_0$ passes through $\tfrac{1}{2}\pi$ as $E$ passes through the resonance energy $\varepsilon_0^Q + \Delta$ where $\Delta$ has the significance of a level shift. The total phase shift is $\eta_0 + \delta_0$, so that the partial cross-section for $l=0$ is

$$q_0 = \frac{4\pi}{k^2}\sin^2(\delta_0 + \eta_0). \qquad (4\text{-}2\text{-}127)$$

If it is assumed that the level shifts $\Delta$ are small, then a calculation of the eigenvalues of $QHQ$ will locate the resonance energies. This eigenvalue problem can be solved by the Rayleigh-Ritz variational method. This formulation can be extended to cover any angular momentum state and also to the discussion of resonances when the energy is such that several inelastic channels are open. All that is necessary is to redefine the projection operator $P$ so that it projects from the total Hilbert space, a sub-space containing all the open channels. The minimum principle can then be shown [55, 58] to apply to the reaction matrix, $K$, rather then to the phase shifts (the diagonal elements of $K$). When different arrangements of the system are possible in the asymptotic region, the construction of the necessary projection operator $P$ can be obtained by working with a multicomponent wave-function as in the method of Hahn discussed earlier.

The particular case discussed by Mittleman [60] was the resonance structure below the $n=2$ excitation threshold of hydrogen. The two channels $e^+ + H(1s)$ and $Ps(1s) + H^+$ are open in this region, and a projection operator due to Mittleman and Chen [62] was employed. The resonances which exist immediately below $n=2$ threshold are similar to those in the corresponding electron–hydrogen atom scattering problem. They occur because the $n=2$ threshold is degenerate and this changes the asymptotic form of the effective potential in the closed channel Hamiltonian from $\alpha/x^4$ to $\beta/x^2$. Attractive potentials of a certain strength behaving like $\beta/x^2$ at large $x$ give rise to an infinite series of bound states [20] which in this case are an infinite series of eigenvalues of $(QHQ)$, converging onto the threshold.

## C. Applications of the Variational Method

### 1. *Positron–hydrogen scattering*

Having surveyed the two main lines of attack on the problem of elastic scattering of positrons by atoms, the polarised orbital and the variational methods, the application of these methods will be discussed in more detail. The first steps in the application of the Kohn variational method to positron scattering by hydrogen atoms were taken by Massey and Moussa [28] in their pioneering paper on positron atom collisions in 1958. It has already been described how, using a phenomenological polarisation potential depending on a parameter, Massey and Moussa were able to show that the influence of polarisation might be sufficient to reduce the cross-sections for positron scattering by the rare gases from the values calculated in the static approximation to the experimental values. To explore the influence of polarisation quantitatively, Massey and Moussa considered positron–hydrogen scattering. They found that the s-wave phase shifts in the static approximation could be accurately reproduced by the Kohn variational method by using a trial function

$$\Psi_t(r, x) = \phi_0(r) F^t(x) \tag{4-2-128}$$

with

$$F^t(x) = x^{-1}[\sin(kx) + (D + C_1 \exp(-x))\cos(kx)]$$

where $D = \tan \eta_0$, and $C_1$ is a variational parameter. Polarisation effects can be included by making the trial function depend on $R = |r - x|$, where $R$ is the distance between the electron and positron. Accordingly Massey and Moussa added to the form (4-2-128) a term

$$G(x, r) = \phi_0(r) x^{-1} C_2 R \exp(-x) \cos(kx). \tag{4-2-129}$$

The phase shifts calculated with this trial function were quite close to those of the static approximation, for example at $k=0.2$, $\delta_0 = -0.098$ compared with the static value $\delta_0 = -0.114$. This result left the basic question in doubt, as it could be explained by supposing that after all polarisation was not important in this system, or alternatively that the trial function did not adequately represent the effect. Further work by Moussa [45] with a different form of trial function also failed to obtain a positive effect.

This doubt was removed by the work of Spruch and Rosenberg [52] in 1960. Again the Kohn variational method was used, but this time at zero energy, in which case a bound is obtained on the scattering length so allowing a choice to be made between the prediction obtained with different trial functions. The boundary condition to be satisfied at zero energy, can be obtained from (4-2-128) by first dividing by $k$ and then taking the limit $k \to 0$

$$\Psi^0(r, x) \underset{x \to \infty}{\sim} \left(1 + \frac{A}{x}\right) \phi_0(r) \tag{4-2-130}$$

where $A$ is the scattering length defined by (4-2-47). The Massey and Moussa trial function in the limit $k \to 0$ becomes

$$\lim_{k \to 0} \frac{1}{k} \Psi_t^0(r, x) = \phi_0(r) \left[1 + \frac{1}{x}(A + C_1' e^{-x}) + C_2' |r - x| x^{-1} e^{-x}\right] \tag{4-2-131}$$

and gives rise to a scattering length [12] $A = -0.487$ which may be compared with the static value $A_s = -0.582$.

The trial function investigated by Spruch and Rosenberg was of the form

$$\Psi_t(r, x) = \phi_0(r)\left[1 + \frac{A}{x}(1 - e^{-x})\right] + C_1 e^{-qx} e^{-sr} + C_2 e^{-tR} e^{-vr}, \tag{4-2-132}$$

Several sets of the non-linear constants $q$, $s$, $t$ and $v$ were investigated, the three linear parameters $A$, $C_1$, $C_2$ being determined variationally. The last term in the trial function represents polarisation, as it depends explicitly on $R$. With a value of $t$ near $\frac{1}{2}$, the factor $\exp(-tR)$ resembles the positronium wave-function and the whole term may be expected to describe virtual positronium formation. While this is not the best function for describing long range polarisation effects, because the polarised hydrogen wave-function in the adiabatic approximation is for $x \gg r$ of the form

$$\{\phi_0(r) + \phi_{\text{Pol}}^0(\boldsymbol{x}, \boldsymbol{r})\} = \frac{1}{\sqrt{\pi}} \left\{1 - \frac{1}{x^2}(r + \tfrac{1}{2}r^2)\cos\theta\right\} e^{-r}$$

some of the long range polarisation is included, because the term in $\exp(-tR - vr)$ is not orthogonal to the p states of hydrogen. Although only three parameters are varied, the same number as in the Massey and Moussa function, the results are quite different. The best set of parameters, $q$, $s$, $t$, $v$, are those which give the greatest value of $A$ and the calculated value of $A$ is a lower bound. With this best set it was found that $A = +1.397$, conclusively proving the importance of distortion of the hydrogen atom. Using a similar form of trial function but at non-zero energies, Spruch and Rosenberg went on to show that at $k = 0.2$, $\delta_0 = 0.156$, showing that even at non-zero energies, the attraction due to distortion was sufficient to change the sign of the phase shift from that predicted by the static approximation.

A further variational calculation of the scattering length, by Allison et al. [69] a year later, confirmed that the positron experiences a net attraction at low energies. Their trial function is designed to estimate the effect of the long range dipole polarisation potential and has the correct form in the asymptotic region. It is (for zero energy scattering)

$$\Psi_t^0(\boldsymbol{r}, \boldsymbol{x}) = \phi_0(r)\{1 + C_1 f(\boldsymbol{r}, \boldsymbol{x})\} F_t(x) \qquad (4\text{-}2\text{-}133)$$

where

$$f(\boldsymbol{r}, \boldsymbol{x}) = x^{-2}\{1 - e^{-\delta r}\}^3 (\tfrac{1}{2}r^2 + r)(\hat{\boldsymbol{r}} \cdot \hat{\boldsymbol{x}})$$

and

$$F_t(x) = x^{-1}\{(\alpha + \beta e^{-r})(1 - e^{-r}) - x\}.$$

The value of the scattering length obtained is a lower bound. It is $A = 0.78$. A comparison of this result with that of Spruch and Rosenberg illustrates that the dipole polarisation potential does not fully account for the effective attraction and that virtual positronium configurations, which lead to a short range attraction, are also important. As already emphasised, the two effects overlap to some extent and the separation is a qualitative one.

From about 1960 onwards, problems of much greater numerical complexity could be tackled because of the availability of fast electronic computers. Schwartz [63] took advantage of this, by using the Kohn variational method in conjunction with an elaborate trial function containing many terms. The form of trial function used was (for the zero angular momentum state)

$$\Psi_t^0(r, x) = \phi_0(r) x^{-1} [\sin(kx) + D(1 - e^{-sx}) \cos(kx)]$$
$$+ \sum_{l, m, n} C_{l, m, n} e^{-s(r+x)} R^l r^m x^n. \quad (4\text{-}2\text{-}134)$$

The first term is an approximation to $P\Psi$ and the second term to $Q\Psi$. The form of the second term is the same as that used successfully in the discussion of the binding energy of helium by Hylleraas. The calculations were carried out by determining the linear parameters $C_{l,m,n}$ by the Kohn variational method, for different values of the non-linear scale parameter $s$. This kind of wave-function for $k \neq 0$ does not give a bound on the phase shift, but the convergence of the procedure can be examined empirically as the number of terms in the trial function is increased, and results were obtained with up to fifty terms. At zero energy, strict bounds could be obtained on the scattering length. In this case convergence could be improved by adding terms, which allow explicitly for the dipole polarisation. The results are included in Table 4-2-1. It is probable that these results are accurate and may be used as a standard to test approximations designed for use in more complicated problems.

A corresponding trial function can be devised for the state with total angular momentum $l=1$, and the phase shifts [64] $\delta_1(k)$ are included in Table 4-2-1. Again it is probable that these results are very accurate and serve as a standard against which other approximations can be tested.

*Bounds on phase shifts*. The trial functions so far described, do not provide bounds on the phase shifts, although a bound is provided at zero energy in the scattering length. To provide a bound on the phase shift (more properly on $\tan \delta_l$) the trial function must be of the form $P\Psi_t + Q\Psi_t$.

Among such functions are the truncated eigenfunction expansions leading to the close coupling equations discussed in the last section. The results for the expansion (4-2-84) in the 1s–2s–2p approximation [65] have already been discussed. The phase shifts, included in Table 4-2-1 are lower bounds, but by comparison with the "exact" results of Schwartz are quite inacurate. McEchran and Fraser [66] have extended this approximation by including the 1s–2s–2p–3s–3p–3d states of hydrogen in the expansion. The results are again lower bounds and are shown in Table 4-2-1 column (d). As required by the minimum principle the phase shifts in the six state approximation are greater than those of the three state approximation, but the difference between the two sets of phase shifts is small, and both sets are far from the exact values. This implies that little is to be gained by adding a further limited number of discrete hydrogenic states. One way of improving the situation is to introduce terms in which the positronium channel appears explicitly. At present, results are available only for the simplest possible case in which the trial function is

$$\Psi_t(r, x) = \phi_0(r) F_0(x) + \psi_0(R) H_0(y). \tag{4-2-135}$$

The results [44] also shown in Table 4-2-1 column (e), are very instructive. For small $k$ the phase shifts (which are again bounds) are more positive than those given by the six state hydrogenic expansion. This clearly indicates the importance of the short range attraction due to positronium formation. The effective attraction is however not nearly large enough, and the overall potential at low energies is still an effective repulsion. Recollecting that the trial function (4-2-132) used by Spruch and Rosenberg, which did lead to an overall attraction at low energies, had a term representing the (positronium + proton) configuration, we see that states beyond the positronium ground state are important. In the Spruch and Rosenberg trial function the parameter $t$, was rather less than 0.5, the value it would have been, if the ground state positronium term alone was important. Clearly a satisfactory close-coupling wave-function, should include the 2s and 2p states of hydrogen (the 2p state giving a sizable fraction of the long range attraction) as well as the 1s, 2s and 2p states of positronium. Such a calculation has not been completed [67].

Fundamentally, trial functions of the mixed type, in which $P\Psi_t$ contains the projection onto all the open, and perhaps some of the closed channels, while $Q\Psi_t$ is represented algebraically, are superior to the purely algebraic and to the pure truncated eigenfunction expansion functions. By using a sufficiently flexible form for $Q\Psi_t$ the effect of the continuum states, which have to be omitted from the eigenfunction expansion for practical reasons, can be determined, and at the same time the bound properties of the phase shifts are preserved. In the important work of Spruch and his collaborators [68], rather than solve the simultaneous integro-differential and algebraic equation (4-2-96) in the manner of Gailitis, the following procedure was adopted. (We shall take the case of s-wave elastic scattering of positrons by hydrogen, and refer the reader to the original papers for the extensions to higher $l$ and to inelastic scattering.)

The exact wave-function $\Psi^0$ can be written as

$$\Psi^0(r, x) = P\tilde{\Psi}^0 + P\chi + Q\Psi^0 \tag{4-2-136}$$

where $P\tilde{\Psi}^0$ is defined to be the solution of the equation

$$P(H - E) P\tilde{\Psi}^0 = 0. \tag{4-2-137}$$

If $P$ is represented as before by eq. (4-2-13) the boundary condition imposed on $P\tilde{\Psi}^0$ may be taken as

$$P\tilde{\Psi} = \phi_0(r) x^{-1} \tilde{f}_0(x)$$
$$\tilde{f}_0(x) \sim \sin(kx) + D \cos(kx). \tag{4-2-138}$$

With this choice, $\tilde{f}_0(x)$ is the radial function in the static approximation. Without approximation the coupled equations (4-2-17) and (4-2-18) become

$$P(H - E) P\chi = - PHQ\Psi^0$$
$$Q(H - E) Q\Psi^0 = - QHP\tilde{\Psi}^0 - QHP\chi. \tag{4-2-139}$$

The first of these equations has the solution

$$P\chi = - G^P(E) PHQ\Psi^0 \tag{4-2-140}$$

where $G^P(E) = [P(H-E)P]^{-1}$. This important point here is that $G^P$ can be constructed explicitly, in fact it is of the form

$$G^P(r, x; r', x') = \phi_0(r) \phi_0(r') g^P(x, x') \tag{4-2-141}$$

where $g^P(x, x')$ is the one-body Green's function associated with scattering by the static potential $U(x)$. Substituting into the second of equations (4-2-139), $Q\Psi^0$ is seen to satisfy

$$(\mathcal{H} - E) Q\Psi^0 = - QHP\tilde{\Psi}^0 \tag{4-2-142}$$

where

$$\mathcal{H} = Q(H - HG^P H) Q. \tag{4-2-143}$$

Writing the exact wave-function as

$$\Psi^0 = P\tilde{\Psi}^0 - G^P(E) PHQ\Psi_t^0 + Q\Psi_t^0 + \Omega \tag{4-2-144}$$

which defines the error in the wave-function $\Omega$ and proceeding as in the proof of the identity (4-2-79), it is possible to show that

$$2\pi k \tan \eta_0 = 2\pi k \tan \tilde{\eta}_0 + 2(Q\Psi^0 |H| P\tilde{\Psi}^0)$$
$$+ (Q\Psi_t^0 |(\mathcal{H} - E)| Q\Psi_t^0) - (Q\Omega |\mathcal{H} - E| Q\Omega) \tag{4-2-145}$$

where $\tilde{\eta}_0$ is the phase shift in the static approximation. Provided that $E$ is less than the lowest eigenvalue of $Q\mathcal{H}Q$, the last term in (4-2-145) is negative, so that a variational bound is obtained for $\tan \eta_0$,

$$2\pi k \tan \eta_0 \geqslant 2\pi k \tan \tilde{\eta}_0 + 2(Q_0 \Psi_t^0 | H | P \tilde{\Psi}^0) + (Q \Psi_t^0 | \mathcal{H} - E | Q \Psi_t^0).$$
(4-2-146)

The best value of $\eta_0$ is obtained by maximising the right-hand side of (4-2-146), with respect to the parameters in $Q\Psi_t^0$. The explicit form of the trial function used by Hahn and Spruch in the case of s-wave scattering was

$$Q\Psi_t^0 = \sum_{i,j} C_{ij} \chi_{ij}(x) w_{ij}(x) (rx)^{-1} P_j(\hat{r} \cdot \hat{x})$$
(4-2-147)

with

$$\chi_{ij}(r) = r^{j+1} \exp(-b_{ij}r) - r\left(\frac{2}{1+b_{ij}}\right) e^{-r} \delta_{0j},$$

$$w_{ij}(x) = x^{j+1} \exp(-d_{ij}x).$$

A predetermined choice was made of the non-linear parameters $b_{ij}$, $d_{ij}$ and the $c_{ij}$ were determined variationally. The results, shown in Table 4-2-1 column (f), were obtained by taking all terms with $j \leqslant 5$ (s-waves). The figures in brackets represent an estimate obtained by extrapolating the results to higher values of $j$. The values representing rigorous bounds are slightly less positive than the values obtained by Schwartz which are in excellent agreement with the extrapolated values.

## 2. Positron–helium scattering

The Kohn variational method has been applied to the zero energy scattering of positrons by helium atoms (Allison et al. [69], Houston and Moisewitsch [70]). If the helium atom ground state wave function were known exactly, the scattering length obtained would be a lower bound on the exact value, but as the helium wave-function is only known approximately, it cannot be certain that the results represent a bound. The general form of trial function is

$$\Psi_t(r_1, r_2, x) = \phi_0(r_1, r_2) F(r_1, r_2, R_1, R_2, x)$$
(4-2-148)

where $r_1$ and $r_2$ are the position vectors of the two electrons, and $x$ is the position vector of the positron, and $R_1$, $R_2$ are defined as

$$R_1 = r_1 - x, \quad R_2 = r_2 - x.$$
(4-2-149)

The ground state wave-function of helium $\phi_0$ was taken to be

$$\phi_0(r_1, r_2) = U(r_1) U(r_2)$$
(4-2-150)

where

$$U(r) = \sqrt{\frac{N}{4\pi}} (e^{-\lambda r} + c e^{-2\lambda r}).$$

The constants in this function (introduced by Green et al. [71]) are found by the Rayleigh-Ritz variational method to have the values $\lambda = 1.4558$, $c = 0.6$, $N = 2.9684$) and the function is a good approximation to the Hartree-Fock wave function. The chosen form of $F$ was

$$F = x^{-1}[A(1 - e^{-\delta x}) - x]$$
$$+ e^{-\alpha} \sum_{l,m,n} C_{l,m,n} \{e^{-\beta r_1} x^l r_1^m R_1^n + e^{-\beta r_2} x^l r_2^m R_2^n\}. \quad (4\text{-}2\text{-}151)$$

In form the trial function is similar to that used by Schwartz in scattering by hydrogen, and to the extent that the helium wave-function is satisfactory, it should produce accurate results. Houston and Moiseiwitsch consider all terms such that $l+m+n \leqslant 1$, 2 or 3 (that is 4, 10 or 20 terms) and in addition to varying the linear parameters, they were able to optimise the non-linear parameters $\alpha$, $\beta$ and $\delta$ for the four and ten term functions. For the twenty term function the same values of $\alpha$, $\beta$ and $\delta$ were used as for the ten term function. The best value of the scattering length was $A = 0.398$, and corresponds to an effective attraction between the positron and helium. As we shall see it is rather smaller than some of the values indicated by the polarised orbital approximation, but it is likely to be accurate.

It will be recalled that measurements exist of the annihilation rate of positrons in helium, and for thermal positrons Falk et al. [22] measured an annihilation rate which corresponded to an effective electron number $Z_{\text{eff}} = 3.9$. With the best 20 term variational function of Houston and Moiseiwitsch, $Z_{\text{eff}}$, was calculated by Houston [72]. The value obtained, $Z_{\text{eff}} = 4.61$, is in reasonable agreement with experiment.

It is also possible to use trial functions of the truncated eigenvalue expansion type to discuss positron–helium scattering. Kraidy and Fraser [73] (see also Kraidy [74]), give some calculated phase shifts over a range of energies in the two state approximation

$$\Psi_t(r_1, r_2, x) = \phi_0(r_1, r_2) F_0(x) + w(r_1) \Psi(R_2) H_0(y_2)$$
$$+ w(r_2) \Psi(R_1) H_0(y_1) \quad (4\text{-}2\text{-}152)$$

where $y_1 = \frac{1}{2}(r_1 + x)$, $y_2 = \frac{1}{2}(r_2 + x)$. As before $\phi_0$ denotes the helium ground state wave-function, and $\psi_0$ the positronium ground state wave-function. The last two terms represent the channel which contains positronium and a helium ion. The ground state wave function of the helium ion is $w(r)$ and $H_0$ is the wave-function of relative motion. The two electrons are in a singlet spin state, so that the trial function is symmetrical in the space coordinates $r_1$ and $r_2$. The variational equations again take the form of coupled integro-differential equations for the functions $F_0(x)$ and $H_0(y)$. The effective potentials depend on the form assumed for $\phi_0$, the helium wave-function, which in this work was taken to be the simple variational form

$$\phi(r_1, r_2) = N \exp(-\lambda r_1 - \lambda r_2). \qquad (4\text{-}2\text{-}153)$$

The phase shifts obtained for s and p waves are shown for a few values of $k$ below the positronium formation threshold ($k^2 = 1.22$, 38 eV) in Table 4-2-4 and the scattering lengths are shown in Table 4-2-5. The values of $Z_{\text{eff}}$ can be calculated from the trial function as a function of energy. These are shown in Table 4-2-6. It is seen for very low energy positrons, $Z_{\text{eff}} = 2.45$, which is some way below the experimental value and about half that found

TABLE 4-2-4

Positron–helium phase shifts at low energies in various approximations

| | Phase shifts in radians | | | | | | |
|---|---|---|---|---|---|---|---|
| $k$ | (a) | (b) | (c) | (d) | (e) | (f) | (g) |

s-waves ($l = 0$)

| $k$ | (a) | (b) | (c) | (d) | (e) | (f) | (g) |
|---|---|---|---|---|---|---|---|
| 0 | | | | | | | |
| 0.1 | −0.038 | 0.0193 | 0.077 | 0.1662 | 0.050 | 0.036 | 0.0069 |
| 0.2 | −0.076 | 0.0196 | 0.102 | 0.2279 | 0.072 | 0.047 | −0.0068 |
| 0.4 | −0.149 | −0.0153 | 0.021 | 0.1626 | 0.056 | 0.020 | −0.0676 |
| 0.6 | −0.218 | −0.0729 | −0.102 | 0.0483 | 0.002 | −0.039 | −0.1454 |
| 0.8 | −0.280 | −0.1367 | −0.208 | −0.0449 | −0.066 | −0.107 | −0.2241 |
| 1.0 | −0.335 | −0.1987 | −0.288 | −0.1125 | −0.134 | −0.176 | −0.2968 |

p-waves ($l = 1$)

| $k$ | (a) | (b) | (c) | (d) | (e) | (f) | (g) |
|---|---|---|---|---|---|---|---|
| 0.2 | −0.004 | 0.007 | 0.0155 | 0.0236 | – | – | 0.0084 |
| 0.4 | −0.003 | 0.0216 | 0.0575 | 0.0830 | – | – | 0.0219 |
| 0.6 | −0.0103 | 0.0335 | 0.0696 | 0.1108 | – | – | 0.0284 |
| 0.8 | −0.0215 | 0.0376 | 0.0523 | 0.1049 | – | – | 0.0238 |
| 1.0 | −0.0364 | 0.0333 | 0.0256 | 0.0855 | – | – | 0.0110 |

(a) Static approximation.
(b) Polarised orbital approximation.
(c) Virtual positronium formation.
(d) Virtual positronium formation (plus dipole polarisation).
(e) Polarised orbital with monopole suppression.
(f) Variational method (Drachman).
(g) Extended polarised orbital method.

TABLE 4-2-5

Scattering lengths for positron–helium collisions in various approximations

| Approximation | Scattering length† $A$ (a.u.) |
|---|---|
| Static | −0.420* |
| Polarised orbital (monopole suppression) | 0.659 |
| Polarised orbital (variational-Drachman) | 0.511* |
| Variational (Houston and Moiseiwitsch) | 0.398* |
| Polarised orbital (monopole + dipole + quadrupole components) | 0.575 |

\* These figures would be lower bounds if the helium ground state wave-function as-summed had been exact.
† $A = \lim_{k \to 0} \eta_0(k)/k$.

TABLE 4-2-6

$Z_{\text{eff}}(v)$ as a function of momentum for positron annihilation in helium

| $k$ | (a) | (b) | (c) | (d) | (e) |
|---|---|---|---|---|---|
| 0.0 | | | | 6.32 | 6.66 |
| 0.1 | 0.733 | 1.247 | 3.703 | | |
| 0.2 | 0.749 | 1.194 | 2.804 | 5.58 | 3.30 |
| 0.4 | 0.805 | 1.127 | 2.000 | 4.86 | 3.02 |
| 0.6 | 0.886 | 1.127 | 1.792 | 4.62 | 2.98 |
| 0.8 | 0.980 | 1.168 | 1.721 | 4.66 | 3.36 |
| 1.0 | 1.077 | 1.229 | 1.718 | 4.88 | 3.36 |

(a) Static approximation.
(b) Polarised orbital (dipole component only).
(c) Virtual positronium formation plus dipole polarisation.
(d) Polarised orbital with monopole suppression.
(e) Variational method (Drachman).

by Moiseiwitsch. This is to be expected as it is known from the case of hydrogen, that this two state approximation is not reliable. Improvements in the two state theory, which allow for polarisation will be considered in the next section.

D. DEVELOPMENTS OF THE POLARISED ORBITAL METHOD

The very elaborate variational calculations that have been applied to positron scattering by hydrogen and helium are hardly feasible for more complicated systems. Because of this, there has been a great deal of interest in the polarised orbital method, which has been shown to be quite effective in electron–atom collisions and which can be applied to the more complicated systems. In the case of positron scattering, the simple approximation in which the terms of longest range, the dipole term $w_1(x)$ is retained in the adiabatic potential can be seen from Table 4-2-1 not to be sufficiently accurate, as it considerably underestimates the effective attraction of the positron to the atom. In this section some improvements of the theory will be examined, some of which have the desirable feature of providing a variational bound on the phase shift.

### 1. *Positron scattering by hydrogen*

One of the defects in the simple polarised orbital method in which only the dipole contribution $w_1(z)$ (or some approximation to $w_1(x)$) is retained, is that it makes too little allowance for the short range attractive forces, which may be thought of as being due to virtual positronium formation. The two channel approximation based on the trial function (4-2-135), which allows for virtual positronium formation, leads to coupled equations for the wave-functions of relative motion $F_0(x)$ and $H_0(y)$, of the general form

$$(\nabla_x^2 + k^2 - U(x))F_0(x) = \int K(x, y) H_0(y) \, dy$$
$$(\nabla_y^2 - \lambda^2) H_0(y) = \int \bar{K}(y, x) F_0(x) \, dx. \qquad (4\text{-}2\text{-}154)$$

Cody et al. [44] modified the first of these equations representing the positron–hydrogen channel, by replacing the static interaction $U(x)$, by $\{U(x) + w_1(x)\}$, thus allowing for the long range polarisation effect. This procedure is not consistent with the variational method. The phase shifts obtained in this way, are greater than those of Schwartz, for example at $k = 0.2$, $\delta_0 = 0.293$, compared with the value of Schwartz $\delta_0 = 0.188$, and while such large values are not ruled out by calculations that give a lower bound, they are likely to be in error.

The polarisability of positronium is eight times that of hydrogen, so that the long range force in the positronium channel, represented by the second of equations (4-2-154), must also be important. If the appropriate adiabatic potential is added in this channel the phase shifts become even greater, showing that the net attraction is seriously overestimated [75]. We shall see in a moment how to overcome these defects by making a fully variational calculation.

Instead of attempting to include the short-range attractive forces by considering virtual positronium formation, the complete adiabatic potential $W^{\text{pol}}(x)$, defined by (4-2-54) can be used in a one-body equation. This potential contains contributions from in the all closed channels and must describe the short-range attractive forces in an approximate way. The consequences of this approach have been studied by Drachman [76], who has solved the radial Schrödinger equation for the partial waves $l=0$, 1 and 2 for scattering by the potential $U(x)+W^{pol}(x)$. The s-wave phase shift obtained is more positive than the accurate results of Schwartz, for example at $k=0.2$, $\delta_0=0.241$ compared with the accurate value of $\delta_0=0.188$. It follows that this simple treatment overestimates the effective attraction and some modification is required. The term in the multiple expansion of the adiabatic potential that dominates the short range part of the interaction, and is mainly responsible for the over-estimate of the effective attraction, is the large exponentially decreasing mono-pole term, which dominates the interior region of the potential. To correct for this overestimate, Drachman introduced a parameter, that suppresses the contribution of the mono-pole potential, which can be adjusted to bring the computed phase shifts in agreement with those of Schwartz. It was hoped that a similar procedure could be carried through for helium and other atoms for which elaborate variational calculations were not feasible.

The parameter $\alpha$ is introduced by writing the effective potential as

$$W(x) = W_A(x) + (\alpha - 1) w_0(x) \tag{4-2-155}$$

so that $\alpha=0$ corresponds to complete suppression and $\alpha=1$ to no suppression of the mono-pole potential $w_0(x)$. It was found that to obtain results in agreement with those of Schwartz nearly all the mono-pole potential must be suppressed, the optimum value of $\alpha$ being $\alpha=0.1$.

The higher order phase shifts do not depend strongly on the inner parts of the potential because the centrifugal barrier prevents the wave-function from penetrating into this region. This illustrated by the fact that the computed phase shifts for $l=1$ and $l=2$ do not vary strongly with $\alpha$ in the range $0 \leqslant \alpha \leqslant 1.0$. The computed phase shifts are smaller than those of Armstead [64] and of Kleinman et al. [68] even when $\alpha=1.0$ (by some 10% at $k=0.4$ for example), and the error must be attributed to the inadequacy of the higher terms in the multipole expansion. In particular, perturbation theory (which should be more accurate for $l>0$ and large $k$), suggests that virtual positronium formation is more effective in determining p waves, than s waves (Bransden [77]) and the adiabatic wave-function poorly represents this effect. Further work (Bransden and Jundi [43]) shows that the monopole, dipole and quadrupole terms in the adiabatic potential provide the major contribution to the phase shift. The difference between the phase shifts calculated with the complete adiabatic potential $W_A$ and those computed retaining only the (monopole + dipole + quadrupole) components is of the order of 15% of the total phase shift in the case of the s waves and 10% in the case of p waves.

The uncertainties in the polarised orbital method can be eliminated by using the adiabatic wave-function as a trial function in a fully variational calculation. There are several ways in which this can be done. One version of the minimum principle (Hahn [59]) can be stated as follows. If $Q\Psi_t^0$ is some approximation to the second of the coupled equations (4-2-18) and $P\Psi^0$ is the *exact* solution of the equation

$$P(H - E) P\Psi^0 = - PHQ\Psi_t^0 \qquad (4\text{-}2\text{-}156)$$

where $P\Psi^0$ has the boundary condition

$$P\Psi^0 \sim \phi_0(r) \, x^{-1} [\sin(kx) + \tan\bar{\eta}_0 \cos(kx)] \qquad (4\text{-}2\text{-}157)$$

then

$$\tan\eta_0 \geq \tan\bar{\eta}_0 - (2\pi k)^{-1} (Q\Psi_t | H - E | \Psi_t). \qquad (4\text{-}2\text{-}158)$$

If we take as our approximation to $Q\Psi_t^0$, the adiabatic wave-function

$$Q\Psi_t = \phi_0^{pol}(r, x) \, x^{-1} f_0(x) \qquad (4\text{-}2\text{-}159)$$

then equation (4-2-156) is identical with equation (4-2-53) defining the polarised orbital method, and the phase shift $\eta_0$ is that given by the polarised orbital approximation. Jundi (unpublished) has evaluated the correction to the polarised orbital phase shifts for the case in which only the dipole term is retained in $\phi_0^{pol}$. The terms, on the right of (4-2-158), envelve the matrix elements of the kinetic energy operator with respect to $\phi_0^{pol}$, thus providing a correction to the adiabatic assumption. As expected, the effect is to moderate the effective attraction. The correction to the dipole term is quite small, for example the phase shifts shown in column (b) of Table 4-2-1 should be modified by the addition of $\Delta\eta_0 = -0.0085$ at $k = 0.2$ and by $\Delta\eta_0 = -0.0044$ at $k = 0.6$. Of more interest would be the modification of the monopole attraction, which appears to be much too large. Such a calculation has not been attempted through equation (4-2-158), but Callaway et al. [78] and Drachman [79] have independently tackled this problem, using a slightly different approach.

The only condition placed on the projection operator $P$, is that $P\Psi$ must satisfy the boundary condition (4-2-11), and include the open channel. No matter what $P$ we select, we know from our prevous discussion, that the phase shifts obtained by setting the effective potential $W = 0$, and solving the equation

$$P(H - E) P\tilde{\Psi} = 0 \qquad (4\text{-}2\text{-}160)$$

satisfy a bound, provided the energy $E$ is less than the lowest eigenvalue of $QHQ$. To obtain a version of the adiabatic method, that satisfies the minimum principle, $P$ may be chosen to be the projector onto the distorted states of hydrogen $\{\phi_0(r) + \phi_0^{\text{pol}}(r, x)\}$ rather then onto the unperturbed state $\phi_0(r)$. We can write

$$P = M(x) \{\phi_0(r) + \phi_0^{\text{pol}}(r, x)\} \int dr' \, M(x) \{\phi_0(r') + \phi_0^{\text{pol}}(r', x)\}. \qquad (4\text{-}2\text{-}161)$$

The distorted wave-function $\{\phi_0(r) + \phi_0^{\text{pol}}(r, x)\}$ is not in general normalised to unity, so that a normalisation factor $M(x)$ is required that satisfies

$$M^2(x) \int dr \, |\phi_0(r) + \phi_0^{\text{pol}}(r, x)|^2 = 1, \quad \text{all } x. \qquad (4\text{-}2\text{-}162)$$

in order that $P^2 = P$. In the usual way a wave-function $F(x)$ is defined by

$$P\Psi(r, x) = \{\phi_0(r) + \phi_0^{\text{pol}}(r, x)\} F(x) \qquad (4\text{-}2\text{-}163)$$

(notice that the normalisation factor, $M(x)$, has been absorbed in the definition of $F(x)$).

If $\phi_0^{\text{pol}}$ is calculated to first order in the adiabatic approximation it is easy to reduce eq. (4-2-160), to the form (Drachman [79])

$$[1 + N(x)][\nabla_x^2 + k^2] F(x)$$
$$= \left\{ U(x) + W_A(x) + W_B(x) + W_C(x) - \frac{dN(x)}{dx} \cdot \frac{d}{dx} \right\} F(x) \qquad (4\text{-}2\text{-}164)$$

where $U(x)$ and $W_A(x)$ are the usual static and adiabatic potentials and

$$N(x) = \int \{\phi_0^{\text{pol}}(r, x)\}^2 \, dr$$

$$W_B(x) = \int \phi_0^{\text{pol}}(r, x) \, V(r, x) \, \phi_0^{\text{pol}}(r, x) \, dr \qquad (4\text{-}2\text{-}165)$$

$$W_C(x) = -\int \phi_0^{\text{pol}}(r, x) \, \nabla_x^2 \phi_0^{\text{pol}}(r, x) \, dr.$$

The most important modifications of the effective potential of the adiabatic method, $U(x) + W_A(x)$, are the addition of a further attraction $W_B(x)$ and a large short range repulsion $W_C(x)$. The additional potentials behave at large $x$ like

$$W_B(x) \sim -\frac{213}{2x^7}, \quad \frac{dN(x)}{dx} \sim \frac{43}{2x^5}, \quad N(x) \sim \frac{43}{8x^4}. \qquad (4\text{-}2\text{-}166)$$

The net effect of the additional potentials, and of the moderating factor $[1+W(x)]^{-1}$ is to lower the phase shifts from those given by the polarised orbital method using the full potential $\{W_A(x)+U(x)\}$. The resulting phase shifts represent lower bounds, but unfortunately the calculated phase shifts are much too small. This can be seen from Table 4-2-2 where the scattering lengths given by various forms of the polarised orbital method are collected together and compared with the value obtained by Schwartz. It is seen that $A=1.0$ compared with the accurate value of 2.10.

A modification of equation (4-2-164) has been introduced by Callaway et al. The repulsive potential $W_C(x)$ can be written as

$$W_C(x) = W_D(x) - \tfrac{1}{2}\nabla_x^2 N(x) \qquad (4\text{-}2\text{-}167)$$

where $W_D(x)$ defined by

$$W_D(x) = \int |\nabla_x \phi_0^{\text{pol}}(r,x)|^2 \, dr \qquad (4\text{-}2\text{-}168)$$

is termed a distortion potential. If all terms in the effective potentials of third or higher order are neglected, eq. (4-2-151) reduces to

$$[\nabla^2 + k^2]F(x)$$
$$= \left[ U(x) + W_A(x) + W_D(x) - \tfrac{1}{2}\nabla_x^2 N(x) - \frac{dN}{dx}\cdot\frac{d}{dx} \right] F(x). \qquad (4\text{-}2\text{-}169)$$

The terms in $N(x)$ can now be eliminated by changing the normalisation of $F(x)$. If $F(x)$ is replaced by $(1-\tfrac{1}{2}N(x))F(x)$ we arrive at the equation of Callaway et al. [78]

$$[\nabla^2 + k^2]F(x) = [U(x) + W_A(x) + W_D(x)]F(x). \qquad (4\text{-}2\text{-}170)$$

As, in this equation, only first or second order terms in the effective potentials have been retained, the phase shifts no longer satisfy the minimum principle. The distortion potential $W_D$ has the interesting property that it exactly cancels $W_A$ as $x \to 0$,

$$\lim_{x \to 0}[W_A(x) + W_B(x)] = 0. \qquad (4\text{-}2\text{-}171)$$

At large $x$, $W_D(x)$ behaves like $\beta/x^6$ and the sum $W_A + W_D$ is correct to order $1/x^6$ in the asymptotic region. In numerical calculations, Callaway et al. did not employ the complete potentials $W_D$, but only the sum of the $l=0$ and $l=1$ multipole components of this potential, but this is not expected to give rise to a significant error. The results of this 'extended' polarised orbital method are in rather good agreement with those of Schwartz. When assessing the extended polarised orbital method it is disturbing that the variationally based equation (4-2-164) to which (4-2-170) is an approximation provided a poor result for the scattering length but, on the other hand, it may be argued that it is preferable to work consistently to second order in constructing the effective potential.

In the polarised orbital method, the trial function

$$P\Psi_t = \phi_0(r) F(x), \quad Q\Psi_t = \phi_0^{\text{pol}}(r, x) F(x) \qquad (4\text{-}2\text{-}172)$$

is severely restricted by presence of the same function $F(x)$ in both $P\Psi_t$ and $Q\Psi_t$. Drachman [79] has considered the more flexible trial function

$$P\Psi_t = \phi_0(r) F(x), \quad Q\Psi_t = \phi_0^{\text{pol}}(r, x) G(x). \qquad (4\text{-}2\text{-}173)$$

By varying $F(x)$ and $G(x)$ independently, the coupled equations

$$[\nabla^2 + k^2 - U(x)] F(x) = W_A(x) G(x)$$

$$\left[ N(x)(\nabla^2 + k^2) + W_A(x) - W_B(x) - W_C(x) + \frac{dN}{dx}\frac{d}{dx} \right] G(x)$$
$$= W_A(x) F(x) \qquad (4\text{-}2\text{-}174)$$

are obtained, and in this case the resulting values of $\tan\delta_l$ satisfy a bound. The results of this procedure are encouraging. Column (h) of Table 4-2-1 shows that the phase shifts for $l=0$ are quite close to those of Schwartz, and like the extended polarised orbital method of Callaway et al. the method can be adapted for scattering by more complex atoms.

## 2. *Positron scattering by the rare gases*

The extension of the polarised orbital method to more complex systems is straight forward, if the ground state wave-functions are expressed in the form of sums of products of single electron wave-functions. The effect of the field of the incident particle, in the adiabatic approximation, on each of the single electron functions can be calculated to first order. For example if the helium ground state wave function is of the form

$$\phi(r_1, r_2) = U(r_1) U(r_2) \qquad (4\text{-}2\text{-}175)$$

the single electron orbital $U(r)$ can be replaced by the polarised orbital

$$U(r) \to U(r) + U^{\text{pol}}(\mathbf{r}_1, \mathbf{r}). \tag{4-2-176}$$

It has been the usual practice to only retain the first order correction to the wave-function which becomes

$$\phi(\mathbf{r}_1, \mathbf{r}_2) = U(r_1) U(r_2) + U^{\text{pol}}(r_1, x) U(r_2) + U^{\text{pol}}(r_2, x) U(r_1). \tag{4-2-177}$$

In the simple approximation (4-2-153) the one electron functions $U(r)$ are hydrogenic functions for an electron moving in a "screened" nuclear field of charge $Z = 1.6879$

$$U(r) = \sqrt{\frac{Z^3}{\pi}} e^{-Zr}. \tag{4-2-178}$$

In this case a reasonable approximation to $U^{\text{pol}}(r, x)$ is to use the perturbed hydrogen function, but again scaled to be consistent with a 'screened' nuclear charge $Z$. The amplitude of $U^{\text{pol}}$ and the effective charge $Z$ can be adjusted to obtain the known polarisability of the atom (in which case $Z = 1.5992$ (Drachman [79])), alternatively $Z$ can be set equal to the effective charge ($Z = 1.6789$) used in the unperturbed function. As in the case of hydrogen, the function $U^{\text{pol}}(r, x)$ can be expanded in a Legendre polynomial series, giving rise to a multipole expansion of the polarisation potential.

Massey et al. [80] have applied the polarised orbital method to positron scattering by helium, argon, neon and krypton, retaining only the dipole term in the effective potential known to given good results for electron scattering was used, but for helium and neon a procedure similar to that outlined in the previous paragraph, was used, starting from Hartree-Fock functions for the unperturbed ground states. The simplifications of setting the perturbed function equal to zero when $x < r$, due to Temkin and Lampkin [40], was used. This will not alter the asymptotic form of the polarisation potential, which is proportional to $\alpha_1/r^4$ as usual.

The calculated total and momentum loss (diffusion) cross-sections are shown in Figs. 4-2-1 and 4-2-2 for helium, neon and argon. The helium and argon cross-sections exhibit a dip at low velocities which is due to the vanishing of the s-wave phase shift. This effect, known as the Ramsauer-Townsend effect is masked to some extent by the non-vanishing p-wave phase shifts. The vanishing of the s-wave phase shift is directly due to the change in sign of the potential which is repulsive in the inner region, dominated by

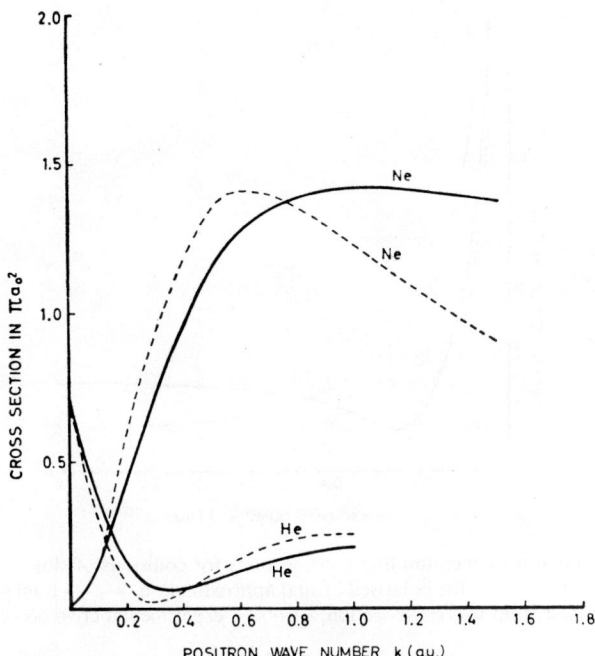

Fig. 4-2-1. Total and momentum loss cross-sections for collisions of slow positrons with helium and neon atoms calculated in the polarised orbital approximation [80].
———— total cross-section, ------ momentum loss cross-section.

the static interaction $U(x)$, and attractive (negative) in the outer region which is dominated by the long range polarisation potential. At low velocities, the outer region of the potential controls the scattering process and the phase shift is positive, but as the energy increases, the inner region becomes of greater and greater importance, until the phase shift passes through zero and becomes negative, corresponding to an effective repulsion. From the results for hydrogen discussed earlier, the simple form of polarised orbital method is expected to underestimate the effective attraction and the predicted position of Ramsauer minimum is probably at too low an energy. In the case of neon, a small increase in the effective attraction would produce a Ramsauer minimum close the threshold, which is absent in the simple calculations.

Although the helium cross-section is small, it is not as small as that indicated by the experiments of Marder et al. [31] which led to the value $0.12\ \pi a_0^2$ at 15 eV. The calculated cross-section is about eight times as large. For argon the total cross-section is of the same order of magnitude as

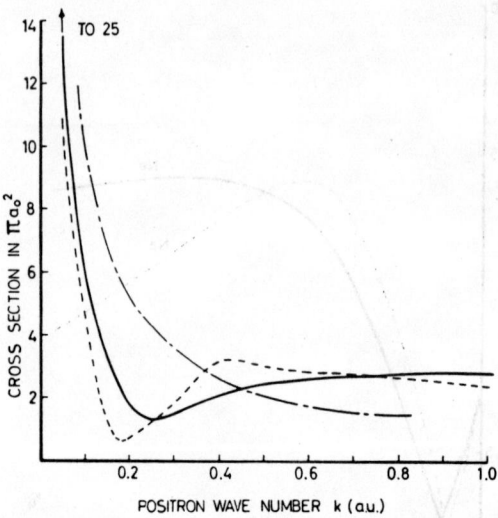

**Fig. 4-2-2.** Total and momentum loss cross-sections for collisions of slow positrons with argon atoms, calculated in the polarised orbital approximation. ──── total cross-section, ------ momentum loss cross-section, –·–·–·– experimental cross-section [22].

estimated by Falk et al. [22] (see Fig. 4-2-2), but the experimental cross-section fails to exhibit the Raumsauer minimum.

Massey et al. [80] tabulated the parameter $Z_{eff}$ as a function of velocity for each of the atoms, He, Ne, A and K. The calculated zero velocity value for helium, $Z_{eff} = 1.22$, is rather smaller then that indicated by experiment, $Z_{eff} = 3.9$, or that obtained by Houston and Moiseiwitsch in variational calculations, which suggests that, as in positron hydrogen scattering, the simple polarised orbital method cannot provide quantitative results.

To allow for virtual positronium formation in addition to polarisation effects, Kraidy and Fraser [73] added a polarisation potential to the equations of the two state approximation. We have seen that the same procedure is the case of hydrogen (cf. p. 226ff) led to an overestimate of the effective attraction, and in this case the phase shifts (Table 4-2-4) also seem to be much larger than those calculated by the variational method. The value of $Z_{eff} = 3.70$ is consistent with the experiment value, but this is likely to be coincidental. The overestimate of the attraction pulls the positron wave-function in towards the atom and this increases the overlap integral leading to $Z_{eff}$, however the same overlap is achieved by the wave-function of Houston and Moiseiwitsch, for a smaller value of the scattering length.

Kestner et al. [81] have also studied the polarised orbital method, using the $l=0$, 1 and 2 multipole components of the effective adiabatic potential and Drachman [76] has employed the complete adiabatic potential with the degree of monopole suppression indicated by his calculation for positron–hydrogen scattering. His results appear in column (e) of Table 4-2-4 and in Table 4-2-6. The calculated wave-function leads to a very high value of $Z_{eff}$, $Z_{eff}=6.9$, nearly twice as large as the experimental value. The scattering length is also rather greater than those of others except that given by Fraser and Kraidy's calculation including virtual positronium formation.

Drachman [79] has extended his variational calculation based on a trial function of the form (4-2-173) to the case of helium. The phase shifts calculated in this way are shown in Table 4-2-4, and the corresponding value of $Z_{eff}$ in Table 4-2-6. If the helium ground state function were exact then results would represent a lower bound on the phase shifts and scattering length. The low energy value of $Z_{eff}$, 3.66, is in good agreement with the experimental value and the scattering length (0.511) is not far from that given by the variational calculation of Houston and Moiseiwitsch (0.938).

The extended polarised orbital method has been applied to the case of helium by Callaway et al. [78], but in this case the method fails rather badly (assuming that the results of Drachman represent lower bounds). It can be seen from Table 4-2-4, the method overestimates the repulsive correction to the adiabatic potential and that the phase shifts are too negative.

The interplay between the repulsive and attractive parts of the effective interaction in positron scattering makes the phase shifts depend critically on the precise form of the wave function and much more work needs to be done before the positron helium phase shifts can be considered known beyond question. More elaborate variational (Moussa [83]) and close coupling calculations (Smith et al. [67]) are underway and should help to remove some of the ambiguities in the existing position.

## § 4-3. The formation and scattering of positronium

In section 4-1 it was seen that the formation of positronium plays an important role in the attenuation of a positron beam in a gas. Once formed, parapositronium decays rapidly, but the quenching rate of orthopositronium is determined by the scattering that it undergoes by the gas atoms. Comparitively little work has been done on either the positronium formation reaction, or on the elastic and inelastic scattering of orthopositronium. Although both processes are important, the positronium formation reaction has a particular interest as it is perhaps the most simple example of a rearrangement collision. We shall start by discussing this reaction in hydrogen and in helium.

## A. THE FORMATION OF POSITRONIUM BY POSITRON IMPACT ON HYDROGEN AND HELIUM

The first general discussion of positronium formation and scattering was given by Massey and Mohr [4] in 1954. The simplest reaction of this kind is the formation of positronium in the (1s) ground state, by positron impact in the ground state of hydrogen,

$$e^+ + H(1s) \rightarrow Ps(1s) + H^+. \tag{4-3-1}$$

The positronium atom may be formed in either the triplet or the singlet state. In the absence of spin-dependent forces the transition matrix element cannot depend on the spin of the positron, so that if the cross-section for the reaction, ignoring the spins of the positron, is $q$, then for an unpolarised positron beam, the cross-section for the formation or orthopositronium will be $\tfrac{3}{4}q$ and for parapositronium $\tfrac{1}{4}q$.

The cross-section for the reaction (4-3-1) has been computed by Massey and Mohr in the Born approximation. Their calculated cross-section rises from the threshold at 6.8 eV to a maximum near 13.5 eV and then decreases. The cross-section is large, being $\sim 4\pi a_0^2$ at maximum and decreasing to about $\pi a_0^2$ at 32 eV. There is some doubt as to the validity of first order perturbation methods in rearrangement collisions (Bransden [84]) at high energies, and at low energies even the order of magnitude of the cross-section can be in error. Because of this, Massey and Mohr discussed a distorted wave approximation in which the wave-function of the initial system of (positron + hydrogen atom) was determined in the static approximation. It is now known that this approximation is inadequate, but the calculation served to show that the formation cross-section was quite sensitive to the form of the approximate wave-functions. The Born approximation cross-section for positronium formation in helium has been discussed by Massey and Moussa [85] in 1961 and more recently by Kraidy [74]. The threshold is at 17.6 eV and the cross-section rises to a maximum of about $0.55\pi a_0^2$ near 28 eV.

Cheshire [86] attempted to obtain a more accurate cross-section for the formation reaction in hydrogen by using the impulse approximation. He obtained a cross-section not disimilar from that of the Born approximation. However, this is essentially a high energy approximation and would not be expected to provide accurate results in the low energy region where the coupling between the channels is strong. At low energies, the cross-section is best investigated by an extension of the variational and coupled channel techniques discussed in connection with elastic scattering, which offer a chance of systematically improving the calculated cross-section, and which have been at least partially successful in explaining the features of inelastic electron scattering by atoms.

### 1. *General formalism*

Let us consider the extension of the formalism, developed in § 4-2 for elastic scattering, to cover the case in which just two channels are open, the incident channel, that composed of a positron moving freely with respect to the ground state of hydrogen, and the rearranged channel, in which positronium in the ground state moves freely with respect to a proton. The energy region of interest is from 6.75 eV, the positronium threshold to 10.2 eV, which is the threshold for excitation of the 2s and 2p states of hydrogen.

The boundary conditions can be imposed by defining projection operators $P_1$ and $P_2$, which project each of the open channel sub-spaces from the complete Hilbert space of the problem. At the same time, we may effect an angular momentum decomposition and isolate that part of the wave-function for which the orbital angular momentum of the scattered particle is $L$. The total wave-function is written as $\Psi = \sum \Psi^L$ where $\Psi^L$ is an eigenfunction of angular momentum belonging to the eigenvalue $L$.

The explicit form of $P_1$ and $P_2$ may be chosen to be

$$P_1 = \phi_0(r) \int dr' \, \phi_0(r') \tag{4-3-2}$$

$$P_2 = \psi_0(R) \int dR' \, \psi_0(R'). \tag{4-3-3}$$

We now would like to find the projection operator $P$ that projects both open channels simultaneously from the complete Hilbert space. The prescription $P = P_1 + P_2$ will not do, because in this case $P^2 \neq P$ since $P_1 P_2 \neq 0$. Various methods of overcoming this difficulty have been proposed [62], but we shall follow the procedure of Hahn [57], that was introduced in section 4-2. The wave-function is again written in two component form

$$\Psi^L = \begin{pmatrix} \Psi^L_1 \\ \Psi^L_2 \end{pmatrix}. \tag{4-3-4}$$

The projection operator $P$ is taken to be the diagonal matrix with components

$$(\boldsymbol{P})_{ij} = \delta_{ij} P_i. \tag{4-3-5}$$

The relationship $\boldsymbol{P}^2 = \boldsymbol{P}$ now follows trivially.

Open channel wave-functions $g_L$ and $f_L$ describing the relative motion of the particles in each channel can be defined by

$$\begin{aligned}P_1 \Psi_1^L(\boldsymbol{r}, \boldsymbol{x}) &= \phi_0(r) P_L(\cos\theta) x^{-1} f_L(x) \\ P_2 \Psi_2^L(\boldsymbol{r}, \boldsymbol{x}) &= \psi_0(R) P_L(\cos\alpha) y^{-1} g_L(y)\end{aligned} \tag{4-3-6}$$

where $(\theta, \phi)$ and $(\alpha, \beta)$ are the polar angles of $\boldsymbol{x}$ and $\boldsymbol{y}$ respectively. It is most convenient to work with real functions, and to ensure this, standing wave boundary conditions can be specified by requiring

$$\begin{aligned}f_L(x) &\sim A_1^L \sin(k_1 x - \tfrac{1}{2} L\pi) + B_1^L \cos(k_1 x - \tfrac{1}{2} L\pi) \\ g_L(x) &\sim A_2^L \sin(k_2 x - \tfrac{1}{2} L\pi) + B_2^L \cos(k_2 x - \tfrac{1}{2} L\pi)\end{aligned} \tag{4-3-7}$$

when $k_1$ is the momentum (in atomic units) of the incident positron and $k_2$ is the momentum of the positronium atom in the final state. Of the four constants $A_i$, $B_i$, two may be specified arbitrarily while two are to be determined from the Schrödinger equation. The $A_i$ and $B_i$ are related by the reaction matrix $K^L$ by*

$$B_i^L = \sum_{j=1}^{2} K_{ij}^L A_j^L \left(\frac{v_j}{v_i}\right)^{\frac{1}{2}} \tag{4-3-8}$$

where $v_1 = k_1$ and $v_2 = \tfrac{1}{2} k_2$ are the relative velocities in each channel. The partial cross-sections of order $L$, for scattering from channel $i$ to channel $j$ are

$$\sigma(i \to j) = \frac{4\pi(2L+1)}{k_i^2} \left| \left(\frac{K}{1 - iK}\right)_{ji} \right|^2. \tag{4-3-9}$$

The projection operator on the closed channels $Q$ is defined as $Q = 1 - P$ and satisfies the conditions

$$Q^2 = Q, \quad QP = PQ = 0. \tag{4-3-10}$$

The Schrödinger equation is then equivalent to the equations, similar in form

---

* The theory of the reaction matrix and its connection with the cross-section may be consulted in Mott and Massey [21] or in Newton [87]. Time-reversal invariance requires $K$ to be real and symmetric, properties that are preserved in the approximations to be discussed.

to (4-2-17) and (4-2-18)

$$P(H - E) P\Psi^L = - PHQ\Psi^L$$
$$Q(H - E) Q\Psi^L = - QHP\Psi^L \qquad (4\text{-}3\text{-}11)$$

where $(H-E)$ is the matrix with elements

$$(H - E)_{ij} \equiv H - E. \qquad (4\text{-}3\text{-}12)$$

The effective (matrix) potential acting in the open channels is $W$ such that

$$P(H - E + W) P\Psi^L = 0 \qquad (4\text{-}3\text{-}13)$$

and is given by an expression similar in form to (4-2-24)

$$W = PHQ \frac{1}{Q(H - E)Q} QHP. \qquad (4\text{-}3\text{-}14)$$

It can be shown that, below the second inelastic threshold, $W$ is a negative definite operator, provided that $QHQ$ has no discrete eigenvalue spectrum. Gailitis [55] and Spruch et al. [58] (see also Blankenbecler and Sugar [36]) have shown that the minimum principle applies in that if a sequence of approximation to $W$, $W^N$ is defined in which $Q\Psi_t^L$ spans progressively more and more of $Q$ space, then

$$W^{N+1} < W^N \qquad (4\text{-}3\text{-}15)$$

from which

$$K^L_{(N+1)} > K^L_{(N)}.$$

A particular form of the minimum principle, is the extension of (4-2-158) to inelastic scattering. It states that if $Q\Psi_A^L$ is any approximation to $Q\Psi^L$ and if $P\tilde{\Psi}^L$ is the exact solution of the equation

$$P(H - E) P\tilde{\Psi}^L = - PHQ\Psi_A^L \qquad (4\text{-}3\text{-}16)$$

giving rise to a reaction matrix $\bar{K}^L$, then

$$-(A|K^L|A) \leqslant -(A|\bar{K}^L|A) + (Q\Psi_A^L|H - E|Q\Psi_A^L + P\Psi^L). \qquad (4\text{-}3\text{-}17)$$

2. *Trial functions and applications*

The close coupling method, that we have described earlier for elastic scattering, is obtained in the two state approximation by the choice $Q\Psi_A^L = 0$. This will be refered to as approximation (A). Trial functions of the polarised orbital variety can be discussed, by taking

$$Q_1 \Psi_{1A}^L = \phi_0^{pol}(r, x) x^{-1} f_L(x) P_L(\cos \theta)$$
$$Q_2 \Psi_{2A}^L = \psi_0^{pol}(R, y) y^{-1} g_L(y) P_L(\cos \alpha). \quad (4\text{-}3\text{-}18)$$

As in the case of elastic scattering, the functions $\phi_0^{pol}$ and $\psi_0^{pol}$ can be calculated in the adiabatic approximation, and either the complete functions or their multipole components can be used. In this approximation, the coupled equations for $f_L$ and $g_L$ reduce to

$$\left[ \frac{d^2}{dx^2} - \frac{L(L+1)}{x^2} + k^2 - U(x) - W_A(x) \right] f_L(x)$$
$$= 2x \int dr \int d(\cos \theta) P_L(\cos \theta) \phi_0(r)$$
$$\times \left[ (H - E)(\psi_0(R) + \psi_0^{pol}(R, y)) y^{-1} g_L(y) P_L(\cos \alpha) \right]$$
$$\left[ \frac{d^2}{dy^2} - \frac{L(L+1)}{y^2} - 4\overline{W}_A(y) \right] g_L(y) \quad (4\text{-}3\text{-}19)$$
$$= 4y \int dR \int d(\cos \alpha) P_L(\cos \alpha) \psi_0(R)$$
$$\times \left[ (H - E)(\phi_0(r) + \phi_0^{pol}(x, r)) x^{-1} f_L(x) P_L(\cos \theta) \right].$$

As usual $U(x)$ is the static potential in the positron–hydrogen atom channel and $W(x)$ is the corresponding adiabatic potential. In the positronium channel, because of the coincidence of the centre of charge and centre of mass of positronium, the static interaction vanishes. For the same reason $\overline{W}_A$, the polarisation potential in the second channel contains only odd multipole components.

Only preliminary numerical calculations based on equation (4-3-19) have been made (by Bransden and Jundi [88]). In this work, the terms envolving $\phi_0^{pol}$ and $\psi_0^{pol}$ appearing in the coupling terms on the right-hand side of (4-3-19) were dropped. This may not be too bad an approximation because these kernels are large and of short range, but more seriously no attempt was made to correct the $K$ matrix by evaluating the term which appears on the right-hand side of (4-3-17). This is equivalent to ignoring the distortion potential of Callaway et al. and we saw in our discussion of elastic scattering that in the absence of this, or similar terms, the effective attraction is too large.

Despite this criticism, the qualitative results are interesting and suggest that the system deserves further study. These results are illustrated in Figs. 4-3-1 and 4-3-2. The two state close coupling approximation (A) has already

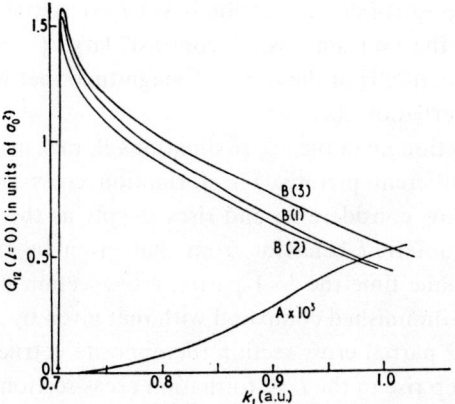

Fig. 4-3-1. Partial cross-sections with $l=0$ for positronium formation by positron impact on hydrogen atoms. A, B(1), B(2) and B(3) refer to the no-polarization and polarization approximations described in the text.

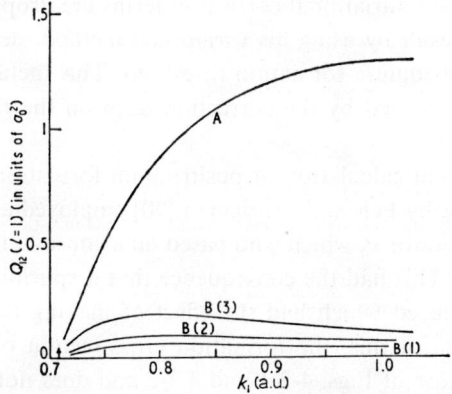

Fig. 4-3-2. Partial cross-sections with $l=1$ for positronium formation by positron impact on hydrogen atoms. A, B(1), B(2) and B(3) refer to the no-polarization and polarization approximations described in the text.

been defined. The other approximations B are found by solving the coupled equations (4-3-19) ignoring $\phi_0^{pol}$ and $\psi_0^{pol}$ on the right-hand side. In approximation B(1) only, the dipole contribution to $W_A$ was retained, in B(2), the (dipole + quadrupole) contributions and in B(3) the (monopole + dipole + quadrupole) contributions were retained. In each case $\bar{W}_A$ was approximated by the dipole contribution.

The most surprising feature of approximation A is the abnormally small cross-section for positronium formation in the $l=0$ partial wave. The partial cross-section for the $l=1$ state is, in contrast, large $(\sim \pi a_0^2)$ and the total formation cross-section is of the order of magnitude that would be expected from simple perturbation theory.

The cross-section in all the approximations B, that include polarisation, present a very different picture. The formation cross-section in the $l=0$ partial wave is now considerable and rises steeply as the energy decreases, which is a very different behavior from that given by the Born approximation. At the same time the $l=1$ partial cross-section in the polarisation approximation is diminished compared with that given by approximation A, while for the $l=2$ partial cross-section the opposite is true.

The very steep rise in the $l=0$ formation cross-section near threshold is apparently due to a pole in the reaction matrix just below threshold, which would give rise to a resonance in positron hydrogen scattering at energy just below the formation threshold. Drachman [89] has however shown that this effect is spurious and arises from the overestimate of the effective attraction that occurs when the variational correction terms are dropped. He was able to establish this result by using his variational method, described in 4-2, at the energy of positronium formation threshold. This included the non-adiabatic effects represented by the correction term on the right-hand side of (4-3-17).

An independent calculation of positronium formation by positron impact on hydrogen, by Fels and Mittleman [90] employed a different form of the projection operator $P$, which who based on a coordinate system centered on the positron.* This had the consequence that a spurious centrifugal potential was introduced which had the effect of making the interaction less attractive. Because of this, the formation cross-section obtained has little resemblence to those of Figs. 4-3-1 and 4-3-2 and does not exhibit the peak close to threshold. The divergence of the results illustrates the sensitivity of the calculations to the particular approximation employed and points the need for a more quantitative investigation in which the minimum principle is preserved.

Fraser and Kraidy [73] (see also Kraidy [74]) have used the (two state + polarisation) model to calculate the positronium formation cross-section in helium, retaining only the dipole part of the polarisation potential. The same approximations were made as in the work of Bransden and Jundi, discussed above. The results are not consistent with the variational principle and

---

* This work has now been extended to helium (private communication from M. F. Fels).

probably overestimate the effective attraction. In the absence of polarisation (model A), the $l=0$ partial cross-section is again remarkably small and in this case the most important contributions arise from the $l=2$ and $l=3$ partial waves. The total cross-section rises linearly and is quite small, $q \approx 0.3\pi a_0^2$ at 27 eV. The addition of the polarisation potential in the positron channel does not lead to much change in the total cross-section. This is understandable because the polarisability of helium ($\alpha_1 = 1.376$) is smaller than that of hydrogen ($\alpha_1 = 4.5$). On the other hand the addition of the polarisation in the positronium channel (the polarisability of positronium is eight times that of hydrogen) leads to a great increase in the magnitude of the cross-section, which increases to a maximum of $\sim 1.4\pi a_0^2$ near 23 eV. The cross-section rises very sharply from threshold to a subsidiary maximum at $k^2 = 1.22$, and this resonance like behavior occurs in the $l=1$ state. The position is very interesting, but these results must be considered to be qualitative in character.

## B. THE SCATTERING OF POSITRONIUM BY HYDROGEN AND HELIUM

To estimate the annihilation rate of positronium in a gas, it is necessary to know the wave-function for scattering of the positronium atoms by the gas atoms. In 1954 Massey and Mohr [4] made some estimates, using the Born approximation, of the elastic scattering cross-sections of orthopositronium in hydrogen and also of the ortho→para conversion cross-section in hydrogen. A few years later, in 1960, Fraser [91] put the theory of positronium scattering by hydrogen on a more quantitative basis. As the chief region of interest is at very low energies below the first excitation threshold of positronium, the problem was treated as a single open channel system and only s wave-scattering was considered. The force of longest range acting between positronium and an atom, is the attractive dipole–dipole interaction which occurs between two neutral polarisable systems. This interaction varies like $-\beta/r^6$ at large distances of separation of the colliding systems. As well as being of shorter range, it is of smaller magnitude than the polarisation force arising between a charged particle and a neutral system that is important in positron scattering. For this reason, it is possible that the approximation in which the closed channel wave-function $Q\Psi$ is ignored may be adequate.

In the positronium–hydrogen atom system, the two electrons may be in a relative spin zero state, in which case by the Pauli principle the spatial part of the wave-function must be symmetrical, or the electrons may be in a relative state of spin one, with an antisymmetrical spatial wave-function. As the symmetry of the electron system is conserved, there are two independent solutions to the problem, $\Psi^+$ and $\Psi^-$ which are symmetrical $(+)$ or antisymmetrical $(-)$ in $r_1$ and $r_2$, the electron coordinates. In the static exchange approximation the open channel wave-function may be represented as (for $l=0$)

$$P\Psi^{\pm}(x, r_1, r_2) = \psi_0(R_1) \phi_0(r_2) y_1^{-1} f_0^{\pm}(y_1) \pm \psi_0(R_2) \phi_0(r_1) y_2^{-1} f_0^{\pm}(y_2) \tag{4-3-20}$$

where $R_i = r_i - x$ and $y_i = \tfrac{1}{2}(r_i + x)$.

The radial functions $f_0^{\pm}(y)$ satisfy the usual boundary condition in terms of phase shifts

$$f_0^{\pm}(y) \sim \sin(ky) + \tan\eta_0^{\pm} \cos(ky) \tag{4-3-21}$$

where $k$ is momentum in atomic units in the centre of mass system.

By analyzing the spin wave-function of the system, Fraser was able to show that the elastic scattering cross-section for scattering of an unpolarised beam of orthopositronium was given by

$$q = \frac{\pi}{k^2} \{3 \sin^2 \eta_0^- + \sin^2 \eta_0^+\} \tag{4-3-22}$$

while the conversion cross-section was

$$q_\alpha = \frac{\pi}{4k^2} \{\sin^2 \eta_0^- + \sin^2 \eta_0^+ - 2 \sin \eta_0^+ \sin \eta_0^- \cos(\eta_0^+ - \eta_0^-)\}. \tag{4-3-23}$$

Application of the usual variational method leads to an integrodifferential equation for each of the functions $f^{\pm}(y)$. It is of the form

$$\left(\frac{d^2}{dy_1^2} + k^2\right) f_0^{\pm}(y_1) = \pm \int_0^\infty dy_2 K(y_1, y_2) y_2^{-1} f_0^{\pm}(y) \tag{4-3-24}$$

where

$$K(y_1, y_2) = y_1 \int_{-1}^{+1} d(\cos\alpha_1) \int_{-1}^{+1} d(\cos\alpha_2) \int dR_1$$
$$\times \int dr_2 \, \psi_0(R_1) \phi_0(r_2) (H - E) \psi_0(R_2) \phi_0(r_1) \tag{4-3-25}$$

and $(\alpha_1, \beta_1)$, $(\alpha_2, \beta_2)$ are the polar angles of $y_1$ and $y_2$.

The values of $\tan\eta_0^{\pm}$ obtained from this equation satisfy the minimum principle and are lower bounds.

As we have mentioned earlier, the direct static interaction between positronium and an atom vanishes, and the effective interaction $K(y_1, y_2)$ is due entirely to electron exchange. The calculated conversion cross-section in this approximation falls rapidly from $0.176\pi a_0^2$ at zero energy to $0.07\pi a_0^2$ at 6.8 eV, while the total cross-section falls from $192\pi a_0^2$ to $2.92\pi a_0^2$ between the same energies. This cross-section is large enough to explain the rapid quenching of orthopositronium by gases with an unpaired electron.

Elastic scattering of orthopositronium by helium has been discussed in the same approximation by Fraser and Kraidy [92] and by Bransden and Barker [93]. As the ground state of helium is a singlet state, the conversion reaction is impossible at low velocities and only elastic scattering occurs. The quenching rate can be worked out, as in the case of positron scattering, and the results expressed in terms of a parameter $Z_{\text{eff}}(v)$. The calculated value of $Z_{\text{eff}}$ at thermal energies is $Z_{\text{eff}} = 0.037$. This result is considerably smaller than that given by the most recent measurements, which give the values $Z_{\text{eff}} = 0.135 \pm 0.068$ (Heymann et al. [94]), $0.118 \pm 0.011$ (Duff and Heymann [95]) and $0.25 \pm 25\%$ (Roellig and Kelly [96]).

The discrepancy between theory and experiment could be due to the neglect of the Van der Waals polarisation potential or to the neglect of short range correlation terms in the wave-function. The influence of the Van der Waals interaction has been estimated by Bransden and Barker, who calculated the long range potential to lowest order and added this interaction to the effective potential. The new value of $Z_{\text{eff}}$ obtained was 0.048. This result shows that the correlation terms in the wave-function must be important and polarisation forces by themselves are insufficient. Using the variational method, it should be possible to investigate the importance of such terms in a systematic manner, and this is the direction in which future work is expected to go.

### Ackowledgement

It is a pleasure to thank Professor R. Geballe and his colleagues for the hospitality offered to the author during his visit, in the Spring and Summer of 1968, to the University of Washington, where the major part of this chapter was written.

### References

1. P. A. M. DIRAC, Proc. Poy. Soc. A117 (1928) 610, 118 (1928) 351.
2. C. D. ANDERSON, Phys. Rev. 43 (1933) 491.
3. P. M. S. BLAKETT and G. P. S. OCCHIALINE, Proc. Roy. Soc. A139 (1933) 699.
4. H. S. W. MASSEY and C. B. O. MOHR, Proc. Phys. Soc. (London) 57 (1954) 695.

5. J. M. JAUCH and F. ROHRLICH, The Theory of Photons and Electrons (Addison-Wesley, 1965).
6. J. D. BJORKEN and S. D. DRELL, Relativistic Quantum Mechanics (McGraw-Hill, 1964).
7. A. ORE and J. L. POWELL, Phys. Rev. **75** (1949) 1696, 1963.
8. A. E. RUARK, Phys. Rev. **68** (1945) 278.
9. M. DEUTSCH, Progr. Nucl. Phys. **3** (1953) 131.
10. S. DE BENEDETTI and H. C. CORBEN, Ann. Rev. Nucl. Sci. **4** (1954) 191.
11. R. A. FERRELL, Phys. Rev. **110** (1958) 1355.
12. J. A. WHEELER, Ann. N.Y. Acad. Sci. **48** (1946) 219.
13. E. HYLLERAAS, Phys. Rev. **71** (1947) 491.
14. R. R. SHARMA, Phys. Rev. **171** (1968) 36.
15. F. H. GERTLER, H. B. SNODGRASS and L. SPRUCH, Phys. Rev. **172** (1968) 110.
16. L. SPRUCH, in: Lectures in Theoretical Physics, XI (Gordon and Breach, 1969).
17. A. A. FROST, M. INOKOTI and J. P. LOWE, J. Chem. Phys. **41** (1964) 482.
18. M. DEUTSCH, Phys. Rev. **82** (1951) 455.
19. E. MERZBACHER, Quantum Mechanics (Wiley, 1960).
20. L. D. LANDAU and E. M. LIFSHITZ, Non-Relativistic Quantum Mechanics (2nd ed., Pergamon Press, 1965).
21. N. F. MOTT and H. S. W. MASSEY, Theory of Atomic Collisions (3rd ed., O.U.P., 1965).
22. W. P. FALK, P. H. R. ORTH and G. JONES, Phys. Rev. Letters **14** (1965) 447.
23. P. A. FRASER, Advan. in Atomic and Molecular Physics **4**, (1968) 63.
24. H. FESHBACH, Ann. Phys. (N.Y.) **5** (1958) 357.
25. P. M. MORSE and H. FESHBACH, Methods of Theoretical Physics (2 vols.) (McGraw-Hill, 1953).
26. A. MESSIAH, Quantum Mechanics (2 vols.) (North-Holland, 1962).
27. A. ORE, Univ. i Bergen Arbok Natur. Rekke **9** and **12**.
28. H. S. W. MASSEY and A. H. A. MOUSSA, Proc. Phys. Soc. (London) **71** (1958) 38.
29. P. G. BURKE and K. SMITH, Rev. Mod. Phys. **34** (1962) 458.
30. R. K. PETERKOP and V. VELDRE, Advan. in Atomic and Molecular Phys. **2** (1966) 263.
31. S. MARDER, V. W. HUGHES, C. S. WU and W. BENNETT, Phys. Rev. **103** (1956) 1258.
32. W. B. TEUSCH and V. W. HUGHES, Phys. Rev. **103** (1956) 1266.
33. P. E. OSMON, Phys. Rev. **138** (1965) B216.
34. L. O. ROELLIG and T. M. KELLY, Phys. Rev. Letters **15** (1965) 746.
35. D. A. L. PAUL, Proc. Phys. Soc. (London) **84** (1964) 563.
36. R. BLANKENBECLER and R. SUGAR, Phys. Rev. **136** (1964) B474.
37. L. CASTELLEJO, I. C. PERCIVAL and M. J. SEATON, Proc. Roy. Soc. **A254** (1960) 259.
38. C. J. KLEINMAN, Y. HAHN and L. SPRUCH, Phys. Rev. **165** (1968) 53.
39. A. TEMKIN, Phys. Rev. **107** (1957) 1007.
40. A. TEMKIN and J. C. LAMKIN, Phys. Rev. **121** (1961) 788.
41. H. REEH, Z. Naturforsch. **15a** (1960) 377.
42. A. DALGARNO and H. LYNN, Proc. Phys. Soc. (London) **A70** (1957) 223.
43. B. H. BRANSDEN and Z. JUNDI, Proc. Phys. Soc. (London) **89** (1966) 7.
44. W. J. CODY, J. LAWSON, H. S. W. MASSEY and K. SMITH, Proc. Roy. Soc. **A278** (1964) 479.
45. Expressions like this have been used by A. H. MOUSSA, Proc. Phys. Soc. (London) (1959).
46. L. HULTHÉN, Kgl. Fysiograf. Sallskap. Lund, Forh. **14** (1944) 1.
47. W. KOHN, Phys. Rev. **74** (1948) 1763.
48. J. SCHWINGER, unpublished lecture notes (1947).
49. L. SPRUCH, in: Lectures in Theoretical Physics, Vol. 4, eds. W. E. Brittin, B. W. Downs and J. Downs (Interscience, 1962).

50. Yu. N. Demkov, Variational Principles in the Theory of Collisions, transl. N. Kemmer (Pergamon, 1963).
51. B. L. Moiseiwitsch, Variational Principles (Interscience, 1966).
52. L. Spruch and L. Rosenberg, Phys. Rev. **117** (1960) 143.
53. L. Rosenberg, L. Spruch and T. F. O'Malley, Phys. Rev. **118** (1960) 184 and **119** (1960) 164.
54. K. Smith, Proc. Phys. Soc. **78** (1961) 549.
55. M. Gailitis, Soviet Phys. JETP **20** (1965) 107.
56. P. G. Burke and A. J. Taylor, Proc. Phys. Soc. (London) **88** (1966) 549.
57. Y. Hahn, Phys. Rev. **142** (1966) 603.
58. Y. Hahn, T. F. O'Malley and L. Spruch, Phys. Rev. **130** (1963) 381 and Phys. Rev. **134** (1964) B911.
59. Y. Hahn, Phys. Rev. **139** (1965) B212.
60. M. H. Mittleman, Phys. Rev. **152** (1956) 76.
61. T. F. O'Malley and S. Geltman, Phys. Rev. **137** (1965) A1344.
62. J. C. Y. Chen and M. H. Mittleman, Ann. Phys. (N.Y.) **37** (1966) 364.
63. C. Schwartz, Phys. Rev. **124** (1961) 1468.
64. R. L. Armstead, Phys. Rev. **171** (1968) 91.
65. P. G. Burke and H. Schey, Phys. Rev. **126** (1962) 147.
66. R. P. McEchran and P. A. Fraser, Proc. Phys. Soc. (London) **86** (1965) 396.
67. A calculation is in progress however, see H. S. W. Massey, K. Smith and C. Wardle, Abstract of the Vth Intern. Conf. on the Physics of Electronic and Atomic Collisions, Leningrad (1967).
68. Y. Hahn, T. F. O'Malley and L. Spruch, Phys. Rev. **134** (1964) B397;
Y. Hahn and L. Spruch, Phys. Rev. **140** (1965) A18;
C. J. Kleinman, Y. Hahn and L. Spruch, Phys. Rev. **140** (1965) A413.
69. D. S. C. Allison, H. A. J. McIntyre and B. L. Moiseiwitsch, Proc. Phys. Soc. **78** (1961) 1169.
70. S. K. Houston and B. L. Moiseiwitsch, J. Phys. B (2), **1** (1968) 29.
71. L. G. Green, M. M. Mulder, M. N. Lewis and J. W. Woll, Phys. Rev. **93** (1954) 757.
72. S. K. Houston, J. Phys. B. (2), **1** (1968) 34.
73. P. A. Fraser and M. Kraidy, Abstracts of the Vth Intern. Conf. on the Physics of Electronic and Atomic Collisions, Leningrad (1967).
74. M. Kraidy, Thesis for University of Western Ontario (1967).
75. B. H. Bransden and Z. Jundi (unpublished).
76. R. J. Drachman, Phys. Rev. **138** (1965) A1582.
77. B. H. Bransden, Proc. Phys. Soc. (London) **79** (1962) 190.
78. J. Callaway, R. W. Labahn, R. T. Pu and W. M. Duxler, Phys. Rev. (1968).
79. D. J. Drachman, Goddard Space Flight Centre preprint X-641-68-75.
80. H. S. W. Massey, J. Lawson and D. G. Thompson, in: Quantum Theory of Atoms, Molecules and the Solid State (Academic Press, 1966) p. 202.
81. N. R. Kestner, J. Jortner, M. H. Cohen and S. A. Rice, Phys. Rev. **140** (1965) A56.
82. R. J. Drachman, Phys. Rev. **144** (1966) 25.
83. A. H. A. Moussa, Abstracts of the Vth Intern. Conf. on the Physics of Electronic and Atomic Collisions, Leningrad (1967).
84. B. H. Bransden, Advan. in Atomic and Molecular Collisions **1** (1965) 85;
B. H. Bransden, in: Lectures in Theoretical Physics, XI (Gordon and Breach, 1969).
85. H. S. W. Massey and A. H. A. Moussa, Proc. Phys. Soc. (London) **77** (1961) 811.
86. I. M. Cheshire, Proc. Phys. Soc. (London) **82** (1964) 113.
87. R. C. Newton, Scattering Theory of Waves and Particles (McGraw-Hill, 1966).
88. B. H. Bransden and Z. Jundi, Proc. Phys. Soc. (London) **92** (1967) 880.

89. R. J. DRACHMAN, Phys. Rev. **171** (1968) 110.
90. M. F. FELS and M. H. MITTLEMAN, Phys. Rev. **163** (1967) 129.
91. P. A. FRASER, Proc. Phys. Soc. (London) **78** (1951) 333.
92. P. A. FRASER and M. KRAIDY, Proc. Phys. Soc. (London) **89** (1966) 533; see also P. A. FRASER, Proc. Phys. Soc. (London) **79** (1962) 721.
93. B. H. BRANSDEN and M. I. BARKER, J. Phys. B (2) **1** (1968) 1109.
94. F. F. HEYMANN, P. E. OSMON, J. J. VEIT and W. F. WIELCAUS, Proc. Phys. Soc. (London) **78** (1961) 1038.
95. B. G. DUFF and F. F. HEYMANN, Proc. Roy. Soc. **A270** (1963) 517.
96. L. O. ROELLIG and T. M. KELLY, Phys. Rev. Letters **18** (1967) 387.

CHAPTER 5

# EXPERIMENTS WITH COLLIDING CHARGED-PARTICLE BEAMS

BY

## K. T. DOLDER

*Department of Atomic Physics, University of Newcastle upon Tyne*

# Contents

|  | Page |
|---|---|
| 5-1. Introduction | 251 |
| 5-2. General principles of experiments with colliding charged–particle beams | 251 |
|    A. Typical experimental arrangements | 251 |
|    B. The calculation of cross sections from experimental data | 252 |
|    C. Form factors | 254 |
|    D. Some effects of space charge on electron and ion beams | 257 |
|    E. The separation of signals from backgrounds | 260 |
|    F. The separation of beams of ions and atoms | 266 |
|    G. Initial states of excitation of parent ions | 269 |
| 5-3. The ionization of positive ions by electron impact | 272 |
|    A. Experiments which have been performed with crossed beams | 272 |
|    B. Four experimental approaches | 272 |
|    C. Results of the experiments | 283 |
| 5-4. The excitation of positive ions by electron impact | 292 |
|    A. Introduction | 292 |
|    B. The measurements of the excitation function $He^+(1S-2S)$ by Dance et al. | 292 |
|    C. Methods which rely upon the direct observation of light emitted by colliding electron and ion beams | 300 |
|    D. The experiment of Bacon and Hooper | 301 |
|    E. The experiment of Lee and Carleton | 304 |
| 5-5. The detachment of electrons from negative ions by electron impact | 306 |
|    A. Introduction | 306 |
|    B. The experiment of Dance, Harrison and Rundel | 306 |
|    C. The experiments of Tisone and Branscomb | 308 |
|    D. Results and discussion | 309 |
| 5-6. Measurements of cross sections for the production of protons by collisions between electrons and $H_2^+$ ions | 311 |
|    A. Introduction | 311 |
|    B. The experiments of Dunn, Van Zyl and Zare | 312 |
|    C. The experiment of Dance et al. | 314 |
|    D. Results and discussion | 315 |
| 5-7. Experiments with colliding ion beams | 317 |
| 5-8. Alternative methods used to study collisions between electrons and ions | 323 |
|    A. Sequential mass spectrometry | 323 |
|    B. Collisions between electron beams and low-density plasmas | 324 |
|    C. Collisions between electrons and highly-charged ions | 325 |
| 5-9. Some experimental techniques | 325 |
|    A. Ion sources | 325 |
|    B. Electron beam sources | 326 |
|    C. Electron and ion optics | 327 |
|    D. Vacuum techniques | 327 |
|    E. Particle detectors | 327 |
|    F. Intensity calibration of UV detectors | 330 |
|    G. Miscellaneous techniques | 330 |
| Acknowledgements | 330 |
| References | 330 |

## § 5-1. Introduction

The study of collisions between charged particles is a fundamental part of plasma physics, and the intense interest in laboratory and astrophysical plasmas has recently stimulated a search for new ways to measure cross-sections for these collisions. The most powerful and promising experimental methods involve the study of products formed when well-defined beams of charged particles collide, and although these methods have only been developed during the last decade, they have already yielded considerable information.

These experiments are particularly fascinating because they pose new and subtle problems and there is considerable scope for further research. The experiments which have already been performed can be classified as follows.

(a) Electron impact ionization of positive ions,

$$A^{n+} + e \rightarrow A^{m+} + (m - n + 1)e. \tag{5-1-1}$$

(b) Electron detachment from negative ions by electron impact,

$$A^- + e \rightarrow A + 2e. \tag{5-1-2}$$

(c) Excitation of positive ions by electron impact,

$$A^+ + e \rightarrow {}^*A^+ + e, \tag{5-1-3}$$

where $^*A^+$ denotes an excited state of the $A^+$ ion.

(d) Dissociative ionization or excitation of positive ions by electron impact, e.g.,

$$e + H_2^+ \rightarrow H^+ + H^+ + 2e. \tag{5-1-4}$$

(e) Collisions between ions, e.g.

$$H^+ + H^- \rightarrow H + H. \tag{5-1-5}$$

Separate sections of this review will be devoted to each of these five groups and further sections will deal briefly with other methods which have been used to study collisions between charged particles and with experimental techniques. But, before proceeding, we must discuss some general principles.

## § 5-2. General principles of experiments with colliding charged-particle beams

A. TYPICAL EXPERIMENTAL ARRANGEMENTS

General features of crossed beam experiments have been discussed by several authors [1–6, 140, 165] and an apparatus which could be used to study inelastic collisions between electrons and ions is represented schematically by Fig. 5-2-1. A collimated beam of ions of given type and energy acts as a target which is bombarded by an electron beam. The ions receive little momentum from these collisions so they are collected as a well-defined beam by collector $C_1$.

If it is required to study ionizing collisions of the types defined by (5-1-1) and (5-1-4) the parent and product ions can be separated by the field of either an electric or magnetic analyzer. Typically, only a small proportion of the target ions ($\lesssim 10^{-7}$) are ionized and these can be collected at $C_2$.

When the collision product is uncharged (reaction (5-1-2)) a neutral particle detector (N) with large aperture is used instead of $C_2$. Alternatively, if a calibrated photon detector (P) is placed near the collision region it is possible to measure excitation cross sections (reaction (5-1-3)). Different systems are used to study collisions between ion beams and these will be described in section 5-7.

## B. The calculation of cross sections from experimental data

We shall now derive expressions which relate collision cross sections to measurable quantities. Figure 5-2-2 represents two parallel monoenergetic beams of charged particles which collide at an angle $\theta$. Suppose that one beam is composed of ions and the other of electrons and, since these beams are usually very tenuous, we need only consider single collisions.

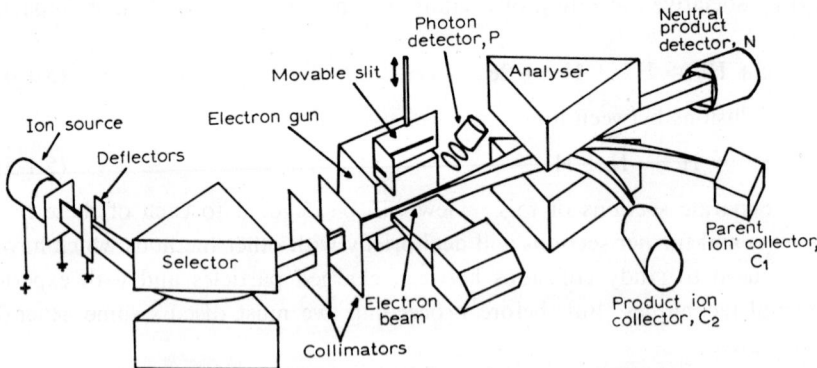

Fig. 5-2-1. An apparatus which could be used to study inelastic collisions between electrons and ions. A momentum-selected ion beam is collimated and bombarded by electrons. Ions which do not collide are collected at $C_1$ and more highly-ionized collision products are collected at $C_2$. Neutral particles and photons formed by the collisions are detected at N and P, respectively. A movable slit is used in determinations of the current density distributions of the colliding beams.

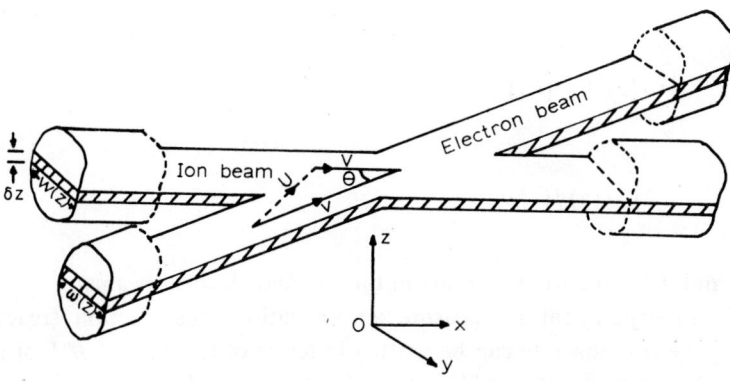

Fig. 5-2-2. Ion and electron beams colliding at an angle $\theta$.

The number of collisions occurring per second in the element of height $\delta z$ at $z$ is,

$$k(z)\,\delta z = \sigma(E)\,N(z)\,n(z)\,U\,\operatorname{cosec}\theta\,W(z)\,w(z)\,\delta z \tag{5-2-1}$$

where $N(z)$ and $n(z)$ are, respectively, the particle number densities in the ion and electron beams at height $z$. $W(z)$ and $w(z)$ represent the respective beam widths and $\sigma(E)$ is the collision cross section for an energy which corresponds to a relative velocity, $U$. This velocity can be expressed in terms of the ion and electron particle velocities, $V$ and $v$ by,

$$U^2 = V^2 + v^2 - 2Vv\cos\theta. \tag{5-2-2}$$

The particle number densities can be expressed in terms of the ion and electron currents $i(z)\,dz$ and $j(z)\,dz$, respectively, which flow through areas which have the widths of the beams and a height $\delta z$. When this is done we obtain,

$$k(z)\,dz = \frac{\sigma(E)\,U\,\operatorname{cosec}\theta\,i(z)\,j(z)\,dz}{ne^2vV} \tag{5-2-3}$$

where $ne$ is the charge of the parent ion.

The total collision rate $(k)$ can be found by integrating over the beam heights which we shall assume to extend between $\pm\infty$. For beams which intersect with $\theta = 90°$,

$$\sigma(E) = \frac{ne^2vV}{(v^2 + V^2)^{\frac{1}{2}}} \cdot \frac{k}{IJ} \cdot F \tag{5-2-4}$$

where

$$F \equiv \frac{\int_{-\infty}^{+\infty} i(z)\,dz \int_{-\infty}^{+\infty} j(z)\,dz}{\int_{-\infty}^{+\infty} i(z)j(z)\,dz} \tag{5-2-5}$$

and $I$ and $J$ are the total currents in the ion and electron beams.

In an experiment to measure an ionization cross section (reaction (5-1-1)) the reaction rate can be written in terms of the current $I^{m+}$ of ions formed by electron impacts. This gives $k = I^{m+}/me$ and,

$$\sigma(E) = \frac{nevV}{m(v^2 + V^2)^{\frac{1}{2}}} \cdot \frac{I^{m+}}{IJ} \cdot F. \tag{5-2-6}$$

In many experiments the ion velocity is negligible compared with that of the electrons so (5-2-6) can be simplified. When this is not the case, one can express the energy of colliding particles in centre of mass co-ordinates rather than laboratory co-ordinates. For electron and ion beams which intersect at right angles, the centre of mass energy, $E_{cm}$, is,

$$E_{cm} = \frac{M}{M+m}\left(E_e + \frac{m}{M}E_i\right) \tag{5-2-7}$$

where $m$, $M$, $E_e$ and $E_i$ are the masses and laboratory energies of the electrons and ions, respectively.

## C. Form Factors

A quantity such as defined by (5-2-5), which takes account of non uniformities in the current densities of the colliding beams, is called a form factor and it is usually computed from measured current density distributions. These distributions are obtained by sliding an L-shaped shutter, driven by a micrometer screw, vertically through the two beams. A typical shutter, illustrated in Fig. 5-2-1, is pierced by a narrow horizontal slit of height $s$ which simultaneously selects currents $I_s(z)$ and $J_s(z)$ from various regions of the beams. If $s$ is much less than the beam heights,

$$I = \int_{-\infty}^{+\infty} i(z)\,dz \approx \frac{1}{s}\int_{-\infty}^{+\infty} I_s(z)\,dz \equiv \frac{A_i}{s} \tag{5-2-8a}$$

$$J = \int_{-\infty}^{+\infty} j(z)\,dz \approx \frac{1}{s}\int_{-\infty}^{+\infty} J_s(z)\,dz \equiv \frac{A_j}{s} \qquad (5\text{-}2\text{-}8b)$$

and

$$\int_{-\infty}^{+\infty} i(z)j(z)\,dz \approx \frac{1}{s^2}\int_{-\infty}^{+\infty} I_s(z)J_s(z)\,dz \equiv \frac{A_{ij}}{s^2} \qquad (5\text{-}2\text{-}8c)$$

where $A_i$, $A_j$ and $A_{ij}$ are the areas under curves obtained by plotting $I_s(z)$, $J_s(z)$ and $I_s(z)J_s(z)$ against $z$. These areas can easily be computed from measured values of $I_s(z)$, $J_s(z)$ and the pitch of the micrometer screw. The heights of the beam and the slit ($s$) are not required since it follows from (5-2-5) that, $F \approx A_i A_j / A_{ij}$.

This determination of $F$ encounters several sources of error. For example, $s$ may be about 0.3 mm whilst the beam heights are about 5 mm. Consequently eqs. (5-2-8) are not exact. Moreover, the shutter usually intersects the beams just *before* they collide and the subsequent beam interaction and spread, caused by space charge forces and imperfect focussing, ensure that the measured profiles are not exactly the same as those in the interaction region. Further errors might arise from the introduction of a metal shutter to the beams because the shutter modifies electric fields within the beams and also introduces secondary electrons.

Fortunately, in spite of these problems, the accurate determination of form factors is not as difficult as it might seem. It follows from (5-2-5) that if the ion beam has uniform current density (i.e. $i(z)$ is constant over the beam height) and the electron beam is contained within the height of the ion beam, then $F$ is independent of the electron beam profile.

In this laboratory the ion source is usually situated some distance from the interaction region so that the ion trajectories correspond closely to a bundle of parallel rays and the ion beam height is defined by an aperture, or preferably a pair of apertures, placed just before the interaction region. Under these conditions the ion beam profile is quite uniform and $F$ is within a few percent of the height of the ion beam defining aperture.

In spite of these precautions there must be some errors in the determination of electron beam profiles. Here again we are helped by the form of eq. (5-2-5). Any error in $j(z)$ or $i(z)$ will appear in both the numerator and denominator of (5-2-5) so that the errors will, to some extent, cancel.

Some authors [13] have preferred a dimensionless form factor,

$$F' \equiv \frac{\int_{-\infty}^{+\infty} i(z)\,dz \int_{-\infty}^{+\infty} j(z)\,dz}{h \int_{-\infty}^{+\infty} i(z)j(z)\,dz}. \qquad (5\text{-}2\text{-}9)$$

This includes the height ($h$) of the ion beam defining slit which must then be included in the numerator of (5-2-4) to retain the correct expression for $\sigma(E)$. The factor $F'$ is unity when the ion beam profile is uniform (provided that the electron beam is contained within the ion beam height) so that the deviation of $F'$ from unity reveals the correction which has been applied for non-uniform current distributions. Frequently $F'$ lies between 0.95 and 1.00. The choice between $F$ and $F'$ is arbitrary and does not affect the accuracy or validity of any experiment.

Lineberger et al. [2] pointed out that errors could arise even with $F'=1$ if, for example, the electron and ion beam heights were equal at the shutter but subsequent spreading caused some electrons to pass above and below the ion beam. These errors need never occur in practice but it is nevertheless advisable to check that measured cross sections are insensitive to beam current density distributions. Several authors [2, 7, 15] have done this; Tisone and Branscomb [15] found no significant change in measured cross sections even when form factors were deliberately varied by a factor of six.

One concludes that the total errors associated with the determination of form factors need not exceed one or two percent but several authors have attempted to reduce these errors further. Wareing and Dolder [7], for example, measured ion beam profiles before and after the ion beam passed through the interaction region. They calculated a mean form factor from the two measurements but the ion beam spread was so small that this refinement was scarcely worthwhile.

Since, in the vicinity of the interaction region, the electron beam spreads more rapidly than the ion beam it might seem more attractive to make two determinations of the electron beam profile. The snag here is the relatively high density of charge in the electron beam. For the reasons previously mentioned the mere introduction of a metal shutter modifies the electron beam profile and it is likely that no real improvement in accuracy would stem from multiple determinations of electron beam profiles. In short, it seems best to ensure that the ion beam is carefully collimated and that its height is accurately defined just before the interaction region; the electron beam should always be contained well within the ion beam height.

The determination of form factors is quite laborious because they must be found for every setting of the colliding beams. The procedure was simplified by Dunn and Van Zyl [5] who measured their ion and electron currents by a digital meter which transferred its readings to punched tape in a form suitable for immediate computation.

### D. SOME EFFECTS OF SPACE CHARGE ON ELECTRON AND ION BEAMS

Two aspects of space charge are particularly relevant to crossed beam experiments. First we must consider the spreading of beams due to repulsion between similarly charged particles, and second we must take account of deflections which occur when two charged beams intersect.

Relevant reviews of space charge effects in electron beams have been given by Spangenberg [8], Klemperer [9] and Pierce [10]. In addition an extensive collection of graphs and equations which relate to the design of both electron and ion beams has been assembled by Von Ardenne [11].

Accurate equations describing the spread of charged beams would have to take account of some eminently intractible processes. Fortunately, the following simple approximate solutions collected by Von Ardenne are sufficient for our purpose because we merely need to know the conditions under which space charge is likely to modify beam geometries. Problems associated with space charge can usually be avoided by limiting beam currents.

The spread of an initially parallel homocentric beam of ions with charge $ne$ is given by,

$$\frac{d}{d_0} \approx \left[ 4.1 \times 10^{10} \frac{\sqrt{M/n}\, I_\rho l^2}{U^{3/2}} \right]^{1/2} \tag{5-2-10}$$

where $d_0$ and $d$ are the initial and final beam diameters and $l$ (metres) is the beam length defined in Fig. 5-2-3. $M$ is the ion mass relative to hydrogen, $I_\rho$ is the initial current density (A cm$^{-2}$) and $U$ is the beam voltage. This solution should only be used when $2 \lesssim d/d_0 \lesssim 5$ but solutions for more extreme cases are given by Klemperer [9], Pierce [10] and Thompson and Headrick [12].

The corresponding solution for an ion beam of rectangular cross section is,

$$\frac{h}{h_0} \approx \frac{2.04 \times 10^{10} \sqrt{M/n}\, I_\rho l^2}{U^{3/2}} + 1 \tag{5-2-11}$$

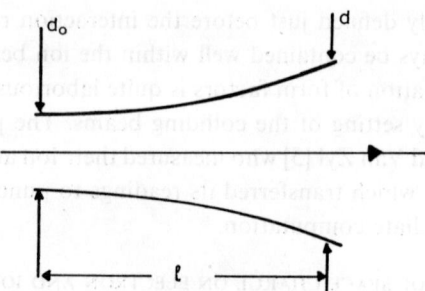

Fig. 5-2-3. Divergence of a homocentric beam of charged particles due to space charge.

where $h_0$ and $h$ are the initial and final beam heights and it has been assumed that the beam width is much greater than $h$.

For homocentric and rectangular electron beams the corresponding solutions are, respectively,

$$\frac{d_l}{d_0} \approx \left(9.6 \times 10^8 \frac{I_p l^2}{U^{\frac{3}{2}}}\right)^{\frac{1}{2}} \tag{5-2-12}$$

when $2 \lesssim d_l/d_0 \lesssim 5$, and,

$$\frac{h_l}{h_0} \approx \frac{4.8 \times 10^8 I_p l^2}{U^{\frac{3}{2}}} + 1. \tag{5-2-13}$$

As an example let us calculate the spread of an initially parallel beam of He$^+$ ions ($M/n = 4$) with rectangular cross section 0.1 cm × 1.0 cm. When the beam current is $10^{-6}$ A and the beam voltage is 5000, equation (5-2-11) predicts that the initial beam width of 0.1 cm will expand to about 0.21 cm in a path of 1 m. A similar calculation shows that the height of an electron beam initially 0.2 cm × 2.0 cm will expand to 0.224 cm in a path of only 1 cm if the beam carries a current of $10^{-3}$ A at an energy of 100 eV.

These examples illustrate conditions under which space charge significantly modifies beam geometries and the perveances* of these beams are close to the maxima which can usefully be employed in crossed beam experiments.

It follows that quite tenuous beams must be used. Typically, the particle densities of ions and electrons are $10^6$ cm$^{-3}$ and $10^7$ cm$^{-3}$, respectively.

The electric fields due to space charge, even in such low-perveance electron beams, are sufficient to deflect the trajectories of ions which pass

---

* Perveance is defined as $I/U^{\frac{3}{2}}$ where $I$ is the beam current and $U$ is the beam voltage.

through, and errors arise if these deflections prevent ions from reaching their collectors. This is illustrated by Fig. 5-2-4 which shows the deflection of an ion beam with initial height $h_i$ by an electron beam of height $h_e$ and width $l$. Some ions fail to enter a collector of height $h_c$ situated a distance $L$ from the electron beam.

If we assume that $l \gg h_e$ it follows from Gauss' law that the vertical component of electric field inside the electron beam due to space charge is,

$$E_z = \frac{\rho h_e}{\varepsilon_0}, \qquad (5\text{-}2\text{-}14)$$

where $\rho$ is the electron charge density (coulomb m$^{-3}$) and $\varepsilon_0$ is the permittivity of free space. Ions which have been accelerated from rest through a potential difference $U_i$ and which initially move in the $x$ direction, are deflected by the field $E_z$ through an angle $\phi(z)$ where,

$$\tan \phi(z) = \frac{lE_z}{2U_i}. \qquad (5\text{-}2\text{-}15)$$

Ions which pass near the electron beam without actually passing through it, will usually suffer smaller deflections.

Ions fail to reach the collector if,

$$\tan \phi(z) \gtrsim \frac{h_c + h_e}{2L} \qquad (5\text{-}2\text{-}16)$$

where it is assumed that $L \gg l$. It follows that ions will be lost if the electron charge density exceeds,

$$\rho_{max} \approx \frac{(h_c + h_e) U_i \varepsilon_0}{Llh_e}. \qquad (5\text{-}2\text{-}17)$$

This condition can be expressed in terms of $J_{max}$ which is the maximum permissible electron current if all ions are to reach their collector. In m.k.s.

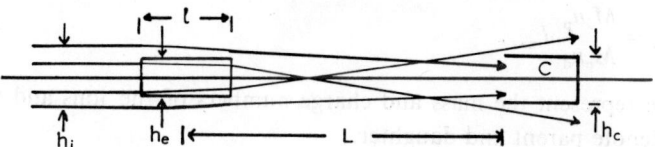

Fig. 5-2-4. Deflection of ion beam of initial height $h_i$ by an electron beam of height $h_e$. Some ions fail to enter the collector (C) of height $h_c$.

units,

$$J_{max} \approx 1.05 \times 10^{-5} U_i U_e^{\frac{1}{2}} \left(\frac{h_c + h_e}{L}\right). \tag{5-2-18}$$

Harrison [1] has measured values of $J_{max}$ for a 5 keV ion beam with $h_c = 1$ cm, $h_e = 0.2$ cm and $L = 50$ cm. Under these conditions (5-2-18) gives $J_{max} = 1.26 \times 10^{-3} U_e^{\frac{1}{2}}$ amperes. This result is shown by the dotted line in Fig. 5-2-5 which also includes Harrison's experimental points. In spite of the crude assumptions made to derive (5-2-18) we see that it predicts the observed functional dependence of $J_{max}$ upon $U_e$ and there is quantitative agreement within a factor of about 4. To some extent this discrepancy must be due to the partial neutralization of electronic space charge by ions formed from residual gas.

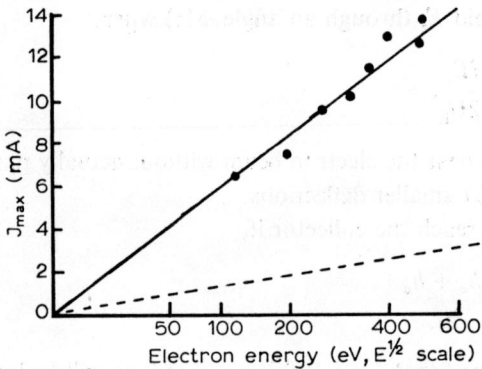

Fig. 5-2-5. Maximum electron currents ($J_{max}$) which can be used if loss of ions by space charge interaction is to be avoided. The broken line is calculated from equation (5-2-18) and the continuous line is drawn through experimental values.

In many reactions the daughter ion is either more highly charged or less massive than the parent and so it is deflected more easily. Smaller electron currents ($J'_{max}$) must then be used where,

$$J'_{max} = \frac{M_d n_p}{M_p n_d} J_{max}. \tag{5-2-19}$$

$M$ and $n$ represent the mass and charge numbers of the ions and the subscripts denote parent and daughter.

E. THE SEPARATION OF SIGNALS FROM BACKGROUNDS

In crossed beam experiments the signal is usually due to a tenuous beam of ions, neutral particles or photons which must be separated from backgrounds of similar particles formed by extraneous processes. Several methods have been devised to effect this separation and it is convenient to divide them into two groups according to whether the signal gives rise to a current greater or less than about $10^{-15}$ A. Electrometer amplifiers can conveniently be used to measure currents greater than $10^{-15}$ A but, when the currents are smaller, it is usual to detect and count individual particles.

An experiment in which ion currents were measured by a vibrating-capacitor electrometer was the determination of cross sections for,

$$He^+ + e \rightarrow He^{2+} + 2e \tag{5-2-20}$$

by Dolder et al. [13]. In their apparatus it was impractical to reduce the pressure much below $10^{-6}$ torr so that $He^{2+}$ was also produced by charge stripping collisions involving residual gas (R),

$$He^+ + R \rightarrow He^{2+} + R + e \tag{5-2-21}$$

and the rates of production of $He^{2+}$ by (5-2-20) and (5-2-21) were roughly equal.

It might seem that the current of $He^{2+}$ produced according to (5-2-20) could be identified because it should increase linearly with electron current. Unfortunately, outgassing of the apparatus also increased with electron current and the consequent pressure rise caused the production of $He^{2+}$ by both reactions to increase simultaneously.

The ion and electron beams were therefore chopped by rectangular pulses with a repetition frequency of 5 kHz, and every 30 sec the pulses were alternated between the two modes illustrated by Fig. 5-2-6. In the anticoincidence mode (Fig. 5-2-6a) the electron and ion beams crossed the interaction region at different times so that no collisions could occur between them. Only currents ($I_A^{2+}$) due to reaction (5-2-21) then reached collector $C_2$ (see Fig. 5-2-1). The same backgrounds were still collected when the ions and

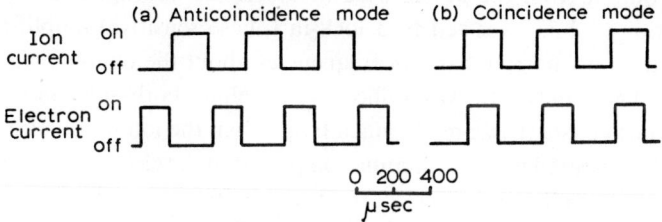

Fig. 5-2-6. Pulsing sequences to separate products of electron ion collisions from extraneous currents.

electrons were pulsed in coincidence (Fig. 5-2-6b) but now, in addition, electrons produced $He^{2+}$ according to (5-2-20). The total collected current was then $I_C^{2+}$ so that the required current of $He^{2+}$, due only to electron ion collisions was,

$$I^{2+} = I_C^{2+} - I_A^{2+}. \qquad (5-2-22)$$

It was essential that the duration of pulses ($10^{-4}$ sec) should be much less than the time in which significant pressure changes could occur in the apparatus. An apparatus of volume $V$ and pumping speed $S$ has a characteristic time $V/S$ and in these experiments $V/S$ was of order $10^{-2}$ sec, which was sufficiently large. To avoid errors due to finite rise and fall times of pulses the duty cycle was 50% for the ions but rather less for the electrons.

This technique works well if the signal to background ratio (SBR) is not too small and if pressure changes do not occur during the 30 sec periods of each mode. The most extreme application of this technique was by Dance et al. [14], who measured cross sections for proton productions in collisions between electrons and $H_2^+$ ions. In this experiment the SBR was only $10^{-2}$ and particular care was necessary to minimise the pressure fluctuations (see section 5-6-B).

An alternative method was used by Tisone and Branscomb [15, 16] in their experiments on electron detachment from negative ions, e.g.,

$$H^- + e \rightarrow H + 2e. \qquad (5-2-23)$$

The electron and $H^-$ beams were chopped to produce symmetric periodic wave currents at repetition frequencies 50 Hz and 20 kHz, respectively. A full description of this method involves Fourier analysis of the wave currents but, for simplicity, we shall illustrate the principles for the special case of sinusoidal modulation.

The signal due to electron ion collisions is proportional to $\sin\omega_j t \sin\omega_i t$ which consists of two sidebands with frequencies 20050 and 19950 Hz. The current output from the device used to detect the hydrogen atoms formed by reaction (5-2-23) was fed to a lock-in (phase sensitive) amplifier which amplified only a narrow band of frequencies about the reference frequency, 20 kHz. The bandwidth was sufficient to include both sidebands and the output time constant was made much larger than the ion modulation period but small compared with the modulation period of the electrons. Consequently the output included a direct current and a 50 Hz signal which was proportional to the amplitudes of the sidebands. This was fed to a second lock-in amplifier which operated at the reference frequency $\omega_j = 50$ Hz. This amplified only the 50 Hz signal and gave a direct current output which was proportional to the magnitude of the side bands and hence to the required signal.

The overall sensitivity of the system was proportional to the product of the detector gain and the efficiency and gain of the two amplifiers. Subsidiary calibration experiments were therefore necessary. The detector was calibrated when the electron beam was replaced by a beam of light of known frequency and its efficiency was found in terms of known photodetachment cross sections. Electronic techniques were used to calibrate the amplifiers. This method is more complicated than that illustrated by Fig. 5-2-9 but it has proved useful for a SBR less than $10^{-2}$.

Similar SBR's were resolved by Dunn and Van Zyl [5] in their experiments with crossed beams of electrons and $H_2^+$ ions. The electrons were modulated at low audio frequencies and the currents of protons formed by electron impacts were amplified and isolated by a single phase sensitive amplifier.

In some experiments the signal is very small indeed and consists of particles or photons which must be counted individually. The backgrounds, which can be due to both colliding beams, may then be 3 or 4 orders of magnitude greater than the signal. A more refined pulsing technique which can then be used was developed by Harrison and his colleagues [1, 17].

The ion and electron beams were chopped as in Fig. 5-2-8 and the counts due to signal and backgrounds were fed simultaneously to two scalers which were gated by the pulse trains shown in the figure.

Assume that in one counting period of a scaler there are on average $B_i$, $B_e$ and $B_0$ background counts which can be attributed, respectively, to the

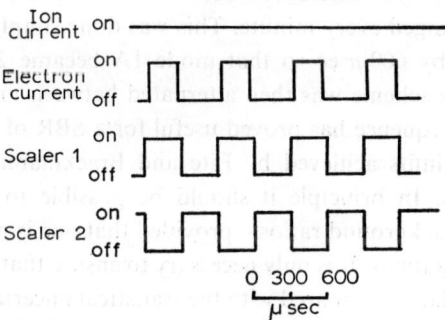

Fig. 5-2-7. Pulsing sequence suitable for crossed beam experiments in which individual collision products are counted and in which only the ion beam gives rise to extraneous signals.

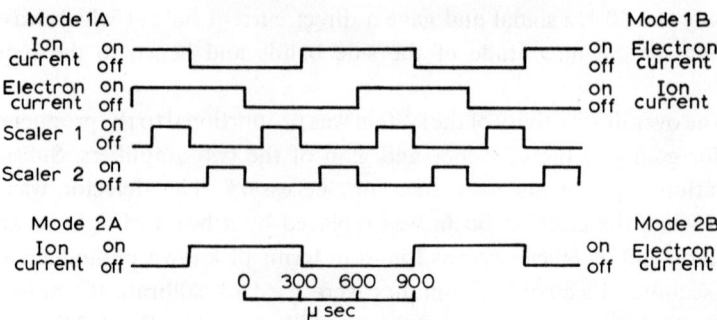

Fig. 5-2-8. Pulsing sequence suitable for crossed beam experiments in which individual collision products are counted and in which extraneous signals arise from both colliding beams.

ion beam, the electron beam and to all other sources of background. Let the average number of signal counts in the same period be $S$. Then, if the pulses are as in mode 1A, scaler 1 will on average record $S + B_e + B_i + 2B_0$ counts in two successive gating periods. In similar periods scaler 2 will record $B_i + B_e + 2B_0$. Thus, the signal is obtained by subtracting the counts registered by scalers 1 and 2 after a sufficient time.

Further refinements were made. Suppose that the ion and electron currents were not constant for the duration of each pulse. Mode 1A would then be in error because, for example, scaler 1 observed the latter half of the ion pulse whilst scaler 2 observed the first half. Every 50 ion pulses the pulsing sequence was therefore automatically alternated between modes 1A and 1B so that the scalers sampled from both ends of the current pulses. To eliminate errors due to differences between the scalers, the scalers were effectively interchanged every minute. This was done simply by delaying the ion beam pulsed by 600 μsec so that mode 1A became 2A. For the next minute the pulsing scheme was then alternated between modes 2A and 2B.

This pulsing sequence has proved useful for a SBR of only $10^{-4}$ which is similar to the limits achieved by Fite and Brackmann [18] with phase sensitive detection. In principle it should be possible to work with even smaller signal to background ratios – provided that one is prepared to count for sufficiently long times. It is only necessary to ensure that the total number of signal counts is large compared with the statistical uncertainty (i.e., 'noise') of the background. Elementary statistical arguments show that the tolerable SBR is almost inversely proportional to the square root of the counting time.

A disadvantage of this pulsing scheme is that the signals are counted for only a fraction ($\approx \frac{1}{6}$th) of the duration of the experiment, but this fraction can be enhanced if no significant part of the background is derived from one of the colliding beams. Assume, for example, that the electrons do not contribute to the background, then the ion beam need not be pulsed and the two scalers and electrons can be pulsed as shown in Fig. 5-2-7. The signal is recorded for almost half the duration of the experiment and it is given by the difference between the counts registered by scalers 1 and 2.

A new pulser, which is constructed from integrated circuits, has recently been developed by Molyneux, Dolder and Peart [19]. It performs any of the sequences illustrated by Figs. 5-2-6, 5-2-7 and 5-2-8. The pulser is smaller and cheaper than those previously used by at least an order of magnitude and it is attractive because all pulses are derived from a single train of clock pulses by strict binary logic.

In certain experiments it would be useful to combine modes 1A and 1B into the single mode and this can easily be arranged with the binary logic system. If (as is frequently the case) there is negligible difference between the efficiencies of the two scalers it should then be possible to dispense with modes 2A and 2B and reduce the whole scheme illustrated by Fig. 5-2-8 to a single mode.

None of the pulsing schemes described avoid errors due to "modulated backgrounds" which, as we shall see, can lead to major experimental difficulties. Modulated backgrounds arise from those *extraneous* contributions to the measured signal which depend in magnitude upon the interaction of the colliding charged beams. They appear in various guises to haunt the experimentalist but their general nature can be illustrated as follows.

Imagine that the apparatus shown by Fig. 5-2-1 is used to study an ionizing collision of the type represented by reaction (5-1-1) and suppose that the parent ion beam also contains some charge-stripped ions, some of which will eventually be collected at $C_2$. When these ions pass through the electron beam their trajectories will be deflected and the number which reach $C_2$ may increase. Consequently, there will be a contribution to the current collected at $C_2$ which depends upon the electron current but is unrelated to reaction (5-1-1).

This unwanted contribution is not removed by any of the pulsing techniques and it follows that particular care is necessary in experimental design to minimize all extraneous processes. In the ensuing sections it will be seen that a number of processes can give rise to modulated backgrounds and the presence of these errors is most probably revealed if the measured cross section fails to fall to zero below threshold.

## F. The separation of beams of ions and atoms

In crossed-beam experiments it is frequently necessary to resolve ion beams which have the same *velocity*. Such beams can be separated either by steady electric or magnetic fields and both methods have been used.

Sector fields of the types used in mass spectrometers (e.g. Fig. 5-2-1) have been employed in several experiments. They were arranged to give first-order focussing and the more complicated pole faces needed for higher-order corrections have not so far been needed, although three dimensional focussing could be valuable in experiments on molecular dissociation (section 5-6) in which some of the collision products are produced with appreciable energies.

Alternatively, the parallel plate electrostatic analysers illustrated in Fig. 5-2-9 can be used. These were developed from the original design of Yarnold and Bolton [20] which was subsequently elaborated by Harrower [21]. Ions incident at 45° to the first plate are deflected through 90° in the analyser and the entrance and exit apertures lie in conjugate planes.

Fig. 5-2-9a shows the analyser used by Lineberger et al. [2, 22] which consisted of two plates $9'' \times 5''$ spaced $\tfrac{15}{16}''$ apart. The aperture sizes were $\tfrac{3}{8}'' \times \tfrac{3}{4}''$ which was substantially larger than the nominal size of the ion beam ($\tfrac{1}{16}'' \times \tfrac{1}{4}''$) in this region. The spaces between centres of adjacent apertures were $2''$.

Lineberger et al. found it necessary to include a baffle plate (B) which was held at the local potential so that it did not perturb the analyser field. This baffle was necessary because the analyser was used to separate currents

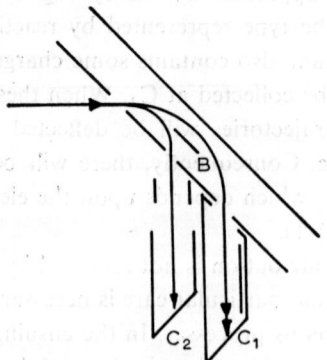

Fig. 5-2-9a. Parallel plate analyser used by Lineberger et al. [22] to separate parent and product ions.

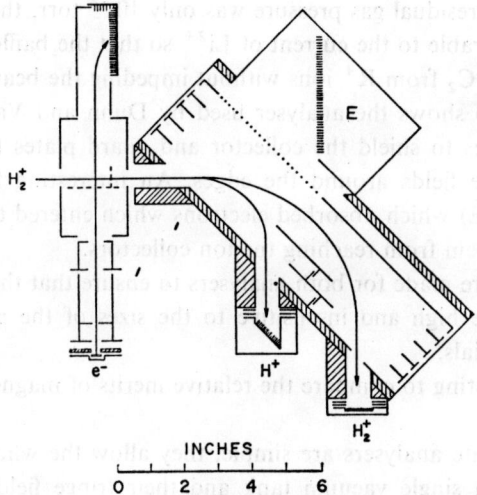

Fig. 5-2-9b. Parallel plate electrostatic analyser with electron trap (E), used by Dunn and Van Zyl [5].

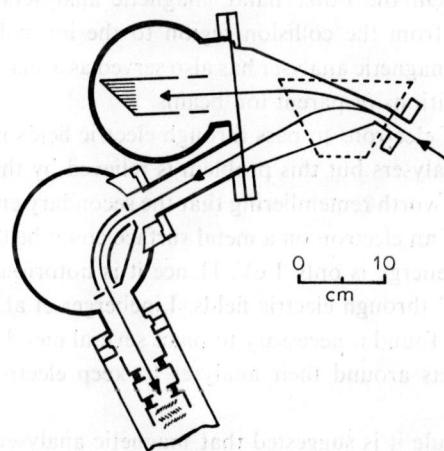

Fig. 5-2-9c. Two stage analyser used by Peart and Dolder [24].

of $Li^+$ and $Li^{2+}$ ions which differed by factors of order $10^8$. As the $Li^+$ beam passed through the analyser, collisions occurred with residual gas (R) which yielded $R^+$ ions by the following processes,

$$Li^+ + R \rightarrow Li^+ + R^+ + e,$$
$$Li^+ + R \rightarrow Li + R^+.$$
(5-2-24)

Even when the residual gas pressure was only $10^{-8}$ torr, the production of $R^+$ was comparable to the current of $Li^{2+}$ so that the baffle was needed to shield collector $C_2$ from $R^+$ ions without impeding the beams.

Fig. 5-2-9b shows the analyser used by Dunn and Van Zyl [5]. This employed baffles to shield the collector and guard plates to minimize the effects of fringe fields around the edges. An interesting feature was the electron trap (E) which absorbed electrons which entered the analyser and so prevented them from reaching the ion collectors.

Checks were made for both analysers to ensure that their transmission efficiencies were high and insensitive to the sizes of the apertures or the analyser potentials.

It is interesting to compare the relative merits of magnetic and parallel plate analysers.

Parallel plate analysers are simple, they allow the whole apparatus to be housed in a single vacuum tank and their fringe fields can easily be localized. The ion paths through them are usually short so that high transmission efficiencies should be obtained even if the beams are not very carefully collimated. On the other hand, magnetic analysers entirely prevent electrons passing from the collision region to the ion collectors. In some experiments [7] a magnetic analyser has also served as a mass spectrometer to check the compositions of parent ion beams.

The ability of electrons to pass through electric fields is the main defect of electrostatic analysers but this problem is relieved by the inclusion of an electron trap. It is worth remembering that the secondary emission (or reflection) coefficient of an electron on a metal surface could be 0.5 even when the incident electron energy is only 1 eV. Hence it is notoriously easy for electrons to "bounce" through electric fields. Lineberger et al. experienced this difficulty and they found it necessary to place several metal shields and small permanent magnets around their analyser to keep electrons from the ion collectors.

As a rough rule it is suggested that magnetic analysers should be considered when it is desired to separate beams which differ by more than 7 or 8 orders of magnitude. A 60° sector field with a 10 cm radius has proved to be adequate. The electrostatic analyser illustrated by Fig. 5-2-9b has, however, successfully resolved beams which differ by a factor $10^9$.

Peart and Dolder [24] recently combined a magnetic and electric analyser as illustrated by Fig. 5-2-9c to achieve very efficient separation of $Li^+$ and $Li^{3+}$ beams which differed by a factor of order $10^{12}$. The shapes of their vacuum tanks made it impossible to place the entrance slit of the electrostatic analyser closer than 10 cm to the focal plane of the magnetic analyser so that the ion beam was quite broad when it entered the electric field. Nevertheless, the 90° sector field generated between the cylindrical electrodes effectively rejected stray ions and gave transmission efficiencies of 99% which were very insensitive to the electrode potentials. The properties of electrostatic sector fields are discussed in several of the books on particle optics [e.g., 11, 25].

During the development of this technique a parallel plate analyser was used in place of the 90° sector field but it frequently gave transmission efficiencies less than 80%. This was due to the fact that the apertures in a simple parallel plate analyser are in regions of strong fields so that there is considerable field penetration through the apertures. In this particular apparatus the ion beam was quite broad when it reached the electrostatic field and it struck the bulging equipotentials obliquely. The large aberrations which arose were responsible for the poor transmission efficiencies.

This criticism does not apply to the experiments by Lineberger et al. and Dunn and Van Zyl in which it was possible to place the analyser close to the interaction region.

## G. Initial States of Excitation of Parent Ions

Measured cross sections of atoms or ions will be ambiguous unless the initial states of excitation are known. This problem is particularly relevant to ions because they are often formed by electron impacts which may simultaneously populate excited states. In most experiments the ions take several microseconds to pass from their source to the interaction region and this time will generally be adequate to allow all but the very long-lived states to decay.

The consequences of excited target ions were strikingly illustrated by Latypov et al. [26] during their measurements of cross sections for

$$Xe^+ + e \rightarrow Xe^{2+} + 2e. \tag{5-2-25}$$

Fig. 5-2-10 shows their measurements of this cross section for a constant (unspecified) incident electron energy. These are plotted against the potential of the anode in the ion source. It can be seen that when the ion source conditions were changed, the measured cross section varied by 300% due to variations in the proportion of metastable $Xe^+$ ions in the target beam.

The problems of initial excitation are even more serious for molecular ions because these may have vibrational as well as electronic excitation.

Another possible source of electronic excitation arises from autoionization states (inner shell electronic excitation) with lifetimes greater than about $10^{-6}$ sec. Autoionization occurs in many atoms and ions and Feldman

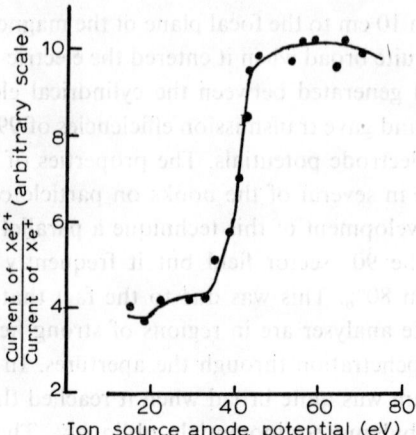

Fig. 5-2-10. Variation of measured (apparent) cross section for the ionization of $Xe^+$ with ion source conditions. The ordinates, which are proportional to the measured cross section, depend sensitively upon the potential applied to the ion source anode.

and Novick [27] recently measured excitation functions and lifetimes for autoionization states of lithium, potassium and rubidium atoms. They found that these lifetimes were 5.1, 90 and 75 μsec, respectively. Autoionization states of $Ca^+$ and $Sr^+$ ions have been studied by Kupriyanov [28] who reported lifetimes greater than $10^{-5}$ sec.

The following methods are suggested to combat these problems.

(a) Many experiments can be performed with ions which have no metastable states or with hydrogenic ions ($He^+$, $Li^{2+}$, etc) in which the metastable 2S state can be "quenched" by electric or magnetic fields.

(b) Thermionic or surface-ionization sources can sometimes be used. These do not generally populate the excited states of ions and beams of unexcited $Li^+$, $Na^+$, $K^+$, $Rb^+$, $Cs^+$, $Ba^+$, etc. have been produced in this way.

(c) When electron bombardment sources must be used, it is sometimes possible to obtain adequate ion beams by chosing an electron energy in the source which is sufficient to produce ionization but insufficient to populate excited states at a single collision [e.g. 17]. A great deal of development of sources of unexcited ions needs to be done.

(d) In experiments with molecular ions it is sometimes possible to design sources which produce beams in calculable states of vibrational excitation. This was demonstrated by Dunn et al. in their experiments with $H_2^+$ ions (section 5-6).

In some experiments it will be impossible to obtain adequate parent beams which are free from metastable excitation and this will certainly be the case when experiments are undertaken with highly-charged ions of astrophysical importance. It would then be necessary to determine the proportion of metastables in the beam because, if this could be measured and controlled, the cross sections for ionization from metastable states could be found – an intriguing prospect!

A promising approach has been suggested by Turner et al. [100] who measured the fractions of excited ions present in beams of $O^+$ and $O_2^+$. Essentially the same method has also been used by Gilbody et al. [104] to determine the metastable populations in beams of fast helium atoms. The principles are as follows.

Consider an ion beam of initial intensity $I_0$ which contains an unknown fraction ($f$) of ions in a metastable state, whilst the remainder of the ions are unexcited. If the ions pass through a cell of length $l$ filled with gas atoms of number density, $n$, then the intensity which emerges is,

$$I = I_0 [(1-f) \exp(-n\sigma l) + f \exp(-n\sigma^* l)]. \quad (5\text{-}2\text{-}26)$$

Here $\sigma$ and $\sigma^*$ respectively represent cross sections for all reactions which lead to the loss of unexcited and metastable ions in their passage through the cell. Frequently, $\sigma^* \gg \sigma$ so that, by plotting $I/I_0$ on a logarithmic scale against $n$, one obtains a straight line at the higher densities as shown in Fig. 5-2-11. An extrapolation of this line gives an intercept on the ordinate equal to $(1-f)$ and it is also possible to deduce $\sigma$ and $\sigma^*$.

An alternative method to determine $f$ is suggested by the metastable ion detector recently described by Daly, McCormick and Powell [151].

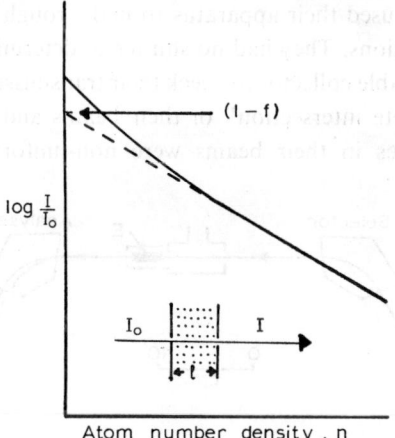

Fig. 5-2-11. Schematic representation of the absorption by a gas target of an ion beam which contains ions in the ground state and one metastable state.

## § 5-3. The ionization of positive ions by electron impact

### A. EXPERIMENTS WHICH HAVE BEEN PERFORMED WITH CROSSED BEAMS

Ionizing collisions of the type defined by eq. (5-1-1) are interesting because they determine the ionization equilibrium in thin plasmas and comparisons of the cross sections of ions and neutral atoms give some indication of the effects of ionic Coulomb fields.

Table 5-3-1 summarizes experiments which have been performed with crossed beams. The first column defines the reaction (i.e. $A^{n+} \to A^{m+}$ denotes reaction (5-1-1)); the highest electron energy used and the estimates of error made by the various authors are also shown. In most experiments the percentage errors are least when the incident electron has several times the ionization energy and it is this minimum error which is given in the table. Not all of these estimates are accurately comparable because various authors assess their errors differently. The symbol "G" in the sixth column indicates that the target beam was composed almost exclusively of ions in the ground state. Experimentalists are identified by a number which appears in the list of references.

### B. FOUR EXPERIMENTAL APPROACHES

The apparatus used by Latypov et al. [26], Lineberger et al. [2, 22] and Peart and Dolder [24] are represented by Fig. 5-3-1 and the apparatus of Dolder et al. [13] resembled that shown in Fig. 5-2-1. Some interesting points arise from a critical discussion of these experiments.

Latypov et al. used their apparatus to make rough estimates of a wide variety of cross sections. They had no shutter to determine form factors (cf. Fig. 5-2-1a) or movable collector to check their transmission efficiencies. They reported "incomplete intersection" of their beams and acknowledged that the current densities in their beams were non-uniform. They also used

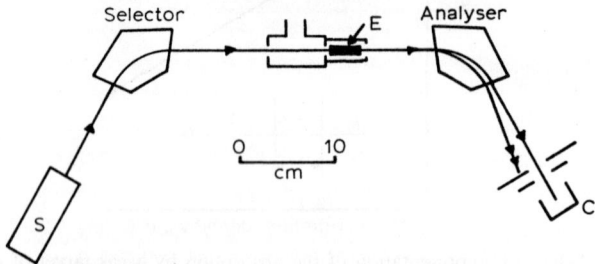

Fig. 5-3-1a. Apparatus used by Latypov et al.

TABLE 5-3-1

Measurements of cross sections for the ionization of ions by electron impact

| Reaction | Highest electron energy (eV) | Errors (%) | Initial state of excitation | Comments | Reference |
|---|---|---|---|---|---|
| $He^+ \to He^{2+}$ | 1000 | 10 | G | See note (m), p. 275 | [13] |
| $Li^+ \to Li^{2+}$ | 800 | 12 | G | | [2, 22] |
| $Li^+ \to Li^{2+}$ | 1000 | 10 | G | Results superceded by [30] | [7] |
| $Li^+ \to Li^{2+}$ | 3000 | 8 | G | | [30] |
| $Li^+ \to Li^{2+}$ | 25000 | 12 | G | See Fig. 5-3-5 | [145] |
| $Li^+ \to Li^{3+}$ | 2500 | 10 | G | | [24] |
| $N^+ \to N^{2+}$ | 500 | 11 | Included $2\,^1D_2$ and $2\,^1S_0$. Also $N_2^{2+}$ in parent beam | | [32] |
| $N^{2+} \to N^{3+}$ | 900 | 10 | — | 10% contribution from metastables | [146,163] |
| $Ne^+ \to Ne^{2+}$ | 1000 | 8 | G | | [31] |
| $Na^+ \to Na^{2+}$ | 1000 | 8 | G | Good agreement | [23] |
| $Na^+ \to Na^{2+}$ | 3500 | 6 | G | | [29] |
| $Mg^+ \to Mg^{2+}$ | 2000 | 7 | G | | [44] |
| $Mg^{2+} \to Mg^{3+}$ | 3000 | 14 | G | | [56] |
| $Ar^+ \to Ar^{2+}$ | 500 | 30 | — | | [26] |
| $Ar^{2+} \to Ar^{3+}$ | 500 | 30 | — | | [26] |
| $K^+ \to K^{2+}$ | 1000 | 8 | G | Good agreement | [23] |
| $K^+ \to K^{2+}$ | 3000 | 6 | G | | [29] |
| $Kr^{2+} \to Kr^{3+}$ | 500 | | — | | [26] |
| $Xe^+ \to Xe^{2+}$ | 500 | 30 | — | | [128] |
| $Xe^+ \to Xe^{6+}$ | 500 | | — | | [128] |
| $Xe^{+2} \to Xe^{3+}$ | 500 | | — | | [26] |
| $Ba^+ \to Ba^{2+}$ | 2000 | 6 | G | Strong autoionization contribution. See also [161] | [30] |
| $Hg^+ \to Hg^{2+}$ | 500 | | — | | [26] |
| $Hg^+ \to Hg^{3+}$ | 500 | 30 | — | Also similar results for $Xe^+$, $Kr^+$ and $Ne^+$ | [128] |
| $Hg^+ \to Hg^{4+}$ | 500 | | — | | [128] |
| $Hg^+ \to Hg^{5+}$ | 500 | | — | | [128] |
| $Hg^{2+} \to Hg^{3+}$ | 500 | | — | | [128] |

*Table* 5-3-2

| Reaction<br>Reference<br>Units<br>Notes | $He^+ \to He^{2+}$<br>[13]<br>$10^{-18}$ cm$^2$<br>(m) | $Li^+ \to Li^{2+}$<br>[22]<br>$10^{-18}$ cm$^2$ | $Li^+ \to Li^{2+}$<br>[30]<br>$10^{-18}$ cm$^2$<br>(a) | $Li^+ \to Li^{3+}$<br>[24]<br>$10^{-21}$ cm$^2$ | $Na^+ \to Na^{2+}$<br>[23]<br>$10^{-17}$ cm$^2$ |
|---|---|---|---|---|---|
| Electron energy (eV) | | | | | |
| 5.0  | –           | –          | –         | –            | –         |
| 9.0  | –           | –          | –         | –            | –         |
| 10   | –           | –          | –         | –            | –         |
| 12.5 | –           | –          | –         | –            | –         |
| 15   | –           | –          | –         | –            | –         |
| 17   | –           | –          | –         | –            | –         |
| 20   | –           | –          | –         | –            | –         |
| 30   | –           | –          | –         | –            | –         |
| 40   | –           | –          | –         | –            | –         |
| 50   | –           | –          | –         | –            | –         |
| 75   | 2.42 (27)*  | –          | –         | –            | –         |
| 100  | 3.74 (16)   | 1.69 (16)  | 1.52 (15) | –            | 1.45 (14) |
| 150  | 4.85 (11)   | 3.54 (14)  | 2.85 (10) | –            | 2.30 (11) |
| 200  | 4.85 (10)   | 4.28 (12)  | 3.92 ( 6) | –            | 2.60 (10) |
| 300  | 4.41 (10)   | 4.50 (12)  | 4.14 ( 6) | 3.50 (15, 9) | 2.67 ( 9) |
| 400  | 3.98 (10)   | 4.25 (12)  | 4.10 ( 6) | 6.50 ( 8, 9) | 2.54 ( 9) |
| 500  | 3.54 (10)   | 3.98 (12)  | 3.90 ( 6) | 8.90 ( 5, 9) | 2.33 ( 9) |
| 750  | 2.87 (10)*  | 3.21 (12)* | 3.32 ( 6) | 10.3 ( 4, 9) | 1.94 ( 8)* |
| 1000 | 2.34 (10)*  | –          | 2.85 ( 6) | 9.40 ( 5, 9) | 1.69 ( 8) |
| 1250 | –           | –          | 2.47 ( 7) | 8.40 ( 5, 9) | –         |
| 1500 | –           | –          | 2.20 ( 7) | 7.20 ( 5, 9) | –         |
| 1750 | –           | –          | 2.01 ( 8) | 6.10 ( 6, 9) | –         |
| 2000 | –           | –          | 1.86 ( 8) | 5.30 ( 6, 9) | –         |
| 2250 | –           | –          | 1.75 ( 9) | 4.80 ( 6, 9) | –         |
| 2500 | –           | –          | 1.62 ( 9) | 4.40 ( 6, 9) | –         |
| 2750 | –           | –          | 1.52 ( 9) | –            | –         |
| 3000 | –           | –          | 1.42 ( 9) | –            | –         |

*Notes from Table 5-3-2*

(a) Results to 25 keV electron energy, see Fig. 5-3-5.
(b) Cross section $0.66 \times 10^{-17}$ cm$^2 \pm 10\%$ at 3500 eV.
(c) Only random errors shown. Normalization errors were not estimated.
(d) Includes contributions from cascading.
(e) Threshold cross section $1.17\pi a_0^2$.
(f) With theoretical correction to remove cascading contribution.
(g) Errors were not estimated.
(h) All cross sections graphically interpolated from published values.

# IONIZATION BY ELECTRON IMPACT

| Na$^+$→Na$^{2+}$ [29] 10$^{-17}$ cm$^2$ (b) | K$^+$→K$^{2+}$ [23] 10$^{-17}$ cm$^2$ | K$^+$→K$^{2+}$ [29] 10$^{-17}$ cm$^2$ | Mg$^+$→Mg$^{2+}$ [44] 10$^{-17}$ cm$^2$ | Mg$^{2+}$→Mg$^{3+}$ [56] 10$^{-17}$ cm$^2$ | Ba$^+$→Ba$^{2+}$ [30] 10$^{-16}$ cm$^2$ (n) |
|---|---|---|---|---|---|
| – | – | – | – | – | – |
| – | – | – | – | – | – |
| – | – | – | – | – | – |
| – | – | – | – | – | – |
| – | – | – | – | – | 1.30 (25) |
| – | – | – | – | – | 1.60 (20) |
| – | – | – | 3.13 (20) | – | 4.10 ( 8) |
| – | – | – | 4.75 (12) | – | 4.25 ( 6) |
| – | 3.50 (16) | – | 4.74 (11) | – | 4.25 ( 6) |
| – | 6.42 (16) | 8.00 (6) | 4.55 (10) | – | 4.20 ( 6) |
| 0.70 (6) | 8.09 (16)* | 9.50 (6) | 4.25 ( 9)* | – | 4.05 ( 6) |
| 1.58 (6) | 8.56 (16) | 9.90 (6) | 4.01 ( 8) | 0.27 (3, 7) | 3.55 ( 6) |
| 2.36 (6) | 8.26 (11) | 9.10 (6) | 3.72 ( 8) | 0.86 (3, 7) | 2.16 ( 6)* |
| 2.55 (6) | 7.52 ( 9) | 8.35 (6) | 3.48 ( 7) | 1.16 (3, 7) | 2.80 ( 6) |
| 2.56 (6) | 6.41 ( 9) | 6.60 (6) | 3.06 ( 7)* | 1.36 (2, 7) | 2.27 ( 6) |
| 2.41 (6) | 5.41 ( 9) | 5.58 (6) | 2.74 ( 7)* | 1.28 (2, 7) | 1.92 ( 6) |
| 2.23 (6) | 4.69 ( 9) | 4.82 (6) | 2.46 ( 7) | 1.25 (2, 7) | 1.67 ( 7) |
| 1.87 (6) | 3.59 ( 9)* | 3.75 (6) | 2.04 ( 7)* | 1.18 (2, 7)* | 1.25 ( 7) |
| 1.58 (6) | 2.93 ( 9) | 3.05 (6) | 1.62 ( 7) | 0.99 (3, 7) | 1.01 ( 8) |
| 1.36 (6) | – | 2.58 (6) | 1.44 ( 8)* | 0.84 (3, 7)* | 0.85 ( 9) |
| 1.22 (6) | – | 2.27 (7) | 1.25 ( 8) | 0.75 (4, 7)* | 0.75 (10) |
| 1.10 (6)* | – | 2.05 (8)* | 1.17 ( 9)* | 0.69 (5, 7)* | 0.65 (10) |
| 1.00 (6) | – | 1.83 (8) | 1.08 ( 9) | 0.62 (5, 7) | 0.60 (10) |
| 0.96 (8)* | – | 1.69 (9)* | – | 0.55 (5, 7) | – |
| 0.92 (8) | – | 1.55 (9) | – | 0.54 (6, 7) | – |
| 0.85 (9)* | – | 1.50 (9)* | – | 0.52 (4, 7) | – |
| 0.75 (9)* | – | 1.45 (9) | – | 0.48 (4, 7) | – |

(i) Systematic errors quoted in paper are slightly assymetric. Mean values given here.
(j) At energies below 30 eV irregularities in ionization function, see Fig. 5-6-1. Cross section $6.82\pi a_0^2$ at 7.0 eV.
(k) Electron energies in laboratory scale. Add 2.72 eV for centre of mass energy.
(l) Electron energies in laboratory scale. Add 1.36 eV for centre of mass energy.
(m) Recent measurement have been made for electron energies between threshold and 10000 eV [160].
(n) See also, ELFORD et al. [161].

\* Interpolated from published data.

*Table 5-3-2 (cont.)*

| Reaction<br>Reference<br>Units<br>Notes | $Ne^+ \to Ne^{2+}$<br>[31]<br>$10^{-17}$ cm$^2$ | $He^+(1S \to 2S)$<br>[17]<br>$10^{-2}\pi a_0^2$<br>(c) (d) (e) | $He^+(1S \to 2S)$<br>[17]<br>$10^{-2}\pi a_0^2$<br>(e) (f) (g) | $H^- \to H$<br>[71]<br>$\pi a_0^2$<br>(h) |
|---|---|---|---|---|
| Electron energy (eV) | | | | |
| 5.0 | – | – | – | – |
| 9.0 | – | – | – | 50.1 (11, 19) |
| 10 | – | – | – | 52.0 (10, 17) |
| 12.5 | – | – | – | 56.0 (7, 16) |
| 15 | – | – | – | 57.0 (6, 16) |
| 17 | – | – | – | 55.4 (6, 16) |
| 20 | – | – | – | 51.8 (5, 15) |
| 30 | – | – | – | 43.7 (3, 14) |
| 40 | – | – | – | 35.8 (3, 14) |
| 50 | 0.37 (10) | 1.47 (3) | 1.38 | 30.6 (3, 14) |
| 75 | 1.42 (10)* | 1.05 (4)* | 0.85* | 24.9 (4, 14) |
| 100 | 2.33 ( 8) | 0.84 (4) | 0.64 | 19.9 (4, 14) |
| 150 | 2.97 ( 8) | 0.74 (4)* | 0.55* | 15.0 (4, 14) |
| 200 | 3.13 ( 8) | 0.68 (4) | 0.48 | 12.0 (5, 14) |
| 300 | 2.93 ( 8) | 0.59 (4) | 0.41 | 8.0 (5, 14) |
| 400 | 2.61 ( 8) | 0.51 (4) | 0.35 | 6.3 (7, 14) |
| 500 | 2.37 ( 8) | 0.44 (4, 5) | 0.29 | 5.4 (9, 14) |
| 750 | 1.96 ( 8)* | 0.32 (4, 5) | 0.20 | – |
| 1000 | 1.65 ( 8) | – | – | – |
| 1250 | – | – | – | – |
| 1500 | – | – | – | – |
| 1750 | – | – | – | – |
| 2000 | – | – | – | – |
| 2250 | – | – | – | – |
| 2500 | – | – | – | – |
| 2750 | – | – | – | – |
| 3000 | – | – | – | – |

*For notes see page* 274–275.

| H⁻→H [15] $\pi a_0^2$ (h) (i) | O⁻→O [15] $\pi a_0^2$ (h) (i) | $H_2^+$→H⁺ [14] $\pi a_0^2$ (i) (j) | $H_2^+$→H⁺ [5] $\pi a_0^2$ (k) | $D_2^+$→D⁺ [5] $\pi a_0^2$ (l) |
|---|---|---|---|---|
| – | – | 6.08 (24, 18) | – | – |
| 41.3 (10, 19) | 6.10 (13, 19) | 5.93 (9, 10) | – | – |
| 43.0 (10, 19) | 6.40 (13, 19) | 5.90 (6, 9)* | 8.76 (18, 18) | – |
| 47.5 (12, 19) | 7.15 (9, 19) | 5.35 (9, 10)* | 6.5 (17, 18) | – |
| 49.5 (12, 19) | 7.90 (9, 19) | 4.95 (8, 9)* | 5.44 (16, 18) | – |
| 50.0 (9, 19) | 8.45 (8, 19) | 4.90 (6, 9)* | 5.28 (12, 16)* | – |
| 49.7 (5, 19) | 8.70 (8, 19) | 4.75 (6, 9)* | 5.02 (10, 14)* | – |
| 43.3 (5, 19) | 8.80 (8, 19) | 4.15 (5, 8)* | 4.36 (10, 10)* | – |
| 36.6 (5, 19) | 9.42 (6, 19) | 3.48 (4, 8)* | 3.97 (10, 10)* | – |
| 34.5 (5, 19) | 8.75 (5, 19) | 3.12 (3, 8)* | 3.68 (10, 10) | 3.44 (4, 9) |
| 27.8 (5, 19) | 7.38 (6, 19) | 2.50 (3, 8)* | 3.02 (6, 9)* | – |
| 23.0 (5, 19) | 6.22 (6, 19) | 2.18 (4, 8)* | 2.49 (1, 8) | 2.40 (2, 8) |
| 18.2 (6, 19) | 5.31 (6, 19) | 1.78 (3, 8)* | 1.95 (5, 8) | – |
| 16.6 (7, 19) | 4.65 (7, 19) | 1.46 (3, 8)* | 1.55 (5, 7) | 1.53 (3, 7) |
| 12.3 (8, 19) | 3.65 (7, 19) | 1.10 (3, 8)* | 1.19 (4, 7) | – |
| 10.6 (9, 19) | 3.25 (7, 19) | 0.91 (2, 8) | 0.90 (3, 8) | 0.90 (1, 8) |
| 8.9 (9, 19) | 2.87 (7, 19) | 0.80 (5, 8)* | 0.80 (4, 7)* | – |
| – | – | 0.58 (4, 8)* | 0.46 (6, 7) | – |
| – | – | 0.47 (4, 8)* | 0.47 (9, 7) | – |
| – | – | – | 0.40 (9, 7)* | – |
| – | – | – | 0.35 (9, 7) | – |

electron currents up to 20 mA which were probably greater than the values of $J'_{max}$ defined by equation (5-2-19).

Several useful improvements could have been made to their apparatus. For example, there was only one diffusion pump between the ion source and the selector magnet so that the production of ions by charge stripping was unnecessarily large, and no modulation technique was used to distinguish these ions from those formed by electron impacts. The magnets ($M_1$ and $M_2$) deflected ion in the same sense whereas Dolder et al. deliberately arranged their magnets in opposite senses to achieve partial cancellation of fringe fields around the electron gun; even so they found it necessary to place a soft iron shield around the electron gun and collector. Latypov et al. used the same collector (C) for their product and target ions. This is undesirable because the two ion currents differed by several orders of magnitude and, if parent ions are allowed to strike the inner walls of an apparatus, they produce scattered ions or electrons which may obscure tenuous beams of product ions.

In spite of these criticisms the Russian experiments were valuable. They demonstrated that a wide variety of cross sections are within a factor 5 of the predictions of Thomson's classical theory. Further, they showed (e.g. Fig. 5-2-10) that when some of the target ions are in metastable states, the measured cross section can vary by factors of 4 or 5, depending on the relative population of these states in the ion source.

The corollary is surely that there is little point in performing further experiments in which target beams contain substantial unknown proportions of excited ions.

The experiments by Lineberger [22, 23] and his colleagues were elegant and this stemmed mainly from their use of thermionic ion sources. These sources are extremely simple and compact and they are not easily "poisoned" by impurities. They require no separate vacuum system and they produce adequate stable beams of alkali metal ions which are entirely in the ground state and in which at least 99% of the ions are of a given type.

The purity of the ion beam obviated the need for a selector, although electrostatic deflectors were used to remove slow ions and neutral particles from the beams. In experiments with lithium, isotopically-pure material was used to produce beams of $^7Li^+$.

These experiments also used the electrostatic analyser illustrated by Fig. 5-2-9a. This analyser is compact and its fringe fields can easily be localized so that it was possible to house the whole apparatus in a single tank which was only 21" in diameter and 6" high. Only one diffusion pump

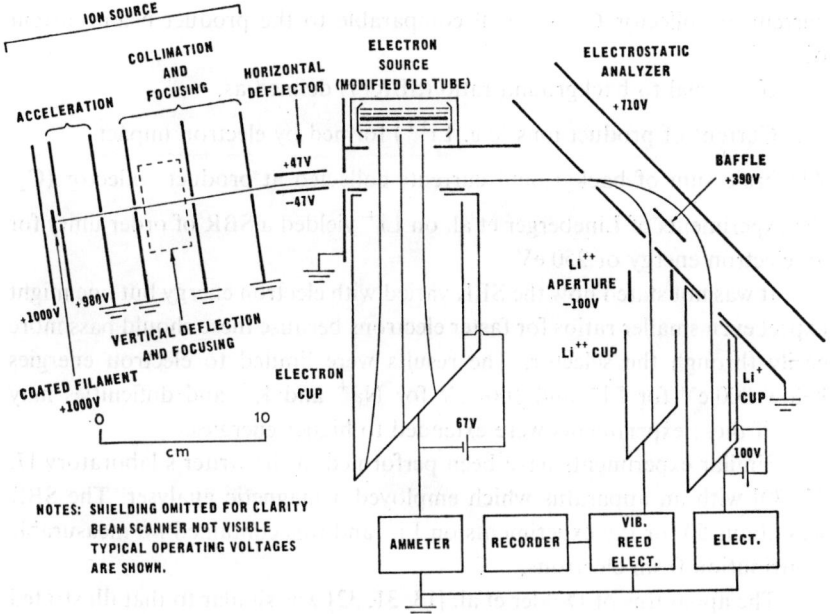

Fig. 5-3-1b. Apparatus used by Lineberger et al.

was needed to evacuate the whole apparatus to $10^{-8}$ torr. At these pressures there was no evidence of the production of ions by charge stripping (this was a surprising feature of the experiments with alkali metal ions) so that the experiments could be performed with unmodulated beams, although it was verified that the measured cross sections were unchanged when the beam modulation technique illustrated by Fig. 5-2-6 was used.

The apparatus included a sliding shutter (Fig. 5-2-1), so that corrections could be made for non-uniform current densities, and checks were made to ensure that the measured cross section did not depend on electron current or ion beam energy. It was estimated that the cross sections measured for $Li^+$, $Na^+$ and $K^+$ ions are accurate to better than $\pm 10\%$ over quite a wide range of electron energies.

The main criticism of these experiments concerns the choice of an electrostatic analyser. It was pointed out in section 5-2-F that electrons can pass through electric fields with surprising ease so that a small proportion of electrons from the electron gun reached the ion collectors. Lineberger et al. largely overcame this by placing metal shields and small magnets around their analyser but, in their experiments on $Li^+$, the stray electron current at collector $C_2$ was still comparable to the product beam current of $Li^{2+}$.

If a signal to background ratio (SBR) is defined as,

$$\frac{\text{Current of product ions (e.g. } Li^{2+}\text{) formed by electron impact}}{\text{Algebraic sum of background currents collected at product collector } (C_2)}$$

the experiments of Lineberger et al. on $Li^+$ yielded a SBR of order unity for an electron energy of 250 eV.

It was not stated how the SBR varied with electron energy but one might expect even smaller ratios for faster electrons because these should pass more easily through the selector. The results were limited to electron energies below 800 eV for $Li^+$ and 1000 eV for $Na^+$ and $K^+$ and difficulties may arise if these experiments were extended to higher energies.

Similar experiments have been performed in the writer's laboratory [7, 29, 30] with an apparatus which employed a magnetic analyser. The SBR was about 50 for the experiments on $Li^+$ and this contained no measurable contribution from electrons.

The apparatus of Dolder et al. [13, 31, 32] was similar to that illustrated by Fig. 5-2-1. It was used to measure cross sections of $He^+$, $Ne^+$ and $N^+$ ions which were obtained from an oscillating electron source, similar to that described by Nielsen [11, 33]. An electrostatic einzel lens (L) and two sets of horizontal and vertical deflector plates ($D_1$ and $D_2$) were used to focus the target beams.

The electron gun included four gridded electrodes one of which was biased to retain slow secondary electrons in the gun. Slow electrons would have insufficient energy to ionize ions but they could have reached the electron collector. If this had occurred the cross section would have been underestimated. It is not clear whether this precaution was worthwhile because Peart and Dolder used one of these guns to study alkali ions whilst Lineberger et al. used a simple triode; both groups obtained very similar results.

Several criticisms can be made of the apparatus. For example, there were only three stages of differential pumping between the ion source and the selector magnet ($M_1$) so that an unnecessarily large number of ions which had been formed by charge stripping were collected at $C_2$. These charged stripped ions were responsible for the modulated backgrounds as described in section 5-2-F so that in the experiments with $He^+$ the measured cross section did not quite fall to zero at the ionization threshold (54.4 eV). Corrections of a few percent were therefore applied to the measured cross sections of He$^+$ but it would have been better to eliminate these problems by the inclusion of more stages of differential pumping between the source and the selector magnet (see very recent experiment, [160]).

In the experiments on N$^+$ even larger signals were obtained below the ionization threshold of N$^+$. These were attributed to the dissociative ionization of N$_2^{2+}$ ions which were not separated from the target beam by the selector magnet. This interpretation could have been checked if isotopically-enriched nitrogen had been fed to the source; beams of $(N_{14}N_{15})^{2+}$ could then have been isolated and their dissociative ionization studied.

Several features of this type of experiment are illustrated by recalling the first abortive attempt to measure a cross section of He$^+$. No modulation technique was used in this early experiment and the ratio of currents $I^{2+}/I^+$ collected at $C_2$ and $C_1$ was plotted against the electron current $(J)$. It was hoped to find a linear relation from which the cross section could be deduced.

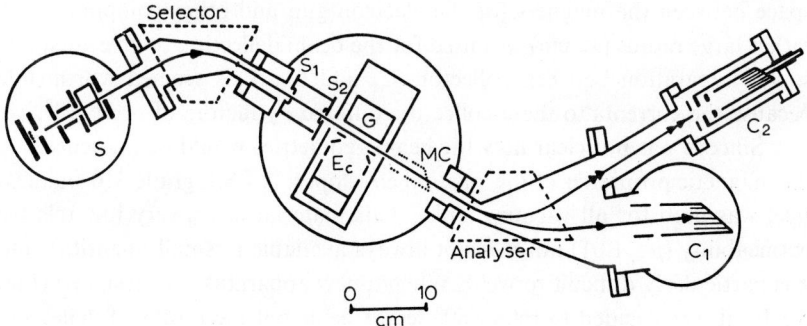

Fig. 5-3-1c. Apparatus used by Peart and Dolder.

The actual result is illustrated schematically by Fig. 5-3-2. For zero electron current there was an intercept (OA) on the ordinate which was attributed to charge-stripped ions. The region AB was non-linear, primarily because increases in electron current caused the electron gun and collector to outgas. Consequently the production of He$^{2+}$ ions by charge stripping and electron impact ionization increased simultaneously. For larger electron currents $(J > J'_{max})$ He$^{2+}$ ions were deflected from the beam by the electronic space charge and when these losses predominated (CD) the cross section was apparently negative!

These difficulties were largely overcome by modulating the beams and using electron currents less than $J'_{max}$.

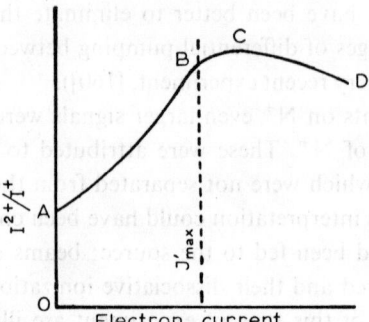

Fig. 5-3-2. Type of result obtained in an early attempt to measure the ionization cross sections of He$^+$ ions by electron impact.

When the apparatus was designed, 60° sector magnetic fields were chosen for the analyser and selector because this geometry leaves ample space between the magnets for the electron gun and other components. A rather large radius (15 cm) was used for the beam deflection to give adequate spatial separation between collectors $C_1$ and $C_2$. This seemed worthwhile because the currents to these collectors differed by factors of order $10^8$.

Since it was not clear how the beam geometries would be influenced by the magnetic properties of the vacuum envelopes, A.I.S.I. grade 310 stainless steel was used for all vacuum tanks. This material has a very low relative permeability ($\mu \lesssim 1.01$) but it is not always available in small quantities and it is particularly difficult to weld. When a new apparatus was designed (Fig. 5-3-1c) it was decided to retain 60° sector fields but the radius of deflection was only 10 cm. This reduced the size of the apparatus by about $\frac{1}{3}$. The new apparatus was largely built from A.I.S.I. grade 321 stainless steel ($\mu \approx 1.5$) although the tanks between the magnet polepieces were made from grade 316/Ti ($\mu \approx 1.1$). These are both titanium-stabilized steels which weld easily to themselves or to each other and, by contrast to grade 310, they require no heat treatment after welding.

Fig. 5-3-1c shows a recent form of this apparatus which was developed from that described by Wareing and Dolder [7]. In the form illustrated, the apparatus has been used with three different types of ion source to measure cross sections of Li$^+$, Mg$^+$ and Ba$^+$. Experiments have also been performed on Na$^+$ and K$^+$ ions but for these the selector magnet was not included.

When the selector was used, ions of a given type were selected and collimated by two slits $S_1$ and $S_2$ which were 3 mm high and, respectively, 4 mm and 1 mm wide. The space between the slits was magnetically shielded by a mumetal tube (B). This collimation ensured that transmission efficiencies* greater than 99% were easily obtained and the electrostatic deflectors near the ion source could be used to align the parent beam so that the form factors were close to unity. The apparatus included a movable collector ($C_m$) and a sliding shutter to check transmission efficiencies and measure form factors. A magnetic analyser was used to separate the target and product ion beams and very high vacuum techniques were employed so that the pressure in the interaction region was about $2 \times 10^{-8}$ torr.

The selector magnet was essential when experiments were made with $Mg^+$ because an electron bombardment source [34, 35] was used which simultaneously produced a variety of ions. A simple surface-ionization source [30] was developed for $Ba^+$ and, since this also produced a few percent of impurity ions, the selector was also necessary.

The thermionic source of $^7Li^+$ produced very few impurities, mainly $Na^+$ and $K^+$, but if these heavier ions were allowed to accompany the target beam they were not greatly deflected by the analyser and collided with the walls of the vacuum tanks. After reflections some of these ions entered collector $C_2$ and contributed to backgrounds. It was therefor worthwhile to retain the selector even for experiments with $^7Li^+$; a signal to background ratio (SBR) of 50 was then obtained.

Fig. 5-3-1c therefore shows an apparatus which attains a good SBR primarily from the employment of a magnetic analyser. Further advantages accrue from mass selection and rigorous collimation of the parent ion beam and the attainment of very high vacua. Beam modulation techniques were not necessary for experiments with $Li^+$, $Na^+$ and $K^+$ ions because charge-stripping was completely negligible. They were, however, required for the experiments with $Mg^+$ and $Ba^+$.

Recently this apparatus has been further developed by the inclusion of the compound analyser and single particle detector shown in Fig. 5-2-9c. In this form it has been used to measure cross sections for, $Mg^{2+} + e \rightarrow Mg^{3+} + 2e$ and $Li^+ + e \rightarrow Li^{3+} + 3e$. In the latter experiment the SBR was of order 5 in spite of the extremely small cross sections.

## C. Results of the Experiments

Perhaps it is worth emphasizing that these experiments give absolute cross sections, without assuming any other calculated or measured cross section. This is not the case for crossed beam experiments with atomic hy-

---

* Transmission efficiency is the percentage of a beam of given ions which passes from the interaction region to the appropriate collector.

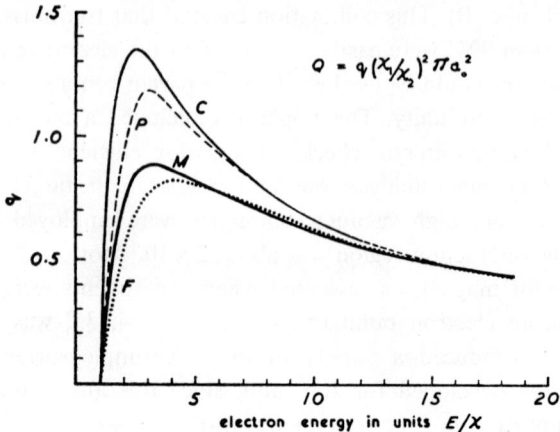

Fig. 5-3-3. Cross sections for $He^+$ and atomic hydrogen which have been reduced by classical scaling: $\chi_2$ and $\chi_1$ are the respective ionization energies. Curves C and P are based on Coulomb-Born and plane-wave Born calculations for hydrogen. Curves M and F represent measurements for $He^+$ and atomic hydrogen, respectively.

drogen or certain other neutral atoms in which the results are frequently normalized to calculations based on Born's approximation. Consequently, some of the results to be described provide useful quantitative tests of theoretical approximations.

The attractive Coulomb field of a positive ion should enhance the cross section for ionization by electrons which have little more than threshold energy, but fast electrons should scarcely be influenced. This view is supported by Fig. 5-3-3 which show calculations by Burgess [36] for, $He^+ + e \rightarrow He^{2+} + 2e$, which are based on the plane wave Born (P) and Coulomb-Born (C) approximations. The Coulomb field is seen to enhance the cross section only near threshold.

Measurements of this cross section [7] are represented by curve M and it can be seen that there is excellent agreement with theory at the higher energies. The disparity nearer threshold is partly due to the neglect of electron exchange and the inadequacy of Born's approximation at these energies. More detailed calculations, which agree well with experiment, have been made by Rudge and Burgess [37], Burke and Taylor [38], Percival [40] and Rudge and Schwartz [see 150].

The effect of the ionic Coulomb field upon the cross section is also illustrated by comparing the measurements for $He^+$ with those made for atomic hydrogen by Fite and Brackmann [18]. Curve F shows the latter

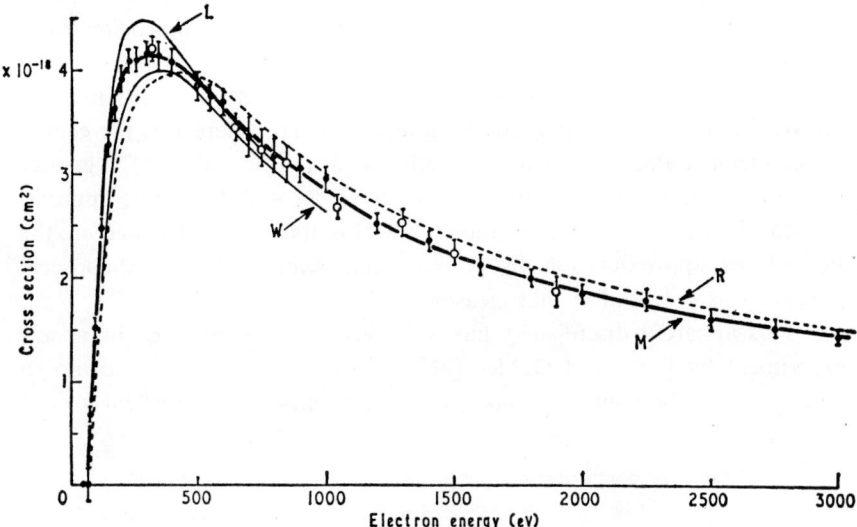

Fig. 5-3-4. Cross sections for the ionization of Li$^+$ ions measured by Lineberger et al. (curve L); Wareing and Dolder (W); Peart and Dolder (M). The open circles represent results of experiments with modulated beams whilst the closed circles were obtained with direct current beams. The broken curve (R) shows ionization cross sections for helium measured by Rapp and Englander-Golden which have been scaled classically.

results after they have been scaled classically and it can be seen that the ionic field only enhances the cross section near threshold. The classical scaling law, which follows from Thomson's theory, implies that the cross section ($\sigma_1$ and $\sigma_2$) of two isoelectronic ions are related by

$$\frac{\sigma_1}{\sigma_2} = \left(\frac{\chi_2}{\chi_1}\right)^2, \tag{5-3-1}$$

provided that the incident electron energies are expressed in terms of the respective ionization energies ($\chi_1$ and $\chi_2$).

The effect of ionic Coulomb fields is further illustrated by Fig. 5-3-4. This shows measurements for Li$^+$ + e → Li$^{2+}$ + 2e by Lineberger et al. [2,22] (curve L), Wareing and Dolder [7] (W) and Peart and Dolder [30] (M). These are compared with cross sections for helium, measured by Rapp and Englander-Golden [41] (R), which have been scaled classically. It can be seen that the curves agree quite well at the higher energies but, near threshold the results for Li$^+$ tend to lie above those for helium.

At energies above about 1000 eV curve M is within a few percent of the results of recent calculations by Moores [147] and McDowell [148] which were respectively based on the Coulomb-Born and Born approximations, without corrections for electron exchange.

A problem arose when the experimental cross sections ($\sigma$) for Li$^+$ (curves L and M) were plotted as $\sigma E$ against $\log E$, where $E$ is the energy of the incident electrons. A linear result was found for values of $E$ greater than 250 eV and this is *qualitatively* in agreement with the Bethe approximation. However, when the gradient of this line was calculated in the Bethe-Born approximation by Inokuti and Kim [149], the theoretical gradient was only half of that measured.

This apparent discrepancy has only recently been resolved by a new experiment by Peart and Dolder [145] who used electron energies up to 25 keV. It can be seen (Fig. 5-3-5) that the results are in excellent accord

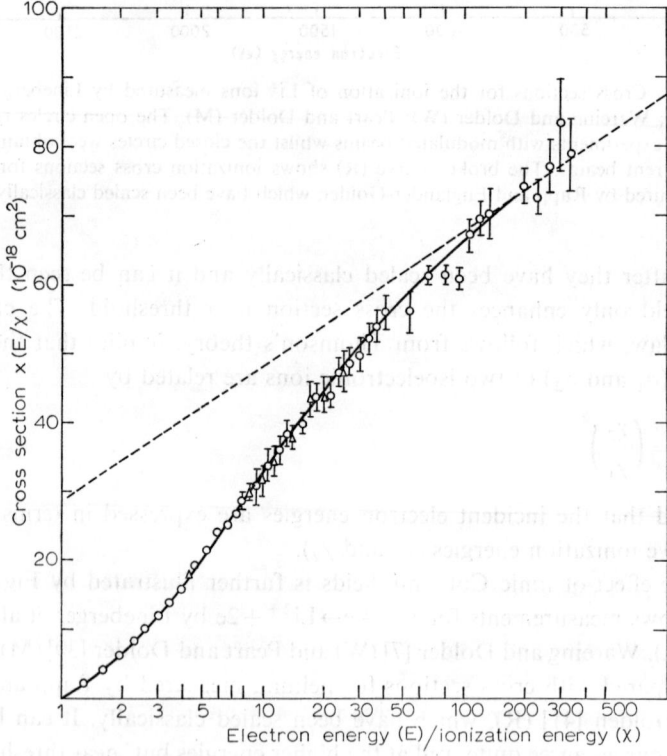

Fig. 5-3-5. Cross sections ($\sigma$) measured by Peart and Dolder [30, 145] for single ionization of Li$^+$ plotted as $\sigma E/\chi$ against $\log(E/\chi)$ where $E$ is the incident electron energy and $\chi$ is the ionization energy. The dashed line represents the gradient calculated by Inokuti and Kim [149] in the Bethe-Born approximation.

with the Bethe approximation for Li$^+$ only when $E \gtrsim 7$ keV. Relativistic corrections should be applied to experimental results at these energies but, when this is done, the agreement between theory and experiment is not significantly changed. It seems that Li$^+$ has a particularly "tight" electronic structure which complies with the assumptions of the Bethe approximation only at very high energies.

Qualitatively similar results also exist for neutral helium but in this case the Bethe approximation is valid when $E \gtrsim 1$ keV. The corresponding situation for H$^-$ is not entirely clear and will be briefly discussed in section 5-5-D.

New measurements have been made on He$^+$ for electron energies to 10 keV [160]. Comparison with calculations by OMIDVAR [162] based on the Born and Bethe approximations shows that the former is valid for $E \gtrsim 0.9$ keV and the latter when $E \gtrsim 3.5$ keV. The new results are in excellent agreement with the earlier measurements by DOLDER et al. [13] but a useful improvement in accuracy at the lower energies has now been attained.

Classical scaling has also been used to compare cross sections of Na$^+$

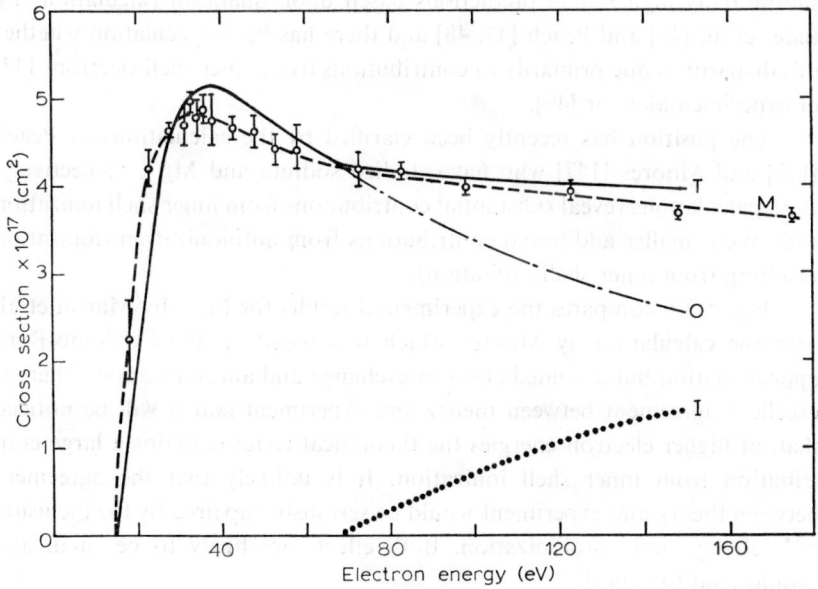

Fig. 5-3-6. The dashed curve (M) locates measurements by Martin et al. [44] for the single ionization of Mg$^+$ ions. The continuous curve (T) is the cross section calculated by Moores [147]. At electron energies greater than the inner shell ionization energy, 68 eV, curve T is the sum of the contributions from inner and outer shell ionization represented by curves I and O, respectively.

with neon and $K^+$ with argon. In these cases it was necessary to apply corrections (based on measurements [42, 43]) to the cross sections of neon and argon to exclude contributions from multiple ionization. Classical scaling worked well for $K^+$ but failed, by a factor of about 2, in the case of $Na^+$ (Fig. 5-3-8).

Cross sections for another isoelectronic pair,

$$Mg^+ + e \to Mg^{2+} + 2e$$

and

$$Na + e \to Na^+ + e$$

have been respectively measured by Martin et al. [44] and McFarland and Kinney [45]. Both experiments employed crossed beams but, to obtain absolute results, one must determine the particle density in the parent beams. This presents no difficulty in the case of $Mg^+$ but a surface ionization detector must be used for sodium and this introduces the possibility of error (see Chapter 2 of this book).

The matter is of interest because the measurements for sodium are substantially higher than predictions based upon quantum calculations by Bates et al. [46] and Peach [47, 48] and there has been speculation whether this disparity is due primarily to contributions from inner shell electrons [47] or experimental error [49].

The position has recently been clarified by the calculations of Peach [152] and Moores [147] who have studied sodium and $Mg^+$, respectively. Both calculations reveal substantial contributions from inner shell ionization and much smaller additional contributions from autoionization (ionization resulting from inner shell excitation).

Fig. 5-3-6 compares the experimental results for $Mg^+$ by Martin et al. with the calculation by Moores which was based on the Coulomb-Born approximation but excluded electron exchange and autoionization. There is excellent agreement between theory and experiment and it will be noticed that, at higher electron energies the theoretical result contains a large contribution from inner shell ionization. It is unlikely that the agreement between theory and experiment would be seriously impaired by the inclusion of exchange and autoionization. Both effects are likely to be small and should tend to cancel.

Bely [50] recently predicted that large contributions from autoionization should occur in the ionization functions of sodium-like ions and he discussed the astrophysical implications. Experiments have so far failed to detect autoionization in sodium or $Mg^+$ and so it seems certain that Bely's theory

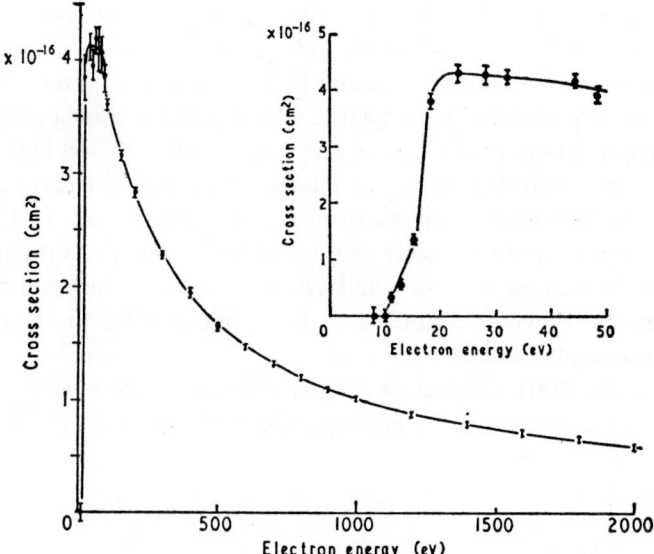

Fig. 5-3-7. Measured ionization cross sections of Ba$^+$. The inset shows the rise in the cross section between 16 and 18 eV due to the onset of autoionization.

considerably overestimates the magnitude of these effects. Nevertheless, autoionization dominates the ionization functions of certain ions. This was illustrated by Peart and Dolder [30] who measured cross sections for Ba$^+$. The results are illustrated by Fig. 5-3-7 in which the inset shows details of the results at lower electron energies. It can be seen that the cross section increases abruptly by a factor of three at the autoionization threshold ($\approx 17$ eV) and more detailed results, which show structure in the ionization function just above the autoionization threshold, are given in the original paper.

Simple classical arguments by Seaton [51] suggest that the excitation cross section of an ion should be finite at threshold and this is supported by quantum calculations (e.g. Burke et al. [52]) and by the experiment of Dance et al. [17] which will be discussed in section 5-4-B. The autoionization cross section for Ba$^+$ (inner-shell excitation) was also found to rise almost to a maximum within 2 eV of threshold. This finite rate of rise could entirely be explained by the energy spread of electrons in the beam and so it is concluded that, within the limits of resolution (2 eV), the autoionization cross section rises discontinuously at threshold to nearly its maximum value.

In spite of the crude underlying assumptions, the simple scaling law expressed by equation (5-3-1) works surprisingly well for measured cross sections of the following isoelectric pairs, H(He$^+$); He(Li$^+$) and Ar(K$^+$) but the law fails for Ne(Na$^+$). Seaton [53] pointed out that this "classical" law can also be deduced from quantum theory and it should hold exactly for hydrogenic ions in the limit of high ionic charge ($Z$) and high electron velocities, provided that relativistic effects are ignored. Evidence of convergence, with increasing $Z$, in an isoelectronic series is seen in Fig. 5-3-8 which compares measured cross sections of Mg$^{2+}$ ions [56] with classically scaled cross sections for Na$^+$ and Ne. Corrections, which are based on measurements, have been applied to the results for Ne to eliminate the contributions of multiple ionization.

Recently, Peart and Dolder [24] have measured cross sections for the double ionization of Li$^+$ by electron impact (Fig. 5-3-9). No theoretical

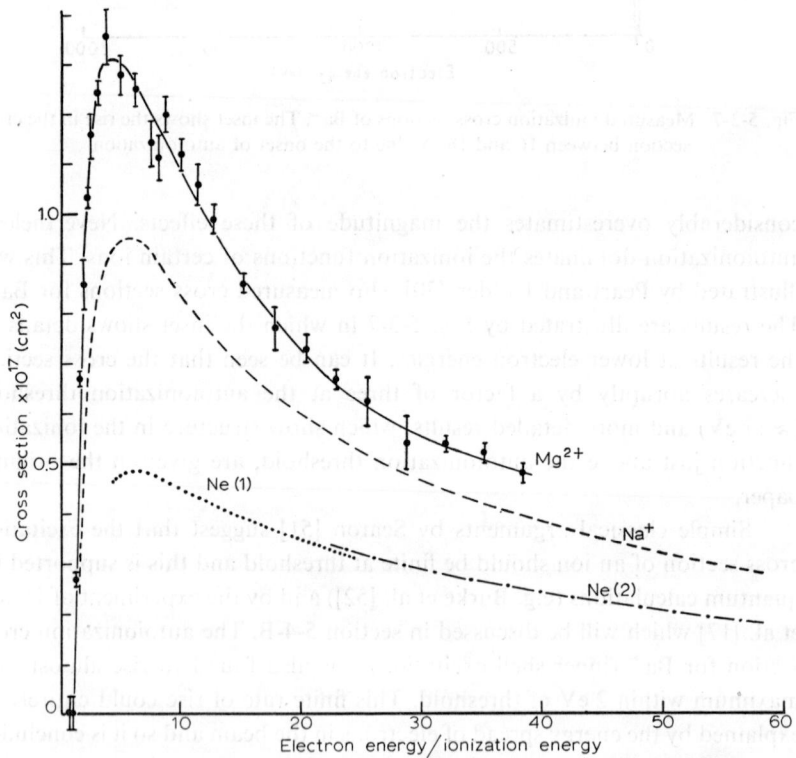

Fig. 5-3-8. The experimental points show cross sections for the ionization of Mg$^{2+}$ to Mg$^{3+}$. Classically scaled measurements for Na$^+$ and neon are also included. The curves for neon represent partial cross sections for single ionization.

work has been performed on this cross section although several papers have discussed the analogous process in helium. In particular, Byron and Joachain [54, 55] suggested that calculations of the double ionization of helium should depend very sensitively upon the choice of wavefunctions. A comparison between theoretical experimental values might therefore provide a stringent check of wavefunctions. Further discussions of the double ionization of helium are given in papers reviewed by Rudge [150] and McDowell and Coleman [153].

Similar theoretical methods should be applicable to $Li^+$, provided the incident electron energies are large, and it is hoped that the results in Fig. 5-3-9 will provide a useful comparison.

The results of a number of selected experiments on electron ion collisions are summarized by Table 5-3-2. In some cases it has been necessary to interpolate the published results to obtain a cross section at the listed electron energy and these interpolated values are marked with an asterisk. Estimates of error (random and systematic) are denoted by the figures in parentheses (these are based on the estimates in the original papers). Thus, for example, the cross section for the ionization of $He^+$ by 100 eV electrons is $3.74 \times 10^{-18}$ cm$^2 \pm 16\%$. Where two numbers are included in the parentheses, the first refers to the estimate of random error and the second to systematic error. More detailed results are, in some cases, given in the various papers and in the report by Kieffer and Dunn [76].

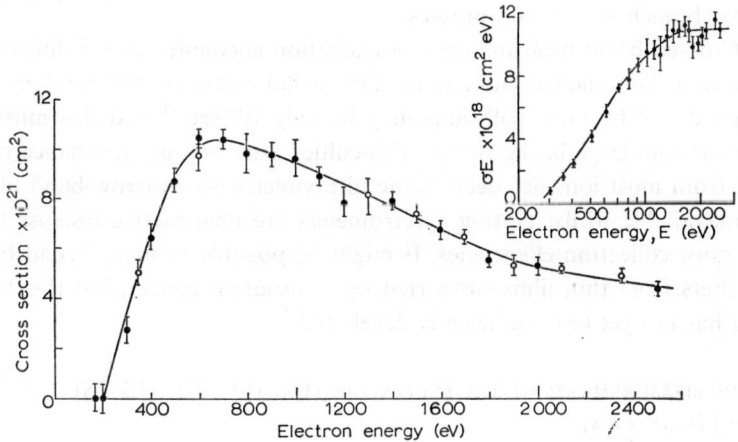

Fig. 5-3-9. Measured cross sections ($\sigma$) for the ionization of $Li^+$ to $Li^{3+}$ plotted against electron energy ($E$). The brackets denote random errors expressed as 90% confidence limits. The inset shows $\sigma E$ plotted against $E$ on a logarithmic scale.

## § 5-4. The excitation of positive ions by electron impact

A. INTRODUCTION

Apart from measurements of autoionization (inner shell excitation) two crossed beam methods have been developed to study the excitation processes defined by equation (5-1-2).

The first of these was described by Dance et al. [17] who studied, $He^+(1S) + e \to He^+(2S) + e$. Their method is restricted to this particular transition in hydrogen-like ions and, although similar principles were used by Fite and Brackmann [57] to investigate the excitation of atomic hydrogen, the two experiments are very different in technique.

The hydrogen-like structure of $He^+$ greatly simplifies the corresponding theory and the experimental results can be made absolute by comparing them, at high electron energies, with theoretical predictions based on the first Born approximation, assuming that for direct excitation the Born series converges to its first term. Dance et al. were the first to show the finite cross section at threshold which is characteristic of ions and their experiment will be discussed in section 5-4-B.

A more versatile approach has been developed simultaneously by Bacon and Hooper [58] and Lee and Carelton [59]. Both groups use a photomultiplier to detect light emitted when electron and ion beams collide and the required frequency is selected by an interference filter. It is already clear that this technique has great potential and points the way to a new, but thorny, branch of collision physics.

Crossed beam measurements of excitation encounter severe difficulties which stem from the tenuous nature of the colliding beams. The total photon flux produced by these collisions may be only $10^4$ $sec^{-1}$ and this must be resolved from large backgrounds. Difficulties arise because resonance radiation from most ions lies deep in the ultraviolet where narrow-band filters are unavailable whilst grating spectrometers are unattractive because they have poor collection efficiencies. It might be possible to make broad band UV filters from thin films supported by transparent gauzes, but this technique has not yet been sufficiently developed.

B. THE MEASUREMENTS OF THE EXCITATION FUNCTION $He^+(1S-2S)$ BY DANCE ET AL.

1. *The method*

The apparatus is illustrated by Fig. 5-4-1. A beam of $He^+$ ions 2 mm

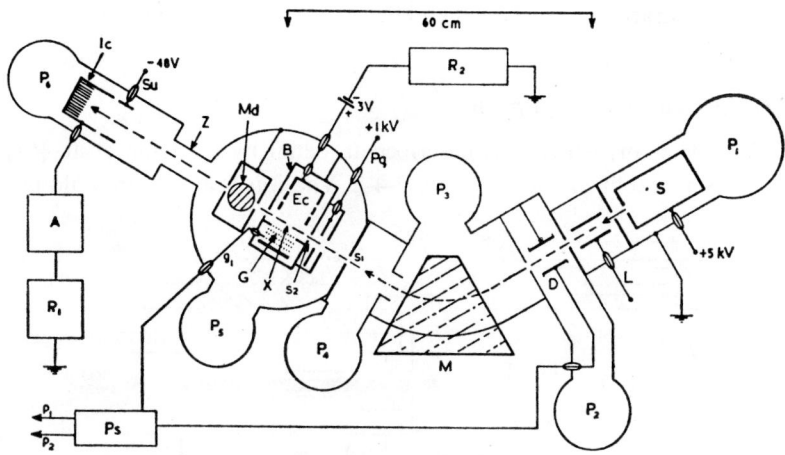

Fig. 5-4-1. Schematic plan view of the apparatus. S, ion source; L, einzel lens; D, deflector plate for pulsing the ion beam; M, electromagnet; $s_1$ and $s_2$, collimating slits; Pq, pre-quench electrode; X, collision region; B, magnetic shield for the electron gun G; $g_1$, control grid of the electron gun; Ec, electron collector; Md, He$^+$ (2S) ion detector; Z, impedance; Ic, ion collector; Su, secondary electron suppressor; A, d.c. amplifier; $R_1$ and $R_2$, recorders; $P_1$ to $P_6$, diffusion pumps; Ps, pulse generator; $p_1$ and $p_2$, gating pulses for scalers.

high and 1 mm wide, which was selected by apertures $S_1$ and $S_2$, was bombarded by electrons which passed from the gun (G) to the electron collector (Ec). Collisional excitation of He$^+$ occurred in the interaction region (X) but virtually all states, except the metastable 2S state, decayed almost immediately. Consequently, the beam of He$^+$ ions which arrived at the photon detector (Md) consisted primarily of ground state ions mixed with a small proportion ($\sim 10^{-9}$) of metastable He$^+$ (2S).

In a field free region the lifetime ($\tau$) of the $2\,^2S_{\frac{1}{2}}$ state of He$^+$ is $2.2 \times 10^{-3}$ sec but Lamb and Skinner [68] showed that, in the presence of an electric field $U$ (V/cm), $\tau$ is only $1.6 \times 10^{-2} U^{-2}$ sec. This follows because the $2\,^2S_{\frac{1}{2}}$ and $2\,^2P$ states become mixed by the field so that the ions can decay.

Dance et al. arranged an electric field of order 2 kV cm$^{-1}$ to act on the He$^+$ beam in the region of the photon detector (Md). This caused about 90% of the He$^+$ ($2^2\,S_{\frac{1}{2}}$) ions to decay within a distance of 0.5 cm with the emission of 304 Å photons. Some of these photons entered the detector where they were converted to electrical pulses which actuated scalers. The efficiency ($\eta$) of the detection system was defined as the counting rate due to the required signal divided by the rate of de-excitation of He$^+$ ($2\,^2S_{\frac{1}{2}}$); it was found that $\eta$ was $(0.83 \pm 0.26)\%$. The signal count was typically 20 sec$^{-1}$ but this was obscured by background counts of order $10^3$ sec$^{-1}$.

## 2. *The detector for 304 Å photons*

The detector, which was described in detail by Harrison et al. [61] is shown in a simplified form by Fig. 5-4-2. The ion beam entered along the

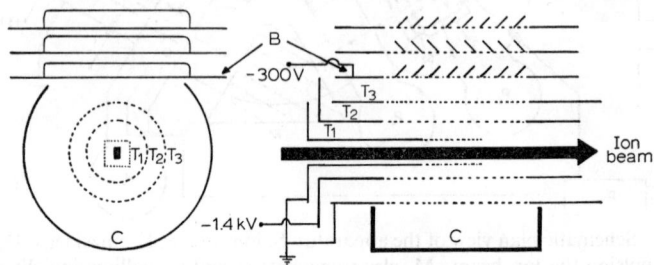

Fig. 5-4-2. Simplified diagrams of the detector for 304 Å photons. The ion beam passes along the axis of tube $T_1$ and is quenched by the field between tubes $T_1$ and $T_2$. The 304 Å photons emitted eject electrons from photocathodes B and C. Tube $T_3$ shields the photocathodes from the potential applied to $T_2$.

axis of the earthed tube $T_1$ and experienced an electric field due to the potential ($-1.4$ kV) of tube $T_2$. Both tubes were perforated so that most of the photons released by the quenching of $He^+$ (2S) could strike the photocathodes B and C, which were screened from the quenching field by an earthed tube $T_3$. The perforations in $T_2$ and $T_3$ were covered by 1000 Å thick aluminium films so that they were each about 45% transparent to 304 Å radiation but impervious to most charged or neutral particles.

The three tubes acted as an einzel lens and this geometry was chosen to preserve the narrow cross section of the ion beam.

The cathode B, which formed the first stage of a 17 stage particle multiplier, was maintained at $+300$ V to attract photoelectrons from C. In an earlier form of the detector, the photocathode C was omitted but the counting rates were then too small for convenience. When cathode C was added it was found that the counting rate due to photoelectrons from C was about 6.4 times greater than that due to B.

The multiplier had AgMg venetian blind electrodes but subsequent experience has shown that BeCu electrodes have more stable secondary emission coefficients in demountable vacuum systems. The cathodes B and C were respectively made from untreated platinum and tungsten which have high ($\approx 0.1$) photoelectric yields for 304 Å radiation.

The output pulses from the multiplier were amplified and fed through a discriminator to two scalers mounted in parallel. The beams and scalers were modulated as illustrated by Fig. 5-2-8.

3. *Backgrounds*

The ion and electron beams each produced backgrounds which were two orders of magnitude larger than the signal. The principal sources of these backgrounds were as follows.

(a) $He^+$ *(2S) ions excited in the ion source.* In preliminary experiments there were many more $He^+$ (2S) ions in the target beam than were formed by electron impacts. This was true in spite of the fact that most of the metastable ions formed in the source were quenched by the fields used to accelerate and deflect the ions. The problem was virtually eliminated by restricting the energy of electrons in the ion source to less than the threshold (65.4 eV) for direct excitation of $He^+$ (2S) from unexcited helium atoms.

(b) *Collisions between* $He^+$ *ions and residual gas.* Collisional excitation produced photons which could actuate the detector and, in addition, the charged particles formed by ionizing collisions led to further background signals. For example, the electrons formed near tube $T_2$ acquired energies up to 1.4 keV so that they might either penetrate the aluminium foils and enter the detector or produce X-rays, whilst ions formed near the detector were attracted to $T_2$ where they produced fast electrons by secondary emission.

(c) *Collisions between* $He^+$ *and residual gas between the prequench electrode (Pq in Fig. 5-4-1) and the detector.* Some of these collisions produced $He^+$ (2S) ions which actuated the detector. The prequench electrode was biased so that its field de-excited most of the $He^+$ (2S) ions in the parent beam before they entered the interaction region.

(d) *X-rays produced in the electron gun and collector.* The electron beam probably produced more than $10^{12}$ X-ray photons per second from the vicinity of the electron gun. Some of these photons entered the detector and this number increased with electron energy. At 750 eV, which was the highest electron energy used, the background associated with the electron beam was 50 times the signal.

Although the beam modulation system successfully separated the numbers of signal and background counts, two serious problems arose.

First, to obtain reasonably accurate results it was necessary to count for about an hour at each electron energy and current used. Only after such long periods was the statistical uncertainty in the large number of background counts small compared with the number of signal counts recorded in the same period.

Second, space charge interaction caused geometrical changes in the colliding beams. For example, the ion beam changed the profile of the electron beam and consequently altered the number and distribution of the X-rays which it produced. Thus the background $B_i$ was not independent of the electron current as assumed in section 5-2-F. Similarly, $B_e$ depended upon the ion current. These "modulated backgrounds" caused the measured excitation function to be considerably distorted and it did not fall to zero at threshold.

Dance et al. therefore devised corrections, which were necessarily very approximate, to take account of this distortion. The magnitude of the correction can be judged from Fig. 5-4-3 in which the rate of signal counts for unit electron and ion beam currents is plotted against electron energy. The points represent corrected count rates and the curve was drawn through the points *before* the correction was made.

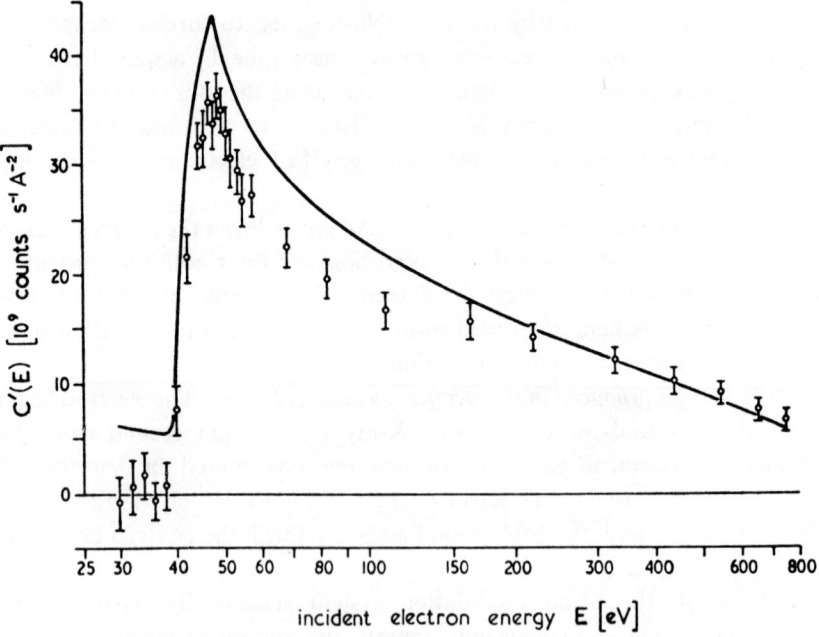

Fig. 5-4-3. The corrected count rate of 304 Å photons, which is proportional to the excitation cross section, plotted against electron energy. The continuous curve shows the measured excitation function which is distorted by modulated backgrounds. The points show experimental results *after* a correction has been applied.

## 4. Results

The results, which are represented by the points in Figs. 5-4-4 and 5-4-5, refer to the population of the 2S state by cascading from higher levels and direct excitation from the ground state. These cross sections were made absolute by normalizing at higher electron energies to the results of Massey's calculation [62] which was based on the plane wave Born approximation. Alternatively, absolute values could be deduced from an estimate of the detector efficiency and the cross in Fig. 5-4-5 shows the result for 700 eV electrons which was obtained in this way.

The measured excitation function did not rise quite vertically at threshold as predicted by theory but this can be attributed to the spread of energies in the electron beams. Rough experimental estimates were made of this spread and a corrected function was deduced which is shown by the continuous curves.

Fig. 5-4-5 also includes several theoretical results. These are based on the close-coupling approximation [52] (curve CC); the Coulomb-Born-Oppenheimer approximation [63] (CBOII); the Coulomb-Born approximation [64] (CBII) and the plane wave Born approximation [62] (PB).

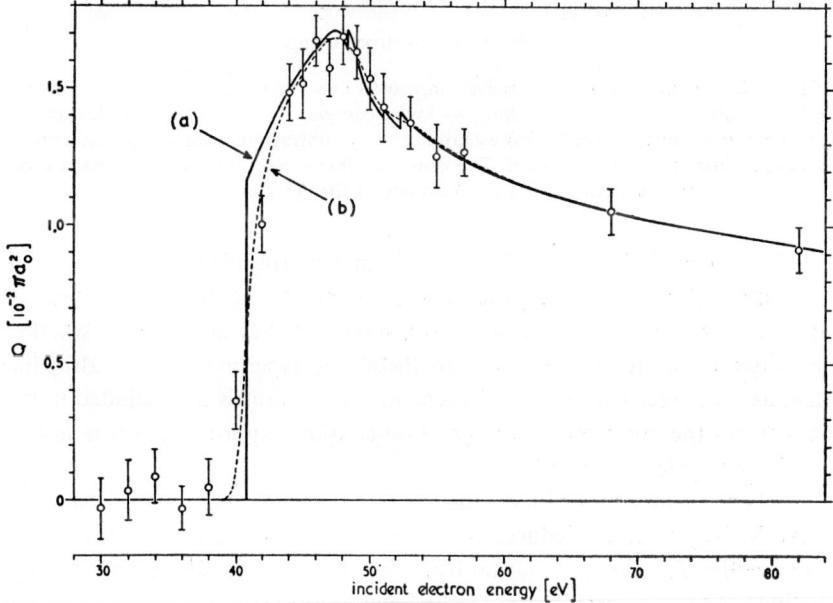

Fig. 5-4-4. The points joined by the broken curve, show measured excitation cross sections for He$^+$(1S–2S). These results include contributions from cascading. The continuous curve has been corrected for the spread of energies in the electron beam.

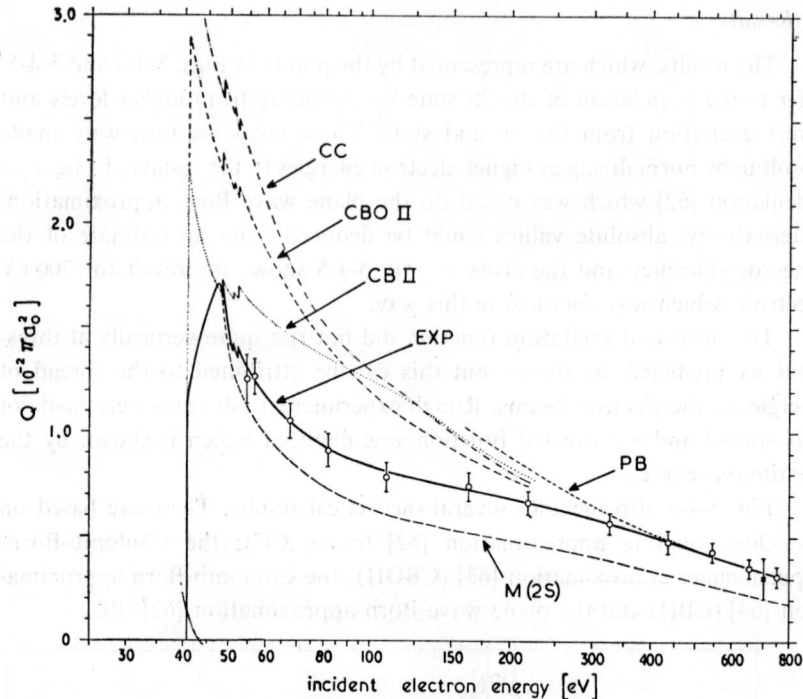

Fig. 5-4-5. The circles show normalized measured cross sections for He$^+$(1S–2S) which include contributions from cascading. At lower energies it lies below theoretical curves identified in the text. Curve M(2S) was obtained by subtracting a theoretical estimate of cascading from the measurements. The cross at 700 eV electron energy shows a cross section deduced from an estimate of the detector efficiency.

Very recently Burke and Taylor [130] have performed new close-coupling calculations for He$^+$ (1S–2S) excitation. They find that their result does not agree well with experiment close to threshold but they are not sure whether this discrepancy is due primarily to theory or experiment. They also find that, at lower electron energies (where all open channels are included in the expansion) the convergence of the close-coupling approximation is not as good for e$^-$–He$^+$ as for e$^-$–H.

Theory can also estimate the contribution of cascading so that the curve M (2S) could be deduced from the measurements. This represents the cross section for direct excitation from the ground state. The cascading correction is of course zero below the 3P excitation threshold (48.4 eV) but the higher thresholds give rise to the saw-tooth irregularities in the theoretical curves.

## 5. Discussion

None of the theories represented in Fig. 5-4-5 agree with the measurements at lower electron energies; in particular the measured excitation function does not fall monotonically with energy from a maximum at threshold.

This disparity may be due to the inadequacy of Born's approximation at low energies and the effects of autoionizing states of He which lie above the $n=2$ threshold of $He^+$. Contributions from these states have recently been included in a close-coupling calculation of the $He^+$ (1S–2S) cross section by Ormonde et al. [65] and this predicts considerable structure in the excitation function for electron energies between 44 and 50 eV. Fortunately, this is just the range of this excitation function that is accessible to measurement by sequential mass spectrometry (see section 5-7-B) and so it was possible for Daly and Powell [66] and Redhead and Feser [67] to obtain experimental confirmation of this structure. Their results are respectively represented by curves D and R in Fig. 5-4-6 which also shows the theoretical result of Ormonde et al. (curve O). The ordinates of the three curves are unrelated.

Information which would be vital to any new crossed-beam study of $He^+$ (1S–2S) recently came to light when Fite et al. [129] started to remeasure cross sections for H (1S–2S). Previously it was believed, from the calculation

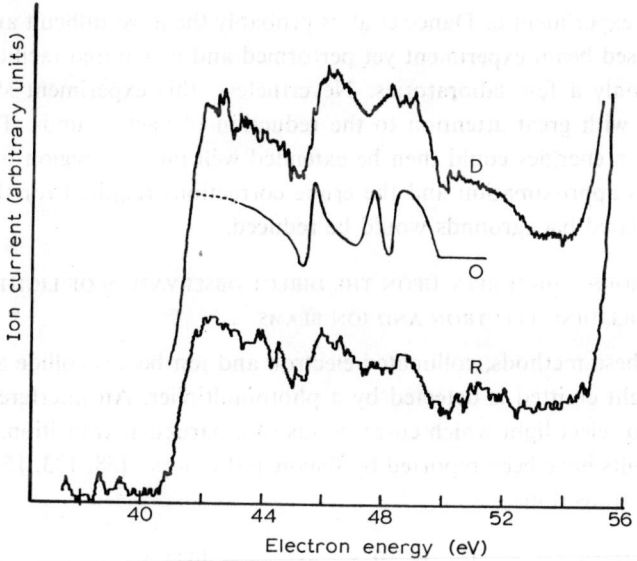

Fig. 5-4-6. Curves D, R and O represent excitation functions for $He^+$(1S–2S) obtained by sequential mass spectrometry (D and R) and a calculation of Ormonde et al. (O). The ordinate scales are unrelated.

by Lichten [60], that the radiation due to quenching the $2\,{}^2S_{\frac{1}{2}}$ state of hydrogen was emitted isotropically. In other words, the polarization factor $(P)$ defined by

$$P = \frac{I_{\parallel} - I_{\perp}}{I_{\parallel} + I_{\perp}}, \qquad (5\text{-}4\text{-}1)$$

was zero. Fite et al. were therefore surprised that their measurements gave $P = -0.30 \pm 0.02$, but it transpired that this is close to the value $-0.329$ very recently calculated by Fano [131].

It seems that Lichten [60] included only quenching via the $2\,{}^2P_{\frac{1}{2}}$ state and excluded the $2\,{}^2P_{\frac{3}{2}}$ state because it is energetically more distant from $2\,{}^2S_{\frac{1}{2}}$ by an order of magnitude. However, Fano's calculation revealed a contribution to the radiation intensity from a cross product of terms involving the $2\,{}^2P_{\frac{1}{2}}$ and $2\,{}^2P_{\frac{3}{2}}$ states which was about 20% of that from the $2\,{}^2P_{\frac{1}{2}}$ state alone. Any new experiment should therefore take account of polarization.

The measurements of Dance et al. should not be too sensitive to polarization because their photon counter had wide angular acceptance and their quenching field had components in three dimensions. Sequential mass spectrometry is, of course, completely unaffected by polarization.

The experiment of Dance et al. is probably the most difficult and elaborate crossed beam experiment yet performed and it required facilities available in only a few laboratories. Nevertheless, this experiment should be repeated with great attention to the reduction of backgrounds. The range of electron energies could then be extended well into the region of validity of Born's approximation and the crude corrections required for the effects of modulated backgrounds would be reduced.

## C. Methods Which Rely Upon the Direct Observation of Light Emitted by Colliding Electron and Ion Beams

In these methods, collimated electron and ion beams collide and some of the light emitted is detected by a photomultiplier. An interference filter is used to select light which corresponds to a particular transition.

Results have been reported by Bacon and Hooper [58, 133, 157] for the resonant transitions,

$$\text{Ba}^+(6\,{}^2S_{\frac{1}{2}}) + e \rightarrow \text{Ba}^+(6\,{}^2P^0_{\frac{1}{2}}) + e; \quad \lambda = 4934\,\text{Å}$$

and

$$\text{Ba}^+(6\,{}^2S_{\frac{1}{2}}) + e \rightarrow \text{Ba}^+(6\,{}^2P^0_{\frac{3}{2}}) + e; \quad \lambda = 4554\,\text{Å}.$$

Lee and Carleton [59] have given a preliminary account of their measurement of cross sections for the emission of 3914 Å radiation by collisions between electrons and $N_2^+$ ions.

### D. THE EXPERIMENT OF BACON AND HOOPER

A beam of unexcited $Ba^+$ ions, which were obtained by surface ionization, was passed along the axis of a light-tight tube and deflected through 90° by an electrostatic analyser. The main purpose of this analyser was to separate the ions from light and neutral particles radiated by the source.

The ions then entered an interaction region where they were bombarded by an electron beam. Some of the light produced by these collisions entered an optical system which was mounted vertically above the interaction region. This system included lenses, an interference filter and a photomultiplier. Either of two filters was used to select one of the resonance lines; the filters each had peak transmission ($\approx 60\%$) close to resonance and their bandwidth (full-width half maximum) was 50 Å. It was verified that only 0.5% of incident radiation was transmitted outside a pass band of 600 Å so that a filter effectively isolated one line.

The photons which actuated the photomultiplier gave signals which were amplified and counted, and the experiment consisted of the determination of the count rate due to the particular transition. The major difficulty was the separation of this signal from backgrounds.

#### 1. *Backgrounds*

The modulation scheme illustrated by Fig. 5-4-7 was used to separate the signal from background currents which were caused mainly as follows.

(a) Some neutral barium vapour escaped from the source and reached the collision region where it was excited by electron impacts. This produced radiation within the passband of the filter and gave a spurious signal ($N_e$ counts $sec^{-1}$) in phase with the electron beam.

(b) Collisions between $Ba^+$ ions and residual gas produced photons ($N_i$ $sec^{-1}$) in phase with the ion beam modulation.

(c) The main background ($N_0$ $sec^{-1}$) was independent of the electron and ion beam intensities and was primarily due to visible light from the ion and electron sources. This was reduced by deflecting the ion beam and by operating the electron gun cathode at the lowest temperature consistent with adequate emission. The photomultiplier was carefully shielded and cooled to $-20\,°C$ so that dark currents were almost entirely eliminated.

Most of the experiments were performed with a residual gas pressure

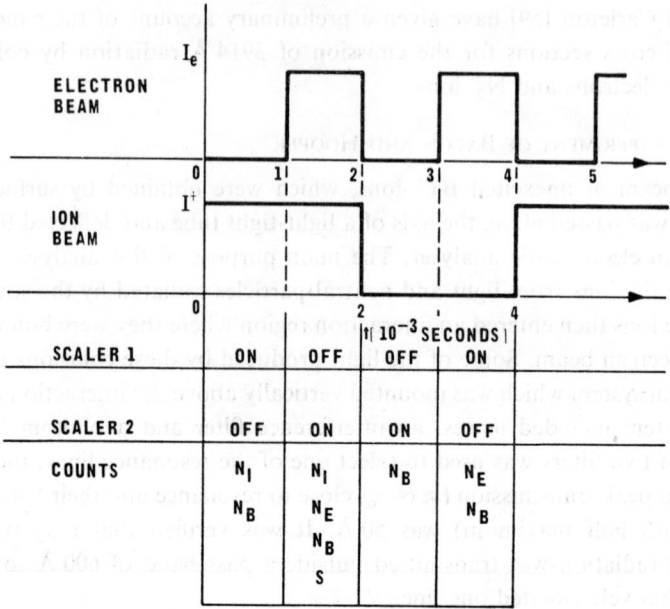

Fig. 5-4-7. Beam modulation scheme used by Bacon and Hooper.

of about $8 \times 10^{-9}$ torr and then the ratio $N_0 : N_e : N_i$ was typically 87:10:3. The SBR ranged from 0.2 to 0.01 so that counting periods of 15 minutes at each electron current and energy were needed to produce signals with statistical errors of only a few percent.

## 2. *Other experimental consideration*

The reports by Bacon and Hooper describe their experiment in detail and so attention will be drawn only to features of particular interest.

The anisotropic distribution of radiation emitted when two beams collide can be deduced if the polarization is measured. This was done by placing a sheet of polaroid in the optical system and a correction (which amounted to only a few percent) was applied.

A most important feature of this type of measurement was inadvertently illustrated by one experiment in which the cross section appeared, unexpectedly, to increase by 30%. This occurred because one end of the electron cathode became poisoned and the position of the interaction region was changed in such a way that the *average* collection efficiency of the optical system was enhanced. The authors suggest that it would therefore be desirable to monitor the extent of the electron beam in both the horizontal and vertical planes.

A refinement was introduced which will probably be adopted quite widely. A small slit was made in the electron beam collector through which about 1% of the electron beam passed. This sample then passed through a 127° electrostatic energy-analyser so that the energy distribution of the electrons could be monitored during the course of the experiments. Corrections for inhomogeneities in the energies (typically 2 eV) could then be applied.

Perhaps the greatest problem that confronts this type of experiment is the determination of the overall efficiency of the optical system; this information is needed to make the measurements absolute. Bacon and Hooper [58] accepted the manufacturers data for the efficiencies of the filters and the photomultiplier to obtain the absolute cross sections represented by the points in Fig. 5-4-8. Fairly crude theoretical estimates of these cross sections were made by a semi-empirical method suggested by Seaton [51] and cascading contributions were estimated to be about 10%, because this ratio was found for caesium (isoelectronic with $Ba^+$) by Zapesochny and Shimon [69]. The final theoretical estimates of these cross sections are represented by the continuous curves in Fig. 5-4-8 and, in view of the assumptions which were made, the agreement between theory and experiment must be partly fortuitous. It should, however, be emphasised that these experiments are still in progress and a calibration of the sensitivity of the optical system with a standard tungsten lamp has recently been described [133, 164].

This type of experiment is capable of many further developments. It should, for example, be possible to mount the ion and electron sources in separate tanks from the interaction region. Backgrounds due to scattered light would be greatly reduced and lower pressures could be attained in the interaction region. There is also a pressing need for accurate low-intensity light standards which could be used to determine the efficiency of optical systems, preferably well into the UV region (see section 5-8-F).

Bacon and Hooper performed several consistency checks on their results but the most revealing check is to ensure that the measured cross section is

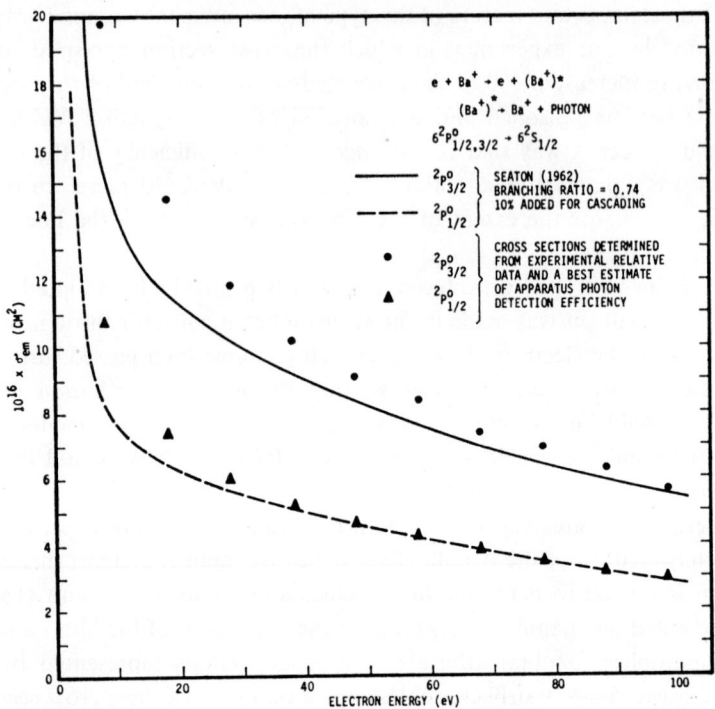

Fig. 5-4-8. Comparisons between measured emission cross sections for Ba⁺ and theoretical estimates.

zero below threshold. Unfortunately, the thresholds were so low in this experiment ($\approx 3$ eV) that this was not possible.

### E. The Experiment of Lee and Carleton

Lee and Carleton [59] have developed similar methods to obtain the results shown in Fig. 5-4-9 (note logarithmic ordinate scale) for the reaction,

$$e + N_2^+ \rightarrow e + (N_2^+)^* \rightarrow N_2^+ + e + h\nu \quad (\lambda = 3914 \text{ Å}).$$

It is proposed subsequently to study the resonant excitation of $Mg^+$, $Ca^+$, $Sr^+$ and $Ba^+$ ions [70].

In this experiment, light from the colliding beams was directed by a lens (f/2.5 aperture) to a shielded photomultiplier which was cooled by liquid nitrogen. The required spectral line was isolated by an interference filter with a pass band of 25 Å.

Although the vacuum in the apparatus was only $10^{-7}$ torr, a SBR of

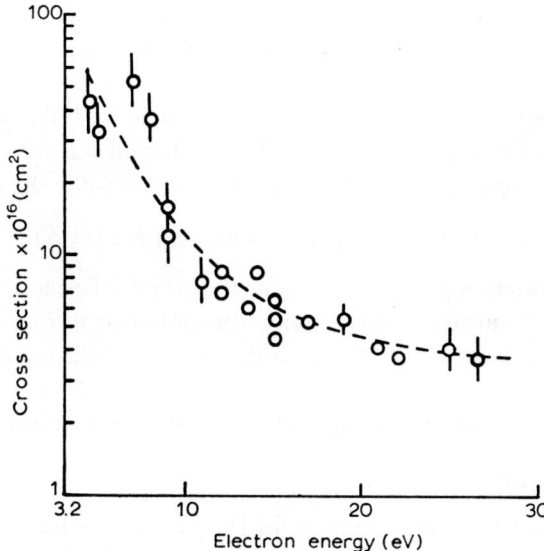

Fig. 5-4-9. Lee and Carleton's measurements of cross sections for the excitation of $\lambda = 3914$ Å radiation by collisions between electrons and $N_2^+$ ions.

order 10 was achieved and a simple modulation technique was sufficient to separate signals from backgrounds.

The apparatus is still under development so that refinements have not yet been included. For example, there is no provision to monitor form factors or the extent of the collision region, and the $N_2^+$ beam has not yet been mass-analyzed so that it contains about 15% of $N^+$ ions. The polarization of the radiation has not been checked, although errors due to anisotropy are probably small, nor has it been possible to observe excitation near threshold because space charge effects made it impracticable to work with electron energies below 5 eV.

The ultimate uncertainty in this type of experiment will probably concern the initial state of excitation of the target beam. The $N_2^+$ ions were formed by electron bombardment (electron energy $\approx 50$ eV) so that many excited states of $N_2^+$ were populated and the measured cross sections must contain contributions from each of these.

Points of particular experimental interest were the methods used to calibrate the electron energy scale and the efficiency of the optical system.

In a subsidiary experiment the apparatus was filled to a low pressure with helium. Measurements of the appearance potentials for electron impact excitation of the $\lambda = 3888$ Å and 5016 Å lines of He I showed that the electron energies were 3 eV ($\pm 1$ eV) lower than the potential difference between the cathode and anode of the electron gun.

The efficiency of the optical system was found by filling the apparatus with nitrogen. The signal due to $\lambda = 3914$ Å radiation was measured and the efficiency was found in terms of known cross sections for the reaction,

$$e + N_2 \rightarrow (N_2^+)^* + 2e \rightarrow N_2^+ + 2e + h\nu \quad (\lambda = 3914 \text{ Å}).$$

This seems a promising way to determine the overall efficiency but it includes the errors in the absolute cross section for molecular nitrogen and further uncertainties associated with the exact extent of the collision region.

## § 5-5. The detachment of electrons from negative ions by electron impact

### A. Introduction

Measurements of cross sections for $H^- + e \rightarrow H + 2e$ have been reported by Dance et al. [71], Tisone [16], Tisone and Branscomb [15, 134] and the latter authors have also described [15] a similar experiment with $O^-$ ions.

These experiments encounter three difficulties which were not experienced in those described by section 5-3-B. First, the main collision products are uncharged so that their flux is more difficult to measure accurately, and second, the cross sections for charge stripping of negative ions by residual gas are so large that considerable backgrounds arise even under conditions of ultra high vacuum. Finally, the ionization energies are too low (0.75 eV for $H^-$) for it to be verified that the measured cross section is zero below threshold. These difficulties are partially mitigated because the electron impact detachment cross sections are so large ($\sim 10^{-14}$ cm$^2$) that adequate signals are easily obtained.

The results for $H^-$ ions are particularly interesting, not only do these ions have a simple structure, but there has been considerable uncertainty about the contribution of $H^-$ ions to the opacity of stellar photospheres. The rate of destruction of $H^-$ by electron impacts has been the subject of several calculations and Tisone [16] collected four theoretical results, all based on the Born-Oppenheimer approximation, which differed by almost four orders of magnitude – a delightful prelude to experiment!

### B. The experiment of Dance, Harrison and Rundel

The apparatus was broadly similar to that represented by Fig. 5-2-1 but collector $C_2$ was removed and a neutral particle detector was placed in line with the beam of $H^-$ parent ions. The analyser magnet deflected $H^-$ ions into collector $C_1$, whilst the hydrogen atoms passed straight through the magnetic field and struck the first dynode of a particle multiplier, which served as an efficient detector of individual neutral particles. Typically, there were between 10 and 30 counts $sec^{-1}$ due to electron ion collisions but these were obscured by about $10^3$ counts $sec^{-1}$ from extraneous processes; the separation of signal from backgrounds was again a major problem.

About 90% of the extraneous counts were due to the formation of atomic hydrogen by charge stripping collisions between $H^-$ ions and residual gas; this gave a background $(B_i)$ in phase with the ion beam modulation. Most of the remaining background was in phase with the electron modulation $(B_e)$ and increased with electron energy. It was probably due to soft X-rays produced near the electron gun.

This sensitivity to electromagnetic radiation is sometimes an undesirable property of particle multipliers and Dance et al. noted that their multiplier was also actuated by light from the ion source, until baffles were fitted inside the apparatus.

Since it was impossible to verify that the measured cross section was zero below threshold, tests were devised to ensure that the results were not invalidated by modulated backgrounds. The detailed description of these tests, which appears in the paper, gives an excellent insight into some subtleties of this type of experiment. Several potential sources of error were discussed including the possibility that the electronic space charge could change the ion beam geometry in phase with the electron modulation. Suppose, for example, that ions which struck collector $C_1$ released any kind of radiation which could actuate the particle counter; any variation of this extraneous signal in phase with the electron beam would then lead to error in the measured cross section. Consider also that the efficiency of the particle multiplier might not be uniform over the whole of its aperture (see section 5-9-E); changes in the ion geometry would then vary the spread of the beam of hydrogen atoms so that non uniformities in the detector efficiency would introduce error.

Tests were therefore made in which the ion beam geometry was deliberately changed by varying, successively, the ion energy, electron current, and the field of the analyzer magnet. These showed no evidence of significant errors from modulated backgrounds.

The neutral particle multiplier used in these experiments was a 17-stage venetian blind type electron multiplier with BeCu electrodes. Its efficiency was measured for beams of $H^+$ and $H^-$ ions by comparing the count rate with currents measured at a Faraday cup. Provided the ion energy exceeded 10 keV the efficiency of detection of these ions differed by only 3% and was always above 90%. The average of these efficiencies was therefore assumed for fast hydrogen atoms.

The results of this experiment will be discussed in section 5-5-D but it can be noted that random errors were estimated to be between $\pm 3\%$ and $\pm 11\%$, depending on electron energy, whilst maximum systematic errors ranged from $\pm 14\%$ to $\pm 19\%$. A detailed discussion of individual sources of systematic error is given in the paper.

## C. The Experiment of Tisone and Branscomb

The apparatus used by Tisone and Branscomb is represented schematically by Fig. 5-5-1. This method is broadly similar to that just described but there are interesting differences.

The colliding beams of electrons and $H^-$ ions were chopped at frequencies of 20 kHz and 50 Hz, respectively and the atomic hydrogen formed was detected by a neutral particle detector of the type described by Daly [72, 73] in which fast atoms released electrons from a BeCu target. The electrons were accelerated to a plastic scintillator and the light produced was converted to electrical signals by a photomultiplier. Signals, proportional to the interaction between the electron and ion beams, were isolated by the modulation scheme described in section 5-2-5. The parent $H^-$ ions were prevented by electrostatic deflection from reaching the particle detector.

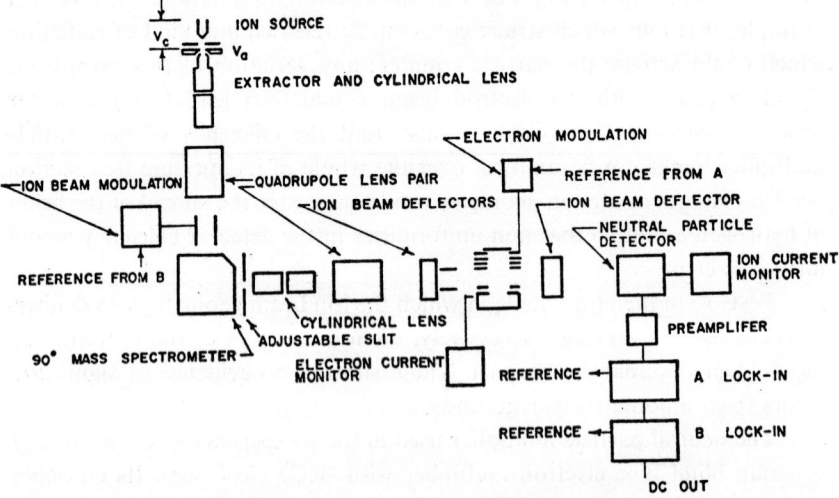

Fig. 5-5-1. Schematic diagram of apparatus used by Tisone and Branscomb.

Large backgrounds were found which depended primarily upon the pressure of residual gas in the apparatus so that test experiments were performed in which this pressure was deliberately varied. This produced little change in the measured cross section and indicated that modulated backgrounds did not lead to serious error.

An interesting refinement was the inclusion of an electrostatic field which deflected the parent ions through 5° just before they passed through the electron beam. Consequently, atomic hydrogen which had already been formed by charge stripping, could not proceed to the detector.

### D. RESULTS AND DISCUSSION

Table 5-5-1 shows typical values of the currents of parent ions ($I^-$) and electrons ($J$), the background pressure ($P$), the rates of formation of H by

TABLE 5-5-1

| | $J$(mA) | $I^-$(A) | $P$(torr) | $n_s$(sec$^{-1}$) | $n$(sec$^{-1}$) | Electron energies (eV) | Ion energies (keV) |
|---|---|---|---|---|---|---|---|
| Dance et al. | 0.03–3.0 | $8 \times 10^{-11}$ | $2 \times 10^{-9}$ | 10–30 | $10^3$ | 9.0–500 | 15 |
| Tisone and Branscomb | 0.7 | $4 \times 10^{-8}$ | $5 \times 10^{-8}$ | $3 \times 10^4$ | $3 \times 10^6$ | 8.4–488 | 2.5 |

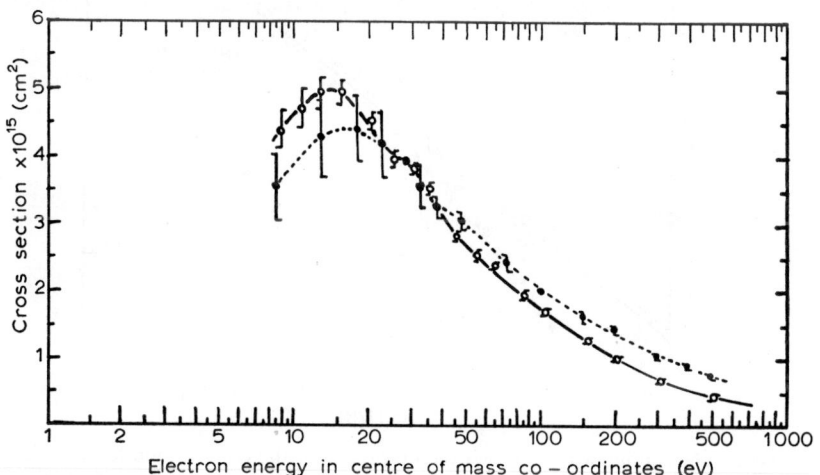

Fig. 5-5-2. Cross sections for the detachment of electrons from H$^-$ ions measured by Dance et al. (○) and Tisone and Branscomb (●). The brackets show standard deviations from the mean of random errors.

electron impacts ($n_s$) and by extraneous processes ($n_b$) and the ranges of electron and ion energies used in the two experiments.

The results of the experiments are shown in Fig. 5-5-2. In the original papers the two groups expressed their random errors and electron energies differently, but to facilitate comparison, the brackets shown in the figure represent the standard deviation of random errors from the mean for both experiments. The results of Dance et al. were expressed for "incident electron energies" ($E_{inc}$) rather than in terms of the centre of mass. These incident energies are given by

$$E_{inc} = E_e + \frac{m}{M} E_i \qquad (5\text{-}5\text{-}1)$$

but, since $m \ll M$, they differ insignificantly from the centre of mass energies defined by equation (5-2-7) and plotted in Fig. 5-5-2.

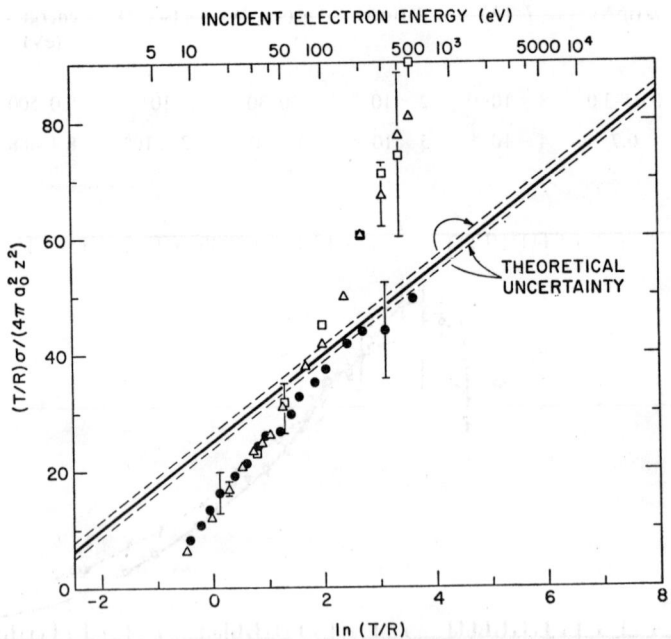

Fig. 5-5-3. The continuous line shows the magnitude of the cross section for single and double ionization of H$^-$, ($\sigma$) calculated in the Bethe-Born approximation by Inokuti and Kim. In the figure, $T/R$ is the incident energy in Rydberg and $a_0$ is the Bohr radius. The hollow and solid symbols respectively represent measurements of single ionization by Tisone and Branscomb and by Dance et al.

It can be seen from Fig. 5-5-2 that the disagreement between the two measurements is particularly large at high energies, and this is surprising because experimental problems are usually less severe under these conditions. Important qualitative differences appear if the results are plotted as $\sigma E$ against $\log E$, where $\sigma$ is the detachment cross section and $E$ is the interaction energy. This is illustrated by Fig. 5-5-3 which also shows the linear relation which Inokuti and Kim [144] have calculated from oscillator strength sum rules in the limit of the Bethe-Born approximation. This theory gives the total cross section for single and double ionization of $H^-$, although the latter process is relatively unimportant.

The results of Dance et al. agree quite closely with this asymptotic theory and with the results of calculations by McDowell and Williamson [74, 75] and Bely and Schwartz [156]. Tisone and Branscomb's results, however, are much larger at the higher energies. These discrepancies can only be resolved by a further experiment in which particular care should be taken to ensure that the two potential sources of error mentioned in section 5-6-C are absent, since they might cause the cross section to be overestimated, particularly at higher energies.

## § 5-6. Measurements of cross sections for the production of protons by collisions between electrons and $H_2^+$ ions

### A. Introduction

The extension of crossed-beam techniques to the study of molecular ions raised difficulties because the parent ions were in various states of vibrational excitation. Studies of $H_2^+$ are particularly interesting because it has a simple structure and its dissociative ionization and excitation are relevant to some of the devices used to explore the possibility of controlled thermonuclear fusion. Moreover, there is very little information about dissociation, even of neutral molecules, so the experiments to be described [5, 78, 14] contribute to a sparsely-covered branch of collision physics.

These experiments measure cross sections for the sum of reactions which lead to the production of protons by collisions between electrons and $H_2^+$ ions. The significant processes are,

$$H_2^+ + e \begin{cases} (H_2^+)^* + e \to H + H^+ + e & \text{(dissociative excitation)} \\ H_2^{2+} + 2e \to H^+ + H^+ + 2e & \text{(dissociative ionization)} \end{cases}$$

where the hydrogen atom may, or may not, be excited. Thus, the measured cross section is $\sigma = \sigma(\text{exc}) + 2\sigma(\text{ion})$ where $\sigma(\text{exc})$ and $\sigma(\text{ion})$ refer respectively, to the two reactions shown above, and $\sigma(\text{exc})$ represents the sum of all dissociative excitation cross sections.

In experiments on molecular dissociation particular attention must be paid to the efficient collection of molecular fragments since these may be formed with considerable initial velocities. Otherwise the experimental techniques are similar to those described in section 5-3 so that the discussion is restricted to topics of particular interest.

### B. THE EXPERIMENTS OF DUNN, VAN ZYL AND ZARE [5, 78, 79]

These were crossed beam experiments in which the electrostatic analyser shown in Fig. 5-2-9b was used to separate parent ($H_2^+$) and product ($H^+$) ions.

To facilitate the interpretation of results, care was taken to produce parent beams in which the population of vibrational states was as close as possible to that given by the Franck-Condon factors between $H_2$ and $H_2^+$.

This distribution will be approached if the ions source conditions approximate to the following.

(a) Electrons in the source should have energies much greater than the highest vibrational energy of $H_2^+$ ($\approx 18$ eV).

(b) Ions formed in the source should be extracted before they have opportunity to change their state, either spontaneously or by collision.

(c) Ions should be formed from $H_2$ molecules which are only in the lowest vibrational state.

Accordingly, an ion source was designed in which the pressure was only $10^{-3}$ torr and the electron energy was usually 200 eV. High fields were used to extract ions soon after their formation and, since the source temperature (except the filament) was only about 50°C, there was little initial excitation of the hydrogen gas.

Tests showed that changes in the pressure ($10^{-3}$ to $2 \times 10^{-2}$ torr), electron energy (100 to 400 eV) and the extraction potential in the source caused the measured cross sections to change by less than the 3% scatter of the measurements. This indicated that the vibrational population was quite constant and probably close to that given by the Franck-Condon factors. Confirmation was obtained by other experiments on the photodissociation of $H_2^+$ in which vibrational populations were measured [158].

Several consistency checks were performed which are described in the paper. In one of these the pressure was deliberately increased by an order of magnitude whilst the signal due to electron impact dissociation was found to change by only 20%. It follows that pressure-dependent modulated backgrounds contribute no more than a few percent to the experimental errors. This is an important check in experiments, such as these, in which it is impractical to verify that the measured cross section is zero below threshold.

A number of refinements were included in Dunn and Van Zyl's experiment and particular attention was paid to the determination of form factors. The electron gun was longer than usual so that the area of the electron beam was fairly constant. Thus little error was introduced by measuring the electron profile *before* the interaction space. The labour associated with the calculation of form factors was reduced by measuring the electron and ion currents with digital meters and the outputs of these were punched on to tape in a form suitable for immediate computation. The ion beam was not modulated but the electrons were pulsed at low audio-frequencies and the proton current due to collisions was resolved by a low-noise phase sensitive amplifier.

TABLE 5-6-1

| Experiment | $I^+$(A) | $J$(A) | $p$(torr) | $I_s^+$(A) | $I_b^+$(A) | Electron energies (eV) | Ion energies (keV) |
|---|---|---|---|---|---|---|---|
| Dunn et. al. [5] | $10^{-6}$ | $10^{-3}$ | $10^{-9}$ | $10^{-15}$ | $10^{-12}$ | 10–1500 | 2–10 |
| Dance et. al. [14] | $10^{-6}$ | $10^{-3}$ | $2 \times 10^{-9}$ | $10^{-14}$ | $10^{-12}$ | 5–1000 | 10–20 |

Some of the experimental conditions are summarized in Table 5-6-1. This shows magnitudes of the parent ion beam current ($I^+$), electron currents ($J$), residual pressure ($p$) and the collected currents of protons formed by electron impacts ($I_s^+$) and other processes ($I_b^+$). The ranges of electron and parent ion energies are also shown and comparisons are made with the experiment of Dance et al.

Similar experiments [79] were subsequently performed with $N_2^+$ and $O_2^+$ ions. The measured cross sections for the formation of $N^+$ and $O^+$ by electron impact can be represented respectively, by,

$$\sigma = \left(\frac{923}{E}\right) \log_{10} E - \left(\frac{1411}{E}\right) \qquad (5\text{-}6\text{-}1)$$

and

$$\sigma = \left(\frac{972}{E}\right) \log_{10} E - \left(\frac{1544}{E}\right) \qquad (5\text{-}6\text{-}2)$$

where $E$ is the interaction energy in electron volts and $\sigma$ is in units of $\pi a_0^2$. These equations are only valid for $E \geqslant 100$ eV.

## C. THE EXPERIMENT OF DANCE ET AL. [14]

This employed an apparatus broadly similar to that represented by Fig. 5-2-1 in which collector $C_1$ received the parent beam whilst the protons were collected by $C_2$.

Dance et al. chose a Nielsen-type [11, 33] source which was not specially designed to have the properties listed in the previous section. However, they demonstrated that there were no significant changes in their results when the ion source pressure was varied between $8 \times 10^{-3}$ and $13 \times 10^{-2}$ torr or when the electron accelerating potential in the source increased from 50 to 200 V. Unfortunately it was only possible to measure dissociation over this range of ion source conditions when the energy of electrons in the interaction region (not the source) was high. These tests would have been more sensitive if the interaction energies had been smaller, because the influence of vibrational excitation upon dissociation is usually more marked when incident electrons have little more than the dissociation energy.

In these experiments the currents at collector $C_2$ were too large ($\sim 10^{-12}$ A) to be measured by particle counters. A vibrating capacitor electrometer was employed and the simple modulation technique illustrated by Fig. 5-2-6 was used to resolve the small differences between $(I_b^+ + I_s^+)$ and $I_b^+$. It was very remarkable that this simple technique sufficed, and since several minutes were needed to take each reading, the currents and background pressure had to be extremely stable over this period. Electronic feedback was used to stabilize the ion beam to 1% over several hours and care was taken to eliminate fissures from the inner walls of the vacuum chamber since it was found that these occasionally released bursts of gas which gave unacceptable fluctuations in pressure.

These problems were not so acute in the experiments by Dunn and his colleagues because they only needed such stability over a modulation period ($\lesssim 10^{-2}$ sec).

Two further problems arose when experiments were performed with fast electrons.

Some slow secondary electrons from the gun and collector tended to drift around the interaction space. Although these electrons had little energy, they produced dissociation with a large cross section when they collided with the fast parent ions and this introduced errors as large as 20%. A negatively-biased box-shaped electrode was therefore fitted around the interaction space to eject these slow electrons and this effectively solved the problem.

A second spurious effect arose from gas ejected from the electron collector by fast electron beams. Normally, pressure fluctuations caused by the electron beam are averaged out if the beam is pulsed at a sufficiently high frequency (see section 5-2-6). In this case, however, the gas formed a crudely collimated molecular beam which passed through the interaction space, roughly in phase with the electron modulation. The solution was deliberately to distort the electron beam geometry inside the collector so that a well-defined molecular beam could not form.

These problems were also considered by Dunn and Van Zyl. Their electron collector was also shaped (Fig. 5-2-9b) so that a well-defined gas beam could not form and the modulation frequency was high enough to average out the effects of pressure fluctuations. Further systematic checks did not reveal significant errors due to slow secondary electrons.

One expects that these errors might arise in experiments with other easily-ionized parent ions and this should be borne in mind during any future experiment with $H^-$ ions (cf. section 5-5-D).

### D. Results and Discussion

For energies above 15 eV the differences between the results of the two experiments are always less than 10%. This may be seen from Figs. 5-6-1 and 5-6-2 in which the triangles and circles respectively represent the results of Dunn and Van Zyl, and Dance et al. plotted against incident electron energy in centre of mass coordinates. The bars show standard deviations of the measurements from the mean.

At lower energies the deviations are quite marked and this must be partly due to the different vibrational populations of parent beams used in the two experiments. Comparison between theory and experiment is complicated because $H_2^+$ ions may exist in all vibrational levels, each of which will have a different cross section. Moreover, from each initial state transitions are possible to several final states.

Following Dunn and Van Zyl one may simplify matters by neglecting all transitions of $H_2^+$ except those from the $1s\,\sigma g$ state to the $2p\,\sigma u$, $2p\,\pi u$, $2s\,\sigma g$ and the ionized states. The cross section for proton production can then be written,

$$\sigma = \sum_v p_v \sigma_v (2p\,\sigma u) + \sum_v p_v \sigma_v (2p\,\pi u) + \sigma(2s\,\sigma g) + 2\sigma(\text{ion}) \qquad (5\text{-}6\text{-}3)$$

where $p_v$ are Franck-Condon factors between $H_2$ and $H_2^+$ and the other

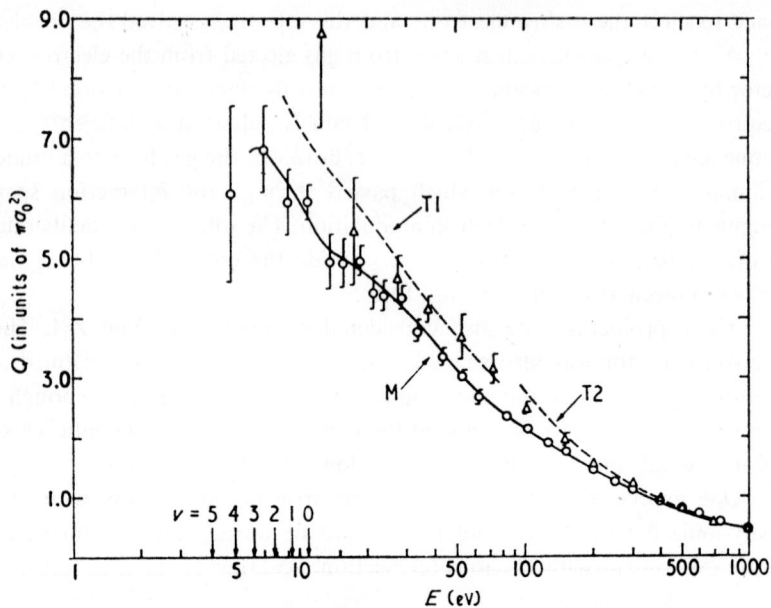

Fig. 5-6-1. Cross sections ($Q$) for proton production in collisions between electrons and $H_2^+$ ions. The circles and triangles respectively refer to the results of Dance et al. and Dunn and Van Zyl. Curves T1 and T2 are based on theories by Peek.

symbols represent excitation and ionization cross sections. The variation of $\sigma(2s\,\sigma g)$ and $\sigma(\text{ion})$ with internuclear spacing is neglected because theoretical data is not available and the dependence is likely to be small.

Theoretical values of the quantities on the right of equation (5-6-3) were available from sources cited in the paper. In particular, values of $\sigma_v(2p\,\sigma u)$, $\sigma_v(2p\,\pi u)$ and $\sigma_v(2s\,\sigma g)$ were taken from the work of Peek [80, 81]. It was then possible to deduce $\sigma$ as a function of the interaction energy and the result is shown by the broken curve T1 in Fig. 5-6-1. Curve T2 represents the results of a calculation by Peek [81] which attempted to include the effects of excitation to all possible final states.

These results confirm that the distribution of vibrational states of $H_2^+$, particularly in the experiment by Dunn and Van Zyl, was close to that calculated from Franck-Condon factors. Moreover, the strong variation with internuclear spacing predicted by Peek for $\sigma_v(2p\,\sigma u)$ is consistent with the experimental results.

Peek [81] also performed a closure calculation which predicts the gradient of the linear relation which should exist between $\sigma E$ and $\log E$

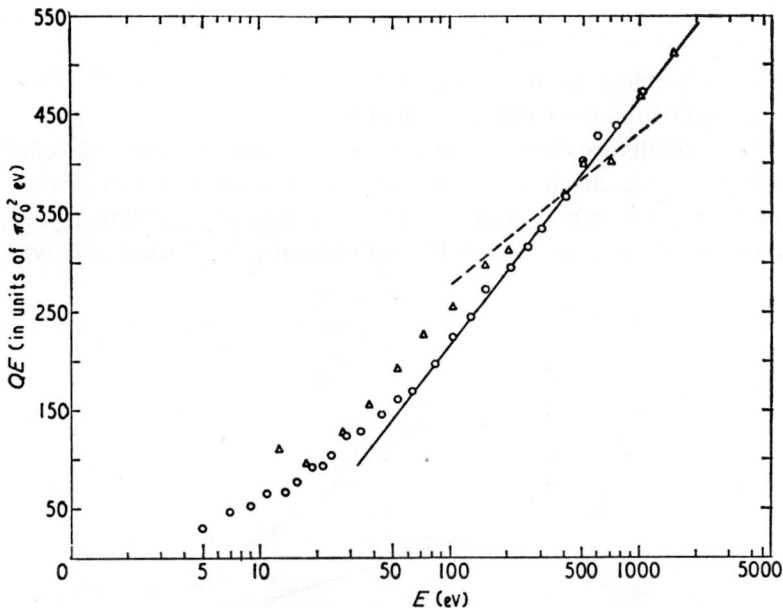

Fig. 5-6-2. Cross sections ($Q$) for proton production in collisions between electrons and $H_2^+$ ions plotted as $QE$ against the electron energy ($E$) on a logarithmic scale. The circles and triangles respectively denote the measurements of Dance et al. and Dunn and Van Zyl. The broken curve is based on theory by Peek.

(cf. section 5-3-C) at high electron energies. It can be seen from Fig. 5-6-2 that his result disagrees with both experiments. The cause of this discrepancy is not yet clear, but better agreement might be found if the experiments were extended to higher energies (cf. Figs. 5-3-5 and 5-5-3).

## § 5-7. Experiments with colliding ion beams

The discussion has hitherto been confined to collisions involving electrons but new methods are being developed to observe interactions between ions. These can be divided into three groups according to whether the angle at which the beams intersect is 90°, 0°, or some intermediate angle.

The first group has been developed mainly by Guidini and his colleagues [82, 83] who built an apparatus in which two ion beams intersect at right angles. Typically, a beam of $H_2^+$ ions with current of order $5 \times 10^{-7}$ A and energy between 20 and 250 keV collides with a target beam of $N_2^+$ which has 2 keV energy and a particle density of $2 \times 10^7$ cm$^{-3}$. The $H^+$, $H_2^+$ ions and hydrogen atoms which emerge from the target are separated by a magnetic field and the collision products are detected by scintillation counters. Modulation techniques are needed to distinguish the products of collisions between ions from those due to residual gas.

Some results are illustrated by Fig. 5-7-1 in which the curves labelled 1 and 3 show, respectively, the measured cross sections for the formation of protons and hydrogen atoms by collisions between $H_2^+$ and slow $N_2^+$ ions. The results are given for various $H_2^+$ ion energies ($E$). The dashed curve (2)

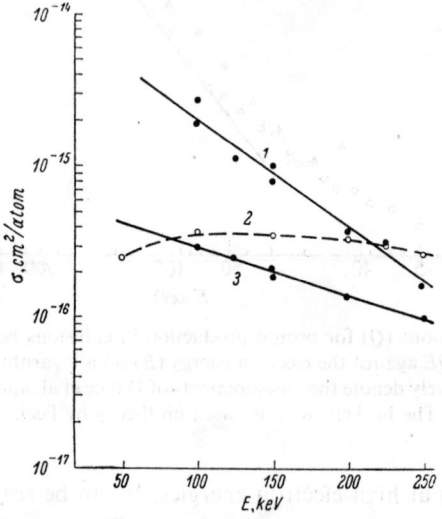

Fig. 5-7-1. Curves 1 and 3 show respectively cross sections for the formation of protons and hydrogen atoms by collisions between fast $H_2^+$ and slow $N_2^+$ ions. Curve 2 shows cross sections for the production of protons by collisions between fast $H_2^+$ ions and molecular nitrogen. The abscissa denotes the energy ($E$) of the $H_2^+$ ions.

locates measured cross sections for the formation of protons by collisions between $H_2^+$ ions and molecular nitrogen.

An interpretation of these results would be complicated by the vibrational excitation of both the projectile and target ions. Nevertheless, the results are particularly relevant to the design of devices intended to produce thermonuclear fusion.

A second, and potentially very powerful method, employs merging beams ($\theta \approx 0°$). These techniques have been described by Trujillo et al. [84] and Belyaev et al. [85] and reviewed by Neynaber [86]. Merging beams have mainly been used to study collisions between ions and neutral particles but the technique is being extended to give valuable information about interactions between ions. This method is uniquely suitable for the study of collisions with exceptionally low interaction energies (i.e. the relative energy of two ions expressed in centre of mass coordinates). This can be shown as follows.

Consider two beams of particles which merge so that they move along parallel lines and, for simplicity, assume that the beams are composed of ions with equal mass (the general features of the method are not changed if the masses are unequal). Then the interaction energy ($W$) can be written in terms of the laboratory energies, $E_1$ and $E_2$ of the particles,

$$W = \tfrac{1}{2}(E_2^{\frac{1}{2}} - E_1^{\frac{1}{2}})^2. \tag{5-7-1}$$

If the difference ($\Delta E$) between these energies is small, and $E$ represents their mean value,

$$W \approx \frac{(\Delta E)^2}{8E}. \tag{5-7-2}$$

It follows that exceptionally small interaction energies can be obtained. Neynaber [86] illustrated this by considering particles of equal mass for which $E_1 = 5000$ eV and $E_2 = 5100$ eV. Equation (5-7-2) then gives an interaction energy of only 0.25 eV.

This "deamplification" applies not only to the difference between $E_1$ and $E_2$ but to any perturbation of this difference; this is a vital factor in experimental design. Consider a perturbation $\delta E$ in the difference $\Delta E$. Neynaber shows that the corresponding perturbation ($\delta W$) in the interaction energy is,

$$\delta W \approx \frac{2W\, \delta E}{\Delta E}. \tag{5-7-3}$$

If, in the example we have just considered, there was a (laboratory) energy spread of 1.5 eV in each of the merging beams, it follows from (5-7-3) that the uncertainty in the interaction energy ($\delta W$) is only 0.0075 eV.

Clearly, these arguments point the way to beautifully refined experiments, particularly on charge changing collisions between ions since these can depend critically upon the interaction energy.

Considerable progress has already been made by Brouillard [87] who has constructed an apparatus in which he proposes to study, $He^+ + He^{2+} \rightarrow He^{2+} + He^+$ and similar reactions. His method is illustrated by Fig. 5-7-2.

Both ionic species ($He^+$ and $He^{2+}$) are produced by the same source (S) and the collimated beam of mixed ions is accelerated by the potential of

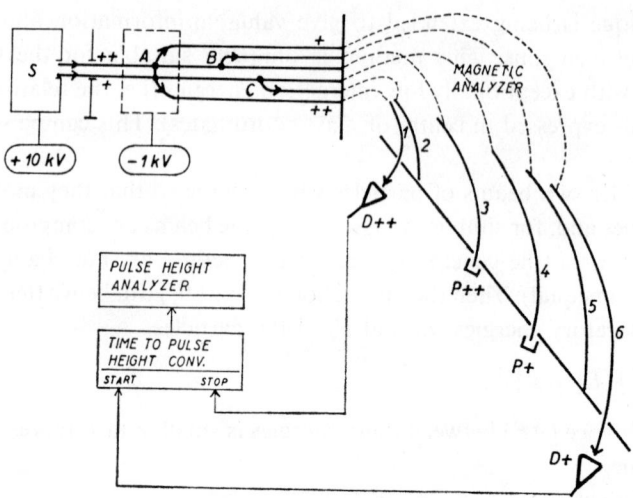

Fig. 5-7-2. Brouillard's apparatus to study collisions between He⁺ and He²⁺ ions.

about $+10$ kV which is applied to the source. The beam then traverses a region, labelled A, of uniform negative potential in which the ions, by virtue of their different charges, acquire different additional velocities. Singly and doubly-charged ions which are formed by charge-changing collisions in region "A" are deflected by a magnetic analyser to the detectors $D^+$ and $D^{++}$ respectively, whilst ions which do not collide in "A" are collected by the Faraday cups, $P^+$ and $P^{++}$. Extraneous signals arise from collisions between ions and residual gas but these are minimized by evacuating the region "A" to about $10^{-10}$ torr. The observation of coincidences between signals at $D^+$ and $D^{++}$ successfully distinguishes products of the required processes from the extraneous signals.

A feature of the design which seems particularly attractive is that the experiment should respond only to collisions in "A". Since this region is small it can easily be baked and differentially pumped to extremely high vacua. Consequently, problems due to collisions with residual gas can be simplified.

Merging beams have also been used by Aberth et al. [136] to measure cross sections for neutralization in collisions between $N^+$ and $O^-$. They performed measurements with (centre of mass) energies in the range 0.1 to 86 eV and achieved energy resolution of about 0.1 eV.

Merging beam techniques are of limited value for investigations of reactions of the type, $H^+ + H^- \rightarrow H + H$ in which the two products are identical. In the reaction cited, one could not distinguish between the atoms produced from $H^+$ or $H^-$. This problem led to the design of experiments with inclined beams in which, for example, beams of $H^+$ and $H^-$ ions collide obliquely. Very little scattering occurs in charge-changing collisions between fast ions so that the product atoms are "labelled" by the direction of their trajectories. If the beams collide obliquely there is still some "deamplification" of the interaction energy but not to the extent obtained with merging beams.

The cross section for $H^+ + H^- \rightarrow H + H$ has recently been measured by Aitken et al. [146] with the apparatus illustrated by Fig. 5-7-3. Well-collimated, mass analyzed beams of $H^+$ and $H^-$ ions intersected at an angle of 20°. The hydrogen atoms formed from $H^+$ in the collision region were detected and counted, whilst the proton beam was deflected electrostatically

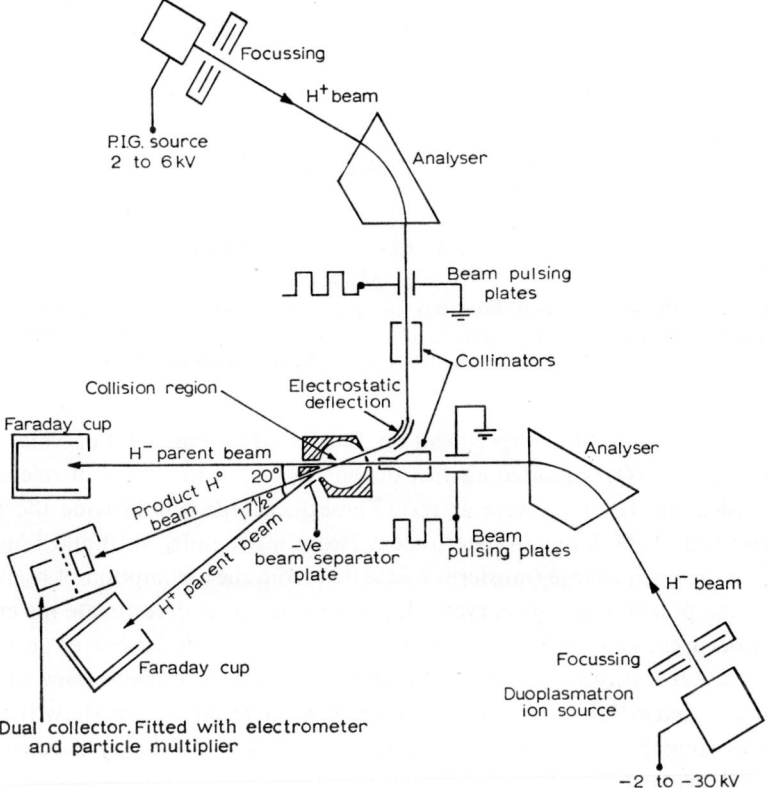

Fig. 5-7-3. Apparatus used by Aitken, Rundel and Harrison to study products of $H^+$ and $H^-$ ion beams which collided at an angle of 20°.

through $17\frac{1}{2}°$ into a Faraday cup. To reduce backgrounds due to the formation of hydrogen atoms by collisions between $H^+$ and residual gas, the proton beam was deflected immediately before and after the collision region. This reduced the path length over which the protons could "see" the neutral particle detector. In spite of this precaution, and a residual pressure of only $10^{-9}$ torr, the SBR ranged from $3 \times 10^{-3}$ to $3 \times 10^{-4}$ so that the modulation scheme illustrated by Fig. 5-2-8 was needed to separate signals from background.

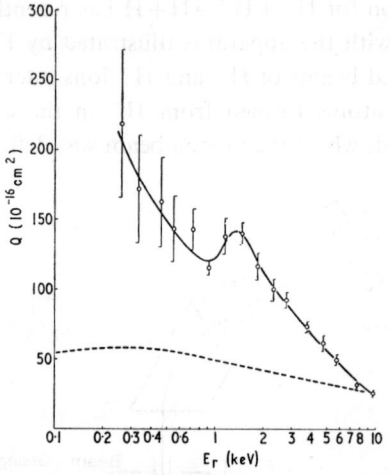

Fig. 5-7-4. Measured cross sections ($Q$) for charge exchange between $H^+$ and $H^-$ ions. The energy scale ($E_r$) is in the frame in which the $H^-$ ions were at rest. The broken curve shows results of a calculation by Bates and Lewis (see text).

The results of this experiment are illustrated by Fig. 5-7-4 in which the cross section ($Q$) is plotted against the energy ($E$) in the frame of reference in which the $H^-$ ions were at rest. These measurements provide the first direct test of the Landau-Zener theory (see, for example, Mott and Massey [159]) in which charge transfer is calculated from the assumption of pseudo-crossing potential energy curves. This theory should underestimate the cross section because it neglects transitions between the potential curves at internuclear separations away from the crossing point. The inadequacy of the Landau-Zener theory is revealed by the present results which show that $Q$ varies roughly as $E_r^{-1}$ at higher energies, whereas the theory predicts an $E_r^{-\frac{1}{2}}$ dependence.

Fig. 5-7-4 also shows the results of a calculation by Bates and Lewis [119] which included only charge transfer to the $n=2$ and $n=3$ states of hydrogen. Since the contribution from higher states should be small, it seems that this theory significantly underestimates the cross section at lower energies and does not predict the structure which was found when $E_r$ was approximately 1 keV.

It has not yet been possible to make measurements at lower energies, but the trend of the present results suggests that charge transfer might be large at thermal energies. If this is so, the reaction could be an important loss mechanism for $H^-$ in stellar atmospheres.

## § 5-8. Alternative methods used to study collisions between electrons and ions

Several other techniques which can be used to study electron ion collisions but do not employ colliding beams. Strictly, these lie outside the scope of the present article but they are of such interest that they will be mentioned briefly.

### A. Sequential Mass Spectrometry

This is the name given to a technique introduced by Baker and Hasted [89, 90] in which ions are trapped primarily by the potential depression due to space charge in an electron beam. This depression is typically a fraction of a volt so that it is sufficient to retain ions with modest thermal energies. Conventional ion sources can easily be modified to provide the basis of a suitable apparatus and the same electron beam is used to trap and bombard the ions. The current of multiply-charged ions formed by these collisions is then resolved by a mass spectrometer and measured.

The advantage of this method is that it shows ionization functions near threshold in much greater detail than has yet been achieved by crossed beam experiments. Baker and Hasted [89], for example, report well-defined structure in the ionization function of $Kr^+ + e \rightarrow Kr^{2+} + 2e$ which can be correlated with known energy levels.

The method is not absolute and without further development it can only be employed over a very limited range of electron energies. For example, the ionization of $Kr^+$ just cited can be studied only between its threshold and the threshold for formation of $Kr^{2+}$ by $Kr + e \rightarrow Kr^{2+} + 3e$; at higher energies it would be impossible to distinguish between the $Kr^{2+}$ formed by the two reactions.

Baker and Hasted studied the single ionization of $Ne^+$, $Ar^+$, $Kr^+$, $Xe^+$, $Ne_2^+$ and $CO^+$ and also the dissociation of $CO^+$ and $CH_4^+$. Subsequently Daly and Powell [91] used similar methods to observe the ionization of

He$^+$ and they have also studied [92] the reactions, $N^+ + e \rightarrow N^{2+} + 2e$; $N_2^+ + e \rightarrow N_2^{2+} + 2e$ and $N_2^{2+} + e \rightarrow N^{2+} + N + e$. Papers by Cuthbert et al. [93–95] have described measurements of the dissociation of several organic molecular ions.

A related, but not identical, technique was developed by Emel'yanov et al. [96] to observe ionization functions for $Cs + e \rightarrow Cs^{3+} + 3e$ and $Cs^+ + e \rightarrow Cs^{2+} + 2e$ within a few eV of threshold. They found a quadratic dependence with energy for the former reaction and a linear dependence for the latter.

The extension of this technique by Daly and Powell [66] and Redhead and Feser [67] to study the (1S–2S) excitation of He$^+$ was mentioned in section 5-4-B.

### B. Collisions between Electron Beams and Low-Density Plasmas

In section 5-4-1 it was pointed out that severe difficulties arise when crossed beams are used to observe the excitation of ions by electron impact. These stem primarily from the tenuous nature of the ion beams which usually have particle densities less than $10^7$ cm$^{-3}$. An alternative approach has been explored by Lucas. A detailed description of these experiments is in preparation [99] and preliminary accounts have already appeared [97, 98].

Lucas employed a radiofrequency discharge in helium to obtain a weakly ionized plasma. The ion density and electron temperature, which were measured by probes, were $10^{11}$ cm$^{-3}$ and 4 eV, respectively. This comparatively dense ion target was bombarded by an electron beam and an evacuated grating monochromator was used to isolate the ensuing 1640 Å radiation formed by the reaction,

$$He^+ (n=1) + e \rightarrow He^+ (n=3) + e \rightarrow He^+ (n=2) + e + h\nu \quad (\lambda = 1640 \text{ Å})$$

where $n$ denotes principal quantum numbers of the states of He$^+$.

From measurements of this line intensity and the ion density, the cross section was deduced for the excitation of $n=3$ state from unexcited He$^+$. These experiments could only be performed for electron energies between threshold and 73 eV since at higher energies $\lambda = 1640$ Å radiation was also produced by the excitation of neutral helium.

Formidable difficulties were encountered. Originally it was intended to observe the $\lambda = 4686$ Å line of He$^+$ but this was obscured by radiation from excited He$_2$ molecules. Even when the $\lambda = 1640$ Å line was studied it was partially obscured by radiation from CO molecules, which were present in minute amounts in spite of stringent precautions. There is no space to review the problems which beset this experiment, but they were finally overcome and a mean value of $1.64 \times 10^{-18}$ cm² was obtained for the cross section over the accessible range of electron energies.

C. COLLISIONS BETWEEN ELECTRONS AND HIGHLY-CHARGED IONS

One expects to see the extension of crossed beam techniques to the study of collisions between electrons and the highly-charged ions which are of direct astrophysical interest. It is not too difficult to obtain adequate parent beams of highly-charged ions [11, 109] but it is usually impossible to ensure that they are all in the ground state (see section 5-2-G). Further difficulties would arise because the cross sections would be small and charge analysers give relatively small dispersion between highly-charged parent and product ions. The compound analyser and single particle detector illustrated by Fig. 5-2-9c seems well suited to this type of experiment.

An alternative approach was proposed by Powell and Fletcher [101] who considered a fully ionized deuterium plasma produced by the collision of two plane shock waves. The plasma was "seeded" with small quantities of oxygen and, from measurements of the temporal growth of intensity of a spectral line of multiply-ionized oxygen, one could in principle compute the ionization cross section of the ion. The method seems feasible because most of the properties of shock-heated plasmas can be calculated fairly accurately from measured shock speeds and the initial conditions in the shock tube. The cross section obtained would, of course, be smeared over the energy distribution of electrons in the plasma.

Plasma spectroscopy can also be employed to study the excitation of highly-charged ions. Boland et al. [102] have, for example, used a high-current toroidal discharge in experiments to observe the electron impact excitation function of $N^{4+}$ ions. Similar methods have also been used by Kunze et al. [132] to measure electron impact ionization and excitation rate coefficients for $C^{4+}$ ions. Their results are of considerable astrophysical interest.

## § 5-9. Some experimental techniques

It is unnecessary to describe the broad range of techniques used in crossed beam experiments because several excellent reviews already exist. It is merely proposed to draw attention to some of these reviews and add a few comments.

A. ION SOURCES

Crossed beam experiments require stable, pure, unexcited and well-collimated ion beams with currents between $10^{-10}$ and $10^{-6}$ A. Although these currents are intermediate between those generally used in mass spectrometers and isotope separators, several suitable sources are available.

For singly-charged ions of metals with low ionization energies (e.g., $Li^+$, $Na^+$ and $K^+$) Kunsman sources are very strongly recommended for their simplicity and stability. Relevant details have been widely published [e.g. 2, 11, 22, 23, 103, 104].

Materials with somewhat higher ionization energies (e.g. barium) can still be surface ionized if the material is made to strike a hot target which has a work function at least comparable to the ionization energy. The golden rule with this type of source is to keep the target so hot ($\approx 2000\,°K$) that it cannot be contaminated by impurities which would lower its work function. Zandberg and Ionov [107] have given a comprehensive review of surface ionization whilst Harrison [108], Elford, Bacon and Hooper [58, 135] have described sources which rely upon these principles. A simple and versatile device which served very well either as a Kunsman or hot target surface ionization source was mentioned earlier [30].

Materials with ionization energies higher than about 6 eV are usually ionized by electron bombardment. Neff [35] described the production of $Ca^+$ and singly-charged ions of more volatile metals by an oscillating-electron source which has been further developed in this laboratory [34]. Here, it produced beams of $Mg^+$ and $Mg^{2+}$ and a pure resolved beam of $10^{-10}$ A of $Li^{2+}$ (very suitable for some interesting experiments on ion atom collisions!). An attractive feature of this source is that it can easily be converted to yield positive and negative gaseous ions.

The sources of negative ions used by Branscomb and his colleagues [16] are quite simple and very successful whilst the low pressure source of $H_2^+$ described by Dunn and Van Zyl [5] achieved impressive stability and yielded adequate currents of ions with a well defined distribution of vibrational states. The Nielsen [33] source used at Culham is unnecessarily large for most of the experiments described here but it is also able to produce beams of multiply-charged ions. In fact it was used by Harrison and Dolder in unpublished experiments on the ionization of $Ar^{2+}$ and it has recently been used [46] to study the ionization of $N^{2+}$.

A number of sources have also been developed especially to produce highly-charged ions and references to these can be found in reviews by Anderson [109] and Von Ardenne [11].

B. ELECTRON BEAM SOURCES

The design of Simpson and Kuyatt [110] has been adopted in some experiments but there is clearly a need for electron guns which give adequate beams uncontaminated by slower electrons. Relevant reviews of electron gun design have recently been given by Simpson [111] and Klemperer [112] and both writers included accounts of the production of nearly mono-energetic electron beams.*

When working with easily-ionized parent ions care must be taken to exclude slow electrons from the collision region. Suitable methods [5, 14] were mentioned in section 6-2.

C. ELECTRON AND ION OPTICS

There is a considerable literature devoted to these topics but the books by Klemperer [9], Von Ardenne [11], Barnard [114] and Duckworth [115] are particularly helpful.

D. VACUUM TECHNIQUES

Again, there is an extensive literature but the books by Dushman [116], Turnbull et al. [117] and Guthrie [118] deserve mention.

Diffusion pumps have generally been used and, in spite of initial misgivings, Apiezon "C" oil proved to be a suitable fluid, provided that it is adequately trapped. Silicone oils have been avoided by most groups because it is sometimes said that they may decompose on electrodes to produce non-conducting layers. One expects to see the wider use of titanium getter sublimation pumps [120, 121] which will have enormous pumping speeds, except for the noble gases.

The experience of Harrison's group at Culham is interesting. They use diffusion pumps (Apiezon "C" fluid) to evacuate a large stainless steel apparatus. A liquid nitrogen-cooled trap and a chevron baffle (cooled by the Peltier effect) are fitted to each diffusion pump and at lower pressures simple sublimation pumps are also used. The apparatus reaches vacua approaching $10^{-10}$ torr *without baking*. They use indium widely as a gasket material, the internal surfaces of the vacuum tanks are electropolished and a number of components are vacuum-baked before they are placed in the vacuum system.

Similar results are acheived by Dunn's group who use "Convalex 10" fluid trapped by zeolite.

E. PARTICLE DETECTORS

Many experiments require the efficient detection of individual atoms or

---

* See also chapter 9 by Kerwin et al. in this book.

ions. Seventeen-stage particle multipliers (E.M.I. type 9643) with venetian blind-shaped dynodes made from BeCu have been successfully used and their efficiency can be greater than 90% for incident particles with more than 10 keV energy.

Curiously, this is rather more than the efficiency claimed by the makers, but it can be attained if the incident particles strike only the central part of the first dynode and no secondary electrons are lost from the sides or front of the multiplier.

Harrison and his colleagues operated their multipliers with the first dynode at or near earth potential so that it had high efficiency only for energetic ions. They fitted an additional electrode in front of the first dynode [71] to give a more uniform potential distribution and to prevent the loss of secondary electrons from the front of the multiplier.

In this laboratory the whole negative potential ($\approx 3.5$ keV) across the dynode chain was applied to the first dynode. Thus an incident positive ion gained very considerable energy from this potential and efficiencies greater than 90% were achieved for $Mg^{2+}$, $Mg^{3+}$ and $Li^{2+}$ ions which originally had energies of only a few keV. It might be thought that the application of large potentials to the front of a multiplier might induce tracking and lead to noisy operation but, provided care was taken to exclude dust, this was not so. Typically, the height of output pulses from the multiplier was 5 V but with one volt discrimination the noise was only one count $sec^{-1}$. A cap-shaped electrode was fitted around the front of the multiplier as shown in Fig. 5-9-1 and this was held a few hundred volts negative with respect to the first dynode to prevent the loss of electrons during the initial stages of multiplication. An additional electrode, maintained at +100 V is shown in the figure. This was intended to repel slow ions but it proved unnecessary and has been removed.

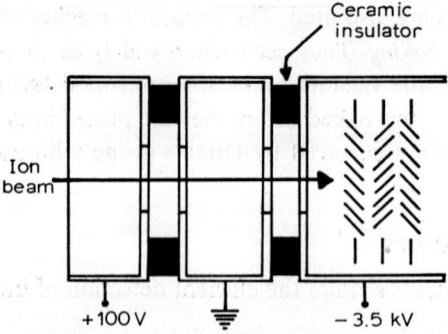

Fig. 5-9-1. Electrode system used to obtain high efficiency from a particle multiplier.

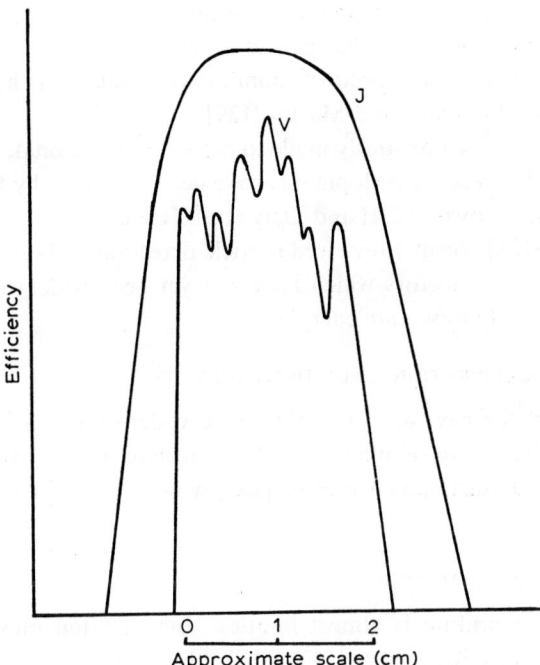

Fig. 5-9-2. Variation of efficiencies of Johnston multiplier (J) and venetian blind multiplier (V) with point of impact of a fine beam of 10 keV protons on first dynode. The ordinate scales are unrelated.

Venetian blind multipliers suffer from the disadvantage that their efficiency depends to some extent on the point at which an incident particle strikes the first dynode. This was clearly illustrated by Rundel [122] who probed the first dynode with a very fine 10 keV proton beam. As the beam was moved across the dynode, the efficiency varied as shown in Fig. 5-9-2. The loss of efficiency at the multiplier edges and the effect of the individual slats can clearly be seen.

An alternative is the Johnston focused mesh multiplier which employs twenty stages and has a circular aperture 1.375" dia. over which the efficiency is fairly uniform. Tests by Rundel [122] showed that the efficiency is about 90% for 2 keV hydrogen ions and, as shown in Fig. 5-9-2, it is uniform over a large fraction of the aperture. The ordinate scales of the two curves in this figure are unrelated and it is not meant to imply that the *peak* efficiencies of the multipliers are significantly different; both can be nearly 100%. The

Johnston multiplier does, however, have more uniform efficiency and it is certainly rugged, compact, efficient and expensive.

Details of statistical aspects of counting with multipliers have been given by Dietz [137], Lombard and Martin [139].

References were previously made to detectors based on designs by Daly. Some interesting recent developments have been described by Freeman et al. [123], Daly and Powell [124] and Daly et al. [151].

Barnett [138] recently reviewed particle detection and described properties of Si barrier detectors which have not yet been widely used to detect particles with only moderate energies.

### F. Intensity calibration of UV detectors

The need for new ways to calibrate UV detectors was mentioned in section 5-4. Recent developments have been described in papers by Saris et al. [154] and Aarts and De Heer [155] which also refer to some earlier work.

### G. Miscellaneous techniques

Here the literature is almost limitless, but mention must be made of excellent books by Rosebury [125] and Kohl [126].

### Acknowledgements

A preliminary draft of this article was read by Professor L. Branscomb, Dr. G. H. Dunn, Mr. M. F. A. Harrison, Mr. B. Peart and Dr. R. D. Rundel. I am grateful to them for their suggestions and to those who provided information in advance of publication.

### References

1. M. F. A. Harrison, Brit. J. App. Phys. **17** (1966) 371.
2. W. C. Lineberger, J. W. Hooper and E. W. McDaniel, Tech. Progr. Rept. No. 8, A.E.C. Contract No. AT-(40-1)-3027, Georgia Inst. of Tech. (1966).
3. M. F. A. Harrison, Methods of Experimental Physics **7B** (revised volume), eds. B. Bederson and W. L. Fite (Academic Press, New York, 1968).
4. S. J. Smith, Methods of Experimental Physics **7A**, eds. B. Bederson and W. L. Fite (Academic Press, New York, 1968) pp. 179–208.
5. G. H. Dunn and B. Van Zyl, Phys. Rev. **154** (1967) 40.
6. Proc. 3rd Intern. Conf. Physics Electronic and Atomic Collisions (London, 1963) p. 387.
7. J. B. Wareing and K. Dolder, Proc. Phys. Soc. **91** (1967) 887.
8. K. R. Spangenberg, Vacuum Tubes (McGraw Hill, New York, 1948).

9. O. KLEMPERER, Electron Optics (second ed., Cambridge University Press, 1953).
10. J. R. PIERCE, Theory and Design of Electron Beams (second ed., Van Nostrand, Princeton, N.J., 1954).
11. M. VON ARDENNE, Tabellen der Electronen Physik, Ionenphysik und Übermikroskopie 1 (Deut. Verlag. Wiss., Berlin, 1954).
12. N. J. THOMPSON and L. B. HEADRICK, Proc. I.R.E. **28** (1940) 319.
13. K. T. DOLDER, M. F. A. HARRISON and P. C. THONEMANN, Proc. Roy. Soc. **A264** (1961) 367.
14. D. F. DANCE, M. F. A. HARRISON, R. D. RUNDEL and A. C. H. SMITH, Proc. Phys. Soc. **92** (1967) 577.
15. G. C. TISONE and L. M. BRANSCOMB, Phys. Rev. **170** (1968) 169.
16. G. C. TISONE, JILA Report No. 73 (University of Colorado, Boulder, 1966).
17. D. F. DANCE, M. F. A. HARRISON and A. C. H. SMITH, Proc. Roy. Soc. **A290** (1966) 74.
18. W. L. FITE and R. T. BRACKMANN, Phys. Rev. **112** (1958) 1141.
19. L. MOLYNEUX, K. DOLDER and B. PEART (to be published).
20. G. D. YARNOLD and H. C. BOLTON, J. Sci. Instr. **26** (1949) 38.
21. G. A. HARROWER, Rev. Sci. Instr. **26** (1955) 850.
22. W. C. LINEBERGER, J. W. HOOPER and E. W. MCDANIEL, Phys. Rev. **141** (1966) 151.
23. J. W. HOOPER, W. C. LINEBERGER and F. M. BACON, Phys. Rev. **141** (1966) 165.
24. B. PEART and K. T. DOLDER (to be published).
25. H. EWALD, Encyclopaedia of Physics **33** (Springer-Verlag, Berlin, 1956).
26. Z. Z. LATYPOV, S. E. KUPRIYANOV and N. N. TUNITSKII, Soviet Phys. JETP **19** (1964) 570.
27. P. FELDMAN and R. NOVICK, Phys. Rev. **160** (1967) 143.
28. S. E. KUPRIYANOV, JETP Letters **4** (1966) 233.
29. B. PEART and K. T. DOLDER, J. Phys. B. (Proc. Phys. Soc.) **1** (1968) 240.
30. B. PEART and K. T. DOLDER, J. Phys. B. (Proc. Phys. Soc.) **1** (1968) 872.
31. K. T. DOLDER, M. F. A. HARRISON and P. C. THONEMANN, Proc. Roy. Soc. **A274** (1963) 546.
32. M. F. A. HARRISON, K. T. DOLDER and P. C. THONEMANN, Proc. Phys. Soc. **82** (1963) 368.
33. K. O. NIELSEN, Nucl. Instr. and Methods **1** (1957) 289.
34. S. O. MARTIN, Ph.D. Thesis, University of Newcastle upon Tyne (1969).
35. S. H. NEFF, Ph.D. Thesis, Harvard University (1963).
36. A. BURGESS, Astrophys. J. **132** (1960) 503.
37. A. BURGESS and M. R. H. RUDGE, Proc. Roy. Soc. **A273** (1963) 372.
38. P. G. BURKE and A. J. TAYLOR, Proc. Roy. Soc. **A287** (1965) 105.
39. A. BURGESS, Astrophys. J. **132** (1960) 503.
40. R. ABRINES, L. C. PERCIVAL and N. A. VALENTINE, Proc. Phys. Soc. **89** (1966) 515.
41. D. RAPP and P. ENGLANDER-GOLDEN, J. Chem. Phys. **43** (1965) 1464.
42. W. BLEAKNEY, Phys. Rev. **36** (1930) 1303.
43. B. L. SCHRAM, A. J. H. BOERBOOM and J. KISTEMAKER, Physica **32** (1966) 185.
44. S. O. MARTIN, B. PEART and K. T. DOLDER, J. Phys. B. (Proc. Phys. Soc.) **1** (1965) 537.
45. R. H. MCFARLAND and J. D. KINNEY, Phys. Rev. **137** (1965) A1058.
46. D. R. BATES, A. H. BOYD and S. S. PRASAD, Proc. Phys. Soc. **85** (1965) 1121.
47. G. PEACH, Proc. Phys. Soc. **87** (1966) 375.
48. G. PEACH, Proc. Phys. Soc. **87** (1966) 381.
49. M. R. C. MCDOWELL, Case Studies in Atomic Collision Physics (North-Holland, Amsterdam, 1969) Chapter 2.
50. O. BELY, J. Phys. B (Proc. Phys. Soc.) **1** (1968) 23.

51. M. J. SEATON, Atomic and Molecular Processes, ed. D. R. Bates (1962) p. 374.
52. P. G. BURKE, D. D. MCVICAR and K. SMITH, Proc. Phys. Soc. **83** (1964) 397.
53. M. J. SEATON (private communication).
54. F. W. BYRON Jr. and C. J. JOACHAIN, Phys. Rev. **146** (1966) 1.
55. F. W. BYRON Jr. and C. J. JOACHAIN, Phys. Rev. **157** (1967) 1.
56. B. PEART, S. O. MARTIN and K. T. DOLDER (to be published).
57. W. FITE and R. T. BRACKMANN, Phys. Rev. **112** (1958) 1151.
58. F. M. BACON and J. W. HOOPER, Report No. ORO-3027-12, Georgia Inst. of Tech. (1968).
59. A. R. LEE and N. P. CARLETON (in course of publication).
60. W. LICHTEN, Phys. Rev. Letters **6** (1961) 12.
61. M. F A HARRISON, D. F. DANCE, K. T. DOLDER and A. C. H. SMITH, Rev. Sci. Inst. **36** (1965) 1443.
62. H. S. W. MASSEY, Encyclopaedia of Physics **36** (Springer-Verlag, Berlin 1956).
63. O. BELY, JILA Report No. 89 (1966).
64. A. BURGESS, D. G. HUMMER and J. A. TULLY (unpublished), see [17, 52, 63].
65. S. ORMONDE, W. WHITAKER and L. LIPSKY, Phys. Rev. Letters **19** (1967) 1161.
66. N. R. DALY and R. E. POWELL, Phys. Rev. Letters **19** (1967) 1165.
67. R. A. REDHEAD and S. FESER, Can. J. Phys. (in course of publication).
68. W. E. LAMB and M. SKINNER, Phys. Rev. **78** (1950) 539.
69. I. P. ZAPESOCHNYI and L. L. SHIMON, Soviet Phys. Doklady **11** (1966) 44.
70. A. R. LEE and N. P. CARLETON (private communication).
71. D. F. DANCE, M. F. A. HARRISON and R. D. RUNDEL, Proc. Roy. Soc. **A299** (1967) 525.
72. N. R. DALY, Rev. Sci. Instr. **31** (1960) 264.
73. N. R. DALY, Rev. Sci. Instr. **31** (1960) 720.
74. M. R. C. MCDOWELL and J. H. WILLIAMSON, Phys. Letters **4** (1966) 159.
75. M. R. C. MCDOWELL and J. H. WILLIAMSON, see [71].
76. L. J. KIEFFER and G. H. DUNN, Rev. Mod. Phys. **38** (1966) 1.
77. M. INOKUTI, Y.-K. KIM and R. L. PLATZMAN, Phys. Rev. **164** (1967) 55.
78. G. H. DUNN, B. VAN ZYL and R. N. ZARE, Phys. Rev. Letters **15** (1965) 610.
79. B. VAN ZYL and G. H. DUNN, Phys. Rev. **163** (1967) 43.
80. J. M. PEEK, Phys. Rev. **134** (1964) A877.
81. J. M. PEEK, Phys. Rev. **154** (1967) 52.
82. J. GUIDINI, C. MANUS, T. SINDA and G. WATEL, Proc. 4th Intern. Conf. Physics Electronic and Atomic Collisions (1965) p. 450.
83. T. SINDA, C. MANUS and J. GUIDINI, Proc. 5th Intern. Conf. Physics Electronic and Atomic Collisions (1967) p. 597.
84. S. M. TRUJILLO, R. H. NEYNABER and E. W. ROTHE, Rev. Sci. Instr. **37** (1966) 1655.
85. V. A. BELYAEV, B. G. BREZHNEV and E. M. ERASTOV, JETP Letters **3** (1966) 207.
86. R. Y. NEYNABER, Methods of Experimental Physics **8**, eds. B. Bederson and W. L. Fite (Academic Press, New York, 1967).
87. F. BROUILLARD, D.Sc. thesis, Université de Louvain (1968).
88. M. F. A. HARRISON (private communication).
89. F. A. BAKER and J. B. HASTED, Phil. Trans. Roy. Soc. London **261** (1966) 33.
90. F. A. BAKER and J. B. HASTED, Proc. 4th Intern. Conf. Physics Electronic and Atomic Collisions (Quebec, 1965) p. 447.
91. N. R. DALY and R. E. POWELL, Proc. Phys. Soc. **89** (1966) 281.
92. N. R. DALY and R. E. POWELL, Proc. Phys. Soc. **89** (1966) 273.
93. J. CUTHBERT, J. FARREN and B. S. PRAHALLADA RAO, Proc. Phys. Soc. **91** (1967) 63.
94. J. CUTHBERT, J. FARREN and B. S. PRAHALLADA RAO, Proc. Phys. Soc. **88** (1966) 91.

95. J. CUTHBERT, J. FARREN and B. S. PRAHALLADA RAO, J. Phys. B. **1** (1968) 62.
96. A. M. EMEL'YANOV, Yu. S. KHODEYEV and L. N. GOROKHOV, Proc. 5th Intern. Conf. Physics Electronic and Atomic Collisions (Leningrad, 1967).
97. C. LUCAS, Proc. 5th Intern. Conf. Physics Electronic and Atomic Collisions (Leningrad, 1967).
98. C. LUCAS, Physikalische Verhandlungen **18** (1967) 27.
99. C. LUCAS (private communication).
100. B. R. TURNER, J. A. RUTHERFORD and D. M. J. COMPTON, J. Chem. Phys. **48** (1968) 1602.
101. A. L. T. POWELL and W. H. W. FLETCHER, Report No. 0-7/61 A.W.R.E., Aldermaston (1961).
102. B. C. BOLAND, F. C. JAHODA, T. J. C. JONES and R. W. P. MCWHIRTER, Proc. 4th Intern. Conf. Physics Electronic and Atomic Collisions (Quebec, 1965).
103. E. W. MCDANIEL, Collision Phenomena in Ionized Gases (John Wiley, New York, 1964).
104. H. B. GILBODY, R. BROWNING and G. LEVY, J. Phys. B. **1** (1968) 230.
105. I. LANGMUIR and H. K. KINGDON, Proc. Roy. Soc. **A107** (1925) 61.
106. I. LANGMUIR and H. K. KINGDON, Phys. Rev. **21** (1923) 380.
107. E. Ya. ZANDBERG and N. I. IONOV, Soviet Phys. Usp. **67** (1959) 255.
108. M. F. A. HARRISON, Report No. GP/R2505, A.E.R.E., Harwell (1958).
109. C. E. ANDERSON, Methods of Experimental Physics **4A**, eds. V. W. Hughes and H. L. Schultz (Academic Press, New York, 1967).
110. J. A. SIMPSON and C. E. KUYATT, Rev. Sci. Instr. **34** (1963) 265.
111. J. A. SIMPSON, Methods of Experimental Physics **4A**, eds. V. W. Hughes and H. L. Schultz (Academic Press, New York, 1967).
112. O. KLEMPERER, Reports on Progress in Physics **28** (1965) 77.
113. M. INOKUTI (private communication).
114. G. A. BARNARD, Modern Mass Spectrometry (The Institute of Physics, London, 1953).
115. H. E. DUCKWORTH, Mass Spectroscopy (Cambridge University Press, 1960).
116. S. DUSHMAN, Scientific Foundations of Vacuum Technique (second ed., John Wiley, New York, 1962).
117. A. H. TURNBULL, R. S. BARTON and J. C. RIVIERE, An Introduction to Vacuum Technique (Geo. Newnes Ltd., London, 1962).
118. A. GUTHRIE, Vacuum Technology (John Wiley, New York, 1963).
119. D. R. BATES and J. T. LEWIS, Proc. Phys. Soc. **A68** (1955) 173.
120. G. M. MCCRACKEN, Vacuum **15** (1965) 433.
121. G. M. MCCRACKEN and N. A. PASHLEY, J. Vac. Sci. and Tech. **3** (1966) 96.
122. R. D. RUNDEL (private communication).
123. N. J. FREEMAN, N. R. DALY and R. E. POWELL, Rev. Sci. Instr. **38** (1967) 945.
124. N. R. DALY and R. E. POWELL, Proc. Conf. Heavy Particle Collisions (Belfast, 1968) p. 261.
125. F. ROSEBURY, Handbook of Electron Tube and Vacuum Techniques (Addison-Wesley, Reading, Mass., 1965).
126. W. H. KOHL, Materials and Techniques for Electron Tubes (Reinhold, New York, 1960).
127. G. HAGEN, Proc. 5th Intern. Conf. Physics Electronic and Atomic Collisions (Leningrad, 1967).
128. S. E. KUPRIYANOV and Z. Z. LATYPOV, J.E.T.P. (U.S.S.R.) **45** (1963) 815.
129. W. L. FITE, W. E. KAUPILLA and W. R. OTT, Phys. Rev. Letters **20** (1968) 409.
130. P. G. BURKE and A. JOANNA TAYLOR, A.E.R.E., Harwell Report, T.P.343 (1968).
131. U. FANO, private communication quoted in [129].

132. H. J. Kunze, A. H. Gabriel and H. R. Griem, Proc. 5th Intern. Conf. Physics Electronic and Atomic Collisions (Leningrad, 1967) p. 362.
133. J. W. Hooper, Report No. ORO-3027-13, Georgia Inst. of Tech. (April 1968).
134. G. Tisone and L. M. Branscomb, Phys. Rev. Letters **17** (1966) 236.
135. M. T. Elford, R. M. Bacon and J. W. Hooper (paper submitted to Rev. Sci. Instr.).
136. W. Aberth, J. R. Peterson, D. C. Lorents and C. J. Cook, Phys. Rev. Letters **20** (1968) 979.
137. L. A. Dietz, Rev. Sci. Instr. **36** (1965) 1763.
138. C. F. Barnett, Proc. Conf. Heavy Particle Collisions (Belfast, 1968) invited review.
139. F. J. Lombard and F. Martin, Rev. Sci. Instr. **32** (1961) 200.
140. G. H. Dunn, Proc. Intern. Atomic Phys. Conf. (New York, 1968) invited review.
141. G. Hagen, Proc. 21st Gaseous Elec. Conf. (Boulder, 1968).
142. L. P. Theard, Proc. 21st Gaseous Elec. Conf. (Boulder, 1968).
143. P. Mahadevan, Proc. 21st Gaseous Elec. Conf. (Boulder, 1968).
144. M. Inokuti and Y.-K. Kim, Phys. Rev. **173** (1968) 154.
145. B. Peart and K. T. Dolder (to be published).
146. K. L. Aitken, M. F. A. Harrison and R. D. Rundel, J. Phys. B (to be published).
147. D. L. Moores (private communication).
148. M. R. C. McDowell (private communication).
149. M. Inokuti and Y.-K. Kim, Proc. 6th. Intern. Conf. Physics Electronic and Atomic Collisions (Boston, 1969).
150. M. R. H. Rudge, Rev. Mod. Phys. **40** (1968) 564.
151. N. R. Daly, A. McCormick and R. E. Powel, Rev. Sci. Instr. **39** (1968) 1163.
152. G. Peach (private communication).
153. M. R. C. McDowell and J. P. Coleman, Introduction to the Theory of Ion-Atom Collisions (North-Holland, Amsterdam, 1969).
154. F. W. Saris, B. F. J. Luyken, F. J. De Heer and J. Aarts, Proc. ESRO Symposium on Calibration Methods in UV and X-ray Regions of the Spectrum, ed. D. D. Clark (Munich, 1968).
155. J. F. M. Aarts and F. J. De Heer, J. Opt. Soc. Am. **58** (1968) 1666.
156. O. Bely and S. B. Schwartz, J. Phys. B **2** (1969) 159.
157. F. M. Bacon and J. W. Hooper, Phys. Rev. **178** (1969) 182.
158. G. H. Dunn, Phys. Rev. **172** (1968) 1.
159. N. F. Mott and H. S. W. Massey, The Theory of Atomic Collisions (The Clarendon Press, Oxford, 3rd edition, 1965).
160. B. Peart, D. S. Walton and K. T. Dolder, to be published.
161. M. T. Elford, R. K. Feeney and J. W. Hooper, Proc. 6th Intern. Conf. Physics Electronic and Atomic Collisions (Boston, 1969).
162. K. Omidvar, Phys. Rev. **177** (1969) 212.
163. K. L. Aitken, M. F. A. Harrison and R. D. Rundel, Proc. 6th Intern. Conf. Physics Electronic and Atomic Collisions (Boston, 1969).
164. M.O. Pace and J. W. Hooper, Proc. 6th Intern. Conf. Physics Electronic and Atomic Collisions (Boston, 1969).
165. G. H. Dunn, Article on "Colliding beams", in: Atomic Physics (Plenum Press, 1969).

CHAPTER 6

# BINARY-ENCOUNTER AND CLASSICAL COLLISION THEORIES

BY

## L. VRIENS

*FOM-Instituut voor Atoom- en Molecuulfysica,
Kruislaan 407, Amsterdam-O.*

# Contents

|  | Page |
|---|---|
| 6-1. Introduction | 337 |
| 6-2. Historical survey | 338 |
|     A. Early work | 338 |
|         1. Scattering by nuclei of atoms | 338 |
|         2. Scattering by atomic electrons | 339 |
|     B. Recent developments | 344 |
| 6-3. Two particle collisions | 346 |
|     A. Basic relations | 346 |
|     B. Some special applications | 349 |
| 6-4. Charged particle–atom collisions | 353 |
|     A. Collision models | 353 |
|     B. Single ionization | 354 |
|     C. Excitation | 357 |
|     D. Double ionization | 358 |
|     E. Charge transfer | 359 |
|     F. Electronic stopping | 361 |
|     G. Velocity distribution of atomic electrons | 362 |
| 6-5. Comparison with the Born approximation | 363 |
|     A. Double differential cross sections | 363 |
|     B. The cross section per unit energy transfer | 366 |
|     C. The total ionization cross section | 368 |
|     D. Stopping power | 369 |
|     E. Sum rules and total cross section | 370 |
|     F. Alternative method of deriving binary-encounter formulae | 372 |
| 6-6. The classical theory | 376 |
|     A. Differential cross sections | 380 |
|     B. Total cross sections | 381 |
| 6-7. Comparison with experiment | 382 |
|     A. Differential cross sections | 382 |
|     B. Total cross sections | 384 |
|     C. Similarity of scaled cross sections | 385 |
| 6-8. Resonances | 390 |
| 6-9. Summary and conclusions | 393 |
| References | 395 |

## § 6-1. Introduction

The main emphasis of this review is on the binary-encounter collision theory which is used to describe ionization, excitation and charge transfer in electron–atom and ion–atom collisions. The ions considered are protons, α-particles and other nuclei. The basic approximations involved are:
(i) the incident particle interacts with only one target particle (electron or nucleus) at a time,
(ii) the mutual interaction between the atomic electrons and nucleus can be disregarded during the collision.

The atomic electrons and the nucleus are thus assumed to be independent scattering centers. This assumption is justified of course only if the effective interaction between the primary particle and the atom takes place in a region small compared to the atomic dimensions. If this is the case, the momentum transfer to one of the atomic particles is large compared to the momenta of the atomic electrons, and if the incident particle collides with an atomic electron the energy transferred to it is much larger than its binding energy.

In the calculation of binary-encounter collision cross sections, the target system of free atoms (for a gaseous target) is replaced by a target system of free nuclei (at rest) and free electrons (which may have the same velocity distribution as in the atom). The first task of the binary-encounter theory is to derive cross section relations for collisions between two free moving structureless charged particles. This may be done using either classical mechanics or quantum-mechanics. Due to the special nature of the Coulomb interaction, classical mechanics and quantum mechanics yield identical cross section formulae for Coulomb scattering of unlike particles, whereas for identical particles the only difference arises from interference between direct and exchange terms. The second task of the binary-encounter theory is to relate these Coulomb cross sections with those for charged particle-atom collisions. Here approximations are involved.

The chief advantage of the binary-encounter theory is that it gives in a direct and simple way reasonable estimates of cross sections for a wide variety of processes. In this "case study" we include a historical survey, a detailed derivation of binary-encounter cross section relations and also give several applications. The range of validity of the theory is studied by comparing with experiment, first Born approximation and three body classical theory.

## § 6-2. Historical survey

A considerable amount of work on the subjects covered by this review was done between 1900 and 1940. After that time, a period of little activity followed until Gryzinski's 1959 paper [1] initiated a new series of investigations on binary-encounter and classical collision theories. In fact, the general interest in electronic and atomic collisions has largely been revived in the last 10 to 15 years, partly owing to the need there exists in other fields of physics (e.g. in astrophysics, plasma physics and upper atmosphere physics) for large numbers of cross sections.

### A. Early work

#### 1. *Scattering by nuclei of atoms*

Experiments on scattering of $\alpha$-particles by numerous targets performed between 1906 and 1920 by Rutherford, Geiger and others, see for a review Rutherford et al. [2], have played a significant role in the development of atomic models. This and later experimental evidence on single "large" angle scattering of nuclei (protons, $\alpha$-particles) by solid or gaseous targets, and of large angle elastic scattering of fast electrons by atoms, can be summarized in the Rutherford formula

$$\sigma_P \, dP = \frac{8\pi e^4 z_1^2 Z^2}{v_1^2 P^3} \, dP. \tag{6-2-1}$$

Here, $\sigma_P$ is the cross section per unit momentum transfer $P$ for scattering of an incident particle with velocity $v_1$ and charge $z_1 e$ by a free target nucleus with charge $Ze$. For a target system containing $N$ atoms per unit volume one has to multiply $\sigma_P$ by $N$.

In the first Born approximation, the corresponding cross section for elastic scattering of a structureless charged particle by an atom is [3, 4]

$$\sigma_P \, dP = \frac{8\pi e^4 z_1^2}{v_1^2 P^3} |Z - F(K)|^2 \, dP \tag{6-2-2}$$

where $F(K)$ is the atomic form factor defined by

$$F(K) = \sum_s (\psi_0 | \exp(i\mathbf{K} \cdot \mathbf{r}_s) | \psi_0) \tag{6-2-3}$$

where $K\hbar = P$, $\psi_0$ is the ground state wave function, ( | | ) stands for an atomic matrix element, and the sum extends over all the atomic electrons, with position vectors $r_s$. Hence, the total elastic scattering amplitude is a sum of amplitudes corresponding to scattering from the individual target particles (nucleus and electrons). For large $K$, $\exp(i\mathbf{K}\cdot\mathbf{r}_s)$ in eq. (6-2-3) oscillates rapidly with $r_s$ and $F(K) \ll Z$; i.e., the amplitudes for scattering by the target electrons vanish and eq. (6-2-2) reduces to eq. (6-2-1). The scattering can then essentially be described as a binary-encounter between the incident particle and the target nucleus. For small $K$, $F(K)$ approaches $Z$ and $F(0) = Z$. The shielding effect of the atomic electrons thus prevents $\sigma_P$ from becoming infinite at $K = 0$ (scattering angle $\Theta = 0$).

The Rutherford formula (6-2-1) is correct both in classical theory and in quantum theory except for scattering of identical particles. For particles having spin 0 ($\alpha$-particles) the total wave function is symmetrical in the coordinates of the two nuclei, and for spin $\frac{1}{2}$ (protons) the total wave function is antisymmetrical. Because of these symmetry properties one has to add an exchange scattering amplitude to the direct scattering amplitude in eq. (6-2-1). Hence one obtains exchange and interference terms in the cross section formula. Mott [5] first derived the relations for these cases. The first experimental verifications of this quantum-mechanical exchange effect were given by Chadwick [6] and Blackett and Champion [7].

Since we are more concerned here with processes like ionization, excitation and charge transfer we now leave the subject of "elastic" scattering of charged particles by (nuclei of) atoms.

## 2. Scattering by atomic electrons

Thomson [8] first used the binary-encounter theory for calculating cross sections for ionization of atoms by electrons. The bound target electrons were assumed to be free and at rest before the collision. Equation (6-2-1) with $Z = -1$ also applies for this case. For target electrons that are initially at rest, the energy transfer $E = P^2/2m$, where $m$ is the electron mass. Equation (6-2-1) can therefore be rewritten as

$$\sigma_E \, dE = \frac{2\pi e^4 z_1^2}{m v_1^2 E^2} \, dE = \frac{4\pi a_0^2 z_1^2 R}{(v_1/v_0)^2 \, E^2} \, dE \tag{6-2-4}$$

where $\sigma_E$ is the differential cross section per unit energy transfer, $R$ the Rydberg energy, $a_0$ the Bohr radius and $v_0 = e^2/\hbar$, hence $\frac{1}{2} m v_0^2 = R$. The Thomson ionization cross section per target electron with binding energy $U_i$ is

$$Q_i = \int_{U_i}^{E_{\max}} \sigma_E \, dE \tag{6-2-5}$$

where, according to conservation laws, $E_{max} = \tfrac{1}{2}mv_1^2$ for incident electrons and $2mv_1^2$ for incident heavy particles to which Thomson's theory can also be applied. For electrons $z_1^2$ in eq. (6-2-4) is equal to unity.

Thomas [9, 10] and Williams [11] refined the theory by partly taking the velocity of the atomic electrons into account. They considered the case that a charged particle moves through a cloud of free target electrons moving in random directions with velocity $v_2$. Thomas [9, 10] derived the differential cross sections $\sigma_{E,P}$ and $\sigma_E$. The first of these is for transfer of energy between $E$ and $E + dE$ and momentum between $P$ and $P + dP$. Because Thomas gave no derivation of the formula for $\sigma_{E,P}$ it is not clear whether he took the recoil of the target electron and the incident particle during the collision into account or not. Williams [11] independently derived $\sigma_E$ for large $v_1$. Both, Thomas and Williams used $\sigma_E$ to calculate the stopping power of different targets.

Thomas [10] further argued that the binary encounter takes place in a region small compared to the atomic dimensions (this is only true for hard collisions). The incident electron must therefore gain kinetic energy from the atomic field before the binary encounter. Thomas took this additional kinetic energy equal to the absolute magnitude of the potential energy $(U_i + \tfrac{1}{2}mv_2^2)$ of the target electron. The cross sections for large $v_1$ are then reduced (see eq. (6-2-4) and also section 6-4-A) by a factor $(\tfrac{1}{2}mv_1^2)/(\tfrac{1}{2}mv_1^2 + U_i + \tfrac{1}{2}mv_2^2)$. Webster et al. [12] later introduced another multiplication factor $(\tfrac{1}{2}mv_1^2 + U_i + \tfrac{1}{2}mv_2^2)/(\tfrac{1}{2}mv_1^2)$ which converts the new relations back to the original ones. This latter multiplication factor takes account of the focussing of the electrons by the atomic (nuclear) field. This effect may be important for inner shell ionization.

New progress was made when Mott [5] derived the quantum mechanical cross section formulae for Coulomb scattering of identical particles in both the center of mass system and the coordinate frame in which one of the particles is initially at rest. An experimental verification of the exchange effect for electrons was given by Williams [13]. He used a cloud chamber to measure large angle inelastic scattering of 20 keV electrons by $N_2$ and $O_2$, by studying the secondary electrons with large energy which appear as branches to the main tracks of the 20 keV electrons. This experiment simultaneously verified the binary-encounter (electron–electron) nature of the collision and the effect of indistinguishability of the electrons. One year later Møller [14] gave the relativistic generalization of Mott's theory. It is interesting to note that the Rutherford, Møller and Mott formulae are still being used in the theory of stopping power and in radiation physics to calculate the energy distribution of fast secondary electrons and the contribution of the close collisions to the stopping power.

In these latter applications one in general disregards the motion of the target (atomic) electrons. To a good approximation this is justified when calculating $\sigma_E$ for large $E$; this cross section per unit energy transfer then directly gives the energy distribution of the ejected electrons. However, the assumption of target electrons initially being at rest leads to completely erroneous results for the double differential cross section $\sigma_{E,P}$. For target electrons at rest, $P^2 = 2mE$ and the energy distribution for one angle (thus one $P$) is a $\delta$-function. In an illustrative series of measurements, Hughes and coworkers [15–17] clearly demonstrated that the energy distribution of initially monoenergetic electrons (with energy between 1000 and 4000 eV) inelastically scattered under 34° from He and $H_2$ is actually rather broad. In the case of 1000, 2000, 3000 and 4000 eV electrons incident on He, the measured most probable (corresponding approximately to scattering by target electrons which are initially at rest) energy loss in inelastic scattering was about 300, 600, 950 and 1300 eV, respectively. The corresponding full width at half maximum of the energy distribution was found to be about 200, 270, 340 and 400 eV, respectively. For electrons incident on $H_2$, the most probable energy loss was the same for the same incident energy, but the half width was smaller by about a factor of 0.85. Apart from the broad inelastic peak also a sharp elastic peak (scattering from the nucleus) was detected.

Hughes et al. [15–17] also gave a theoretical interpretation of their results. They have shown that the shape of the experimentally measured distribution of energies among the inelastically scattered electrons can directly be related to the distribution of component velocities of the atomic electrons. This theoretical relationship originated from earlier work on Compton scattering, in which case the internal motion of the atomic electrons is also directly responsible for the (experimentally found) spread in wave-length of scattered photons. Jauncey [18–20], Kirkpatrick et al. [21] and Hicks [22] made calculations of this effect.

Some of these older experimental (Hughes et al.) and theoretical (Kirkpatrick et al. and Hicks) results for helium are shown here in Figs. 6-2-1 and 6-2-2.

The theory behind Hughes' work and the Thomas [9, 10] and Williams [11] theories are only at one point closely related. The target electrons are not assumed to be at rest. Thomas and Williams used their theory to calculate $\sigma_E$ and to calculate the stopping power. In these applications only the average kinetic energy of the target electrons needs to be known (see section

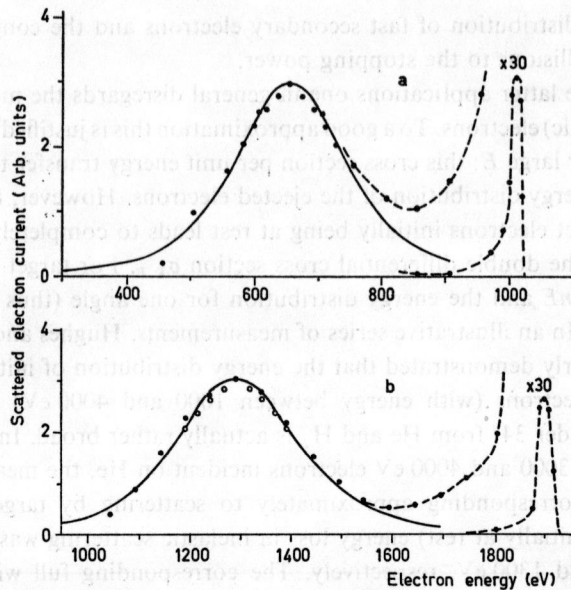

Fig. 6-2-1. Experimental [16] (points and broken curves) and theoretical (full curves [22] and open circles [21]) energy distributions of 1010 eV (a) and 1900 eV (b) electrons inelastically and elastically scattered through 34.2° by helium.

6-4-G) and the approximation that the target electrons have only one velocity $v_2$ does not lead to errors. On the contrary, in Hughes' work the velocity distribution of the target electrons is of vital importance (see section 6-5-A).

Another application of the binary-encounter theory is given by Thomas [23] for charge transfer. The basic assumption made is that an atomic electron is transferred to the passing nucleus as a result of two successive binary encounters. If the velocity $v_1$ of the incident nucleus is much greater than the (average) velocities of the atomic electrons, one may in first approximation disregard these latter velocities. The atomic electron which is to be captured must gain a momentum $|P| \approx m|v_1|$ in the first binary encounter of the incident nucleus with this electron. One can easily show [23] that the angle between $P$ and $v_1$ then must be approximately equal to 60°. In order to be captured, the atomic electron must next undergo a second (elastic) binary-encounter with the atomic nucleus in such a way that its final velocity $v_2'$, with $|v_2'| \approx |v_1|$, is directed approximately parallel to $v_1$. Thus the electron must be scattered by the nucleus also through an angle of about 60°. It is then assumed to be captured if $\frac{1}{2}m|v_2' - v_1|^2$ is smaller than the final potential energy between the passing nucleus and the ejected electron.

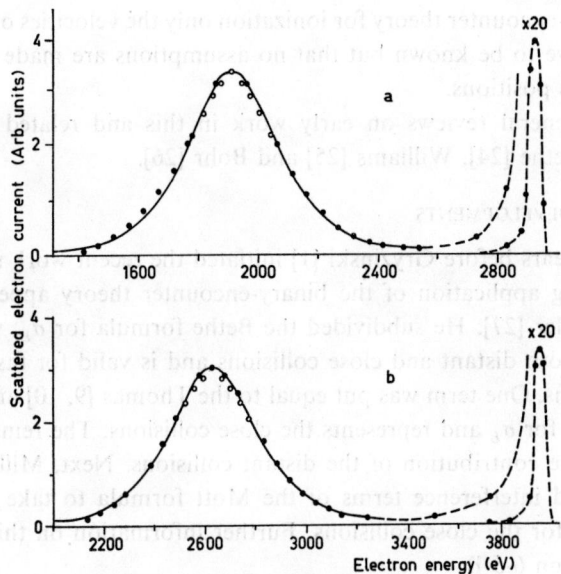

Fig. 6-2-2. As in Fig. 6-2-1 with 2930 eV (a) and 3922 eV (b) incident electrons.

Some general remarks on the range of validity of the theories summarized above are now in order. More details are given in following sections. The basic assumption of the binary-encounter theory is that the atomic electrons and nuclei are independent scattering centers. This requires that the interaction region is small compared to the atomic dimensions. This requirement is fulfilled in large angle inelastic and elastic scattering of fast charged particles by atoms, for example in the measurements of Rutherford, Geiger, Chadwick, Blackett and Champion, Williams and Hughes et al. The theories of Thomas [9, 10] and Williams [11] are incomplete insofar as they don't take account of distant collisions. These are of major importance for the differential cross sections $\sigma_{E,P}$ and $\sigma_E$ for small $E$ and small $P$ (see § 6-5).

In the theory of Thomas [23] on charge transfer other difficulties arise. An electron is assumed to be captured if its final kinetic energy relative to the passing nucleus is smaller than the final potential energy between the two particles. One must therefore simultaneously know relative velocity and relative position, and thus ignore the uncertainty principle in the calculation. Furthermore, the electron is captured in a discrete state which cannot correctly be described in the binary-encounter and classical theories. Note that in the binary-encounter theory for ionization only the velocities of the atomic electrons have to be known but that no assumptions are made about their simultaneous positions.

More general reviews on early work in this and related fields were written by Bethe [24], Williams [25] and Bohr [26].

## B. Recent Developments

A few years before Gryzinski [1] initiated the recent work in this field, an interesting application of the binary-encounter theory appeared in the thesis of Miller [27]. He subdivided the Bethe formula for $\sigma_E$, which takes account of both distant and close collisions and is valid for fast electrons, into two terms. One term was put equal to the Thomas [9, 10] and Williams [11] formula for $\sigma_E$ and represents the close collisions. The remaining term then gives the contribution of the distant collisions. Next, Miller used the exchange and interference terms of the Mott formula to take account of these effects for the close collisions. Further information on this subject is given in section 6-5-B.

Gryzinski [1, 28–30] first derived, in very great detail, classical relations for Coulomb collisions of two moving particles. His calculations were based on results of Chandrasekhar [31] and Chandrasekhar and Williams [32] for collisions of gravitational masses. Next, Gryzinski applied his relations in an amazing amount of different ways to charged particle–atom collisions. Later, it was shown that many of Gryzinski's derivations and relations could be simplified by using different procedures and different variables. It was also shown that in his applications some unrealistic approximations were made. For these reasons we do not reproduce here Gryzinski's derivations. Nevertheless, we emphasize that they were extremely helpful for, and initiated, later investigations in this field.

We briefly summarize the various cases to which Gryzinski [30] applied his theory. These are:

(i) single ionization of atoms (and also $H_2$) by electrons and protons;
(ii) innershell ionization of atoms (e.g. Ni, Ag, Li and Ar) by electrons;
(iii) charge transfer (non-resonant and resonant);
(iv) multiple ionization (e.g. double ionization of He by electrons);
(v) excitation of atoms by electrons and protons;
(vi) stopping power; and
(vii) scattering of electrons by atoms (i.e. angular dependences).

In these applications Gryzinski introduced several new features. For charge transfer, for example, the only previously available classical theories were those of Thomas [10] and Bohr [26]. Gryzinski's simple theory for this process (see section 6-4-E) is quite different from these previous theories. Gryzinski's application of the binary-encounter or classical theory to multiple ionization is also new. His treatment (section 6-4-D) of double ionization is similar to Thomas' treatment [23] of capture, in that both processes are assumed to proceed via a double binary-encounter.

Corrections and extensions of Gryzinski's theory and simpler derivations are given by Ochkur and Petrunkin [33], Stabler [34], Vriens [35–39], McDowell [40], Gerjuoy [41] and Garcia, Gerjuoy and Welker [42, 43]. Bauer and Bartky [44] further extended the theory to excitation and ionization of molecules and Robinson [45] suggested the use of Slater's rules to obtain estimates of the average kinetic energies of the atomic electrons (see also Catlow and McDowell [46]).

For ionization and excitation of atoms by electrons, Burgess [47, 48] followed a more elegant approach, which allows for both close and distant collisions. For the close collisions he assumed (Thomas [10]) that the incident electron gains a kinetic energy $U_i + \frac{1}{2}mv_2^2$ and simultaneously loses the same amount of potential energy before the binary-encounter with an atomic electron. In this symmetrical model (section 6-4-A) both electrons are in the same potential field during the interaction. Burgess further used quantum mechanics to derive the binary-encounter cross section formulae. Exchange and interference are therefore included. Burgess combined the above binary-encounter theory with the impact parameter method for the distant encounters. Burgess' theory is in some aspects similar to the theory of Miller [27]. In the derivation of his cross section formulae, Burgess made the unrealistic assumption that the collision cross section is invariant for transformation from center of mass to laboratory coordinates. In reality only the collision rate is invariant. Vriens [38] obtained the correct formulae for the symmetrical collision model. These formulae are in fact simpler than those of Burgess and in general give better agreement with experiment. Vriens [37] also suggested a combined method in which collisions involving large momentum transfers $P$ should be treated with the binary-encounter theory, and collisions involving small $P$ with for example the Bethe theory.

Further investigations in this field have been reported by Prasad and Prasad [49], Kingston [50–54], Vriens [55], McFarland [56, 57], Sheldon and Dugan [58], Percival and Valentine [59], Garcia [60] and Vriens and Bonsen [61].

Quite independent approaches to the problems discussed here have been made by Bates and coworkers and by Percival and coworkers. The main emphasis of the work of Bates, Cook and Smith [62], Bates and Mapleton [63–65] and Bates and Walker [66, 67], is on charge transfer. Bates et al. further refined and extended the early work of Thomas [23]. Abrines and Percival [68–70], Abrines, Percival and Valentine [71] and Valentine [72] made a detailed study of the classical three-body problem for ionization and charge transfer in collisions of protons, electrons and positrons with the hydrogen atom.

In the binary-encounter theory it is not necessary to make any assumptions about classical orbits of electrons in atoms. In this theory only knowledge about the velocity (velocity-distribution) of the target electrons is required. No assumptions are made about the positions of the target electrons. These favourable circumstances do not apply to the theories of Bates and Percival. Percival used a statistical theory for the hydrogen atom. The microcanonical velocity distribution of the atomic electrons (Mapleton [73]) in this theory is identical to the quantum mechanical velocity distribution for the electron in the hydrogen atom. Percival and coworkers further used Monte Carlo methods to obtain collision cross sections.

## § 6-3. Two particle collisions

### A. Basic relations

In the center of mass system, the Schrödinger equation of the interaction of two structureless particles with masses $m_1$ and $m_2$ and charges $z_1 e$ and $z_2 e$ is given by (see for example Messiah [74] or Mott and Massey [75])

$$\left[ -\frac{\hbar^2}{2m^*} \nabla^2 + \frac{z_1 z_2 e^2}{r} \right] \psi(\mathbf{r}) = \varepsilon \psi(\mathbf{r}) \tag{6-3-1}$$

where $m^* = m_1 m_2/(m_1 + m_2)$, $\varepsilon = \hbar^2 k^2/2m^* = \frac{1}{2} m^* v_r^2$ is the energy in the center of mass system and $v_r$ is the relative velocity. Hence, this problem is identical to the one in which a particle with mass $m^*$, charge $z_1 e$ and velocity $v_r$ is incident on an infinitely heavy nucleus with charge $z_2 e$ and velocity zero. Equation (6-3-1) has the solution (Messiah, Mott and Massey)

$$\psi(\mathbf{r}) = \psi(r, \Theta) = e^{-\frac{1}{2}\pi\gamma} \Gamma(1 + i\gamma) e^{ikz} {}_1F_1(-i\gamma; 1; ik\zeta) \tag{6-3-2}$$

where

$$\gamma = z_1 z_2 e^2/\hbar v_r \quad \text{and} \quad \zeta = r - z = r(1 - \cos\Theta).$$

This wave function has the asymptotic form $\psi \sim I_n + S_c f(\Theta)$, where

$$I_n = [1 + \gamma^2/ik(r - z) + \cdots] \exp[ikz + i\gamma \ln k(r - z)], \tag{6-3-3}$$

$$S_c = r^{-1} \exp[ikr - i\gamma \ln 2kr], \tag{6-3-4}$$

and

$$f(\Theta) = \frac{z_1 z_2 e^2}{2m^* v_r^2} \operatorname{cosec}^2 \tfrac{1}{2}\Theta \exp[-i\gamma \ln \tfrac{1}{2}(1 - \cos\Theta) + 2i\eta_0 + i\pi] \tag{6-3-5}$$

where

$$\exp(2i\eta_0) = \Gamma(1 + i\gamma)/\Gamma(1 - i\gamma). \tag{6-3-6}$$

The incident wave $I_n$ is thus normalized to unit amplitude. The scattering cross section $\sigma(\Theta, \varphi) \, d\Omega$ for scattering through an angle $\Theta$ into solid angle $d\Omega = \sin\Theta \, d\Theta \, d\varphi$, is then given by

$$\sigma(\Theta, \varphi) = |f(\Theta)|^2 = \left[\frac{z_1 z_2 e^2}{2m^* v_r^2}\right]^2 \operatorname{cosec}^4 \tfrac{1}{2}\Theta \tag{6-3-7}$$

for non-identical particles and by

$$\sigma^{\pm}(\Theta, \varphi) = |f(\Theta) \pm f(\pi - \Theta)|^2 \tag{6-3-8}$$

for identical particles. In the latter case $+$ and $-$ stand for the case of symmetrical $(+)$ and antisymmetrical $(-)$ space wave functions.

We now make a transformation of variables; we replace the scattering angle $\Theta$ by the momentum transfer $P$. The following relations are used

$$\sin\tfrac{1}{2}\Theta = \tfrac{1}{2}|v_r - v_r'|/v_r = |v_1 - v_2 - v_1' + v_2'|/2v_r$$
$$= (P/m_1 + P/m_2)/2v_r = P/2m^* v_r, \tag{6-3-9}$$

$$P \, dP = m^{*2} v_r^2 \sin\Theta \, d\Theta, \tag{6-3-10}$$

and

$$\sigma_P = \frac{2\pi P}{m^{*2} v_r^2} \sigma(\Theta, \varphi) \tag{6-3-11}$$

where $\sigma_P$ is, as before, differential per unit momentum transfer. For non-identical particles eqs. (6-3-5), (6-3-7), (6-3-9) and (6-3-11) yield

$$\sigma_P = \frac{8\pi z_1^2 z_2^2 e^4 P}{v_r^2} \left|\frac{\exp(i\eta_P)}{P^2}\right|^2 \tag{6-3-12}$$

where the phase

$$\eta_P = -2\gamma \ln(P/2m^* v_r) + 2\eta_0 + \pi. \tag{6-3-13}$$

For identical particles, $m^* = \frac{1}{2}m_1 = \frac{1}{2}m_2 = \frac{1}{2}m$, and

$$\cos \tfrac{1}{2}\Theta = \tfrac{1}{2}|\boldsymbol{v}_r + \boldsymbol{v}_r'|/v_r = |\boldsymbol{v}_1 - \boldsymbol{v}_2 + \boldsymbol{v}_1' - \boldsymbol{v}_2'|/2v_r$$
$$= S/2m^*v_r = S/mv_r \qquad (6\text{-}3\text{-}14)$$

where $S = m(\boldsymbol{v}_1 - \boldsymbol{v}_2') = m(\boldsymbol{v}_1' - \boldsymbol{v}_2)$ is the exchange momentum transfer which is obtained from $P = m(\boldsymbol{v}_1 - \boldsymbol{v}_1') = m(\boldsymbol{v}_2' - \boldsymbol{v}_2)$ by interchanging $\boldsymbol{v}_1'$ and $\boldsymbol{v}_2'$. Equations (6-3-5), (6-3-8), (6-3-9) and (6-3-14) yield

$$\sigma_P^{\pm} = \frac{8\pi z_1^2 z_2^2 e^4 P}{v_r^2} \left| \frac{\exp(i\eta_P)}{P^2} \pm \frac{\exp(i\eta_S)}{S^2} \right|^2 \qquad (6\text{-}3\text{-}15)$$

where in this case $z_1 = z_2$ and

$$\eta_S = -2\gamma \ln(S/2m^*v_r) + 2\eta_0 + \pi$$
$$= -2\gamma \ln(S/mv_r) + 2\eta_0 + \pi. \qquad (6\text{-}3\text{-}16)$$

It is important to note that all variables occurring in eqs. (6-3-12), (6-3-13), (6-3-15) and (6-3-16) are invariant to coordinate transformations. Further, the collision rate $\alpha_P = v_1 \sigma_P$ or $\alpha_P^{\pm} = v_1 \sigma_P^{\pm}$ should be invariant to coordinate transformations. The special case considered above (scattering in the center of mass system) was equivalent to the problem of a particle with mass $m^*$, charge $z_1 e$ and velocity $v_r$ incident on an infinitely heavy nucleus with charge $z_2 e$ and velocity zero. Thus for eqs. (6-3-12) and (6-3-15) $\alpha_P = v_r \sigma_P$ and $\alpha_P^{\pm} = v_r \sigma_P^{\pm}$. We now simultaneously generalize eqs. (6-3-12) and (6-3-15) to an arbitrary coordinate system in which the particles have velocities $v_1$ and $v_2$ respectively, and we reintroduce the c.m.s. azimuthal scattering angle $\varphi$ as a variable, via

$$\alpha_P = v_1 \sigma_P = 2\pi v_1 \sigma_{P,\varphi} = 2\pi \alpha_{P,\varphi}. \qquad (6\text{-}3\text{-}17)$$

Equation (6-3-12) then gives

$$\alpha_{P,\varphi} = \frac{4z_1^2 z_2^2 e^4 P}{v_r} \left| \frac{\exp(i\eta_P)}{P^2} \right|^2 \qquad (6\text{-}3\text{-}18)$$

and eq. (6-3-15) gives

$$\alpha_{P,\varphi}^{\pm} = \frac{4z_1^2 z_2^2 e^4 P}{v_r} \left| \frac{\exp(i\eta_P)}{P^2} \pm \frac{\exp(i\eta_S)}{S^2} \right|^2. \qquad (6\text{-}3\text{-}19)$$

The c.m.s. variable $\varphi$ can now be transformed to the variable $E$, the energy loss of the incident particle with velocity $v_1$ and mass $m_1$, via

$$\alpha_{E,P} = \frac{d^2\alpha}{dP\,dE} = \frac{d^2\alpha}{dP\,d\varphi} \times \frac{d\varphi}{dE} = \alpha_{P,\varphi} \frac{d\varphi}{dE}. \tag{6-3-20}$$

The calculation of $d\varphi/dE$ is lengthy but straightforward and will not be given here*. As a result we get

$$\alpha_{E,P} = v_1 \sigma_{E,P} = \frac{8z_1^2 z_2^2 e^4}{v_1 v_2 X^{\frac{1}{2}}} \left| \frac{\exp(i\eta_P)}{P^2} \right|^2 \tag{6-3-21}$$

and

$$\alpha_{E,P}^{\pm} = v_1 \sigma_{E,P}^{\pm} = \frac{8z_1^2 z_2^2 e^4}{v_1 v_2 X^{\frac{1}{2}}} \left| \frac{\exp(i\eta_P)}{P^2} \pm \frac{\exp(i\eta_S)}{S^2} \right|^2 \tag{6-3-22}$$

where

$$\begin{aligned} X &= -\cos^2\phi + 2(\hat{v}_1 \cdot \hat{P})(\hat{v}_2 \cdot \hat{P})\cos\phi + 1 - (\hat{v}_1 \cdot \hat{P})^2 - (\hat{v}_2 \cdot \hat{P})^2 \\ &= (\cos\phi_{\min} - \cos\phi)(\cos\phi - \cos\phi_{\max}) \\ &= \left(\frac{v_r}{v_1 v_2 P}\right)^2 (E_{\max} - E)(E - E_{\min}). \end{aligned} \tag{6-3-23}$$

Here, $\phi$ is the angle between $v_1$ and $v_2$, and $\hat{v}_1$, $\hat{v}_2$ and $\hat{P}$ are the unit vectors along $v_1$, $v_2$ and $P$. One can show that

$$\hat{v}_1 \cdot \hat{P} = \frac{E}{v_1 P} + \frac{P}{2m_1 v_1} \quad \text{and} \quad \hat{v}_2 \cdot \hat{P} = \frac{E}{v_2 P} - \frac{P}{2m_2 v_2}. \tag{6-3-24}$$

In eqs. (6-3-21) and (6-3-22), $X \geq 0$. Integrating $\alpha_{E,P}$ and $\alpha_{E,P}^{\pm}$ over $E$ using

$$\int_{E_{\min}}^{E_{\max}} \{(E_{\max} - E)(E - E_{\min})\}^{-\frac{1}{2}} dE = \pi \tag{6-3-25}$$

directly yields the $\sigma_P$ and $\sigma_P^{\pm}$ of eqs. (6-3-12), (6-3-15) and (6-3-17).

## B. Some Special Applications

In most applications of the binary-encounter theory, for example for ionization, excitation and charge transfer, the target particles are electrons ($m_2 = m$ and $z_2 = -1$). The incident particles may be heavy (protons, $\alpha$-particles) in which case $m_1 \gg m$ and $m^* = m$, or may be electrons.

We now confine ourselves to target systems which only contain electrons and consider both cases of incident heavy particles and of incident electrons.

---

* I am indebted to D. Banks for showing me this method of calculating $\sigma_{E,P}$.

We assume that the target electrons act as independent scattering centers. We may integrate therefore over scattering cross sections instead of over scattering amplitudes.

Equation (6-3-21) which applies for incident heavy particles then simplifies to

$$\sigma_{E,P}(\phi) = \frac{8z_1^2 e^4}{v_1^2 v_2 P^4 X^{\frac{1}{2}}} \tag{6-3-26}$$

and eq. (6-3-22) which applies for incident electrons becomes

$$\sigma_{E,P}^{\pm}(\phi) = \frac{8e^4}{v_1^2 v_2 X^{\frac{1}{2}}} \left[ \frac{1}{P^4} + \frac{1}{S^4} \pm \frac{2\cos(\eta_P - \eta_S)}{P^2 S^2} \right] \tag{6-3-27}$$

where

$$\eta_P - \eta_S = -2\gamma \ln(P/S) = (2e^2/\hbar v_r) \ln(S/P) \tag{6-3-28}$$

and $\phi$ is, as before, the angle between $v_1$ and $v_2$. The first, second and third term in eq. (6-3-27) are the direct, the exchange and the interference term respectively. For large $v_1$ (and thus for large $v_r$) $\eta_P$, $\eta_S$ and $\eta_P - \eta_S$ approach zero and $\cos(\eta_P - \eta_S)$ tends to unity.

If we consider ionization or excitation of atoms, these atoms in general have no specific orientation with respect to the incident beam. The target electrons then have an isotropic velocity distribution. The corresponding cross sections are given by [38, 39]

$$\sigma_{E,P} = \int_0^\pi \sigma_{E,P}(\phi) \tfrac{1}{2} \sin\phi \, d\phi = \frac{4\pi z_1^2 e^4}{v_1^2 v_2 P^4} \tag{6-3-29}$$

for incident heavy particles, and by

$$\sigma_{E,P}^{\pm} = \int_0^\pi \sigma_{E,P}^{\pm}(\phi) \tfrac{1}{2} \sin\phi \, d\phi$$

$$= \frac{4\pi e^4}{v_1^2 v_2} \left[ \frac{1}{P^4} + \frac{v_1^2 + v_2^2 - P^2/2m^2 - 2E^2/P^2}{m^4 |v_1^2 - v_2^2 - 2E/m|^3} \right.$$

$$\left. \pm \frac{2\Phi}{m^2 P^2 |v_1^2 - v_2^2 - 2E/m|} \right] \tag{6-3-30}$$

for incident electrons. Here, $\Phi$ is a function of $v_1$, $v_2$, $P$ and $E$; its values lie between zero and unity and approach unity for large $v_1$. An approximation for $\Phi$ is given in [38].

The condition $X \geq 0$ determines the integration limits of eqs. (6-3-26) and (6-3-27), where the integration variable may be $\phi$, $P$, $E$, $v_2$ or $v_1$. Equations (6-3-29) and (6-3-30) have one common integration limit (for the variables $P$, $E$ or $v_2$):

$$m(v_2' - v_2) \leq P \leq m(v_2' + v_2) \qquad (6\text{-}3\text{-}31)$$

where $\tfrac{1}{2}m(v_2'^2 - v_2^2) = E$. Consequently,

$$\frac{P^2}{2m} - v_2 P \leq E \leq \frac{P^2}{2m} + v_2 P \qquad (6\text{-}3\text{-}32)$$

and

$$v_2 \geq \left| \frac{E}{P} - \frac{P}{2m} \right|. \qquad (6\text{-}3\text{-}33)$$

Here we assumed $E$ to be positive since we only consider inelastic collisions in which the incident particle loses energy. One further limitation for incident heavy particles (eq. (6-3-29)) is that

$$M(v_1 - v_1') \leq P \leq M(v_1 + v_1') \qquad (6\text{-}3\text{-}34)$$

where $M = m_1$ and $\tfrac{1}{2}M(v_1^2 - v_1'^2) = E$. From eq. (6-3-34) it follows that $E \leq v_1 P - P^2/2M \approx v_1 P$ and $v_1 \geq E/P + P/2M \approx E/P$. The corresponding relation for incident electrons is

$$m(v_1 - v_1') \leq P \leq m(v_1 + v_1') \qquad (6\text{-}3\text{-}35)$$

and consequently $E \leq v_1 P - P^2/2m$ and $v_1 \geq E/P + P/2m$.

To evaluate the formulae for $\sigma_E$ for incident heavy particles one has to distinguish between three cases – these are direct consequences of eqs. (6-3-31) and (6-3-34). In the first case, $m(v_2' - v_2) \geq M(v_1 - v_1') \approx E/v_1$ and therefore $2v_1 \geq v_2' + v_2$ and $E \leq 2mv_1(v_1 - v_2)$. In this case eq. (6-3-31) applies and integrating over $P$ yields

$$\sigma_E = \frac{2\pi z_1^2 e^4}{mv_1^2} \left( \frac{1}{E^2} + \frac{2mv_2^2}{3E^3} \right). \qquad (6\text{-}3\text{-}36)$$

In the second case $m(v_2' - v_2) \leq E/v_1 \leq m(v_2' + v_2)$, which implies that $v_2' - v_2 \leq 2v_1 \leq v_2' + v_2$ and $2mv_1(v_1 - v_2) \leq E \leq 2mv_1(v_1 + v_2)$. In this case integrating over $P$ between $E/v_1$ and $m(v_2' + v_2)$ gives

$$\sigma_E = \frac{\pi z_1^2 e^4}{3v_1^2 v_2 E^3} \left\{ 4v_1^3 - \tfrac{1}{2}(v_2' - v_2)^3 \right\}. \qquad (6\text{-}3\text{-}37)$$

In the third case, $E/v_1 \geq m(v_2' + v_2)$, no integration region is left and $\sigma_E = 0$.

For incident electrons either eq. (6-3-31) or eq. (6-3-35) applies. In the first case $m(v_1 - v_1') \leq m(v_2' - v_2)$, simultaneously $m(v_2' + v_2) \leq m(v_1 + v_1')$ and integrating eq. (6-3-30) over $P$ gives

$$\sigma_E^{\pm} = \frac{2\pi e^4}{mv_1^2} \left[ \left( \frac{1}{E^2} + \frac{2mv_2^2}{3E^3} \right) + \left( \frac{1}{D^2} + \frac{2mv_2^2}{3D^3} \right) \pm \frac{2\Phi}{ED} \right] \quad (6\text{-}3\text{-}38)$$

where $D = \tfrac{1}{2}mv_1^2 - \tfrac{1}{2}mv_2'^2 (= \tfrac{1}{2}mv_1^2 - \tfrac{1}{2}mv_2^2 - E)$ is the "exchange energy transfer" which may be obtained from $E = \tfrac{1}{2}mv_1^2 - \tfrac{1}{2}mv_1'^2$ by interchanging the final kinetic energies of the interacting electrons. In this first case $D \geq 0$. In the second case $m(v_2' - v_2) \leq m(v_1 - v_1')$, simultaneously $m(v_1 + v_1') \leq m(v_2' + v_2)$ and integrating eq. (6-3-30) over $P$ gives

$$\sigma_E^{\pm} = \frac{2\pi e^4}{mv_1^2} \left[ \left( \frac{1}{E^2} + \frac{2mv_1'^2}{3E^3} \right) + \left( \frac{1}{D^2} + \frac{2mv_1'^2}{3|D|^3} \right) \pm \frac{2\Phi}{E|D|} \right] \frac{v_1'}{v_2}. \quad (6\text{-}3\text{-}39)$$

In the latter case $D \leq 0$. Note that eqs. (6-3-38) and (6-3-39) have the correct symmetry with respect to interchange of the final kinetic energies of the electrons. If in the direct term $v_1'$ in $E = \tfrac{1}{2}mv_1^2 - \tfrac{1}{2}mv_1'^2$ is replaced by $v_2'$, one obtains $\tfrac{1}{2}mv_1^2 - \tfrac{1}{2}mv_2'^2 = D$ of the exchange term. This replacement leaves the interference term unchanged.

All previous relations of this section were derived starting from first principles, i.e. from the Schrödinger equation. When using classical mechanics instead of quantum mechanics, all conservation laws and therefore all integration limits remain unchanged. Furthermore, owing to the special nature of the Coulomb interaction all cross section relations also remain unchanged except that no information is obtained about phases ($\eta_P$ and $\eta_S$). No interference terms can therefore be obtained for identical particles.

Finally, it is interesting to note that if eq. (6-3-12) is multiplied by $\tfrac{1}{2} \sin \phi \, d\phi$ and integrated over $\phi$ between the proper integration limits, or if eq. (6-3-29) is integrated over $E$ one obtains for incident heavy particles

$$\sigma_P = \frac{8\pi z_1^2 e^4}{v_1^2 P^3}, \quad (6\text{-}3\text{-}40)$$

provided that $P^2/2m + v_2 P \leq v_1 P - P^2/2M$. Similarly, eqs. (6-3-15) or (6-3-30) yield for the direct term in the case of incident electrons

$$\sigma_P^{\pm} = \frac{8\pi e^4}{v_1^2 P^3}, \quad (6\text{-}3\text{-}41)$$

for $P^2/2m + v_2 P \leq v_1 P - P^2/2m$. Equations (6-3-40) and (6-3-41) are identical to the Rutherford or Thomson formula for scattering by target electrons which are initially at rest.

## § 6-4. Charged particle–atom collisions

### A. COLLISION MODELS

For incident electrons two models are being used for calculating ionization and excitation cross sections. The first one is the unsymmetrical collision model of Thomson and Gryzinski. In this model the long range interaction between the incident electron and the atom is not taken into account. For example, if, for the hydrogen atom the incident and the atomic electron have the same distance $r$ from the nucleus, it is assumed that because of the interaction with the nucleus the atomic electron has potential energy $-e^2/r$, whereas the incident electron has potential energy 0. Because the incident electron does not gain any kinetic energy before the binary-encounter with the atomic electron, it cannot lose more than its initial kinetic energy $T = \frac{1}{2}mv_1^2$. Exchange excitation, for example triplet excitation of helium, cannot occur in this model. The second model used for electron–atom collisions is the symmetrical model of Thomas and Burgess. In this model one assumes that the incident electron with initial kinetic energy $T$ gains a kinetic energy $U_i + E_2 = U_i + \frac{1}{2}mv_2^2$ and simultaneously loses the same amount of potential energy, before it interacts with the atomic electron which is assumed to be bounded with this potential energy. For example for the hydrogen atom, the total energy of the atomic electron is $-U_i = \langle V \rangle + \langle E_2 \rangle$ and the expectation value of the potential energy $\langle V \rangle = -U_i - \langle E_2 \rangle$ where $\langle E_2 \rangle$ is the expectation value of the kinetic energy.

In § 6-3 we derived cross section formulae using wave functions that are symmetrical $(+)$ or antisymmetrical $(-)$ in the space coordinates of the electrons. If $\chi$ is the angle between the spin directions of the interacting electrons, then the actual cross section is (Mott and Massey [75])

$$\sigma = \tfrac{1}{4}(1 - \cos \chi)\, \sigma^+ + \tfrac{1}{4}(3 + \cos \chi)\, \sigma^-. \tag{6-4-1}$$

We now further assume that $\chi$ is arbitrary, which is the case if the incident electron beam or (and) the target atom is unpolarized. Then we must average over $\chi$ and

$$\sigma = \tfrac{1}{4}\sigma^+ + \tfrac{3}{4}\sigma^-. \tag{6-4-2}$$

For the *unsymmetrical* model eqs. (6-3-38), (6-3-39) and (6-4-2) then yield

$$\sigma_E = \frac{\pi e^4}{T}\left[\left(\frac{1}{E^2} + \frac{4E_2}{3E^3}\right) + \left(\frac{1}{D^2} + \frac{4E_2}{3D^3}\right) - \frac{\Phi}{ED}\right] \quad (6\text{-}4\text{-}3)$$

for $D = T - E_2 - E \geq 0$, and

$$\sigma_E = \frac{\pi e^4}{T}\left[\left(\frac{1}{E^2} + \frac{4(T-E)}{3E^3}\right) + \left(\frac{1}{D^2} + \frac{4(T-E)}{3|D|^3}\right) - \frac{\Phi}{E|D|}\right]\left(\frac{T-E}{E_2}\right)^{\frac{1}{2}} \quad (6\text{-}4\text{-}4)$$

for $D \leq 0$ and $T \geq E$.

In the *symmetrical* model $T$ must be replaced by $T + U_i + E_2$, which gives

$$\sigma_E = \frac{\pi e^4}{T + U_i + E_2}\left[\left(\frac{1}{E^2} + \frac{4E_2}{3E^3}\right) + \left(\frac{1}{(T+U_i-E)^2}\right.\right.$$
$$\left.\left. + \frac{4E_2}{3(T+U_i-E)^3}\right) - \frac{\Phi}{E(T+U_i-E)}\right] \quad (6\text{-}4\text{-}5)$$

for $T + U_i - E \geq 0$, and

$$\sigma_E = \frac{\pi e^4}{T + U_i + E_2}\left[\left(\frac{1}{E^2} + \frac{4(T+U_i+E_2-E)}{3E^3}\right) + \left(\frac{1}{(T+U_i-E)^2}\right.\right.$$
$$\left.\left. + \frac{4(T+U_i+E_2-E)}{3|T+U_i-E|^3}\right) - \frac{\Phi}{E|T+U_i-E|}\right]\left(\frac{T+U_i+E_2-E}{E_2}\right)^{\frac{1}{2}} \quad (6\text{-}4\text{-}6)$$

for $-E_2 \leq T + U_i - E \leq 0$ and $T \geq 0$.

For incident heavy particles the amount of additional kinetic energy $U_i + E_2$ is negligible compared to the values of $\frac{1}{2}Mv_1^2$ considered in the binary-encounter theory. It is therefore irrelevant to introduce a symmetrical collision model for incident heavy particles and we can directly use eqs. (6-3-36) and (6-3-37).

### B. SINGLE IONIZATION

An atomic electron is assumed to be ejected into the continuum if the energy transferred to it in the binary-encounter exceeds its binding energy $U_i$.

The total ionization cross section per atomic electron for incident heavy particles is then given by [39]

$$Q_i = \int_{U_i}^{2mv_1(v_1+v_2)} \sigma_E \, dE = \frac{2\pi z_1^2 e^4}{mv_1^2}\left\{\frac{1}{U_i} + \frac{mv_2^2}{3U_i^2} - \frac{1}{2m(v_1^2 - v_2^2)}\right\} \quad (6\text{-}4\text{-}7)$$

for $U_i \leq 2mv_1(v_1-v_2)$, and by

$$Q_i = \int_{U_i}^{2mv_1(v_1+v_2)} \sigma_E \, dE = \frac{\pi z_1^2 e^4}{mv_1^2} \left\{ \frac{1}{2mv_2(v_1+v_2)} + \frac{1}{U_i} \right. $$
$$\left. + \frac{m}{3v_2 U_i^2} \left[ 2v_1^3 + v_2^3 - \left(\frac{2U_i}{m} + v_2^2\right)^{\frac{3}{2}} \right] \right\} \quad (6\text{-}4\text{-}8)$$

for $2mv_1(v_1-v_2) \leq U_i \leq 2mv_1(v_1+v_2)$. For $U_i \geq 2mv_1(v_1+v_2)$, $Q_i$ is equal to zero.

The total ionization cross section per atomic electron for incident electrons is given by

$$Q_i = \tfrac{1}{2} \{Q_{i,\,\text{dir}} + Q_{i,\,\text{exc}} + Q_{i,\,\text{int}}\} \quad (6\text{-}4\text{-}9)$$

if

$$Q_{i,\,x} = \int_{U_i}^{T} \sigma_{E,\,x} \, dE \quad (6\text{-}4\text{-}10)$$

where $x$ may stand for direct (dir), exchange (exc) and interference (int).

For the unsymmetrical model it follows from eqs. (6-4-3) and (6-4-4) that [34]

$$Q_{i,\,\text{dir}} = \frac{\pi e^4}{T} \left[ \frac{1}{U_i} + \frac{2E_2}{3U_i^2} - \frac{1}{T-E_2} \right] \quad (6\text{-}4\text{-}11)$$

for $T \geq E_2 + U_i$, and

$$Q_{i,\,\text{dir}} = \frac{2\pi e^4 (T-U_i)^{\frac{3}{2}}}{3T U_i^2 E_2^{\frac{1}{2}}} \quad (6\text{-}4\text{-}12)$$

for $U_i \leq T \leq E_2 + U_i$. However, it follows directly from eqs. (6-4-3), (6-4-4) and (6-4-10) that $Q_{i,\,\text{exc}}$ diverges. In any correct theory $Q_{i,\,\text{exc}}$ must be equal to $Q_{i,\,\text{dir}}$. One cannot therefore include exchange and interference terms in the unsymmetrical collision model. Instead one directly puts $Q_i$ equal to $Q_{i,\,\text{dir}}$ and eq. (6-4-9) is not used.

In the symmetrical model it follows from eq. (6-4-5) that [38]

$$Q_{i,\,\text{dir}} = Q_{i,\,\text{exc}} = \frac{\pi e^4}{T + U_i + E_2} \left[ \left(\frac{1}{U_i} - \frac{1}{T}\right) + \frac{2}{3} E_2 \left(\frac{1}{U_i^2} - \frac{1}{T^2}\right) \right] \quad (6\text{-}4\text{-}13)$$

and

$$Q_{i,\,\text{int}} = -\frac{\pi e^4}{T + U_i + E_2} \left[ \frac{2\Phi'}{T + U_i} \ln \frac{T}{U_i} \right]. \quad (6\text{-}4\text{-}14)$$

An approximation for $\Phi'$ is given in [38]. For $U_i \gg R$, $\Phi' \approx 1$ and for $U_i \ll R$, $\Phi'$ can best be put equal to zero. Eqs. (6-4-9), (6-4-13) and (6-4-14) yield

$$Q_i = \frac{\pi e^4}{T + U_i + E_2} \left[ \left( \frac{1}{U_i} - \frac{1}{T} \right) + \frac{2}{3} E_2 \left( \frac{1}{U_i^2} - \frac{1}{T^2} \right) - \frac{\Phi'}{T + U_i} \ln \frac{T}{U_i} \right]. \tag{6-4-15}$$

This result is also obtained from the formula

$$Q_i = \int_{U_i}^{\frac{1}{2}(T+U_i)} \sigma_E \, dE \tag{6-4-16}$$

where $\sigma_E$ includes the direct, exchange and interference terms. The interpretation of eq. (6-4-16) is that one cannot distinguish between scattered and ejected electron and the faster of the two is considered to be the scattered one. The maximum kinetic energy that the ejected electron may have [27] is $\frac{1}{2}(T-U_i)$ and the maximum energy transfer $\frac{1}{2}(T+U_i)$. In integrating up to $T$ instead of $\frac{1}{2}(T+U_i)$, all events are counted twice, which explains the factor $\frac{1}{2}$ in eq. (6-4-9).

The ionization cross section per shell or sub-shell of an atom is obtained by multiplying the above formulae by the number of electrons in that shell. The cross section for single ionization of argon, for example, is then equal to six times the cross section for ejection of a 3p electron plus two times the cross section for ejection of a 3s electron. In a more detailed treatment one even subdivides the 3p-shell into a $3p_{\frac{1}{2}}$ and a $3p_{\frac{3}{2}}$ subshell. The K and L shells in argon hardly contribute to single ionization and need not be taken into account.

The chief arguments in favor of using the binary-encounter theory for calculating ionization cross sections are:
(i) the great simplicity and wide applicability of eq. (6-4-15) and related formulae,
(ii) such formulae give reasonable estimates (often within a factor of 2) of ionization cross sections over a significant range of energies of the incident particles, and
(iii) the binary-encounter theory gives a correct description of the close collisions.

The major argument against application of the binary-encounter or classical theory is that the distant collisions are not described correctly. In any correct (many-body quantum) theory for ionization, and even in the first

Born approximation which is binary-encounter in the sense that the incident particle is assumed to interact with only one target particle (electron or nucleus) but is many-body in that the mutual interaction of the atomic particles is taken into account, $\sigma_{E,P} \sim v_1^{-2} P^{-1}$ for small $P$. Consequently, $\sigma_E \sim v_1^{-2} \ln P_{\min}^{-1} \sim v_1^{-2} \ln v_1$ and therefore $Q_i \sim v_1^{-2} \ln v_1$ for large $v_1$. In the binary-encounter theory and in any purely classical theory $\sigma_{E,P}$ decreases faster than $P^{-1}$ for small $P$ and $\sigma_E$ and $Q_i$ are proportional to $v_1^{-2}$. In other words: the (continuum) optical oscillator strength is finite in many-body quantum theory, but is zero in binary-encounter and many-body classical theory (see section 6-5-A).

## C. Excitation

In the binary-encounter and in classical collision theory energy and angular momentum quantization cannot properly be taken into account. One usually assumes that excitation to a level with excitation energy $U_n$ occurs if $U_n \leq E < U_{n+1}$ where $U_{n+1}$ is the next higher level. Furthermore, one does not distinguish between different final $l$ (azimuthal quantum number) states. These difficulties in $E$- and $l$-quantization do not occur for ionization. Finally, excitation processes generally are less violent than ionization processes ($E$ is smaller and on the average $P$ is smaller). For these reasons, it is not considered worthwhile to use the binary-encounter or classical theories to calculate excitation cross sections.

With one exception, the above arguments also apply for exchange excitation, for example triplet excitation of helium. The exception is that the incident electron must transfer more than its initial kinetic energy to one of the target electrons, provided that spin-flip is forbidden. For large velocity of the incident electron exchange excitation then occurs via violent binary-encounters with one atomic electron. The exchange excitation cross section per target electron with the correct spin direction is thus given by (see eq. (6-4-5))

$$Q_{e,\,\text{exc}} = \int_{U_n}^{U_{n+1}} \sigma_{E,\,\text{exc}}\, dE = \frac{\pi e^4}{T + U_i + E_2} \left[ \left\{ \frac{1}{T + U_i - U_{n+1}} - \frac{1}{T + U_i - U_n} \right\} + \frac{2}{3} E_2 \left\{ \frac{1}{(T + U_i - U_{n+1})^2} - \frac{1}{(T + U_i - U_n)^2} \right\} \right]$$

(6-4-17)

for $T \geq U_{n+1}$, and

$$Q_{e,\,exc} = \int_{U_n}^{T} \sigma_{E,\,exc}\, dE = \frac{\pi e^4}{T + U_i + E_2}\left[\left\{\frac{1}{U_i} - \frac{1}{T + U_i - U_n}\right\}\right.$$
$$\left. + \frac{2}{3} E_2 \left\{\frac{1}{U_i^2} - \frac{1}{(T + U_i - U_n)^2}\right\}\right] \quad (6\text{-}4\text{-}18)$$

for $U_n \leq T \leq U_{n+1}$. For large $T$, $Q_{e,\,exc} \sim T^{-3}$ in accordance with, for example, the Ochkur approximation. In the calculation of eqs. (6-4-17) and (6-4-18) we used the $\sigma_{E,\,exc}$ of eq. (6-4-5). For simple atoms where spin-orbit coupling need not be taken into account, eqs. (6-4-17) and (6-4-18) give in a simple way useful order of magnitude estimates of exchange cross sections.

## D. Double Ionization

In the binary-encounter theory double electron ejection must involve a double binary-encounter in which two electrons, initially bound in the atom, acquire sufficient energy to escape. Hence, the incident charged particle, can in one encounter with an atom collide with only one of the atomic electrons which in turn has a binary encounter with another atomic electron, transfers part of its energy, and causes both to be ejected. The primary particle can also subsequently collide with two atomic electrons with the result that both are ejected. This model [1], due to Gryzinski [30], contains many unrealistic features and will not be discussed in more detail. Instead we briefly discuss some general aspects of double ionization, insofar as these clarify relationships with the binary-encounter theory.

Especially for large $v_1$, a crucial feature of double ionization is that it can occur because the atom is not composed of separable electrons. If one uses correlated many-electron wave functions, $Q_{2i}$ is non-zero even in the first Born approximation (this corresponds to the first of the above mentioned binary-encounter processes). There is also a contribution to $Q_{2i}$ from the second Born approximation, which includes double processes (like the second binary-encounter process). Further, single electron ejection, can lead, via Auger or shake-off processes, to multiple ionization. A significant part of multiple ionization cross sections can thus be due to single and even "distant" collisions. Measurements of Schram [76] clearly show that the cross sections for formation of $Ar^{3+}$, $Kr^{2+}$, $Kr^{3+}$, $Kr^{4+}$ and $Xe^{2+}$ up to $Xe^{5+}$ have a strong dipole character. On the other hand, measurements of Schram, Boerboom and Kistemaker [77] show that the dipole contribution to the cross sections for formation of $He^{2+}$, $Ne^{2+}$ and $Ne^{3+}$ is very small. The latter processes may proceed therefore predominantly via violent binary-encounters.

## E. Charge Transfer

The elaborate treatments of electron capture given by Thomas [23] and Bates and coworkers [63–67] have been mentioned already in § 6-2. The basic assumption of their theories is that an atomic electron is captured by the passing nucleus as a result of two successive binary-encounters. The first binary-encounter is between the incident nucleus and one atomic electron; the second is between that atomic electron and the atomic nucleus. The major difficulties and disadvantages of this classical theory are:

(i) the electron is captured in some discrete quantum state which cannot be described classically;
(ii) the uncertainty principle is ignored in the calculation because assumptions are made about simultaneous velocity and position of the target electron in the final (and initial) state, and
(iii) the theory is rather complicated.

The major argument in favor of the theory of Thomas and Bates et al. is its success. Further, the capture process may, for large $v_1$, for a significant part indeed proceed via two successive violent binary-encounters. It is therefore especially for the larger $v_1$ that the theory of Thomas and Bates should be applied. For small $v_1 (\approx v_2)$ the capture process cannot be considered to proceed via two successive binary-encounters. In this case no argument in favor of such a complicated theory is left.

For small $v_1$, the simple theory of Gryzinski [30] becomes more attractive. In this theory an electron can be captured if it gains an energy corresponding to the velocity of the incident nucleus plus the difference $U_i^a - U_i^b$ of the binding energies of the two systems. Here, we assumed that ion $b^+$ was incident on atom a, resulting in atom b and ion $a^+$. Also, the acquired translational energy cannot be greater than the sum of translational energy of $b^+$ plus ionization energy of the level in which the electron is captured. Hence, capture can occur if

$$\tfrac{1}{2}mv_1^2 + U_i^a - U_i^b \leqq E \leqq \tfrac{1}{2}mv_1^2 + U_i^a + U_i^b \qquad (6\text{-}4\text{-}19)$$

where $v_1 = v_{b^+} = v_b$. Equation (6-4-19) gives, as noted by Gryzinski, a necessary but not a sufficient condition for capture. The velocity vectors of ejected electron and ion $b^+$ must be contained in a sufficiently small angle, this angle rapidly decreases with increase of the velocity of the incident ion. Gryzinski simply took the capture cross section equal to

$$Q_c = \int_{\frac{1}{2}mv_1^2 + U_i^a - U_i^b}^{\frac{1}{2}mv_1^2 + U_i^a + U_i^b} \sigma_E \, dE. \qquad (6\text{-}4\text{-}20)$$

For large $v_1$ this formula gives too large cross sections. For small $v_1$, the velocity vectors of ejected electron and ion $b^+$ need not be contained in a small angle, and eq. (6-4-20) may well give reasonable estimates of capture cross sections. Equations (6-3-36) and (6-3-37) may be used in eq. (6-4-20) to yield $Q_c$. In eq. (6-4-20) we have to require that $\frac{1}{2}mv_1^2 + U_i^a - U_i^b > 0$. For $\frac{1}{2}mv_1^2 + U_i^a + U_i^b \leq 2mv_1(v_1 - v_2)$ we find

$$Q_c = \frac{4\pi z_1^2 e^4}{mv_1^2} \left[ \frac{U_i^b}{(\frac{1}{2}mv_1^2 + U_i^a)^2 - (U_i^b)^2} + \frac{mv_2^2 U_i^b (\frac{1}{2}mv_1^2 + U_i^a)}{3\{(\frac{1}{2}mv_1^2 + U_i^a)^2 - (U_i^b)^2\}^2} \right]. \qquad (6\text{-}4\text{-}21)$$

One may proceed similarly for the other integration regions. Detailed calculations of this sort have been made by Garcia, Gerjuoy and Welker [43].

One should note that the integration limits of eqs. (6-4-7), (6-4-8) and (6-4-20) overlap. For sufficiently large $v_1$, eqs. (6-4-7) and (6-4-8) give the total electron removal instead of the total ionization cross section. For large $v_1$, however, $Q_c$ according to eqs. (6-4-20) and (6-4-21) is proportional to $v_1^{-6}$ whereas $Q_i \sim v_1^{-2}$. Hence $Q_c \ll Q_i$ for large $v_1$. For not too large $v_1$, $Q_i$ may be overestimated by eqs. (6-4-7) and (6-4-8) due to the overlap of capture and ionization cross sections.

One further unrealistic feature of eq. (6-4-20) is, for example for incident protons, that capture in the ground state $(n=1)$ occurs if $\frac{1}{2}mv_1^2 + U_i^a - U_{i,1} \leq E \leq \frac{1}{2}mv_1^2 + U_i^a + U_{i,1}$, that capture in the $n=2$ state occurs if $\frac{1}{2}mv_1^2 + U_i^a - U_{i,2} \leq E \leq \frac{1}{2}mv_1^2 + U_i^a + U_{i,2}$ and so on. Here $U_{i,n} = R/n^2$. Hence, all these energy ranges completely overlap. A more logic approach would be to assume that capture to state $n$ occurs if $\frac{1}{2}mv_1^2 + U_i^a - U_{i,n} \leq E \leq \frac{1}{2}mv_1^2 + U_i^a - U_{i,n+1}$ and if $\frac{1}{2}mv_1^2 + U_i^a + U_{i,n+1} \leq E \leq \frac{1}{2}mv_1^2 + U_i^a + U_{i,n}$. The latter approach, which has as far as we know not yet been used, leads to capture cross sections for the $n$th hydrogenic level that are proportional to $n^{-3}$ in agreement with the quantum mechanical Born and Kramers-Brinkman approximations. On the contrary, the integration region $\frac{1}{2}mv_1^2 + U_i^a - U_{i,n} \leq E \leq \frac{1}{2}mv_1^2 + U_i^a + U_{i,n}$ predicts $Q_c$ to be proportional to $n^{-2}$ which is quite unrealistic. Garcia, Gerjuoy and Welker [43] pointed out that Gryzinski's theory is not consistent with detailed balancing. Because the theory works better when the initial binding energy $U_i^a$ exceeds the final binding energy $U_i^b$ and because $Q_c$ becomes infinite if $\frac{1}{2}mv_1^2 + U_i^a - U_i^b \leq 0$, which can occur only if $U_i^a < U_i^b$, they suggest using Gryzinski's theory only if $U_i^a \geq U_i^b$. For $U_i^a < U_i^b$ they suggest calculating the cross section via detailed balancing using the Gryzinski cross section for the reversed reaction.

Most difficulties in the theory arise in the region of very small $v_1$, where the interactions do not take place in regions small compared to the atomic dimensions. Here the interaction time is not short compared to the revolution time of the atomic electron. For very small $v_1$ there is therefore little reason to use the binary-encounter theory.

## F. Electronic Stopping

If we only consider atomic excitation and ionization, the average energy loss per unit pathlength of a *fast* charged particle traversing a gaseous medium is

$$-\frac{dE}{ds} = \sum_i N_{ai} \, \mathbf{S}_n E_{ni} \sigma_{ni} \qquad (6\text{-}4\text{-}22)$$

where the summation extends over all different atoms or molecules $i$, with density $N_{ai}$, of the target. $\mathbf{S}_n$ indicates a summation over all discrete and an integration over the continuum states with excitation energies $E_{ni}$ and cross sections $\sigma_{ni}$. We now confine ourselves to one sort of target atoms, which removes the summation over $i$. In the binary-encounter theory $-dE/ds$ then becomes

$$-\frac{dE}{ds} = N_a \sum_j N_j \int_{U_j}^{E_{max}} E \sigma_E \, dE \qquad (6\text{-}4\text{-}23)$$

where $N_j$ is the number of electrons in the $j$th shell, $U_j$ is the lowest inelastic (excitation) threshold energy for shell $j$ and $E_{max}$ is the same as for ionization.

Upon using eqs. (6-3-36) and (6-3-37) for incident heavy particles we find

$$\begin{aligned}-\frac{dE}{ds} &= \frac{2\pi z_1^2 e^4}{mv_1^2} N_a \sum_j N_j \left[ \ln\left\{ \frac{2m(v_1^2 - v_{2j}^2)}{U_j} \right\} + \frac{2mv_{2j}^2}{3U_j} + \frac{4}{3} \right] \\ &= \frac{2\pi z_1^2 e^4}{mv_1^2} N_a N \left[ \ln\left\{ \frac{2mv_1^2}{I} \right\} + C + O\left(\frac{1}{v_1^2}\right) \right] \end{aligned} \qquad (6\text{-}4\text{-}24)$$

where $N$ is the total number of electrons in atom, $I$ a constant which may be taken equal to the average excitation energy (see section 6-5-D) and $C$ is also a constant. Equation (6-4-24) holds for sufficiently large $v_1$ (per shell for $2mv_1(v_1-v_2) \geq U_j$).

For incident electrons we obtain

$$-\frac{dE}{ds} = \frac{\pi e^4}{T} N_a \sum_j N_j \left[ \ln\left\{\frac{T}{8U_j}\right\} + \frac{4E_{2j}}{3U_j} + 1 + O\left(\frac{1}{T}\right) \right] \quad (6\text{-}4\text{-}25)$$

if eq. (6-4-5) with $E_{\max} = \frac{1}{2}(T+U_i)$ and $\Phi = 1$ is used in eq. (6-4-23).

The main reason for paying attention to the stopping power is that much weight is given to the close (violent) collisions, because before integrating over $E$ we multiply $\sigma_E$ with $E$.

For ionization we directly integrate $\sigma_E$ over $E$. A disadvantage in calculating $-dE/ds$ compared to $Q_i$ is that transitions to discrete excited states must be included. For further information see section 6-5-D.

### G. Velocity distribution of atomic electrons

In all previous sections we assumed that the target electrons have only one velocity $v_2$. Equation (6-3-29) and all further relations are derived on the assumption that the target electrons have an isotropic velocity distribution, thus the direction of $v_2$ is arbitrary with respect to the direction of $v_1$.

The velocity (momentum) distribution of atomic electrons is given by the square of the wave function in momentum space. This wave function is exactly known only for hydrogenic atoms and ions. However, only the square of the wave function of the initial state and thus a diagonal matrix element needs to be known. Reasonably accurate velocity distributions may therefore be obtained for more complicated atoms.

From the special structure of eqs. (6-3-36), (6-3-38), (6-4-3), (6-4-5), (6-4-7), (6-4-11), (6-4-15), (6-4-17), (6-4-21), (6-4-24) and (6-4-25) it follows that for sufficiently large $v_1$ (and thus $T$) averaging over $v_2$ can simply be carried out by replacing $\frac{1}{2}mv_2^2 = E_2$ by its expectation value $\langle E_2 \rangle$. In these cases no knowledge of the correct velocity distribution is required. Slater's [78] rules can be used to obtain, for complex atoms rough estimates of, values for $\langle E_2 \rangle$. Better estimates can be obtained using Hartree-Fock wave functions.

For small $v_1$, properly averaging over $E_2$ or simply replacing $E_2$ by $\langle E_2 \rangle$ can lead to very different results. For the cross section $\sigma_{E,P}$ this is also the case for large $v_1$ (see following sections).

## § 6-5. Comparison with the Born approximation

### A. DOUBLE DIFFERENTIAL CROSS SECTIONS

At sufficiently large velocity of the incident particle (electron, proton, α-particle) the Bethe (first Born) theory gives for ionization an adequate description of the collision process. The essential features of the cross section $\sigma_{E,P}$ are then conveniently represented by the generalized oscillator strength (density) $f_E(K)$ of the atomic transition involved (the notation $R\,df(K)/dE$ instead of $f_E(K)$ is also frequently used). In the Bethe theory [3], the relation between $\sigma_{E,P}$ and $f_E(K)$ is

$$\sigma_{E,P} = \frac{8\pi a_0^2 z_1^2 v_0^2}{PEv_1^2} f_E(K) \tag{6-5-1}$$

where as before $v_0 = e^2/\hbar$ and $\hbar K = P$. The general definition of $f_E(K)$ is [3]

$$f_E(K) = (E/R)(Ka_0)^{-2} \left|\sum_s (\psi_f|\exp(i\mathbf{K}\cdot\mathbf{r}_s)|\psi_i)\right|^2 \tag{6-5-2}$$

where $\psi_i$ and $\psi_f$ are the initial and final state wave functions. The sum extends over all the atomic electrons with position vectors $\mathbf{r}_s$ and the atomic matrix element ( | | ) stands for an integration over all coordinate space.

If $K$ is sufficiently large, $\exp(i\mathbf{K}\cdot\mathbf{r}_s)$ oscillates rapidly over a distance where $\psi_i$ varies slowly. The matrix element in eq. (6-5-2) then becomes very small unless $\psi_f$ oscillates equally rapidly. Thus $f_E(K)$ takes appreciable values only if $\psi_f \sim \exp(i\mathbf{k}\cdot\mathbf{r}_s)$ and $k \approx K$, where $\hbar k$ is the momentum of the ejected electron; $(ka_0)^2 = (E - U_i)/R$. This situation corresponds with a binary-encounter of the incident charged particle with a single electron.

For ionization of the *hydrogen atom* from the ground state, Bethe [3] derived

$$f_E(K) = 2^7(E/R)\left[(Ka_0)^2 + \tfrac{1}{3}E/R\right]\left[(E/R - (Ka_0)^2)^2 + 4(Ka_0)^2\right]^{-3}$$
$$\times \left[1 - \exp(-2\pi/ka_0)\right]^{-1}$$
$$\times \exp\left[-\frac{2}{ka_0}\arctan\frac{2ka_0}{(Ka_0)^2 - (ka_0)^2 + 1}\right] \tag{6-5-3}$$

where for the multivalued function arctan one has to use the value that lies between 0 and $\pi$.

The range of validity of the binary-encounter theory can for this case be tested by comparing with the exact eq. (6-5-3). We use eq. (6-5-1) with binary-encounter $\sigma_{E,P}$ as a phenomenological definition of "binary-encounter" generalized oscillator strength. We first compare three binary-encounter models, in all of which we neglect possible exchange and interference effects as is justified for large $v_1$. The simplest model is that of target electrons which are initially at rest. From the direct term of the Mott formula and from eq. (6-5-1) it follows that for this model $f_E(K) = (E/R)(Ka_0)^{-2} \delta(E/R - (Ka_0)^2)$. The second model is that of target electrons which all have velocity $v_2 = v_0$ (as in the Bohr model) but are moving in random directions. Equations (6-3-29) and (6-3-30) for this case yield $f_E(K) = \frac{1}{4}(E/R)(Ka_0)^{-3}$ for $m(v_2' - v_0) \leq K\hbar \leq m(v_2' + v_0)$ and zero otherwise. The third and most sophisticated model is that in which eqs. (6-3-29) and (6-3-30) are integrated over the correct velocity distribution

$$f(v_2) \, dv_2 = \frac{32 v_2^2 v_0^5}{\pi (v_2^2 + v_0^2)^4} \, dv_2 \tag{6-5-4}$$

which yields

$$f_E(K) = \frac{2^8 (E/R)(Ka_0)^3}{3\pi \left[(E/R - (Ka_0)^2)^2 + 4(Ka_0)^2\right]^3}. \tag{6.5.5.}$$

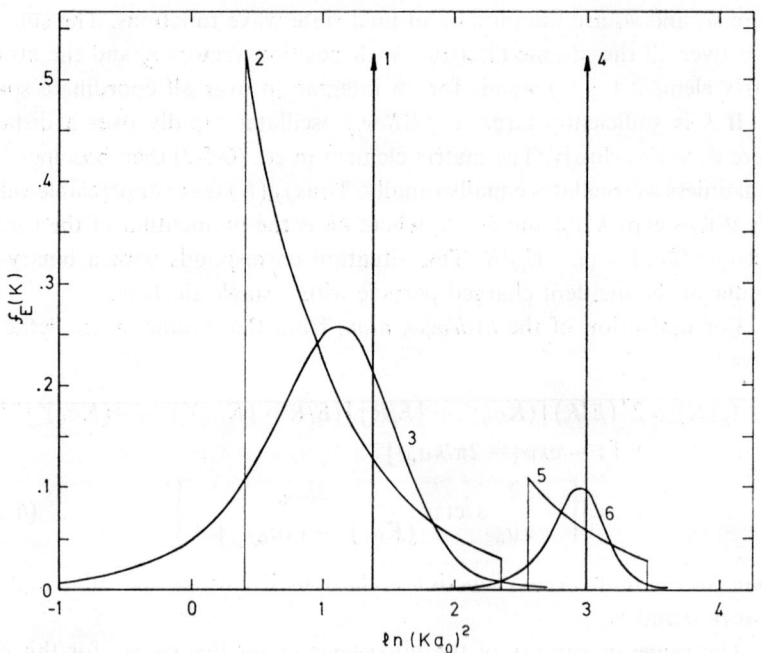

Fig. 6-5-1. Generalized oscillator strengths according to the binary-encounter theory for $E/R = 4$ (curves 1, 2, 3) and $E/R = 20$ (curves 4, 5, 6). "Curves" 1 and 4: target electrons at rest; curves 2 and 5: target electrons which are moving in random directions with one velocity $v_0$; curves 3 and 6: according to eq. (6-5-5), i.e. with the correct velocity distribution.

## 6-5 COMPARISON WITH THE BORN APPROXIMATION

The three binary-encounter models are compared with each other in Fig. 6-5-1. They clearly lead to very different results. In Figs. 6-5-2 and 6-5-3 we compare the $f_E(K)$ according to eqs. (6-5-3) and (6-5-5). The binary-encounter and the Bethe theories excellently agree with each other for large $K$ and $E$. A similar test for ionization from the 2s and 2p states of the hydrogen atom has been made by Vriens and Bonsen [61]. General conclusions are:

(i) Good agreement between binary-encounter and Bethe theory is found for $(Ka_0)^2 > E/2R$ and $E > 4R/n^2$, where $R/n^2$ is the ionization energy.

(ii) For $E > 4R/n^2$, the $f_E(K)$ are peaked around $(Ka_0)^2 = E/R$, that is $P^2/2m = E$. From the analytical expressions for $f_E(K)$ it follows that binary-encounter and Bethe theory tend to the same limit for $E \to \infty$ and $P^2/2m \approx E$.

(iii) The width and shape of the function $f_E(K)$ for large $E$ (the Bethe ridge, see Inokuti and Platzman [79]) is determined essentially by the initial velocity distribution of the atomic electron. This arises because the binary-encounter theory contains no information about the interaction of the ejected electron with the nucleus but still gives the same results as the Bethe theory, provided that the correct velocity distribution of the atomic electron is used.

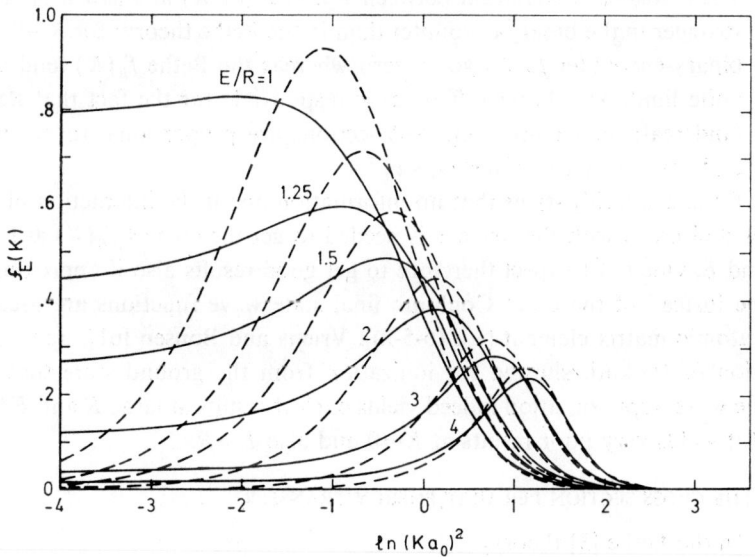

Fig. 6-5-2. Bethe (full curves) and binary-encounter (broken curves) generalized oscillator strengths for different values of the energy transfer for ionization of the hydrogen atom from the ground state.

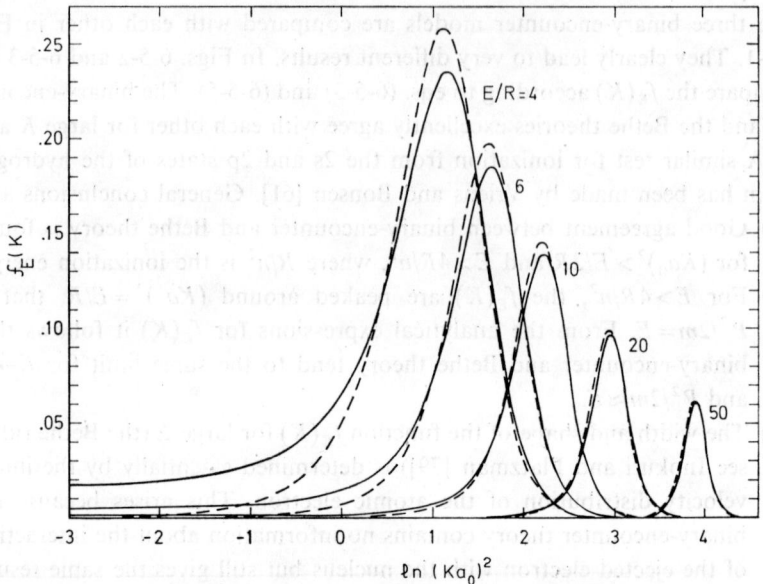

Fig. 6-5-3. As in Fig. 6-5-2 with larger values of $E/R$.

(iv) The statistical correlation between $E$ and $P$ (or $K$) at small $E$ is much stronger in the binary-encounter than in the Bethe theory; for $K \to 0$, the binary-encounter $f_E(K)$ go to zero whereas the Bethe $f_E(K)$ tend to a finite limit. This latter difference is responsible for the fact that Bethe (and real) total ionization cross sections are proportional to $v_1^{-2} \ln v_1$ and the binary-encounter $Q_i \sim v_1^{-2}$.

Conclusion (iii) states that no information about the interaction of the ejected electron with the nucleus is needed to get the correct $f_E(K)$ at large $K$ and $E$. One must expect therefore to get good results also if approximate plane instead of the exact Coulomb final state wave functions are used in the atomic matrix element (eq. (6-5-2)). Vriens and Bonsen [61], upon suggestion of Inokuti, showed for ionization from the ground state that the plane wave approximation indeed yields correct results at large $K$ and $E$, but that it yields very poor results at $K \to 0$ and also $E \to R$.

B. THE CROSS SECTION PER UNIT ENERGY TRANSFER

In the Bethe [3] theory,

$$\sigma_E = \frac{4\pi a_0^2 z_1^2 v_0^2}{v_1^2} \left[ \frac{f_E(0)}{E} \ln\left(\frac{4c_E v_1^2}{v_0^2}\right) + O\left(\frac{1}{v_1^2}\right) \right] \tag{6-5-6}$$

where $f_E(0)$ is the (density of the) optical oscillator strength, which is also often denoted by $R\,df(0)/dE$. Thus $f_E(0)$ is dimensionless as is $f_E(K)$. Miller and Platzman [80] showed that

$$\ln c_E = \int_0^\infty [f_E(K)/f_E(0)]\,d\ln(Ka_0)^2$$

$$-\int_{-\infty}^0 [1 - f_E(K)/f_E(0)]\,d\ln(Ka_0)^2 - 2\ln(E/R). \quad (6\text{-}5\text{-}7)$$

The binary-encounter formula corresponding to eq. (6-5-6) is (see eqs. (6-3-36), (6-4-3) and (6-4-5))

$$\sigma_E = \sum_j \frac{4\pi a_0^2 z_1^2 v_0^2}{v_1^2} N_j \left[\frac{R}{E^2} + \frac{4R\langle E_{2j}\rangle}{3E^3} + O\left(\frac{1}{v_1^2}\right)\right]. \quad (6\text{-}5\text{-}8)$$

Here, $N_j$ is the number of electrons in the $j$th shell, and the summation extends over those shells for which $E \geq U_{ij}$.

For ionization of H1s, Vriens and Bonsen [61] compared values of the functions within the square brackets of eqs. (6-5-6) and (6-5-8). In this case, $N_j = 1$ for $j=1$ and zero for $j \neq 1$. Increasingly good agreement is found the larger $E$.

For a more interesting comparison, the leading term within the square brackets of eq. (6-5-6) is subdivided into

$$\frac{f_E(0)}{E}\ln\frac{v_1^2}{v_0^2} + \frac{f_E(0)}{E}\ln 4c_E. \quad (6\text{-}5\text{-}9)$$

The first term rapidly decreases with increasing $E$. Rau and Fano [81] showed that $f_E(0) \sim E^{-l-\frac{7}{2}}$ for large $E$, where $l$ is the azimuthal quantum number of the initial state. For ionization of the hydrogen atom from s or p states, the first term of eq. (6-5-9) is, consequently, proportional to $E^{-\frac{9}{2}}$ and $E^{-\frac{11}{2}}$, respectively. For large $E$, the difference between eqs. (6-5-6) and (6-5-8) can therefore be studied by calculating

$$E^3 \left[\frac{f_E(0)}{E}\ln 4c_E - \frac{R}{E^2} - \frac{4R\langle E_2\rangle}{3E^3}\right]. \quad (6\text{-}5\text{-}10)$$

Bonsen (unpublished) numerically calculated values of this function for a large number of $E$ values for ionization of the hydrogen atom from the 1s, 2s, 2p$_0$, 2p$_{\pm 1}$ and 2p states. In all cases this function decreases with increasing $E$. For the 1s, 2s and 2p states it does so very rapidly. Omidvar [82] later proved analytically that this function approaches zero for $E \to \infty$, for ionization of the hydrogen atom from the ground state.

A meaningful subdivision of eq. (6-5-6) is therefore

$$\sigma_E = \frac{4\pi a_0^2 z_1^2 v_0^2}{v_1^2} \left[ \frac{f_E(0)}{E} \ln\left(\frac{\kappa_E v_1^2}{v_0^2}\right) + \sum_j N_j \left( \frac{R}{E^2} + \frac{4R\langle E_{2j}\rangle}{3E^3} \right) + O\left(\frac{1}{v_1^2}\right) \right]. \tag{6-5-11}$$

Here, the function $\kappa_E$ varies much more slowly with $E$ than $c_E$. The first term in eq. (6-5-11) takes account of the distant collisions and is important only for small $E$. The second and third term take account of the close collisions. Equation (6-5-11) has first been given by Miller [27]. He also introduced an elegant way of taking account of exchange in close collisions for incident electrons. He replaced the term $R/E^2$ in eq. (6-5-11), which represents the Thomson formula $(E_2=0)$, by $R/E^2 + R/(T-E)^2 - R/E(T-E)$ of the Mott formula. These latter terms do not satisfy the correct symmetry conditions (see eqs. (6-4-9), (6-4-10), (6-4-13) and (6-4-16)). By using eq. (6-4-5) we simultaneously satisfy these symmetry conditions and generalize Miller's treatment. The resulting formula for $\sigma_E$ is

$$\sigma_E = \frac{4\pi a_0^2 v_0^2}{v_1^2} \left[ \frac{f_E(0)}{E} \ln\left(\frac{\kappa_E v_1^2}{v_0^2}\right) + \sum_j \frac{RTN_j}{T + U_{ij} + E_{2j}} \left\{ \frac{1}{E^2} + \frac{4E_{2j}}{3E^3} \right.\right.$$
$$\left.\left. + \frac{1}{(T + U_{ij} - E)^2} + \frac{4E_{2j}}{3(T + U_{ij} - E)^3} - \frac{1}{E(T + U_{ij} - E)} \right\} \right] \tag{6-5-12}$$

where as before $T = \tfrac{1}{2}mv_1^2$ and $\Phi$ of eq. (6-4-5) is taken equal to unity; $E_{2j}$ can conveniently be replaced by $\langle E_{2j}\rangle$. It must be emphasized that in eq. (6-5-12), as in eq. (6-5-11), another term of the order $1/v_1^4$ is disregarded. Equation (6-5-12) should further be used in combination with eq. (6-4-16), whereas eq. (6-5-11) should be integrated to $E_{\max}$ ($=T$ for incident electrons).

C. THE TOTAL IONIZATION CROSS SECTION

Integrating eq. (6-5-6) over $E$ yields the Bethe formula

$$Q_i = \frac{4\pi a_0^2 z_1^2 v_0^2}{v_1^2} \left[ M_i^2 \ln\left(\frac{4c_i v_1^2}{v_0^2}\right) + O\left(\frac{1}{v_1^2}\right) \right] \tag{6-5-13}$$

where

$$M_i^2 = \int_{U_i}^{\infty} E^{-1} f_E(0) \, dE \qquad (6\text{-}5\text{-}14)$$

and

$$M_i^2 \ln c_i = \int_{U_i}^{\infty} E^{-1} f_E(0) \ln c_E \, dE. \qquad (6\text{-}5\text{-}15)$$

The binary-encounter formula corresponding to eq. (6-5-13) is

$$Q_i = \frac{4\pi a_0^2 z_1^2 v_0^2}{v_1^2} \left[ \sum_j N_j \left\{ \frac{R}{U_{ij}} + \frac{2R \langle E_{2j} \rangle}{3 U_{ij}^2} \right\} + O\left(\frac{1}{v_1^2}\right) \right]. \qquad (6\text{-}5\text{-}16)$$

Values of $M_i^2$ and $\ln c_i$ have been calculated by Bethe [3] and Inokuti [83] for H(1s), and by Vriens and Bonsen [61] for H(2s), (2p$_0$), (2p$_{\pm 1}$) and (2p). Experimental values of $M_i^2$ and $\ln c_i$ have been determined for a large number of gases by Schram et al. [77, 84, 85] and by Schram [76]. In all cases the dipole ($M_i^2$) contribution to $Q_i$ is quite significant for which reason eq. (6-5-16) cannot be used for large $v_1$.

As in section 6-5-B one may subdivide $Q_i$ into a distant and a (binary-encounter) close collision term. The binary-encounter theory therefore yields estimates of the exchange and interference terms for the close collisions (see eqs. (6-4-9), (6-4-13), (6-4-14) and (6-4-15)).

D. STOPPING POWER

Elaborate treatments of the theory of stopping power are given by Lindhard, Bethe [86], Bethe and Askin [87] and Fano [88]. If we again confine ourselves to one type of target atoms and to electronic stopping,

$$-\frac{dE}{ds} = N_a \, \underset{n}{S} E_n \sigma_n = N_a \, \underset{n}{S} E_n \int_{P_{min}}^{P_{max}} \sigma_{E_n, P} \, dP. \qquad (6\text{-}5\text{-}17)$$

The integration over $P$ can be subdivided in two (Bethe) or three (Fano) regions.

For incident heavy particles the Bethe theory yields for the interval $(P_{min}, P_x)$

$$-\frac{dE}{ds}(P_{min}, P_x) = \frac{2\pi z_1^2 e^4}{m v_1^2} N_a N \left[ \ln \left\{ \frac{v_1^2 P_x^2}{I^2} \right\} + O\left(\frac{1}{v_1^2}\right) \right] \qquad (6\text{-}5\text{-}18)$$

and for the interval $(P_x, P_{max})$,

$$-\frac{dE}{ds}(P_x, P_{max}) = \frac{2\pi z_1^2 e^4}{mv_1^2} N_a N \left[\ln\left\{\frac{4m^2 v_1^2}{P_x^2}\right\} + O\left(\frac{1}{v_1^2}\right)\right]. \qquad (6\text{-}5\text{-}19)$$

The total stopping power is thus given by

$$-\frac{dE}{ds} = \frac{4\pi z_1^2 e^4}{mv_1^2} N_a N \left[\ln\left\{\frac{2mv_1^2}{I}\right\} + O\left(\frac{1}{v_1^2}\right)\right], \qquad (6\text{-}5\text{-}20)$$

where the definition of the average excitation energy $I$ is given by

$$\ln I = \mathop{S}_{n} f_n(0) \ln E_n. \qquad (6\text{-}5\text{-}21)$$

The relativistic equations corresponding to eqs. (6-5-18), (6-5-19) and (6-5-20) are given by Fano [88]. Comparison of eqs. (6-5-18) and (6-5-19) shows that the small and large $P$ regions give essentially the same contribution to the stopping power for $P_x^2 = 2mI$. On the contrary, the small $P$ region is of major importance for the ionization cross section $Q_i$. Further essential features of the theory are that $P_{max}$ in eqs. (6-5-17) and (6-5-19) is taken equal to $2mv_1$ (this holds for scattering of incident heavy particles by free target electrons that are initially at rest) and that the Bethe and binary-encounter theories give the same result for eq. (6-5-19). Comparison with eq. (6-4-24) shows that the leading terms of eqs. (6-4-24) and (6-5-19) are the same. The discrepancy of a factor of two between the leading terms of eqs. (6-4-24) and (6-5-20) is thus due to the fact that the close encounters (large $P$) are correctly described in the binary-encounter theory, but that the distant (small $P$) collisions which are equally important are not taken into account.

The same discrepancy of a factor of two exists for incident electrons. The binary-encounter theory is used in the same way as in section 6-4-F to obtain the contributions of exchange and interference terms to $-dE/ds$. Bethe [86] used the Mott formula for this purpose. To take the indistinguishability of the electrons into account he further used $P_{max}^2/2m = \frac{1}{4}mv_1^2$ instead of $\frac{1}{2}mv_1^2$.

### E. Sum Rules and Total Cross Section

In the Bethe (first Born) theory the total cross section for momentum transfer $P$ is obtained as a sum of the elastic (eq. (6-2-2)) and total inelastic (Inokuti, Kim and Platzman [89]) cross sections. The result is

$$\sigma_P = \frac{8\pi z_1^2 e^4}{v_1^2 P^3} \left[|Z - F(K)|^2 + ZS_{inc}(K) + O\left(\frac{1}{v_1^2}\right)\right] \qquad (6\text{-}5\text{-}22)$$

where the number of electrons $N$ in the atom has been taken equal to the nuclear charge $Z$. As before, $F(K)$ is the atomic form factor (eq. (6-2-3)) and now, $S_{\text{inc}}(K)$ is the incoherent scattering function [90]

$$S_{\text{inc}}(K) = Z^{-1}\left[\sum_{j,s=1}^{Z} \langle\psi_i|\exp[i\mathbf{K}\cdot(\mathbf{r}_j - \mathbf{r}_s)]|\psi_i\rangle - |F(K)|^2\right]. \quad (6\text{-}5\text{-}23)$$

The binary-encounter formula corresponding to eq. (6-5-22) is directly obtained as a sum of eq. (6-2-1) for scattering by the nucleus and $Z$ (the number of electrons) times eq. (6-3-40) for scattering by the atomic electrons,

$$\sigma_P = \frac{8\pi z_1^2 e^4}{v_1^2 P^3}\left[Z^2 + Z + O\left(\frac{1}{v_1^2}\right)\right]. \quad (6\text{-}5\text{-}24)$$

For large $K$, $F(K)$ tends to zero, $S_{\text{inc}}(K)$ tends to unity and eq. (6-5-22) reduces to eq. (6-5-24) confirming that the atomic electrons and nucleus act as independent scattering centers. For small $K$, $|Z - F(K)|^2$ is proportional to $K^4$ and $S_{\text{inc}}(K)$ is proportional to $K^2$. Owing to the mutual shielding of atomic electrons and nucleus eq. (6-5-22) thus leads to much smaller cross sections for small $K$ than eq. (6-5-24). The binary-encounter theory leads again to serious errors for small $P$.

In the derivation of eq. (6-5-22), where the total cross section is expressed in terms of ground state expectation values, use was made of the sum rule [89] involving the incoherent scattering function $S_{\text{inc}}(K)$,

$$\mathbf{S}_n \frac{R}{E_n} f_n(K) = Z S_{\text{inc}}(K)/(Ka_0)^2. \quad (6\text{-}5\text{-}25)$$

Two other sum rules are of direct interest for the comparison of binary-encounter and Bethe theories. The first one is the Thomas-Reiche-Kuhn rule,

$$\mathbf{S}_n f_n(K) = N \quad (6\text{-}5\text{-}26)$$

where $N = Z$ for neutral atoms, and the second one (see Fano [88]) is,

$$\mathbf{S}_n \frac{E_n}{R} f_n(K) = N(Ka_0)^2 + \frac{4}{3R}\left\langle \sum_{j=1}^{N} \tfrac{1}{2}mv_j^2 \right\rangle + \text{correlation terms}.$$

$$(6\text{-}5\text{-}27)$$

The second term in eq. (6-5-27) thus gives $4/3R$ times the expectation value of the total kinetic energy of the atomic electrons. The summation over $j$ extends over all the atomic electrons.

The binary-encounter formulae (6-3-29) and (6-3-30), and eq. (6-5-1) yield

$$f_E(K) = \frac{m^2 E v_0^2}{2v_2 P^3} \qquad (6\text{-}5\text{-}28)$$

per target electron with velocity $v_2$. The binary-encounter formula corresponding to eq. (6-5-26) is

$$\sum_{j=1}^{N} \int_{E_{\min}}^{E_{\max}} f_E(K) \frac{dE}{R} = N \qquad (6\text{-}5\text{-}29)$$

where we inserted the $f_E(K)$ of eq. (6-5-28) and used the $E_{\min}$ and $E_{\max}$ of eq. (6-3-32). The binary-encounter formula corresponding to eq. (6-5-27) is

$$\sum_{j=1}^{N} \int_{E_{\min}}^{E_{\max}} \frac{E}{R} f_E(K) \frac{dE}{R} = \sum_{j=1}^{N} \left\{ \frac{P^2}{2mR} + \frac{2mv_{2,j}^2}{3R} \right\}. \qquad (6\text{-}5\text{-}30)$$

Equations (6-5-29) and (6-5-26) give the same result. After averaging over the velocity distribution of the atomic electrons, eq. (6-5-30) reduces to eq. (6-5-27) except that no correlation terms are obtained in the binary-encounter theory. For hydrogenic atoms these are absent also in the Bethe theory.

The above sum rules provide us with a new relationship between binary-encounter and Bethe theories. The Thomas-Reiche-Kuhn rule is used in the theory of stopping power and also for photo-absorption (for $K=0$). Sum rule (6-5-27) is also used in the theory of stopping power to calculate fluctuations of energy loss.

F. ALTERNATIVE METHOD OF DERIVING BINARY-ENCOUNTER FORMULAE

In the previous sections, binary-encounter formulae were derived starting from relations for two particle collisions. The alternative method described below is due to Nijboer [91], the mathematical techniques employed have been used before, for example for neutron scattering by Nijboer and Rahman [92].

The Bethe (first Born) definition (eq. (6-5-2)) of generalized oscillator strength may be rewritten as

$$f_E(K) = \frac{E/R}{(Ka_0)^2} \mathbf{S}_f \left| \sum_s \langle \psi_f | \exp(i\mathbf{K}\cdot\mathbf{r}_s) | \psi_i \rangle \right|^2 \delta\left(\frac{E + E_i - E_f}{R}\right) \qquad (6\text{-}5\text{-}31)$$

where $E_i$ and $E_f$ are the initial and final state energy eigenvalues of the atom and $E$ is, as before, the energy loss of the scattered particle. The $\delta$-function imposes energy conservation, when integrated (summed) over f it yields unity and converts eq. (6-5-31) back to eq. (6-5-2), except that eq. (6-5-31) contains all final states $\psi_f$ possible at energy $E_f$, whereas eq. (6-5-2) contains only one. However, when using eq. (6-5-2), as in section 6-5-A, one often also sums over final states. The $\delta$-function in eq. (6-5-31) may be replaced by

$$\delta\left(\frac{E + E_i - E_f}{R}\right) = \frac{R}{2\pi\hbar} \int_{-\infty}^{\infty} \exp\left[\frac{iEt}{\hbar} + \frac{i(E_i - E_f)t}{\hbar}\right] dt \qquad (6\text{-}5\text{-}32)$$

where the integration variable $t$ may be interpreted as time. Using the complex conjugate of

$$\exp\left[\frac{i(E_f - E_i)t}{\hbar}\right] (\psi_f |T_K| \psi_i) = (\psi_f |T_K(t)| \psi_i) \qquad (6\text{-}5\text{-}33)$$

where the time dependent Heisenberg operator

$$T_K(t) = \exp(iHt/\hbar) \, T_K \exp(-iHt/\hbar), \qquad (6\text{-}5\text{-}34)$$

and summing (integrating) over all final states, yields

$$f_E(K) = \frac{E}{2\pi\hbar(Ka_0)^2} \int_{-\infty}^{\infty} \exp\left[\frac{iEt}{\hbar}\right] \langle T_K^*(t) \, T_K(0) \rangle \, dt. \qquad (6\text{-}5\text{-}35)$$

Here

$$\langle T_K^*(t) \, T_K(0) \rangle = \sum_{s,j=1}^{N} (\psi_i |e^{-i\mathbf{K} \cdot \mathbf{r}_s(t)} e^{i\mathbf{K} \cdot \mathbf{r}_j(0)}| \psi_i) \qquad (6\text{-}5\text{-}36)$$

where we inserted already that in the present case

$$T_K(t) = \sum_s \exp[i\mathbf{K} \cdot \mathbf{r}_s(t)]. \qquad (6\text{-}5\text{-}37)$$

The previous equations are exact; binary-encounter formulae with correlation terms still included are now obtained by putting $r_s(t)$ in eq. (6-5-36) equal to $v_s t + r_s(0)$, which holds only if the target electrons are free. Further, one uses

$$\exp\{-i\mathbf{K} \cdot (\mathbf{v}_s t + \mathbf{r}_s(0))\} = \exp(-i\mathbf{K} \cdot \mathbf{v}_s t) \exp(-i\mathbf{K} \cdot \mathbf{r}_s(0))$$
$$\times \{\exp - \tfrac{1}{2}[-i\mathbf{K} \cdot \mathbf{v}_s t, -i\mathbf{K} \cdot \mathbf{r}_s(0)]\} \qquad (6\text{-}5\text{-}38)$$

where the commutator

$$-\tfrac{1}{2}[-i\mathbf{K}\cdot\mathbf{v}_s t, -i\mathbf{K}\cdot\mathbf{r}_s(0)] = -i\hbar K^2 t/2m. \tag{6-5-39}$$

Equation (6-5-36) then becomes

$$\langle T_K^*(t)\, T_K(0)\rangle = \Big\langle \sum_{s,j} \exp\{-i\mathbf{K}\cdot\mathbf{v}_s t\}\exp\{-i\hbar K^2 t/2m\} \\ \times \exp\{i\mathbf{K}\cdot(\mathbf{r}_j(0)-\mathbf{r}_s(0))\}\Big\rangle. \tag{6-5-40}$$

Inserting eq. (6-5-40) into eq. (6-5-35) and integrating over $t$ yields

$$f_E(K) = \frac{E/R}{(Ka_0)^2}\Big\langle \sum_{s,j}\delta\!\left(\frac{E-\hbar\mathbf{K}\cdot\mathbf{v}_s-\hbar^2 K^2/2m}{R}\right)\exp\{i\mathbf{K}\cdot(\mathbf{r}_j(0)-\mathbf{r}_s(0))\}\Big\rangle. \tag{6-5-41}$$

For hydrogenic atoms this formula directly reduces to the binary-encounter formula

$$f_E(K) = \frac{E/R}{(Ka_0)^2}\Big\langle \delta\!\left(\frac{E-\hbar\mathbf{K}\cdot\mathbf{v}-\hbar^2 K^2/2m}{R}\right)\Big\rangle \tag{6-5-42}$$

where $v=v_2$. Because the only variable in the $\delta$-function is $v$, one can calculate the expectation value of this function by integrating it over the velocity distribution of the atomic electron. Hence, instead of using the wave function in coordinate space, one uses the wave function in velocity (momentum) space. As an example we consider ionization of the hydrogen atom from the ground state. The velocity distribution, eq. (6-5-4) for this case, may be rewritten as

$$f(\mathbf{v})\, dv_x\, dv_y\, dv_z = \frac{8v_0^5}{\pi^2(v_x^2+v_y^2+v_z^2+v_0^2)^4}\, dv_x\, dv_y\, dv_z. \tag{6-5-43}$$

If we take the $z$-axis along $\mathbf{K}$, eqs. (6-5-42) and (6-5-43) give after integrating over $v_x$ and $v_y$,

$$f_E(K) = \frac{8v_0^5(E/R)}{3\pi(Ka_0)^2}\int_{-\infty}^{\infty}\frac{1}{(v_z^2+v_0^2)^3}\delta\!\left(\frac{E-\hbar Kv_z-\hbar^2 K^2/2m}{R}\right)dv_z \\ = \frac{2^8(E/R)(Ka_0)^3}{3\pi[(E/R-(Ka_0)^2)^2+4(Ka_0)^2]^3}. \tag{6-5-44}$$

The latter result is the same as eq. (6-5-5) which was obtained in an entirely different way. Equation (6-5-42) can also be used for other initial states of the hydrogen atom. For more complicated atoms, eq. (6-5-41) differs from binary-encounter relations, in that it contains correlation terms.

Alternatively, one may use polar coordinates. In the case of an isotropic velocity distribution of the atomic electron,

$$f(v)\,dv = f(v)\,dv\frac{1}{4\pi}\sin\Theta\,d\Theta\,d\varphi. \tag{6-5-45}$$

Equation (6-5-42) yields after integrating over $\varphi$ and putting $K \cdot v = Kv\cos\Theta$

$$f_E(K) = \frac{E/R}{2(Ka_0)^2}\int_0^\infty dv\,I_1 \tag{6-5-46}$$

where

$$I_1 = f(v)\int_{-1}^{1}\delta\!\left(\frac{E - \hbar Kv\cos\Theta - \hbar^2 K^2/2m}{R}\right)d(\cos\Theta). \tag{6-5-47}$$

Using that

$$\int_a^b f(x)\,\delta[g(x)]\,dx = \sum_i \frac{f(x_i)}{|g'(x_i)|}, \tag{6-5-48}$$

where the sum extends over all roots of the equation $g(x)=0$ in the interval $(a, b)$, gives

$$I_1 = f(v)\frac{R}{\hbar Kv} \tag{6-5-49}$$

and

$$f_E(K) = \int_{v_{\min}}^\infty f(v)\frac{E}{2(Ka_0)^2\,\hbar Kv}\,dv. \tag{6-5-50}$$

From the relations

$$g(x) = \frac{E - \hbar Kv\cos\Theta - \hbar^2 K^2/2m}{R} = 0$$

and $-1 \le \cos\Theta \le 1$ it follows that

$$v_{\min} = \left|\frac{E}{\hbar K} - \frac{\hbar K}{2m}\right| \tag{6-5-51}$$

which reproduces eq. (6-3-33). Equations (6-5-1) and (6-5-50) result in

$$\sigma_{E,P} = \frac{4\pi z_1^2 e^4}{v_1^2 P^4} \int_{v_{min}}^{\infty} \frac{f(v)}{v} dv \qquad (6-5-52)$$

and comparison with eqs. (6-3-29) and (6-3-30) shows the relation between the two methods of deriving binary-encounter formulae.

One further difference between the two derivations of binary-encounter formulae is that in sections 6-2 up to 6-5-E the exact relations for two particle collisions are used, whereas in section 6-5-F we started with the first Born approximation (eq. (6-5-31)). It is special property of the Coulomb potential that, except for the phase factor, the Born approximation gives the same scattering amplitudes as an exact treatment.

## § 6-6. The classical theory

Abrines and Percival [68–70], Abrines, Percival and Valentine [71] and Valentine [72] studied the classical three-body problem for ionization and charge transfer in collisions of protons, electrons and positrons with the hydrogen atom. Percival et al. used statistical (Monte Carlo) methods to obtain cross sections. The atomic electrons are assumed to be moving around the nucleus in elliptical orbits. The probability distribution of different elliptical orbits is chosen in such a way that the "angular momentum parameter" $l^2/l_{max}^2$ is distributed uniformly in the interval (0, 1). Here $l^2 = (r \times mv)^2$ and $l_{max}$ is the angular momentum of a circular orbit. The dimensions of the orbits are further determined by the condition that the total energy of the atom is $-R$. This microcanonical distribution gives the same velocity distribution for the atomic electron as quantum mechanics. To obtain one cross section, for example a total ionization cross section for one incident energy, the results of typically 1000 to 7000 scattering events have to be calculated. For each collision Percival et al. use step-by-step integration of Newton's equations of motion. Monte Carlo methods are used to obtain an arbitrary set of initial conditions. Within the statistical errors one obtains in this way the exact classical cross sections for the three-body problem considered.

Once the classical cross sections for the hydrogen atom in the ground state are known, these are directly scalable to other levels because, as was shown by Fock [93], the form of the velocity distribution is the same for each level ($n$) provided that all states ($l$) are equally populated. Exact relations for the expectation values of potential and kinetic energy give

$\langle 1/r_n \rangle = 1/n^2 a_0$ and $\langle p_n^2 \rangle = \hbar^2/n^2 a_0^2$ (see Bethe and Salpeter [94]). Hence for large $n$, $r_n$ and $p_n$ are proportional to $n^2 a_0$ and $\hbar/na_0$, respectively. Therefore, $\hbar/p_n r_n \sim 1/n$ and $\lambda_n = \hbar/p_n \sim r_n/n$, so $\hbar/p_n r_n$ approaches zero and $\lambda_n \ll r_n$ for large $n$. According to the Bohr correspondence principle, the classical description of the atom (with elliptical orbits of the electrons) becomes valid for large $n$. If we now consider a transition from level $n$ to level $m$ (with $m > n$), $\Delta r = r_m - r_n \sim (m^2 - n^2) a_0$ and $\Delta p = p_n - p_m \sim (\hbar/a_0)(1/n - 1/m)$. Hence, $\hbar/\Delta r \Delta p \sim mn/(m-n)^2 (m+n)$. If we approach the ionization limit ($m \to \infty$),

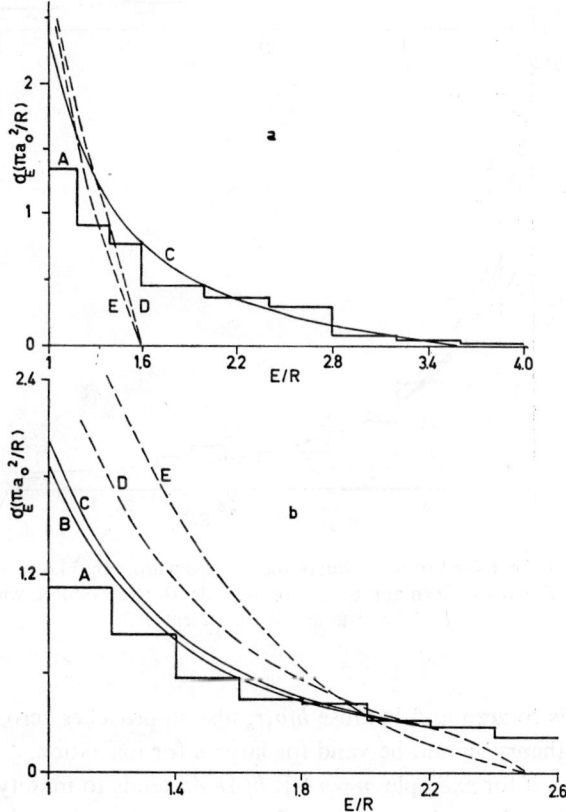

Fig. 6-6-1. The differential cross section $\sigma_E$ for energy transfer $E$ in ionization (and exchange excitation) of the hydrogen atom by 21.8 eV (a) and 34 eV (b) electrons. A. Classical three-body theory; B. symmetrical binary-encounter theory, averaged over the velocity distribution of the atomic electron, interference not included; C. symmetrical binary-encounter theory, unaveraged, no interference; D. unsymmetrical binary-encounter theory, averaged over velocity distribution; E. as D., unaveraged. (Reproduced with courtesy of I. C. Percival and N. A. Valentine.)

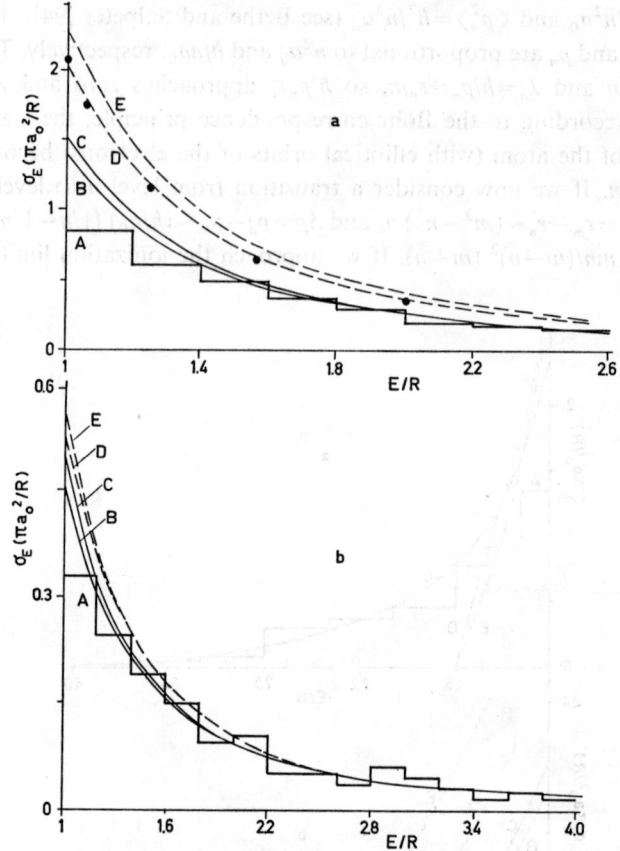

Fig. 6-6-2. As in Fig. 6-6-1 for ionization of the hydrogen atom by 54.4 eV (a) and 218 eV (b) electrons. Full circles: Born approximation (Omidvar). (Reproduced with courtesy of I. C. Percival and N. A. Valentine.)

$\hbar/\Delta r\, \Delta p$ tends to zero and because $\hbar/p_n r_n$ also approaches zero for large $n$, the classical theory should be valid for large $n$ for ionization.

However, if for example $m = n+1$, $\hbar/\Delta r\, \Delta p$ tends to infinity for large $n$ and the classical theory is not expected to give good results at large $n$. Thus even for macroscopic (hydrogen) atoms one should use quantum mechanics to calculate excitation cross sections for levels which are not too far apart. The classical model of orbiting electrons indeed becomes correct at large $n$, but in six dimensional phase-space the orbits are coming closer together the larger $n$ and transitions between them are not classical. Percival et al. confine

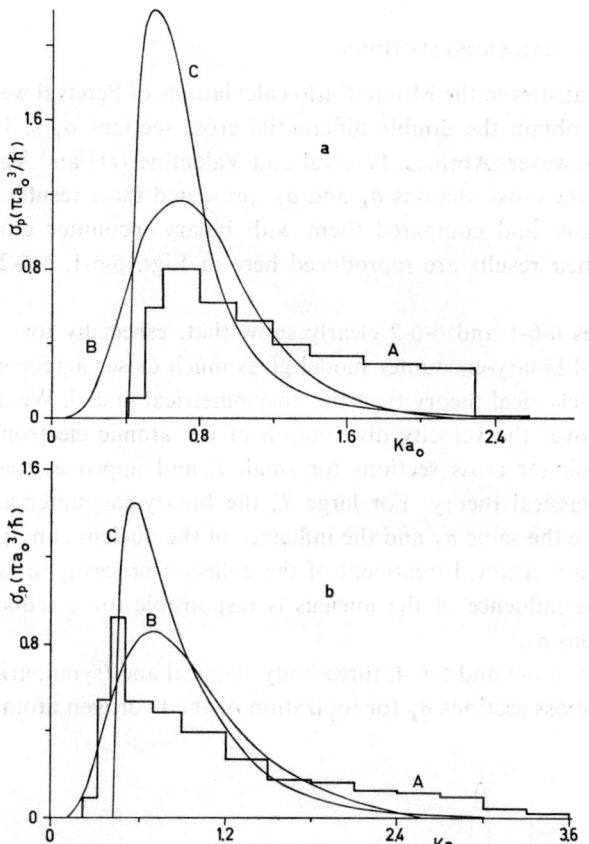

Fig. 6-6-3. The differential cross section $\sigma_P$ for momentum transfer $P = \hbar K$ for ionization of the hydrogen atom by 27.2 eV (a) and 54.4 eV (b) electrons. A, B and C have the same meaning as in Fig. 6-6-1. (Reproduced with courtesy of I. C. Percival and N. A. Valentine.)

themselves to ionization and charge transfer in which cases the classical theory should be valid for large $n$.

The classical scaling law which applies for ionization is that the normalized ionization cross section $\tilde{Q}_i = U_i^2 Q_i$ is a universal function of the normalized energy $\tilde{T} = T/U_i$ of the incident particle (see also section 6-7-C).

The classical theory does not apply for small $n$, but it is useful to have results for small $n$, as comparisons with quantum theory and with experiment give information about where quantum effects are becoming important. Comparisons with the binary-encounter theory further yield information about the influence of the nucleus in the collision process.

## A. Differential Cross Sections

The statistics in the Monte Carlo calculations of Percival were not good enough to obtain the double differential cross sections $\sigma_{E,P}$. For incident electrons however, Abrines, Percival and Valentine [71] and Valentine [72] calculated the cross sections $\sigma_E$ and $\sigma_P$, presented these results in the form of histograms and compared them with binary-encounter cross sections. Some of their results are reproduced here in Figs. 6-6-1, 6-6-2, 6-6-3 and 6-6-4.

Figures 6-6-1 and 6-6-2 clearly show that, especially for small $T$, the symmetrical binary-encounter model gives much closer agreement with the three-body classical theory than the unsymmetrical model. We also see that averaging over the velocity distribution of the atomic electron lowers the binary-encounter cross sections for small $E$ and improves the agreement with the classical theory. For large $T$, the binary-encounter and classical theories give the same $\sigma_E$ and the influence of the nucleus can apparently be neglected in a classical treatment of the collision process. For small $T$ and small $E$ the influence of the nucleus is responsible for a reduction of the cross sections $\sigma_E$.

In Figs. 6-6-3 and 6-6-4, three-body classical and (symmetrical) binary-encounter cross sections $\sigma_P$ for ionization of the hydrogen atom are plotted

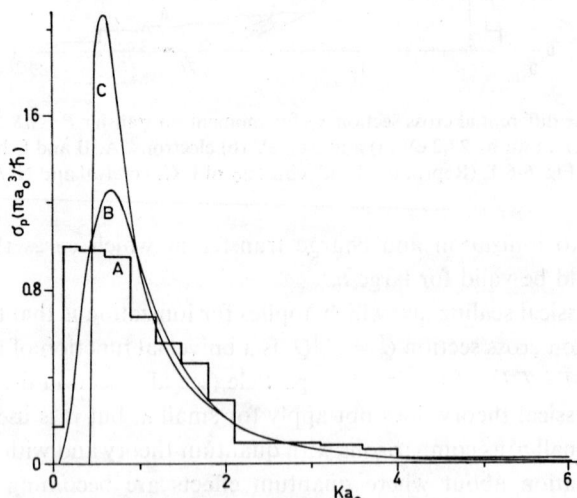

Fig. 6-6-4. As in Fig. 6-6-3 for 544 eV incident electrons. (Reproduced with courtesy of I. C. Percival and N. A. Valentine.)

versus $Ka_0$. Averaging in the binary-encounter theory over the velocity-distribution of the atomic electron greatly improves the agreement with the classical theory. Further, the binary-encounter and classical curves shift with respect to each other with varying $T$. For small $T$ the average value of $P$ according to the classical three-body theory is larger than given by the binary-encounter theory. For large $T$ just the opposite is true and the classical theory gives larger $\sigma_P$ for small $P$.

From Fig. 6-6-2 we know that the $\sigma_E$ and therefore also the $Q_i$ are the same for large $T$. The $\sigma_P$ for transfer of momentum $P$ to the atomic electron must therefore also be the same, but in the classical three-body theory $\sigma_P$ stands for transfer of momentum $P$ to the atom. For distant encounters, the momentum transfer to the atom $|\boldsymbol{P}_a| = |\boldsymbol{P}_e + \boldsymbol{P}_p|$ is small and the momentum transfers to the atomic electron $\boldsymbol{P}_e$ and to the nucleus $\boldsymbol{P}_p$ will have on the average opposite directions. Consequently, $P_e$ is on the average larger than $P_a$, which explains the difference between the theories for large $T$. This effect also explains a part of the difference between the binary-encounter and Bethe theories (see Figs. 6-5-2 and 6-5-3). For very large $T$ classical and binary-encounter theories are expected to give the same results, because in order to transfer a momentum $P$, the collision must be closer the larger $T$. For small $T$, the dynamics of the collision process is more complicated and no such simple explanations of the difference between the theories can be given.

## B. Total cross sections

The classical three-body ionization cross sections are proportional to $v_1^{-2}$ and reduce to the binary-encounter $Q_i$ for large $v_1$, in disagreement with the Born approximation.

For incident protons, classical ionization and charge transfer cross sections are given by Abrines and Percival [70], see their figures 1 and 2. They compare their results with experiment,* Born approximation and other theories. The corresponding binary-encounter cross sections for ionization (averaged as well as unaveraged over the velocity distribution of the atomic electron) are given in [39]. For proton energies between 100 and 250 keV, classical, binary-encounter, Born and experimental ionization cross sections agree very well with each other. Below 100 keV, the binary-encounter cross sections are too large. This difference arises because the binary-encounter theory gives the total electron removal cross sections for $\frac{1}{2}mv_1^2 \geqq U_i^b$ (see section 6-4-E). In this case one should compare these cross sections with

* The disagreement between classical and experimental results for charge transfer in [70] was caused by an error in the experiment. Later experiments give much closer agreement.

those for ionization plus charge transfer. For $\frac{1}{2}mv_1^2 < U_i^b$, the binary-encounter cross sections correspond to the total ionization plus part of the charge transfer cross sections.

For incident electrons, three-body classical, binary-encounter and experimental cross sections are compared with each other in Fig. 6-6-5. Again we find that the symmetrical collision model of the binary-encounter theory

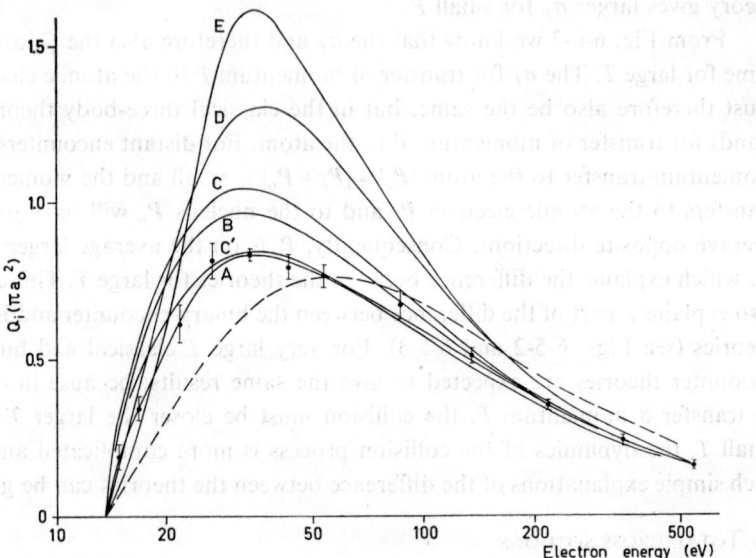

Fig. 6-6-5. The total cross section $Q_i$ for ionization of the hydrogen atom by electrons. A. (points with error bars) classical three-body theory; B. symmetrical binary-encounter theory, averaged over the velocity distribution of the atomic electron, interference not included; C. symmetrical binary-encounter theory, unaveraged, no interference; C'. as C with interference included (eq. (6-4-15)); D. unsymmetrical binary-encounter theory, averaged over velocity distribution; E. as D, unaveraged; broken curve: experiment [95]. (Curves A, B, C, D and E reproduced with courtesy of I. C. Percival and N. A. Valentine.)

yields much better results than the unsymmetrical model. Averaging over the velocity distribution of the atomic electron also gives better results at small $T$. Finally, curves C and C' show that including interference in the binary-encounter theory significantly lowers the cross sections at small $T$, also in better agreement with experiment.

## § 6-7. Comparison with experiment

### A. Differential cross sections

Experimental results on differential cross sections, pertinent to this review, are rather scarce. The older work is summarized in section 6-2-A, where special attention was given to the measurements of Hughes et al. [15–17]. Recently, Rudd, Sautter and Bailey [96] measured the energy and angular distributions of electrons ejected from He and $H_2$ by 100 to 300 keV protons. Their experimental results are very interesting since they measured for energies of the ejected electron up to 700 eV. The binary-encounter theory should, for large $v_1$, work very well for such violent collisions. Rudd et al. also calculated binary-encounter cross sections using an incorrect formula of Gryzinski and taking $\langle E_2 \rangle$ equal to $U_i$. For helium it follows directly from the virial theorem that $\langle E_2 \rangle = 1.6 U_i$. Their comparison with the binary-encounter theory is for these two reasons not conclusive. More recently, Rudd and Gregoire [97] made improved calculations which give better agreement with experiment and with the Born approximation as is shown in Figs. 6-7-1 and 6-7-2. The comparison with experiment is not yet entirely conclusive because the binary-encounter and Born cross sections $\sigma_E$, shown in these figures, are scaled hydrogenic ones. Hence in the binary-encounter calculation Rudd and Gregoire used the correct formulae for $\sigma_E$,

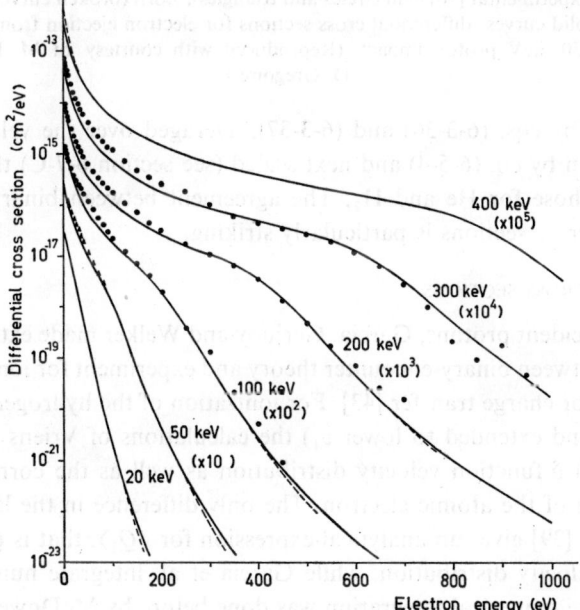

Fig. 6-7-1. Experimental [96] (full circles), Born (broken curves) and binary-encounter (solid curves) differential cross sections for electron ejection from $H_2$ resulting from 20 to 400 keV proton impact, plotted versus the energy of the ejected electron. (Reproduced with courtesy of M. E. Rudd and D. Gregoire.)

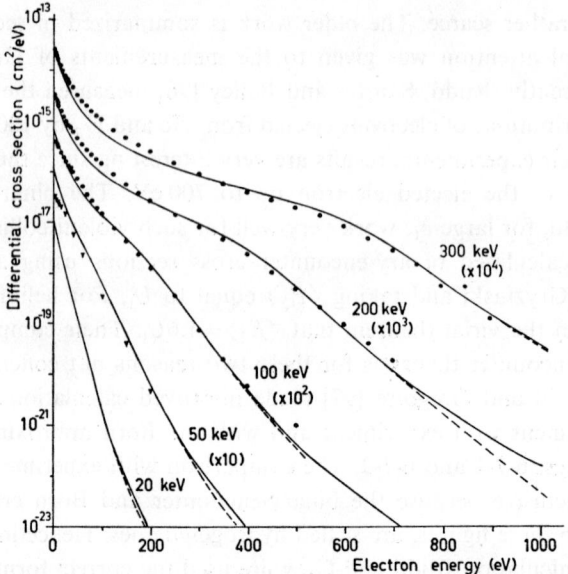

Fig. 6-7-2. Experimental [96] (full circles and triangles), Born (broken curves) and binary-encounter (solid curves) differential cross sections for electron ejection from He resulting from 20 to 300 keV proton impact. (Reproduced with courtesy of M. E. Rudd and D. Gregoire.)

given here by eqs. (6-3-36) and (6-3-37), averaged over the velocity distribution given by eq. (6-5-4) and next scaled (see section 6-7-C) the obtained $\sigma_E$ to get those for He and $H_2$. The agreement between binary-encounter and Born cross sections is particularly striking.

B. TOTAL CROSS SECTIONS

For incident protons, Garcia, Gerjuoy and Welker made extensive comparisons between binary-encounter theory and experiment for ionization [42] as well as for charge transfer [43]. For ionization of the hydrogen atom they repeated (and extended to lower $v_1$) the calculations of Vriens [39] in that they used a $\delta$-function velocity distribution as well as the correct velocity distribution of the atomic electron. The only difference in the latter case is that Vriens [39] gives an analytical expression for $\langle Q_i \rangle$, that is $Q_i$ averaged over the velocity distribution, while Garcia et al. integrate numerically. A less accurate numerical integration was done before by McDowell [40], who also gave results for Li. Garcia's calculations for the rare gases, Li, Na, K and Rb were all done with a $\delta$-function velocity distribution of the atomic electrons and with $E_2 = U_i = U_i^a$. The binary-encounter theory in all cases overestimates the ionization cross sections for proton energies between 5 and about 200 keV. As mentioned before, part of the discrepancy must be ascribed to the inclusion of the total or of part of the charge transfer cross section in the binary-encounter calculation. Catlow and McDowell [46] calculated proton impact ionization cross sections for He, Li, N and O by averaging over velocity-distributions obtained from Fourier transforms of Hartree-Fock density distributions. They could only compare with experiment for He, and the agreement is good. For N and O they only considered contributions from the outer 2p electrons.

For incident electrons, most investigaters (Ochkur and Petrunkin [33], Stabler [34], McDowell [40], Catlow and McDowell [46], Garcia et al. [42]) used the unsymmetrical collision model. As we have seen already in § 6-6, this model is inferior to the symmetrical model and overestimates the cross sections for small $v_1$ often by a factor of 2 or more (see also Fig. 6-6-5). Other investigators (Prasad and Prasad [49], Vriens [55], Kingston [50, 52, 54], McFarland [56, 57], Bauer and Bartky [44]) used Gryzinski's [1, 30], rather empirical relations. The theoretical basis of the latter work is rather poor, but in general reasonable (within 50%) agreement with experiment is obtained. Similar good agreement is obtained with the empirical method of Drawin [98]. The symmetrical collision model has been used before only for the hydrogen atom (Vriens [38, 39], Abrines et al. [71], Valentine [72]). An application to other atoms is shown in Fig. 6-7-4. The comparison with classical theory and experiment is given already in § 6-6.

## C. SIMILARITY OF SCALED CROSS SECTIONS

Classical scaling laws can conveniently be used to compare ionization and charge transfer cross sections for different atoms. Energies are scaled by dividing them by the ionization energy. Hence, the scaled ionization energy $\tilde{U}_i$ is equal to unity. $\tilde{E}_2 = E_2/U_i$, $\tilde{v}_2^2 = v_2^2/U_i$, $\tilde{T} = T/U_i$, $\tilde{v}_1^2 = v_1^2/U_i$ and $\tilde{E} = E/U_i$. The scaled energies are dimensionless. The scaled ionization cross section $\tilde{Q}_i = U_i^2 Q_i / N z_1^2 R^2$, where $Q_i$ is the total cross section for an atom and $N$ the number of electrons responsible for the process. The scaling laws do not directly apply for atoms where different shells with different ionization energies give comparable contributions to the total cross section. The scaled capture cross section $\tilde{Q}_c = U_i^2 Q_c / N z_1^2 R^2$, where $U_i = U_i^a$.

Upon scaling, eq. (6-4-7) becomes

$$\tilde{Q}_i = \frac{8\pi a_0^2}{m\tilde{v}_1^2} \left\{ 1 + \tfrac{1}{3} m \tilde{v}_2^2 - \frac{1}{2m(\tilde{v}_1^2 - \tilde{v}_2^2)} \right\} \qquad (6\text{-}7\text{-}1)$$

and eq. (6-4-8) yields

$$\tilde{Q}_i = \frac{4\pi a_0^2}{m\tilde{v}_1^2} \left\{ \frac{1}{2m\tilde{v}_2(\tilde{v}_1 + \tilde{v}_2)} + 1 + \frac{m}{3\tilde{v}_2} \left[ 2\tilde{v}_1^3 + \tilde{v}_2^3 - \left(\frac{2}{m} + \tilde{v}_2^2\right)^{\frac{3}{2}} \right] \right\}. \tag{6-7-2}$$

For incident electrons eq. (6-4-15) gives

$$\tilde{Q}_i = \frac{4\pi a_0^2}{\tilde{T} + 1 + \tilde{E}_2} \left[ 1 - \frac{1}{\tilde{T}} + \frac{2}{3} \tilde{E}_2 \left(1 - \frac{1}{\tilde{T}^2}\right) - \frac{\Phi'}{\tilde{T} + 1} \ln \tilde{T} \right]. \tag{6-7-3}$$

For capture eq. (6-4-21) becomes

$$\tilde{Q}_c = \frac{8\pi a_0^2 \tilde{U}_i^b}{(\frac{1}{2}m\tilde{v}_1^2)^3} \left[ \frac{1}{(1 + 2/m\tilde{v}_1^2)^2 - (2\tilde{U}_i^b/m\tilde{v}_1^2)^2} \right.$$
$$\left. + \frac{2\tilde{v}_2^2(1 + 2/m\tilde{v}_1^2)}{3\tilde{v}_1^2 \{(1 + 2/m\tilde{v}_1^2)^2 - (2\tilde{U}_i^b/m\tilde{v}_1^2)^2\}^2} \right] \tag{6-7-4}$$

and the term within the square brackets of eq. (6-7-4) tends to unity for large $v_1$.

In applications of the binary-encounter theory one often takes $E_2$ equal to $U_i$, which is correct only for hydrogenic atoms. Then $\tilde{E}_2 = \frac{1}{2}m\tilde{v}_2^2 = 1$ and eqs. (6-7-1) to (6-7-4) do not depend upon any properties of the target atom. Three problems will now be studied:

(i) are the real (experimental) cross sections indeed scalable in this way,
(ii) does the binary-encounter-theory (eqs. (6-7-1) to (6-7-4)) give a good description of the scaled cross sections, and
(iii) are better results obtained with more realistic values of $E_2$?

Scaled experimental and binary-encounter ionization cross sections for incident protons are given in Fig. 6-7-3. The experimental data were taken from Fite et al. [99] and Ireland and Gilbody [100] for hydrogen and from De Heer et al. [101] for He, Ne, Ar, Kr and $H_2$. More experimental data are available in literature, but it would be confusing to include them in Fig. 6-7-3. The binary-encounter curves are obtained: A. from eqs. (6-7-1) and (6-7-2) with $\frac{1}{2}m\tilde{v}_2^2 = 1$ ($\frac{1}{2}mv_2^2 = U_i$), B. from eqs. (6-7-1) and (6-7-2) with $\frac{1}{2}m\tilde{v}_2^2 = 4$, and C. from eq. (15) of Vriens [39] which gives the ionization cross section averaged over the velocity distribution of the atomic electron for H1s atoms. Conclusions to be drawn from Fig. 6-7-3 are:

(i) The experimental ionization cross sections are, except for Ne, very well scalable in a classical (binary-encounter) way.
(ii) Because the binary-encounter ionization cross sections as given by eqs.

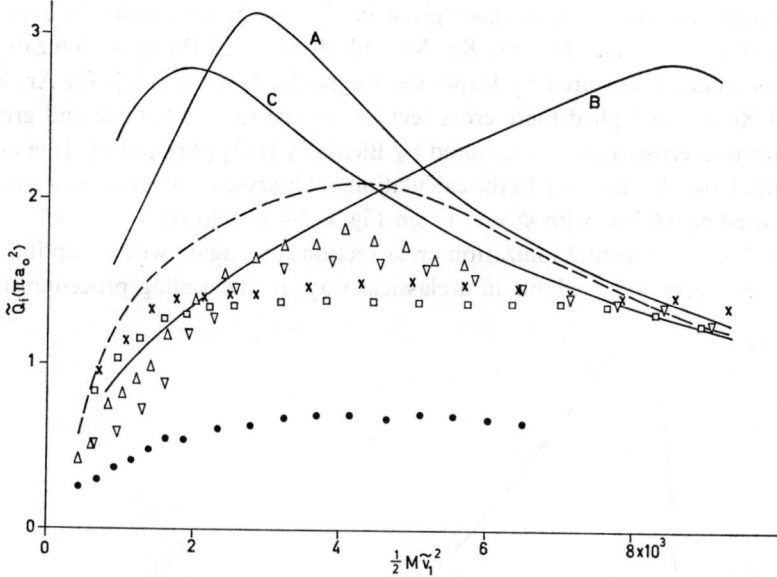

Fig. 6-7-3. Scaled ionization cross sections $\tilde{Q}_i = U_i^2 Q_i / N z_1^2 R^2$ for incident protons plotted versus the scaled proton energy $\frac{1}{2} M \tilde{v}_1^2 = \frac{1}{2} M v_1^2 / U_i$. Full curves: binary-encounter theory, A. with $\frac{1}{2} m v_2^2 = U_i$, B. with $\frac{1}{2} m v_2^2 = 4 U_i$ and C. averaged over the velocity distribution for the hydrogen atom. Broken curve: experimental results for ionization of the hydrogen atom [99, 100]. The experimental data of De Heer et al. [101] are given by △(He), ▽(H₂), ●(Ne), □(Ar) and ×(Kr).

(6-7-1) and (6-7-2) contain a part of the charge transfer cross sections, they must be too large for small $v_1$. Curves A and C give therefore satisfactory agreement with experiment.

(iii) Taking a higher, and thus for Ne, Ar and Kr a more realistic, value for $\tilde{v}_2$ (curve B) does not improve the agreement with experiment but makes it even worse.

Scaled experimental capture cross sections are plotted in Fig. 10 of Garcia et al. [43], for protons incident on He, Ne, Ar, Kr, Xe and K. This illustrative figure shows that classical (binary-encounter) scaling works even better for capture than for ionization. The other figures of their paper [43] show that the binary-encounter cross sections are correct within about a factor of 2. The variation of binary-encounter capture cross sections with $\tilde{v}_2$ has not yet been studied.

Scaled experimental and binary-encounter ionization cross sections for incident electrons are given in Fig. 6-7-4. The experimental data for atomic hydrogen are the same as those given in Fig. 6-6-5. To obtain the experimental data for He, Ne, Ar, Kr, Xe and $H_2$ we used the gross ionization cross sections measured by Rapp and Englander-Golden [102]; for Ar, Kr and Xe we multiplied these cross sections by the ratios of single and gross ionization cross sections measured by Bleakney [103] (Ar) and by Tate and Smith [104] (Kr and Xe). In the calculation of binary-encounter cross sections we used eq. (6-7-3) with $\Phi' = 1$. From Fig. 6-7-4 it follows:

(i) The experimental ionization cross sections are, again with exception of Ne, very well scalable in a classical way. In the scaling procedure for

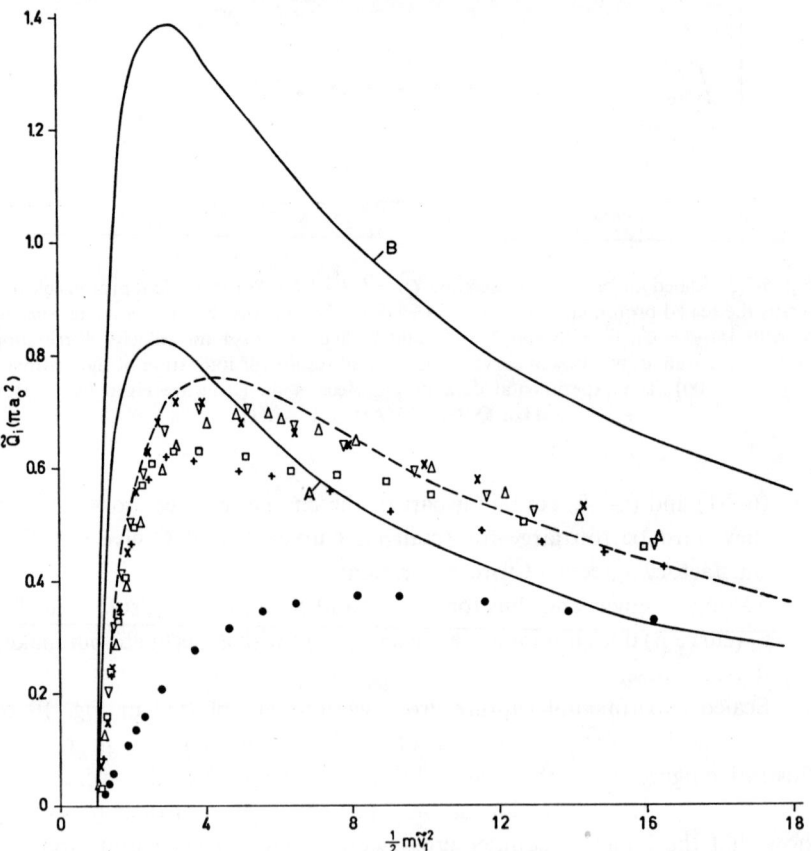

Fig. 6-7-4. Scaled ionization cross sections for incident electrons plotted versus the scaled electron energy $\tilde{T} = T/U_1 = \frac{1}{2}m\tilde{v}_1^2$. Full curves: binary-encounter theory (eq. (6-7-3)), A. with $\tilde{E}_2 = 1$ (curve C' of Fig. 6-6-5), B. with $\tilde{E}_2 = 4$. Broken curve: experimental results for ionization of the hydrogen atom [95]. The experimental data of Rapp and Englander-Golden [102] are given by △ (He), ▽ ($H_2$), ● (Ne), □ (Ar), × (Kr) and + (Xe).

Ne, Ar, Kr and Xe only the 6 outer-shell p electrons were taken into account; hence for Ar, Kr and Xe we neglected the inner-shell contribution to single ionization.

(ii) The binary-encounter curve A ($\tilde{E}_2 = 1$) gives too large cross sections for small $T$ and too small ones for large $T$.

(iii) Taking a higher, and thus for Ne, Ar, Kr and Xe a more realistic, value for $\tilde{E}_2$ (curve B) again (see Fig. 6-7-3) results in more unrealistic cross sections for small $T$.

In Fig. 6-7-4 we confined ourselves to the region of small and intermediate electron energies. For large $T$, the region where the Born approximation holds, we separately investigate whether classical scaling applies. Equation (6-5-13) may for this purpose be rewritten as

$$\tilde{Q}_i = \frac{4\pi a_0^2}{\tilde{T}} \left[ \tilde{M}_i^2 \ln \tilde{T} + \tilde{C}_i + O\left(\frac{1}{\tilde{T}}\right) \right] \qquad (6\text{-}7\text{-}5)$$

where $\tilde{M}_i^2 = U_i M_i^2 / RN$ and $\tilde{C}_i = \tilde{M}_i^2 \ln(4c_i U_i / R)$. For a large number of gases $M_i^2$ and $c_i$ or $\ln c_i$ values have accurately been calculated or measured. To obtain the $\tilde{M}_i^2$ and $\tilde{C}_i$ values listed in Table 6-7-1, we used the theoretical $M_i^2$ and $c_i$ values of Inokuti [83] for H(1s) and of Vriens and Bonsen [61] for H(2s) and H(2p). We further used the experimental $M_i^2$ and $\ln c_i$ values of Van der Wiel et al. [105] for single ionization of He, Ne and Ar and the experimental $M_i^2$ values of Schram [76] for single ionization of Kr and Xe. No $\tilde{C}_i$ values for single ionization of Kr and Xe were calculated because no suf-

TABLE 6-7-1

Values of $\tilde{M}_1{}^2 = U_1 M_1{}^2 / RN$ and $\tilde{C}_1 = \tilde{M}_1{}^2 \ln(4c_1 U_1/R)$ for single ionization of different atoms

| atom | $\tilde{M}_1{}^2$ | $\tilde{C}_1$ |
|---|---|---|
| H(1s) | 0.2834 | 1.257 |
| H(2s) | 0.2058 | 1.124 |
| H(2p) | 0.1328 | 1.459 |
| He | 0.443 | 0.912 |
| Ne | 0.489 (0.426) | −0.393 (−0.298) |
| Ar | 0.718 (0.608) | −0.397 (−0.279) |
| Kr | 0.678 (0.579) | − |
| Xe | 0.780 (0.665) | − |

The values in parenthesis are obtained by taking not only the outer $np^6$ but also the $ns^2$ electrons into account and assuming similarity in behaviour for s and p electrons. Hence we used $\tilde{M}_1{}^2 = M_1{}^2/(N_p R/U_{1p} + N_s R/U_{1s})$ and $\tilde{C}_1 = [\tilde{M}_1{}^2/(1 + N_s U_{1p}/N_p U_{1s})] \times [\ln(4c_1 U_{1p}/R) + (N_s U_{1p}/N_p U_{1s}) \ln(4c_1 U_{1s}/R)]$. For Ne the similarity assumption may be rather poor.

ficiently accurate values of $c_i$ were available. In the calculation of the $\tilde{C}_i$ value for He we used the renormalized value of $\ln c_i$ (see [105]).

From Table 6-7-1 it follows that the $\tilde{M}_i^2$ and $\tilde{C}_i$ values may be rather different for different atoms and different initial states. Hence, classical scaling does not apply at large $T(v_1)$. If, for single ionization of Ne, Ar, Kr and Xe, the inner-shell $(ns^2)$ contributions to $M_i^2$ are substracted from the values given in the table, these are reduced by about 15 to 20% (see the table). For ionization from highly excited initial states $\tilde{M}_i^2$ approaches zero and Percival's classical theory applies.

## § 6-8. Resonances

We introduce this section by an anecdote. At the 1965 Quebec conference on Electronic and Atomic Collisions, in the session on classical collision theories, Gryzinski was asked whether he thought it possible to describe resonances classically, a question which produced some hilarity. Although it seems quite unrealistic at first sight, we will show nevertheless that some resonance phenomena can be understood (at least qualitatively) with classical models.

Resonances occur in scattering of slow electrons by atoms or molecules at those particular energies at which (negative) compound ion states can be formed. At these energies, one finds sharp resonance structures in measured [106] and calculated [107] elastic and inelastic scattering cross sections. From the (energy) widths of these structures and the uncertainty relation $\Delta E \Delta \tau \approx h$ it follows that the lifetimes of the compound ions exceed the time an incident electron needs to traverse a distance comparable to the atomic dimensions by a factor of 100 to 1000. To condemn further classical considerations on resonances one then argued that such long lifetimes imply that the incident electron must revolve around the nucleus by a few hundred times which indeed seems extremely unrealistic.

We now replace this above qualitative argumentation by the following one. If for example a 19.3 eV electron is incident on a helium atom, it can excite one of the 1s electrons to a 2s orbital and itself be captured also in a 2s orbital. The binding energy of the excited 2s electrons is then 0.52 eV, because the threshold energy for $2^3S$ excitation is equal to 19.82 eV (the binding energy of the captured electron is 1.3 eV). The expectation value of the kinetic energy of a 2s electron is then about 0.8 to 1.5 eV and the ratio of the velocities of the captured (2s) and incident electron is about $(1.5/19.3)^{\frac{1}{2}}$. Furthermore, in a hydrogenic model, the size of a classical orbit in an atom is inversely proportional to the binding energy of the electron. The ratio of the sizes of orbits in the He$^-$ 1s2s$^2$ ion and the He 1s$^2$ atom is thus approximately equal to 24.58/0.52 and 24.58/1.3. Hence the revolution times of the 2s electrons in the compound ion exceed the time a 19.3 eV electron needs to pass a He atom by factors of about 300 to 700 and half or one or two revolutions already suffice to reach the aforementioned long lifetimes. Moreover, one can think of other simple classical models. For example, if the 19.3 eV electron transfers in a collision a significant part of its kinetic energy to an atomic electron, both will move away from the nucleus in general in different directions. Consequently both will be subject to Coulomb and to larger and larger polarization attractions, slow down to zero velocity and move back to the nucleus. One or two of such oscillations will again be sufficient to reach long lifetimes. Long lifetimes are therefore no argument disfavoring application of a classical model with orbiting and oscillating electrons. The only real argument against application of classical theory is that compound ions are formed at very discrete energies. Discrete states are indeed a phenomenon which can be described only by quantum theory. Nevertheless it may be interesting to extend Percival's theory for electron–hydrogen atom collisions to lower energies, to study scattering times, and to find out whether the maximum probability for long lifetimes occurs near the energy ($\approx 9.7$ eV) where in reality a H$^-$ 2s$^2$ ion can be formed. Such pictorial models as given above may be useful in improving our qualitative understanding on some scattering phenomena. Two examples are given below.

In measurements on excitation of atoms by electrons several investigators found a typical threshold behaviour of the polarization of emitted radiation. The (orbital) angular momentum transferred to the atom is $\Delta J = m(r_1 \times v_1 - r_1' \times v_1')$, where $r_1$ and $r_1'$ are the initial and final "impact" parameters respectively*. At the threshold $v_1' = 0$, hence the outgoing electron cannot take away any angular momentum and $\Delta J \perp v_1$ which exactly determines the theoretical threshold values. Among others, Heideman, Smit and Smit [108] found for example for the emission line $4^1D \to 2^1P$ in He that the theoretical polarization of 60% is reduced already to 30% at an electron energy of 0.5 eV above threshold. Such a drastic reduction can only occur if the scattered electron can take away a significant amount of angular momentum**. Because $v_1'$ is still very small at 0.5 eV, $r_1'$ must be very large

---

* In quantum theory $r_1$ and $r_1'$ are meaningless and $m(r_1 \times v_1)$ and $m(r_1' \times v_1')$ have to be replaced by $j_1$ and $j_1'$.
** I am indebted to H. G. M. Heideman, Chr. Smit and J. A. Smit for informing me about this interpretation. See also [108].

and indeed this fits very nicely in the above pictorial model where the size of orbits in a He$^-$ ion exceeds those of a He atom by a factor of about 20 to 50. Also, the mentioned (and some other) polarization function does not show one dip near threshold, but gradually increases above 0.5 eV above threshold. Consequently, not only one but a whole series of resonances must be responsible for the shape of the measured polarization functions*. In some recent close coupling calculations on scattering of electrons by Li, Burke and Taylor** also found that the shape of the polarization functions near threshold is determined predominantly by resonances.

We now proceed to the second example. Reichert and Deichsel [109] measured spin-polarization in 90° elastic electron scattering from neon***. Electrons with energies in the region 16.0 to 16.3 eV were found to be significantly polarized after scattering and just in this region a double resonance due to the $P_{\frac{3}{2}}$ (16.04 eV) and $P_{\frac{1}{2}}$ (16.135 eV) compound ion states occurs. These compound ion states both give enlarged cross sections for 90° scattering. Around 16.0 eV a positive spin-polarization of the scattered electrons is found. Because the corresponding compound ion state is $P_{\frac{3}{2}}$, more electrons with positive orbital angular momentum $r_1 \times p_1$ than with negative $r_1 \times p_1$ are scattered. Around 16.1 eV a negative spin-polarization is found. The corresponding ion state is $P_{\frac{1}{2}}$, and again more electrons with positive angular momentum than with negative $r_1 \times p_1$ are detected. Because elastic scattering through 90° is measured, the nucleus must play a significant role. The experiments [109] now tell us that the incident electron strongly prefers to make a $\frac{1}{4}$ (90°) turn around the nucleus instead of making a $\frac{3}{4}$ (270° measured as 90°) turn. Here the $\frac{1}{4}$ turn corresponds (as do $\frac{5}{4}$, $\frac{9}{4}$ etc. turns) with positive $r_1 \times p_1$ whereas a $\frac{3}{4}$ (and $\frac{7}{4}$, $\frac{11}{4}$ etc.) turn corresponds to negative $r_1 \times p_1$. The attraction of the incident electron by the nucleus excludes a $-\frac{1}{4}$ turn (repulsion by the nucleus) which also gives negative $r_1 \times p_1$. In reality complications may arise for example because exchange of electrons can occur. However, in elastic scattering the outgoing electron must have the same orbital angular momentum and spin directions as the incident one owing to conservation of angular momentum and spin, for which reason the situation will not largely change. The experimental results can again be understood qualitatively from our pictorial model.

---

\* This was suggested by J. A. Smit.
\*\* I am indebted to P. G. Burke for informing me about this matter.
\*\*\* I am indebted to E. Reichert for explaining the relation between the sketched pictorial model and his experimental results.

## § 6-9. Summary and conclusions

In this review we were concerned primarily with those collisions of electrons or protons with atoms in which the individual atomic particles, electrons or nucleus, act as independent scattering centers.

We have seen that these binary collisions always involve large momentum transfers (hence large scattering angles), and for scattering by the atomic electrons large energy transfers.

In the historical survey we summarized the information available before 1940 on elastic (Rutherford) scattering by nuclei of atoms and on large angle inelastic scattering by the atomic electrons. The measurements of Hughes et al. [15–17] (Figs. 6-2-1 and 6-2-2) are particularly interesting in that they clearly show the relationship between the velocity distribution of the atomic electrons and the (broad) energy distribution of electrons scattered through a given angle.

To describe binary collisions of electrons or protons with atomic electrons a set of basic relations for Coulomb collisions between two moving particles must be known. In § 6-3 we derived such a set of relations starting from the Schrödinger equation. In that section we also derived the relevant cross section relations for the case of a charged particle incident on a cloud of free-electrons moving in arbitrary directions (as in an atom).

In the next section we described the binary-encounter models which are used to describe electron–atom and proton–atom collisions and derived the cross section relations for these models.

In § 6-5 a detailed comparison with the Born approximation was made. Here we confined ourselves to ionization by fast charged particles, for which case the Born approximation holds. Figures 6-5-1, 6-5-2 and 6-5-3 clearly show the influence of the motion of the atomic electron on the angular distribution of electrons or protons scattered with a particular energy-loss. Of particular interest is section 6-5-F, where a method is described to derive binary-encounter formulae starting from the Born transition matrix element, and not from Coulomb cross section relations. For two and more electron atoms this method also gives information on correlation effects.

In § 6-6 we compared the binary-encounter theory with Percival's [68–72] classical three-body theory. Figures 6-6-1 to 6-6-5 show that, if interference is not included in the binary-encounter theory, the classical three-body cross sections are smaller than the binary-encounter cross sections for small energy transfer and not too large energies of the incident electron. Hence, the influence of the nucleus is responsible for a reduction of the cross sections.

Curves A, C and C' of Fig. 6-6-5 are interesting in that they suggest that if interference could be included in the classical three-body-theory, as in the binary-encounter theory, the resulting cross sections for small electron energies would almost coincide with the experimental ones.

In § 6-7 we compared experimental and binary-encounter cross sections for ionization and charge transfer. Figures 6-7-1 and 6-7-2 show that the binary-encounter theory can well be used to calculate the energy distribution of ejected electrons resulting from impact of, also not very fast, protons. Figures 6-7-3 and 6-7-4 and also Fig. 10 of [43] are illustrative in another aspect. The binary-encounter theory predicts that the collision cross sections, as a function of the electron or proton energy in units of the ionization energy, are simply proportional to the number of electrons $N$ in the atomic shell involved in the interaction and are inversely proportional to the square of the ionization energy $U_i$. Hence in a plot of scaled cross sections $U_i^2 Q/N$ versus the scaled energy $\frac{1}{2}mv_1^2/U_i$ or $\frac{1}{2}Mv_1^2/U_i$ of the incident particle a universal function should be obtained for different atoms. The mentioned figures show that such a binary-encounter scaling procedure indeed gives very good results for small and intermediate velocities of the incident particles. Table 6-7-1 shows that this scaling procedure does not work at large velocities. In literature one usually refers to this procedure as classical scaling and indeed it applies to Percival's theory for ionization of the hydrogen atom from any $n$-level. However, a classical more than three-body theory (e.g. electrons on He) may well give scaled cross sections different from those obtained in the classical three-body theory. Further, one should note that the binary-encounter theory indeed gives reasonable cross sections, but that better agreement with experiment is obtained by just drawing by hand a curve through the middle of the experimental data shown in Figs. 6-7-3, 6-7-4 and Fig. 10 of [43]. Hence for predicting unknown total ionization or charge transfer cross sections, for not too large velocities of the incident particles, such empirical methods must still be preferred above other more elaborate theoretical methods. Finally, we mention the interesting fact that the measured ionization functions for K (ref. [10]), Rb (ref. [104]) and Cs (refs. [104] and [111]) have two maxima. The binary-encounter theory gives a direct qualitative explanation of the two maxima. The first one near threshold is caused by the outer shell electron and the second one comes from the inner shell electrons (see ref. [56] and also [55]).

In conclusion: the binary-encounter and classical theory are useful in improving our understanding of close collisions between charged particles and atoms. The binary-encounter theory provides us in a simple way with the differential cross sections $\sigma_E$ and $\sigma_{E,P}$ (or $\sigma_{E,\Theta}$) which are rather accurate for sufficiently large values of the energy transfer $E$ and momentum transfer $P$. Some characteristic features of total ionization and charge transfer cross sections are well explained by the binary-encounter theory, which can be and is used as a basis for empirical calculations. The theory is not applicable for distant collisions and should not be used to describe excitation to bound states.

Finally, I wish to thank Professors I. C. Percival, B. R. A. Nijboer and J. A. Smit and Drs. M. Inokuti, M. R. C. McDowell, N. A. Valentine, F. J. De Heer, D. A. Vroom, T. F. M. Bonsen and M. E. Rudd for valuable contributions and/or comments and Dr. M. Gryzinski for initially raising my interest in the problems discussed here. I also wish to acknowledge that a preprint of a review by Burgess and Percival [112] has been very helpful in preparing the present one. Burgess and Percival's review is less detailed on the main subjects discussed here but covers a broader field.

### References

1. M. GRYZINSKI, Phys. Rev. **115** (1959) 374.
2. E. RUTHERFORD, J. CHADWICK and C. D. ELLIS, Radiations from Radioactive Substances (Cambridge University Press, 1930 and 2nd edition 1951).
3. H. BETHE, Ann. Physik **5** (1930) 325.
4. N. F. MOTT, Proc. Roy. Soc. **A127** (1930) 658.
5. N. F. MOTT, Proc. Roy. Soc. **A126** (1930) 259.
6. J. CHADWICK, Proc. Roy. Soc. **A130** (1931) 380.
7. P. M. S. BLACKETT and F. C. CHAMPION, Proc. Roy. Soc. **A130** (1931) 380.
8. J. J. THOMSON, Phil. Mag. **23** (1912) 449.
9. L. H. THOMAS, Proc. Camb. Phil. Soc. **23** (1927) 713.
10. L. H. THOMAS, Proc. Camb. Phil. Soc. **23** (1927) 829.
11. E. J. WILLIAMS, Nature (London) **119** (1927) 489.
12. D. L. WEBSTER, W. W. HANSEN and F. B. DUVENECK, Phys. Rev. **43** (1933) 839.
13. E. J. WILLIAMS, Proc. Roy. Soc. **A128** (1930) 459.
14. C. MØLLER, Z. Physik **70** (1931) 786; see also
    H. KOLBENSTVEDT, Phys. Rev. **163** (1967) 112.
15. A. L. HUGHES and S. S. WEST, Phys. Rev. **50** (1936) 320.
16. A. L. HUGHES and M. M. MANN, Phys. Rev. **53** (1937) 50.
17. A. L. HUGHES and M. A. STARR, Phys. Rev. **54** (1938) 189.
18. G. E. M. JAUNCEY, Phil. Mag. **49** (1925) 427; Phys. Rev. **25** (1925) 723.
19. G. E. M. JAUNCEY, Phys. Rev. **46** (1934) 667.
20. G. E. M. JAUNCEY, Phys. Rev. **50** (1936) 326.
21. P. KIRKPATRICK, P. A. ROSS and R. O. RITLAND, Phys. Rev. **50** (1936) 928.
22. B. HICKS, Phys. Rev. **52** (1937) 436.
23. L. H. THOMAS, Proc. Roy. Soc. **A114** (1927) 561.
24. H. BETHE, Handbuch der Physik, Vol. **24** (Springer Verlag, Berlin, 1933).
25. E. J. WILLIAMS, Rev. Mod. Phys. **17** (1945) 217.
26. N. BOHR, Kgl. Danske Videnskab. Selskab., Mat. Fys. Medd. **18**, No. 8 (1948).

27. W. F. MILLER, Ph.D. thesis, Purdue University (1956).
28. M. GRYZINSKI, Phys. Rev. **138** (1965) A305.
29. M. GRYZINSKI, Phys. Rev. **138** (1965) A327.
30. M. GRYZINSKI, Phys. Rev. **138** (1965) A336.
31. S. CHANDRASEKHAR, Astrophys. J. **93** (1941) 285.
32. R. E. WILLIAMSON and S. CHANDRASEKHAR, Astrophys. J. **93** (1941) 305.
33. V. I. OCHKUR and A. M. PETRUNKIN, Opt. Spectry. (USSR) **14** (1963) 245.
34. R. C. STABLER, Phys. Rev. **133** (1964) A1268.
35. L. VRIENS, Phys. Letters **9** (1964) 295.
36. L. VRIENS, Phys. Letters **10** (1964) 170.
37. L. VRIENS, Phys. Rev. **141** (1966) 88.
38. L. VRIENS, Proc. Phys. Soc. **89** (1966) 13; for a relativistic generalization see H. KOLBENSTVEDT, Phys. Rev. **163** (1967) 112.
39. L. VRIENS, Proc. Phys. Soc. **90** (1967) 935.
40. M. R. C. MCDOWELL, Proc. Phys. Soc. **89** (1966) 23.
41. E. GERJUOY, Phys. Rev. **148** (1966) 54.
42. J. D. GARCIA, E. GERJUOY and J. E. WELKER, Phys. Rev. **165** (1968) 66.
43. J. D. GARCIA, E. GERJUOY and J. E. WELKER, Phys. Rev. **165** (1968) 72.
44. E. BAUER and C. D. BARTKY, J. Chem. Phys. **43** (1965) 2466.
45. B. B. ROBINSON, Phys. Rev. **140** (1965) A764.
46. G. W. CATLOW and M. R. C. MCDOWELL, Proc. Phys. Soc. **92** (1967) 875.
47. A. BURGESS, Proc. 3rd Intern. Conf. on Electronic and Atomic Collisions, London, 1963, ed. M. R. C. McDowell (North-Holland Publ. Co., Amsterdam, 1964) p. 237.
48. A. BURGESS, Proc. Symp. on Atomic Collision Processes in Plasmas, Culham, 1964 (A.E.R.E. Rept. 4818) p. 63.
49. S. S. PRASAD and K. PRASAD, Proc. Phys. Soc. **82** (1963) 655.
50. A. E. KINGSTON, Phys. Rev. **135** (1964) A1529.
51. A. E. KINGSTON, Phys. Rev. **135** (1964) A1537.
52. A. E. KINGSTON, Proc. Phys. Soc. **87** (1966) 193.
53. A. E. KINGSTON, Proc. Phys. Soc. **89** (1966) 177.
54. A. E. KINGSTON, J. Phys. B. (Proc. Phys. Soc.) **1** (1968) 559.
55. L. VRIENS, Phys. Letters **8** (1964) 260.
56. R. H. MCFARLAND, Phys. Rev. **139** (1965) A40.
57. R. H. MCFARLAND, Phys. Rev. **159** (1967) 20.
58. J. W. SHELDON and J. V. DUGAN, J. Appl. Phys. **36** (1965) 650.
59. I. C. PERCIVAL and N. A. VALENTINE, Proc. Phys. Soc. **88** (1966) 885.
60. J. D. GARCIA, Phys. Rev. **159** (1967) 39.
61. L. VRIENS and T. F. M. BONSEN, J. Phys. B. (Proc. Phys. Soc.) **1** (1968) 1123.
62. D. R. BATES, C. J. COOK and F. J. SMITH, Proc. Phys. Soc. **83** (1964) 49.
63. D. R. BATES and R. A. MAPLETON, Proc. Phys. Soc. **85** (1965) 605.
64. D. R. BATES and R. A. MAPLETON, Proc. Phys. Soc. **87** (1966) 657.
65. D. R. BATES and R. A. MAPLETON, Proc. Phys. Soc. **90** (1967) 909.
66. D. R. BATES and J. C. G. WALKER, Proc. Phys. Soc. **90** (1967) 333.
67. D. R. BATES and J. C. G. WALKER, Planetary Space Sci. **14** (1966) 1367.
68. R. ABRINES and I. C. PERCIVAL, Phys. Letters **13** (1964) 216.
69. R. ABRINES and I. C. PERCIVAL, Proc. Phys. Soc. **88** (1966) 861.
70. R. ABRINES and I. C. PERCIVAL, Proc. Phys. Soc. **88** (1966) 873.
71. R. ABRINES, I. C. PERCIVAL and N. A. VALENTINE, Proc. Phys. Soc. **88** (1966) 885.
72. N. A. VALENTINE, Ph.D. thesis, University of London (1968).
73. R. A. MAPLETON, Proc. Phys. Soc. **87** (1966) 219; see also refs. [69] and [93].
74. A. MESSIAH, Quantum Mechanics, Vol. **1** (North-Holland Publ. Co., Amsterdam, 1964) p. 421.

75. N. F. MOTT and H. S. W. MASSEY, Theory of Atomic Collisions (Clarendon Press, Oxford, 3rd ed., 1965) p. 53.
76. B. L. SCHRAM, Physica 32 (1966) 197; see also ref. [105].
77. B. L. SCHRAM, A. J. H. BOERBOOM and J. KISTEMAKER, Physica 32 (1966) 185.
78. J. C. SLATER, Quantum Theory of Atomic Structure, Vol. 1 (McGraw-Hill Book Co., New York, 1960) p. 369.
79. M. INOKUTI and R. L. PLATZMAN, Abstr. 4th Intern. Conf. on Electronic and Atomic Collisions, Quebec, 1965 (Science Bookcrafters Inc., New York) p. 408.
80. W. F. MILLER and R. L. PLATZMAN, Proc. Phys. Soc. 70 (1957) 299.
81. A. R. P. RAU and U. FANO, Phys. Rev. 162 (1967) 68.
82. K. OMIDVAR, private communication.
83. M. INOKUTI, Argonne National Laboratory Rept. 6769 (1963).
84. B. L. SCHRAM, F. J. DE HEER, M. J. VAN DER WIEL and J. KISTEMAKER, Physica 31 (1965) 94.
85. B. L. SCHRAM, M. J. VAN DER WIEL, F. J. DE HEER and H. R. MOUSTAFA, J. Chem. Phys. 44 (1966) 49.
86. Ref. [24] and M. S. LIVINGSTON and H. A. BETHE, Rev. Mod. Phys. 9 (1937) 245.
87. H. A. BETHE and J. ASKIN, Experimental Nuclear Physics, Vol. 1, ed. E. Segré (John Wiley, New York, 1953) p. 166.
88. U. FANO, Ann. Rev. Nucl. Sci. 13 (1963) 1.
89. M. INOKUTI, Y. K. KIM and R. L. PLATZMAN, Phys. Rev. 164 (1967) 55.
90. Y. K. KIM and M. INOKUTI, Phys. Rev. 165 (1968) 39.
91. B. R. A. NIJBOER, private communication.
92. B. R. A. NIJBOER and A. RAHMAN, Physica 32 (1966) 419.
93. V. FOCK, Z. Physik 98 (1935) 145.
94. H. A. BETHE and E. E. SALPETER, Handbuch der Physik, Vol. 35 (Springer-Verlag, Berlin, 1957) p. 88.
95. W. L. FITE and R. T. BRACKMANN, Phys. Rev. 112 (1958) 1141;
    A. BOKSENBERG, Ph.D. thesis, University of London (1961);
    E. W. ROTHE, L. L. MARINO, R. H. NEYNABER and S. M. TRUJILLO, Phys. Rev. 125 (1962) 582.
96. M. E. RUDD, C. A. SAUTTER and C. L. BAILEY, Phys. Rev. 151 (1966) 20.
97. M. E. RUDD and D. GREGOIRE, private communication.
98. H. W. DRAWIN, Z. Physik 164 (1961) 513.
99. W. L. FITE, R. F. STEBBINGS, D. G. HUMMER and R. T. BRACKMANN, Phys. Rev. 119 (1960) 663.
100. H. B. GILBODY and J. V. IRELAND, Proc. Roy. Soc. 277 (1964) 137.
101. F. J. DE HEER, J. SCHUTTEN and H. R. MOUSTAFA, Physica 32 (1966) 1766.
102. D. RAPP and P. ENGLANDER-GOLDEN, J. Chem. Phys. 43 (1965) 1464.
103. W. BLEAKNEY, Phys. Rev. 36 (1930) 1303.
104. J. T. TATE and P. T. SMITH, Phys. Rev. 46 (1934) 773.
105. M. J. VAN DER WIEL, Th. M. EL-SHERBINI and L. VRIENS, Physica 42 (1969) 411.
106. See for example H. EHRHARDT, L. LANGHAUS and F. LINDER, Z. Physik 214 (1968) 179;
    G. E. CHAMBERLAIN and H. G. M. HEIDEMAN, Phys. Rev. Letters 15 (1965) 337.
107. See for example P. G. BURKE, S. ORMONDE and W. WHITAKER, Proc. Phys. Soc. 92 (1967) 319;
    P. G. BURKE and A. J. TAYLOR, Proc. Phys. Soc. 92 (1967) 336;
    P. G. BURKE, A. J. TAYLOR and S. ORMONDE, Proc. Phys. Soc. 92 (1967) 345.
108. H. G. M. HEIDEMAN, CHR. SMIT and J. A. SMIT, Physica, in press.
109. E. REICHERT and H. DEICHSEL, Phys. Letters 25A (1967) 560.
110. G. O. BRINK, Phys. Rev. 127 (1962) 1204.

111. H. HEIL and B. SCOTT, Phys. Rev. **145** (1966) 279;
 Yu. P. KORCHEVOI and A. M. PRONSKI, Soviet Physics, JETP **24** (1967) 1089;
 K. J. NYGAARD, J. Chem. Phys. **49** (1968) 1995.
112. A. BURGESS and I. C. PERCIVAL, Advances in Atomic and Molecular Physics, Vol. 4, eds. D. R. Bates and I. Estermann (Academic Press, New York, 1968) p. 109.

CHAPTER 7

# COINCIDENCE MEASUREMENTS

BY

## QUENTIN C. KESSEL

*Robert J. Van de Graaff Laboratory,*
*High Voltage Engineering Corporation, Burlington, Massachusetts, U.S.A.*

## Contents

|  | Page |
|---|---|
| 7-1. Introduction | 401 |
| 7-2. The theory of measurement | 404 |
|     A. The inelastic energy loss | 404 |
|     B. Charge state analysis | 405 |
| 7-3. The experiment | 407 |
|     A. Apparatus | 407 |
|     B. Procedure | 410 |
| 7-4. Coincidence scattering data | 412 |
|     A. Argon–argon | 413 |
|         1. Inelastic energy loss | 413 |
|         2. Ionization states | 419 |
|         3. Correlation between scattered and recoil charge states | 423 |
|         4. Correlation of the triple-peak region | 426 |
|     B. Neon–neon | 430 |
|         1. Inelastic energy loss | 430 |
|         2. Ionization states | 431 |
|         3. Correlation between scattered and recoil charge states | 434 |
|     C. Krypton–krypton | 436 |
|     D. Neon–argon | 438 |
|     E. Oxygen–argon | 440 |
|     F. Related, non-coincident data | 440 |
|         1. Anomalies in the total differential cross-section | 440 |
|         2. Energies of ionized electrons | 441 |
|     G. Summary | 445 |
| 7-5. Models concerning ionization in heavy ion collisions | 445 |
|     A. Collective oscillations | 446 |
|     B. Energy level crossings of molecular wave functions | 447 |
|         1. The $Ar^+$–$Ar$ collision | 450 |
|         2. The $Ne^+$–$Ne$ collision | 451 |
|     C. Statistical models for ionization | 451 |
| 7-6. Additional coincidence measurements | 452 |
|     A. Total scattering cross sections | 453 |
|     B. Electron–ion coincidence experiment | 456 |
| 7-7. Appendix | 456 |
|     A. Calculation of $T_1$, $T_2$ and $Q$ | 456 |
|     B. Some experimental details | 458 |
|         1. Charge analysis of coincident ions | 458 |
|         2. Coincidence resolving time | 459 |
| Acknowledgments | 460 |
| References | 460 |

## § 7-1. Introduction

The development of new experimental techniques, due in part to technological advances, is making possible atomic scattering experiments of greater and greater complexity. The increased sophistication of differential measurements of large angle scattering events is an example. These scattering reactions are commonly investigated by passing a beam of particles through a thin gas target. Such a single collision reaction, illustrated in Fig. 7-1-1, may be specified by the equation

$$A^{+i} + B \rightarrow A^{+m} + B^{+n} + (m+n-i)e \qquad (7\text{-}1\text{-}1)$$

where $i$ represents the initial charge of the incident ion, $m$ the final charge state of the scattered incident ion, and $n$ that of the recoil target ion.

Many investigators have made large angle scattering studies of reactions represented by eq. (7-1-1). During the 1930's Ramsauer and Kollath [1] studied the angular distributions of protons scattered by gas targets, and Frische [2] and Rouse [3] studied alkali ions scattered by gas targets. More recently laboratories under the direction of Fedorenko in the USSR and Everhart in the USA have made extensive investigations of colliding atomic systems. These laboratories have studied not only the angular distributions of scattered ions but also the ionization states within these angular distributions [4, 5]. Furthermore, they have made similar studies of the low energy recoil ions resulting from these collisions [6, 7]. These latter investigations [6, 7] have also measured the inelastic energy losses associated with atomic collisions; the inelastic energy loss represents the internal energy absorbed by the atoms during the collision.

These experiments, though providing important information on atomic

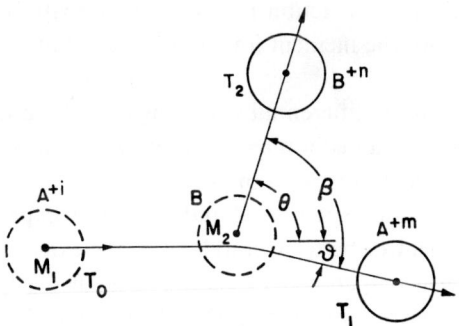

Fig. 7-1-1. Collision schematic.

scattering, have a common deficiency. The measurements made on the scattered ions are actually averages over all possible charge states of the unseen recoil ion, and there is similar averaging in the experiments making measurements using just the recoil ion; it is impossible to know if significant information is being concealed by this averaging effect. Consider some reactions that may contribute to an ordinary electron capture cross section:

$$A^+ + B \rightarrow A + B^+ \quad \text{single electron capture,}$$
$$A^+ + B \rightarrow A^* + B^{+*} \quad \text{electron capture and excitation,} \quad (7\text{-}1\text{-}2)$$
$$A^+ + B \rightarrow A + B^{+2} + e \quad \text{electron capture and ionization.}$$

The electron capture cross section measured by detecting only atom A after collision is not a pure electron capture cross section but is a sum of several different cross sections. An experiment detecting and determining the charge of both ions after collision removes this ambiguity. The identification of both the ions resulting from a single collision is possible through the use of single particle detectors, a delay line, and a coincidence circuit. An added advantage of a delayed coincidence experiment is that it automatically determines the inelastic energy loss of the event being measured. Hasted [8], in 1958, was the first to consider seriously such a coincidence counting experiment. He proposed the notation $(i0/mn)$ for the reaction of eq. (7-1-1). Here the "$i0$" refers to the charge states of the incident ion and target atom before collision and the "$mn$" to their charge states after collision. The cross section for such a collision is written $_{i0}q_{mn}$. With this notation the examples of single charge transfer, and charge transfer together with single ionization, would be written (10/01) and (10/02), respectively. In general the reaction $(10/0n)$ would correspond to electron capture plus, for $n > 1$, ionization of the target atom. The reaction $(10/m0)$ would correspond to stripping of just the incident ion while $(10/1n)$ would indicate ionization of just the target atom. The most common reaction for large angle scattering is $(10/mn)$; in violent collisions both the incident ion and the target atom are often highly ionized.

The measurement of differential scattering events is difficult because the large angle scattering of a fast ion from a thin gas target is a rare event. The detection of these rare events is an important part of any scattering experiment, and more accurate experiments and often new experiments have followed the development of better methods of particle detection. Until about 1955 atomic scattering experiments used the same methods of charged particle detection that were used during the 1930's: namely, a Faraday cage combined with a galvanometer or an electrometer. With the refinements of the 1950's these techniques were reliable for measuring (with difficulty) currents as small as $10^{-16}$ amperes. During this time the nuclear and cosmic ray physicists were using Geiger tubes to detect particles with much higher energies. In so doing, they developed the techniques of pulse counting and pulse analysis that are in common use today. Unfortunately these counting techniques were not applicable to atomic scattering experiments without an appropriate detector that would detect single atoms and ions with energies as low as a few keV. Although Allen [9] had used secondary electron multipliers for the detection of low energy, positive ions in 1939, the practice did not become common until the early 1950's. At this time mass spectroscopists recognized the value of electron multipliers and began to fabricate their own; soon the use of commercial electron multipliers [10] was commonplace. Finally atomic physicists began to use the counting techniques that nuclear physicists had long enjoyed. The counting of individual events allows "current measurements" of $10^{-19}$ amperes to be fairly simple; this represents an increase in detection sensitivity over conventional electrometers of several orders of magnitude. More important, the ability to obtain scattering information in the form of pulses rather than d.c. electrical currents made possible a delayed coincidence experiment, in which both ions from a single encounter could be recognized.

Preliminary results of the first delayed coincidence experiment were reported in 1963 by Afrosimov, Gordeev, Panov and Fedorenko [11], who published their full report in 1964 [12]. They studied the reactions $Ar^{0,+1} + Ar \rightarrow Ar^{+m} + Ar^{+n}$ at 12.5 and 50 keV for scattering angles from 3 to 35 degrees, and measured the inelastic energy loss for several $(00/mn)$ and $(10/mn)$ reactions. Their report also included an analysis of some relative differential cross section data for $(10/mn)$ reactions. An extensive study of large angle (8 to 40 degrees) argon ion–atom scattering for energies from 3 to 400 keV was later reported by Kessel and Everhart [13–16]. These latter papers detail the dependence upon incident ion energy and scattering angle of the inelastic energy loss of many $(10/mn)$ events, and also determine the relationship between $m$ and $n$, the charges of the scattered and recoil ions after collision. More recently these laboratories have made delayed coincidence measurements of the $Ne^+$–Ne [17–19], $Kr^+$–Kr [19, 20] and $Ne^+$–Ar [19] reactions at keV energies. In other laboratories, Bingham [21] has made similar measurements using keV $O^+$–Ar collisions, and Kessel [22] has investigated the $Ar^+$–Ar reaction for energies between 0.5 to 1.5 MeV. The first coincidence measurements of total cross sections have been made by Afrosimov and co-workers [23, 24]; for the combination $H^+$–Ar they have measured the total cross section for $(10/mn)$ reactions with $m=0$, $\pm 1$ and $1 \leq n \leq 3$.

The data from these first coincidence experiments put the early models of heavy ion collisions to their strictest test, and phenomena that were not readily explained gave birth to several new ideas. The experiments give partial justification to a statistical theory of ionization proposed earlier by Russek [25]. At the same time collisions in which inner atomic shells of electrons are made to interpenetrate seem best explained by a molecular orbital level crossing theory proposed by Fano and Lichten [26, 27]. Collective oscillations of electron shells have also been discussed; first by Afrosimov et al. [11, 12] and later by Amusia [28]. With the accumulation of additional data on heavy ion collisions, including electron spectroscopy data by Rudd and co-workers [29], the strengths and weaknesses of these ideas are becoming evident. These early coincidence data have opened new avenues of thought toward the understanding of heavy ion collisions, and indicate the necessity of additional work, both experimental and theoretical.

## § 7-2. The theory of measurement

In concept, a coincidence experiment measuring both ions after a single collision is straightforward. The properties of a two-body, central force interaction are well known, and one of these properties is that such an interaction takes place in a single plane. Experimentally, the axis of the incident beam of ions together with the chosen axis of the scattered beam of ions define this plane; all target particles recoiling *elastically* from these scattered ions are found in the same plane at an angle determined by known collision parameters. However, large angle collisions are seldom elastic, and during the collision some kinetic energy is usually lost to excitation of the electrons surrounding the colliding atoms. This excitation energy, known as the inelastic energy, is removed from the system through ionization of electrons and radiation of photons. The effect of this inelastic energy loss on the trajectories of the ions must be understood before their ionization states can be measured. In fact, the ability to measure this inelastic energy loss together with the final ionization states of the two ions has made the coincidence experiment doubly valuable.

### A. THE INELASTIC ENERGY LOSS

The inelastic energy loss, $Q$, equals the sum of the excitation energies absorbed by the two atoms during the collision. The $Q$ experimentally measured may be considered to be the sum of several excitation and subsequent energy loss processes:

$$Q = \sum V_{m+n-i} + \sum T_{m+n-i} + \sum T_p + \sum T^*, \tag{7-2-1}$$

where $\sum V_{m+n-i}$ represents the ionization energy required to remove the $m+n-i$ electrons of eq. (7-1-1). The term $\sum T_{m+n-i}$ represents the final kinetic energy of these ionized electrons; $\sum T_p$ represents the energy removed as a result of photon emission, and $\sum T^*$ represents the energy removed from the system by the excitation of long-lived metastable states. The terms $\sum V_{m+n-i}$ and $\sum T_{m+n-i}$, pertaining to electron removal, are thought to account for most of the inelastic energy loss.

In Fig. 7-1-1 the incident ion, $A^{+i}$ having energy $T_0$ and mass $M_1$, is scattered through the angle $\vartheta$ by atom B and leaves the collision with an energy $T_1$. Atom B, with mass $M_2$, recoils from ion A and moves away at an angle $\theta$ with energy $T_2$. The inelastic energy loss represents the decrease in kinetic energy of the system and is given by

$$Q = T_0 - T_1 - T_2. \tag{7-2-2}$$

The effect of a finite $Q$ on the ion trajectories is to make the angle $\beta = \vartheta + \theta$ smaller than it would be if $Q$ were equal to zero.

The energies $T_1$ and $T_2$ may be derived, in the classical approximation [30], from the equations for the conservation of energy and momentum. Neglecting the thermal motion of the target atoms and the momenta of the ionized electrons, one may write (see section 7-7-A)

$$T_1/T_0 = \sin^2(\beta - \vartheta)/\sin^2 \beta \tag{7-2-3}$$

and

$$T_2/T_0 = \gamma \sin^2 \vartheta/\sin^2 \beta \tag{7-2-4}$$

where $\gamma = M_1/M_2$. These equations, together with eq. (7-2-2), show that the simultaneous measurement of $T_0$, $\vartheta$ and $\beta$ for a single collision is sufficient for the determination of $Q$ for that collision. If there is no analysis of the charge of the ions appearing at $\vartheta$ and $\beta$ the quantity determined is $\bar{Q}$, which represents the average inelastic energy loss for all the scattering events having a known $T_0$ and $\vartheta$.

B. CHARGE STATE ANALYSIS

The delayed coincidence experiment is unique in its ability to determine the charge state of each ion following a single collision. The average inelastic energy loss $\bar{Q}$, which does not depend upon the charges of the ions after collision, can sometimes be determined by analysis of just the recoil ions [6, 7]. The value of knowing $\bar{Q}$ is diminished, however, when the ionization energies contributing to it are unknown. When in addition to $T_0$, $\vartheta$ and $\beta$, the charge states $m$ and $n$ of the scattered and recoil ions are measured, $\bar{Q}_{mn}$ can be determined by using equations (7-2-2) through (7-2-4). The quantity $\bar{Q}_{mn}$ is the average inelastic energy loss for a collision where not only $T_0$ and $\vartheta$ are specified, but $m$ and $n$ are also specified. The correct notation for the energy loss in eq. (7-1-1) would be $_{i0}\bar{Q}_{mn}$; however, most experiments use a singly ionized beam and the notation $\bar{Q}_{mn}$, and later $\bar{p}_{mn}$, will be used to signify properties of the $(10/mn)$ reaction. Substituting an experimental value for $\bar{Q}_{mn}$ and the corresponding ionization potentials into eq. (7-2-1), one may obtain the remaining energy available for conversion to electron kinetic energy, radiation, and excitation of metastable states.

The relationship of the scattered ion charge $m$ to the recoil ion charge $n$ after a single collision can be determined by a delayed coincidence experiment. Using eq. (7-1-1) with $i=1$ for an example, it is necessary to measure the relative number of $(10/mn)$ events for each value of $m$ and $n$. If, for a specified combination of $T_0$ and $\vartheta$, $N_{mn}$ represents the relative number of events having a scattered charge state $m$ and a recoil charge state $n$, then

$$\bar{p}_{mn} = N_{mn} / \sum_{m,n} N_{mn} \qquad (7\text{-}2\text{-}5)$$

where $\bar{p}_{mn}$ is the probability of the $(10/mn)$ event occurring. Furthermore the relative probability $\bar{P}'_i$ of the scattered ion having charge $i$ is given by

$$\bar{P}'_i = \sum_n \bar{p}_{in}. \qquad (7\text{-}2\text{-}6)$$

The relative probability $\bar{P}''_i$ of the recoil particle having charge state $i$ is given by

$$\bar{P}''_i = \sum_m \bar{p}_{mi}. \qquad (7\text{-}2\text{-}7)$$

The average charge state of the scattered component is given by

$$\bar{m} = \sum_i i \bar{P}'_i, \qquad (7\text{-}2\text{-}8)$$

with a similar equation for $\bar{n}$. Another useful quantity is $\bar{n}^{(m)}$ given by

$$\bar{n}^{(m)} = \sum_n n \bar{p}_{mn} \qquad (7\text{-}2\text{-}9)$$

which represents the average recoil ionization state found in coincidence with scattered ions of charge state $m$. The bars have been placed over many experimental quantities because either they are, or may be, averages. The quantity $\bar{Q}$ is actually a weighted average of the $\bar{Q}_{mn}$'s. Even the inelastic energy loss $\bar{Q}_{mn}$ with $m$ and $n$ specified may be an average; there may be several inelastic processes resulting in the same final charge state.

## § 7-3. The experiment

### A. Apparatus

The study of fast ions scattered to large angles by stationary gas atoms requires an apparatus with two basic features. The target gas must be isolated from the high vacuum of the ion analysis region and the scattered ion analyzer must be capable of rotary motion about a center located within the target gas. A coincidence scattering chamber must provide rotary motion for two analyzers about the same center. This allows the apparatus to analyze both the scattered ions and the recoiling target ions simultaneously; furthermore, the use of single particle detection and delayed coincidence techniques allows the specific identification of which scattered ion struck which recoil ion.

The first coincidence apparatus was described by Afrosimov et al. [12] and is outlined in Fig. 7-3-1. The incident ion beam enters the target gas chamber C, after being collimated by slits $S_1$ and $S_2$. A high vacuum is maintained in this collimating region by differential pumping, i.e. any target gas atoms passing through slit $S_2$, are rapidly pumped away by pump $P_1$. The target gas is maintained at a low pressure (about $5 \times 10^{-4}$ Torr) to ensure that an incident ion rarely suffers more than one collision. A fast ion scattered through the angle $\vartheta$ passes through collimator $K_1$ and into the magnetic analyzer $A_1$. By varying the magnetic field of $A_1$, the different scattered charge states may be identified and deflected into detector $D_2$, where they are counted. Neutral particles are not deflected and enter detector $D_1$. The charge states of the slower recoil ions passing through collimator $K_2$ are identified in the same way by analyzer $A_2$. Detectors $D_1$–$D_4$ post-accelerate the ions through 10–20 kV to an electron emitter, the secondary electrons produced are accelerated by the same voltage, and this pulse is detected with a scintillation counter. The bellows attached to the target gas chamber allow collimator $K_1$ and analyzer $A_1$ to cover the range of $\vartheta$ from 0° to 50°. The recoil collimator $K_2$ and analyzer $A_2$ can be rotated from 40° to 100° on their side of the incident ion beam.

The coincidence scattering apparatus built by Kessel and Everhart [13]

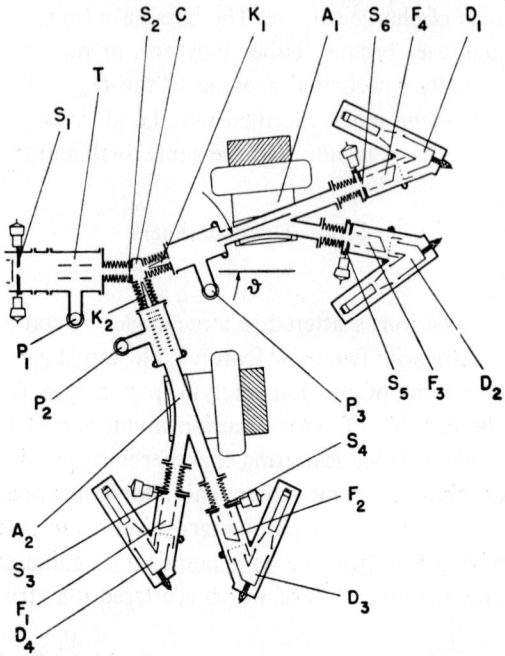

Fig. 7-3-1. Apparatus used by Afrosimov et al. [12] for coincidence measurements of differential scattering. $S_1$–$S_6$, adjustable collimation slits; T, deflection plates for the removal of ions from atomic beams; C, target gas chamber; $K_1$ and $K_2$, scattered and recoil beam collimators; $A_1$ and $A_2$, magnetic analyzers for the scattered and recoil beams; $F_1$–$F_4$, detectors for measuring the intensity of beams of charged and neutral particles; $D_1$–$D_4$, single particle detectors; $P_1$–$P_3$, diffusion pumps.

was constructed without knowledge of the concurrent (coincident) effort by Afrosimov and co-workers and is mechanically quite different. A schematic of this apparatus is shown in Fig. 7-3-2. The construction is unusual; a 0.74 m diameter vacuum chamber serves as the "target gas cell" and within this chamber are three separate vacuum enclosures. The incident ion beam enters the target gas after passing through the rigidly mounted enclosure containing collimation apertures a, b and c. Each scattered and recoil vacuum enclosure houses collimation apertures, an electrostatic analyzer, and a movable secondary electron multiplier. An ion scattered to angle $\vartheta$ in the plane of the apparatus passes through apertures d and e. These ions may be made to follow either of two paths. If the plates of the analyzer are grounded, the ions will pass directly into the detector in position "A". If the electric field in the analyzer is varied, the different charge states of the ions may be

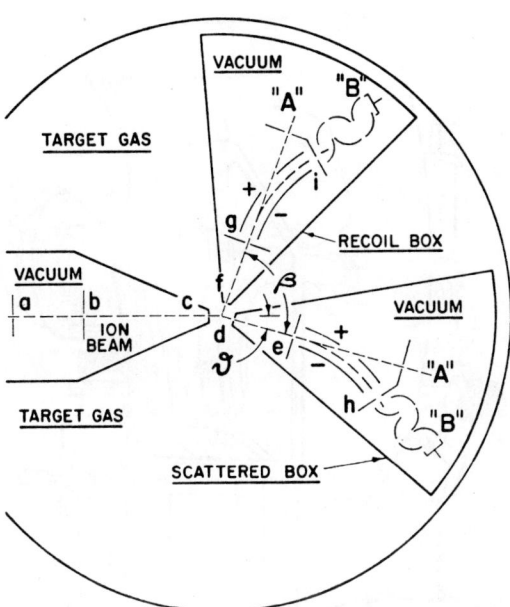

Fig. 7-3-2. Schematic of the coincidence scattering apparatus used by Kessel and Everhart [13].

identified and deflected into the detector in position "B". For the detection of neutral atoms electric fields are applied and the detector is put in position "A". The slower recoiling target gas atoms and ions passing through apertures f and g may be analyzed in the same way. The design of this apparatus has the advantage that there are no bellows to limit the angular motion of either detecting chamber. These 45° wedge-shaped chambers may come to within 45° of each other or within 45° of the incident ion collimator. Another coincidence chamber, using the same construction technique, has been built by Kessel [31] for the investigation of 0.5 to 10.0 MeV ion–atom collisions. A cutaway view of this chamber, 1.32 m in diameter, is shown in Fig. 7-3-3. The high vacuum enclosures for the incident ion beam collimator C and for the rotating recoil ion chamber R are indicated. The inside of the scattered ion enclosure S is open and shows the electrostatic analyzer $A_1$, and a detector $D_1$, for counting unanalyzed ions. Another scattered ion detector, not shown, counts the analyzed ions. The vacuum and the motion for the rotating vacuum enclosures are provided through the pumping manifold M.

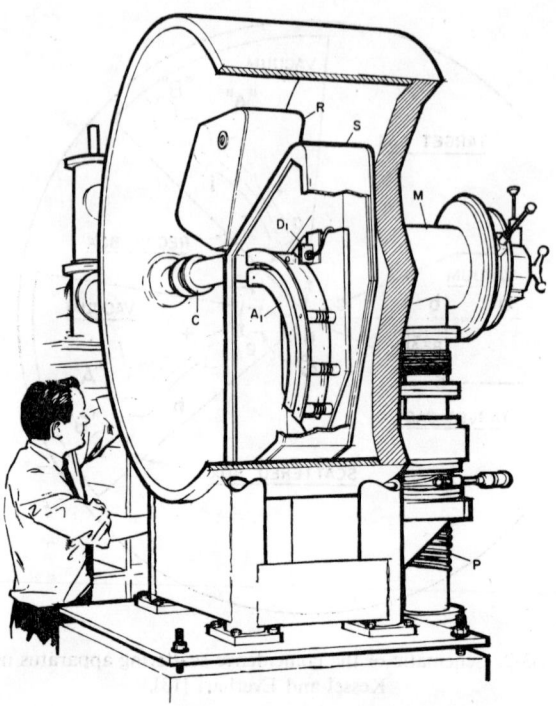

Fig. 7-3-3. Cutaway view of the coincidence apparatus used by Kessel [31]. C, vacuum chamber for incident ion beam; R and S, vacuum chambers for recoil and scattered beams; $A_1$, electrostatic analyzer; $D_1$, solid state detector; M, vacuum manifold for chambers R and S; P, diffusion pump.

## B. Procedure

The calculation of the inelastic energy loss for a collision of two particles of known mass requires the knowledge of three experimental parameters: the incident ion energy $T_0$, the angle $\vartheta$, to which the incident ion was scattered, and the angle $\beta$ (measured with respect to $\vartheta$) at which the recoil ion emerged. These may be substituted into equations (7-2-2) to (7-2-4) in order to calculate $Q$, the inelastic energy loss. Because of their fundamental importance, the variables $T_0$ and $\vartheta$ are held constant for a given measurement and the angle $\beta$ is determined experimentally. Consider Fig. 7-3-2 to see how this is accomplished. An ion beam of known energy $T_0$ passes through aperture c into the target gas region. The scattered box is positioned to accept ions scattered through the desired angle $\vartheta$, and its detector is placed in position "A" behind grounded analyzer plates. With this arrangement the detector will record all ions scattered through the angle $\vartheta$ regardless of their charge state. Similarly the recoil box, with its detector in position "A" and its analyzer grounded, can detect all ions emerging from the collision at the angle $\beta$. In order to establish that an observed recoil ion resulted from the same collision as an ion detected in the scattered box, the pulses from their detectors (after the scattered pulse has been delayed by an amount corresponding to the difference in the time of flight of the two ions) are passed through a coincidence circuit. No coincidences will be recorded, however, unless the angle $\beta$ is such that the set of recoil ions corresponding to the ions scattered to angle $\vartheta$ passes through apertures f and g. The correct setting for $\beta$ is found by rotating the recoil ion chamber until coincidences are recorded. The value of $\beta$ for which the maximum number of coincidences are recorded is the value used to calculate the inelastic energy loss $Q$. The $Q$ determined in this way does not depend upon the charge states of the ions, and is denoted by $\bar{Q}$ or $\bar{Q}_{TT}$, the average inelastic energy loss for all collisions of incident ions with energy $T_0$ scattering to the angle $\vartheta$. The (T, T) notation, for "total-total", refers to the coincidences being obtained by using the total beam passing through the scattered collimator and the total beam passing through recoil collimator without regard to charge states. The curve labeled (T, T) in Fig. 7-3-4 is a plot of the number of coincidence recorded versus $\beta$ for 50 keV Ar$^+$ ions being scattered to 15°. The maximum number of coincidences occurs for $\bar{\beta}$ equal to about 88.2°, which corresponds to an average inelastic energy loss $\bar{Q}$, of 779 eV.

The parameters $\vartheta$ and $\beta$ associated with a particular $(10/mn)$ reaction are more difficult to determine. In order to adjust the analyzers to pass a specific set of charge states $m$ and $n$, the energies $T_1$ and $T_2$ of these ions

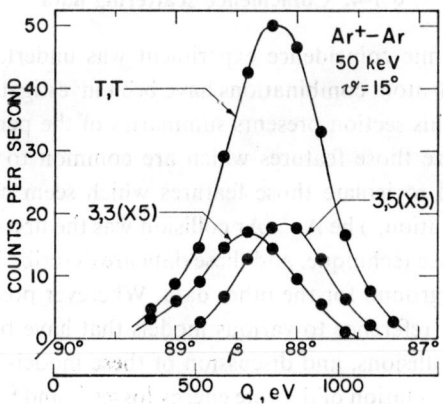

Fig. 7-3-4. Profiles of coincidence counts versus recoil angle $\beta$ (Kessel and Everhart [13]).

must be known. These energies, it would seem from equations (7-2-3) and (7-2-4), cannot be determined except by experiment. In general, the kinetic energies of scattered particles vary as the cosine-squared of the angle between the incident beam and the ions in question. However, this is not strictly true for the events observed by the coincidence technique. The coincidence experiment selects a certain subset of scattered ions, i.e. those ions scattered to $\vartheta$ whose recoiling target ions appear at the angle $\beta$. The energies of the ions in this subset are the energies given by equations (7-2-3) and (7-2-4). These equations show that for $\gamma=1$, for which $\beta\approx 90°$ and $\sin^2\beta\approx 1$, $T_2$ depends only on $\vartheta$ whereas $T_1$ depends on $\vartheta$ and $\beta$. This means that if $\vartheta$ is being held constant, then $T_2$ may be considered constant while $\beta$ is varied over a small range of angles near 90°. However as $\beta$ is changed, the energy $T_1$ of the coincident scattered particle varies; this variation, in $\sin^2(\beta-\vartheta)$, for values of $\vartheta$ up to 15° is less than one percent for a one degree variation in $\beta$. As a practical matter this variation in $T_1$ is small enough so that the analyzer may be set for the value of $T_1$ (eq. (7-2-3)) used in the calculation of $\bar{Q}$, i.e. the value of $T_1$ obtained without using the analyzers. These complications, discussed further in section 7-7-B, become more serious for values of $\gamma$ less than unity, that result in values of $\beta$ not close to 90°.

Fig. 7-3-4 shows the coincidence counts plotted versus $\beta$ for the (10/33) and (10/35) reactions recorded following 50 keV, $\vartheta=15°$, Ar$^+$–Ar collisions. These correspond to $\bar{Q}_{mn}$ values of $685\pm 25$ eV and $814\pm 18$ eV; the 129 eV increase for the (10/35) event reflects the additional energy required to remove the fourth and fifth electrons from the recoil ion.

## § 7-4. Coincidence scattering data

The first atomic coincidence experiment was undertaken in 1963 and fewer than ten ion–atom combinations have been investigated in the six years since that time. This section presents summaries of the pertinent findings in order to emphasize those features which are common to all the ion–atom combinations and to isolate those features which seem characteristic of a particular combination. The Ar$^+$–Ar collision was the first to be investigated with the coincidence technique, and these data are described in detail in order to provide a background for the other data. Wherever possible the data are presented without reference to various models that have been suggested for these heavy ion collisions, and discussion of these models will be presented in § 7-5. The interpretation of discrete energy losses found for certain Ar$^+$–Ar collisions, however, has prompted some investigators to obtain and present their data within the framework of certain interpretations, and these interpretations are briefly discussed to clarify the data when necessary. In particular, many of the data of Afrosimov and co-workers [12, 19] are presented in terms of $Q^*$, a reduced and averaged form of the inelastic energy loss.

## A. ARGON–ARGON

### 1. *Inelastic energy loss*

The $Ar^+$–Ar collision had been studied extensively, and therefore represented a logical collision combination for the first coincidence investigation. Among the earlier non-coincidence investigations, a paper by Morgan and Everhart [7] is of special interest. Following 3 to 100 keV, $Ar^+$–Ar collisions, they measured the target atom's energy and its angle of recoil and calculated the inelastic energy loss of the collisions. In making these measurements they discovered an anomalous behavior for collisions whose distance of closest approach $r_a$,* was approximately 0.23 Å. Near this value of $r_a$, the inelastic energy losses increased rapidly with decreasing $r_a$ and in some instances were multiple-valued. This double and sometimes triple structure in the inelastic energy loss spectrum was found to be consistent with a model that assumes electrons leave the collision in pairs. This unusual idea could not be verified without observing the ionization state of both ions following an $Ar^+$–Ar collision whose distance of closest approach is about 0.23 Å. When the first coincidence experiment was performed by Afrosimov et al. [11, 12] their data showed that the electrons did not leave in pairs, but these data did suggest a new model for the collision. This model and other ideas postulated after further data were available are summarized in § 7-5.

Using incident ion energies of 12.5 and 50 keV Afrosimov et al. [11, 12] investigated this anomalous region in great detail. They measured the inelastic energy losses for many $(00/mn)$ and $(10/mn)$ argon reactions. Some of their data curves are shown in Fig 7-4-1 through 7-4-4 The curves shown in 7-4-1 and 7-4-2 are analogous to those in Fig. 7-3-4, but show a triple structure in the inelastic energy loss. Fig. 7-4-1 shows data for the $(10/22)$ process. The position of each peak corresponds to a different value of $\bar{Q}_{22}$ with $\bar{Q}_{22}^{I} < \bar{Q}_{22}^{II} < \bar{Q}_{22}^{III}$; the relative heights of these three peaks vary for different values of $r_a$, and similar variations are shown in Fig. 7-4-2 for the $(10/23)$ process. It is seen that while the relative heights vary, their positions along the $Q$ axis do not; in the limited region for which these data were

---

* The calculation of $r_a$, the distance of closest approach, generally makes use of a screened-Coulomb potential, see Everhart et al. [32].

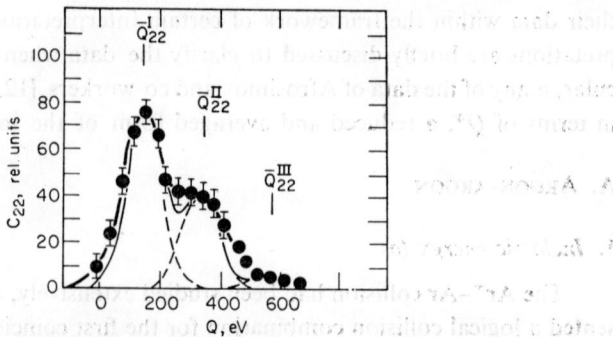

Fig. 7-4-1. Profile of coincidence counts versus the energy loss $Q$ for Ar(10/22) events following 50 keV, $\vartheta = 7.5°$ scattering. The distance of closest approach is $r_a = 0.25$ Å (Afrosimov et al. [12]).

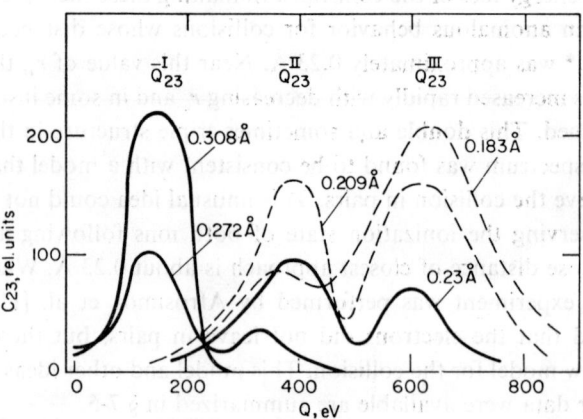

Fig. 7-4-2. Variation of the coincidence profiles versus $Q$ for different values of the distance of closest approach $r_a$. The values of $r_a$ are indicated; $T_0 = 50$ keV (Afrosimov et al. [12]).

taken, the inelastic energy loss for the (10/23) event takes on three discrete values, $\bar{Q}_{23}^I$, $\bar{Q}_{23}^{II}$ and $\bar{Q}_{23}^{III}$, which do not vary greatly for different values of $r_a$. Fig. 7-4-3 shows the relation between the distance of closest approach and the probability for excitation of these three lines. For gentle collisions having large values of $r_a$, only the lowest $\bar{Q}_{23}^I$ value is found; as $r_a$ becomes less, the other two values, $\bar{Q}_{23}^{II}$ and $\bar{Q}_{23}^{III}$, appear.

Table 7-4-1 gives some $\bar{Q}_{mn}$ values measured by Afrosimov et al. [12]; the data shown were taken with incident ion energies of 50 keV, and the 0.330 Å value of $r_a$ corresponds to approximately 3° scattering and the 0.183 Å values to 15° scattering. These data, taken over a relatively small

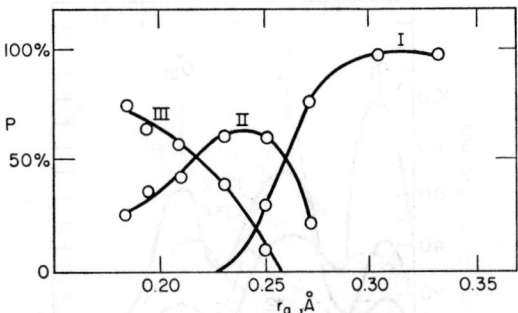

Fig. 7-4-3. Relative probability $P$ of exciting $Q_{23}{}^{\rm I}$, $Q_{23}{}^{\rm II}$ and $Q_{23}{}^{\rm III}$ plotted versus the distance of closest approach. I, $Q_{23}{}^{\rm I}$; II, $Q_{23}{}^{\rm II}$; III, $Q_{23}{}^{\rm III}$; $T_0 = 50$ keV (Afrosimov et al. [12]).

range of $r_a$, exhibit an interesting feature. The values of $\bar{Q}_{mn}^{\rm I}$, $\bar{Q}_{mn}^{\rm II}$ and $\bar{Q}_{mn}^{\rm III}$ take on different values for different combinations of $m$ and $n$. However, if the sum of the ionization energies $V_{mn} = \sum V_{m+n-1}$ of the $(m+n-1)$ electrons removed during the $(10/mn)$ process is subtracted from its corresponding $\bar{Q}_{mn}$ as shown in eq. (7-4-1),

$$Q_{mn}^{*} = \bar{Q}_{mn} - V_{mn} \qquad (7\text{-}4\text{-}1)$$

the resulting $Q_{mn}^{*}$ seems to be largely independent of $m$ and $n$. This surprising result is illustrated in Fig. 7-4-4, where the peaks of different $(10/mn)$ events are seen to coincide after the subtraction of $V_{mn}$ from the corresponding values of $\bar{Q}_{mn}$. Using the data in Table 7-4-1 Afrosimov et al. [12] averaged the values of $Q_{mn}^{*}$ in each group and arrived at three values of $Q^{*}$. These three values, $Q_{\rm I}^{*} = 53 \pm 14$ eV, $Q_{\rm II}^{*} = 263 \pm 16$ eV and $Q_{\rm III}^{*} = 475 \pm 22$ eV, were

TABLE 7-4-1

Some $\bar{Q}_{mn}$ values for 50 keV, argon–argon collisions measured by Afrosimov et al. [12] are shown. The range of $r_a$, the distance of closest approach, includes values for which three groups of energy losses, $Q^{\rm I}$, $Q^{\rm II}$ and $Q^{\rm III}$ are observed.

| $r_a$, Å | $(i0/mn)$ | $Q^{\rm I}$ | $Q^{\rm II}$ | $Q^{\rm III}$ | $r_a$, Å | $(i0/mn)$ | $Q^{\rm I}$ | $Q^{\rm II}$ | $Q^{\rm III}$ |
|---|---|---|---|---|---|---|---|---|---|
| 0.30 | (10/22) | 109 | | | 0.25 | (10/33) | 200 | 430 | 635 |
| 0.30 | (10/23) | 160 | | | 0.25 | (10/34) | | 505 | 680 |
| 0.30 | (10/33) | 207 | | | 0.25 | (10/44) | | 517 | 722 |
| 0.25 | (10/22) | 150 | 340 | 565 | 0.18 | (10/23) | | 377 | 626 |
| 0.25 | (00/22) | 185 | 360 | | 0.18 | (10/33) | | 405 | 633 |
| 0.25 | (10/13) | 125 | 340 | 570 | 0.18 | (10/34) | | 462 | 690 |
| 0.25 | (10/23) | 199 | 385 | 610 | 0.18 | (10/44) | | | 735 |
| 0.25 | (00/23) | 189 | 412 | 635 | 0.18 | (10/55) | | | 842 |

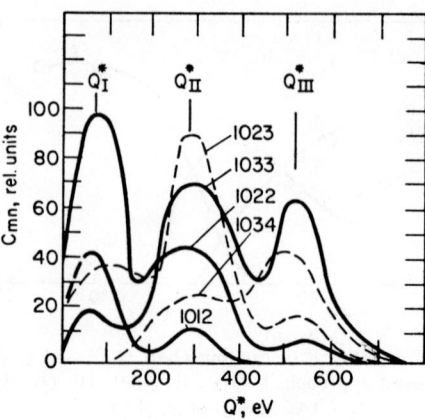

Fig. 7-4-4. Profiles of coincidence counts versus the reduced energy loss $Q^*$ for several $(10/mn)$ processes. $T_0 = 50$ keV, $r_a = 0.25$ Å (Afrosimov et al. [12]).

designated "characteristic excess energy losses" and a model of collective electron oscillations, discussed in section 7-5-A, was proposed to explain them.

The $Ar^+$–Ar investigation by Kessel and Everhart [13] used ions with energies from 3 to 400 keV and included scattering angles from 8° to 40°. The energy losses measured include not only $\bar{Q}_{mn}$ values for specific $(10/mn)$ events but also $\bar{Q}$ (or $\bar{Q}_{TT}$) values, where $\bar{Q}$ represents the average inelastic energy loss for all events. Sets of data were obtained for many combinations of incident ion energy $T_0$ and scattering angle $\vartheta$. A set of data included the measurement of $\bar{Q}$ and also several representative $\bar{Q}_{mn}$ values for the $(T_0, \vartheta)$ combination chosen. Some of these data [33] are shown in Table 7-4-2 where the (T, T) notation indicates the value of the average inelastic energy loss $\bar{Q}_{TT} = \bar{Q}$. This average inelastic energy loss, ranging from 57 to 2430 eV, generally increases as either $T_0$ or $\vartheta$ increases. This is shown in Fig. 7-4-5 where $\bar{Q}$ is plotted versus $T_0$. The discontinuity in these curves is due to the triple-peak region where for 25 keV, 16° there exist three separate values for $\bar{Q}$. Below this, $\bar{Q}$ increases gradually with increasing $T_0$, but does not vary appreciably with changes in $\vartheta$. Above the triple-peak region $\bar{Q}$ increases smoothly from 700 eV to 2500 eV and exhibits some variation for different scattering angles. Inelastic energy losses approaching 6000 eV have been reported by Kessel [22], who has measured $\bar{Q}$ for $Ar^+$–Ar collisions with energies from 0.5 to 1.5 MeV, using the coincidence technique. These data, together with those of Kessel and Everhart [13], are plotted versus the distance of closest approach $r_a$, in Fig. 7-4-6. Contours are shown for several

## TABLE 7-4-2

Inelastic-energy-loss values $\bar{Q}_{mn}$ for Ar$^+$-Ar collision for reactions where charge states $m$ and $n$ after collision are both specified. The notation T, T (or total-total) refers to measurements of average $Q$ wherein all particles are counted, irrespective of charge. Thus $\bar{Q}_{TT} \equiv \bar{Q}$. The incident ion energy $T_0$ and angle $\vartheta$ are given for each data set (Kessel and Everhart [13]).

| $T_0, \vartheta$ | $m, n$ | $\bar{Q}_{mn}$(eV) | $m, n$ | $\bar{Q}_{mn}$(eV) |
|---|---|---|---|---|
| 6 keV, 8° | T, T | 57 ± 3 | 1, 1 | 55 ± 3 |
| | 0, 1 | 36 ± 3 | 2, 1 | 79 ± 4 |
| | 1, 0 | 30 ± 4 | | |
| 25 keV, 8° | T, T | 94 ± 2 | 1, 2 | 91 ± 5 |
| | 0, 1 | 27 ± 5 | 1, 3 | 127 ± 5 |
| | 0, 2 | 70 ± 7 | 2, 2 | 123 ± 5 |
| | 1, 0 | 29 ± 5 | 2, 3 | 167 ± 7 |
| | 1, 1 | 60 ± 5 | | |
| 25 keV, 16° | T, T[a] | 90 ± 17 | 2, 2[b] | 353 ± 7 |
| | T, T[b] | 379 ± 10 | 2, 3[b] | 362 ± 9 |
| | T, T[c] | 613 ± 14 | 2, 3[c] | 636 ± 14 |
| | 1, 1[a] | 62 ± 6 | 3, 3[b] | 468 ± 6 |
| | 2, 2[a] | 160 ± 7 | 3, 3[c] | 647 ± 10 |
| 50 keV, 15° | T, T | 779 ± 9 | 4, 4 | 805 ± 16 |
| | 3, 3 | 685 ± 25 | 4, 5 | 861 ± 17 |
| | 3, 5 | 814 ± 18 | 5, 5 | 915 ± 25 |
| | 4, 3 | 745 ± 20 | | |
| 100 keV, 10° | T, T | 905 ± 12 | 4, 4 | 845 ± 25 |
| | 3, 3 | 700 ± 25 | 4, 5 | 925 ± 25 |
| | 4, 3 | 790 ± 25 | 5, 5 | 995 ± 25 |
| 200 keV, 10° | T, T | 1430 ± 30 | 5, 6 | 1400 ± 60 |
| | T, T[d] | 1340 ± 40 | 5, 7 | 1600 ± 60 |
| | 4, 4 | 1030 ± 70 | 6, 6 | 1560 ± 60 |
| | 5, 4 | 1100 ± 50 | 7, 7 | 1900 ± 60 |
| | 5, 5 | 1260 ± 60 | | |
| 200 keV, 20° | T, T | 2030 ± 70 | 6, 6 | 1850 ± 110 |
| | T, T[d] | 1920 ± 70 | 7, 7 | 2160 ± 90 |
| | 5, 5 | 1470 ± 180 | 8, 8 | 2340 ± 180 |
| 300 keV, 20° | T, T[d] | 2180 ± 140 | | |
| 300 keV, 10° | T, T[d] | 2430 ± 200 | | |

[a] Data correspond to the low value, $Q^I$, in cases where more than one value is found.
[b] Data correspond to the intermediate value $Q^{II}$.
[c] Data correspond to the high value $Q^{III}$.
[d] Data taken using Ar$^{2+}$ as incident ion.

incident ion energies, with the larger distances of closest approach for a given contour corresponding to the more gentle, small angle collisions. There are several regions of interest here: The step increases in $\bar{Q}$ of the triple-peak region are evident for $r_a$ of about 0.23 Å. Between 0.11 and 0.08 Å there is another rise of about 1000 eV in the data, and a sudden increase of nearly

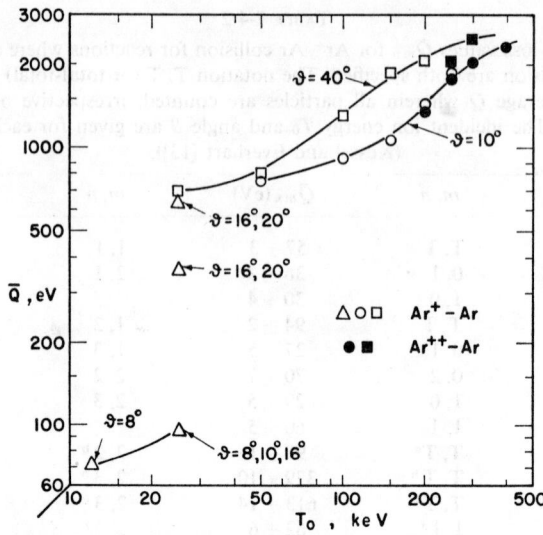

Fig. 7-4-5. For Ar$^+$–Ar collisions, the average inelastic energy loss $\bar{Q}$ is plotted versus the incident ion energy $T_0$ with contours of constant scattering angle $\vartheta$ (Kessel and Everhart [13]).

3000 eV is found when $r_a$ decreases from 0.02 to 0.01 Å. These distances of closest approach [32] indicate that the nuclei pass well within the outer shells of their collision partners; the $r_a$ of 0.01 Å (1.5 MeV, 20°) corresponds to the nucleus of one atom passing well within the K shell (about 0.031 Å radius for argon) of the other atom. It must be remembered that the values of $r_a$ in Fig. 7-4-6 are calculated with the use of a screened Coulomb potential; the calculation does not take shell effects into account.

The $\bar{Q}_{mn}$ data in Table 7-4-2 show an interesting feature: different $\bar{Q}_{mn}$ values for the same $(10/mn)$ event are obtained for different combinations of $T_0$ and $\vartheta$; for example, the $\bar{Q}_{55}$ for 200 keV, 20° scattering is 565 eV larger than the $\bar{Q}_{55}$ measured for 50 keV, 15° scattering. In general, for fixed values of $m$ and $n$, $\bar{Q}_{mn}$ is larger for the more violent collisions. This behavior is shown in Fig. 7-4-7 where $\bar{Q}_{mn}$ for equal $m$ and $n$ are plotted versus $\bar{Q}$. Significantly, $\bar{Q}_{mn}$ increases as the violence of the collision $(\bar{Q})$ increases and a particular $(10/mn)$ event does not have a fixed inelastic energy loss. If the sum of the $m+n-1$ ionization energies is subtracted from $\bar{Q}_{mn}$, as in eq. (7-4-1), the result will be a $\bar{Q}_{mn}^*$ value no longer discrete as was found to be true in the narrower range of phenomena studied by Afrosimov et al. [12].

The 6 keV, 8° and the 25 keV, 8° data indicate the presence of $(10/10)$ and $(10/01)$ events. These events, representing "elastic" collisions and simple

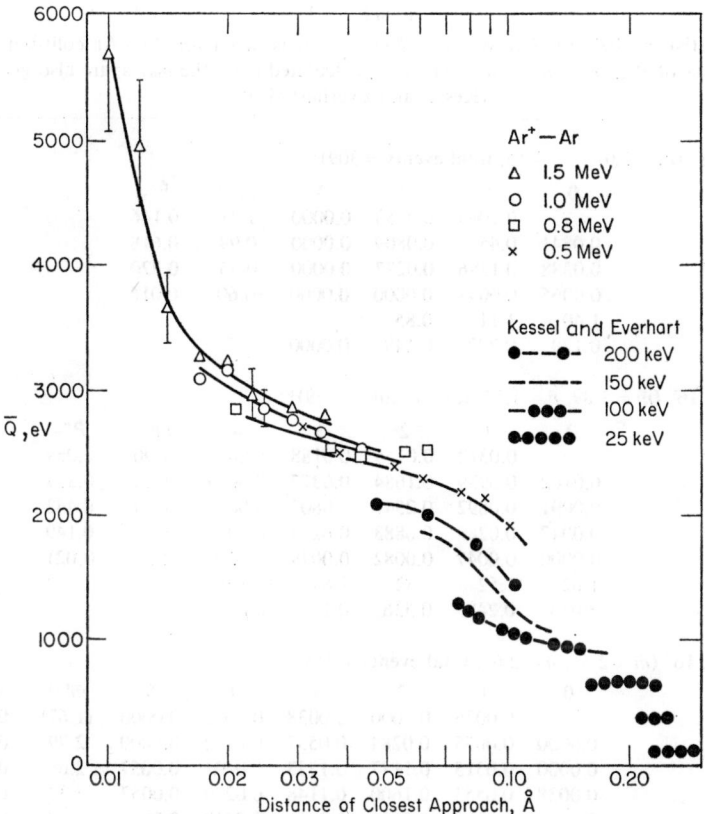

Fig. 7-4-6. The average inelastic energy loss $\bar{Q}$ is plotted versus the distance of closest approach, with the incident energy $T_0$ as a parameter. The data of Kessel and Everhart [13] are indicated (Kessel [22]).

charge capture, are seen to have inelastic energy losses exceeding the first (15.7 eV) and second (27.8 eV) ionization potentials of argon. These atoms and ions evidently leave the collision in excited states. Although not indicated in Table 7-4-2, the $(10/mn)$ collisions seem to be symmetric with respect to energy loss within a given data set, i.e. the loss for the $(10/mn)$ event equals that for the $(10/nm)$ event.

## 2. *Ionization states*

Tables 7-4-1 and 7-4-2 give the inelastic energy losses of many $(10/mn)$ events, but give no indication about the relative probability of these events occurring. The $(10/55)$ event, for example, is observed with an energy loss

## TABLE 7-4-3

The relative probability $\bar{p}_{mn}$ of the (10/$mn$) event is given for Ar$^+$–Ar collisions. The values of $\bar{P}'_m$, $\bar{P}''_n$, $\bar{n}^{(m)}$, $\bar{m}^{(n)}$, $\bar{m}$ and $\bar{n}$, calculated from the $\bar{p}_{mn}$'s, are also given (Kessel and Everhart [13]).

6 keV, 8° ($\bar{m} = 1.02$, $\bar{n} = 1.15$, total events = 309):

| $m =$ | 0 | 1 | 2 | 3 | $\bar{m}^{(n)}$ | $\bar{P}''_n$ |
|---|---|---|---|---|---|---|
| $n = 0$ |        | 0.1003 | 0.0453 | 0.0000 | 1.31 | 0.146 |
| $n = 1$ | 0.0841 | 0.4531 | 0.0809 | 0.0000 | 0.99 | 0.618 |
| $n = 2$ | 0.0388 | 0.1586 | 0.0227 | 0.0000 | 0.93 | 0.220 |
| $n = 3$ | 0.0065 | 0.0097 | 0.0000 | 0.0000 | (0.60) | 0.016 |
| $\bar{n}^{(m)}$ | 1.40 | 1.11 | 0.85 | | | |
| $\bar{P}'_m$ | 0.129 | 0.722 | 0.149 | 0.0000 | | |

25 keV, 16° ($\bar{m} = 1.89$, $\bar{n} = 1.73$, total events = 379)[a]:

| $m =$ | 0 | 1 | 2 | 3 | 4 | $\bar{m}^{(n)}$ | $\bar{P}''_n$ |
|---|---|---|---|---|---|---|---|
| $n = 0$ |        | 0.0312 | 0.0165 | 0.0188 | 0.0029 | 1.90 | 0.069 |
| $n = 1$ | 0.0162 | 0.0939 | 0.1684 | 0.0377 | 0.0029 | 1.74 | 0.319 |
| $n = 2$ | 0.0091 | 0.0892 | 0.2541 | 0.0807 | 0.0085 | 1.99 | 0.442 |
| $n = 3$ | 0.0047 | 0.0289 | 0.0883 | 0.0268 | 0.0000 | 1.92 | 0.149 |
| $n = 4$ | 0.0000 | 0.0047 | 0.0082 | 0.0079 | 0.0000 | (2.15) | 0.021 |
| $\bar{n}^{(m)}$ | 1.62 | 1.52 | 1.82 | 1.81 | (1.39) | | |
| $\bar{P}'_m$ | 0.030 | 0.248 | 0.536 | 0.172 | 0.14 | | |

25 keV, 16° ($\bar{m} = 2.51$, $\bar{n} = 2.63$, total events = 318)[b]:

| $m =$ | 0 | 1 | 2 | 3 | 4 | 5 | $\bar{m}^{(n)}$ | $\bar{P}''_n$ |
|---|---|---|---|---|---|---|---|---|
| $n = 0$ |        | 0.0076 | 0.0000 | 0.0038 | 0.0000 | 0.0000 | (1.67) | 0.011 |
| $n = 1$ | 0.0000 | 0.0055 | 0.0284 | 0.0537 | 0.0172 | 0.0000 | 2.79 | 0.105 |
| $n = 2$ | 0.0000 | 0.0315 | 0.1177 | 0.1533 | 0.0431 | 0.0057 | 2.64 | 0.351 |
| $n = 3$ | 0.0038 | 0.0553 | 0.1600 | 0.1148 | 0.0229 | 0.0057 | 2.32 | 0.363 |
| $n = 4$ | 0.0038 | 0.0138 | 0.0618 | 0.0464 | 0.0219 | 0.0000 | 2.47 | 0.148 |
| $n = 5$ | 0.0000 | 0.0076 | 0.0080 | 0.0093 | 0.0038 | 0.0000 | (2.32) | 0.029 |
| $\bar{n}^{(m)}$ | (3.50) | 2.70 | 2.74 | 2.46 | 2.56 | (2.50) | | |
| $\bar{P}'_m$ | 0.008 | 0.121 | 0.376 | 0.381 | 0.109 | 0.011 | | |

25 keV, 16° ($\bar{m} = 3.12$, $\bar{n} = 3.12$, total events = 209)[c]:

| $m =$ | 1 | 2 | 3 | 4 | 5 | $\bar{m}^{(n)}$ | $\bar{P}''_n$ |
|---|---|---|---|---|---|---|---|
| $n = 1$ | 0.0000 | 0.0000 | 0.0192 | 0.0099 | 0.0000 | (3.34) | 0.029 |
| $n = 2$ | 0.0083 | 0.0052 | 0.0950 | 0.0447 | 0.0114 | 3.28 | 0.165 |
| $n = 3$ | 0.0000 | 0.1070 | 0.2513 | 0.1194 | 0.0036 | 3.04 | 0.481 |
| $n = 4$ | 0.0000 | 0.0488 | 0.1708 | 0.0748 | 0.0114 | 3.16 | 0.306 |
| $n = 5$ | 0.0000 | 0.0052 | 0.0140 | 0.0000 | 0.0000 | (2.73) | 0.019 |
| $\bar{n}^{(m)}$ | (2.00) | 3.32 | 3.12 | 3.04 | (3.00) | | |
| $\bar{P}'_m$ | 0.008 | 0.166 | 0.550 | 0.249 | 0.026 | | |

50 keV, 15° ($\bar{m} = 3.84$, $\bar{n} = 3.86$, total events = 8859):

| $m =$ | 1 | 2 | 3 | 4 | 5 | 6 | $\bar{m}^{(n)}$ | $\bar{P}''_n$ |
|---|---|---|---|---|---|---|---|---|
| $n = 1$ | 0.0000 | 0.0002 | 0.0013 | 0.0025 | 0.0027 | 0.0002 | (4.20) | 0.007 |
| $n = 2$ | 0.0004 | 0.0011 | 0.0116 | 0.0234 | 0.0159 | 0.0033 | 4.13 | 0.056 |
| $n = 3$ | 0.0014 | 0.0089 | 0.0651 | 0.1246 | 0.0638 | 0.0084 | 3.98 | 0.272 |

Table 7-4-3 (cont.)

| | | | | | | | | |
|---|---|---|---|---|---|---|---|---|
| $n=4$ | 0.0025 | 0.0230 | 0.1259 | 0.1916 | 0.0797 | 0.0089 | 3.81 | 0.432 |
| $n=5$ | 0.0012 | 0.0161 | 0.0708 | 0.0891 | 0.0307 | 0.0022 | 3.66 | 0.210 |
| $n=6$ | 0.0008 | 0.0035 | 0.0108 | 0.0071 | 0.0021 | 0.0002 | (3.28) | 0.025 |
| $\bar{n}(m)$ | (4.10) | 4.22 | 4.00 | 3.83 | 3.65 | (3.44) | | |
| $\bar{P}'_m$ | 0.006 | 0.053 | 0.286 | 0.438 | 0.195 | 0.023 | | |

of $915\pm25$ eV following 50 keV, 15° scattering and with an energy loss of $1470\pm180$ eV following 200 keV, 20° scattering. It is also important to know the relative probability $\bar{p}_{mn}$ of these events occurring; Table 7-4-3 [33] gives these probabilities for several scattering combinations of $T_0$ and $\vartheta$. This table shows that for 6 keV, 8° scattering, simple ionization of the target predominates; the probability for the (10/11) event occurring is over 45 percent. For the more violent collisions of 50 keV, 15°, higher charge states predominate and the probability of the (10/11) event is zero. The last row of each entry is the sum of the $\bar{p}_{mn}$'s of that column and represents $\bar{P}'_m$, the probability of the scattered ion having charge state $m$ as given by eq. (7-2-6). Similarly the

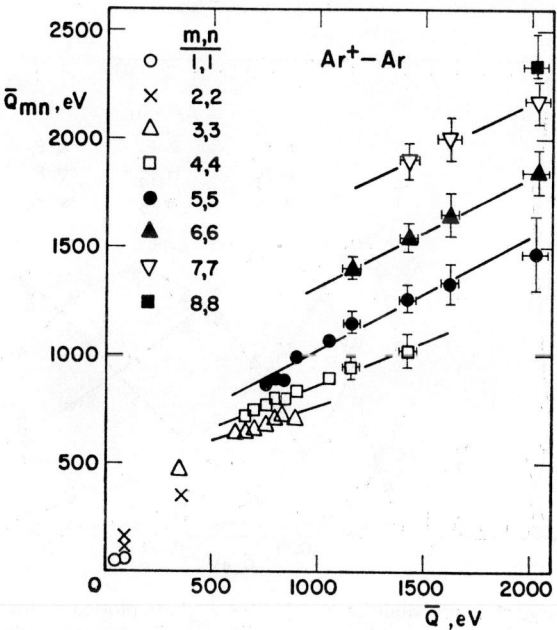

Fig. 7-4-7. For Ar+–Ar collisions, the $\bar{Q}_{mn}$ values for (10/$mn$) processes are plotted versus the average inelastic energy loss $\bar{Q}$ (Kessel and Everhart [13]).

last column of each row gives $\bar{P}_n''$, the probability of the recoil ion having charge state $n$. The quantity $\bar{n}^{(m)}$ is the average recoil charge state corresponding to a scattered charge state of $m$, as described in eq. (7-2-9). Similarly $\bar{m}^{(n)}$ is the average scattered charge state seen in coincidence with the recoil charge state $n$. Statistically poor values, representing less than 3 percent of the data, are shown in parentheses. Also shown are $\bar{m}$ and $\bar{n}$, the overall average charge states of the scattered and recoil ions; the total number of events contributing to each data set are included in order to indicate the statistics involved.

A great deal of symmetry is evident in the data sets of Table 7-4-3; to a first approximation the matrices are symmetric with $\bar{p}_{mn} \approx \bar{p}_{nm}$ and $\bar{n}^{(i)} \approx \bar{m}^{(i)}$. A comparison of $\bar{P}_m'$ and $\bar{P}_n''$ for $m=n$ for any given $T_0$, $\vartheta$ entry shows them to be approximately equal. It follows (eq. (7-2-8)) that $\bar{m}$ and $\bar{n}$ must also be nearly equal; the final charge states do not reflect the one electron asymmetry of the initial system. The larger the inelastic energy loss of a collision, the higher are the final ionization states: Fig. 7-4-8 plots $\bar{P}_i = \frac{1}{2}(\bar{P}_i' + \bar{P}_i'')$ versus $\bar{Q}$. Except for the non-continuous character of $\bar{Q}$ in the triple peak region the different $\bar{P}_i$ curves are remarkably smooth functions of $\bar{Q}$.

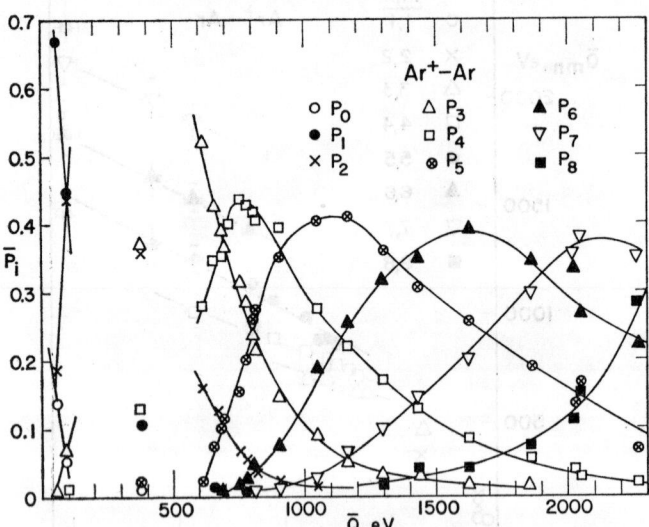

Fig. 7-4-8. Charge state probabilities $\bar{P}_i = \frac{1}{2}(\bar{P}'_i + \bar{P}''_i)$ are plotted versus the average inelastic energy loss $\bar{Q}$ for large angle Ar$^+$–Ar collisions. The data points between 100 and 600 eV are not connected by empirical lines because the data are discontinuous in that region (Kessel and Everhart [13]).

Fig. 7-4-9 plots $\bar{Q}$, the average inelastic energy loss for a data set, versus $\bar{m}+\bar{n}-1$, the average number of electrons lost for that set. From the data of Table 7-4-3 and the sums of the spectroscopic ionization energies for each $(10/mn)$ event $V_{mn}$, the average spectroscopic energy deficit $\bar{V}$,

$$\bar{V} = \sum_{m,n} \bar{p}_{mn} V_{mn},$$

may be calculated for each data set. This energy deficit is shown by the dashed line in Fig. 7-4-9. The energy difference between this line and the data curve represents the kinetic energy of the ionized electrons plus any contribution made by photon emission and the excitation of metastable states. Below the triple-peak discontinuity, ionization energies comprise most of the inelastic energy loss; for the more violent collisions, they contribute less than half the energy loss. This figure sheds some light on the character of this triple-peak discontinuity. When $\bar{m}+\bar{n}-1$ increases from about 3 to 4 or from 4 to 5 the energy loss in each case increases in steps of about 200 eV. As photon emission and the excitation of metastable states probably represent a small amount of the total loss, these 200 eV increases correspond to an increase in the total kinetic energy of the ionized electrons.

### 3. Correlation between scattered and recoil charge states

As Table 7-4-3 shows, there are many possible combinations of $m$ and $n$. The $\bar{p}_{mn}$ values of Table 7-4-3 give the relative probability of the $(10/mn)$ event occurring for certain values of $T_0$ and $\vartheta$; from these values, empirical

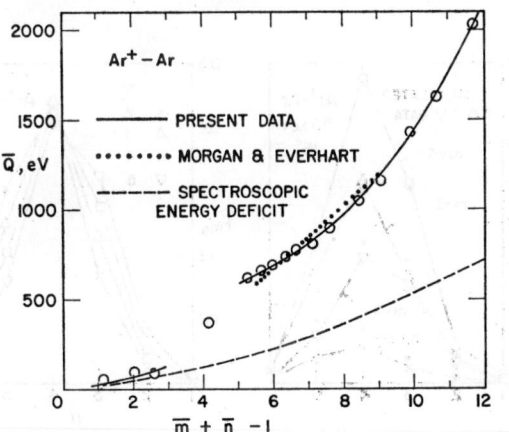

Fig. 7-4-9. The average inelastic energy losses $\bar{Q}$, each one corresponding to collisions of known $T_0$ and $\vartheta$, are plotted versus $\bar{m}+\bar{n}-1$, the average number of electrons lost during the collisions (Kessel and Everhart [13]).

relations between $m$ and $n$ may be determined. It is useful to plot these $\bar{p}_{mn}$ values versus $n$ (or sometimes $m$), and Fig. 7-4-10 shows such a plot. In (a) the $\bar{p}_{mn}$ values for various values of $m$ are plotted along the vertical axis and of $n$ along the absissa for the 50 keV, 15° data of Table 7-4-3. The contours are for constant values of $m$; for example, the uppermost data point gives a probability of about 19 percent for the (10/44) event occurring, and along the same curve are the probabilities of the (10/41), (10/42), (10/43), (10/45) and the (10/46) events occurring. The existence of any correlation between $m$ and $n$ is not obvious from these curves, for they all have their maximum value at $n=4$. In fact, if the contours of different $m$ were identical in shape, one would have to conclude that $m$ did not depend upon $n$, i.e. that there is no correlation between the final charge states of the two ions. In Fig. 7-4-10b these data are normalized to allow an easier comparison of the relative proportions of the contours. It is seen that these contours are essentially the same, although systematic deviations for $n=2$, 3, 5 and 6 do suggest a small amount of correlation between $m$ and $n$. Fig. 7-4-10 does show to a first approximation, however, that the final charge states of the two ions are independent of one another. If the two ions do reach their final charge states independently, the probability of the (10/$mn$) event occurring should be predictable from the $\bar{P}'_m$ and $\bar{P}''_n$, the average final charges of each ion, through the laws governing the probabilities of independent events occurring together; i.e. the equation

$$\bar{p}_{mn} = \bar{P}'_m \bar{P}''_n \tag{7-4-2}$$

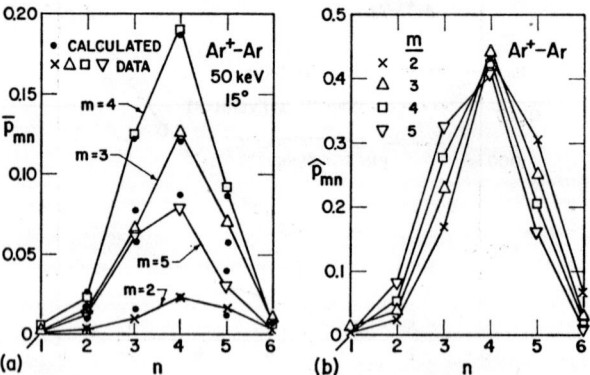

Fig. 7-4-10. (a) The probability $\bar{p}_{mn}$ of the (10/$mn$) reaction is plotted versus $n$ with contours of constant $m$, for 50 keV, 15° scattering. (b) Same, except that renormalized values, $\hat{p}_{mn}$, are plotted (Kessel and Everhart [13]).

should hold true for the data of Table 7-4-3. The results of such a calculation for the 50 keV, 15° data are shown as solid circles in Fig. 7-4-10a. The agreement is good, and for this individual case the relation is seen to be qualitatively correct.

In searching for a correlation between the final charge states $m$ and $n$, Afrosimov, Gordeev, Polyanskii and Shergin [34] made use of the quantities $\bar{n}^{(m)}$ and $\eta_{(nm)}$, where $\eta^2_{(nm)}$ is given by

$$\eta^2_{(nm)} = \sum_m (\sum_n \bar{p}_{mn}) (\bar{n}^{(m)} - \bar{n})^2 / \sum_{mn} \bar{p}_{mn}(n - \bar{n})^2 \tag{7-4-3}$$

and called a correlation ratio. Although Afrosimov et al. [34] applied these quantities to the correlation of the triple-peak region, it is instructive to apply them first to the simple, 50 keV, 15° scattering of Fig. 7-4-10 for comparison with the triple peak data later. Fig. 7-4-11a shows $\bar{n}^{(m)}$ plotted versus $m$ for these 50 keV data: the average slope of this line is about $-0.2$ and shows that there is a small tendency for a higher charge state on one ion to appear in coincidence with a lower charge state on the other ion. This is a quantitative evaluation of the gradual displacement of the curves in Fig. 7-4-10b.

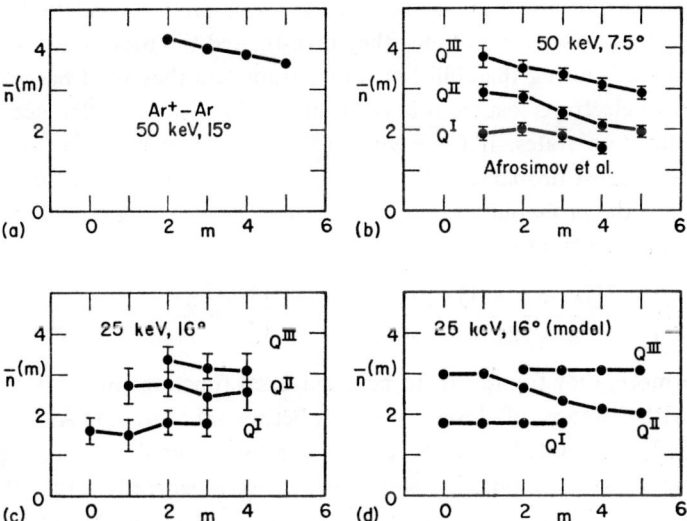

Fig. 7-4-11. The average recoil charge $\bar{n}^{(m)}$ associated with scattered ions of charge $m$ is plotted versus $m$. (a) The 50 keV, 15° data of Fig. 7-4-10; (b) 50 keV, 7.5° triple peak data of Afrosimov et al. [34]; (c) 25 keV, 16° triple peak data of Kessel and Everhart [13]; (d) 25 keV, 16° prediction assuming $Q^I$ and $Q^{III}$ are uncorrelated, but that the $Q^{II}$ data is given by the $P_m$ values in Table 7-4-4 and correlated according to eq. (7-4-5).

The correlation ratio is more difficult to understand; it characterizes the decrease in the width of the charge distribution for one ion when a specific value is assigned to the charge of the other. If there were a one to one correlation between the charge states $m$ and $n$ there would be only one entry in each row or column of Table 7-4-3, and $\eta_{(nm)}$ would be equal to unity. Conversely, if the charge distributions on each ion were identical and there were no correlation, $\bar{n}^{(m)} = \bar{n}$ and $\eta_{(nm)}$ would be equal to zero. For the 50 keV, 15° data of Fig. 7-4-10 $\eta_{(nm)} = \eta_{(mn)} = 0.18$. Care must be taken when comparing $\eta_{(nm)}$ values for different data sets. For example, $\eta_{(nm)}$ cannot be less than zero, and the effect of scatter in the data being analyzed will result in an $\eta_{(nm)} > 0$ for data that might be uncorrelated. Furthermore, if correlation is present, the maximum value of $\eta_{(nm)}$ may not be unity but will depend upon the type of correlation involved.

### 4. Correlation of the triple-peak region

The correlations found in examining the data of the anomalous triple-peak region have been useful in determining the phenomena that give rise to this discontinuity in the data. Afrosimov et al. [12] were the first to study these $(10/mn)$ ionization states at 50 keV using the coincidence techniques. However, at that time, rather than concentrating on the correlation between $m$ and $n$ for the separate atoms, they investigated the total number of electrons ionized during the collision. For parameters they used $m+n-1$, the number of electrons lost from both atoms, and $m-n$, the difference in their final ionization states. If for a given $(T_0, \vartheta)$ data set near the triple peak region, the most probable number of electrons removed is $(\bar{m}+\bar{n}-1)$, they found that the probability of an individual event losing $(m+n-1)$ electrons could be approximated by

$$\exp\left\{-\frac{3[(m+n-1)-(\bar{m}+\bar{n}-1)]^2}{2(\bar{m}+\bar{n}-1)}\right\} \qquad (7\text{-}4\text{-}4)$$

Furthermore, they found this to be a universal function for all sets of data taken with distances of closest approach between 0.18 to 0.33 Å. Afrosimov et al. [12] also noted that the cross sections for collisions having small differences in final ionization states, $m-n$, were generally larger than the cross sections for those events with large values of $m-n$. At the time they interpreted this to mean that symmetrical ionization processes with a low value of $m-n$ predominate. This would infer a correlation of final charge states, i.e. ions with high values of $m$ would appear in coincidence with ions having high values of $n$. Their later data have shown this inference to be incorrect [34]; for in general, there seems to be a tendency for the opposite to hold true, as shown in Fig. 7-4-11b where $\bar{n}^{(m)}$ is plotted versus $m$ for their 50 keV, 7.5° triple-peak data.

Kessel et al. [13, 16] approached the triple-peak correlation in a very different way and asked if it were possible for the $Q^{II}$ peak to be some combination of the $Q^I$ and $Q^{III}$ peaks. This is possible if the 25 keV, 16° collision is exciting an appreciably higher level in the atoms during some but not all of the collisions. Fig. 7-4-12 illustrates how this might happen. In (a) both atoms receive their usual excitation A, and in (c) each atom is excited additionally by the amount B. If only the scattered or the recoil ion received this additional amount, as shown in (b), the inelastic loss for this collision would be A+A+B. For 25 keV, 16° scattering, the average inelastic energy losses for the three peaks are $\bar{Q}^I = 90 \pm 17$ eV, $\bar{Q}^{II} = 379 \pm 10$ eV, and $\bar{Q}^{III} = 613 \pm 14$ eV. If the second peak does represent the suggested combination of first and third peaks, then one would expect $\bar{Q}^{II} \approx \frac{1}{2}(\bar{Q}^I + \bar{Q}^{III})$; this calculation predicts a $\bar{Q}^{II}$ of 352 eV which is within 10 percent of the actual value of $379 \pm 10$ eV. It also follows that if the $Q^{II}$ peak is the result of an asymmetric collision, where one ion leaves the collision more highly excited than the other ion, this should be reflected in their final ionization states.

Fig. 7-4-13a shows the data points (from Table 7-4-3) for the $Q^I$ set corresponding to the lowest of the three $\bar{Q}$ values found for 25 keV, 16° scattering. As indicated, this figure plots the $\bar{p}_{mn}$ data points for $m$ versus $n$ and $n$ versus $m$. The solid contours in this figure pass through the average of these points, i.e. these curves assume $\bar{p}_{mn} = \bar{p}_{nm}$. Fig. 7-4-13c plots the $\bar{p}_{mn}$'s for the higher $Q^{III}$ peak, and to a first approximation the $m$ and $n$ values for

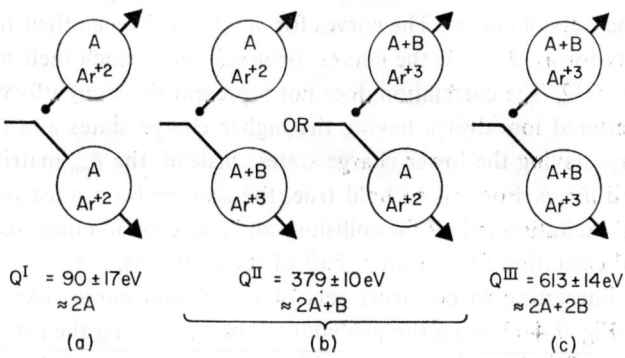

Fig. 7-4-12. A possible explanation [16] for the three discrete energy loss values found for 25 keV, 16°, Ar⁺–Ar collisions. (a) Each ion excited to state A; (b) one ion excited to state A and other ion excited to state A+B; (c) each ion excited to state A+B.

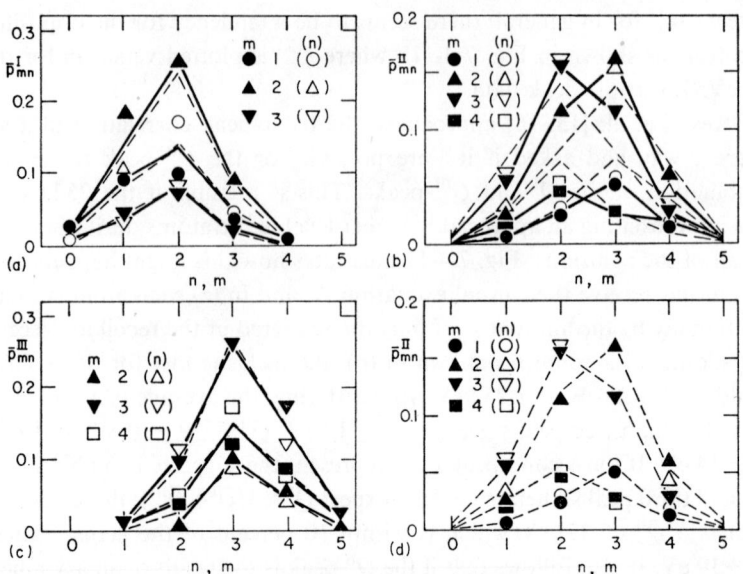

Fig. 7-4-13. For 25 keV, 16°, Ar$^+$–Ar collisions, the probabilities $\bar{p}_{mn}$ of the (10/$mn$) events occurring are plotted versus $n$ with contours of constant $m$ (solid points) and versus $m$ with contours of constant $n$ (open points). The solid lines connect the data points; the dashed lines are calculated. (a) $Q^{\mathrm{I}}$ data; (b) $Q^{\mathrm{II}}$ data; (c) $Q^{\mathrm{III}}$ data; (d) $Q^{\mathrm{III}}$ data compared with calculation using eq. (7-4-4) and the $P_m$ values from Table 7-4-4.

these figures are uncorrelated; however, the $\bar{p}_{mn}$'s corresponding to the $Q^{\mathrm{II}}$ peak and shown in Fig. 7-4-13b do show correlation. This is unlike the slight correlation of the 50 keV, 15° data illustrated by the gradual displacement of the curves in Fig. 7-4-10b. Instead, the contours of Fig. 7-4-13b represent two distinct sets of curves. The curves for $m=1$ and 2 reach their maximum probability for $n=3$, while the curves for $m=3$ and 4 reach their maximum values for $n=2$. The correlation does not represent the straightforward case of the scattered ion always having the higher charge states and the recoil ions always having the lower charge states; instead, the $\bar{p}_{mn}$ matrix is symmetric and $\bar{m} \approx \bar{n}$. For this to hold true, the scattered ion must receive the additional excitation in half the collisions and the recoil ion must receive this additional excitation for the other half of the collisions.

It is interesting to construct sets of correlation curves like the solid curves in Fig. 7-4-13 using the probability laws governing the simultaneous occurrence of independent events. If, for example, you take values of $P'_i = P''_i = \bar{P}_i$, where $\bar{P}_i$ is obtained from the data by averaging the $\bar{P}'_i$ and $\bar{P}''_i$ from the $Q^{\mathrm{I}}$ data set, and multiply them together as in eq. (7-4-2), you can generate an idealized $\bar{p}^{I}_{mn}$ matrix. This idealized matrix is thus generated under the assumption that the final charge states $m$ and $n$ are reached independently, i.e. that there is no correlation between $m$ and $n$. The dashed lines in Fig. 7-4-13a pass through these idealized $\bar{p}^{I}_{mn}$ values for the $Q^I$ data. These dashed lines are nearly identical with the solid data curves. Fig. 7-4-13c plots a similar reconstruction of the $Q^{III}$ data, and again the dashed lines are a good approximation to the data. This procedure cannot produce two families of curves as would be required to reconstruct the solid data curves in Fig. 7-4-13b.

If the situation illustrated in Fig. 7-4-12b is correct, the $\bar{p}^{II}_{mn}$ matrix might be predictable from the average $\bar{P}_n$'s of the $Q^I$ and $Q^{III}$ peaks. Denoting these average $\bar{P}_n$'s by $\bar{P}^I_n$ and $\bar{P}^{III}_n$ the $\bar{p}^{II}_{mn}$'s would be predicted by

$$\bar{p}^{II}_{mn} = \tfrac{1}{2}(\bar{P}^I_m \bar{P}^{III}_n + \bar{P}^I_n \bar{P}^{III}_m) \qquad (7\text{-}4\text{-}5)$$

if each ion received the additional excitation in half the collisions. Values of $\bar{p}^{II}_{mn}$ calculated this way are shown by the dashed lines in Fig. 7-4-13b. This reconstruction of the data has produced two sets of contours that are a fair approximation of the actual data. This qualitative agreement is significant' for it suggests that eq. (7-4-5) does represent the particular type of correlation present in the $Q^{II}$ data. This allows a further refinement to be made: using the $\bar{P}^I_n$ and $\bar{P}^{III}_n$ obtained from the data in Table 7-4-3 as initial guesses and eq. (7-4-5) as a restraint, a least squares fit to the $\bar{p}^{II}_{mn}$ data points can be made. This generates new sets of percentages, $P^{I*}_n$ and $P^{III*}_n$, and a symmetric $p^{II*}_{mn}$ matrix with the correlation of eq. (7-4-3) built into it. The values of $P^{I*}_n$ and $P^{III*}_n$ obtained this way are given in Table 7-4-4; interestingly, the average charge $\bar{m}$ for the more highly excited set of $P^{III*}_m$'s is just one charge higher than that for the $P^{I*}_m$'s. The $p^{II*}_{mn}$'s generated from $P^{I*}_m$ and $P^{III*}_m$ are indicated by the dashed lines in Fig. 7-4-13d. These are an excellent fit to the actual data points, which are also shown in that figure.

The quantities $\bar{n}^{(m)}$ and $\eta_{(nm)}$ are rather insensitive to the specific type of correlation described by eq. (7-4-5). Values of $\bar{n}^{(m)}$ are plotted versus $m$

TABLE 7-4-4

The percentages $P_m{}^{I*}$ and $P_m{}^{III*}$ and the corresponding average charge states $\bar{m}^I$ and $\bar{m}^{III}$ are given. These values were calculated using eq. (7-4-5) as a restraint and making a least squares fit to the correlated 25 keV, 16° data of Kessel and Everhart [13].

| $m$ | 0 | 1 | 2 | 3 | 4 | 5 | |
|---|---|---|---|---|---|---|---|
| $P_m{}^{I*}$ | 0.019 | 0.216 | 0.507 | 0.234 | 0.024 | 0.000 | ($\bar{m}^I = 2.03$) |
| $P_m{}^{III*}$ | 0.000 | 0.005 | 0.243 | 0.520 | 0.209 | 0.023 | ($\bar{m}^{III} = 3.00$) |

for several data sets in Fig. 7-4-11. Fig. 7-4-11d shows what these curves would look like if the $Q^I$ and $Q^{III}$ data exhibited no correlation and the $Q^{II}$ data were completely correlated according to eq. (7-4-5). Fig. 7-4-11a shows the curve for the relatively uncorrelated 50 keV, 15° data; unfortunately its slope is approximately the same as the average slope of the $Q^{II}$ curve in Fig. 7-4-11d, which is the calculated result of a completely different type of correlation. The $\eta_{(nm)}(=\eta_{(mn)})$ value for the idealized $p_{mn}^{II*}$ data, shown in Fig. 7-4-13d by dashed lines, is only 0.29. It does not equal unity even though the $p_{mn}^{II*}$'s are completely correlated in the sense of eq. (7-4-5). The corresponding data points for the $Q^{II}$ peak give values of $\eta_{(nm)}^{II}=0.15$ and $\eta_{(mn)}^{II}=0.21$ while the $Q^I$ and $Q^{III}$ data have correlation ratios of $\eta_{(nm)}^{I}=0.15$, $\eta_{(mn)}^{I}=0.14$, $\eta_{(nm)}^{III}=0.17$, and $\eta_{(mn)}^{III}=0.15$.

## B. NEON–NEON

### 1. *Inelastic energy loss*

The $Ne^+$–Ne collision represents an ion–atom combination that in many ways is similar to the $Ar^+$–Ar combination. Neon and argon are both noble gas atoms with full outer shells of eight electrons. Neon is a "harder" atom with higher ionization energies than argon; therefore in the range of collision energies from 10 to 200 keV, the $Ne^+$–Ne collisions may be similar in nature to the lower energy $Ar^+$–Ar collisions of section 7-4-A. Kessel, McCaughey and Everhart [18] investigated the $Ne^+$–Ne combination in a study that closely parallels the previous $Ar^+$–Ar work. They found that the 200 keV, $Ne^+$–Ne collision does exhibit a behavior that may be analogous to the triple-peak region in the $Ar^+$–Ar data. The major portion of the $Ne^+$–Ne data, however, is not complicated by any unusual activity; these data describe one of the simpler heavy ion collisions. Table 7-4-5 gives the inelastic energy loss for several combinations of $T_0$ and $\vartheta$. As shown in Fig. 7-4-14, $\bar{Q}(=\bar{Q}_{TT})$ increases smoothly as either $T_0$ or $\vartheta$ is increased for incident ion energies from 10 to 150 keV. Within this range of energies there is no evidence of structure of the type found for 25 keV, 16° $Ar^+$–Ar collisions; the $\bar{Q}_{mn}$ values for the $Ne^+$–Ne combination are not constant but increase as the $\bar{Q}$ of the collision increases. Fig. 7-4-16 shows $\bar{Q}$ plotted versus $r_a$, the distance of closest approach of the two neon nuclei. Except for a slight velocity (energy) dependence, $\bar{Q}$ is seen to depend upon $r_a$ and to increase gradually as $r_a$ becomes smaller. For the high energy scattering where $r_a$ is less than 0.05 Å, the $\bar{Q}$ data indicate the presence of a second, unresolved peak at about 1500 eV in the energy loss spectrum. The second

## TABLE 7-4-5

The inelastic energy $\bar{Q}_{mn}$ is given for Ne$^+$–Ne reactions where charge states $m$ and $n$ after collision are specified. The notation T, T refers to an overall average. Thus $\bar{Q}_{TT} \equiv \bar{Q}$. The incident ion energy $T_0$ and angle $\vartheta$ are given for each data set (Kessel, McCaughey and Everhart [18]).

| $T_0$(keV), $\vartheta$ | $m, n$ | $\bar{Q}_{mn}$(eV) | $m, n$ | $\bar{Q}_{mn}$(eV) |
|---|---|---|---|---|
| 6.4, 10° | T, T | 85 ± 5 | 1, 1 | 70 ± 5 |
|  | 0, 1 | 60 ± 15 | 1, 2 | 100 ± 10 |
|  | 1, 0 | 45 ± 20 | | |
| 12, 10° | T, T | 120 ± 10 | 2, 1 | 120 ± 5 |
|  | 1, 1 | 90 ± 5 | 2, 2 | 170 ± 10 |
| 25, 10° | T, T | 245 ± 10 | 2, 2 | 230 ± 10 |
|  | 1, 2 | 185 ± 10 | 2, 3 | 290 ± 10 |
| 50, 10° | T, T | 335 ± 10 | 2, 3 | 345 ± 15 |
|  | 2, 1 | 195 ± 15 | 3, 3 | 415 ± 15 |
|  | 2, 2 | 260 ± 15 | | |
| 100, 10° | T, T | 440 ± 15 | 2, 4 | 510 ± 40 |
|  | 2, 2 | 305 ± 25 | 3, 3 | 475 ± 25 |
|  | 2, 3 | 390 ± 25 | 3, 4 | 575 ± 25 |
| 200, 8° | T, T [a] | 570 ± 20 | 5, 3 [a] | 710 ± 80 |
|  | 5, 4 [a] | 770 ± 100 | 5, 3 [b] | 1550 ± 100 |
|  | 5, 4 [b] | 1630 ± 100 | 4, 3 [a] | 670 ± 50 |

[a] Data correspond to first peak.
[b] Data correspond to second peak.

peak is well resolved for the (10/53) and (10/54) events at 200 keV, 8° and the losses for these reactions are given in Table 7-4-5.

Afrosimov et al. [19] report several discrete $Q^*$ values for Ne$^+$–Ne scattering at 12.5 and 50 keV. In particular, at 50 keV they find $Q^* \approx 160$ eV for the (10/22), (10/23), (10/33) and (10/43) reactions and $Q^* \approx 110$ eV for the (10/12) reaction. For 12 keV scattering between 4° and 40° they found $Q^*$ to vary from 50 to 110 eV. These values are compatible with $Q^*$ values that can be derived from the data in Table 7-4-5; however, the theory that $Q^*$ takes on only discrete values is not compatible with the smooth variations of $\bar{Q}_{mn}$ shown in Fig. 7-4-15.

### 2. Ionization states

The $\bar{p}_{mn}$ values for several combinations of $T_0$ and $\vartheta$ are given in Table 7-4-6. As in Table 7-4-3, values for $\bar{n}^{(m)}$, $\bar{m}^{(n)}$, $\bar{P}'_m$, $\bar{P}''_n$, $\bar{m}$ and $\bar{n}$ are also given. The 25 keV, 10° and the 50 keV, 10° data sets are uncorrelated, with $\bar{p}_{mn} \approx \bar{p}_{nm}$, $P'_i \approx P''_i$, and $\bar{m} \approx \bar{n}$. This does not hold true for the lower energy collisions: the 6.4 keV, 10° data exhibit little symmetry and show most of the electrons

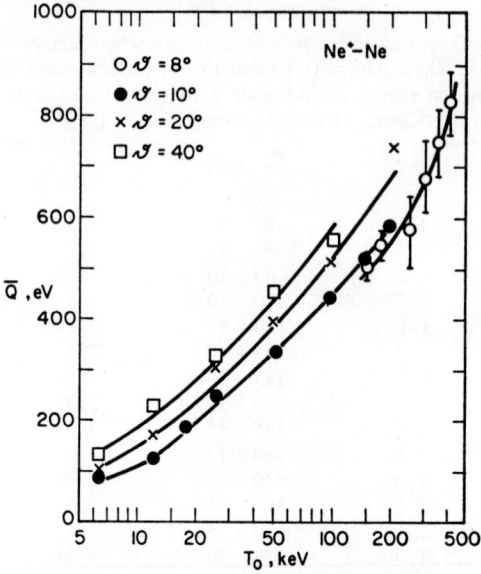

Fig. 7-4-14. Average inelastic energy $\bar{Q}$ is plotted versus incident energy $T_0$ for several scattering angles $\vartheta$. The values above 200 keV were obtained using an incident Ne$^{++}$ ion beam (Kessel, McCaughey and Everhart [18]).

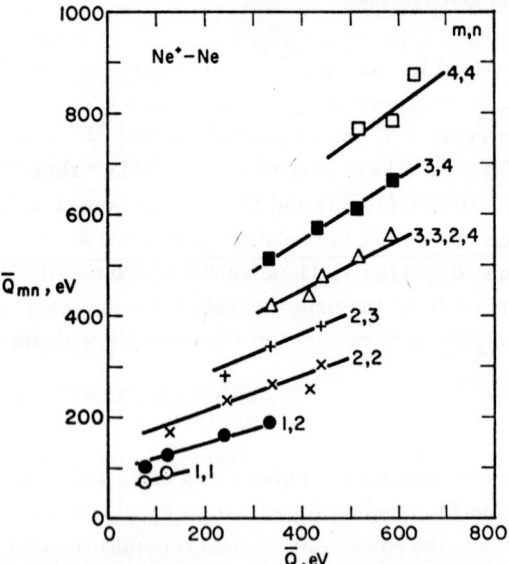

Fig. 7-4-15. Values of $\bar{Q}_{mn}$ are plotted versus $\bar{Q}$ for several (10/$mn$) events (Kessel, McCaughey and Everhart [18]).

### TABLE 7-4-6

The relative probability $\bar{p}_{mn}$ of the (10/$mn$) event is given for Ne$^+$–Ne collisions. The values of $P'_m$, $P''_n$, $\bar{n}^{(m)}$, $\bar{m}^{(n)}$, $\bar{m}$ and $\bar{n}$, calculated from the $\bar{p}_{mn}$'s are also given (Kessel, McCaughey and Everhart [18]).

6.4 keV, 10° ($\bar{m}=1.09$, $\bar{n}=1.45$, total events $=547$)

| $m=$ | 0 | 1 | 2 | $\bar{m}^{(n)}$ | $P''_n$ |
|---|---|---|---|---|---|
| $n=0$ |  | 0.050 | 0.029 | 1.37 | 0.079 |
| $n=1$ | 0.026 | 0.271 | 0.097 | 1.18 | 0.394 |
| $n=2$ | 0.068 | 0.413 | 0.024 | 0.91 | 0.505 |
| $\bar{n}^{(m)}$ | 1.72 | 1.49 | 0.97 |  |  |
| $P'_m$ | 0.094 | 0.734 | 0.150 |  |  |

12 keV, 10° ($\bar{m}=1.32$, $\bar{n}=1.49$, total events $=871$)

| $m=$ | 0 | 1 | 2 | 3 | $\bar{m}^{(n)}$ | $P''_n$ |
|---|---|---|---|---|---|---|
| $n=0$ |  | 0.018 | 0.031 | 0.007 | 1.80 | 0.560 |
| $n=1$ | 0.008 | 0.209 | 0.187 | 0.005 | 1.49 | 0.402 |
| $n=2$ | 0.051 | 0.330 | 0.142 | 0.000 | 1.17 | 0.523 |
| $n=3$ | 0.003 | 0.008 | 0.000 | 0.000 | (0.73) | 0.011 |
| $\bar{n}^{(m)}$ | 2.04 | 1.58 | 1.30 | (0.42) |  |  |
| $P'_m$ | 0.055 | 0.565 | 0.360 | 0.012 |  |  |

25 keV, 10° ($\bar{m}=1.98$, $\bar{n}=2.03$, total events $=4583$)

| $m=$ | 0 | 1 | 2 | 3 | $\bar{m}^{(n)}$ | $P''_n$ |
|---|---|---|---|---|---|---|
| $n=0$ |  | 0.000 | 0.018 | 0.002 | (2.10) | 0.020 |
| $n=1$ | 0.001 | 0.024 | 0.084 | 0.022 | 1.97 | 0.131 |
| $n=2$ | 0.005 | 0.117 | 0.409 | 0.109 | 1.97 | 0.640 |
| $n=3$ | 0.002 | 0.037 | 0.128 | 0.034 | 1.97 | 0.201 |
| $\bar{n}^{(m)}$ | (2.13) | 2.07 | 2.01 | 2.05 |  |  |
| $P'_m$ | 0.008 | 0.178 | 0.639 | 0.167 |  |  |

50 keV, 10° ($\bar{m}=2.42$, $\bar{n}=2.43$, total events $=57867$)

| $m=$ | 1 | 2 | 3 | 4 | $\bar{m}^{(n)}$ | $P''_n$ |
|---|---|---|---|---|---|---|
| $n=1$ | 0.005 | 0.034 | 0.031 | 0.006 | 2.50 | 0.076 |
| $n=2$ | 0.036 | 0.208 | 0.189 | 0.021 | 2.43 | 0.454 |
| $n=3$ | 0.037 | 0.197 | 0.183 | 0.015 | 2.41 | 0.432 |
| $n=4$ | 0.006 | 0.019 | 0.013 | 0.001 | 2.23 | 0.039 |
| $\bar{n}^{(m)}$ | 2.52 | 2.44 | 2.43 | 2.26 |  |  |
| $P'_m$ | 0.084 | 0.458 | 0.416 | 0.043 |  |  |

being removed from the recoiling target ion. Seventy-three percent of the collisions are of the (10/1$n$) type with the scattered ion leaving the collision with its original charge. The reaction (10/12), is the most likely event and occurs with a probability of 41 percent, while the (10/11) event (single ionization of the target atom) results from 27 percent of the collisions. Fig. 7-4-17 shows $\bar{P}_i$ plotted versus $\bar{Q}$. The $\bar{P}_i$'s associated with the lowest $\bar{Q}$ values have primes to indicate they refer to scattered incident ions or double

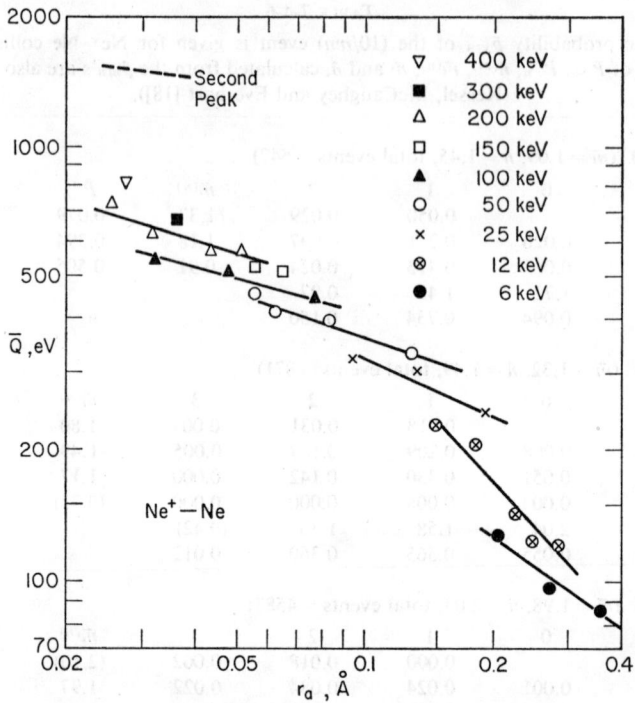

Fig. 7-4-16. The average inelastic energy $\bar{Q}$ is plotted versus $r_a$, the distance of closest approach. The dashed line indicates the presence of a higher energy loss that was not fully resolved (Kessel, McCaughey and Everhart [18]).

primes to indicate they are probabilities for the recoil target ion. The curves are the result of a model, to be discussed in section 7-5-C, that assumes $\bar{P}'_m = \bar{P}''_m$. The initial ion–atom asymmetry is shown clearly by the single and double primed data points; for no interaction $\bar{Q}=0$, and $\bar{P}'_1$ would be unity while $\bar{P}''_1$ would be zero; the data points near $\bar{Q}=100$ eV clearly reflect this asymmetry. For the more violent collisions $\bar{P}'_m \approx \bar{P}''_m$, and these data are represented by unprimed points.

### 3. Correlation between scattered and recoil charge states

The correlation data for the Ne$^+$–Ne collisions fall into three categories. The very low energy data, as indicated in Fig. 7-4-17, reflect the initial ion–atom asymmetry of the collision. The predominance of the (10/11) and (10/12) events in the 6.4 keV, 8° scattering represents a correlation unlike any discussed for argon collisions. However, for incident ion energies from

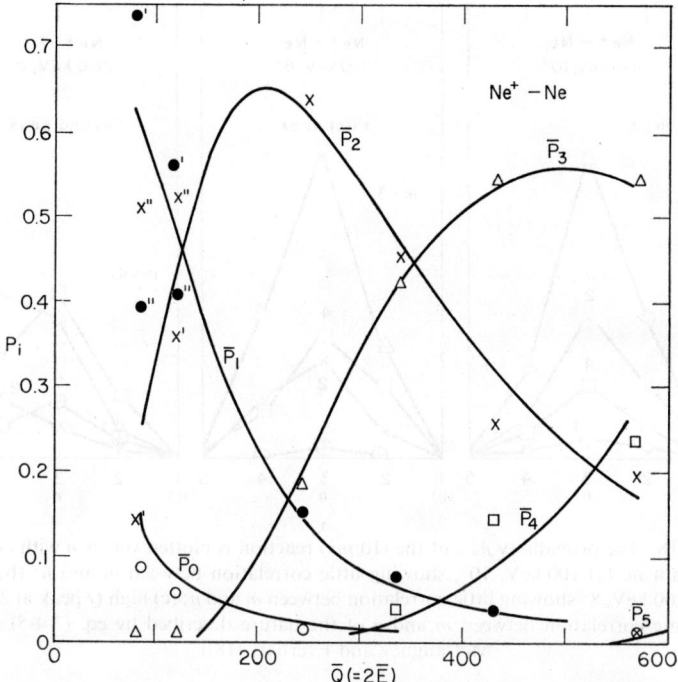

Fig. 7-4-17. Average ionization probabilities $\bar{P}_i$ are plotted versus average inelastic energy $\bar{Q}$. The two lower sets have primes to refer to scattered incident particles and double primes to indicate probabilities for the recoil target particle. The solid lines are computed through the use of a statistical model discussed in section 7-5-C (Kessel, McCaughey and Everhart [18]).

25 to 150 keV the $\bar{p}_{mn}$ matrices are symmetrical and exhibit almost no correlation. This is illustrated in Fig. 7-4-18a where $\bar{p}_{mn}$ values are plotted versus $n$ for the 100 keV, 10° data. For incident ion energies above 150 keV the collisions produce some events having an energy loss about 850 eV greater than the average loss. These events appear as a partly resolved second peak in the energy loss spectrum and the correlation between $m$ and $n$ for these events may be studied separately as was done for the three peaks found for 25 keV, 16° $Ar^+$–Ar collisions.

The $\bar{p}_{mn}$ values for this peak, obtained with 8° scattering at 200 keV, are plotted in Fig. 7-4-18c. The $m=1$ and $m=2$ curves of Fig. 7-4-18c have their maxima at $n=4$ while the $m=3$ and $m=4$ curves resemble those in Fig. 7-4-18b and have their maxima at $n=3$. For this to be completely analogous to the $Ar^+$–Ar structure the presence of a third peak in the energy loss

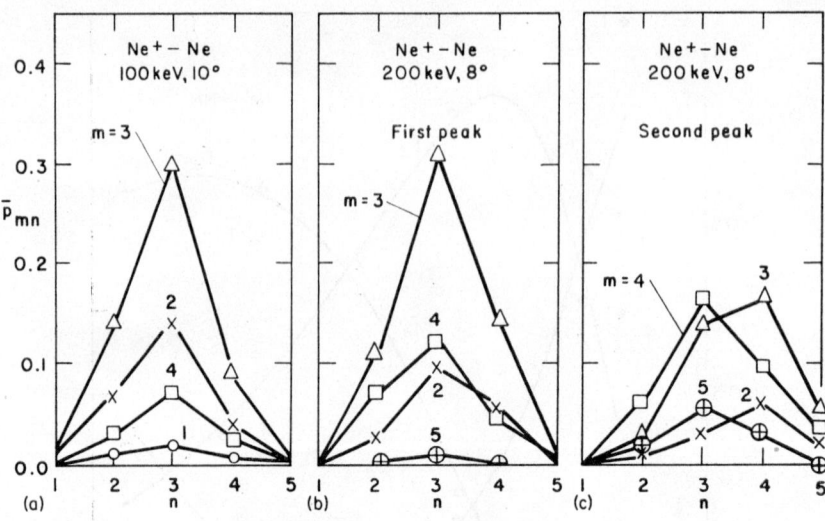

Fig. 7-4-18. The probability $\bar{p}_{mn}$ of the $(10/mn)$ reaction is plotted versus $n$ with contours of constant $m$. (a) 100 keV, 10°, showing little correlation between $m$ and $n$; (b) low $Q$ peak at 200 keV, 8° showing little correlation between $m$ and $n$; (c) high $Q$ peak at 200 keV, 8° showing correlation between $m$ and $n$ of the nature described by eq. (7-4-5) (Kessel, McCaughey and Everhart [18]).

spectrum would be necessary. It is possible that higher incident ion energies would excite a third peak, or this second peak might represent a wholly or partly new mode of excitation.

### C. KRYPTON–KRYPTON

The $Kr^+$–Kr combination has been investigated for 25 and 50 keV collisions with scattering to angles between 5 and 40 degrees by Afrosimov et al. [19]. Their data show several plateaus in the inelastic energy loss spectra for these collisions. Fig. 7-4-19 plots the excess energy loss $Q^*$ for the (10/33) event at 25 keV and the (10/44) event at 50 keV versus the distance of closest approach for these collisions. By adding 140 eV to the $Q^*$ values for the (10/33) events and 276 eV to the $Q^*$ values for the (10/44) events, the actual inelastic energy loss $\bar{Q}_{mm}$ would be obtained for each of the data points in this figure. The M and N shells of krypton have approximate radii of 0.25 Å and 0.80 Å, respectively; therefore the range of $r_a$ in Fig. 7-4-19 corresponds to the range for which the M shells of each atom begin to interpenetrate. The larger angles of scattering correspond to the smaller distances of closest approach, and this figure indicates that at 50 keV the (10/44) event may be

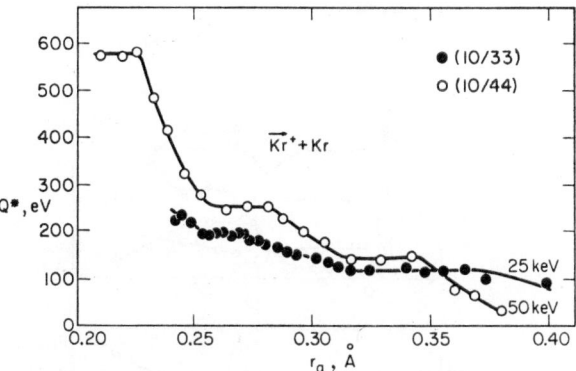

Fig. 7-4-19. The reduced inelastic energy loss $Q^*$ is plotted versus the distance of closest approach $r_a$, for the (10/33) event at 25 keV and the (10/44) event at 50 keV (Afrosimov et al. [19]).

the end result of collisions having widely different energy losses. For 5° scattering, the (10/44) event is produced in collisions having $Q^*$ values close to zero; while for 40° scattering, the (10/44) event has a $Q^*$ of nearly 600 eV. There is no multiple structure as is observed with the $Ar^+$–Ar and $Ne^+$–Ne combinations, but the plateaus in Fig. 7-4-19 do show the existence of levels that seem to become excited at certain distances of closest approach. The effect of these excitations on the final charge states of the scattered ions is shown in Fig. 7-4-20a where $P'_m$, the probability of the scattered ion having the charge $m$, is plotted versus the scattering angle $\vartheta$ for 25 keV collisions. These curves exhibit sudden changes for scattering angles of approximately 9, 23 and 38 degrees and have nearly constant values between these angles. Fig. 7-4-20b plots the average scattered charge $\bar{m}$ and the average excess energy loss $Q^*$ versus $\vartheta$ for 25 keV scattering. The curve for $\bar{m}$ reflects the sudden changes in $P'_m$ seen in Fig. 7-4-20a, and the excess energy loss $Q^*$ exhibits similar variations with $\vartheta$. Fig. 7-4-20c shows $Q^*$ and $\bar{m}$ plotted versus $\vartheta$ for 50 keV scattering; these curves show three distinct plateaus. The sharp increases preceding these plateaus occur for the same distances of closest approach as do the increases in the 25 keV data.

The rise in the inelastic energy loss found for 25 keV scattering near 9° was studied further by McCaughey et al. [20] and their data are shown in Fig. 7-4-21. In (a) the change of $\bar{m}$ with angle for the 25 keV $Kr^+$–Kr data is shown to be similar to that found for $Ar^+$–Ar at the same energy. The inelastic energy loss profiles for the (10/11), (10/22) and (10/33) events are shown in Figs. 7-4-21b, c and d for 8°, 9° and 10° scattering. The heights of

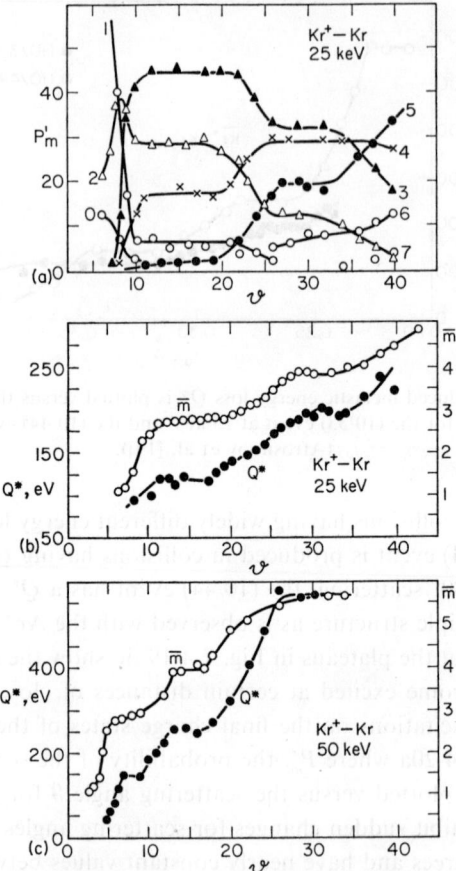

Fig. 7-4-20. (a) The charge state probabilities $P'_m$, of the scattered ion versus the scattering angle $\vartheta$ for 25 keV, Kr$^+$–Kr collisions. (b) The reduced inelastic energy $Q^*$ and the average scattered charge $\bar{m}$ plotted versus $\vartheta$ for 25 keV collisions; (c) same, for 50 keV collisions (Afrosimov et al. [19]).

these peaks represent the probability $P_{nn}$ of the (10/$nn$) event occurring. There is no triple structure of the type found for the Ar$^+$–Ar combination with 25 keV, 16° scattering. There is, however, a rapid change in the probabilities of the different (10/$mn$) events occurring. At 8° the (01/11) event is most common, while for 10° scattering the (10/33) event is prevalent.

## D. Neon–Argon

The Ne$^+$–Ar collision was studied by Afrosimov et al. [19] for incident

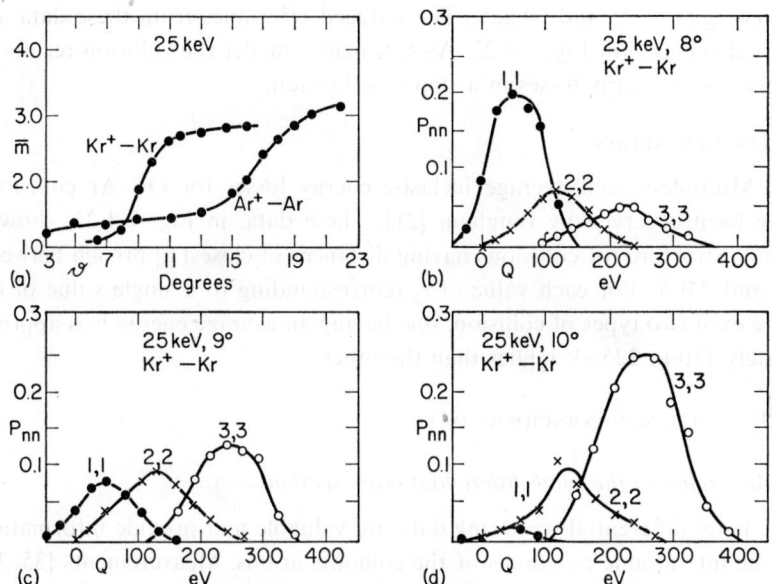

Fig. 7-4-21. (a) Average charge state $\bar{m}$ for $Kr^+$–$Kr$ and $Ar^+$–$Ar$ collisions at 25 keV as a function of scattering angle $\vartheta$. (b) Relative charge state populations after scattering for the $Kr^+$–$Kr$ collision as a function of inelastic energy $Q$, for a scattering angle of 8°; (c) same for 9°; (d) same for 10° (McCaughey et al. [20]).

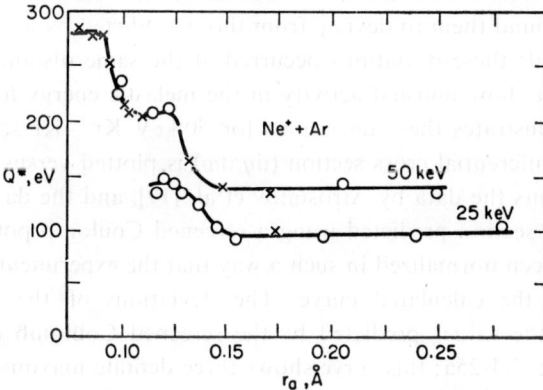

Fig. 7-4-22. The reduced inelastic energy loss $Q^*$ is plotted versus the distance of closest approach $r_a$, for 25 and 50 keV $Ne^+$–$Ar$ collisions. The circles represent $Q^*$ for the (10/23) events and the crosses represent an average of the $Q^*$'s for the (10/12), (10/23), (10/33) and the (10/43) events (Afrosimov et al. [19]).

ion energies of 25 and 50 keV. The reduced $Q^*$ values from these data are plotted versus $r_a$ in Fig. 7-4-22. As $r_a$ is made smaller the collision results in larger excess energy losses in a step-wise fashion.

### E. Oxygen–Argon

Multiple-valued average inelastic energy losses for $O^+$–Ar collisions have been observed by Bingham [21]. These data, in Fig. 7-4-23, show a double structure for collisions having distances of closest approach between 1.0 and 2.0 Å. For each value of $r_a$ (corresponding to a single value of $\vartheta$), there exist two types of collision, one having an average energy loss approximately 110 to 225 eV higher than the other.

### F. Related, non-coincident data

#### 1. *Anomalies in the total differential cross section*

Total differential scattering data are valuable and provide information on the interatomic potentials of the colliding atoms. Measurements [35–37] of the total differential cross section have shown that the interatomic potential for heavy ion collisions can be approximated by a screened Coulomb potential [32, 38]. Such a potential, of the form $(Ce^2/r)\exp(-r/a)$, does not take shell structure into account, yet shell structure can be expected to have an effect in these violent collisions where the nucleus of the incident ion passes through one or more shells of the target atom. Afrosimov et al. [19] measured the total differential cross sections for $Kr^+$–Kr collisions at 25 and 50 keV and found them to deviate from those predicted by a screened Coulomb potential; these deviations occurred at the same distances of closest approach that show unusual activity in the inelastic energy loss spectrum. Fig. 7-4-24 illustrates these deviations for 50 keV $Kr^+$–Kr scattering; the log of total differential cross section $(dq/d\omega)$ is plotted versus $\vartheta$. The solid curve represents the data by Afrosimov et al. [19], and the dashed line is a smooth cross section predicted using a screened Coulomb potential; these curves have been normalized in such a way that the experimental points fall on or above the calculated curve. The deviations of the experimental points from the values predicted by the screened Coulomb potential are plotted in Fig. 7-4-25a; this curve shows three definite maxima. The arrows indicate the scattering angles at which the inelastic energy loss for the collision increases sharply. Fig. 7-4-25b is a similar plot for the 25 keV data, and like behavior is found for the $Ar^+$–Ar and $Ar^+$–Ne ion–atom collisions.

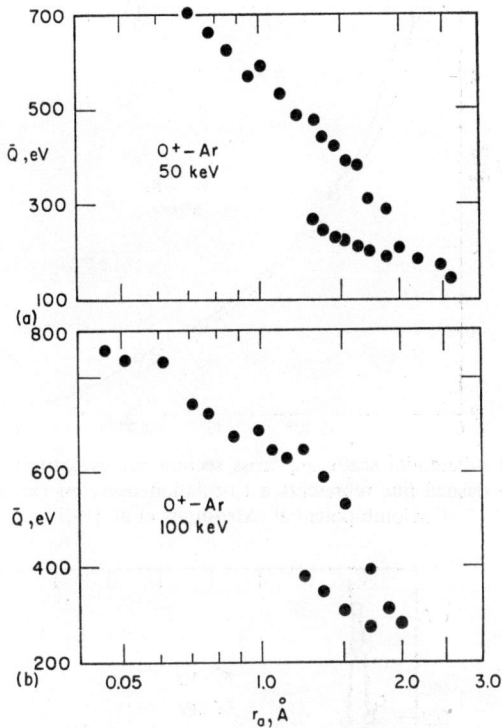

Fig. 7-4-23. The average inelastic energy $\bar{Q}$ is plotted versus the distance of closest approach for O$^+$–Ar collisions. These data show the existence of two values of the energy loss occurring for single values of $r_a$ between 1.0 and 2.0 Å. (a) 50 keV; (b) 100 keV (Bingham [21]).

## 2. Energies of ionized electrons

The inelastic energy loss measured for an ion–atom collision is the sum of several energy losses that may be considered separately. The contributions made by photon emission and the excitation of metastable states are thought to be minor; most of the energy loss results through ionization. Spectroscopic ionization energies may be used to calculate the change in potential energy when the final charge states are known. Therefore the measurement of the inelastic energy loss for the $(10/mn)$ event allows one to estimate (see eq. (7-2-1)) the sum of the final kinetic energies of the ionized electrons. Unfortunately the measurement does not give any information as to the distribution of this energy among these free electrons.

The energy spectra of electrons ejected from these heavy ion collisions have been the subject of several investigations [29, 39, 40]. Such an experi-

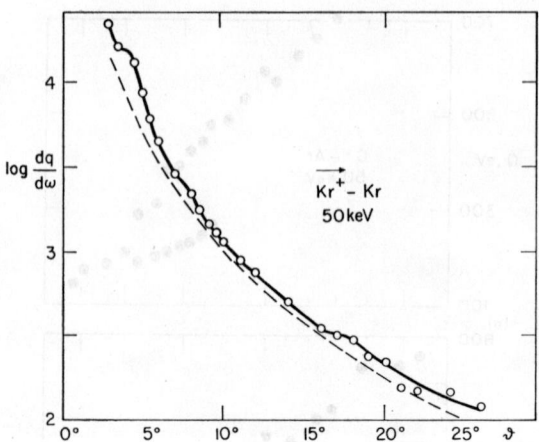

Fig. 7-4-24. Total differential scattering cross section versus scattering angle, Kr⁺–Kr, $T_0 = 50$ keV. The dashed line represents a calculation using an exponentially screened Coulomb potential (Afrosimov et al. [19]).

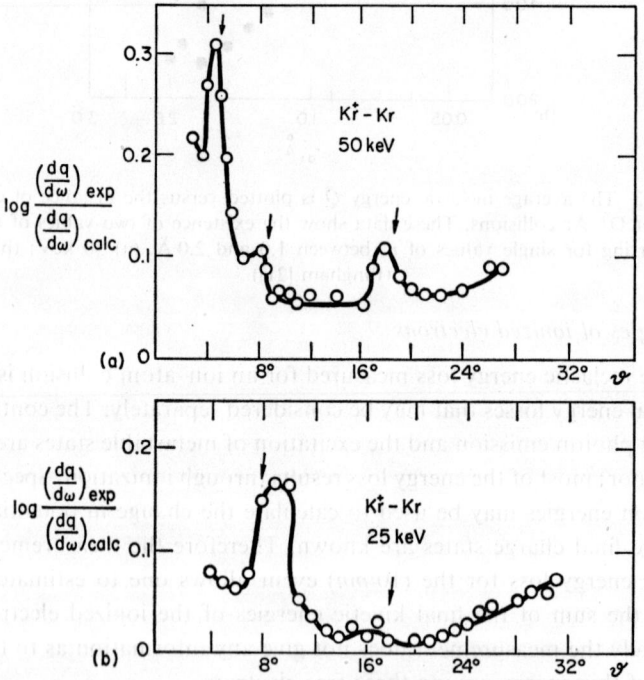

Fig. 7-4-25. Deviation of the total differential cross section from that calculated using a screened Coulomb potential, as a function of the scattering angle for Kr⁺–Kr collisions, (a) 50 keV; (b) 25 keV (Afrosimov et al. [19]).

ment consists of passing an ion beam through a gaseous target and measuring the energies of the electrons ejected from the resulting collisions. Such a measurement is "total" with respect to the ion–atom collision and not differential; the electron spectra represent the sum of all electrons being removed from all the collisions. The measurement of an electron's energy, by itself, does not indicate if it came from a gentle collision having a large impact parameter or a violent, large angle collision having a small impact parameter. In spite of this, these measurements give considerable insight into heavy ion collision processes. Of particular interest are the investigations by Rudd and co-workers [29] on the electron energy spectrum resulting from the $Ar^+$–Ar collisions. Fig. 7-4-26 shows the absolute differential (differential with respect to electron energy) cross sections for the ejection of electrons from 100 keV $H^+$–Ar and $Ar^+$–Ar collisions. These curves indicate the existence of a continuum of electron energies upon which is superimposed a certain amount of structure. An important question concerns the nature of this continuum. Does an ion–atom collision, having a fixed impact parameter, eject only electrons with certain, discrete energies or does it eject electrons having a continuous spectrum of energies? If the former were true, the continuum observed might be the result of detecting electrons from collisions having a continuous range of impact parameters. If the latter were true, one would not expect the structure seen in Fig. 7-4-26. The actual situation probably falls between these extremes, with some groups of electron energies being sharply defined and others less well defined. Only an electron–ion coincidence experiment can answer this question; such an experiment would study the electron energies resulting from ion–atom collisions having a fixed impact parameter.

The $Ar^+$–Ar structure between 120 and 220 eV in Fig. 7-4-26 is an example of a well defined peak in the energy distribution whose presence is attributable to the relatively few (less than 2 percent) [41] 100 keV $Ar^+$–Ar collisions whose distances of closest approach are less than 0.24 Å (this corresponds to collisions where the incident ions are scattered to angles greater than 4°). Fig. 7-4-6 shows that $Ar^+$–Ar collisions with values of $r_a$ greater than 0.24 Å have energy losses of less than 100 eV. Obviously 200 eV electrons cannot result from these collisions; but must result from the rare, large angle collisions that correspond to values of $r_a$ of less than 0.24 Å and whose energy losses exceed 200 eV. A model (section 7-5-B) formulated by Fano and Lichten [26] to explain the triple energy loss structure in $Ar^+$–Ar collisions does predict that these large angle collisions will eject Auger electrons having energies of about 200 eV. Rudd and co-workers [29] have also

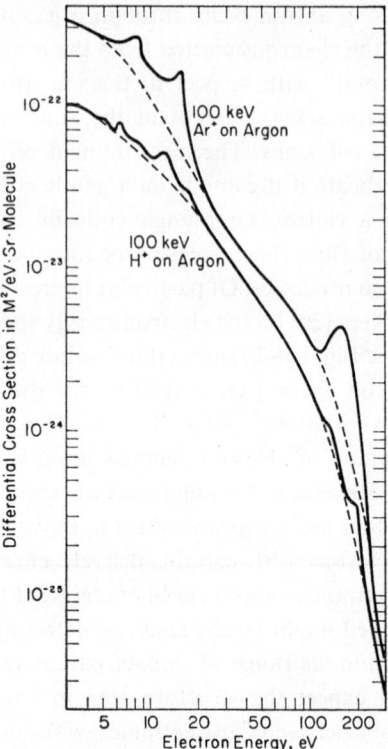

Fig. 7-4-26. Doubly-differential cross sections for ejection of electrons from argon by protons and argon ions at 100 keV. Electrons are detected at 160° from the beam direction (Rudd et al. [29]).

investigated electron energy spectra resulting from 200 keV and 300 keV $Ar^+$–Ar collisions and have resolved three peaks in the electron spectra between 160 and 200 eV. A peak for 181 eV electrons predominates the 100 keV high energy spectrum; this peak is less prominent following the 200 keV collisions and nearly vanishes at 300 keV. A lower peak at 170 eV predominates the 300 keV spectrum. Other peaks, not so clearly resolved, are thought to exist throughout the range of electron energies from 120 to 220 eV. The peaks corresponding to energies between 120 and 181 appear to be due to Auger transitions from multivacancy states. The high energy electron spectra resulting from $H^+$–Ar collisions are not so complicated. Following 200 keV $H^+$–Ar collisions, four high energy peaks are clearly resolved and identified as Auger transitions from specific single-vacancy states.

Auger electrons having about 750 eV energy have been detected following 200 keV $Ne^+$–Ne collisions [18, 42]. As with the $Ar^+$–Ar collisions, the energy removed by these electrons results in discontinuities in the inelastic energy loss spectrum (Fig. 7-4-16) [17, 18].

## G. SUMMARY

The inelastic energy losses measured with these coincidence experiments vary from about 20 eV to 6000 eV, and a wide range of phenomena occur with this latitude. These experiments have generally used noble gas ions and show many similarities. The most striking of these similarities is the presence of structure in all the inelastic energy loss spectra. This structure seems to be related to the shell structure of the colliding atoms: it appears only in collisions whose distances of closest approach allow the inner shells of these atoms to interpenetrate. The multiple-valued losses found with the $Ar^+$–Ar and $Ne^+$–Ne combinations have been associated with the ejection of fast electrons following the collisions. The structure (though not necessarily multiple) found with the $Kr^+$–Kr, $Ne^+$–Ar and $O^+$–Ar combinations may, or may not, be similar in nature.

The correlation between the final charge states $m$ and $n$ for these collisions is complex and the nature of these correlations is often unclear. The 50 keV, 10° $Ne^+$–Ne data show no correlation between $m$ and $n$; the distribution of charges is the same for each ion and the final charge on one ion is independent of the final charge on the other ion. To a first approximation the 50 keV, 15° $Ar^+$–Ar data are also uncorrelated; however, a careful analysis shows a slight tendency for an ion with high charge to have a collision partner with a lower charge state. The 25 keV, 16° $Ar^+$–Ar collisions exhibit a type of correlation that may be approximated by eq. (7-4-5); this type of correlation was also observed in the 200 keV, 8° $Ne^+$–Ne data. The correlation data for 6.4 keV, 10° $Ne^+$–Ne scattering are simpler in nature: the (10/11) and (10/12) reactions clearly represent the preferred channels of ionization.

## § 7-5. Models concerning ionization in heavy ion collisions

The coincidence scattering data have stimulated discussion about violent, heavy ion collisions during which the electron shells are forced to interpenetrate thoroughly. Little is known about the actual mechanisms that give rise to ionization during the collisions of heavy ions. As the body of data (not just coincidence data) describing these collisions has grown, certain concepts seem less valid than initially; however, insight may be gained by discussing the development of these ideas. It seems fitting that a case history of coincidence experimentation should indicate how certain data suggested certain ideas. The first coincidence data, obtained by Afrosimov et al. [12], concerned the triple energy loss structure found with $Ar^+$–Ar collisions having distances of closest approach of about 0.23 Å; they interpreted their data to indicate the presence of levels of collective electron oscillations which become excited during the collision. Amusia further refined this approach by calculating energies that might be associated with collective oscillations of the electron shells of the colliding atoms [28]. The $Ar^+$–Ar region of triple structure was also studied by Kessel, Russek and Everhart [16], whose results seem to verify a model suggested by Fano and Lichten [26], who assume the structure can be explained on the basis of molecular orbitals and energy level crossings. Before coincidence experiments were undertaken, a statistical model for ionization in heavy ion collisions was proposed by Russek [25]. These experiments have verified the validity of a statistical approach in describing some of the data, this is demonstrated by a phenomenological model developed by Everhart and Kessel [14] that is based, in part, on Russek's assumptions.

## A. Collective Oscillations

Collective oscillations, or plasma oscillations, of electron shells in atoms were discussed by Bloch [43] in 1933. A plasma oscillation in an atomic shell might occur if the shell is distorted during an ion–atom collision. If the electron density of a shell is momentarily made to deviate from its equilibrium value, Coulomb forces will act to restore equilibrium and oscillations may result. If such an oscillation were produced in the inner shell of an atom, outer shell electrons would be in the field of this oscillation; in this way energy of oscillation might be transferred to the outer electrons and result in ionization. Kirzhnits and Lozovik [44], in reviewing the possibility of plasma oscillations, note that in considering these oscillations one must evaluate first, the probability of exciting these plasma oscillations and second, the characteristics of these oscillations, such as their natural frequencies, their damping, and their mode of oscillation (radial, or dipole, etc.). These properties are not well known, nor are they easy to calculate; quantitative comparison with atomic scattering data is not possible at present. However, qualitative comparisons have shown [28, 45] that the multiple-valued energy loss data of the $Ar^+$–Ar collisions might be consistent with such a theory.

Afrosimov et al. [12] point out the following features of their $Ar^+$–Ar data, which were taken in the narrow range of $r_a$ from 0.18 Å to 0.33 Å:

(1) The inelastic energy losses can have several discrete values for the same (10/*mn*) event. (2) The excess energy loss, i.e. the total energy loss minus the loss due to the change in ionization potentials, has discrete, characteristic values. These values are independent of the (10/*mn*) reaction chosen, the impact parameter, the distance of closest approach, or the relative velocity of the ions. (3) These lines in the spectrum of characteristic excess energy losses are relatively narrow. On the basis of these observations, they suggest that the characteristic excess energy losses are due to the colliding particles having a system of narrow, discrete excitation levels situated in a continuous energy spectrum; and that the excitation of one of these levels is followed by subsequent autoionizing transitions. They assume that such levels are related to collective excitations of electrons.

### B. Energy level crossings of molecular wave functions

An alternative to collective oscillations is suggested by Fano and Lichten [26], who use phenomena familiar to atomic physics to explain the triple structure found with $Ar^+$–Ar collisions. With the use of Born-Oppenheimer type molecular orbitals and the Landau-Zener theory of level crossings, they suggest that the second and third peaks result from the emission, respectively, of one or two 200 eV Auger electrons. This hypothesis was originally rejected by Afrosimov et al. [12], who reasoned that if the third peak ($Q^{III}$) was the result of two Auger transitions, at least four electrons would have to be removed in the process; the presence of (10/22) and (10/13) events in the $Q^{III}$ peak (Table 7-4-1) seemed to exclude this possibility. By reintroducing an older idea about promotion of molecular orbitals [46, 47], Fano and Lichten have circumvented this difficulty*.

Lichten [27] has published a full account of this technique and interprets both the $Ar^+$–Ar and $Ne^+$–Ne collisions within the framework of molecular orbital theory. Lichten views the collision as a molecule living through a single, aperiodic vibration whose ionization is governed by the crossing of energy levels that occur during the collision. Figures 7-5-1 and 7-5-2 show semi-quantitative molecular orbitals for the Ar–Ar and Ne–Ne systems. In Fig. 7-5-1 the energies corresponding to the K, L and M shells of two separated argon atoms are indicated. The argon molecule, as $r$ goes to zero,

---

* Afrosimov et al. [19] continue to reject the idea of promotion, and point to the existence (even though rare) of events of the (10/1*n*) type in the $Q^{III}$ peak (Table 7-4-1). However, the presence of these events can be attributed to charge transfer and promotion both occurring. There is enough charge transfer during the collision to make the initial one electron asymmetry of the collision undetectable in the final charge states.

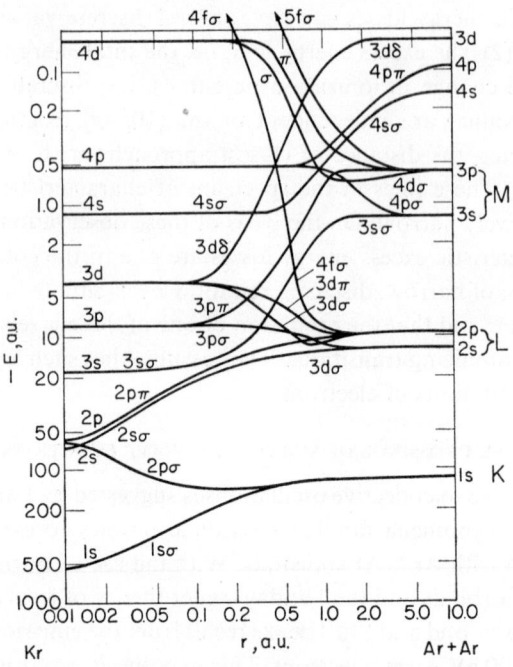

Fig. 7-5-1. Semi-quantitative energy levels of diabatic ($H_2^+$-like) molecular orbitals of the Ar–Ar system (Lichten [27]).

resembles the krypton atom. During a collision between two atoms the internuclear distance $r$ varies, first from infinity to a minimum distance $r_a$ (the distance of closest approach), and then back to infinity. The ionization phenomena occurring during and after the collision are governed primarily by $r_a$; this quantity controls the maximum degree of distortion in the atomic energy levels. There are numerous energy level crossings, and the behavior of an electron at an energy level crossing is governed by several factors. Electron penetration effects, within the framework of an independent electron molecular orbital model, can cause transitions between diabatic ($H_2^+$-like) molecular orbitals of the same parity (s↔s, s↔d, p↔p, p↔f) and of equal orbital angular momentum (σ↔σ, π↔π). Additionally, the effects of nuclear motion and the rotation of the internuclear axis can allow transitions between orbitals of the same parity and unequal orbital angular momentum. At relatively slow collision velocities, configuration interaction may allow transitions between some orbitals of different parity in an outer shell. In general, Lichten notes that: (1) energy levels from the same atomic shell and

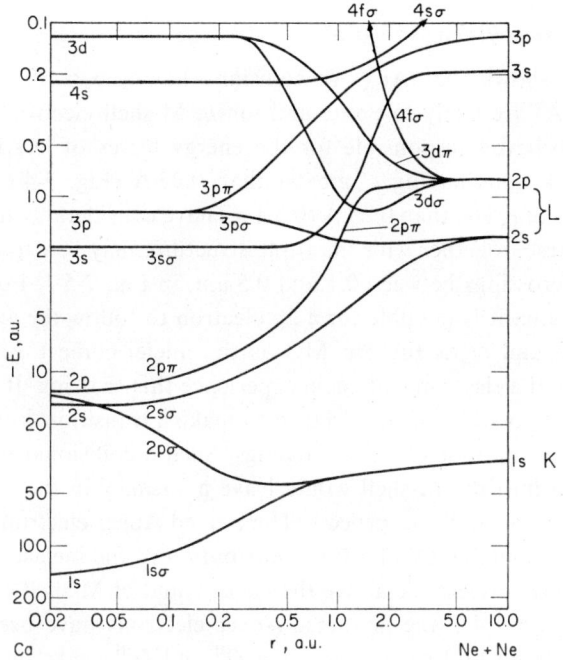

Fig. 7-5-2. Semi-quantitative energy levels of diabatic ($H_2^+$-like) molecular orbitals of the Ne–Ne system (Lichten [27]).

of comparable energy before collision can be widely separated after promotion, (2) there are many crossings of energy levels, and (3) transitions between molecular orbitals at crossings can leave a promoted electron stranded in a higher level after the collision. The separated atoms may rid themselves of this excitation energy through several autoionizing processes, and these processes fall into two general groups. Autoionizing transitions taking place within a single shell occur quickly; in fact, multiple excitations in the outer shells can have decay times that are barely longer than the collision time for 25 keV $Ar^+$–Ar collisions. Ionizations from these transitions will occur immediately after, and possibly even during, the collision. Autoionizing transitions between shells takes longer, typically about $10^{-14}$ or $10^{-15}$ seconds; and presumably occur from excited, discrete ionic states after the atoms are well separated. The energies of the Auger electrons ejected during these transitions may depend upon the ionization state of the ion at the time the inner shell vacancy is filled.

## 1. The $Ar^+-Ar$ collision

Lichten shows that $Ar^+$–Ar collisions having $r_a \approx 1.0$ atomic units (1 a.u. $\approx 0.5$ Å) are likely to excite and ionize M-shell electrons. Such excitations are believed responsible for the energy losses of less than 100 eV found for collisions having $r_a$ greater than 0.25 Å (Fig. 7-4-6). Collisions having values of $r_a$ less than 0.25 Å (0.5 a.u.) have considerably higher energy losses and these, together with the triple structure, may be explained by the energy level crossings between 0.1 and 0.5 a.u. in Fig. 7-5-1. For a collision with $r_a \approx 0.5$ a.u., it is possible for a 2p electron to follow the $4f\sigma$ molecular orbital (MO) and cross the $3s\sigma$ MO as the nuclei come together; as the nuclei recede, the electron will again experience this crossing. If the $3s\sigma$ MO is unfilled it is possible for the electron to make a transition from the $4f\sigma$ to the $3s\sigma$ MO during one of these crossings. Such a collisional promotion of a 2p electron into the M-shell would leave a vacancy in the L-shell which would be filled by an Auger process. The ejected Auger electron would have an energy of about 200 eV and this contribution to the inelastic energy loss would be clearly discernible above the background of M-shell excitation; as noted in section 7-4-F the predicted Auger electrons have been observed. Fano and Lichten [26] suggest that the $Q^{II}$ and $Q^{III}$ peaks in the inelastic energy loss structure are due, respectively, to the promotion of one or two $4f\sigma$ electrons.

The additional rise in the inelastic energy loss found between 0.10 and 0.20 a.u. (0.05 and 0.10 Å in Fig. 7-4-6) may be explained by the promotion of $3d\sigma$ and $3d\pi$ electrons to the $3d\delta$ state or $3p\sigma$ electrons to the $3p\pi$ state. Such interpenetration would also allow the $4f\sigma$ level to cross several more M-shell levels. The net effect of these many different promotion possibilities would be a general increase in the inelastic energy loss of the collision, and the individual promotions would not necessarily be resolved as separate peaks in the inelastic energy loss spectrum. For values of $r$ near 0.01 a.u. the K-shell $2p\sigma$ state crosses the $2s\sigma$ and $2p\pi$ states. Ordinarily one would not expect a K-shell promotion, because both the $2s\sigma$ and the $2p\pi$ states are already filled with electrons. Additionally the $2p\sigma$ to $2s\sigma$ transition is unlikely because of their difference in parity. However, for MeV energy collisions the relative velocity of the atoms is greater than one atomic unit and some of the assumptions made by Lichten are less valid at these velocities. In particular, the energy uncertainty at these velocities might allow an electron transition from the $2p\pi$ level to another level, as the nuclei come together. In this case a $2p\sigma$ electron could cross over into the $2p\pi$ level; such a promotion of a K-shell electron would result in approximately a 3000 eV increase in the inelastic energy loss. Losses of this order are reported by Kessel [22] for 1.5 MeV $Ar^+$–Ar collisions having distances of closest approach of 0.02 to 0.04 a.u. (0.01 to 0.02 Å in Fig. 7-4-6).

## 2. The $Ne^+$–Ne collision

Except for changes in scale, the energy levels for the $Ne_2$ molecule shown in Fig. 7-5-2 are similar to those for the inner K and L-shells of the $Ar_2$ molecule. The same processes that were responsible for L-shell promotion in the $Ar_2$ molecule may excite the outer L-shells of the Ne atoms. There is a good possibility of promoting a K-shell electron during a $Ne^+$–Ne collision; for in this case, the single ionization of one of the collision partners means there may be a vacancy in the $2p\pi$ state. If there is such a vacancy, rotational excitation would allow the $2p\sigma \rightarrow 2p\pi$ promotion to occur. The energy required for this promotion would be the difference in energy between the 2p and the 1s levels, or about 850 eV, and the kinetic energy of the resulting Auger electron would be about 750 eV. Table 7-4-5 shows that where a second energy loss peak is resolved in 200 keV $Ne^+$–Ne collisions, it is about 850 eV higher than the lower peak. As noted in section 7-4-F, 750 eV electrons have been detected following these collisions.

For a collision of the type $Ne^{++}$–Ne there would be twice as many p-shell electrons missing; the 1s electrons should find twice as many vacancies in the $2p\pi$ MO, and twice as many Auger electrons should be emitted. Conversely for a ground state Ne–Ne collision there will be no vacancies in the $2p\pi$ MO and there should be no promotion and no Auger electrons. McCaughey et al. [20] tested this prediction and did find twice as many 750 eV electrons following $Ne^{++}$–Ne collisions as they found following $Ne^+$–Ne collisions. However, following Ne–Ne collisions they also detected some 750 eV electrons. The presence of these electrons could be due to metastable atoms in the incident Ne beam; or a $2p\pi$ vacancy might, at these velocities, be produced by the same means as was discussed for the $Ar^+$–Ar collisions.

## C. Statistical Models for Ionization

As indicated by the Ar–Ar and Ne–Ne energy level diagrams of Lichten, there are many level crossings and countless ways by which an atom may accept varying amounts of excitation energy. Furthermore, following a collision, an excited atom may rid itself of this energy through many channels. The complexity of the situation lends itself to statistical analysis. A statistical model for collisional excitation was developed by Russek [25] before coincidence experiments were undertaken, and he viewed the collision as a two step process. First, when two atoms suffer a violent atomic collision, each atom or ion receives an excitation energy which is a reasonably well defined function of the collision parameters; and second, after separation of the colliding atoms, each atom rids itself of most of this energy through autoionization. Russek further assumed that an atom's excitation energy would be statistically distributed among the outer electrons. Using these assumptions, ionization probability curves, agreeing well with experiment and similar in nature to those in Figures 7-4-8 and 7-4-17, may be derived. This statistical theory was developed before $\approx 200$ eV electrons were observed following $Ar^+$–Ar collisions. This recent finding requires a modification of the idea that the excitation energy is statistically distributed among the outer electrons all the time; for in this case, one of the outer electrons is receiving an energy considerably greater than the average. Improved ionization probability curves may be derived by taking this into account [48].

The ionization of an ion suffering a collision must depend largely upon the excitation energy received by that atom. A coincidence experiment measures the average inelastic energy loss for the collision as a whole, i.e. the sum of the excitation energies of both collision partners. Everhart and Kessel [14] consider various ways of distributing the total energy loss between two atoms, and reject the idea that each atom always receives just half of the total energy loss. Instead they assume that even for fixed collision conditions an individual atom may receive varying amounts of excitation energy, and that independently, its collision partner will do the same. By fitting the widths of energy distributions to the data and assuming that an atom receiving an excitation energy $E$ will have a probability $P_i(E)$ of losing $i$ electrons (a concept introduced by Russek [25]) the actual values for $P_i(E)$ may be calculated. The solid curves in Fig. 7-4-17 are the result of such a calculation. The values of $\bar{Q}_{mn}$ and $\bar{p}_{mn}$ for the $Ar^+$–Ar and $Ne^+$–Ne collisions predicted by this phenomenological model agree well with experimental data [14, 18].

### § 7-6. Additional coincidence measurements

The data reported in § 7-4 have contributed to our understanding of violent heavy ion collisions, but they represent only the first of many possible experiments. The differential scattering measurements primarily used symmetric ion–atom combinations; though slightly more difficult, more asymmetric combinations must be thoroughly investigated. The lighter ions represent another avenue of investigation; collision combinations such as $He^+$–He, $He^{++}$–He and $H^+$–He are two and three electron systems which are more amenable to present theoretical techniques than are the heavy ion systems. Ten and hundred keV collisions of heavy ions show great complexity in their charge state spectra; for energies below 5 keV, fewer electrons are ionized and it may be possible to observe the thresholds for certain ionization processes. The first high energy (MeV) coincidence measurements [22] indicate the possibility of observing energy losses greater than 6 keV in single collisions. Interesting as these differential scattering experiments have already proven to be, they represent only the rare, large angle scattering events. In 1967 Afrosimov and co-workers [23, 24] reported the first coincidence measurements of total scattering cross sections. Another difficult, but very important, experiment that is being considered is an electron–ion coincidence experiment where electron–ion coincidences will be sought following large angle ion-atom collisions.

## A. Total Scattering Cross Sections

Non-coincident total electron capture cross section measurements are unable to discriminate between electron capture and electron capture plus ionization of the target atom. Afrosimov and co-workers [23, 24] have constructed a coincidence scattering apparatus capable of distinguishing between electron capture, ionization of either the projectile or target atom, or any combination of these processes. They have used this apparatus to measure the total cross sections for 5 to 50 keV, $H^+$–Ar collisions for $m=0$, $\pm 1$ and $1 \leq n \leq 3$. The apparatus used for the $H^+$–Ar study by Afrosimov and co-workers is similar in nature to their differential scattering coincidence chamber and is shown in Fig. 7-6-1. The incident ion beam B enters the target gas chamber $C_1$ after being collimated by slits $S_1$ and $S_2$. This beam, after passing through the target gas, continues through slit $S_3$ and is magnetically analyzed by analyzer $A_1$. The charged components of this beam are counted by detector $D_1$ and the neutral component by $D_2$. For incident proton energies of less than 100 keV, most recoiling argon ions will have energies of less than 1 eV. These low energy ions are extracted from the collision region by applying a voltage to the capacitor plate K. These extracted ions pass through slit $S_6$, analyzer $A_2$, and into detector $D_3$. The deflection of the incident beam by the field in K is taken into account by rotating $A_1$ about O, the center of capacitor K.

Afrosimov et al. [24] measured relative ($10/mn$) cross sections with this apparatus and normalized these results to absolute values of the total cross

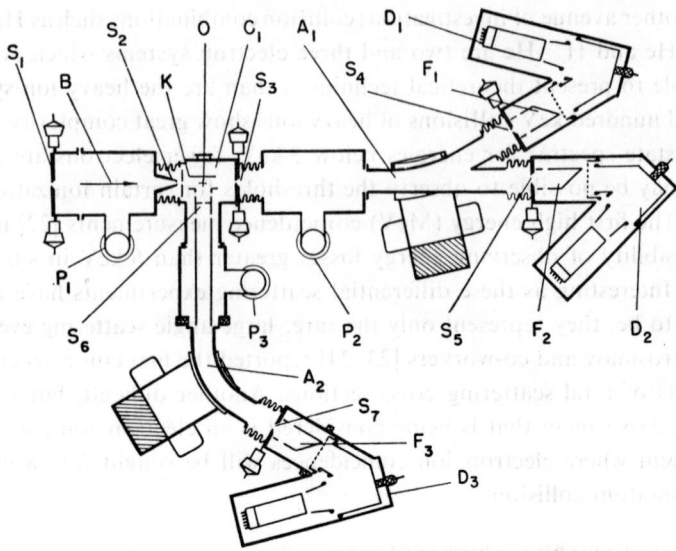

Fig. 7-6-1. Apparatus used by Afrosimov et al. [23] for coincidence measurements of total cross sections. B, incident ion beam; $S_1$–$S_7$, collimation slits; K, capacitor for the extraction of slow ions; O, axis of rotation for the analyzer of fast particles $A_1$ and the chamber C together with the slow ion analyzer $A_2$; $A_1$ and $A_2$, magnetic analyzers; $F_1$–$F_3$, detectors used to measure beam intensities; $D_1$–$D_3$, detectors of fast ions, fast atoms, and slow ions, respectively; $P_1$–$P_3$, high vacuum pumps.

section measured by non-coincident means. Fig. 7-6-2 shows the H$^+$–Ar cross sections for (10/0$n$) type events. At 5 keV electron capture by the incident ion, the (10/01) process, dominates the three cross sections measured. However, at higher energies additional ionization of the target atom becomes important; $_{10}q_{01}$ decreases monotonically and $_{10}q_{02}$ and $_{10}q_{03}$ increase, reaching their maxima between 20 and 40 keV. These data may also be plotted to show the individual components of the total cross sections for the formation of Ar$^+$, Ar$^{+2}$ and Ar$^{+3}$ as shown in Fig. 7-6-3. In (a) the processes producing Ar$^+$ ions at 5 keV are seen to differ from those at 50 keV. At 5 keV simple electron capture is the primary process producing Ar$^+$ ions while at 50 eV over half of the Ar$^+$ ions are produced through simple ionization of the target with no electron capture. Three processes are seen to contribute to the formation of Ar$^{+2}$ ions (Fig. 7-6-3b); most of these result from the (10/02) process, i.e. the combination process of electron capture plus ionization of the target; however, the (10/12) process, double ionization of the target atom with no electron capture, also contributes significantly to the formation of Ar$^{+2}$ ions. A smaller contribution is made by

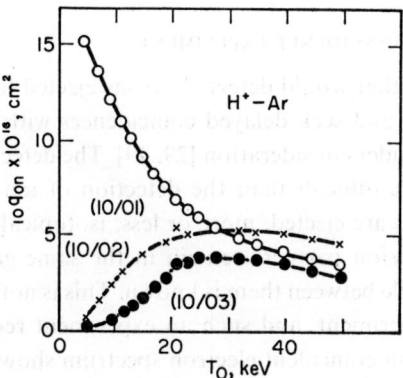

Fig. 7-6-2. Total cross section for elementary processes of the type (10/0n) related to the capture of one electron by a proton. (10/01), pure capture process; (10/02) and (10/03), capture with ionization (Afrosimov et al. [24]).

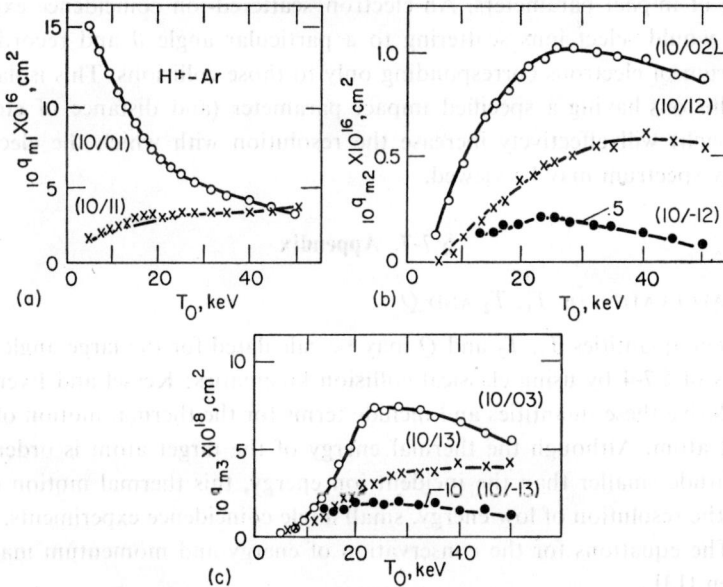

Fig. 7-6-3. Components of the total cross sections of events having a fixed recoil argon charge state following H+–Ar collisions. (a) Cross section for the (10/m1) reactions; (b) for the (10/m2) reactions; (c) for the (10/m3) reactions (Afrosimov et al. [24]).

the $(10/-12)$ process, which represents double electron capture by the proton. The cross sections in Fig. 7-6-3c for the creation of $Ar^{+3}$ all require the production of at least one free electron (ionization); but the cross sections still have the same general nature as do the $Ar^{+2}$ cross sections.

B. Electron–Ion Coincidence Experiment

An experiment that would detect electrons ejected following large angle ion–atom collisions and seek delayed coincidences with either the scattered or the recoil ion is under consideration [29, 49]. The detection of electron–ion coincidences is more difficult than the detection of ion–atom coincidences because the electrons are ejected, more or less, isotopically. In the ion–atom experiment the collision partners remain in the same geometric plane and the approximate angle between them is known. This is not true in an electron–ion coincidence experiment, and such an experiment requires a greater sophistication. The non-coincident electron spectrum shown in Fig. 7-4-26 for $Ar^+$–Ar collisions represents a total cross section with respect to the ion–atom collisions; it is not the spectrum of $Ar^+$–Ar collisions having a specific impact parameter but the sum of electrons from collisions having a wide range of impact parameters. An electron–scattered-ion coincidence experiment would select ions scattering to a particular angle $\vartheta$ and record the spectrum of electrons corresponding only to those collisions. This isolation of collisions having a specified impact parameter (and distance of closest approach) will effectively increase the resolution with which the electron energy spectrum may be viewed.

## § 7-7. Appendix

A. Calculation of $T_1$, $T_2$ and $Q$

The quantities $T_1$, $T_2$ and $Q$ may be calculated for the large angle collisions of § 7-4 by using classical collision kinematics; Kessel and Everhart [13] derive these quantities and include terms for the thermal motion of the target atom. Although the thermal energy of the target atom is orders of magnitude smaller than the incident ion energy, this thermal motion does limit the resolution of low energy, small angle coincidence experiments.

The equations for the conservation of energy and momentum may be written [13]

$$T_0 = T_1 + T_2 + Q, \qquad (7\text{-}7\text{-}1)$$
$$(\gamma T_0)^{\frac{1}{2}}(1 + \varepsilon_x) = (\gamma T_1)^{\frac{1}{2}} \cos \vartheta + T_2^{\frac{1}{2}} \cos(\beta - \vartheta) \qquad (7\text{-}7\text{-}2)$$

and

$$(\gamma T_1)^{\frac{1}{2}} \sin \vartheta + \varepsilon_y (\gamma T_0)^{\frac{1}{2}} = T_2^{\frac{1}{2}} \sin(\beta - \vartheta). \qquad (7\text{-}7\text{-}3)$$

Here $\varepsilon_x$ is the ratio of the initial $x$ momentum of the target atom to the momentum of the incident particle and $\varepsilon_y$ refers to the $y$ momentum of the target. From (7-7-2) and (7-7-3) one solves for

$$(T_1/T_0)^{\frac{1}{2}} = [(1 + \varepsilon_x) \sin(\beta - \vartheta) - \varepsilon_y \cos(\beta - \vartheta)]/\sin\beta \qquad (7\text{-}7\text{-}4)$$

and

$$(T_2/T_0)^{\frac{1}{2}} = \gamma^{\frac{1}{2}} [(1 + \varepsilon_x) \sin\vartheta + \varepsilon_y \cos\vartheta]/\sin\beta. \qquad (7\text{-}7\text{-}5)$$

Setting $\varepsilon_x = \varepsilon_y = 0$ and squaring (7-7-4) and (7-7-5) leads to (7-2-3) and (7-2-4). Eq. (7-7-1) may now be written

$$Q(\beta) = T_0 - T_0 [\sin^2(\beta - \vartheta) + \gamma \sin^2\vartheta]/\sin^2\beta. \qquad (7\text{-}7\text{-}6)$$

Setting $Q(\beta) = 0$ one obtains

$$\begin{aligned}\sin^2(\beta_0 - \vartheta) + \gamma \sin^2\vartheta &= \sin^2\beta_0, \quad &\gamma \ne 1 \\ \beta_0 &= \tfrac{1}{2}\pi, \quad &\gamma = 1\end{aligned} \qquad (7\text{-}7\text{-}7)$$

where $\beta_0$ is the angle for which the recoil particles will emerge following an elastic collision.

Thermal motion in the target gas will have the same effect on both elastic and inelastic collisions, and this effect may be derived by considering only elastic collisions and assuming, for the moment, that the apparatus has infinitely narrow resolving power. Under these conditions one could perform an experiment where $\vartheta$ is held constant and $\beta$ is varied. Near $\beta = \beta_0$ one would find coincident events that could be plotted as in Fig. 7-3-4; only in this case, the width of the curve would be due solely to thermal effects. This thermal effect curve is gaussian and if one defines $\delta\beta_t$ as its half-width at $1/e$ of the peak height, (7-7-2) and (7-7-3) may be used to derive $\delta\beta_t$. The result for $\gamma = 1$ is

$$\delta\beta_t = 2(kt)^{\frac{1}{2}}/T_0^{\frac{1}{2}} \sin 2\vartheta \qquad (7\text{-}7\text{-}8)$$

where $k$ is Boltzmann's constant, and $t$ is the temperature measured in °K. The corresponding half-width in the inelastic energy loss is [13]

$$\delta Q_t = \delta\beta_t\, T_0 \sin 2\vartheta = 2(T_0 kt)^{\frac{1}{2}}. \qquad (7\text{-}7\text{-}9)$$

Another factor contributing to the finite width of data curves, such as those shown in Fig. 7-3-4, is the instrumental broadening $\delta Q_a$ which is due to the finite apertures of the apparatus. The apparatus of Kessel and Everhart has an instrumental half-width equal to 0.07 radian, which contributed a $\delta Q_a = (0.07) T_0 \sin 2\vartheta$ to the experimental half widths [13]. Their experimental half-width $\delta Q$ for average energy loss measurements and $\delta Q_{mn}$ for $(10/mn)$ energy loss measurements together with $\delta Q_t$ and $\delta Q_a$ are shown in Fig. 7-7-1. For low energy, small angle coincidence measurements, $\delta Q_{mn}$ is due almost entirely to thermal broadening; for higher energy, large angle collisions $\delta Q_a$, the instrumental half-width, is primarily responsible for the broadening. The average inelastic energy loss measurements represent a sum of all $(10/mn)$ type events for a particular $(T_0, \vartheta)$ collision, and therefore $\delta Q$ is larger than $\delta Q_{mn}$.

## B. SOME EXPERIMENTAL DETAILS

### 1. *Charge analysis of coincident ions*

There is a considerable spread in the energies of the recoil ions leaving the collision region at a given angle $\theta$. This is demonstrated in Fig. 7-7-2, which shows an analyzer spectrum (curve 1) for recoil $Ar^+$ ions at $\theta = 81°30'$

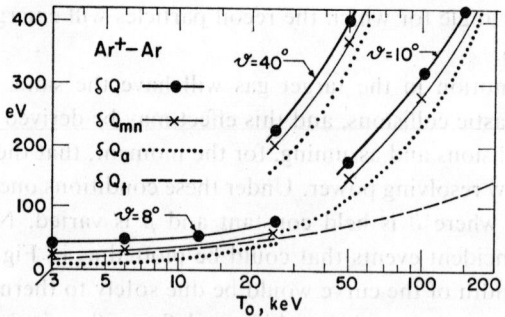

Fig. 7-7-1. Half-widths of the energy loss profiles plotted versus incident ion energy. Here $\delta Q_t$ and the contribution $\delta Q_a$ due to instrumental effects are also shown (Kessel and Everhart [13]).

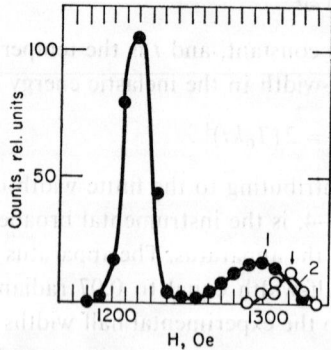

Fig. 7-7-2. Recoil analyzer spectrum for $Ar^+$ ions at $\theta = 81°30'$, following 50 keV $Ar^+$–Ar collisions. Curve 1 represents all $Ar^+$ ions at $\theta = 81°30'$ and curve 2 represents those found in coincidence with scattered $Ar^{+2}$ ions at $\vartheta = 7.5°$ (Afrosimov et al. [12]).

following 50 keV Ar$^+$–Ar collisions. The recoil ions fall into two general groups: the taller peak is due to "soft" scattering and the other to "hard" scattering [6]. Collision partners for these recoil ions may emerge from the collision over a range of scattering angles. Coincidences will not be observed with ions at a given angle $\beta$, unless the recoil analyzer is set to pass ions of the proper energy; a similar difficulty exists with setting the scattered ion analyzer. A coincidence experiment selects a certain subset of events, i.e. those events whose scattered ions pass through angle $\vartheta$ and whose recoil ions pass through the angle $\beta$. Equations (7-2-3) and (7-2-4):

$$T_1/T_0 = \sin^2(\beta - \vartheta)/\sin^2\beta \qquad (7\text{-}2\text{-}3)$$

$$T_2/T_0 = \gamma \sin^2\vartheta/\sin^2\beta \qquad (7\text{-}2\text{-}4)$$

give the appropriate energies for the ions in this subset. For the case of $\gamma = 1$ ($\beta \approx 90°$ and $\sin^2\beta \approx 1$), $T_2$ is a function only of the scattering angle $\vartheta$; so for a fixed value of $\vartheta$, $T_2$ may be considered constant while $\beta$ is varied over a small range of angles near 90°. However, (7-2-3) shows that as $\beta$ is changed, the energy $T_1$, of the coincident scattered particle, changes even though $\vartheta$ is being held constant. This variation is less than one percent for a one degree change in $\beta$ if $\vartheta$ is less than 15° and $\gamma = 1$. For $\gamma \neq 1$, the $\beta$ corresponding to large angle scattering may not be so near 90°; and, for a constant value of $\vartheta$, a change in $\beta$ will produce a greater change in both $T_1$ and $T_2$ than it will for a $\gamma = 1$ collision.

## 2. Coincidence resolving time

The resolving time of the coincidence circuit must meet two criteria. First, it must be short enough to avoid an excess number of random coincidences (the number of random coincidences is directly proportional to the length of the resolving time) [50]; and second, it must be longer than the spread in the transit times of the slow moving recoil ions. Fig. 7-7-2 shows that coincident recoil ions (curve 2) do have a finite energy spread; a very short coincidence resolving time would effectively select only a portion of these coincident events and might lead to fallacious results. An appropriate resolving time is best found by experiment, and Fig. 7-7-3 shows typical data, plotting the number of true coincidences as a function of delay time for three values of the resolving time, $\rho_1 < \rho_2 < \rho_3$, of the coincidence circuit. All three values of $\rho$ are suitable and result in curves of equal height. However, a resolving time less than $\rho_1$ would incorrectly discriminate against some events, and a resolving time greater than $\rho_3$ would result in a needless

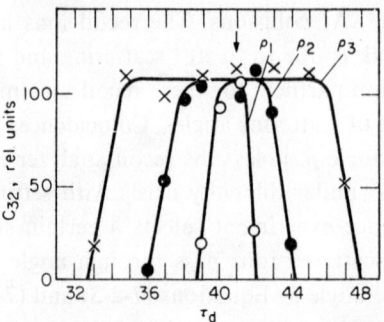

Fig. 7-7-3. The number of coincidences as a function of delay time for three values, $\rho_1 < \rho_2 < \rho_3$, of the coincidence resolving time. The events being observed are the (10/32) argon reactions for 50 keV, $\vartheta = 7.5°$ scattering with $\beta = 87.5°$ (Afrosimov et al. [12]).

number of random coincidences. The value of $\rho_2$, resulting in a narrow plateau, would be a good compromise in this case. Typically, resolving times from 0.1–0.6 μs have been found satisfactory [12, 13].

## Acknowledgements

The author is grateful to Professor Edgar Everhart for his warm interest and encouragement. High Voltage Engineering Corporation has very generously enabled the author to extend the range of coincidence measurements to higher ion energies by providing both a coincidence scattering apparatus and the use of a 3 MV tandem Van de Graaff accelerator.

## References

1. C. RAMSAUER and R. KOLLATH, Ann. Physik **16** (1933) 570.
2. C. A. FRISCHE, Phys. Rev. **43** (1933) 160.
3. A. G. ROUSE, Phys. Rev. **52** (1937) 1238.
4. N. V. FEDORENKO, Zh. Tekhn. Fiz. [Sov. Phys. – Tech. Phys.] **24** (1954) 784;
   D. M. KAMINKER and N. V. FEDORENKO, Zh. Tekhn. Fiz. [Sov. Phys. – Tech. Phys.] **25** (1955) 1843 2239;
   Additional references to atomic collisions before 1960 may be found in a review article by N. V. FEDORENKO, Usp. Fiz. Nauk **68** (1959) 481 [English transl.: Sov. Phys. – Usp. **2** (1959) 526].
5. E. EVERHART, R. J. CARBONE and G. STONE, Phys. Rev. **98** (1955) 1045;
   R. J. CARBONE, E. N. FULS and E. EVERHART, Phys. Rev. **102** (1956) 1524;
   E. N. FULS, P. R. JONES, F. P. ZIEMBA and E. EVERHART, Phys. Rev. **107** (1957) 704.
6. V. V. AFROSIMOV and N. V. FEDORENKO, Zh. Tekhn. Fiz. **27** (1957) 2557 [English transl.: Sov. Phys. – Tech. Phys. **2** (1957) 2391].
7. G. H. MORGAN and E. EVERHART, Phys. Rev. **128** (1962) 667.

8. J. B. HASTED, Penetration of Charged Particles in Matter, ed. E. A. Uehling (Publication 752, 1960, USA National Academy of Sciences, Nat. Res. Council) pp. 82–89.
9. J. S. ALLEN, Phys. Rev. **55** (1939) 966.
10. K. KOYAMA and R. E. CONNALLY, Rev. Sci. Instr. **28** (1957) 833.
11. V. V. AFROSIMOV, YU. S. GORDEEV, M. N. PANOV and N. V. FEDORENKO, Proc. Sixth Intern. Conf. on Ionization Phenomena in Gases, eds. P. Hubert and E. Cremieu-Alcau (Paris, 1963) pp. 111–115.
12. V. V. AFROSIMOV, YU. S. GORDEEV, M. N. PANOV and N. V. FEDORENKO, Zh. Tekhn. Fiz. **34** (1964) 1613, 1624, 1637 [English transl.: Soviet Phys. – Tech. Phys. **9** (1965) 1248, 1256, 1265].
13. Q. C. KESSEL and E. EVERHART, Phys. Rev. **146** (1966) 16.
14. E. EVERHART and Q. C. KESSEL, Phys. Rev. **146** (1966) 27.
15. E. EVERHART and Q. C. KESSEL, Phys. Rev. Letters **14** (1965) 247.
16. Q. C. KESSEL, A. RUSSEK and E. EVERHART, Phys. Rev. Letters **14** (1965) 484.
17. Q. C. KESSEL, M. P. MCCAUGHEY and E. EVERHART, Phys. Rev. Letters **16** (1966) 1189; **17** (1966) 1170.
18. Q. C. KESSEL, M. P. MCCAUGHEY and E. EVERHART, Phys. Rev. **153** (1967) 57.
19. V. V. AFROSIMOV, YU. S. GORDEEV, M. N. PANOV and N. V. FEDORENKO, Zh. Tekhn. Fiz. **36** (1966) 123 [English transl.: Soviet Phys. – Tech. Phys. **11** (1966) 89].
20. M. P. MCCAUGHEY, E. J. KNYSTAUTAS, H. C. HAYDEN and E. EVERHART, Phys. Rev. Letters **21** (1968) 65;
M. P. MCCAUGHEY, E. J. KNYSTAUTAS and E. EVERHART, Phys. Rev. **175** (1968) 14.
21. F. W. BINGHAM, Bull. Am. Phys. Soc. **13** (1968) 613.
22. Q. C. KESSEL, Bull. Am. Phys. Soc. **13** (1968) 68.
23. V. V. AFROSIMOV, YU. A. MAMAEV, M. N. PANOV, V. UROSHEVICH and N. V. FEDORENKO, Zh. Tekhn. Fiz. **37** (1967) 550 [English transl.: Soviet Phys. – Tech. Phys. **12** (1967) 394].
24. V. V. AFROSIMOV, YU. A. MAMAEV, M. N. PANOV and V. UROSHEVICH, Zh. Tekhn. Fiz. **37** (1967) 717 [English transl.: Soviet Phys. – Tech. Phys. **12** (1967) 512];
Additional total cross section, for the $H^+$–Xe collision, have been reported by V. V. AFROSIMOV, YU. A. A. MAMAEV, M. N. PANOV and N. V. FEDORENKO, Zh. Eksperim. i Teor. Fiz. **55** (1968) 97.
25. A. RUSSEK, Phys. Rev. **132** (1963) 246;
A. RUSSEK and M. T. THOMAS, Phys. Rev. **109** (1958) 2015; **114** (1959) 1538; **122** (1961) 506.
26. U. FANO and W. LICHTEN, Phys. Rev. Letters **14** (1965) 627.
27. W. LICHTEN, Phys. Rev. **164** (1967) 131.
28. M. Ya. AMUSIA, Phys. Letters **14** (1965) 36; Zh. Tekhn. Fiz. **36** (1966) 1409 [English transl.: Soviet Phys. – Tech. Phys. **11** (1967) 1053].
29. M. E. RUDD, T. JORGENSEN and D. J. VOLZ, Phys. Rev. **151** (1966) 28.
30. N. F. MOTT and H. S. W. MASSEY, The Theory of Atomic Collisions (Oxford University Press, London, second ed., 1952) pp. 124–126.
31. Q. C. KESSEL, Rev. Sci. Instr. **40** (1969) 68.
32. E. EVERHART, G. STONE and R. J. CARBONE, Phys. Rev. **99** (1955) 1287.
33. The complete set of tables has been deposited as Document No. 8798 with the ADI Auxiliary Publications Project, Photoduplication Service, Library of Congress, Washington, D. C. 20036. A copy may be secured by citing the Document number and by remitting $2.50 for photoprints. Advance payment is required. Make checks or money orders payable to Chief, Photoduplication Service, Library of Congress.
34. V. V. AFROSIMOV, YU. S. GORDEEV, A. M. POLYANSKII and A. P. SHERGIN, Zh. ETF Pis. Red. **6** (1967) 461 [English transl.: JETP Letters **6** (1967) 3].

35. D. M. Kaminker and N. V. Fedorenko, Zh. Tekhn. Fiz. **24** (1955) 2239.
36. E. N. Fuls, P. R. Jones, F. P. Ziemba and E. Everhart, Phys. Rev. **107** (1957) 704.
37. R. J. Carbone, E. N. Fuls and E. Everhart, Phys. Rev. **102** (1956) 1524.
38. O. B. Firsov, Zh. Eksperim. i Teor. Fiz. **33** (1957) 696 [English transl.: Soviet Phys. JETP **6** (1958) 534].
39. E. Blauth, Z. Physik **147** (1957) 228.
40. H. W. Berry, Phys. Rev. **121** (1961) 1714; **127** (1962) 1634.
41. P. R. Jones, F. P. Ziemba, H. A. Moses and E. Everhart, Phys. Rev. **113** (1959) 182.
42. A. K. Edwards and M. E. Rudd, Phys. Rev. **170** (1968) 140.
43. F. Bloch, Z. Physik **8** (1933) 363.
44. D. A. Kirzhnits and Yu. E. Lozovik, Usp. Fiz. Nauk **89** (1966) 39 [English transl.: Soviet Phys.-Usp. **9** (1966) 340].
45. M. Ya. Amusia, V. V. Afrosimov, Yu. S. Gordeev, N. A. Cherepkov and S. I. Sheftel, Phys. Letters **24A** (1967) 394.
46. F. Hund, Z. Physik **40** (1927) 742.
47. R. S. Mulliken, Phys. Rev. **32** (1928) 186.
48. A. Russek and J. A. Meli, private communication.
49. E. Everhart, private communication.
50. See for example, R. D. Evans, The Atomic Nucleus (McGraw Hill Book Company, Inc., New York, 1955) pp. 791–793.

CHAPTER 8

# RECOMBINATION OF RARE GAS IONS WITH ELECTRONS

BY

## H. J. OSKAM

*Department of Electrical Engineering, University of Minnesota,
Minneapolis, USA*

## Contents

|  | Page |
|---|---|
| 8-1. Introduction | 465 |
| 8-2. Mechanisms of electron–ion recombination | 468 |
|    A. Dissociative recombination | 468 |
|       1. Dependence of the recombination coefficient on electron temperature | 469 |
|       2. The direct dissociative recombination process | 471 |
|       3. The indirect dissociative recombination process | 472 |
|    B. Collisional-radiative recombination | 473 |
| 8-3. Experimental methods | 476 |
|    A. Microwave cavity method | 476 |
|       1. Theory of the method | 476 |
|       2. Experimental technique | 484 |
|    B. Light spectrometer techniques | 487 |
|       1. Light emission | 488 |
|       2. Spectral line shape | 489 |
|    C. Mass spectrometer techniques | 494 |
| 8-4. Recombination processes involving helium ions | 503 |
| 8-5. Recombination processes involving rare gas ions other than helium ions | 512 |
|    A. Neon ions | 512 |
|       1. Electron density studies | 512 |
|       2. Light emission studies | 514 |
|       3. Spectral line-shape studies | 516 |
|       4. Dependence of $\alpha\,(N_2{}^+)$ on the electron temperature | 517 |
|    B. Argon ions | 518 |
|    C. Krypton and xenon ions | 520 |
| Acknowledgement | 521 |
| Referenecs | 521 |

## § 8–1. Introduction

An electron colliding with a positive ion can be captured by the latter, forming a neutral particle. This process, called electron–ion recombination, can be effected by various mechanisms, the probability of which greatly depends on the associated possibility of absorbing the neutralization energy. For instance, a great difference exists between recombination of electrons and ions on the walls of a plasma container and in the volume of the plasma. At the walls the electron–ion recombination process is very effective since the molecules or atoms of the surface of the walls are always present as third bodies which can absorb the liberated energy of neutralization. The wall-recombination process thus prevents the presence of a noticeable concentration of charged particles at the walls since the charged particles arriving there are neutralized very rapidly. Hence, the characteristics of the disappearance of the electrons and ions from a plasma can be discussed by assuming a charged particle density close to zero at the walls. The capture of a free electron by a positive ion inside the plasma can occur by means of various mechanisms. Each recombination mechanism has its own numerical value of the coefficient describing the recombination. This coefficient, moreover, may depend in a different way on the various plasma parameters.

The efficiency of the volume electron–ion recombination process is described by the coefficient $\alpha$, which is defined as

$$R = \alpha n^+ n_e, \tag{8-1-1}$$

where $R$ is the number of recombination events per unit volume and unit time, and $n^+$ and $n_e$ are the number densities of the participating positive ions and electrons, respectively. The value of $\alpha$ is in general given in cm$^3$/sec.

The recombination coefficient $\alpha$ can be derived from the recombination cross section $q_r(v)$ of the ion through the relation

$$\alpha = \int_0^\infty v q_r(v) f(v) \, dv, \tag{8-1-2}$$

where $f(v) \, dv$ is the fraction of the electrons having a velocity between $v$ and $v+dv$. Here, the velocity of the ions is considered to be small compared to that of the electrons.

During the second part of the nineteen forties the M.I.T. gaseous electronics group developed the microwave cavity technique for measuring the number density of electrons during the decay period of a plasma [1, 2]. This method made it possible to obtain data concerning collision processes relating to electrons and ions having a Maxwellian energy distribution with a temperature equal to the temperature of the gas in which the plasma is produced. The early microwave cavity studies of decaying plasmas performed in rare gases seemed to result in establishing the occurrence of the dissociative recombination process

$$X_2^+ + e \to X^* + X + \text{kinetic energy} \tag{8-1-3}$$

for all homonuclear diatomic rare gas ions [3, 4]. (The asterisk indicates an excited state of X.) For the rare gases only one of the produced atoms can be in an excited state.

Subsequent studies performed by Oskam [5, 6] and by Kerr et al. [7, 8] however, showed that the recombination coefficient for $He_2^+$ was at least anomalously low. It was even doubted that process (8-1-3) was the process involved in the recombination of $He_2^+$ with electrons. Around 1950 Herman [9] and Johnson et al. [10] had already observed a predominance of the molecular bands in the helium afterglow for pressures larger than about 10 Torr. This is in conflict with the emission spectrum expected from process (8-1-3), since this process should result in the emission of the atomic spectrum.

Studies performed on high-intensity and high-temperature helium afterglows led d'Angelo [11] in 1961 to believe that the three-body recombination process

$$He^+ + 2e \to He^* + e \tag{8-1-4}$$

was responsible for the emission of spectral lines during the plasma decay period. This process was discussed previously by Fowler [12] and Giovanelli [13] in connection with stellar atmospheres.

Although there existed during the early nineteen sixties disagreement among the various authors with respect to the processes responsible for the spectral emission during the decay period of helium plasmas, it was generally agreed upon that the large recombination coefficients measured for the other molecular rare gas ions related to the dissociative recombination process (8-1-3). During the last five years various diagnostic methods have been applied to the study of decaying plasmas in an effort to solve the existing disagreements. Moreover, the significance of data obtained using the microwave cavity method has been carefully evaluated by several authors. These recent studies have also firmly established the occurrence of the dissociative recombination process in plasmas produced in the rare gases other than helium.

## 8-1 INTRODUCTION

The recent progress with respect to the understanding of processes occurring in decaying rare gas plasmas has also had a considerable influence on the study of plasmas produced in other gases. For instance, the improved and new diagnostic techniques to be discussed in this chapter have already been applied successfully to the study of recombination processes involving ions such as $N_2^+$, $NO^+$, $CO_2^+$, etc. [14–16]. Due to space and time limitation the present discussion has to be confined to the recombination processes occurring in rare gas plasmas. For the same reason mainly recent studies will be discussed, while no effort has been made to include references to all the studies which have been reported. A choice has been made of studies the combination of which appears to have contributed most significantly to the understanding of recombination processes active in rare gas plasmas.

The microwave cavity method is the most accurate way of determining the value of the recombination coefficient, provided this value is not much smaller than about $10^{-7}$ cm$^3$/sec. Therefore, this method and the reported numerical calculations related to the effect of loss of electrons by ambipolar diffusion towards the walls on the reliability and accuracy of the value of $\alpha$ measured by means of the cavity method are discussed in section 8-3-A.

To establish the type of recombination process active during the plasma decay period, studies of the emission spectra and the time dependence of the intensity of lines, bands and continua is necessary. Moreover, Biondi et al. [17, 18] designed an experiment for measuring the energy of dissociation liberated during the dissociative recombination process (8-1-3). Light spectrometer techniques as applied to the study of recombination mechanisms are presented in section 8-3-B.

Electron density studies during the decay period of plasmas produced in rare-gas mixtures showed that the recombination coefficients measured by means of the microwave cavity method in neon and argon plasmas related to the recombination of electrons with $Ne_2^+$ and $Ar_2^+$. The experimental condition had to be such that these ions could be expected to be the dominant ions during the plasma decay period. For conditions such that more than one type of ion can be expected to contribute to the disappearance of electrons and/or the light emission during the decay period, the interpretation of the data became extremely difficult. The mass-spectrometric probing of decaying plasmas resolves this difficulty. This method is discussed in section 8-3-C.

Several authors have applied more than one of these diagnostic methods simultaneously. The main results of these studies and their contribution to the determination of recombination mechanisms and coefficients are summarized in sections 8-4 and 8-5. The two recombination processes which appear to be active in rare gas plasmas are the dissociative and the collisional-radiative recombination processes. The main features of these processes are given in section 8-2.

## § 8-2. Mechanisms of electron–ion recombination

There are several mechanisms by which the neutralization energy can be absorbed during the recombination process. A discussion and relevant calculation of the various recombination processes are available in the literature. In this section only those mechanisms and related properties which are needed to understand the studies of recombination processes in decaying rare gas plasmas will be discussed.

### A. Dissociative Recombination

In order to explain the night time decay of the electron density in the E layer of the ionosphere Bates and Massey [19] suggested in 1947 the occurrence of the dissociative recombination process

$$(AB)^+ + e \to A^* + B^* + \text{kinetic energy}. \qquad (8\text{-}2\text{-}1)$$

The asterisks indicate that A, B, or A as well as B can be excited atoms. This neutralization process is possible only if a repulsive potential curve crosses the potential curve of the molecular ion (Fig. 8-2-1). The difference between the potential energy of the ionic state $(AB)^+$ and that of the produced neutral atoms $A^*$ and $B^*$ appears as kinetic energy of these atoms. This energy will be called the dissociation energy $E_d$ and depends on the vibrational state of $(AB)^+$ as shown in Fig. 8-2-1. The total kinetic energy received by $A^*$ and $B^*$ is equal to the sum of the appropriate $E_d$ value and the kinetic energy of the captured electron. In general more than one repulsive potential curve will cross the ionic potential curve. Each curve will result in a different set of dissociation energies and spectral lines emitted as a consequence of the recombination process (8-2-1).

In 1950 Bates [20] used process (8-2-1) to explain the surprisingly large value of the recombination coefficient of $He_2^+$ reported by Biondi and Brown [3]. It is interesting to note that at present it appears that $He_2^+$ is the only homonuclear diatomic rare gas ion for which the electron–ion recombination process is not the dissociative process.

Two mechanisms have been postulated which result in dissociative recombination. Both will be explained briefly, but first the relation between

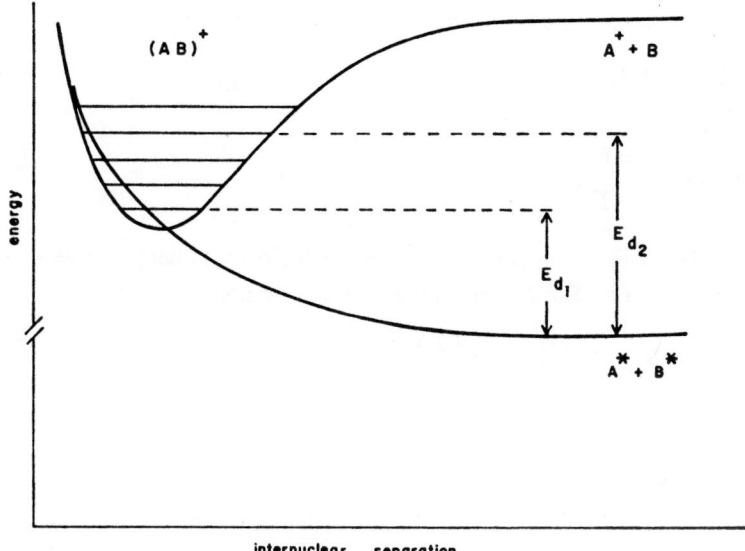

Fig. 8-2-1. Schematic representation of the dissociative recombination process $(AB)^+ + e \rightarrow A^* + B^* +$ kinetic energy.

the dependence of the recombination coefficient on electron energy and the properties of the recombination cross section $q_r(v)$ will be discussed.

1. *Dependence of the recombination coefficient on electron temperature*

For a Maxwellian velocity distribution of the electrons, the relation (8-1-2) between the recombination coefficient $\alpha$ and the cross section $q_r(v)$ can be written as

$$\alpha(T_e) = \int_0^\infty \frac{8\pi m_e \varepsilon}{(2\pi m_e k T_e)^{\frac{3}{2}}} q_r(\varepsilon) \exp\left(-\frac{\varepsilon}{kT_e}\right) d\varepsilon, \qquad (8\text{-}2\text{-}2)$$

where $\varepsilon = \tfrac{1}{2} m_e v^2$. If the dependence of $q_r(\varepsilon)$ on $\varepsilon$ is known, the function $\alpha(T_e)$ can be calculated.

In discussing the function $\alpha(T_e)$ it is convenient to consider $\varepsilon q_r(\varepsilon)$ since $q(\varepsilon)$ contains a factor $\varepsilon^{-1}$. It is possible to consider various limiting conditions such that the dependence of $\alpha$ on $T_e$ can easily be determined.

(a) If $\varepsilon q_r(\varepsilon)$ is a slowly varying function of $\varepsilon$ compared to $\exp(-\varepsilon/kT_e)$ this condition can be approximated by assuming $\varepsilon q_r(\varepsilon)$ to be constant (Fig.

8-2-2a). The temperature dependence of $\alpha$ is then given by

$$\alpha(T_e) = CT_e^{-0.5}, \tag{8-2-3}$$

with

$$C = \frac{4\varepsilon q(\varepsilon)}{(2\pi m_e k)^{\frac{1}{2}}}. \tag{8-2-4}$$

(b) If $\varepsilon q_r(\varepsilon)$ is non-zero only for a small electron energy range $\Delta\varepsilon$ such that $\Delta\varepsilon \ll kT_e$ (Fig. 8-2-2b) the value of $\alpha$ can be approximated by

$$\alpha(T_e) = C' T_e^{-\frac{3}{2}} \exp(-\varepsilon_s/kT_e), \tag{8-2-5}$$

with

$$C' = \frac{8\pi m_e \varepsilon_s}{(2\pi m_e k)^{\frac{3}{2}}} \int_0^\infty q_r(\varepsilon) \, d\varepsilon, \tag{8-2-6}$$

where $\varepsilon_s$ is the average value of $\varepsilon$ over $\Delta\varepsilon$.

The function (8-2-5) has for fixed $\varepsilon_s$ a maximum value at $kT_e = \frac{2}{3}\varepsilon_s$, so that for $\varepsilon_s > \frac{3}{2}kT_e$ the value of $\alpha$ will increase with $T_e$. If $\varepsilon_s \ll kT_e$ the temperature dependence of $\alpha$ will be as $T_e^{-1.5}$.

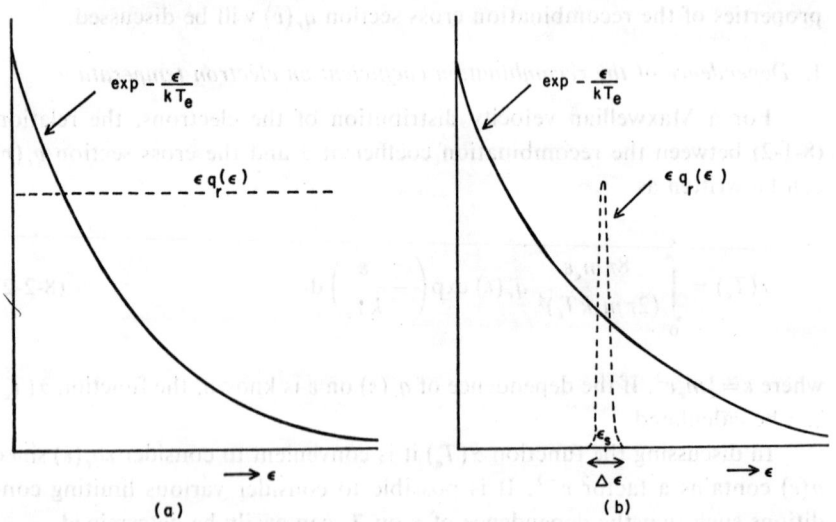

Fig. 8-2-2. The functions $\exp(-\varepsilon/kT_e)$ and $\varepsilon q_r(\varepsilon)$. (a) $\varepsilon q_r(\varepsilon)$ is a slowly varying function of $\varepsilon$ compared to $\exp(-\varepsilon/kT_e)$. (b) $\varepsilon q_r(\varepsilon)$ is non-zero only for a small energy range $\Delta\varepsilon$ with $\Delta\varepsilon \ll kT_e$.

The dependence of $\alpha$ on $T_e$ is in general expressed as

$$\alpha(T_e) = \alpha(300\,°K)\left(\frac{T_e}{300\,°K}\right)^\gamma. \tag{8-2-7}$$

From the previous discussion follows that $\gamma$ can vary from a positive value to $-1.5$.

## 2. The direct dissociative recombination process

Recently, Bardsley [21] has discussed the dissociative recombination process in detail. This author reported two mechanisms leading to dissociative recombination. The first process is analogous to that suggested by Bates [20] and involves only one intermediate electronically excited state, i.e.,

$$(AB)^+ + e \leftrightarrows (AB)_{\text{resonance}} \rightarrow A^* + B^*. \tag{8-2-8}$$

Here, $(AB)_{\text{resonance}}$ is a repulsive intermediate electronically excited state, which has a strong configuration interaction with the continuous configuration of the ion including the free electron. This interaction is strong in the crossing area of the ionic and the repulsive potential energy curve (Fig. 8-2-3). When the system is in the resonance state auto-ionization (formation of $(AB)^+$ and an electron) or pre-dissociation (the two nuclei are separated due to the repulsive forces) can occur. The system is called "stabilized" during the latter process and a survival factor $s_f$ can be defined, which gives the probability for this stabilization. The difference between the more classical theory of Bates and the theory advanced by Bardsley is in the expression for $s_f$. In general, however, $s_f$ is close to one (if $(AB)_{\text{resonance}}$ is produced predissociation occurs), so that both theories lead to the same result.

When the ion is in its ground vibrational energy state it can be assumed that the probability $f(\varepsilon)\,d\varepsilon$ of electrons in the energy range $d\varepsilon$ for a transition from the ionic plus free electron state to the resonance state will not change appreciably over the energy distribution of the electrons, provided this distribution is narrow compared to the interaction width of the two configurations.* This assumption is believed to be valid for electron temperatures up to $1000\,°K$. Bardsley obtained for the recombination cross section

$$q_r(\varepsilon) = \frac{\pi^2 \hbar^2}{2m\varepsilon}\, r\Gamma_c f(\varepsilon)\, s_f, \tag{8-2-9}$$

where $r$ is the ratio of the degeneracy of ionic and resonance states, while

---

* $f(\varepsilon)$ is the oscillator strength.

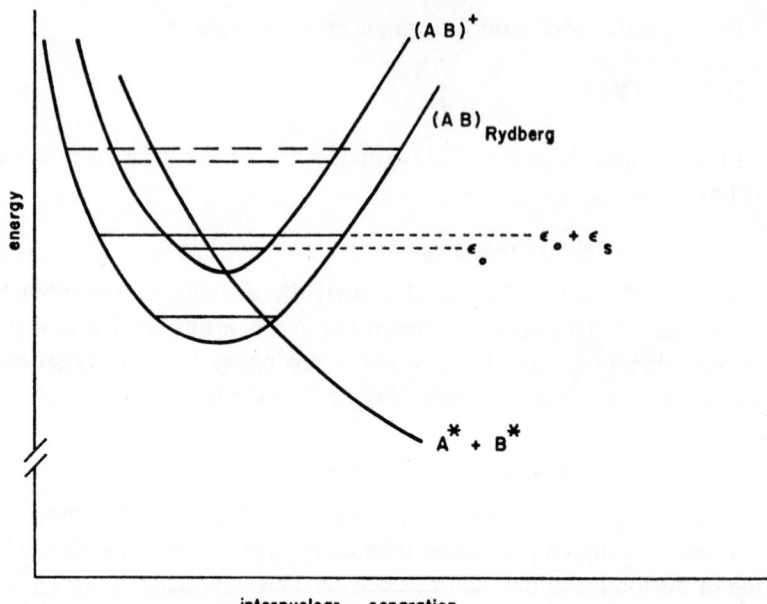

Fig. 8-2-3. Schematic representation of proposed dissociative recombination processes. The direct process involves electron capture in a repulsive resonant state after which this state dissociates into A* and B*. In the indirect process the electron is first captured in a Rydberg state and then dissociates through the resonance state.

$\Gamma_c$ is the transition energy width from ionic plus free electron state to resonance state.

As mentioned, for electron temperatures up to about 1000 °K the transition probability $f(\varepsilon)$ can be considered constant over the thermal energy range of the electrons. Then $\varepsilon q_r(\varepsilon)$ can be approximated as a constant in eq. (8-2-2), which implies that $\alpha$ varies as $T_e^{-0.5}$. If ions are present in excited vibrational energy states the recombination coefficient is obtained by summing over the modes involved. The transition probability $f(\varepsilon)$ will have a stronger dependence on $\varepsilon$ for the higher vibrational modes.* For $T_e = 300$ °K this effect will probably be important only for very high vibrational states.

3. *The indirect dissociative recombination process*

An indirect dissociative recombination process was recently suggested

---

* Winans-Stueckelberg approximation (G. Herzberg, Molecular Spectra and Molecular Structure I: Spectra of Diatomic Molecules (second ed., D. van Nostrand Company, Inc., New York, 1959) p. 393).

by Chen and Mittleman [22] and by Bardsley [21, 23]. This process involves two intermediate electronically excited states, i.e.,

$$(AB)^+ + e \leftrightarrows (AB)_{Rydberg} \leftrightarrows (AB)_{resonance} \to A^* + B^*. \qquad (8\text{-}2\text{-}10)$$

The $(AB)_{Rydberg}$ state is an intermediate metastable state where the free electron is captured due to transfer of the electron energy to the vibrational or rotational energy of the nuclei. The capture occurs in a Rydberg orbital with high principal quantum number. The potential curve of the molecular ion and this Rydberg state will be very similar (Fig. 8-2-3). This means that indirect and direct dissociative recombination processes will always be linked processes since any repulsive state that causes the direct process will also cross the Rydberg state and contribute to the indirect process.

Bardsley gives for the cross section $q_r(\varepsilon)$ of the indirect process the expression

$$q_r(\varepsilon) = \sum_s \pi \frac{h^2}{2m_e \varepsilon} \frac{r}{2} \frac{\Gamma_{sa}\Gamma_{sd}}{(\varepsilon - \varepsilon_s)^2 + \tfrac{1}{4}\Gamma_s^2}, \qquad (8\text{-}2\text{-}11)$$

where $s$ is the number of Rydberg states involved in the process; $\Gamma_{sa}$ and $\Gamma_{sd}$ are partial energy widths related to transitions from ionic to the Rydberg state and that of the latter to the repulsive state, respectively; $\Gamma_s$ is the total width and $\varepsilon_s$ is the energy of the molecule in the Rydberg state with respect of that of the ionic state (Fig. 8-2-3).

From eq. (8-2-11) follows that $\varepsilon q_r(\varepsilon)$ will be narrow with respect to the function $\exp(-\varepsilon/kT_e)$, as shown in Fig. 8-2-2, if $\Gamma_s \ll kT_e$, since the half-width of $\varepsilon q_r(\varepsilon)$ is equal to $\Gamma_s$. From Fig. 8-2-3 it follows also that $\varepsilon_s$ must lie inside the range of electron energies for the indirect process to contribute to dissociative recombination. If these conditions are satisfied the dependence of $\alpha$ on $T_e$ is given by eq. (8-2-5) and if, moreover, $\varepsilon_s \ll kT_e$ the $T_e^{-1.5}$ dependence can be expected.

Bardsley [21] considered the relative importance of the direct and indirect processes. He finds that the indirect process can only be significant for electron temperature below $1000\,°K$ if the electron energy is transferred to the vibrational motion of the nuclei. If this transfer is to the rotational nuclei energy the indirect process is of importance only for $T_e < 10\,°K$. The latter energy transfer has also been discussed by Stabler [24].

B. COLLISIONAL-RADIATIVE RECOMBINATION

The stabilization in the dissociative recombination process occurs through dissociation of the intermediate resonance state as discussed in the previous subsection. Another mechanism for electron-recombination is the direct free-bound transition of an electron. The energy liberated during this transition can appear as radiation resulting in the radiative recombination process

$$X^+ + e \rightarrow X(p) + h\nu, \qquad (8\text{-}2\text{-}12)$$

where $p$ denotes the energy level of the produced excited state.* It may also appear as kinetic energy of a third participating particle, for instance another electron.** This leads to the three-body recombination process***

$$X^+ + e + e \rightarrow X(p) + e. \qquad (8\text{-}2\text{-}13)$$

After the initial capture many of the excited states formed may initially be in high quantum number energy levels. They, therefore, could be re-ionized by the reverse process of (8-2-13), i.e.,

$$X(p) + e \rightarrow X^+ + e + e. \qquad (8\text{-}2\text{-}14)$$

They may also suffer collisional excitation and/or de-excitation

$$X(p) + e \rightleftarrows X(q) + e, \qquad (8\text{-}2\text{-}15)$$

or take part in spectral line absorption and/or emission

$$X(p) + h\nu \rightleftarrows X(q). \qquad (8\text{-}2\text{-}16)$$

The combination of the processes (8-2-12) through (8-2-16) has been called collisional-radiative recombination by Bates et al. [25]. Several calculations have been performed of the magnitude and dependence of the recombination coefficient, defined by (8-1-1), on the density and temperature of the electrons [25, 26]. The calculations involve solving the infinite set of equations describing the population density of the energy states participating in the processes (8-2-12) through (8-2-16). The calculations are simplified by grouping together energy states higher than a given energy level, determined by the temperature of the electrons. These states are supposed to be in Saha equilibrium with the free electrons and the ions. For an optically thin $H^+$ ion plasma the reverse process of (8-2-12) is of no importance and Fig. 8-2-4 shows for this type of plasma the dependence of $\alpha$ on the electron density for three different electron temperatures [26].

* If $X^+$ is an hydrogenic ion the symbol $p$ indicates the principal quantum number of the excited state.
** Three-body recombination processes where the third particle is a neutral particle or an ion are believed to be of considerably less importance.
*** This process will be called the collisional recombination process.

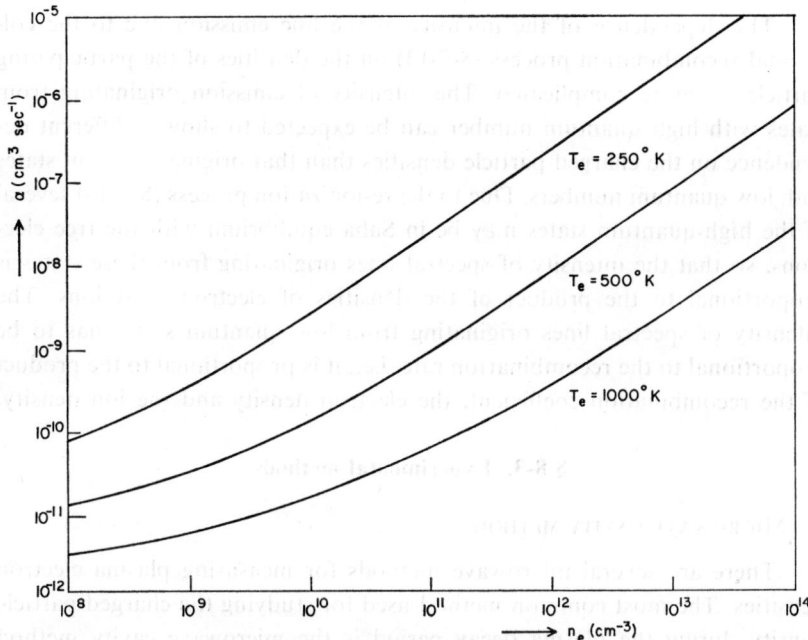

Fig. 8-2-4. Collisional-radiative recombination coefficient $\alpha$ for an optically thin H$^+$ ion plasma as a function of electron density for three electron temperatures (reference 25).

For tenuous plasmas the limit of the collisional-radiative recombination process is the radiative recombination process (8-2-12), since then the effect of processes (8-2-13) and (8-2-14) can be neglected. The recombination coefficient is then independent of the electron density, while the dependence of $\alpha$ on $T_e$ for H$^+$ ions has been calculated to be as $T_e^{-\frac{3}{4}}$. If the electron density is so high that the contribution of the radiative processes (8-2-12) and (8-2-16) can be neglected the recombination is governed exclusively by collision processes. Then the recombination coefficient is proportional to the electron density since the initial electron capture process (8-2-13) is a process involving two electrons. The calculated dependence of $\alpha$ on $T_e$ is for this electron density region about as $T_e^{-4.5}$.

The intensity of the emitted radiation due to the radiative recombination process (8-2-12) is for an optically thin plasma proportional to the product of the densities of the participating electrons and ions. The distribution in the electron energy produces continua, which end at the long wavelength sides corresponding to transitions from the ionization limit to the discrete energy levels.

The dependence of the intensity of the line emission due to the collisional recombination process (8-2-13) on the densities of the participating particles is more complicated. The intensity of emission originating from states with high quantum number can be expected to show a different dependence on the charged particle densities than that originating from states with low quantum numbers. Due to the re-ionization process (8-2-14) several of the high-quantum states may be in Saha equilibrium with the free electrons, so that the intensity of spectral lines originating from these states is proportional to the product of the densities of electrons and ions. The intensity of spectral lines originating from low quantum states has to be proportional to the recombination rate, i.e., it is proportional to the product of the recombination coefficient, the electron density and the ion density.

## § 8-3. Experimental methods

### A. Microwave cavity method

There are several microwave methods for measuring plasma electron densities. The most common method used for studying the charged particle density during the plasma decay period is the microwave cavity method [1, 2, 5]. This method consists of inserting the plasma into a resonant cavity and measuring the resonant frequency shift caused by the plasma.

#### 1. *Theory of the method*

Slater [27] has evaluated the influence of a rarefied electron gas on the properties of a cavity at microwave frequencies. His formula (III.82), reduces to the following first-order approximation, which includes the changes in both the quality and resonant frequency

$$\left(\frac{1}{Q(t)} - \frac{1}{Q_0}\right) - 2j\frac{\Delta f(t)}{f_0} = \frac{1}{\varepsilon_0 \omega_0} \frac{\int \sigma_c(r, t) E^2(r) \, dV}{\int E^2(r) \, dV}. \tag{8-3-1}$$

Here $Q(t)$ and $Q_0$ are the qualities of the cavity with and without the electron gas, respectively; $\varepsilon_0$ is the permittivity of free space; $\omega_0 = 2\pi f_0$ is the angular-resonant frequency of the empty cavity; $dV$ is a volume element of the cavity; $E(r)$ is the amplitude of the electric field [time factor $\exp(j\omega t)$] applied to measure the frequency shift $\Delta f$ and the change in the quality, while the conductivity $\sigma_c$ is the complex ratio of current density and field strength.

The complex conductivity of an electron gas for a high-frequency electric field has been shown to be [28]

$$\sigma_c = -\frac{4}{3}\pi \frac{e^2}{m_e} \int_0^\infty \frac{1}{v_m + j\omega} \frac{\partial f_0^0}{\partial v} v^3 \, dv, \qquad (8\text{-}3\text{-}2)$$

where $e$ and $m_e$ are the electron charge and mass, respectively; $v_m$ is the collision frequency for momentum transfer of the electrons with the gas particles, $f_0^0$ is the steady-state isotropic part of the velocity distribution of the electrons, and $\omega$ is the angular frequency of the electric field.

When it is assumed that $v_m$ is independent of the electron velocity $v$, the complex conductivity is given by*

$$\sigma(r, t) = \frac{e^2 n_e(r, t)}{m_e} \frac{v_m}{v_m^2 + \omega^2} - j \frac{\omega}{v_m^2 + \omega^2}, \qquad (8\text{-}3\text{-}3)$$

independent of the form of $f_0^0$. Substitution of (8-3-3) into (8-3-1) leads to

$$\frac{1}{Q(t)} - \frac{1}{Q_0} = \frac{e^2}{\varepsilon_0 m_e \omega_0^2} \frac{b p_0}{[1 + (b p_0)^2]} \bar{n}_{\mu w}(t) \qquad (8\text{-}3\text{-}4)$$

and

$$\frac{\Delta f(t)}{f_0} = \frac{e^2}{2\varepsilon_0 m_e \omega_0^2} \frac{1}{[1 + (b p_0)^2]} \bar{n}_{\mu w}(t), \qquad (8\text{-}3\text{-}5)$$

where

$$b \equiv v_{m,0}/\omega_0 \qquad (8\text{-}3\text{-}6)$$

and

$$\bar{n}_{\mu w}(t) \equiv \frac{\int n_e(r, t) E^2(r) \, dV}{\int E^2(r) \, dV}. \qquad (8\text{-}3\text{-}7)$$

Here, $p_0$ is the gas pressure (reduced to 273°K) and in the definition (8-3-6) the quantity $v_{m,0}$ is the collision frequency for momentum transfer of electrons in a gas at a pressure of 1 Torr.

The quantity $\bar{n}_{\mu w}(t)$ can be considered as "microwave average" of the electron density $n_e(r, t)$ and has been calculated by Oskam [5] for various types of microwave cavities for a uniform spatial electron density distri-

---

* When $v_m$ depends on $v$, the symbol $v_m$ in (8-3-3) refers to a weighted average.

bution and for distributions resulting from electron diffusion towards the plasma container walls.

Equation (8-3-5) is valid only if $\Delta f/f_0$ is very small. The limits of validity of eq. (8-3-5) have been discussed by Persson [29] and by Buchsbaum and Brown [30]. One of the approximations made in deriving eq. (8-3-1) is that $E$, the field in the cavity with the plasma present, can be replaced by $E_0$, the field without the plasma. When the electron density is low ($\eta = n_e e^2/m_e \varepsilon_0 \omega_0^2 \ll 1$ or $n_e \ll 10^{-8} f_0^2$ electrons/cm$^3$), the approximation is good since the plasma does not appreciably disturb the electric field. This is the region in which the microwave cavity method for measuring the electron density is mostly used. As the electron density is increased $E$ can become appreciably different from $E_0$. There are three reasons for this deviation, i.e., (a) a.c. space charge (plasma resonance) effects become important as $\eta$ approaches unity; (b) as $\eta$ becomes larger than unity the plasma starts to shield its interior from the applied field, and (c) when both $\eta$ and the pressure are high enough so that the quality of the cavity is lowered appreciably, the overlapping of higher cavity modes may also cause $E$ to be different from $E_0$ by adding to $E_0$ some of the higher mode fields. The a.c. space charge effect usually enters first. Its influence on the measurements can be eliminated by using a plasma-microwave cavity mode configuration such that the applied field $E$ is normal to the plasma density gradients. An example of such a configuration is a cylindrical cavity oscillating in the TE$_{011}$ mode with a cylindrical plasma column placed along the axis of the cavity.

The application of the microwave cavity method to the measurement of recombination coefficients is limited to those recombination processes which have a large coefficient (about $10^{-7}$ cm$^3$/sec or larger). This is due to competition of ambipolar diffusion as an electron loss process from the plasma, combined with cavity size and gas pressure limitation. The dissociative recombination process (8-1-3) has been found to be a process of which the coefficient can be measured by means of the microwave cavity method.

When it is assumed that the only process by which electrons disappear from the plasma during the afterglow period is the dissociative recombination process, the rate of change of the electron density is given by

$$\frac{dn_e(t)}{dt} = -\alpha n_e(t) n^+(t), \qquad (8\text{-}3\text{-}8)$$

provided that only one type of ion is present, and that all electron production processes can be neglected. For a quasineutral plasma $(n_e(t) \approx n^+(t))$ the solution of eq. (8-3-8) is

$$1/n_e(t) = \frac{1}{n_e(t_0)} + \alpha(t - t_0). \tag{8-3-9}$$

The curve representing the reciprocal of the electron density as a function of time during the afterglow period should be a straight line as long as the assumptions leading to the solution (8-3-9) are valid. The slope of this line gives the value of the recombination coefficient.

The coefficient $\alpha$ related to the dissociative recombination process should be independent of the gas pressure, since only an electron and an ion are involved in the collision process. If a dependence of the measured $\alpha_m$-value on gas pressure is found, it is either due to a different type of recombination process or is of experimental origin. For instance, the presence of the factor $[1+(bp_0)^2]^{-1}$ in formula (8-3-5) can result in

$$\alpha_m = [1 + (bp_0)^2]\alpha, \tag{8-3-10}$$

where $\alpha_m$ refers to the recombination coefficient calculated from the $1/n_e(t)$ versus time curve when the influence of $v_m$ on the measured frequency shift is neglected. For electrons having an average energy of 0.04 electron volts, the mean free paths have been measured with the aid of microwave techniques [13, 32]. This makes it possible to estimate the gas pressures at which $\alpha_m$ is 10% larger than the actual value $\alpha$. For a frequency of the microwave measuring signal of 9000 Mc/sec these pressures are 100 Torr for helium, 550 Torr for neon, 870 Torr for argon, 36 Torr for krypton, and 10 Torr for xenon. For lower frequencies and/or higher electron energies these pressures are lower.

Various authors have reported a dependence of the measured $\alpha_m$ on the magnitude and duration of the plasma excitation pulse [33–36]. It is evident that the value of $\alpha$, relating to the same recombination process and the same ion, should be independent of these variables, so that the origin of the measured dependences must be due to other phenomena. The type of excitation pulse may influence the initial spatial density distribution of the electrons, or, when impurity particles are present, may influence the number of impurity ions produced.

The influence of electron production processes during the plasma decay period and loss processes different from dissociative recombination have been neglected in eq. (8-3-8). Moreover, it was assumed that the molecular ion studied was the only ion present and that its number density was closely equal to that of the electrons. For plasmas produced in the rare gases it is possible to achieve these experimental conditions with the exception of a possible influence of the ambipolar loss process on the determination of $\alpha$. The effect of ambipolar diffusion on recombination studies has been discussed by Oskam [37] for an infinite plane parallel geometry, by Gray and Kerr [38] for spheres and infinitely long cylinders and recently by Frommhold and Biondi [39] for plasma containers which are finite cylinders or rectangular parallelepipeds.

During the early measurements of the recombination coefficients of molecular rare gas ions it was assumed that the electrons maintain a uniform distribution during the plasma decay period. This assumption was used to calculate the relation between the electron density and the measured resonant frequency shift. The slope of the resulting $1/n_e(t)$ versus time plot was considered to be the value of the recombination coefficient. An important conclusion following from the numerical analysis of the effect of ambipolar diffusion on the measurements is that the linearity of the $1/n_e(t)$ versus time plot has to satisfy certain conditions in order to obtain a meaningful value for the recombination coefficient.

A brief summary of the recent analysis reported by Frommhold and Biondi [39] for finite plasma containers will be given here. These authors followed the method used by Gray and Kerr [38]. The relation between the electron density and the measured frequency shift can be written as

$$\bar{n}(t) = CG\,\Delta f(t), \qquad (8\text{-}3\text{-}11)$$

where $C$ and $G$ are constants of the experiment and are equal to

$$C = \frac{4\pi\varepsilon_0 m_e \omega_0}{e^2}\left[1 + (bp_0)^2\right] \qquad (8\text{-}3\text{-}12)$$

and

$$G = \int_{\text{cavity}} E^2(r)\,dV \Big/ \int_{\text{plasma}} E^2(r)\,dV. \qquad (8\text{-}3\text{-}13)$$

The "average" electron density $\bar{n}(t)$ is

$$\bar{n}(t) = \int_{\text{plasma}} n_e(r,t)\,E^2(r)\,dV \Big/ \int_{\text{plasma}} E^2(r)\,dV. \qquad (8\text{-}3\text{-}14)$$

If the effect of ambipolar diffusion loss is neglected, i.e., the electron density distribution is uniform and eq. (8-3-11) describes the time dependence of the electron density, $\bar{n}(t)$ is equal to $n_e(t)$ and the slope of the $[CG\,\Delta f(t)]^{-1}$ versus time curve gives the recombination coefficient $\alpha$. Thus $\bar{n}(t)$ is equivalent to the quantity often used for presenting the experimental data in recombination analyses. However, it is unrealistic to make the general assumption that the effect of ambipolar diffusion, especially close to the container walls, on the spatial distribution of the electrons can be neglected. When including the ambipolar diffusion loss the continuity equation becomes

$$\frac{\partial n_e(r, t)}{\partial t} = D_a \nabla^2 n_e(r, t) - \alpha n_e^2(r, t), \qquad (8\text{-}3\text{-}15)$$

in which $D_a$ is the ambipolar diffusion coefficient. This equation has to be solved with the appropriate initial and boundary conditions. The solution substituted in (8-3-14) gives $\bar{n}(t)$.

The measured $1/\bar{n}(t) = [CG \Delta f(t)]^{-1}$ versus time plot could exhibit a linear behavior for certain measured ranges of $\Delta f$. The resulting quantity $\alpha_m$ in

$$1/\bar{n}(t) = \frac{1}{\bar{n}(t_0)} + \alpha_m(t - t_0) \qquad (8\text{-}3\text{-}16)$$

is to be related to the recombination coefficient $\alpha$ in eq. (8-3-15). The calculations showed that the relevant relation becomes meaningful only if the "linear range" of eq. (8-3-16), defined as the ratio $f$ between the largest and smallest value of $1/\bar{n}(t)$ for which the data points lie sufficiently close to the best-fitting straight line, is larger than about ten. The correction $\Delta$ defined by

$$\alpha \equiv \alpha_m(1 - \Delta), \qquad (8\text{-}3\text{-}17)$$

is found to be directly related to the value of $f$ and to depend on plasma container and cavity shapes and dimensions, on the microwave probing field distribution, on the spatial electron distribution at $t = t_0$ and on the quantity

$$\beta = \frac{\alpha n_e^2(0, t_0)}{D_a n_e(0, t_0)/\Lambda^2}, \qquad (8\text{-}3\text{-}18)$$

where $n_e(0, t_0)$ is the initial central electron density and $\Lambda$ is the characteristic fundamental diffusion length. The quantity $\beta$ is thus the ratio at $t = t_0$ of the central electron loss rate that would prevail in the absence of diffusion to the corresponding loss rate from only diffusion in the fundamental mode, so that $\beta$ is a measure of the degree to which the plasma is initially recombination controlled ($\beta \gg 1$) or diffusion controlled ($\beta \ll 1$).

The numerical calculations were performed using a "normalized" continuity equation, in which the variables and function are dimensionless, i.e.,

$$\frac{\partial N}{\partial \tau} = \nabla^2 N - \beta N^2, \qquad (8\text{-}3\text{-}19)$$

where $N = n_e(r, t)/n_e(0, t_0)$ and $\tau = (D_a/\Lambda^2) t$. The normalized Cartesian coordinates $\xi, \eta, \zeta$, and cylindrical coordinates $\rho$ and $\zeta$ are obtained by dividing the respective spatial variables $x, y, z, r$ and $z$ by $\Lambda$. The boundary condition for $N$ is $N = 0$ at the walls for all values of $\tau$. The initial normalized electron density distributions were taken as

$$N(r, z, 0) = [J_0(2.4 r/r_0) \cos(\pi z/z_0)]^k \qquad (8\text{-}3\text{-}20)$$

for a finite cylinder (radius $r_0$ and height $z_0$) and as

$$N(x, y, z, 0) = [\cos(\pi x/x_0) \cos(\pi y/y_0) \cos(\pi z/z_0)]^k \qquad (8\text{-}3\text{-}21)$$

for a rectangular parallelepiped. In order to include the variety of initial

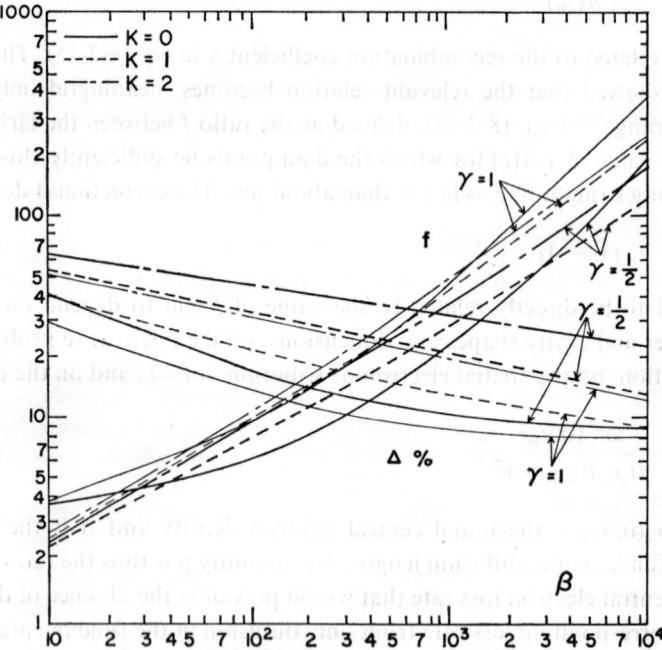

Fig. 8-3-1. $\Delta$ and $f$ values as function of $\beta$ for a finite cylinder with $z_0 = r_0 = 3.96\, \Lambda$ and a TM$_{010}$ mode probing field; $\gamma = 1$ indicates a completely filled microwave cavity; $\gamma = \frac{1}{2}$ refers to $z_0 = z_1$ and $r_0 = \frac{1}{2} r_1$, where $z_1$ and $r_1$ are cavity dimensions (reference 39).

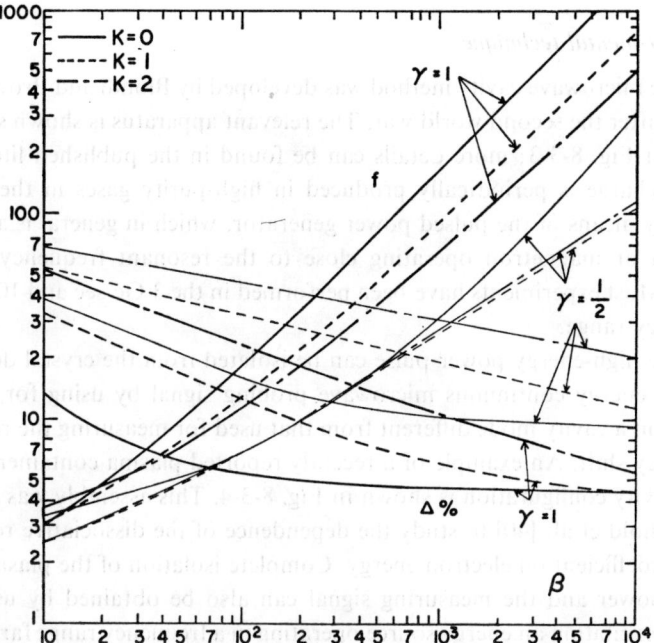

Fig. 8-3-2. $\Delta$ and $f$ values as function of $\beta$ for the same conditions as those of Fig. 8-3-1, with the exception that a TE$_{011}$ mode probing field is used (reference 39).

distributions encountered in experiments $k$ was chosen to be 0, 1 and 2. The distribution $k=0$ corresponds with the uniform "recombination controlled" distribution; the $k=1$ distributions are the fundamental diffusion distributions, while the squared distributions have sometimes been observed at higher gas pressures.

Starting from the selected initial distribution the normalized distribution was computed as a function of time $\tau$ for given plasma container dimensions and value of the normalized recombination coefficient $\beta$. Then the average $1/\bar{N}(\tau)$ is computed for a given plasma container–cavity combination and microwave probing field distribution. Finally, a straight line $y=y_0+s\tau$ is sought to optimally fit the linear portion of the function $1/\bar{N}(\tau)$. An upper and a lower time limit $\tau_u$ and $\tau_l$ are then found such that the $1/\bar{N}(\tau)$ curve deviates by maximal 2% from the straight line. The linear range $f$ mentioned previously is then $f=\bar{N}(\tau_l)/\bar{N}(\tau_u)$ and the correction $\Delta$ defined by (8-3-17) is given by $\Delta=(\alpha_m-\alpha)/\alpha_m=(s-\beta)/s$. Examples of $f$ and $\Delta$ as a function of $\beta$ for a few geometries and microwave probing fields are given in Figs. 8-3-1 and 8-3-2.

## 2. Experimental technique

The microwave cavity method was developed by Biondi and Brown [1, 2] shortly after the second world war. The relevant apparatus is shown schematically in Fig. 8-3-3; more details can be found in the published literature. The discharge is periodically produced in high-purity gases in the quartz bottle by means of the pulsed power generator, which in general is a pulsed klystron or magnetron operating close to the resonant frequency of the cavity. Most experiments have been performed in the 3 Gc/sec and 10 Gc/sec frequency range.

The high-energy power pulse can be isolated from the crystal detecting the low energy continuous microwave probing signal by using for plasma excitation a cavity mode different from that used for measuring the resonant frequency shift. An example of a recently reported plasma container-microwave cavity configuration is shown in Fig. 8-3-4. This assembly was used by Frommhold et al. [40] to study the dependence of the dissociative recombination coefficient on electron energy. Complete isolation of the plasma excitation power and the measuring signal can also be obtained by using for plasma excitation an energy source operating in a frequency range far outside the microwave frequency region [41].

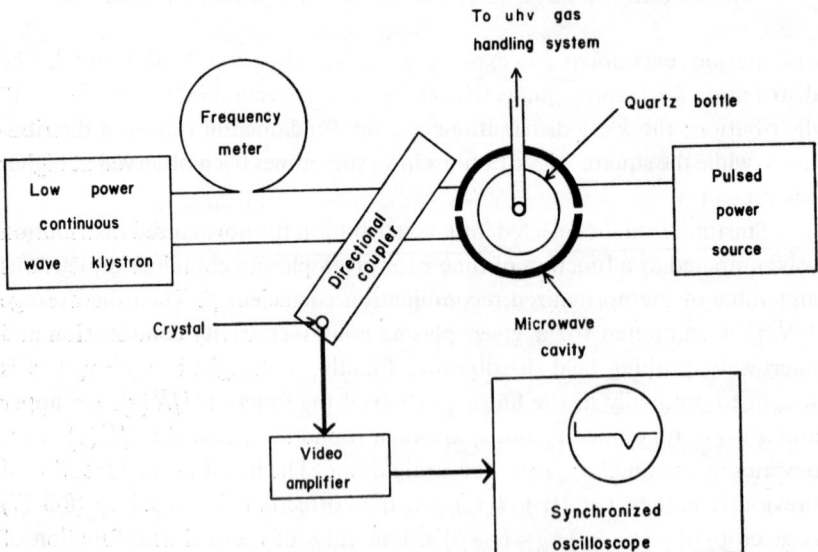

Fig. 8-3-3. Simplified block diagram of the microwave cavity apparatus used in afterglow recombination studies (reference 69).

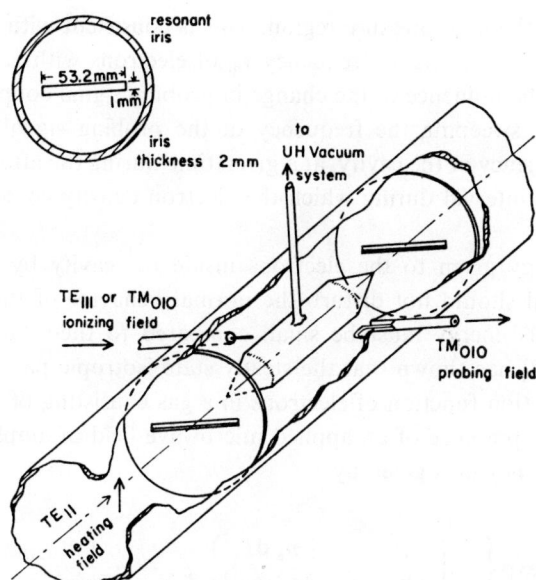

Fig. 8-3-4. Cutaway view drawing of cylindrical cavity-waveguide section bounded by resonant irises (reference 40).

The resonant frequency shift is measured by displaying the reflected energy signal of the continuous microwave probing signal of fixed frequency on a synchronized oscilloscope. At the time during the decay period at which maximum signal absorption occurs the resonant frequency of the cavity is equal to the frequency of the probing signal. By varying this frequency the time dependent function $\Delta f(t)$ is obtained. The change in the quality of the microwave cavity can distort the reflected probing signal. According to formula (8-3-4), the change in the quality of the microwave cavity due to the presence of the plasma depends, for a given gas and pressure, on the electron density. This change in quality causes a change in coupling of the probing signal to the cavity during the afterglow period. When this coupling changes appreciably during the period that the cavity absorbs part of the probing signal, the reflected signal may be distorted so that the moment of maximum absorption (minimum reflection) does not coincide with the time at which the resonant frequency of the cavity is equal to the probing signal frequency. For example, if the coupling increases during the afterglow period, the minimum of the reflected signal may occur at a time appreciably later than the time at which the two frequencies are equal. This phenomenon was observed by Oskam and Mittelstadt [6] during studies in helium, krypton and xenon in the high-pressure region. This is consistent with the relatively large values of the collision frequency $v_m$ of electrons with gas particles in these gases. The influence of the change in probing signal coupling could be eliminated by sweeping the frequency of the probing signal through the resonant frequency of the cavity, at a given time during the afterglow period, within a time interval during which the electron density could be assumed to be constant.

The energy given to the electrons inside the cavity by the low-level probing signal should not disturb the normal behavior of these electrons. Therefore, this energy must be small compared to their thermal energy. Margenau [42] has shown that the steady state isotropic part $f_0^0$ of the velocity distribution function of electrons in a gas consisting of particles with mass $m$ in the presence of an applied microwave field of amplitude $E_0$ and angular frequency $\omega$ is given by

$$f_0^0(v) = \exp\left\{-\int_0^{v^2} \frac{\tfrac{1}{2}m_e\, d(v^2)}{kT_e + [me^2 E_0^2/6m_e^2(v_m^2 + \omega^2)]}\right\}, \tag{8-3-22}$$

provided that the collisions are dominantly elastic. In the absence of the microwave field the electron velocity distribution is Maxwellian with temperature $T_e$. When it is assumed that $v_m$ is independent of electron energy this energy is not increased by the probing field provided

$$E_0^2 \ll \frac{6m_e^2 kT_e(\omega^2 + v_m^2)}{me^2}. \tag{8-3-23}$$

It is possible to express the electric field inside the cavity, due to the absorbed power, in measurable quantities. Let $P$ be the part of the energy of the probing signal absorbed per second by the cavity at resonance. Then inside the cavity an electric-field amplitude $E(r)$ is built up which satisfies the relation

$$Q = \frac{\tfrac{1}{2}\varepsilon_0 \int E^2(r)\, dV}{P/\omega_0}, \tag{8-3-24}$$

or introducing the maximum amplitude $E_m$ of the field inside the cavity,

$$E_m^2 = \frac{2PQ}{\varepsilon_0 \omega_0 V \delta}, \tag{8-3-25}$$

in which

$$\delta = \frac{1}{VE_m^2} \int E^2(r) \, dV \, ; \qquad (8\text{-}3\text{-}26)$$

The combination of (8-3-23) and (8-3-25) yields the condition for the absorbed power $P$ of the probing signal by the cavity:

$$P \ll \frac{3\varepsilon_0 m_e^2 k \omega_0 (\omega_0^2 + v_m^2) V \, \delta T_e}{e^2 Q m}, \qquad (8\text{-}3\text{-}27)$$

or, evaluating the constants,

$$P \ll 7.2 \times 10^{-30} \frac{\omega_0 (\omega_0^2 + v_m^2) V \, \delta T_e}{QM} \, W. \qquad (8\text{-}3\text{-}28)$$

Here, $M$ represents the molecular weight of the gas particles, while $V$ is expressed in (meter)$^3$. Condition (8-3-28) limits the energy of the probing signal to a few microwatts.

Recently, Frommhold et al. [40] and Mehr and Biondi [43] used the 3-mode microwave cavity (resonant frequencies in the 3 Gc/sec region) shown in Fig. 8-3-4 for the study of the dependence of the dissociative recombination coefficient of $Ne_2^+$ and $Ar_2^+$ on the electron energy under the experimental condition that the ion energy is equal to that of the gas particles at 300°K. A high-quality $TE_{111}$ cavity mode was used to produce the pulsed plasma, a high-quality $TM_{010}$ mode was used to measure the cavity detuning by the electrons, while the microwave electron heating field was applied by means of the non-resonant $TE_{11}$ circular waveguide mode. The mean energy of the plasma electrons during the decay period resulting from the heating field was calculated using eq. (8-3-22). In order to correct for the spatial variation in the electron density $E_0^2$ in eq. (8-3-20) was replaced by the average

$$\langle E^2 \rangle_{av} = \int n_e(r, t) E_0^2(r) \, dV \Big/ \int n_e(r, t) \, dV. \qquad (8\text{-}3\text{-}29)$$

The dissociative recombination coefficient could be measured for electron temperatures varying from 300°K to 11 000°K.

B. LIGHT SPECTROMETER TECHNIQUES

Two experimental methods involving light spectrometer techniques have been found very useful in the study of electron–ion recombination processes. The first relates to the determination of the spectral distribution of the light emitted by the decaying plasma and the measurement of the time dependence of the intensity of various spectra lines and bands during the plasma decay period. The second method measures the shape of atomic spectral lines emitted due to the recombination process. The latter method was developed by Biondi et al. [17, 18] specifically in an effort to establish the dissociative recombination processes in plasmas produced in helium, neon and argon. Sometimes, light absorption techniques are employed for determining the possible influence of the presence of metastable excited atoms and/or molecules on recombination studies [40, 44].

1. *Light emission*

The time dependence of the intensity of spectral lines and bands during the decay period of plasmas produced in rare gases has been studied by various investigators [7, 8, 44–48]. The light emitted by the decaying plasma is separated into its various constituents by means of a monochromator. Interference filters can be used for lines and/or bands located sufficiently isolated from other lines (bands). The intensity of the selected line leaving the exit of the spectrometer is detected by an appropriate detector. The time dependence of signal magnitude can be measured by various techniques. One method consists of sensitizing a photomultiplier for a short period of time at a given time $t$ during the decay period. The sensitizing voltage pulse has the same frequency as the discharge pulse and its occurrence is determined by a delay circuit. The photomultiplier current pulses are sampled by applying them to a RC network. The voltage across this network is measured by an electrometer and displayed on a strip chart recorder. By changing the delay time $t$ the time dependence of the intensity of the spectral line is obtained. This sampling technique makes it possible to measure the intensity of spectral lines and bands over a large intensity range [44].

Recently, pulse counting detection systems are becoming more widely used. An example of such a system is the combination of a pulse counter and a multi-channel analyzer. The number of current pulses due to the photons incident on the photocathode of the multiplier are counted at equidistant time intervals during each decay period. The number of pulses at each time interval is stored in the multi-channel analyzer such that each channel corresponds to a fixed time in the decay period. The intensity of the spectral line studied and the desired intensity range determine the number of decay periods to be sampled. This detection method is considerably faster than the previous method and results in at least the same measurable spectral line intensity range. The increased speed at which data can be collected makes the requirements concerning the stability of the experimental system less severe. Moreover, it can be used with photon detectors which exhibit an increased noise signal under pulsed operation.

Cooling of the photomultiplier tube can be used to improve the signal to noise ratio. If only a small part of the photocathode is used the noise signal due to off-center secondary electrons can be strongly reduced by magnetically defocusing these electrons [49].

In order to obtain a complete description of the light emission properties of decaying plasmas it would be desirable to measure the time variation of the intensity of each individual spectral line and band. The time required for this is large; moreover, it is known that many lines (bands) show the same afterglow behavior. One of the methods to find the groups which decay at different rates consists of recording a complete spectral distribution at a few different times in the afterglow by scanning the light spectrometer at a very slow rate at the desired times in the afterglow. By comparing the several spectra the relative decay rates can be determined.

## 2. *Spectral line shape*

The dissociative recombination processes (8-2-8) and (8-2-10) can be effective only if it is exothermic, so that the transition from the bound state $(X_2)_{resonance}$ to the dissociated state $X^*+X$ results in the conversion of potential energy equal to $E_d$ (Fig. 8-2-1) into kinetic energy of the dissociated particles. For a recombining homonuclear diatomic ion this energy of dissociation is shared equally by the two identical atoms. Biondi et al. [17, 18] designed an experiment for detecting and measuring the kinetic energy of dissociation of excited atoms formed by the recombination process.

The spectral line shape due to Doppler effects arising from the thermal motion of the molecular ions and from dissociation of the unstable intermediate excited molecule formed due to the radiationless capture of an electron by the molecular ion was calculated under the following assumptions:

(a) The velocity distribution $f_0(v)$ of the homonuclear diatomic ions is Maxwellian at gas temperature, i.e.,

$$f_0(v) = (\beta/\pi)^{\frac{3}{2}} \exp(-\beta v^2), \qquad (8\text{-}3\text{-}30)$$

where $\beta = m/kT$, with $m$ being the mass of the single atom.

(b) All dissociations leading to a given excited atomic state originate from a single specific molecular energy state, and, therefore, release a definite total kinetic energy $E_D$, which is equally shared by the two dissociating atoms. This means that only one vibrational level of the ions is considered and that the spread in energy of the captured electrons is neglected.

(c) The direction taken by the excited atom upon dissociation is randomly oriented. Combined with assumption (b) this means that the distribution $f_1(v)$ in velocity of the excited atoms relative to the center of mass of the diatomic ions is the isotropic unit impulse function

$$f_1(v') = \delta(v_1), \tag{8-3-31}$$

with $v_1 = (E_D/m)^{\frac{1}{2}}$.

For the calculation of the Doppler shift only the distribution in speed along the observer's line of sight (the $z$-axis) is needed. Integration of the distributions (8-3-30) and (8-3-31) over the $x-y$ plane gives

$$f_0(v_z) = (\beta/\pi)^{\frac{1}{2}} \exp(-\beta v_z^2) \tag{8-3-32}$$

and

$$\begin{aligned} f_1(v'_z) &= \tfrac{1}{2} v_1 \quad \text{for} \quad -v_1 \leqslant v'_z \leqslant v_1 \\ f_1(v'_z) &= 0 \quad \text{for} \quad |v'_z| > v_1. \end{aligned} \tag{8-3-33}$$

The distribution $f_0(v_z)$ is the thermal velocity distribution along the $z$-axis of the centers of mass of the diatomic ions, while $f_1(v'_z)$ is this distribution of the excited atoms with respect to the centers of mass. For every $v'_z$ there is thus a complete distribution of velocities $f_0(v_z)$ of these centers of mass. For the observer this results in a group of excited atoms having a Maxwellian distribution centered on the value $v'_z$, so that the numbers of atoms in the speed range $dv'_z$ at $v_z - v'_z$ is

$$dG = f_0(v_z - v'_z) f_1(v'_z) dv'_z. \tag{8-3-34}$$

The total resultant distribution in $v_z$ is obtained by integrating (8-3-34) over all values of $v'_z$. Biondi et al. first substituted in eq. (8-3-34) for $v_z$ the corresponding Doppler wave number shift $\bar{v} = \bar{v}_0 (v_z/c)$, where $\bar{v}_0$ is the frequency (in wave numbers) of the center of the spectral line. The resultant intensity contour of the line (normalized to unit area) is given by

$$G(\bar{v}) = \frac{a}{4b} [\operatorname{erf}(a\bar{v} + b) - \operatorname{erf}(a\bar{v} - b)], \tag{8-3-35}$$

where $a = (mc^2/kT)^{\frac{1}{2}}/\bar{v}_0$ and $b = (E_D/kT)^{\frac{1}{2}}$. Analysis of the function $G(\bar{v})$ shows that the line contour has a trapezoidal form, while the width at half-intensity gives $E_d$ directly, provided $E_d \gg kT$, as is usually expected. An example of such a predicted line shape calculated by Connor and Biondi at an ion temperature of 300 °K for $\lambda 5852$ radiation $(2p_1 \rightarrow 1s_2)$ emitted from excited neon atoms produced by dissociative recombination of $Ne_2^+$ with electrons is shown in Fig. 8-3-5; a dissociation kinetic energy $E_d = 1.3$ eV was assumed [18].

The simple dissociative line shape shown has been obtained by making several simplifying assumptions and by neglecting collisions between the produced fast excited atoms and other plasma constituents. Biondi et al. [17, 18] have considered various processes which may influence the dissociative line shape. If different vibrational and/or electronic states of the diatomic ion recombine and dissociate through the same and/or different potential energy curves, while resulting in excited atoms in the same electronic state, several values of kinetic dissociation energy $E_d$ have to be considered. The resulting spectral line contour is then a composite of contours shown in Fig. 8-3-5. Each half-width of the sub-trapezoid yields the $E_d$ value involved.

A collision process which can strongly influence the spectral line shape is that of an excitation transfer between the fast excited atom and a slow normal atom before radiation occurs, i.e.,

$$X_f^* + X_s \rightarrow X_s^* + X_f, \tag{8-3-36}$$

where the subscripts f and s refer to fast and slow atoms, respectively. This process thus converts fast-excited atoms into slow-excited atoms, which radiate the thermal Doppler line. When dissociative recombination and excitation transfer are the only significant processes the emitted line consists of

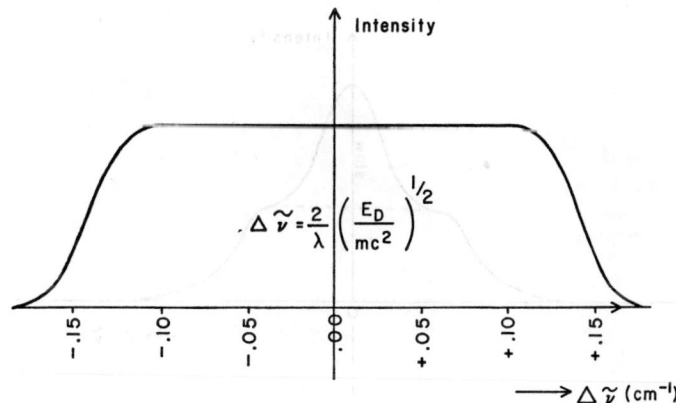

Fig. 8-3-5. Predicted line shape at 300 °K for $\lambda = 5852$ Å radiation emitted from excited neon atoms produced by dissociative recombination, with $E_d = 1.3$ eV (reference 18).

the broad and flat-topped dissociative profile representing those atoms which radiate before excitation transfer occurs, surmounted by a narrower Gaussian shaped core representing slow excited atoms which are produced by the excitation transfer process. Such a composite line profile is shown in Fig. 8-3-6.

Biondi et al. also discussed the influence on the spectral line shape of the isotope effect, Stark effect, gas pressure, and momentum transfer collisions of the fast-excited atoms with slow atoms. The experimental conditions, however, could be chosen such that the line shape was mainly determined by dissociation and excitation transfer.

The apparatus used by Biondi et al. [17, 18] for measuring the line shape of spectral lines emitted during the discharge and the decay period of plasmas produced in helium, neon and argon is shown schematically in Fig. 8-3-7. A dual-mode microwave system is used to produce the pulsed plasma inside a quartz bottle enclosed by a microwave cavity and to measure the average electron density during the decay period. In order to obtain the intensity contour of a given spectral line, that line is first isolated from the rest of the spectrum by a narrow band (about 17 Å half-width) Spectrolab interference filter. During later studies this isolation was improved by replacing the filter by a monochromator with a resolution of about 6 Å. The line then passes through the Fabry-Perot (F-P) interferometer. The concentric rings of the F-P pattern are imaged by the output lens (focus length 20 cm) on a plate containing a hole (0.36 mm diameter) which is aligned with the center of the pattern. Thus a small portion of the radiation in the central spot passes through and falls on the photocathode of a gated photomultiplier. The

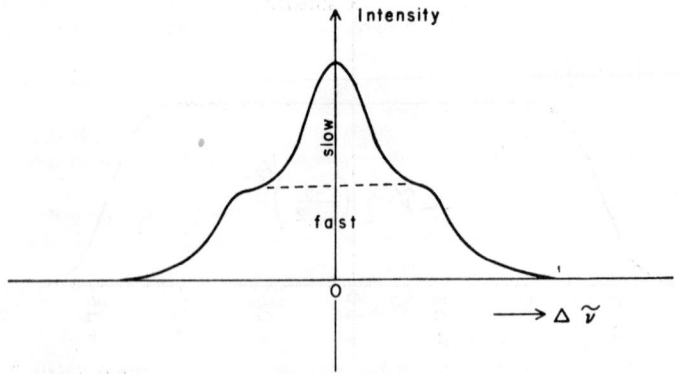

Fig. 8-3-6. The predicted spectral line shape due to the dissociative recombination process and the excitation transfer process $X_f{}^* + X_s \rightarrow X_s{}^* + X_f$ (reference 18).

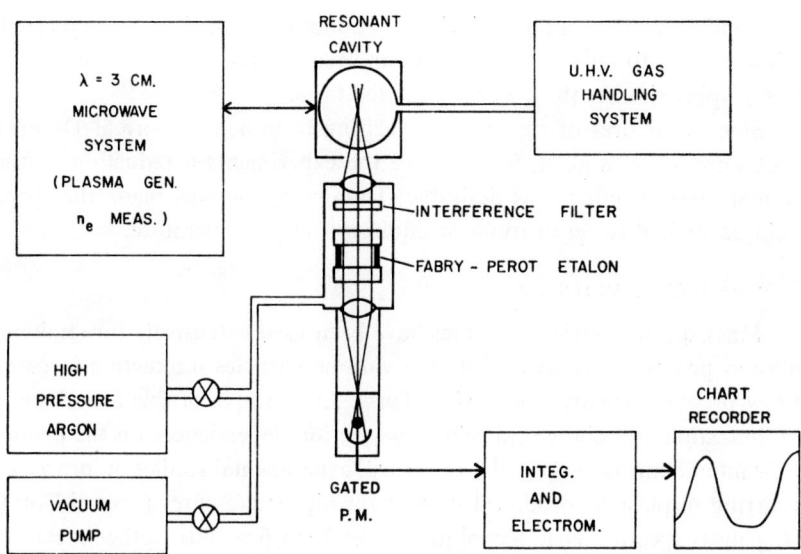

Fig. 8-3-7. Highly simplified block diagram of the experimental apparatus used by Connor and Biondi for the study of spectral line shapes (reference 18).

output signal is fed to an integrating circuit (time constant about 4 sec) and electrometer whose output is displayed on the chart recorder.

The wavelength for constructive interference of beams making up the center of the F-P pattern is given by

$$\lambda = 2\mu d/n, \qquad (8\text{-}3\text{-}37)$$

where $\lambda$ is the vacuum wavelength, $\mu$ the refractive index of the medium between the plates, $d$ the plate spacing, and $n$ the interference order. This wavelength is made to change linearly with time by introducing argon gas at a constant rate to the sealed, initially evacuated housing containing the F-P plates; thus the refractive index between the plates increases linearly with time. The chart paper is driven by a synchronous motor, so that a linear presentation of intensity versus frequency over the line profile is obtained. The frequency scale is calibrated by tuning through several interference orders, since the pattern repeats itself after a wave number interval $\Delta \bar{v}_{\text{fs}} = (2\mu d)^{-1} \approx (2d)^{-1}$. The F-P plates were flat to 1/100 wavelength and coated with dielectric films of $\sim 97\%$ reflectivity. The resulting finesse (the ratio of the free spectral range, $\Delta \bar{v}_{\text{fs}}$, between interference orders to the instrumental half-width) was 42. For a 1.0 cm spacer, $\Delta \bar{v}_{\text{fs}} = (2\mu d)^{-1} \approx 0.50$

$cm^{-1}$ and an instrumental half-width of about 0.01 $cm^{-1}$ can be expected. This width was found to be sufficiently small so that the observed line profiles closely approximates those of the radiated lines.

Since structures of spectral lines arising from a non-thermal Doppler effect were to be studied, for some of the experiments a reduction of the thermal Doppler effect was desirable. Thus provision was made that lines could be studied either at room or liquid nitrogen temperature.

### C. Mass Spectrometer Techniques

Mass spectrometric techniques have been used extensively for studying collision processes occurring between various particles interacting in gases at low pressures (below about $10^{-3}$ Torr). An example of this technique is the ionization and charge transfer cross-section dependences on the energy of the interacting particles [50]. In contrast, experimental studies on processes occurring in plasmas produced in gases at higher pressures (several Torr), using mass spectrometric techniques, have been few until rather recently.

In 1952 Phelps and Brown [51] reported the first studies of decaying helium plasmas by means of a 60° magnetic deflecting type of mass spectrometer with a resolution of about 100. It was found that the dominant ion during the decay period changed from $He^+$ to $He_2^+$ as the gas pressure was increased from 1 to 5 Torr. The number density range of the ions reported was rather small (about 10 to 1) and no further mass spectrometer studies of decaying rare gas plasmas were reported by these authors. About nine years later Kasner et al. [52] published afterglow studies in helium and neon containing small concentrations of nitrogen and oxygen. The time dependence of the number density of $N_2^+$ was measured by means of a Boyd type r.f. mass spectrometer. In 1962 Fite et al. published preliminary mass spectrometer studies of decaying plasmas produced in commercial reagent grade helium, oxygen, nitrogen, oxygen–nitrogen and nitrogen–hydrogen mixtures; a magnetic-sector mass spectrometer was used.

During 1960 Mosharrafa and Oskam [53–55] assembled a set of requirements of a mass spectrometer to be used for the study of collision processes occurring in plasmas produced in gases over wide pressure ranges above 0.1 Torr. The required mass spectrometer features were believed to be:

1. Adequate resolving power. High resolution techniques are generally not required for plasma studies, since only identification of the ionic species is required and could be achieved through isotopic abundance measurements.

2. Very high sensitivity. This feature is particularly desirable in studies of the properties of decaying plasmas, where it is important to measure the

time decay of the density of ionic species over several orders of magnitude.

3. No energy discrimination in the mass analyzing system. This makes it possible to obtain true quantitative representations of the various ion densities.

4. High ion-source pressure capabilities. The mass spectrometer should maintain efficient operation throughout a large range of pressure variation in the gaseous discharge ion source. This enables the study of the pressure dependence of collision processes occurring in the plasma.

5. Bakeable system. It is well known that traces of impurities in the discharge tube as small as 1 part in $10^7$ can have an appreciable influence on plasma properties.

6. Minimal influence of the mass spectrometer system on the plasma behavior. This is important in studies related to the natural state of the plasma which may be very sensitive to external magnetic and electric fields.

After a study had been made of different existing types of mass spectrometers, the quadrupole "mass filter" proposed and developed by Paul [56–58] was believed to be the most suitable type of mass spectrometer for the study of the properties of plasma ions. The main considerations which led to the choice of this particular type of mass spectrometer are:

(a) The "mass filter" has a very high transmission efficiency especially at low resolving powers where the efficiency approaches 100%.

(b) The strong focusing action of the quadrupole analyzing field. This feature permits the mass spectrometer to operate at high residual gas pressures, of the order of $10^{-3}$ Torr, without seriously affecting its characteristics. Hence, it offers a distinct advantage for flow-through systems, as well as eliminating differential pumping requirements for studies employing ion sources operating in this pressure range.

(c) The cylindrical symmetry of the analyzing field. The most practical way of extracting ions from a plasma is via a small circular orifice drilled through the plasma container. The resulting ion beam of cylindrical symmetry therefore matches the optimum injection requirements of the quadrupole field. This maximizes the coupling efficiency between the ion source and the analyzing field, without the use of special lens systems.

The quadrupole mass spectrometer is a mass filter, which transmits only those ions whose mass-to-charge ratio lies within a band whose width is easily variable electronically. The theory of the quadrupole mass filter has been discussed in detail in the literature. Therefore, only the essentials of its operation will be given here. A voltage $(U + V \cos \omega t)$ is applied to the hyperbolic electrode system as shown in Fig. 8-3-8. The charged particles

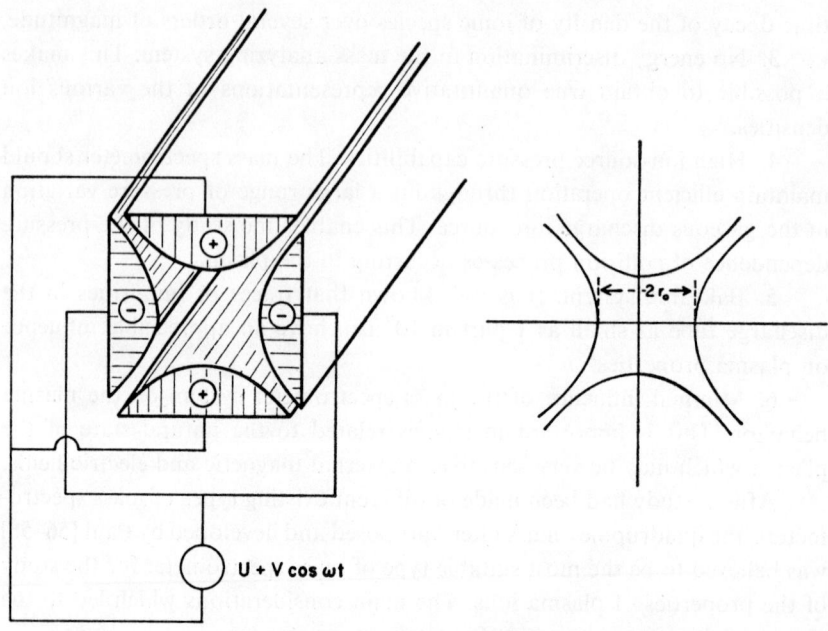

Fig. 8-3-8. Schematic of quadrupole mass spectrometer electrode geometry.

(charge $e$ and mass $m$) are injected axially and move in the potential field

$$\varphi(x, y) = (U + V \cos \omega t) \frac{x^2 - y^2}{r_0^2}. \tag{8-3-38}$$

Substitution of $\omega t = 2\xi$ leads to the following equations of motion of the ion in the $x$–$y$ plane

$$\frac{d^2 x}{d\xi^2} + (a + 2q \cos 2\xi) x = 0 \tag{8-3-39}$$

$$\frac{d^2 y}{d\xi^2} - (a + 2q \cos 2\xi) y = 0, \tag{8-3-40}$$

with

$$a = \frac{8eU}{m\omega^2 r_0^2} \tag{8-3-41}$$

and

$$q = \frac{4eV}{m\omega^2 r_0^2}. \tag{8-3-42}$$

Equations (8-3-39) and (8-3-40) are of the form of the Mathieu differential equation [59]. An extensive discussion of this equation in connection with mass analyzers has been published [57]. It is found that there are several regions in the $a$–$q$ plane for which the amplitudes of both $x$ and $y$ remain smaller than the "radius" $r_0$ of the quadrupole electric field. These are the so-called stable regions of the stability diagram. The largest stable region is used for the operation of the quadrupole mass spectrometer and is shown in Fig. 8-3-9. It is bounded by the $q$-axis and by the lines

$$a_1(q) = +\tfrac{1}{2}q^2 - \tfrac{7}{128}q^4 + \tfrac{29}{2304}q^6 \tag{8-3-43}$$

and

$$a_2(q) = 1 - q - \tfrac{1}{8}q^2 + \tfrac{1}{64}q^3 - \tfrac{1}{1536}q^4. \tag{8-3-44}$$

If the parameters ($U$, $V$, $m$, $\omega$, $r_0$) for a specific ion are such that the operating point lies within this region the motion of the ion in the $x$–$y$ plane is stable and hence it will pass through the quadrupole and can be detected at the exit of the mass spectrometer. If the operating point lies outside the stable region, the ion will oscillate with exponentially increasing amplitude and will eventually be lost at the quadrupole field electrodes. This is the

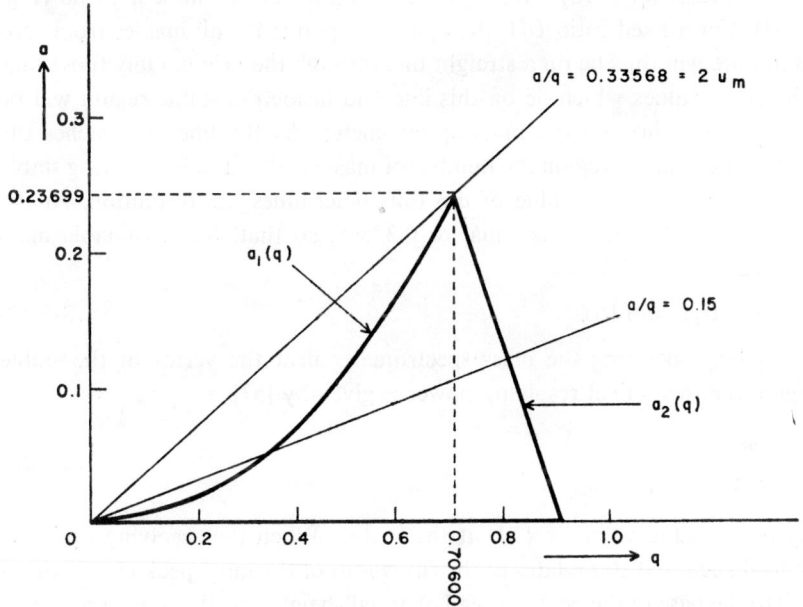

Fig. 8-3-9. Stability diagram of quadrupole mass spectrometer (reference 57).

principle of "mass filtering" upon which the quadrupole spectrometer operates.

In order to ensure complete loss of ions having unstable orbits at the field electrodes these ions should remain inside the quadrupole region during a minimum number $n$ of high-frequency periods, so that its amplitude can exceed $r_0$. It was experimentally found that, for an operation point close to the vertex of the stable region [57],

$$n \geqslant 3.5 \left(\frac{m}{\Delta m}\right)^{\frac{1}{2}}. \qquad (8\text{-}3\text{-}45)$$

Here $m/\Delta m$ is the resolving power where $\Delta m$ is defined as the full width at half-height of the mass peak. Condition (8-3-45) leads to a maximum allowable ion injection voltage $V_z$ given by

$$V_z \approx 4.3 \times 10^{-2} L^2 f^2 M \frac{\Delta m}{m} \text{ V}, \qquad (8\text{-}3\text{-}46)$$

where $L$ is the length of the quadrupole field in cm, $f$ the field frequency in Mc/second and $M$ the molecular weight of the ion in atomic mass units. These units will be used throughout this section.

The ratio $a/q = 2U/V$ represents a straight line in the $a$–$q$ plane (Fig. 8-3-9). For a fixed ratio $U/V$ the operating points for all masses from zero to infinite will thus lie on a straight line through the origin. Only those ions with mass values which lie on this line and inside the stable region will be able to pass through the mass spectrometer. As the line approaches the vertex of the stable region the number of masses related to ions having stable orbits decreases. The value of $a/q$ thus determines the resolution. At the vertex the value of $a/q$ is equal to 0.33568, so that, for reasonable mass resolution

$$u = U/V \approx 0.167. \qquad (8\text{-}3\text{-}47)$$

When operating the mass spectrometer near the vertex of the stable region the theoretical resolving power is given by [57]

$$\frac{m}{\Delta m} \approx 1.5 \left(1 - \frac{u}{u_m}\right)^{-1}, \qquad (8\text{-}3\text{-}48)$$

where $u_m$ is the value of $U/V$ at the vertex. When the resolving power is defined such that $\Delta m$ relates to the full-width of the mass peak close (within 1%) to the base of the peak, instead of at half-height, the theoretical resolving power is about half of that given by eq. (8-3-48).

The operating values of $U$ and $V$ follow from eqs. (8-3-41) and (8-3-42) and are for operation close to the vertex of the stable region given by

$$U \approx 1.22 r_0^2 f^2 M \text{ V} \qquad (8\text{-}3\text{-}49)$$

and

$$V \approx 7.30 r_0^2 f^2 M \text{ V}, \qquad (8\text{-}3\text{-}50)$$

where $r_0$ is expressed in cm. A mass scan with constant $m/\Delta m$ is obtained by keeping $U/V$ constant and changing the frequency with constant voltages (frequency scanning) or changing the voltages with constant frequency (voltage scanning).

The preceding discussion related to ions injected along the quadrupole field axis with zero radial energy. It was found that the maximum radial injection energy $E_r$ a stable ion is allowed to have is equal to [57]

$$E_r \approx \frac{V}{15(m/\Delta m)} \text{ eV}. \qquad (8\text{-}3\text{-}51)$$

The transmission efficiency will be the same for different ions (mass-independent transmission) provided $E_r$ is independent of the selected mass. This can be achieved by constant $V$ and $\Delta m/m$, which in turn requires a constant $U/V$. Since both voltages must be held constant, a mass spectrum is obtained by changing the frequency. An alternate method involves a constant $\Delta M$ and constant $f$, since substitution of eq. (8-3-50) into (8-3-51) leads to

$$E_r \approx 0.48 f^2 r_0^2 \Delta M \text{ eV}. \qquad (8\text{-}3\text{-}52)$$

In order to achieve a value of $\Delta M$ independent of selected ion mass a relation

$$U = \gamma V - \delta \qquad (8\text{-}3\text{-}53)$$

between the r.f. and d.c. voltages is used instead of the relation (8-3-47) [60]. Substitution of eq. (8-3-53) into eq. (8-3-48) yields

$$\Delta M = 3.97(0.16874 - \gamma) M + \frac{0.545 \, \delta}{f^2 r_0^2}. \qquad (8\text{-}3\text{-}54)$$

For $\gamma = 0.16874$, $\Delta M$ is independent of $M$ and is then given by

$$\Delta M = \frac{0.545 \, \delta}{f^2 r_0^2}, \qquad (8\text{-}3\text{-}55)$$

so that, for fixed values of $f$ and $r_0$, its value can be varied by changing $\delta$. A mass scan is obtained by changing at fixed frequency the voltage amplitudes according to the relation (8-3-53).

The maximum ion injection-aperture diameter, so that all stable ions injected with zero radial energy will be contained in a circle of radius $r_0$ (the condition for transmission) is [57]

$$d \approx r_0 \left(\frac{\Delta m}{m}\right)^{\frac{1}{2}}. \tag{8-3-56}$$

The transmission efficiency due to this condition is thus independent of mass for constant resolving power. The limit on radial injection energy (8-3-51) is usually the most restrictive ion injection condition.

The high-frequency power requirement for the quadrupole system is given by [57]

$$P = 6.5 \times 10^{-4} \frac{CM^2 f^5 r_0^2}{Q} \text{ W}, \tag{8-3-57}$$

where $C$ is the system capacitance in pF, and $Q$ is the quality factor of the output circuit.

The highest resolving powers, smallest power requirements, etc. have been obtained with accurately machined hyperbolic field electrode systems. These systems, however, are rather costly and resolving powers of a few hundred can also be obtained with the low cost circular-rod electrode system. The best approximation to a quadrupole field is obtained when the ratio of the rod radius to the field radius $r_0$ is 1.16. The bakeable rod-type quadrupole mass spectrometer constructed by Sauter et al. [61] is shown in Fig. 8-3-10. The field radius is 0.5 cm and the analyzing field lengths used

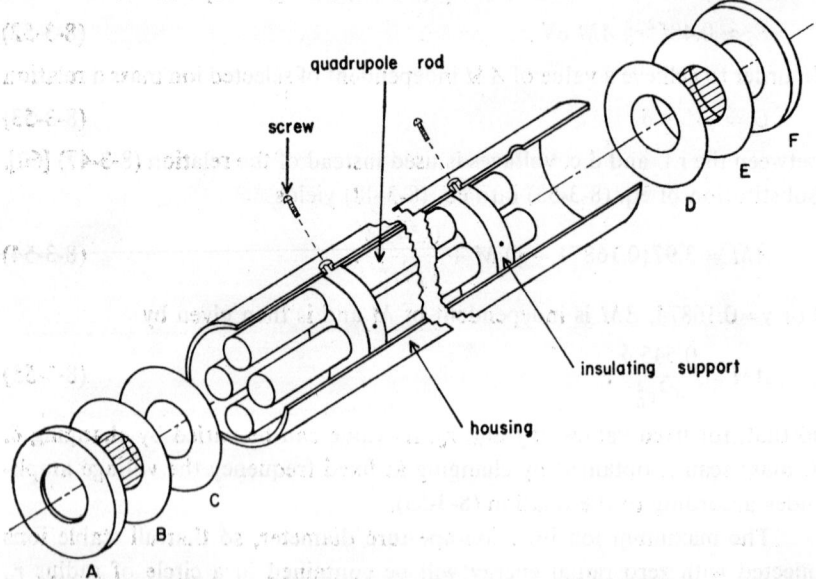

Fig. 8-3-10. Cutaway view drawing of rod-type quadrupole mass spectrometer (reference 61).

varied from 10 to 20 cm. The four electrodes are type 304 stainless steel tubes and are held in proper alignment by two insulating supports made from "Supramica 500". This assembly is placed inside a section of stainless steel tubing. The electrodes are secured in position with stainless steel machine screws which pass through the outside housing and insulating support and are threaded into the electrodes. The screws are insulated from the outside housing by Supramica insulating washers. The entrance grid (B) is separated from the electrode system by insulator (C) and is held close to it by the insulator cap (A). This grid is used to accelerate the ion beam effusing through the hole in the wall of the plasma container. The exit grid (E) makes it possible to avoid or achieve trapping of ions inside the quadrupole field region. This grid is kept in place inside the outside housing by the two insulators (D) and (F). The ion multiplier used for measuring the number of ions passing through the mass spectrometer is placed immediately behind the exit grid. The exit grid also decreases the influence of the ion multiplier accelerating field on the operation of the mass spectrometer.

Various plasma container–mass spectrometer combinations have been reported [14, 15, 51]. The arrangement used by Oskam et al. [61, 62] for the study of the time dependence of the densities of ions during the plasma decay period is shown in Fig. 8-3-11. This configuration eliminates any influence of the quadrupole field on the properties of the plasma. The plasma could be generated by applying a r.f. or d.c. voltage between the molybdenum electrode and the metal plate containing the effusion hole. The r.f. power for producing the plasma can also be applied through metal rings placed around the glass cylinder. The molybdenum electrode serves also as a boundary of the plasma establishing a well-defined characteristic diffusion length of the plasma container. Moreover, it can be used to cover the glass wall with a thin molybdenum layer, which traps non-rare-gas impurities present in the gas or evolving from the walls.

The electronic equipment needed for the operation of the quadrupole mass spectrometer can be kept very simple or made very elaborate and automatic. An example of a very simple and low cost electronic system is shown in Fig. 8-3-12. It employs a U.S. Government surplus BC-375-E transmitter together with a rectifying circuit. The frequency ranges of the transmitter are from 0.2 to 0.8 Mc/sec and from 1.5 to 12.5 Mc/sec. A mass scan can be made by continuously varying the high voltage (0 to 1500 V) to the transmitter. This is accomplished by a motor drive connected to the power supply. If high quadrupole mass spectrometer operating stability is required a more elaborate electronic system is needed.

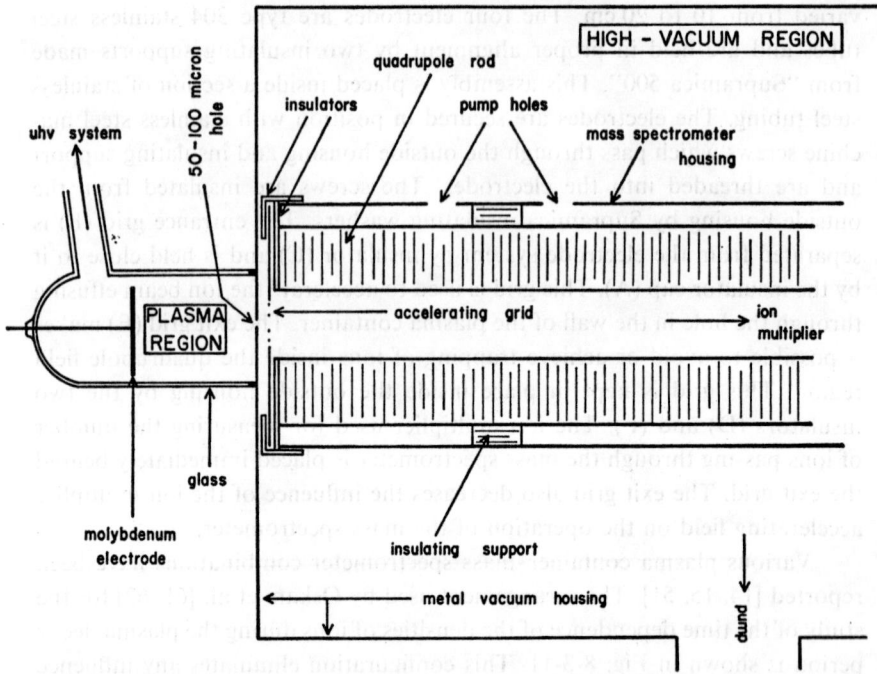

Fig. 8-3-11. The plasma container-quadrupole mass spectrometer combination used by Oskam et al. (references 61, 62, 66).

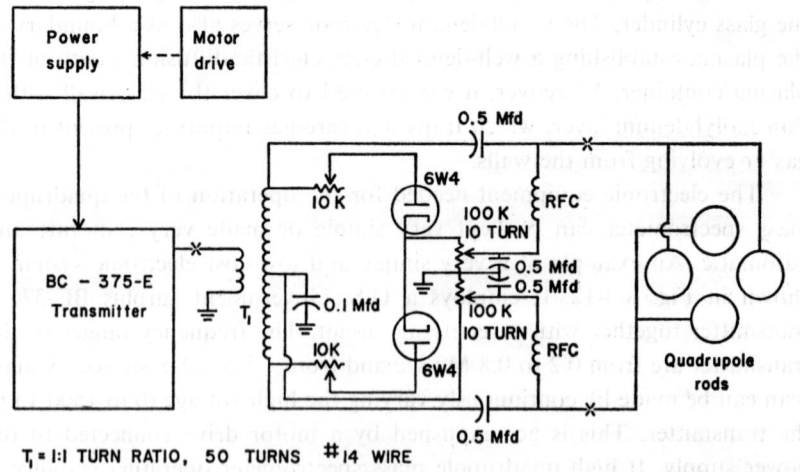

Fig. 8-3-12. Low cost electronic equipment for quadrupole mass spectrometer (reference 61).

The time dependence of the ion density during the plasma decay period can be obtained by detection electronics identical to those used for the light-emission studies.

## § 8-4. Recombination processes involving helium ions

In 1949 Biondi and Brown [3] reported microwave cavity measurements of the time dependence of the electron density during the decay period of a helium plasma. Studies at pressures larger than 20 Torr resulted in the first reported value of $\alpha(\text{He}_2^+) = 1.7 \times 10^{-8}$ cm$^3$/sec. Shortly after the publication of these studies Bates [20] suggested that, since molecular positive ions of the noble gases were known to exist, the unexpected large value of $\alpha(\text{He}_2^+)$ was the result of the dissociative recombination process. Analogous studies reported by Oskam [5] in 1958 and by Kerr [7] in 1960, however showed that the value of $\alpha(\text{He}_2^+)$ was smaller than that reported by Biondi and Brown. The authors were unable to determine a value of $\alpha(\text{He}_2^+)$ from afterglow studies by means of the microwave cavity method. In 1963 Oskam [6] reported another effort in measuring a value for $\alpha(\text{He}_2^+)$. At a helium pressure of 60 Torr the density range over which the $1/n_e(t)$ versus time curves appeared to be a straight line was only 4. The value of the recombination coefficient which could be calculated from this curve was $7 \times 10^{-9}$ cm$^3$/sec. The value of $f$ is too small, however, to assign much significance to this value. The correction following from Gray and Kerr's [38] as well as from Frommhold and Biondi's [39] analysis is very large, and the only conclusion of this study was that, if dissociative recombination is active at all, $\alpha(\text{He}_2^+) \leqslant 4 \times 10^{-9}$ cm$^3$/sec. An analysis of analogous studies reported by other authors leads to the unavoidable conclusion that the microwave cavity method is not suitable to determine the type of recombination process related to $\text{He}_2^+$.

In an effort to solve the discrepancy between the recombination coefficients reported, Kerr [7, 8] conducted careful simultaneous studies of the time dependence of the electron density and that of the intensities of spectral lines and bands emitted during the helium afterglow period. Kerr found that the intensities of spectral lines originating from energy levels having principal quantum number as high as six had the same time dependence. The same property was observed for the molecular bands studies, although the time dependence was different from that of the lines. Kerr did not obtain a consistent interpretation of the recombination processes active during the decay period and concluded that a single recombination process could not explain the results.

Studies of the relative population densities of highly excited atomic states in high-intensity and high-temperature helium afterglows led d'Angelo [11] in 1961 to believe that the collisional recombination process

$$He^+ + 2e \rightarrow He^* + e \qquad (8\text{-}4\text{-}1)$$

was responsible for the emission of spectral lines during the decay period studied. D'Angelo's main reason for assuming the occurrence of process (8-4-1) was the measurement of thermal equilibrium between electrons and He* for large principal quantum numbers. The equilibrium was assumed to be maintained via the processes

$$He^+ + 2e \leftrightarrows He^* + e. \qquad (8\text{-}4\text{-}2)$$

Definite proof of the occurrence of process (8-4-2) could not be obtained from these studies since the change in the electron energy during the decay period studied dominated all other time dependences.

Niles and Robertson [47] showed in 1964 that the afterglow emission of a plasma produced in helium converted for increasing pressure rapidly from line emission to band emission. This indicated that the line emission is mainly related to collision processes involving atomic ions, while the emission of spectral bands is mainly a consequence of recombination processes involving $He_2^+$. These studies thus confirmed previously reported observations by Herman [9], Johnson et al. [10] and Kerr [7, 8].

In 1964 Rogers and Biondi [17] reported results obtained from spectral line broadening measurements during the helium afterglow period. The purpose of these studies was to establish the occurrence of the dissociative recombination of $He_2^+$ with an electron by detecting and measuring the kinetic energy of dissociation of the excited helium atoms resulting from this neutralization process. The results obtained, however, were inconclusive.

Although various collision processes had been proposed and questioned the origin of afterglow emission and the type of recombination processes occurring was not established by the previous studies. One of the difficulties was the uncertainty about the types of ions present during the decay period. At low pressures ($\sim 1$ Torr) it could be assumed that $He^+$ was the dominant ion, while at pressures above about 10 Torr $He_2^+$ should determine the afterglow properties. It was suspected that line emission resulted from a recombination process involving $He^+$, while $He_2^+$ was believed to be responsible for the band emission. In order to remove the uncertainties concerning the relation between ion type and spectral emission Gerber et al. [45] studied the helium afterglow by simultaneously measuring the time dependences of the number density of the various ions present during the decay period and that of the intensities of spectral lines and bands during the same period. Mass spectrometer studies can also be expected to give information about the influence of impurities on the properties of decaying plasmas.

Gerber et al. first determined the validity and significance of mass-spectrometric probing of a decaying plasma by comparing the data obtained for the ambipolar diffusion coefficients of $He^+$ and $He_2^+$ in helium with those obtained by means of other measuring techniques. Excellent agreement was found. Since several of the early studies of the helium afterglow were performed on plasmas produced in commercial reagent grade helium without any further purification, the ionic composition of afterglows of this type of plasma was also determined. The measurements revealed the large influence on the plasma decay properties of neon impurity atoms present in these gases (about 50 parts per million). Fig. 8-4-1 demonstrates that at a helium pressure of about 7 Torr the dominant ion is $Ne^+$ instead of the expected $He_2^+$. At pressure above 10 Torr the dominant ions during the decay period were $Ne^+$ and $(HeNe)^+$. The reason for the large effect of a very small concentration of neon atoms is the occurrence of the following two processes postulated by Oskam [5]:

$$He_2^+ + Ne \rightarrow Ne^+ + 2He \qquad (8\text{-}4\text{-}3)$$

$$Ne^+ + 2He \rightarrow (HeNe)^+ + He. \qquad (8\text{-}4\text{-}4)$$

It was previously believed that the presence of a small number of neon atoms would have a negligible influence on the properties of plasmas produced in helium, since a helium-neon mixture does not constitute a Penning-mixture. The large cross section ($\approx 10^{-15}$ cm$^2$) associated with process (8-4-3), however, makes the influence of a small neon concentration in helium on the plasma properties as large as if it would have been a Penning mixture. Process (8-4-3) is also a major process in the cataphoretic segregation of neon atoms in helium-neon mixtures [63].

In order to be able to interpret the mass spectrometer studies of decaying helium plasmas the time dependences of the densities of $He^+$ and $He_2^+$ have to be calculated. The measured decay curves should exhibit the theoretically expected dependences on the plasma parameters, and the correlation with the time dependences of the intensities of spectral lines will depend on the type of recombination processes active during the plasma decay period. The interpretation of the data is facilitated by choosing experimental conditions such that the various measured time dependencies can

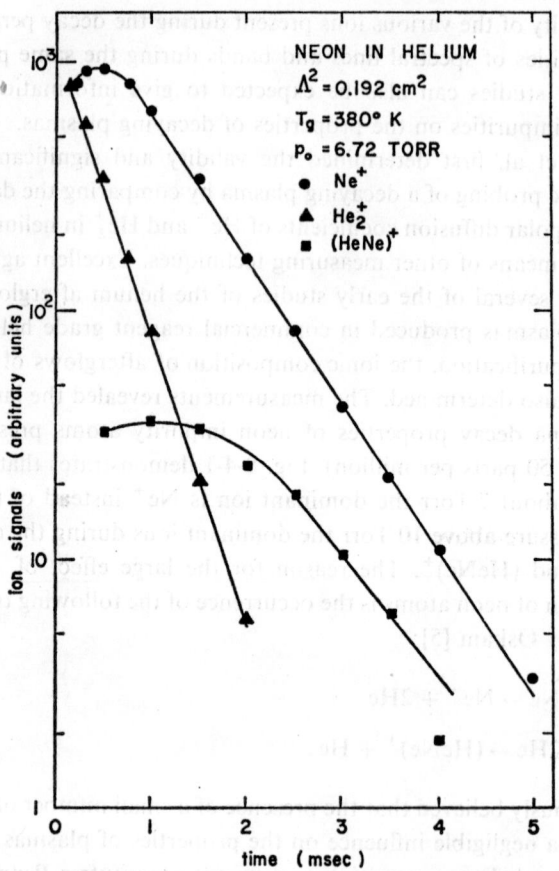

Fig. 8-4-1. Time dependence of atomic and molecular ion signals during the decay period of a plasma produced in a helium-neon mixture (reference 46).

be described by exponential functions during the major part of the decay period. This is rather easily realized for decaying helium plasmas since the recombination processes have for a wide range of conditions a negligible influence on the rate of loss of the helium ions, although the emission during the decay period is a consequence of the electron–ion recombination processes.

It will be assumed that the production of $He^+$ during the decay period by mutual collisions between metastable helium atoms can be neglected. Moreover, the various quantities relating to the efficiency of loss and production processes will be assumed to be independent of time and space coordinates. This condition is satisfied provided the plasma constituents are in thermal equilibrium with their environment, i.e., provided the data relate to measurements conducted late enough during the decay period. The equations describing the rate of change of $He^+$ and $He_2^+$ during the afterglow period under the influence of diffusion, recombination and ion conversion are then given by

$$\frac{\partial n_1(r, t)}{\partial t} = D_{a1} \nabla^2 n_1(r, t) - v_{conv} n_1(r, t) - \alpha_1 n_1(r, t) n_e(r, t) \quad (8\text{-}4\text{-}5)$$

$$\frac{\partial n_2(r, t)}{\partial t} = D_{a2} \nabla^2 n_2(r, t) + v_{conv} n_1(r, t) - \alpha_2 n_2(r, t) n_e(r, t). \quad (8\text{-}4\text{-}6)$$

Here $n_1$, $n_2$ and $n_e$ are the number densities of $He^+$, $He_2^+$ and electrons, respectively. $D_a$ is the ambipolar diffusion coefficient, $v_{conv}$ is the frequency of conversion of $He^+$ into $He_2^+$, and $\alpha$ is the recombination coefficient.

During the late afterglow period the last term in eqs. (8-4-5) and (8-4-6) can be neglected. When the amplitudes of the higher diffusion mode solutions have become small compared with that of the first order diffusion mode solution, the time dependent part of the solutions are given by

$$n_1(t) = n_1(0) e^{-t/\tau_1} \quad (8\text{-}4\text{-}7)$$

$$n_2(t) = \left[ n_2(0) + \frac{v_{conv} n_1(0)}{1/\tau_1 - 1/\tau_2} \right] e^{-t/\tau_2} - \frac{v_{conv} n_1(0)}{1/\tau_1 - 1/\tau_2} e^{-t/\tau_1}, \quad (8\text{-}4\text{-}8)$$

where

$$p_0/\tau_1 = \frac{D_{a1} p_0}{\Lambda^2} + p_0 v_{conv}, \quad (8\text{-}4\text{-}9)$$

$$p_0/\tau_2 = \frac{D_{a2} p_0}{\Lambda^2}. \quad (8\text{-}4\text{-}10)$$

Here $\Lambda$ is the characteristic diffusion length of the plasma container and $p_0$ is the helium pressure in Torr reduced to 273 °K.

Equation (8-4-7) shows that the time dependence of the number density of $He^+$ is exponential with a time constant $\tau_1$. The final time dependence of the density of $He_2^+$ is exponential according to eq. (8-4-8) with a time constant of either $\tau_1$ or $\tau_2$. The conversion frequency $v_{conv}$ for the three-body process

$$He^+ + 2He \rightarrow He_2^+ + He \quad (8\text{-}4\text{-}11)$$

can be written as

$$v_{conv} = C p_0^2. \quad (8\text{-}4\text{-}12)$$

Substitution of (8-4-12) into (8-4-9) and equating to (8-4-10) yields the critical pressure $p_c$ at which $\tau_1 = \tau_2$, i.e.,

$$p_c^3 = \frac{D_{a2}p_0 - D_{a1}p_0}{C\Lambda^2}.\qquad(8\text{-}4\text{-}13)$$

For $p_0 > p_c$ the final decay of the density of $He_2^+$ is characterized by $\tau_2$, while for $p_0 < p_c$ the final decay of the density of $He_2^+$ is the same as that of $He^+$ (determined by $\tau_1$).

When plotting $p_0/\tau$ for $He^+$ versus $p_0^3$ a straight line should result. The slope of this line gives the value of $C$, while the intercept yields the value of $D_{a1}p_0/\Lambda^2$. The value of $p_0/\tau_2$ should be independent of pressure, i.e., the value of $p_0/\tau$ for $He_2^+$ for $p_0 > p_c$ should be constant. The measured time constants related to the decay of the number densities of $He^+$ and $He_2^+$ are shown in Fig. 8-4-2 and their dependencies on pressure are in agreement with theory, while the values of the relevant constants are the same as those obtained through other measuring methods.

The time dependencies of the intensity of spectral lines originating from atomic energy levels with small principal quantum number were found to be identical in agreement with previous studies performed by Kerr et al. [7, 8].

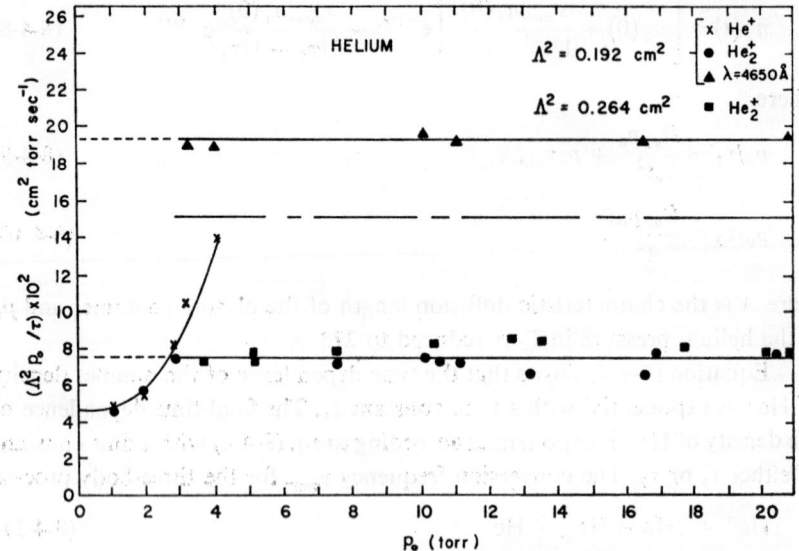

Fig. 8-4-2. Dependence on pressure of $\Lambda^2 p_0/\tau$ for the ion densities and the molecular light intensity (gas temperature 345°K) (reference 45).

The spectral line mainly studied by Gerber et al. [45] was the 3889 Å line and comparison of the time dependence of its intensity during the decay period with that of the density of $He^+$ and $He_2^+$ showed that no relation existed between the intensity of the spectral line and the number density of $He_2^+$. This shows that the dissociative recombination process

$$He_2^+ + e \rightarrow He^* + He \rightarrow 2He + h\nu \qquad (8\text{-}4\text{-}14)$$

does not occur during the decay period or that its "probability" is very small. This is in agreement with Mulliken's [64] discussion of potential energy curves of $He_2$. This author states *"Thus it appears that dissociative recombination to give He\* may be able to compete seriously with collisional-radiative $He_2^*$ formation under conditions where $He_2^+$ ions in high vibrational levels are abundant, but that for $He_2^+$ ions which are in their low or lowest states of vibration, electron recombination should be almost entirely nondissociative,"* and *"The theoretical reasoning on potential curves which leads to the conclusion that dissociative recombination of electrons should, as is observed, be unimportant in helium gas if the vibrational excitation of $He_2^+$ ions is low, is not applicable to the other rare gases, because of the more complicated outer shells and excited states of their atoms".*

Gerber et al. [45] found that for low helium pressures the time constant $\tau_a$ of the intensity of the atomic lines studied was close to one-third that of the $He^+$ ions ($\tau_1$). This is demonstrated in Fig. 8-4-3 where $1/\tau_a$ is exactly three times $1/\tau_1$. At this pressure the time dependencies of the densities of $He^+$ and $He_2^+$ are equal during the same time interval, so that the electron density also has the same time dependence during that time interval. At a helium pressure larger than about 4 Torr $He_2^+$ is the dominant ion during the decay period and for this condition the intensity of the 3889 Å line was observed to be proportional to the product of the number density of $He^+$ and the square of the $He_2^+$ density. These observations thus show that for the lines studied their intensities are proportional to the product of the $He^+$ density and the square of the electron density, i.e., these lines are emitted as a consequence of the collisional recombination process

$$He^+ + 2e \rightarrow He + e + h\nu, \qquad (8\text{-}4\text{-}15)$$

with a recombination coefficient proportional to the electron density. The significance of these studies is that they directly relate the $He^+$ ions with the emission of spectral lines, while moreover they refer to a constant and well-defined electron energy, so that the relevant time dependencies are not dominated by a time dependent recombination coefficient as in previous studies [11, 65].

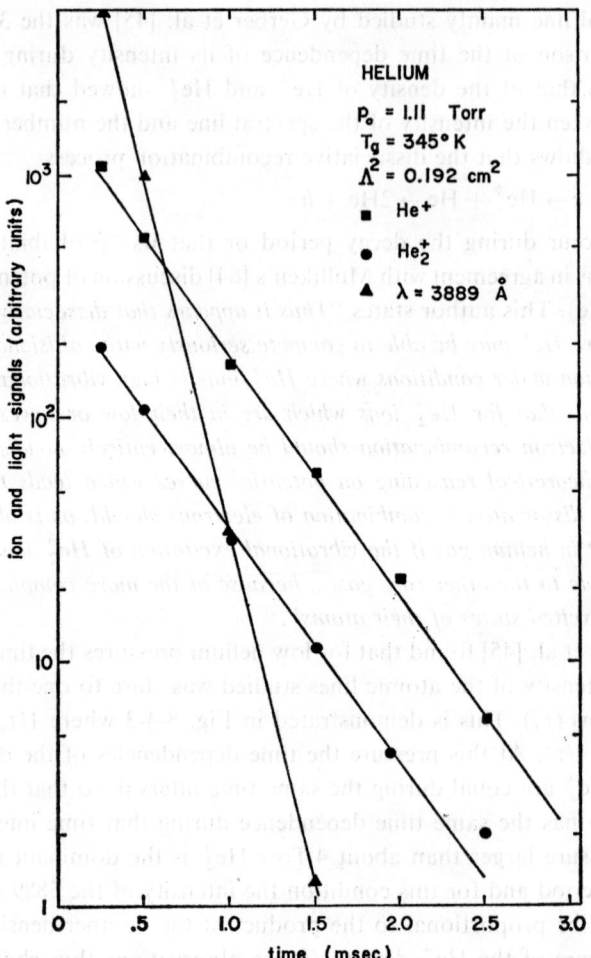

Fig. 8-4-3. Time dependence of helium ion signals and atomic light intensity during the decay period of a plasma produced in helium (reference 45).

Subsequent studies reported by Collins and Hurt [48] confirmed the results obtained by Gerber et al. They measured simultaneously the dependence of the emission intensity of spectral lines and that of the electron density during the decay period of a plasma produced in helium at a pressure of 1 Torr. Lines originating from energy levels with principal quantum number smaller than ten had the same time dependence and their intensity dependence on the charged particle densities was consistent with that determined by Gerber et al. [45]. The intensity of spectral lines originating from energy levels with quantum numbers 21 through 25 was proportional to the product of the measured electron density and the calculated density of $He^+$ ions. This indicates that the relevant excited helium atom group is in collisional equilibrium with the free electrons as can be expected for the recombination process (8-4-15).

Kerr et al. [7, 8] found that the intensity of spectral bands studied during the decay period showed the same time dependence. Gerber et al. [45] studied the time dependence of the intensity of the 4650 Å band ($3^3\Pi_g \to 2^3\Sigma_u^+$) simultaneously with that of the densities of $He^+$ and $He_2^+$. No relation was found between the intensity of this band and the number density of $He^+$. However, a clear relation between the band intensity and the density of $He_2^+$ was observed. Moreover, the intensity of the molecular emission decreased at low pressure, where the relative density of $He_2^+$ also decreased. Values of the ratio $p_0/\tau_m$ of the pressure and the time constant of the band intensity measured during the afterglow period where $He_2^+$ is the dominant ion, so that the time dependence of the electron density can be assumed to be the same as that of $He_2^+$, are shown in Fig. 8-4-2. The value of $p_0/\tau_m$ is larger than $2p_0/\tau_2$ (indicated by the dashed line), so that the radiative recombination process

$$He_2^+ + e \to He_2^* \to 2He + h\nu \qquad (8\text{-}4\text{-}16)$$

alone cannot explain the observations, since this would lead to $p_0/\tau_m = 2p_0/\tau_2$. Most probably the three-body neutralization process

$$He_2^+ + 2e \to He_2^* + e \to 2He + e + h\nu \qquad (8\text{-}4\text{-}17)$$

contributes significantly to molecular light emission in the electron density range studied. The effective recombination coefficient, derived from the relation between the time dependence of the band intensity and those of the charged particles, was found to be proportional to the square root of the electron density. Recently, Veatch and Oskam [66] found the same relation between the 4650 Å band emission and the densities of $He_2^+$ and electrons during the decay period of plasmas produced in helium containing a small concentration of neon atoms.

From the previous discussion it can be concluded that studies performed during the last few years have shown that the line emission during the decay period of helium plasmas is due to collisional recombination of $He^+$ ions with electrons, while the band emission is due to recombination of $He_2^+$ ions with electrons. The process involved is most probably also the collisional recombination process (8-4-17). No experimental evidence is available that the neutralization of $He_2^+$ occurs through dissociative recombination.

## § 8-5. Recombination processes involving rare gas ions other than helium ions

The microwave technique has been very successful in determining the recombination coefficient of molecular rare gas ions other than $He_2^+$. The recombination process involved has been shown to be the dissociative recombination process. It appears that $He_2^+$ is the only molecular rare gas for which this recombination process is improbable if not impossible. This is consistent with the observation that band emission during the decay period is observed in helium and not in plasmas produced in the other rare gases.

### A. NEON IONS

The recombination of $Ne_2^+$ ions with electrons has been studied in detail by means of the measuring techniques described in section 8-3. The results will be discussed rather extensively since they demonstrate the various methods of establishing the type of recombination process occurring in decaying plasmas.

#### 1. *Electron density studies*

In 1949 Biondi and Brown [4] reported studies of the electron density during the decay period of neon plasmas by means of the microwave cavity method. The authors obtained a value of $\alpha = 2 \times 10^{-7}$ cm$^3$/sec for the coefficient describing the disappearance of electrons by volume recombination with ions. The experimental conditions were such that it could be assumed that $Ne_2^+$ determined the plasma decay properties. At the gas pressures studied atomic neon ions produced during the discharge period were converted very rapidly into $Ne_2^+$ ions by the conversion process

$$Ne^+ + 2Ne \rightarrow Ne_2^+ + Ne. \qquad (8\text{-}5\text{-}1)$$

The conversion frequency relating to this process is about $v_{conv} = 60p_0^2$, so that at a pressure of 10 Torr the time constant relating to the disappearance of $Ne^+$ due to process (8-5-1) is about $1.6 \times 10^{-4}$ sec. The afterglow studies extend over several milliseconds. Therefore, even if the discharge would contain a mixture of $Ne^+$ and $Ne_2^+$, it could be assumed that the dominant afterglow ion rapidly becomes the molecular ion.

The large majority of subsequent analogous studies performed on room temperature neon plasmas resulted in values of $\alpha(\text{Ne}_2^+)$ which are, within the experimental errors, independent of pressure and in excellent agreement with the value first reported by Biondi and Brown [4, 6, 40, 67]. The correction of the reported values for the influence of ambipolar diffusion on the data according to the numerical analysis by Frommhold and Biondi leads to $\alpha(\text{Ne}_2^+) = (1.7 \pm 0.2) \times 10^{-7}$ cm$^3$/sec.

Biondi [68] and Oskam [5] reported studies of the decay properties of plasmas produced in rare gas mixtures. The results showed that large values of recombination coefficients are obtained only under conditions during which molecular ions can be assumed to determine the plasma decay properties. The disappearance of electrons during the decay period of a plasma produced in neon at a pressure of 20 Torr was described by a straight $1/n_e(t)$ versus time curve characteristic for electron disappearance by volume recombination with ions, which could be assumed to be $\text{Ne}_2^+$ ions. When adding about $10^{-3}$% argon to the neon gas the electron density decayed considerably slower and the decay curve showed an exponential time dependence characteristic for electron disappearance by ambipolar diffusion towards the container walls. The ambipolar diffusion coefficient calculated from the data agreed with that expected for Ar$^+$ ions in neon. These ions are produced during the discharge and afterglow periods by the Penning process

$$\text{Ne}^m + \text{Ar} \to \text{Ne} + \text{Ar}^+ + e, \qquad (8\text{-}5\text{-}2)$$

where the superscript m indicates an atom in a metastable excited state. If $\text{Ne}_2^+$ ions are initially present they will rapidly disappear by recombining with electrons, so that during the major part of the afterglow period Ar$^+$ ions are the dominant ions. When the concentration of argon atoms in neon was increased the rate of electron disappearance increased and obtained the characteristics of electron disappearance by recombination with ions at argon concentrations of about 1%. This transition from electron disappearance by ambipolar diffusion to electron disappearance by volume recombination is due to the change from Ar$^+$ being the dominant ion type during the decay period to $\text{Ar}_2^+$ ions dominating this period. The latter ions are produced by the conversion process [5]

$$\text{Ar}^+ + \text{Ne} + \text{Ar} \to \text{Ar}_2^+ + \text{Ne}, \qquad (8\text{-}5\text{-}3)$$

which has a conversion frequency $v_{\text{conv}} = 315 p_0(\text{Ar}) p_0(\text{Ne})$. The contribution of a process analogous to (8-5-1) is very small at the argon concentrations studied. These studies thus showed that for conditions during which $\text{Ne}_2^+$

or $Ar_2^+$ can be assumed to determine the electron loss rate, the recombination process became, at the pressure studied, the main electron loss process, while if $Ar^+$ is the major ion type the electron loss rate is considerably smaller and is governed by ambipolar diffusion. This gives the correlation between large recombination coefficients and molecular ions.

## 2. *Light emission studies*

In 1950 Holt et al. [34] observed that the spectrum emitted by a decaying neon plasma was virtually a pure line spectrum. The studies were performed at a neon pressure of 20 Torr, where $Ne_2^+$ should be the dominant ion type during the decay period. The total intensity of emitted light was found to be proportional to the square of the electron density measured by means of the microwave cavity method. The same relation was obtained by Biondi [69] at a neon pressure of about 10 Torr.

In 1965 Sauter et al. [46] reported detailed studies of the emission spectrum during the neon afterglow period. Only spectral lines were observed in agreement with previous studies. The time dependence of the intensity of a large number of spectral lines in the spectral range of 3000 Å to 7000 Å was studied at a neon pressure of 3.85 Torr. The data showed that two distinct groups of spectral lines, each group having its own time dependence, are emitted during the decay period. An example of this phenomenon is shown in Fig. 8-5-1 for a few lines from each group. From studies performed at various pressures it became apparent that two different processes can contribute to the line emission during the decay period. One process populated mainly the d-levels, while the other was mainly responsible for the emission of lines originating at p-levels. The time dependence of the intensity of spectral lines originating from the p-levels can be influenced by both processes as a consequence of electron transitions from the d-levels to the p-levels. This contribution of the "faster" process to the intensity of the "slower" process is apparent in Fig. 8-5-1 in the dependence of the intensity of the 5974 Å line during the early afterglow period. In order to determine which collision processes were responsible for the emission of the two spectral line groups, Sauter et al. measured simultaneously the time dependence of the number density of $Ne^+$ and $Ne_2^+$. It was found that the process populating mainly the p-levels is the dissociative recombination of $Ne_2^+$ with electrons. Sauter et al. observed also a close analogy between the type of time dependence of the intensity of lines originating from the d-levels and that of the density of $Ne^+$. This strongly indicated that these levels are populated by the recombination of $Ne^+$ with an electron. The measurements

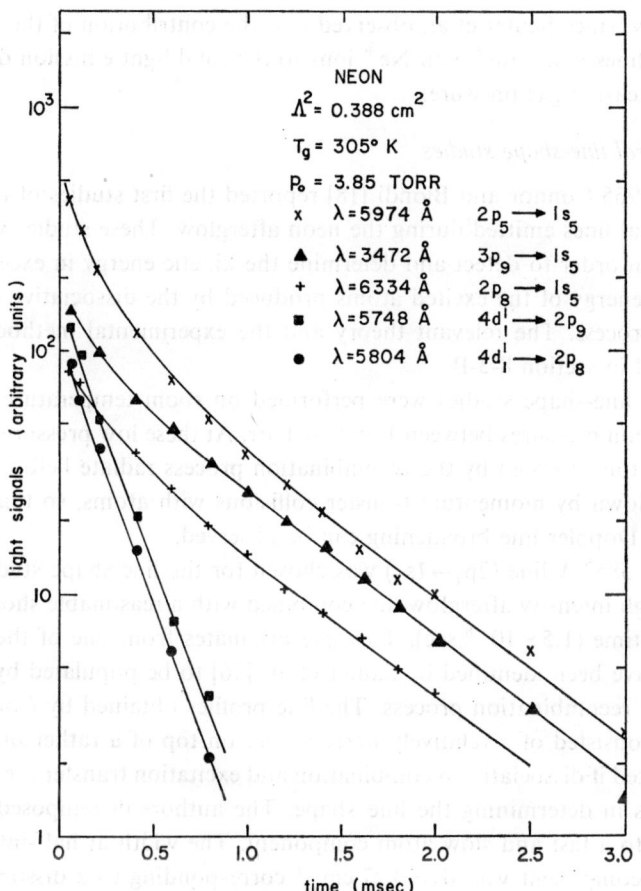

Fig. 8-5-1. Time dependence of atomic light intensity during the decay period of a plasma produced in neon (reference 46).

did not lead to a definite conclusion about the dependence of the line intensities on the electron density, since this had to be inferred from the ion densities. Indications were found for the occurrence of both the radiative and the collisional recombination processes. Subsequent studies performed with helium-neon mixtures resulted in a strong indication that the latter process contributes significantly to emission of lines originating from the d-levels [66].

The studies of Sauter et al. are in agreement with the quadratic dependence on electron density of the total light emission as reported by Holt et al. [34] and Biondi [69] at neon pressures of 20 Torr and 10 Torr, respectively, since Sauter et al. observed that the contribution of the group of spectral lines associated with $Ne^+$ ions to the total light emission decreased with increasing gas pressure.

## 3. *Spectral line-shape studies*

In 1965 Connor and Biondi [18] reported the first studies of the shape of spectral lines emitted during the neon afterglow. These studies were performed in order to detect and determine the kinetic energy in excess of the thermal energy of the excited atoms produced by the dissociative recombination process. The relevant theory and the experimental method used is discussed in section 8-3-B.

The line-shape studies were performed on room temperature plasmas and at neon pressures between 1 and 10 Torr. At these low pressures the fast excited atoms formed by the recombination process radiate before they are slowed down by momentum transfer collisions with atoms, so that the associated Doppler line broadening can be observed.

The 5852 Å line ($2p_1 \rightarrow 1s_2$) was chosen for the line shape studies since it is a high intensity afterglow line combined with a reasonable short upper-state lifetime ($1.5 \times 10^{-8}$ sec). This line originates from one of the p-levels which have been identified by Sauter et al. [46] to be populated by the dissociative recombination process. The line profiles obtained by Connor and Biondi consisted of a relatively narrow core on top of a rather broad base as expected if dissociative recombination and excitation transfer are the main processes in determining the line shape. The authors decomposed the line shape into a fast and slow atom component. The width at half-intensity of the fast component was $W_f = 0.27$ cm$^{-1}$ corresponding to a dissociative kinetic energy $E_d \approx 1.2$ eV. The value of $W_f$ was independent both of time in the afterglow period and neon pressure as it should. The half-intensity linewidth ($\approx 0.07$ cm$^{-1}$) of the slow component during the decay period was equal to the width measured during the discharge period and, moreover, was in good agreement with the value resulting from thermal and pressure line broadening in addition to instrumental linewidth.

Recently, Biondi and Frommhold [70] reported more detailed studies of the shape of spectral line originating from p-levels of neon. A gas temperature variation from 300 to 77°K hardly affected the afterglow line shape, whereas discharge linewidths were reduced at the lower temperature. This is in agreement with the postulate that the width of the spectral line during the decay period is determined by the dissociative recombination process. These authors also observed that the intensity of all the strong spectral lines due to 2p→1s transitions decayed at the same rate, which is in agreement with previous studies by Sauter et al. [46]. The high sensitivity of the experimental apparatus used by Biondi and Frommhold revealed an interesting fine structure in the profiles of the mentioned spectral lines. Instead of interpreting the line shapes as the superposition of one trapezoid and one narrow core, as was done previously, it was possible to interpret the data as a composite of several trapezoids and one narrow core. This means that several values for the dissociation kinetic energy could be obtained from the same spectral line shape, which would imply that molecular neon ions in different vibrational (and possibly electronic) states participate in the dissociative recombination process. This leads also to the significant conclusion that even during the decay period of a neon plasma molecular neon ions in different energy states are present.

## 4. Dependence of $\alpha(Ne_2^+)$ on the electron temperature

Frommhold et al. [40] used the multimode microwave cavity shown in Fig. 8-3-4 to study the dependence of $\alpha(Ne_2^+)$ on the electron temperature. The results are shown in Fig. 8-5-2. The solid line through the experimental data shows a variation of $\alpha(Ne_2^+)$ as $T_e^{-0.43}$.

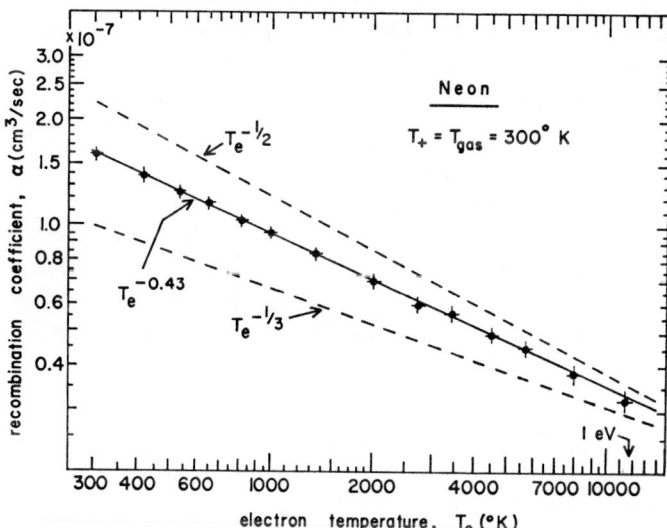

Fig. 8-5-2. Variation of the recombination coefficient, $\alpha(Ne_2^+)$, as a function of electron temperature. The solid line through the points represents a $T_e^{-0.43}$ variation, while the dashed lines indicate the slopes of $T_e^{-\frac{1}{3}}$ and $T_e^{-\frac{1}{2}}$ dependences (reference 40).

Kasner [67] has reported a similar study over a gas temperature range of 295 to 503°K. He obtained sufficiently large linear ranges of $1/n_e(t)$ versus time curves, so that his values are quite accurate and, moreover, observed that the time dependence of the $Ne_2^+$ wall current followed the electron-density decay over the major part of the decay period. Kasner obtained a variation of $\alpha(Ne_2^+)$ as $T_{gas}^{-0.42}$, where $T_+ = T_e = T_{gas}$, in excellent agreement with the results shown in Fig. 8-5-2.

These studies suggest that the dissociative recombination of $Ne_2^+$ with an electron is of the direct type as discussed in section 8-2-A.

## B. Argon Ions

The first measurement of $\alpha(Ar_2^+)$ by means of the microwave cavity method was reported by Biondi and Brown [4]. They obtained $\alpha(Ar_2^+) = 3.0 \times 10^{-7}$ cm$^3$/sec, but suspected the purity of the argon used. The differences between subsequently reported values of $\alpha(Ar_2^+)$ obtained from analogous studies were larger than those for $\alpha(Ne_2^+)$, although it was believed that $\alpha(Ar_2^+)$ was larger than $\alpha(Ne_2^+)$. For instance, an influence of the length and magnitude of the plasma excitation pulse on the value of $\alpha$ has been reported [33–36]. However, when $\alpha$ is related to dissociative recombination of $Ar_2^+$ with an electron its value should be independent of plasma excitation conditions.

In order to investigate the influence of plasma excitation on the determination of $\alpha$, Oskam and Mittelstadt [6] performed studies in which the duration of the plasma-excitation pulse was varied. From observations of the spatial distribution of the visible light emitted by the plasma during the discharge pulse it was concluded that inadequate pulse lengths do not permit the plasma, which is initially created at regions of high excitation-field strength, to fill the container by the ambipolar diffusion process before the excitation ceases and that the subsequent rapid cooling of the electrons during the very early afterglow period "freezes in" the spatial electron density distribution. Consequently, only part of the plasma container will then be filled with the plasma and the value of the recombination coefficient calculated from the measured curve under the false assumption of a completely filled plasma container, will lead to a possibly severe overestimation of $\alpha$. The linear range of such $1/n_e(t)$ versus time curves does not satisfy the condition following from the calculated value of $\alpha$ as given by Grey and Kerr [38] and by Frommhold and Biondi [39]. The time needed by the active plasma to diffuse through the container increased with increasing pressure, and was even more pronounced in krypton and xenon than in argon, as a consequence of the slower ambipolar diffusion process.

A critical survey of reported microwave cavity measurements shows that the recombination coefficient of $Ar_2^+$ is independent of gas pressure and is given by $\alpha(Ar_2^+) = (7 \pm 0.7) \times 10^{-7}$ cm$^3$/sec.

In 1951 Redfield and Holt [71] reported preliminary studies of the emission spectrum of decaying argon plasmas. These authors could not arrive at any definite conclusion about the origin of the emitted light. Very recently, Biondi and Frommhold [70] studied the shape of spectral due to 2p→1s electron transitions during the afterglow period. The lines had a definitely broader base during the afterglow period than during the discharge period, which shows that the emission of these lines is due to the dissociative recombination process.

Another very recent study concerning the recombination of $Ar_2^+$ is the study of the dependence of $\alpha(Ar_2^+)$ on the electron temperature by Mehr and Biondi [43]. The results are shown in Fig. 8-5-3 and refer to data obtained in

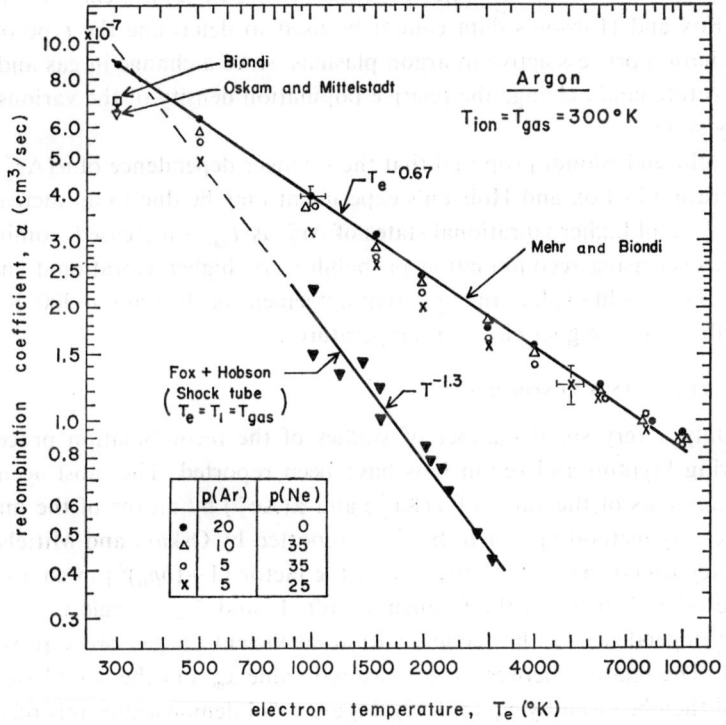

Fig. 8-5-3. Comparison of the various measurements of the recombination coefficient, $\alpha(Ar_2^+)$, and its dependence on electron temperature (reference 43).

argon and neon–argon mixtures. The variation of $\alpha(Ar_2^+)$ as $T_e^{-0.67}$ may indicate the importance of both the direct and indirect dissociative recombination mechanisms. Fox and Hobson [72] studied recombination processes in argon using shock tube and electrostatic double probe techniques. It was assumed that the electrons, ions and atoms were in thermal equilibrium. The dependence of $\alpha(Ar_2^+)$ on this equilibrium temperature is also shown in Fig. 8-5-3. Their results disagree with the data of Mehr and Biondi [43] both as to the magnitude of the recombination coefficient and its dependence on temperature, although a rather daring and extensive extrapolation of the data to 300°K yields a value of $\alpha(Ar_2^+)$ in good agreement with previously reported room temperature values [6, 69]. Mehr and Biondi emphasize that during their studies the ion temperature remains constant at 300°K as $T_e$ is varied, while in Fox and Hobson's studies $T_i$ and $T_e$ vary together. The theories relating to the dependence of the dissociative recombination coefficient on the electron temperature as discussed in section 8-2-A include the assumption that the energy state of the ion involved does not change. Therefore, Fox and Hobson's data cannot be used to determine the type of recombination process active in argon plasmas since a change in gas and ion temperature could change the relative population density of the various ion energy states.

Mehr and Biondi proposed that the stronger dependence of $\alpha(Ar_2^+)$ on temperature in Fox and Hobson's experiment may be due to an increasing population of higher vibrational states of $Ar_2^+$ as $T_{gas}$ is increased combined with a decreasing recombination probability for higher vibrational energy states. This could explain the apparent agreement of the data at 300°K and the difference at higher electron temperatures.

## C. Krypton and Xenon Ions

Only a very small number of studies of the recombination processes involving krypton and xenon ions have been reported. The most accurate measurements of the value of $\alpha(Kr_2^+)$ and $\alpha(Xe_2^+)$ by means of the microwave cavity method appear to be those reported by Oskam and Mittelstadt [6]. They found that if the influence of the factor $[1+(bp_0)^2]$ in eq. (8-3-5) on the relation between the frequency shift $\Delta f$ and $\bar{n}_{\mu w}$ is neglected an apparent dependence of the recombination coefficient on gas pressure is obtained. The relation between the measured value $\alpha_m$ and the actual value $\alpha$ should then be given by eq. (8-3-10). Figure 8-5-4 demonstrates this relation for $\alpha(Xe_2^+)$.

Oskam and Mittelstadt obtained $\alpha(Kr_2^+) = (1.2 \pm 0.1) \times 10^{-6}$ cm³/sec

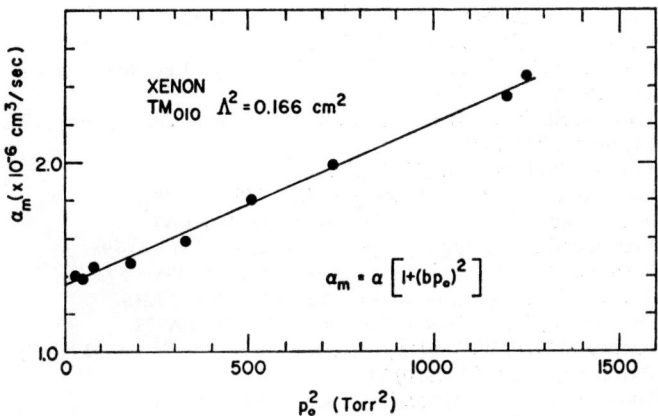

Fig. 8-5-4. The measured recombination coefficient $\alpha_m$ in xenon as a function of the square of the gas pressure (reference 6).

and $\alpha(Xe_2^+) = (1.4 \pm 0.1) \times 10^{-6}$ cm$^3$/sec from $1/\bar{n}_{\mu w}(t)$ versus time curves which were straight lines within 2% over an electron density range of more than 200 to 1.

## Acknowledgement

The author is indebted to the Air Force Office of Scientific Research, the Air Force Cambridge Research Laboratories, the National Science Foundation and the Office of Naval Research for supplying funds to his research group during the past ten years for electron–ion recombination studies. The valuable contributions to these studies made by R. A. Gerber, R. E. Lund, V. R. Mittelstadt, G. F. Sauter and G. E. Veatch during this period have made it possible for the author to write this chapter. The help received from A. R. De Monchy in writing section 8-2 is very much appreciated. The many suggestions made by my colleagues L. M. Chanin and J. Freudenthal have been very helpful. The author is also grateful to M. A. Biondi and L. Frommhold for permission to use unpublished data.

## References

1. M. A. BIONDI, Rev. Sci. Instr. **22** (1951) 500.
2. D. J. ROSE and S. C. BROWN, J. Appl. Phys. **23** (1952) 1028.
3. M. A. BIONDI and S. C. BROWN, Phys. Rev. **75** (1949) 1700.
4. M. A. BIONDI and S. C. BROWN, Phys. Rev. **76** (1949) 1697.
5. H. J. OSKAM, Philips Res. Rept. **13** (1958) 401.
6. H. J. OSKAM and V. R. MITTELSTADT, Phys. Rev. **132** (1963) 1445.
7. E. P. GRAY and D. E. KERR, Bull. Am. Phys. Soc. **5** (1960) 273.

8. D. E. KERR and C. S. LEFFEL, Bull. Am. Phys. Soc. **7** (1962) 131.
9. L. HERMAN, Compt. Rend. **228** (1949) 2016.
10. R. A. JOHNSON, B. T. MCCLURE and R. B. HOLT, Phys. Rev. **80** (1950) 376.
11. N. D'ANGELO, Phys. Rev. **121** (1961) 505.
12. R. H. FOWLER, Statistical Mechanics (second ed., Cambridge University Press, Cambridge, 1936) pp. 726–727;
    R. H. FOWLER, Phil. Mag. **47** (1924) 257.
13. R. G. GIOVANELLI, Australian J. Sci. Res. **A1** (1948) 275, 289.
14. W. H. KASNER and M. A. BIONDI, Phys. Rev. **137** (1965) A317.
15. C. S. WELLER and M. A. BIONDI, Phys. Rev. **172**, No. 1 (1968) 198.
16. C. S. WELLER and M. A. BIONDI, Phys. Rev. Letters **19** (1967) 59.
17. W. A. ROGERS and M. A. BIONDI, Phys. Rev. **134** (1964) A1215.
18. T. R. CONNOR and M. A. BIONDI, Phys. Rev. **140** (1965) A778.
19. D. R. BATES and H. S. W. MASSEY, Proc. Roy. Soc. (London) **A192** (1947) 1.
20. D. R. BATES, Phys. Rev. **77** (1950) 718.
21. J. N. BARDSLEY, J. Phys. B (Proc. Phys. Soc.) [2] **1** (1968) 365.
22. J. C. Y. CHEN and M. H. MITTLEMAN, Abstracts of the Contributed Papers of the Fifth Intern. Conf. on the Physics of Electronic and Atomic Collisions (Nauka, Leningrad, 1968) p. 329.
23. J. N. BARDSLEY, Abstracts of the Contributed Papers of the Fifth Intern. Conf. on the Physics of Electronic and Atomic Collision (Nauka, Leningrad, 1968) p. 265.
24. R. C. STABLER, Phys. Rev. **131** (1963) 1578.
25. D. R. BATES and A. DALGARNO, in: Electronic Recombination in Atomic and Molecular Processes, ed. D. R. Bates (Academic Press, New York, 1962) pp. 608–613. See also D. R. BATES, A. E. KINGSTON and R. W. P. MCWHIRTER, Proc. Roy. Soc. (London) **A267** (1962) 297;
    D. R. BATES and A. E. KINGSTON, Planetary Space Sci. **11** (1963) 1;
    D. R. BATES, Third Intern. Conf. on the Physics of Electronic and Atomic Collisions, London, 1963;
    R. W. P. MCWHIRTER and A. G. HEARN, Proc. Phys. Soc. (London) **82** (1963) 641;
    F. ROBBEN, W. B. KUNKEL and L. TALBOT, Phys. Rev. **132** (1963) 2363;
    D. R. BATES and A. E. KINGSTON, Proc. Phys. Soc. (London) **83** (1964) 43.
26. D. R. BATES, A. E. KINGSTON and R. W. P. MCWHIRTER, Proc. Roy. Soc. (London) **A270** (1962) 155.
27. J. C. SLATER, Rev. Mod. Phys. **18** (1946) 480.
28. W. P. ALLIS, Motions of Ions and Electrons, in: Encyclopedia of Physics, ed. S. Flügge (Springer Verlag, Berlin, 1956) pp. 383–445.
29. K. B. PERSSON, Phys. Rev. **106** (1957) 191.
30. S. J. BUCHSBAUM and S. C. BROWN, Phys. Rev. **106** (1957) 196.
31. A. V. PHELPS, O. T. FUNDINGSLAND and S. C. BROWN, Phys. Rev. **84** (1951) 559.
32. L. GOULD and S. C. BROWN, Phys. Rev. **95** (1954) 897.
33. J. J. LENNON and M. C. SEXTON, J. Electron. Control **7** (1959) 123.
34. R. B. HOLT, J. M. RICHARDSON, B. HOWLAND and B. T. MCCLURE, Phys. Rev. **77** (1950) 239.
35. J. M. RICHARDSON, Phys. Rev. **88** (1952) 895.
36. M. C. SEXTON and J. D. GRAGGS, J. Electron. Control **4** (1948) 493.
37. H. J. OSKAM, Philips Res. Rept. **13** (1958) 335.
38. E. P. GRAY and D. E. KERR, Ann. Phys. (N.Y.) **17** (1962) 276.
39. L. FROMMHOLD and M. A. BIONDI, Ann. Phys. (USA) **48** (1968) 407.
40. L. FROMMHOLD, M. A. BIONDI and F. J. MEHR, Phys. Rev. **165**, No. 1 (1968) 44.
41. H. J. OSKAM and V. R. MITTELSTADT, Phys. Rev. **132** (1963) 1435.
42. H. MARGENAU, Phys. Rev. **69** (1946) 508.

43. F. J. MEHR and M. A. BIONDI, Phys. Rev. **176**, No. 1 (1968) 322.
44. A. V. PHELPS, Phys. Rev. **99** (1955) 1307.
45. R. A. GERBER, G. F. SAUTER and H. J. OSKAM, Physica **32** (1966) 2173.
46. G. F. SAUTER, R. A. GERBER and H. J. OSKAM, Physica **32** (1966) 1921.
47. F. E. NILES and W. W. ROBERTSON, J. Chem. Phys. **40** (1964) 2909.
48. C. B. COLLINS and W. B. HURT, Phys. Rev. **167**, No. 1 (1968) 166.
49. Gy. FARKAS and P. VARGA, J. Sci. Instr. **41** (1964) 704.
50. J. H. BEYNON, Mass Spectrometry and Its Applications to Organic Chemistry (Elsevier Publishing Co., 1960).
51. A. V. PHELPS and S. C. BROWN, Phys. Rev. **86** (1952) 102.
52. W. H. KASNER, W. A. ROGERS and M. A. BIONDI, Phys. Rev. Letters **7** (1961) 321.
53. M. A. MOSHARRAFA and H. J. OSKAM, Design and Construction of a Mass Spectrometer for the Study of Basic Processes in Plasma Physics, ONR Techn. Rept. No. 2, 1961.
54. M. A. MOSHARRAFA and H. J. OSKAM, Proc. ASTM-E14 Conf. on Mass Spectroscopy and Allied Topics, Montreal, 1964, pp. 473–478.
55. M. A. MOSHARRAFA and H. J. OSKAM, Physica **32** (1966) 1759.
56. W. PAUL and H. STEINWEDEL, Z. Naturforsch. **8a** (1953) 448.
57. W. PAUL, H. P. REINHARD and U. VON ZAHN, Z. Physik **152** (1958) 143.
58. W. M. BRUBAKER and J. TUUL, Rev. Sci. Instr. **35** (1964) 1007.
59. N. W. MCLACHLAN, Theory and Application of Mathieu Functions (Oxford, 1951).
60. C. E. WOODWARD and C. K. CRAWFORD, Development of a Quadrupole Mass Spectrometer, Techn. Report 194 Lab. for Insulation Res., Massachusetts Institute of Technology, Cambridge, 1964.
61. G. F. SAUTER, R. A. GERBER and H. J. OSKAM, Rev. Sci. Instr. **37** (1966) 572.
62. R. E. LUND and H. J. OSKAM, Z. Physik **219** (1969) 131.
63. H. J. OSKAM, J. Appl. Phys. **34** (1963) 711.
64. R. S. MULLIKEN, Phys. Rev. **136** (1964) A962.
65. E. HINNOV and J. G. HIRSCHBERG, Phys. Rev. **125** (1962) 795.
66. G. E. VEATCH and H. J. OSKAM, Phys. Rev., to be published.
67. W. H. KASNER, Phys. Rev. **167** (1968) 148.
68. M. A. BIONDI, Phys. Rev. **83** (1951) 1078.
69. M. A. BIONDI, Phys. Rev. **129** (1963) 1181.
70. M. A. BIONDI and L. FROMMHOLD, to be published.
71. A. REDFIELD and R. B. HOLT, Phys. Rev. **82** (1951) 874.
72. J. N. FOX and R. M. HOBSON, Phys. Rev. Letters **17** (1966) 161.

CHAPTER 9

# HIGH RESOLUTION ELECTRON BEAMS AND THEIR APPLICATION

BY

## L. KERWIN, P. MARMET and J. D. CARETTE

*Centre de Recherches sur les Atomes et les Molécules,
et Département de Physique,
Université Laval, Canada*

## Contents

| | Page |
|---|---|
| 9-1. Introduction | 527 |
| 9-2. Some history | 528 |
| 9-3. Instruments | 529 |
|     A. The RPD method | 529 |
|     B. The 127° electrostatic selector | 530 |
|     C. The spherical electrostatic selector | 536 |
|     D. The parallel-plane electrostatic selector | 537 |
|     E. The axial-source cylindrical electrostatic selector | 538 |
|     F. The monokinetron | 539 |
|     G. Magnetic selectors | 540 |
|     H. Time-of-flight selectors | 541 |
|     I. Miscellany | 542 |
| 9-4. Electron spectroscopy | 544 |
|     A. The electron spectroscope | 544 |
|     B. Elastic scattering, resonances | 546 |
|     C. Inelastic scattering | 555 |
|         1. Second mode | 555 |
|         2. Third mode | 561 |
|         3. Rotational levels | 562 |
| 9-5. Ionization | 562 |
|     A. Atoms | 563 |
|         1. Hydrogen | 563 |
|         2. Helium | 566 |
|         3. Ne, Ar, Kr, Xe | 568 |
|         4. Double ionization of inert gases | 571 |
|         5. Higher degree of ionization of inert gases | 572 |
|     B. Diatomic molecules | 573 |
|         1. Hydrogen | 573 |
|         2. Nitrogen | 574 |
|         3. Oxygen | 575 |
|     C. Triatomic molecules | 575 |
|     D. The negative ion $SF_6^-$ | 576 |
| 9-6. Conclusion | 576 |
| References | 577 |

## § 9-1. Introduction

In the scientist's never-ending quest for the complete and detailed description of the universe, the structure of the atom and the simple molecule has occupied much of his time. Complicated, orderly and beautiful in their own right, atoms and simple molecules are also the basic units of most physical and biochemical activity: to understand these phenomena the descriptions of the atoms and molecules must be to hand. Sufficient is now known to cause atomic and molecular science to be the broad base for most of our current technology and much of the economy. Yet so little is known that research projects on atomic hydrogen are still valid and timely.

During the 19th century the work of many spectroscopists revealed the existence of discrete energy levels in the atom. The turn of the century witnessed the discovery of the electron as the probable vehicle for the energy under consideration, the discovery of the atomic nucleus which relegated the electron to the vast, empty outer regions of the atom, and the discovery of the quantum nature of energy. Combined in the Bohr model of the atom these basic notions found full flower in the realm of quantum mechanics which also provided the first satisfactory account of how simple homopolar molecules stayed together. Even simple molecules provide a rich wealth of physical phenomena all by themselves: to the various possibilities of excitation to levels characterized by fine structure, ultra-fine structure, nuclear effects and their attendant probabilities must be overlaid the pattern of various vibrations and quantized rotations, the blind alleys of metastable states and the cheating of autoionizing levels, the complexities of curve-crossing and the various protocols of becoming an ion: direct, impact, photo-ionization, field ionization, etc., not to mention the myriad extensions of these possibilities in the presence of electric and magnetic fields or of even one other molecule.

Clearly such a miniature universe requires many theoretical and experimental techniques for its elucidation. At the moment the experimental ones are the more fruitful, since the difficulty of the quantum mechanics equations has prevented their being successfully applied in detail to the majority of atoms and molecules as yet. In the experimental field spectroscopy still provides most of the data. However within the last 15 years a new tool has been honed: the monoenergetic electron beam. An electron of known energy which penetrates an atom or molecule can provoke most of the phenomena that can be attributed to the penetration of a light quantum of known energy, and in some cases can result in data which escape the classical spectroscopist: either because of selection rules or because of the differences inherent in photon and electron reactions. Basic facts about atoms and simple molecules such as the dissociation energy of $N_2$, the resonance levels in He, the vibrational structure of diatomic molecules, the electron affinity of O have all been obtained with the use of electron beams as probes of atomic structure. In his recent review of Twenty Years of Atomic Physics, Lewis Branscomb [1] lists the development of monoenergetic electron beams as one of the important contributions of the period.

## § 9-2. Some history

Lenard [2] appears to have been the first to bombard atoms with beams of low-energy electrons. He observed, in 1903, inelastic collisions which implied a discrete energy structure of the atoms. This interpretation is a bit a posteriori, for the experiment was not all that clear. A decade later Franck and Hertz [3] performed the elegant experiment which is so well known and which helped to establish the Bohr model of the atom. Similar work was taken up by others and by 1930 Smyth could write a major paper [4] on results obtained by the electron bombardment of hydrogen which included consideration of the vibrational levels of the molecular ion, the nature of the ionization probability curve, the value of the threshold of ionization, the protons resulting from direct ionization according to the Franck-Condon principle, and other matters which are still of much import. This work was done using an electron beam obtained from a thermal emitter. The effect of the energy distribution characteristic of such a beam on the data obtained was a matter of some concern, and for the next interval efforts were made to correct for its effect and various recipes for handling appearance-potential data were evolved. Clarke [5] gives a brief review of these. However the more direct approach of eliminating the electron energy spread from the beam was being studied and around the early 1950's a breakthrough was achieved by the development of the RPD method – to be discussed – and the electron selector. Immediately much greater precision in data became routine, and new phenomena ("close" doublet levels, autoionizing levels, vibrational structure, atomic resonances) were made evident. And, of course, new problems. The effects of surface absorption began to interfere with measurements of the free atoms, exotic molecular ions began to show up and previously unknown metastable levels – sometimes close to ionization limits – were discovered to influence data.

The last decade has seen a flourishing of the instrumentation associated with the production of monoenergetic electron beams and it is now highly reliable. Hundreds of selectors of various kinds are now in use as well as the RPD method, and various techniques for peeling the effects of electron energy spread from measured data are still being tried [6], among which those using on-line computers have a certain vogue. A list of current problems would include much that interested Smyth in 1930, but the knowledge acquired since his day is vast, and the electron beam technique is now firmly established as a major one, rivalling spectroscopy in the field of atomic physics.

## § 9-3. Instruments

In this section we shall briefly consider some of the techniques currently used for high resolution electron beam investigations.

### A. THE RPD METHOD

Fox [7] and his associates first described the "retarding potential difference" method in 1951 at Westinghouse in Pittsburgh. It does not produce a beam of monoenergetic electrons, but effectively gives results as though it did. In Fig. 9-3-1 the simple apparatus used is sketched. The electron beam from a filament is directed to electrode 4 through an opening in which it continues to traverse an ionization chamber 5 before being collected on a measuring electrode. The beam is collimated along its path by a magnetic field parallel to it. It has the energy distribution characteristic of thermal emission ($\sim\frac{1}{2}$ eV) and this is indicated by the dashed curve shown between the filament and electrode 4. Electrode 4 is biased negatively, so that the electrons in the bunch with insufficient energy are cut off. The amputated

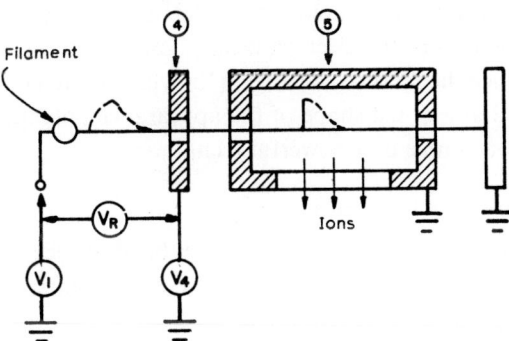

Fig. 9-3-1. Schema of the RPD method [7]. In practice, further electrodes are used to increase beam intensity and sharpen the effective resolution [11].

bunch is then accelerated to the desired mean energy by a potential difference between 4 and 5. Thus – as seen by the dashed curve within the ionization box – only electrons of energy greater than a sharp minimum pass through the box. Here they may produce ionization, excitation, etc. of gas atoms or molecules within the box, and the ions so produced may be detected and measured or otherwise identified by a mass spectrometer or other device. The ion current so measured will have been produced, of course, by the still quite wide spread of energies in the electron beam. This having been done, the potential of electrode 4 is then changed by a small amount $\Delta V_R$. This means that a further group of electrons is now prevented from entering the ionization box and so the amount of ionization or other phenomenon produced is reduced, being caused only by the further amputated beam which also has a considerable energy spread. However the difference between the two ion currents produced at the two voltage settings of electrode 4 may be attributed to the band of electrons with energies varying only by $\Delta V_R$. In practice this band may be as narrow as 0.06 eV, and thus the difference ion current is effectively measured as being produced by a beam of electrons of this energy spread. Pulsed electron beam and ion-withdrawal techniques are used to avoid interference of the fields associated with these two operations. Fox and his associates [8] have done much work with this technique which has been widely used [9]. It was the first of the electron-impact methods to resolve the doublet structure of the rare-gas ions and has made many contributions to the identification of excited levels. It is incorporated in a commercial instrument [10].

The RPD is subject to two difficulties. Its efficiency for producing ions (e.g. the slopes of its ionization efficiency curves) may be adjusted by varying the potentials of the electron gun [11]. This prevents its use for certain applications, such as the measurement of relative cross-sections of succeeding states. The reduction of the electron beam to near-zero energy (by electrode 4) could introduce the relaxation* effect [12] which both increases the energy distribution and affects the shape of the appearance potential curves. However it remains a simple and powerful technique.

B. THE 127° ELECTROSTATIC SELECTOR

The use of this high-resolution electron beam device is probably most widespread. It has undergone a series of developments, mostly at Laval

---

* This is a change in the energy distribution of a beam of charged particles caused by the particles' field interaction.

University. The first instrument was built by Clarke [5] according to the theory of Hughes and Rojanski [13]. The operation of the device is illustrated in Fig. 9-3-2. Thermally emitted electrons form a slightly divergent beam which enters the radial electrostatic field between two concentric cylindrical electrodes. Electrons of a suitable energy are deviated through 127°, at which point a focusing action of the field causes convergence of the beam at an exit slit. Electrons of different energy are deviated elsewhere. The beam is then accelerated to the desired energy and introduced into an ionization box. Clarke obtained an electron beam of $10^{-7}$ ampère with an energy distribution of about 0.25 eV. This sufficed to reveal new structure in the ionization efficiency curve of $N_2^+$, discovered the square-law for the production of doubly-charged ions, and measured the dissociation energy of $N_2$ as 9.7 eV rather than the 7.4 value then held.

The next selector (Marmet and Kerwin [14]) overcame some of the limitations of the first model which could not operate below about 10 eV because of space charge problems. The cylindrical plate electrodes were replaced by grid structures permitting the escape of unwanted electrons, thus avoiding space charge build-up and reducing surface charging. The grids

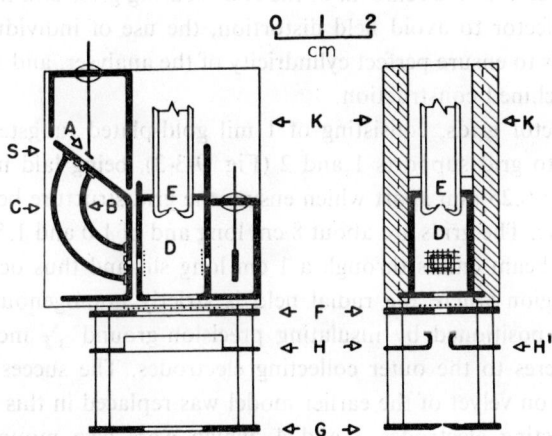

Fig. 9-3-2. The 127° cylindrical electrostatic selector [5]. Electrons from filament A are accelerated by a few volts to an entrance slit in the selector S where a radial electrostatic field between concentric electrodes B and C analyses and focuses a beam of electrons onto an exit slit. The beam is further accelerated to a desired energy by a potential difference between S and the ionization chamber D, where they may collide with gas molecules entering at E. The resultant ions are withdrawn and analysed by a mass spectrometer beam system F-G-H.

were supplemented by low-reflective "electron velvet"* surfaced outer electrodes which absorbed these electrons. The ionization box was similarly treated. An electron beam of FWHM energy dispersion of only 0.04 eV at $10^{-7}$ ampère resulted. This sufficed to reveal for the first time certain structure of molecular ions as well as further details in appearance potential curves including the often-observed anomalous breaks in the argon curve [15]. Schulz [16] used two of these selectors to perform an elegant electron spectroscopy experiment in which the inelastically scattered electrons from nitrogen molecules were observed, revealing the vibrational structure of the neutral molecule.

The most extensively duplicated model is the result of a Laval group effort headed by Marmet and Carette. About 50 instruments have been constructed and sent to various laboratories or have been copied [17]. Technical drawings have been available upon request. The model has been proven over the past five years to be very reliable and gives reproducible results. However its description has not heretofore been published.

In Fig. 9-3-3 may be seen a photograph of the instrument, and one of the assembly drawings. In Fig. 9-4-1 is given an outline of a selector incorporated into an electron spectroscope. Its improvements over the previous model are threefold: the removal of the end focusing grids and the lengthening of the selector to avoid field distortion, the use of individually-placed fine-wire grids to ensure perfect cylindricity of the analyser, and a very solid, precision-machined construction.

The selector grids, consisting of 1 mil gold-plated tungsten wires are spot-welded to grid supports 1 and 2 (Fig. 9-3-3), being laid in machined grooves about 0.25 mm apart which ensure the grid structure being parallel and concentric. The grids are about 8 cm long and of 1.0 and 1.5 cm radius. The electron beam enters through a 1 cm long slit and thus occupies only the center region where the radial field is strictly homogenous. The grid supports are positioned by insulating precision-ground $\frac{3}{32}$ inch pyrex or sapphire spheres to the outer collecting electrodes. The successful but laborious electron velvet of the earlier model was replaced in this instance by smooth collecting electrodes 3 and 4, which were also mounted on the selector box by means of the insulating glass spheres. Electron velvet or similar low-reflectivity surfaces are still used in the ionization chamber. It was found important that the selector box be tightly closed except for the beam entrance and exit slits; otherwise a low background of diffuse energy

* A surface consisting of tightly joined gold-plated copper tubes, each about 0.5 mm diameter by 1 mm long. The appearance is not unlike that of a miniature honeycomb.

Fig. 9-3-3 (a). Photograph of a recent 127° selector.

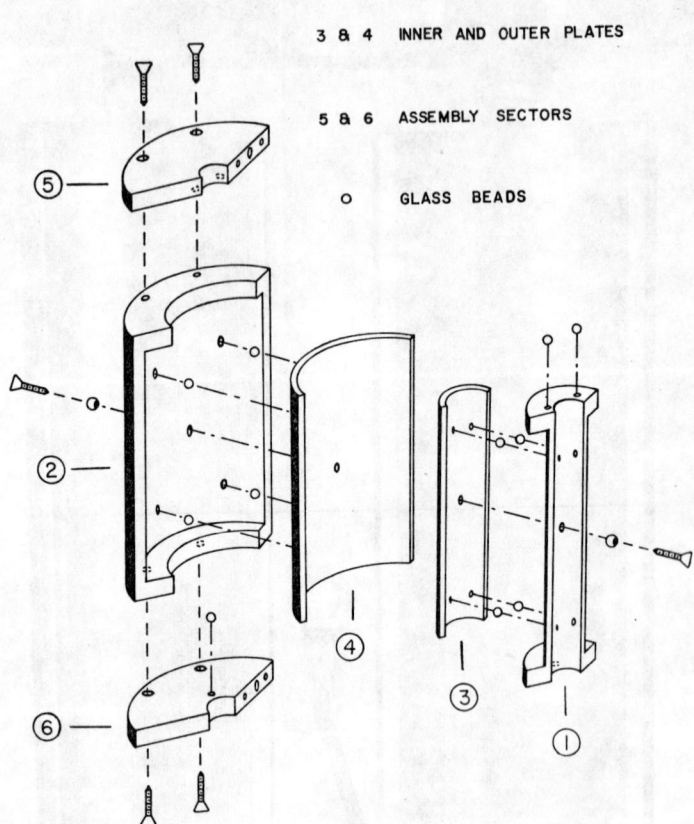

Fig. 9-3-3 (b). One assembly drawing. Overall length is 10 cm. (1) and (2) inner and outer grid wire supports: (3) and (4) inner and outer electron collectors; (5) and (6) end plates – at potential of 2.

electron current was observable. The 1 cm by 0.3 mm slits were machined in soft iron plates forming part of the box. This was necessary to shield the beam from stray magnetic fields. Kovar seals are used to apply the various potentials to the electrodes.

Carette has made an extensive investigation of the operating characteristics of this selector, and has analysed the beam produced by means of a second instrument. In Fig. 9-3-4 is shown a measured energy distribution of FWHM value about 70 mV and a beam current of about $10^{-9}$ A at the output of the analyser. The current from the selector is then about $10^{-7}$ A.

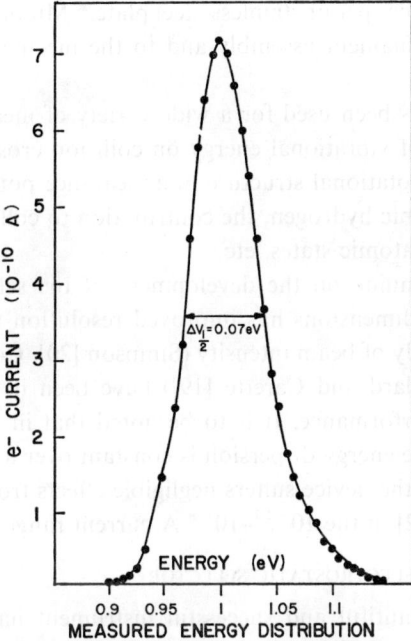

Fig. 9-3-4. Electron current reaching the detector of one analysing 127° selector used to measure the energy distribution of the beam produced by another. The two devices were mounted in "S" formation. Half-height width shown is about 70 mV. Peak current is about $10^{-9}$ A.

However better performance may be attained, by choosing the electrons contained in the center part of the beam solid angle. FWHM energy spreads of 0.060 eV for beam currents of $10^{-7}$ A are very reproducible under the following typical conditions:

| | |
|---|---|
| Filament potential | 0.0 V |
| Box (slit) potential | 0.0 V |
| Inner grid potential | 0.5 V |
| Outer grid potential | −0.5 V. |

The electron beam emerging from the selector is accelerated to the desired energy before entering the ionization chamber. The beam current is found to be constant ($\pm 1\%$) as the beam energy is varied from 5 to 40 eV. Better beam stability ($\pm 0.01\%$) has been achieved by using feedback from a virtual collimator [112].

Several modifications have been found convenient in this model. The fine wire grids have been successfully replaced in some instruments by 90%

transparent nickel, copper or stainless-steel plate.* Minor modifications have been made to the filament assembly and to the mechanical mounting and insulating.

This model has been used for a wide variety of measurements [17] including the effect of vibrational energy on collision cross-sections, preliminary evidence for rotational structure in appearance potential curves, resonance levels in atomic hydrogen, the contribution to collision cross-sections by various excited atomic states, etc.

Work is continuing on the development of the selector. A device of three times larger dimensions has improved resolution to 25 mV FWHM. The theoretical study of beam intensity (Simpson [20], Cotte [18]) and beam profile (Delage, Allard and Carette [19]) have been in general agreement with instrument performance. It is to be noted that in the case of the cylindrical selector the energy dispersion is constant over a considerable beam current range, and the device suffers negligible effects from the problems of beam relaxation [12] in the $10^{-14}$–$10^{-8}$ A current range.

## C. The spherical electrostatic selector

This rather beautiful and successful instrument has been largely developed by Simpson and his associates at the U.S. National Bureau of Standards [20–23], reported in 1963. Based on the analysing properties of the 180° concentric spherical condenser (Purcell [24]), it also incorporates double-aperture beam lenses (Spangenberg and Field [25]) which eliminate the need for slits in the analyser sections. In Fig. 9-3-5 is seen a sketch of a double analyser constructed by Simpson.

In the monochromator half of the instrument, electrons from a heated filament are focused by axially symmetrical lenses into the 1" radius spherical condenser which focuses its analysed beam on a second lens system which delivers it to the experimental region. The monochromator half may be rotated about an axis vertical to the drawing. This permits the electrons scattered by the experiment at various angles to be analysed by the second half of the device which is essentially similar to the first. An energy resolution of about 20 mV for a beam of $10^{-8}$ A was obtained. The energy resolution varies with the output current, going from about 10 mV at $10^{-9}$ A to about 100 mV at $10^{-6}$ A. The theoretical operating conditions have been extensively investigated by Kuyatt and Simpson [21] who find that they have been able to construct instruments which approach closely the theoretical performance limits. Because this type of instrument focuses the electron beam to

---

* Available from Buckbee Mears Company, 245 E. Sixth St., St. Paul, Minnesota 55101.

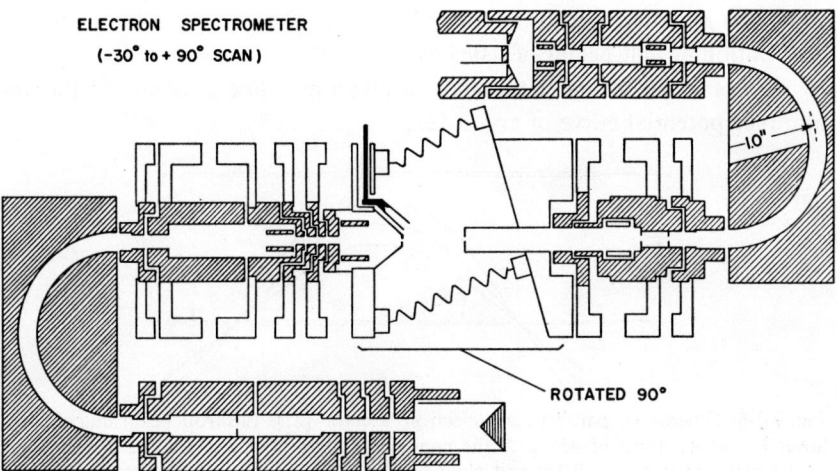

Fig. 9-3-5. A cross-section of the prototype spherical electrostatic electron selector and beam analyser [21]. Filament source is at top, experimental region in center. Both selector and analyser are preceeded and followed by beam lens system. A bellows permits the selector to be rotated about the experimental region.

a "point", it is more subject to beam relaxation than line-focusing devices.

The spherical selector has been used for electron spectroscopy, revealing vibrational structure in HD and resonance levels in He [23].

D. THE PARALLEL-PLANE ELECTROSTATIC SELECTOR

In 1956 Hutchison [26] at Argonne developed a parallel-plane electron selector suitable for low-energy investigations of atoms and molecules. This was based on a higher-energy version of Harrower [27] following considerations by Yarnold and Bolton [28]. The operation of the instrument is indicated in Fig. 9-3-6.

A beam of electrons is focused onto a slit in the lower of two plane parallel electrodes between which is maintained a uniform electric field. Under the influence of the constant opposing field the electrons describe parabolic paths, an entrance angle of 45° giving the greatest "range" which corresponds to optimum conditions. Electrons of a given energy follow a given parabolic path which may be adjusted by varying the electric field so as to impinge on the exit slit. Hutchison used carefully-designed electron gun and beam lenses to produce his beam, and a second parallel plate selector for analysing it. Ions produced at mid-point of his twin-selector device were analysed in turn by a mass spectrometer. The electron beam analyser in the apparatus recorded a beam of $1.4 \times 10^{-10}$ A at 10 eV energy with an energy distribution at half-height of 0.063 eV.

This instrument has been used to investigate fine structure in the appearance potential curve of argon [29].

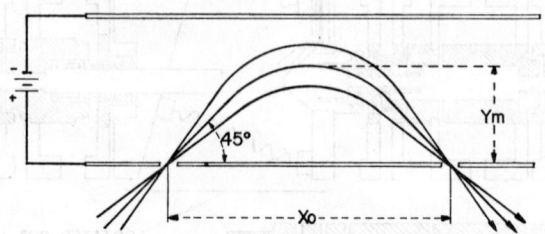

Fig. 9-3-6. Schema of parallel-plate electron selector [27]. Electron beam enters from lower left at an angle of 45° with the condenser plates. For a small beam divergence ($\pm 3°$) plate separation is $0.3x_0$ and plate voltage 0.6 times electron accelerating voltage ("beam energy").

### E. THE AXIAL-SOURCE CYLINDRICAL ELECTROSTATIC SELECTOR

The parallel-plate condenser used in the selector mentioned above may be considered as a small section of a cylindrical condenser. A cylindrical selector has been devised by Blauth [30] and studied by Zashkvara et al. [31]. Sar-el [32] has recently given a theoretical treatment and description of an apparatus used mostly for relatively high energy (100–600 eV) electrons. The apparatus is shown in Fig. 9-3-7. Electrons from a gun B are directed by a hemispherical capacitor system devised by Simpson [33] (A–C–D) through the slit in the inner plate of the inner cylinder. A radial field is maintained

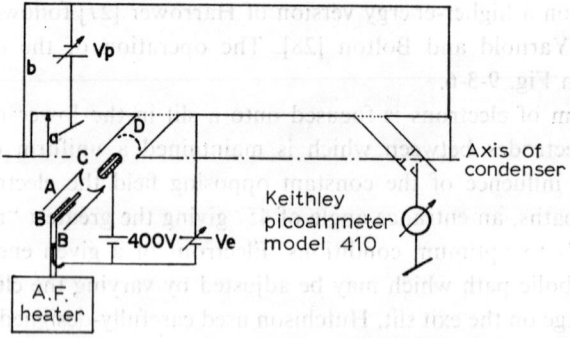

Fig. 9-3-7. Schema of the axial-source cylindrical electrostatic selector [32]. Electrons from the source B are directed by the lens system A-C-D into the cylindrical-slot opening in the inner of two concentric cylinders of radii $a$ and $b$. The angle of entry is 42.3°.

between this and the outer cylinder, the two being concentric about the broken axis shown in the drawing. For a beam-entry angle of 42.3° second-order focusing of electrons of the appropriate energy for being deflected through a second cylindrical slit takes place. At energies of from 100 to 600 eV, the instrument delivered about $10^{-6}$ A with an energy resolution of 1%. Operated at an electron energy of about 1 eV, the current was about $10^{-8}$ A.

Hafner, Simpson and Kuyatt [34] have made a detailed theoretical comparison of the spherical and axial-source cylindrical selectors, and conclude that the cylindrical device is potentially superior to the other. Their estimate of an energy dispersion of only 1% for the cylinders was in fact achieved by Sar-el. No results of investigations with the instrument have been reported as yet.

## F. The Monokinetron

A high-current cousin to the 127° cylindrical selector has recently been proposed by Marmet [35]. The principal of operation is indicated in Fig. 9-3-8. In part (a) of the figure we see a divergent beam of electrons of various energies entering a 127° cylindrical analyser as discussed in section 9-3-B. Those electrons of an appropriate energy $E_0$ will be focused into a convergent beam at $X_0$ and at this point would ordinarily be selected by an exit slit and used for collision studies. However note that the electrons of another energy $E^+$ are *also* focused into a convergent beam at point $X^+$, while those of yet a different energy $E^-$ are focussed at $X^-$. In other words, the line

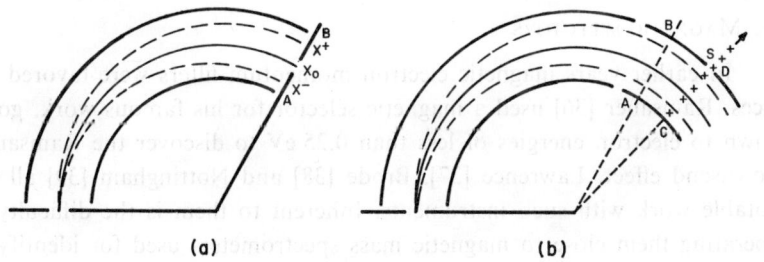

Fig. 9-3-8. (a) Focusing action of the 127° selector. Electrons in the entering beam which have the appropriate "design" energy $E_0$ will focus at $X_0$; those with greater and lesser energies will also be focused – at points $X^+$ and $X^-$, say.
(b) Energy correction of the monokinetron. The field gradient between the correcting plates slows up faster electrons, speeds up slower ones, focuses all electrons to the same kinetic energy along line C-D. Ions formed by collision with electrons will drift along field lines; those formed at line of focus are expelled through orifice S.

A–B is an image line, at any point of which all incident electrons have the same energy. The variation in electron energy from one end of line A–B to the other is a simple, known function.

The next step is to make all of these energies the same. This means slowing up the electrons on one side of $X_0$ and speeding up those on the other. Marmet has shown that this may be accomplished – for example – by changing the field gradient past a certain point in the beam path. In part (b) of Fig. 9-3-8 this is shown accomplished by an insulated extension to the selector. By using the proper field gradients, all electrons impinging on line C–D have become monokinetic. The first advantage of the device is that most electrons entering it are brought to the same kinetic energy, while the usual selector eliminates the greater part of them. It may be shown that the edge effects of the selector extension do not affect the first-order performance of the monokinetron. A second advantage is also indicated in Fig. 9-3-8b. In order to withdraw the ions formed by electron bombardment from the ionization box, an extraction potential is used in many applications. This can seriously interfere with the selector performance (for example, in the RPD method, the difficulty is avoided by pulsing the electron beam and ion extraction voltage alternatively). In the monokinetron we see that the line of monoenergetic electrons C–D is distributed close to a field line, so that ions formed in this region will be accelerated perpendicularly to the cylinder walls. A small opening at S will permit them to be removed without disturbing the electron beams. Marmet has worked out several versions of the monokinetron, and in a typical application expects to obtain a gain in electron current for the same energy dispersion as in a similar electron selector.

### G. Magnetic selectors

In earlier years magnetic electron momentum filters were favored devices. Ramsauer [36] used a magnetic selector for his famous work, going down to electron energies of less than 0.25 eV to discover the Ramsauer-Townsend effect. Lawrence [37], Brode [38] and Nottingham [39] all did notable work with such instruments. Inherent to them is the difficulty of operating them close to magnetic mass spectrometers used for identifying ions. The success of electrostatic selectors has deviated attention from magnetic instruments, but Marmet [35] has proposed a magnetic version of the monokinetron. A sketch of the proposed device is shown in Fig. 9-3-9. The principle of operation is similar to that discussed in connection with Fig. 9-3-8. A divergent beam of electrons entering the magnetic field is deviated through 180°, when all electrons of a given energy are focused at $X_0$.

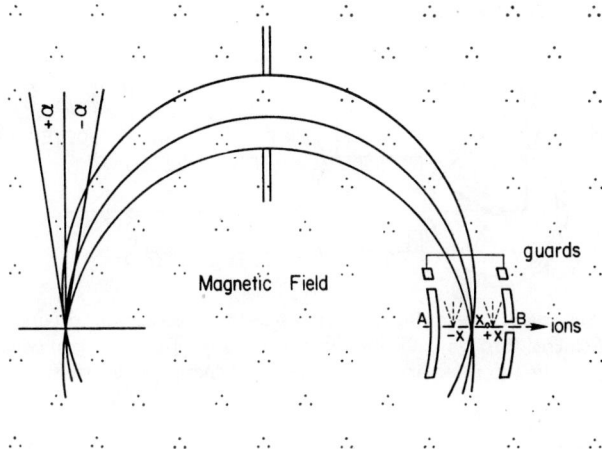

Fig. 9-3-9. Magnetic monokinetron. Magnetic field is normal to drawing. Action of correcting electrostatic field is similar to that of Fig. 9-3-8. Ions formed along line of focus for monoenergetic electrons A–B are expelled through orifice.

Electrons of a higher energy are focused at $X^+$, those of lesser energy at $X^-$. Thus the line A–B is the image line for electrons of energies varying in a known way from A to B. These may all be brought to the same energy by being passed through an appropriate electrostatic field between the curved electrodes. This field would also permit the withdrawal of the ions formed between them along the line of constant electron energy through an aperture in the outer electrode. One advantage of the magnetic instrument would be the relative absence of surfaces. These are now known to produce three deleterious effects: space charge, wall phenomena and spurious electrostatic fields. These will be discussed briefly below.

## H. Time-of-flight selectors

Although long used successfully in mass spectrometry [40], the time-of-flight instrument has only recently been applied to low energy electron beam work. Chanin, Phelps and Biondi [41] have used time-of-flight techniques to measure electron attachment for electrons of energy between about 0.05 and 3 eV. An instrument devised by Baldwin and Friedman [42] is more comparable to the usual high resolution beam device. Its operating principle is indicated in Fig. 9-3-10. Time-calibrated bursts of electrons are emitted from a source, and a collimated beam of them – with a wide energy distribution – enters a drift tube. Consider those electrons with a precise energy

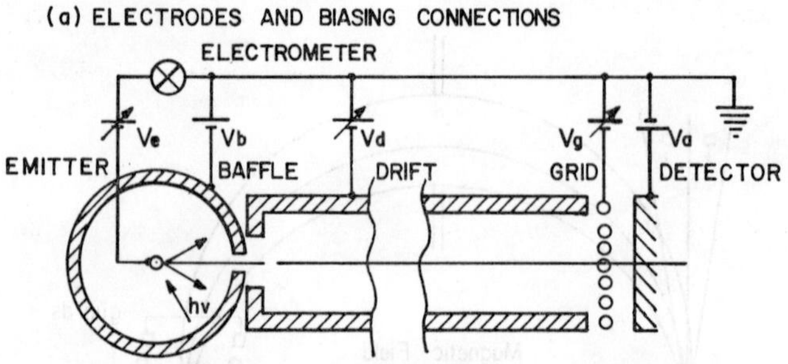

**(a) ELECTRODES AND BIASING CONNECTIONS**

Fig. 9-3-10. Schema of time-of-flight electron selector. Precisely timed bursts of electrons from emitter emerge through baffle into drift tube. Time-of-flight to grid (whence accelerating to detector) gives velocity and energy of electrons.

$E_0$ corresponding to a precise velocity $v_0$. They will arrive at the other end of the drift tube in a precise time, whose measurement is a measure of the electron energy. The number of electrons arriving at this precise interval after each burst is measured. Then the gas under investigation is admitted to the drift tube, and the number of electrons detected under the new conditions is measured. The difference is the number which has been scattered by the gas molecules, and from the drift tube length and gas pressure, the collision cross section may be determined for this electron energy. The cross sections for other energies are similarly determined by simply changing the time of flight interval being measured. By suitable electronic techniques a complete spectrum of time intervals may be measured simultaneously. The slot grid at the end of the drift region in the figure serves to separate the field-free drift tube from the relatively high-field region necessary to accelerate the electrons for detection by an electron multiplier.

The time-of-flight instrument is very complex, and subject to many special conditions which have been studied in detail by Baldwin and Friedman. However since it is specially adapted for work at electron energies below 1 eV it should be particularly valuable. Indeed, its energy resolution improves at low electron velocities in view of the improved precision in longer time interval measurements. It has been used to measure Ramsauer scattering in argon, the Ramsauer transparency at 0.25 eV being beautifully brought out. The instrument has been operated down to 150 mV, with experimental points spaced by about 80 mV.

I. MISCELLANY

Mention was made of the problems introduced into high resolution electron beam work by the presence of surfaces. Space charge plagues most low-energy electron devices, causing a broadening of the energy resolution by relaxation and other effects. It is usually reduced by using very small currents, grid electrodes and electron absorbing surfaces. It is most aggravating in devices which focus the electron beam to a point rather than to a line. Its effects would be minimised in time-of-flight devices.

Wall phenomena are also troublesome, although they offer a potential research field of much interest. Molecules adsorbed on metal surfaces, or on surfaces which are oxidized or covered with a gas layer will have their physical characteristics modified (ionization potential, dissociation energy, etc.), and if struck by an electron beam may furnish ions at energies which do not correspond to those of the free molecule. This effect (see, for example, Marmet and Morrison [43]) can give rise to "spurious" detail in energy levels and no doubt accounts for some of the mysterious levels now abounding [29]. If these interesting effects are to be avoided, then wall surfaces must be kept to a minimum by means of jet and similar techniques. Experiments in which the wall surface is varied systematically are also useful [44].

Surface potentials are a third source of much difficulty. It has recently been shown by Petit-Clerc and Carette [45] that ordinary surfaces (physically "clean" and evacuated to $10^{-8}$ Torr) when bombarded by electron and ion beams may build up local surface potentials of up to 1 volt which decrease slowly, persisting for several hours. Such effects have long been qualitatively known, and cause distortions of field patterns resulting in poor instrument performance. Instrument design, which avoids slits [21], may help. This problem is critical in the ionization chamber.

High resolution electron beam devices are usually sensitive to extraneous fields of various kinds. Television and other H.F. fields are distressingly noxious, and should be shielded out by appropriate filters in leads [20]. The earth's magnetic field is usually shielded out or nullified by mumetal shields or Helmholtz coils. Operating electron selectors in combination with mass spectrometers is a common problem, and a small fringe-field-free spectrometer has been designed for this purpose [46].

Although not involving high resolution electron beams, two other atomic physics techniques should be mentioned. One is the use of photoelectrons as ionizing and exciting agents. A typical reference would be the paper by Ditchburn and Opik [47]. The second is the use of thin foils for stripping and exciting fast ions, resulting in radiation from highly ionized species. This "beam-foil" technique has been reviewed by Bashkin [48].

## § 9-4. Electron spectroscopy

The energies of atomic and molecular electronic levels are generally well known as a result of optical spectroscopy. Its experimental techniques usually attain a precision of one part per million and often better. Such precision from electron spectroscopy is illusory at the present time. As a matter of fact, the actual state of art of this method provides under the most favorable conditions a precision of about one part per ten thousand. This has been reported by many papers (e.g. [14–17, 20]).

The major interest in the measurements now being made by means of electronic spectroscopy is therefore not the precise measurement of levels, but rather the verification of laws and theories which attempt to describe the electron-target interaction during their collision. All experimental results reported in this section have been done for this purpose. The particular objectives sought in electron scattering experiments at the moment are to resolve close (0.1 eV) excited electronic or vibrational states, to measure the excitation probability or cross-section as a function of projectile energy, to analyse the profiles of the lines in the energy spectrum of the scattered projectiles and to determine their angular distribution.

To reach these goals one needs a monokinetic and well-collimated electron beam, 0.05 eV FWHM or better, and it is desirable that this beam has as high an intensity as possible to facilitate detection problems. Recent developments of electron selectors, described previously, have allowed scientists to satisfy these imperatives. They have used these instruments to build up what we may call an electron spectroscope.

A. THE ELECTRON SPECTROSCOPE

Many types of spectroscope have been constructed and used by different research groups [16, 20, 49, 50, 51]. Their common feature is that they all have three main parts: a monokinetic electron source which is an electron selector, a target which is either a cell containing a high pressure gas or a molecular beam, and finally an analyser which is usually similar to the selector. A schematic diagram of such an apparatus is shown in Fig. 9-4-1 and a cut view drawing to scale in Fig. 9-3-5. This set-up allows three main modes of operation, according to the various settings of the potentials $V_1$ and $V_2$.

In the first mode it is mainly the geometry of the apparatus which permits the study of elastic scattering of an electron beam by the gas target as a function of the incident electron energy $eV_1$. The size of the cell, the

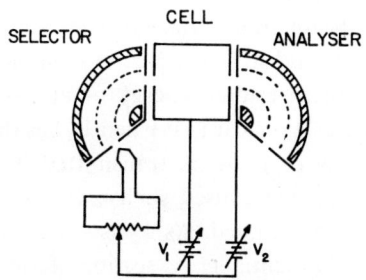

Fig. 9-4-1. Schematic diagram of electron spectrometer. Selector: monoenergetic electron source; cell: chamber containing the gas target or, instead of a chamber, a molecular beam; analyser: may be rotated.

selector-cell and cell-analyser distances and the slit width are chosen such that any electron having been scattered outside a given angle, usually by one degree or less, is eliminated from the incident beam. If the analysis is made in the forward direction then one detects the electrons of the incident beam minus those which have been scattered. If the cell and the analyser are properly oriented then one can detect the scattered electrons at a given angle.

In the case of elastic scattering the analyser could be replaced by a simple electron collector in front of the exit slit of the cell. But as the signal is usually extremely weak, one wishes to eliminate from the transmitted beam all spurious electrons having lost energy by inelastic collision or by collision on surfaces. This elimination is performed by the analyser. In this type of measurement the potential $V_2$ is set to the same value as that of the exit slit of the selector and $V_1$ is varied in the desired energy range. The electrons are thus accelerated up to an energy $eV_1$, scattered, and decelerated to about one eV. This deceleration process is necessary because the energy spread of the analyser (like that of the selector) is proportional to the energy of electrons.

The other two modes of operation concern the inelastic scattering of electrons by a target. In the second mode the potential $V_1$ is held constant throughout the measurement, for example $V_1 = 200$ V. Potential $V_2$ is the parameter which is varied. Electrons emerging from the selector are first accelerated to an energy $eV_1$; some are then elastically scattered in the cell; then in the cell-analyser interval they lose energy by an amount proportional to $V_1 - V_2$. When no target is present in the cell and $V_2$ is equal to the potential of the exit slit of the selector, the electrons at the entrance slit of the analyser are back at their original selector energy $eV_0$, about one eV, and the beam is transmitted by the analyser. However if a gas is contained in the cell or a molecular beam crosses the electron beam, many electrons will excite atoms or molecules and so lose a part of their energy. Those electrons will find themselves with a residual energy of $e(V_1-V_e)$, where $V_e$ is the energy of the excited state. It is clear that when $V_2$ has the value $V_2 = V_e + \Delta V$, where $e \Delta V$ is the energy of particles transmitted by the analyser (about one eV) only electrons having caused such excitation will be transmitted, while any others will be eliminated. As $V_2$ is varied, this experiment gives the value of the relative differential cross-section of excitation for the various electronic levels of the target for electron impact at a given energy.

It is the dependence on the electron energy for the excitation cross-section of a given electronic state which is studied in the third mode of operation of the electron spectroscope. In this case the apparatus is operated in such a way that only the electrons which have lost a given amount of energy $eV_e$ are transmitted. This is done by varying $V_1$ and setting $V_2$ in accordance with $V_2 = V_e + \Delta V$, where again $\Delta V$ determines the energy of electrons transmitted by the analyser (it is equal to $kV_g$, where $k$ is a constant depending on the geometry of the analyser and $V_g$ is the difference of potential between the focussing electrodes of the analyser). Thus when $V_1$ is swept in the desired energy range the transmitted current is proportional to the differential cross-section of the studied electronic state.

In résumé, the main data available from the electron spectroscope are: a) The value of the relative differential elastic cross-section as a function of the electron energy and angle obtained by the measurement of the current transmitted across the cell. b) The value of the relative differential inelastic cross-section for the various electronic states of the target at a given incident electron energy and angle, obtained by sweeping the energy spectrum of electrons having suffered an inelastic collision. c) The value of the relative differential inelastic cross-section for a given electronic state as a function of incident electron energy and angle obtained by sorting out of the emerging beam from the cell those electrons which have lost a given amount of energy.

## B. Elastic Scattering, Resonances

Elastic scattering of low energy electrons in gases was studied as early as 1921 with a monokinetic electron source described by Ramsauer [36]. The early reported results [52, 53] although they permitted the discovery of the important Ramsauer and Townsend effects, which could be explained by quantum treatment of scattering, did not show any evidence of resonance in the elastic scattering. However Golden and Bandel [54] have recently reported the observation of such resonance phenomena with a similar apparatus. Resonance in elastic scattering, already well known in nuclear physics, has been the object of fruitful experimental studies in atomic physics during the last five years. The first measurements were by Schulz [55, 57], Simpson and Fano [56] who reported the existence of a resonance in the elastic scattering of electrons by helium. On the continuous curve of the scattered electron current it appears as a sharp increase of the transmitted current in the forward direction of the beam or an equivalent decrease in the angularly scattered electron current. After many calibrations [20, 22, 50, 55, 56] this resonance has been fixed in the energy scale at $19.3 \pm 0.02$ eV. This is about 0.5 eV below the first excited state of helium, the metastable

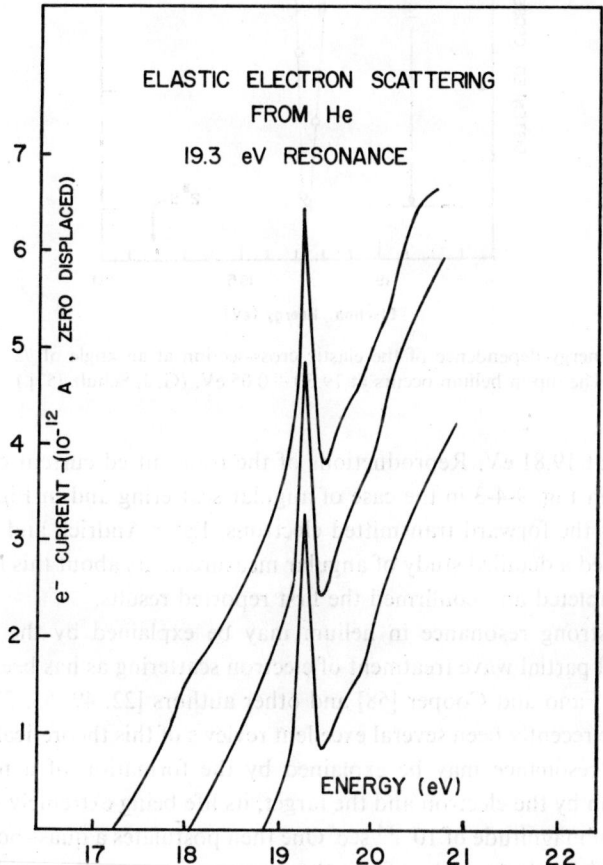

Fig. 9-4-2. Value of the electron current transmitted in the forward direction in an elastic scattering experiment on helium. Three runs are shown. The $y$-axis zero has been displaced for each. (A. Jacob and J. D. Carette, unpublished.)

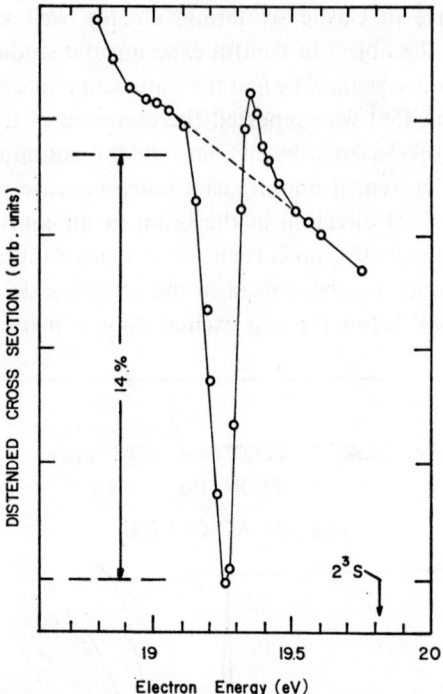

Fig. 9-4-3. Energy dependence of the elastic cross-section at an angle of 72° in helium. The dip in helium occurs at 19.30 ± 0.05 eV. (G. J. Schulz [57].)

state $2\,^3S$ at 19.81 eV. Reproductions of the transmitted current curves are presented in Fig. 9-4-3 in the case of angular scattering and in Fig. 9-4-2 in the case of the forward transmitted electrons. Later Andrick and Ehrhardt [49] reported a detailed study of angular measurements about this resonance which completed and confirmed the first reported results.

This strong resonance in helium may be explained by the quantum mechanical partial wave treatment of electron scattering as has been demonstrated by Fano and Cooper [58] and other authors [22, 49, 55, 56, 59–61]. There have recently been several excellent reviews of this theoretical problem [62]. This resonance may be explained by the formation of a temporary negative ion by the electron and the target, its life being extremely short – of the order of magnitude of $10^{-14}$ sec. One then postulates a quasi-bound state of the incident electron. Far from the resonance energy the incident wave associated with the electron is scattered as on a hard sphere, and there is no important phase shift in the scattered wave. In the neighbourhood of the resonance energy the incident wave interacts deeply with inner parts of the target during a time of the order $\hbar/\Gamma$ and is reemitted as a scattered wave. The density probability of the electron around the target is then very strong and the scattered wave has an important phase lag which provokes a large change in the scattering cross-section. This particularly strong density probability of the electron around the target is somewhat similar to a bound state, except that its life is extremely short while that of a bound state is infinite.

The main data which may be obtained characterizing these resonances are the energy $E_r$ at which they occur, the type of partial wave involved in the scattering, the dispersion of the wave (particularly its width at half

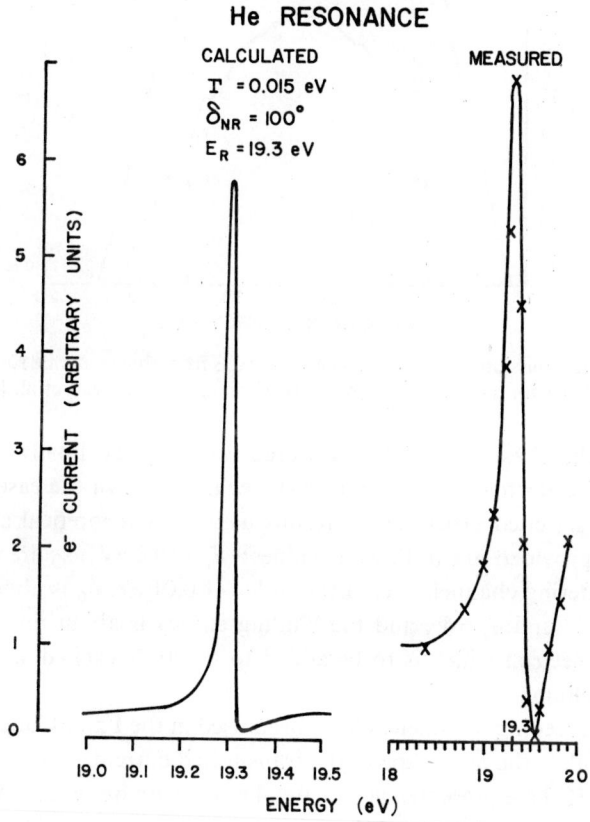

Fig. 9-4-4. Calculated and measured electron current transmitted in the forward direction in helium as a function of incident electron energy around the 19.3 eV resonance. (A. Jacob and J. D. Carette, unpublished.)

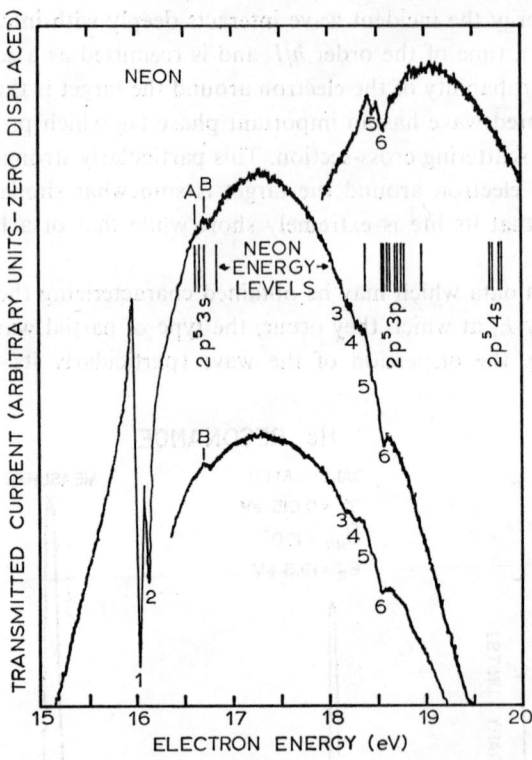

Fig. 9-4-5. Transmission of electrons by neon. Results from three separate runs are shown, together with the lowest energy levels of neon. (C. E. Kuyatt et al. [22].)

height $\Gamma$), the phase shift of the scattered waves $\delta_l$ and finally the binding energy of the electron of the temporary negative ion. In the case of helium the previously cited experimental results as well as theoretical calculations [62] have provided the following values: $E_r = 19.3$ eV as already stated; s-wave scattering channel, $\Gamma$ is of the order of 0.01 eV, $\delta_0$ is about equal to 100°, $\delta_1 = 25°$ and $\delta_2 = 4°$, and the binding energy is about one eV for the additional electron which is to be added to the 1s 2s excited state configuration of helium.

The preceeding parameters have been used in the Fano-Cooper formula [58] to calculate the line shape of the transmitted current. The results of this calculation [63] are presented at Fig. 9-4-4 along with our experimental data. Considering the dispersion of the electron beam and the widening due to Doppler broadening effect the agreement between calculated and measured values is quite good as far as the shape of the curve is concerned.

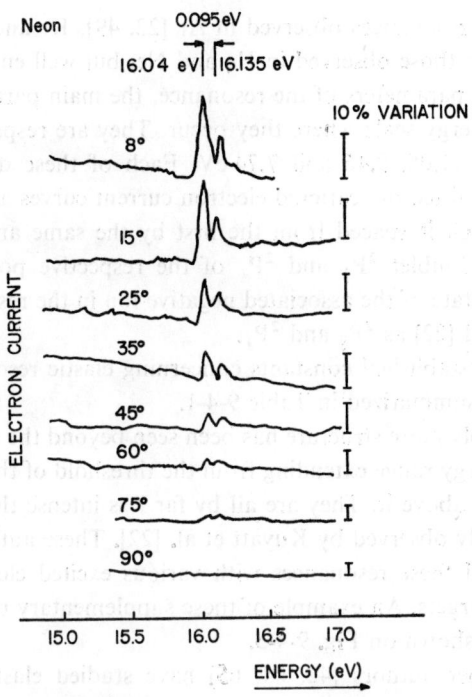

Fig. 9-4-6. Intensity dependence of elastically scattered electrons in neon as a function of energy of incident electrons, for various angles. (D. Andrick and H. Ehrhardt [49].)

Similar resonance phenomena in the elastic scattering of electrons have also been encountered for other noble gas atoms: Ne, Ar and Xe [22, 49, 56, 57]. They are always located at approximately half a volt below the first excited state of the atom. The neon target shows two strong resonances at 16.04 and 16.135 eV; they are thus separated by 0.095 eV. They appear as two sharp decreases in the forward transmitted electrons, and corresponding increases in the angularly scattered electrons. These two features are illustrated in Figs. 9-4-5 and 9-4-6. The shape of the transmitted and scattered electron current curves enables one to determine that the resonance occurs by the intermediary of a p partial wave. Andrick and Ehrhardt [49] have determined the potential phases $\delta_l$ for $l=0$ and 1; they are $\delta_0 = 300°$ and $\delta_1 = 160°$. It is interesting to note that the energy difference between the two first elastic peaks, 0.095 eV, is closely the same as the interval between the $^2P_{\frac{3}{2}}$ and $^2P_{\frac{1}{2}}$ doublet levels of the neon ion $Ne^+$. The relative intensity 2:1 of the two peaks permits deducing the configuration $^2P_{\frac{3}{2}}$ and $^2P_{\frac{1}{2}}$ for the state of the negative ion $Ne^-$ associated with the resonance.

The elastic resonances observed in Ar [22, 49], Kr and Xe [22] are all less intense than those observed in He and Ne, but well enough profiled to determine some parameters of the resonance, the main parameter being the values in the energy scale where they occur. They are respectively for each of these atoms 11.06, 9.46 and 7.74 eV. Each of these discontinuities in either the transmitted or scattered electron current curves are followed by a second one which is spaced from the first by the same amount of energy separating the doublet $^2P_{\frac{3}{2}}$ and $^2P_{\frac{1}{2}}$ of the respective positive ions. The corresponding state of the associated negative ion in the resonance has thus been determined [22] as $^2P_{\frac{3}{2}}$ and $^2P_{\frac{1}{2}}$.

The main established constants concerning elastic resonances in noble gases has been summarized in Table 9-4-1.

Considerably more structure has been seen beyond the strong first resonance in an energy range extending from the threshold of the first one up to about five volts above it. They are all by far less intense than the first, but have been clearly observed by Kuyatt et al. [22]. These authors have tentatively associated these resonances with various excited electronic configurations of the targets. An example of these supplementary resonances in the case of neon is shown on Fig. 9-4-5.

Finally other authors [42, 64, 65] have studied elastic scattering of

TABLE 9-4-1

|    | $E_r$ (eV) | $\Gamma$ (eV) | $\delta_l$ | Wave | Configuration of negative ions |
|----|-----------|---------------|-----------|------|-------------------------------|
| He | 19.30 | 0.015 | $\delta_0 = 100°$<br>$\delta_1 = 25°$<br>$\delta_2 = 4°$ | s | $(1s\,2s^2)\,^2S_{\frac{1}{2}}$ |
| Ne | 16.04 | 0.01 | $\delta_0 = 300°$ | p | $(1s^2\,2s^2\,2p^5\,3s^2)\,^2P_{\frac{3}{2},\frac{1}{2}}$ |
|    | 16.13 | 0.01 | $\delta_1 = 160°$ |   |   |
| Ar | 11.06 | 0.02 | – | p | $(KL\,3s^2\,3p^5\,4s^2)\,^2P_{\frac{3}{2},\frac{1}{2}}$ |
|    | 11.23 | 0.02 | – |   |   |
| Kr | 9.46 | 0.02 | – | p | $(KLM\,4p^5\,5s^2)\,^2P_{\frac{3}{2},\frac{1}{2}}$ |
|    | 10.10 | 0.02 | – |   |   |
| Xe | 7.74 | 0.02 | – | p | $(KLMN\,5p^5\,6s^2)\,^2P_{\frac{3}{2},\frac{1}{2}}$ |
|    | 9.02 | 0.02 | – |   |   |

electrons by helium using newly developed techniques like the time-of-flight electron monochromator [42, 65]. As mentioned in section 9-3-H, this method has enabled a remarkably accurate measurement of the Ramsauer effect in elastic electron scattering through helium at low energy of incident electrons, between 0 and 1.5 eV.

Following the discovery of resonances in the elastic scattering of electrons in noble gases, many searches have been undertaken to find similar processes in other atoms or molecules. Although always harder to find, several successful achievements have been made in both scattering channels: the elastic and the inelastic ones. The particular species which have been investigated are atomic hydrogen [54, 66–71], lithium [72], mercury [22], hydrogen and deuterium molecules [73, 74], the nitrogen molecule [49, 73, 75]. Helium transmission type of resonances have rarely been found; one of them is seen, however, in $N_2$ at 11.48 eV. The curve of the transmitted current obtained by Heideman et al. [75] in the range from 11 to 13 eV is presented in Fig. 9-4-7; a sharp and narrow peak is observed at 11.48 eV. The authors have made very high energy resolution measurements of this peak's half-width and found a measured value as low as 0.02 eV, which they reported was apparently still limited by instrumental resolution and Doppler

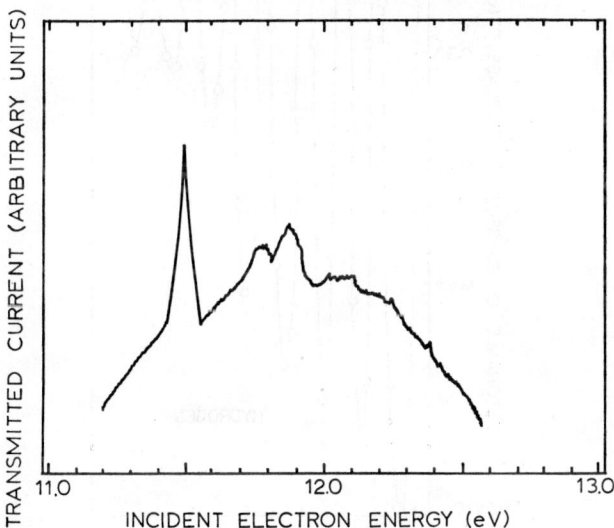

Fig. 9-4-7. Transmission of electrons by $N_2$, showing a sharp "window" type resonance at 11.48 ± 0.05 eV. The zero of current has been displaced. Additional structure occurs at 11.75 and 11.87 eV. The latter is partly due to an inelastic threshold (E $^3\Sigma_g^+$). The nitrogen pressure was ~0.04 Torr. (H. G. M. Heideman et al. [75].)

broadening. This resonance has been interpreted to be caused by a compound $N_2^-$ electronic state associated with either an unidentified state at 12.26 eV or the E state at 11.87 eV.

One of the most interesting and important species related to resonance in elastic scattering is atomic hydrogen, because this is the simplest known system and the easiest to calculate. In particular, resonance associated with the 2S and 2P states have been observed and compared with predicted values [66–71], the agreement being extremely good.

Resonance processes in elastic scattering of low energy electrons from diatomic molecules offers a peculiar feature. The energy spectrum of the transmitted or angularly scattered electrons sometimes contains a structure associated with the vibrational levels of the temporary negative ion associated with the resonance. Such a behaviour has been particularly evidenced for the diatomic molecules $H_2$ and $N_2$ [54, 68, 73, 75]. For each of these molecules many series of peaks have been detected either at very low energy (approximately two eV above threshold) or at higher energy: around 15 eV. An example of results presenting the value of the cross section for $H_2$ [54]

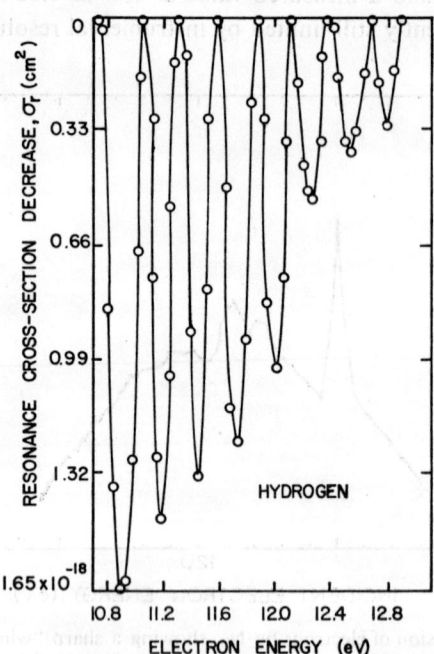

Fig. 9-4-8. Resonance cross section in hydrogen versus electron energy. (D. E. Golden and H. W. Bandel, Phys. Rev. Letters **14** (1965) 1010.)

between 10.8 and 12.8 eV of incident electron energy is shown at Fig. 9-4-8. One observes clearly the vibrational structure represented by peaks in the curve of the cross-section. Previously cited authors have observed many similar vibrational spectra.

## C. Inelastic scattering

The numerous new results and data obtained by experimental studies of electron elastic scattering from atoms and molecules with the aid of the improved electron spectrometers were accompanied by even more new results in the inelastic scattering channel. As stated earlier, these results are divided in two parts according to the method of using the electron spectrometer. The first method consists in the measurement of the relative excitation cross-section of various electronic states as a function of a fixed incident electron energy and the second the excitation of a determined excited state as a function of an energy varying electron beam. Considering the transmission type of experiment as the first, we may refer to them as the second and third operating modes of the electron spectrometer.

### 1. *Second mode*

Relative to the energy range, this mode may be divided in two: the high and the low energy. This subdivision is very arbitrary and relative but in the field of electron scattering we may fix the limits as follows: low energy between 0 and 100 eV, high energy starting from 100 eV and up.

The main purpose of these measurements is the verification of various theories predicting the behaviour of electrons colliding with a gas target and the discovery of new, unpredicted ways of excitation, like the resonance processes. Quantum mechanical treatment of collisions between an electron and an atom or molecule predicts that at sufficiently high energy the excitation laws will be the same as those governing the excitation by ultra-violet photon impact. Explicitly the conclusions are [61]:

a) The probability of excitation to a state implying an allowed optical transition is by far larger than the probability of a transition forbidden by the same rules.

b) The excitation probability decreases rapidly as the scattering angle increases.

In fact, if in the high energy range electron impact the Born approximation is usually valid and the transition selection rules are the same as those which govern optical excitation, such is not the case in the low energy range (0 to 100 eV), where many forbidden transitions take place. In the last ten years such low energy processes have been reported by many authors concerning various atoms and simple molecules. These new data and results have of course been made possible by the newly developed techniques of electron spectroscopy using either the various electrostatic analysers or the electron trap method.

In the low energy range the most studied atoms and molecules are He [22, 51, 76–80], $H_2$ [73, 81], $N_2$ [51, 73, 75, 82], $O_2$ [83], CO [73, 82], $H_2O$ [84], COS [63], ethylene, benzene and many other organic molecules [51, 77, 78]. We may see on Fig. 9-4-9 three aspects of the helium electron scattered spectrum [80] for incident electrons of 50 eV. Each spectrum has been measured at a different scattering angle. At this electron energy the dominant peaks in the spectrum correspond to the levels $n\,^1P$ and $2\,^1S$. According to

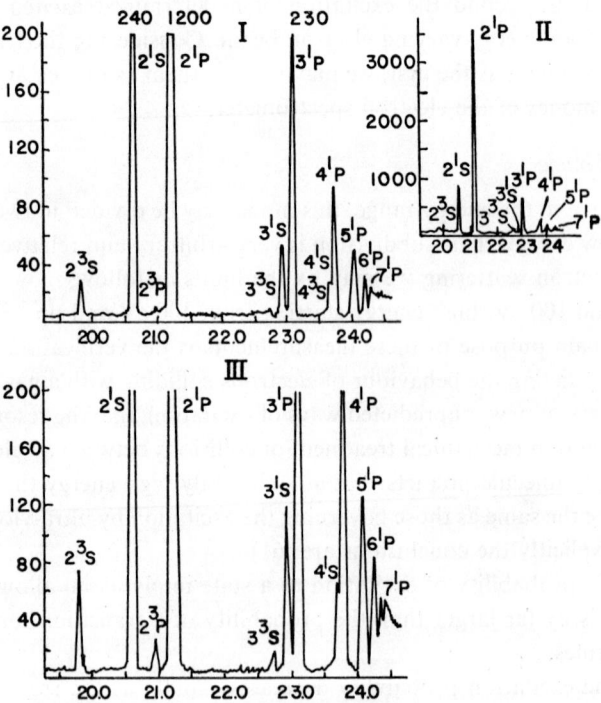

Fig. 9-4-9. Electron impact spectra of helium. The inset at upper right shows the complete spectrum; in the other spectra, the high peaks have been left off scale to show the weaker transitions more clearly. Intensities should be compared only in single spectra due to change in focussing and discriminators settings. Electron energy 50 eV; on the axes counts/sec versus energy loss eV. I: scattering angle $\theta \sim 3°$; II: scattering angle $\theta \sim 0°$; III: scattering angle $\theta \sim 8°$. (A. Skerbele and E. N. Lassettre [80].)

the rules previously stated, we may expect that the excitation from ground level to the $n\,^1P$ levels will be the most probable. However, at low electron energy we observe an appreciable amount of scattering due to excitation ending at $2\,^3S$, $2\,^1S$, $3\,^3S$, $2\,^1S$ levels. The contribution of these levels is enhanced as the electron energy is decreased. An example of this behaviour is illustrated by the spectra [51] shown at Fig. 9-4-10. The salient and most remarkable feature of these spectra is the increasing intensity of the $2\,^3S$, up to the point where it is the dominant peak of the spectrum. In a different

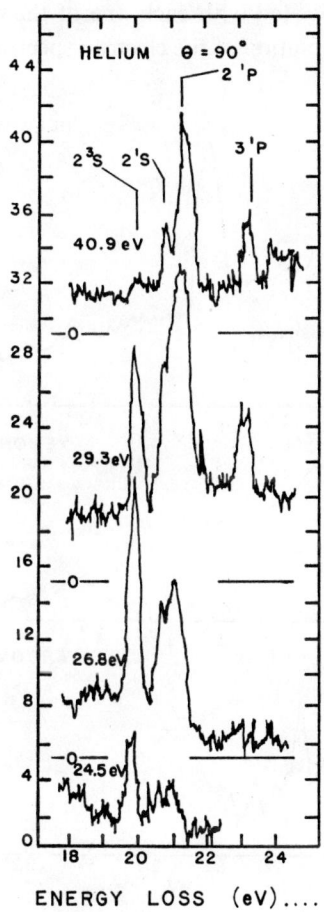

Fig. 9-4-10. 90° electron-impact spectra of helium at 24.5, 26.8, 29.3 and 40.9 eV. The pressure in the collision chamber for all these runs was pproximately 0.040 Torr. The zero level is displaced as shown. (J. P. Doering and A. J. Williams [51].)

energy range, 20 to 70 eV, another example of optically forbidden transitions concerning rare gases is illustrated by the spectra [76] presented at Fig. 9-4-11. These results illustrate clearly the fact that at low energy the conclusions relative to the excitation cross-section by electron impact do not in general agree with the conclusions obtained through the Born approximation, at least in the case of rare gases.

Studies similar to those performed with noble atoms have been conducted in the case of diatomic and triatomic molecules. The resulting electron energy loss spectra contain in this instance a new feature: it is the vibrational spectrum generally associated with each one of the electronic levels of the molecules. When the resolution of the electron spectrometer is poor this new

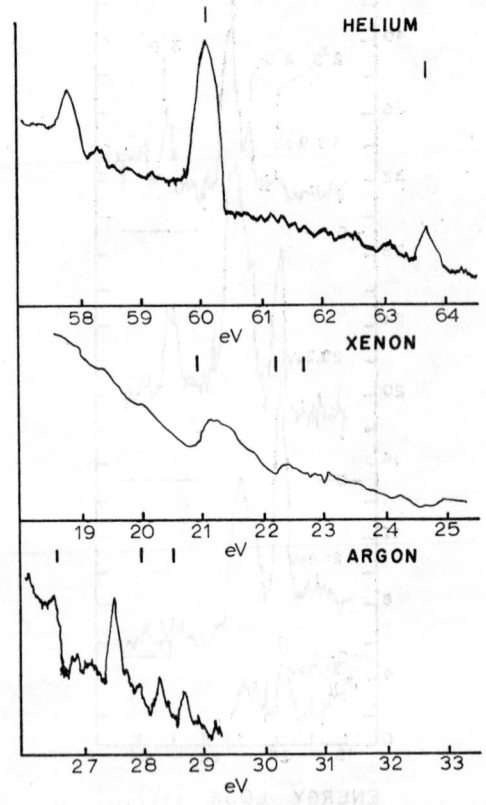

Fig. 9-4-11. The intensity of inelastically scattered electrons as a function of energy loss for He, Xe and Ar. The strongest observed optical lines are indicated above. Incident energies were 90 eV (except He: 100 eV). (J. A. Simpson et al., J. Opt. Soc. Am. **54** (1964) 269.)

structure may complicate the spectrum giving broad peaks or bands which make the interpretation of a spectrum very difficult. But if the lines of the vibrational levels are well enough resolved it may often facilitate the interpretation of a spectrum. A typical illustration of such a situation is offered by results given at Fig. 9-4-12. One may see on this figure four spectra of electron energy loss in hydrogen for various nominal incident electron energy. The profile and the sequence of the lines of these various spectra are due to vibrational states, and enables us to note four electronic states of this molecule. According to the authors [81], the most likely electronic states are $B\,^1\Sigma_u^+$, $C\,^1\Pi_u$, $c\,^3\Pi_u$ and $a\,^3\Sigma_g^+$. Earlier measurements [73] with the trapped electron method had shown a strong and wide peak around 12 eV.

Many atomic and molecular species have also been used as targets to

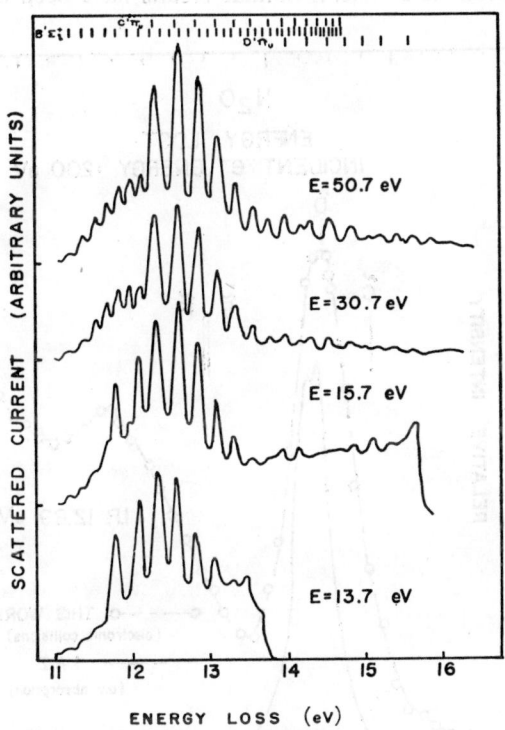

Fig. 9-4-12. Energy-loss spectra of $H_2$ for different incident electron energies $E$. For the two incident energies of 13.7 and 15.7 eV there is a smoothly rising background due to a contribution of low-energy secondary electrons generated inside the collision chamber. Although within one energy-loss spectrum the peak heights are a good measure of the relative cross-sections, the peak intensities in different spectra are not comparable. The $H_2$ pressure in the collision chamber was $\sim 0.02$ Torr. (H. G. Heideman et al. [81].)

study the inelastic scattering with high energy incident electrons. The main species considered are He [63, 85, 86], $N_2$ [85–87], NO [85], CO [86, 88], $CO_2$ [85, 89, 90], $H_2O$ [86], $NH_3$ [86], $N_2O$ [63], COS [63] and benzene [86]. This enumeration is not necessarily complete but represents typical measurements and data about scattering in the energy range 100 eV to about 500 eV. All the results, with few exceptions, have indicated the fact that the most probable states to be excited are those which correspond to optically allowed transitions. An example of an energy loss spectra in the case of the triatomic molecule $N_2O$ is presented on Fig. 9-4-13. On the same figure are shown the data concerning the absorption of ultra-violet radiation [91] for the same energy range. Within the commonly explored energy range the agreement between the probability of excitation by electron impact and by ultra-violet photons is excellent. Similar results have been obtained in the

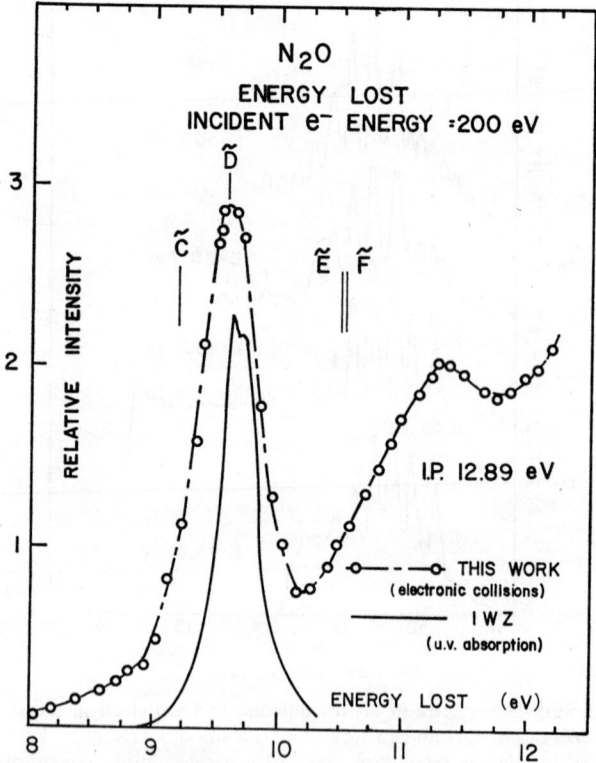

Fig. 9-4-13. Energy loss of forward scattered electrons in $N_2O$ for 200 eV incident electrons and comparison with absorption in ultra-violet radiation as measured by Watanabe and Zelikoff. (A. Jacob and J. D. Carette, unpublished.)

case of $CO_2$ [63, 90] and COS [63]. So far only one exception to this rule has been reported [85–87]. It concerns an intense peak observed at 12.91 eV in nitrogen, while in ultra-violet absorption spectrum the maximum appears at 12.71 eV. In general the results concerning the atoms and molecules listed above confirm the excitation rules by electron impact derived from the Born approximation.

## 2. *Third mode*

There are relatively very few reported results obtained by the third mode of operating the electron spectrometer. Typical results of such measurements are available concerning inelastic scattering in helium [92] and in hydrogen [93]. The curve representing the number of electrons that have lost a fixed amount of energy as a function of incident electron energy in helium is presented on Fig. 9-4-14. These data indicate many resonance processes in the inelastic scattering channels as pointed out by the authors [92].

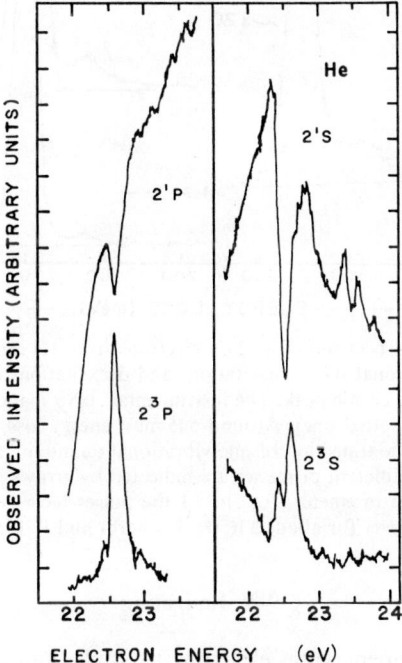

Fig. 9-4-14. Forward inelastic electron scattering intensity (zero displaced) versus incident electron energy in helium. The number of electrons that have lost a fixed amount of energy is shown as a function of the incident energy. The final-state helium levels are indicated by the labels on the curves. (G. E. Chamberlain, Phys. Rev. Letters **14** (1965) 581.)

## 3. Rotational levels

To illustrate the great value and efficiency of electron spectrometers in the field of atomic physics research let us mention the recently reported results concerning the rotational levels of $H_2$ [94]. It is the very first time that unambiguous evidence of rotational levels effects in electron scattering experiment has been obtained. The electron energy spectrum obtained in this experiment is presented in Fig. 9-4-15. The figure and its caption amply convey the elegant nature of the experiment and its considerable significance.

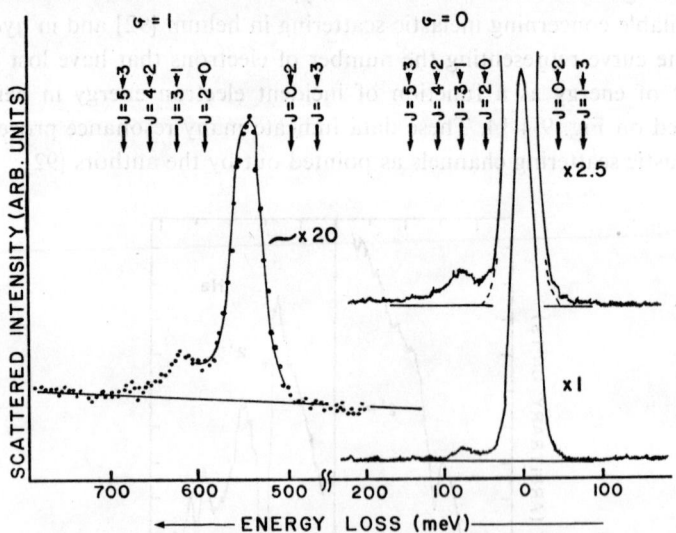

Fig. 9-4-15. Energy-loss spectrum of 4.42 eV electrons from $H_2$ molecules at a scattering angle of 20°. Pure rotational $\Delta J = 2$ excitations and deexcitations appear to the left and right, respectively, of the elastic peak. The instrumental curve shape of this peak has been measured using argon (dotted line). Around 545 meV energy loss, rotational transitions occur accompanied by the transition of one vibrational quantum. The expected positions of energy losses for the different processes are indicated by arrows. The curves $\Delta v = 0$ are recorder traces using the ratemeter; for $\Delta v = 1$ the pulses have been stored in a multichannel analyzer for about 3 h. (H. Ehrhardt and F. Linder [94].)

## § 9-5. Ionization

The use of monoenergetic electrons to study ionization is a valuable contribution to the improvement of ionization probability data when fine structure is present. Above the ground state, which is marked by the threshold of the ionization efficiency curve, monoenergetic electrons are essential for the study of the threshold laws and for the detection of excited states just above the threshold of ionization. Still further up (several volts) above threshold we expect that the appearance of new excited states would add new cross-sections for ionization which would make smaller and smaller relative contributions with respect to the total ionization cross-section as the energy increased. Consequently the use of monoenergetic electrons is expected to give best results at or near threshold. The situation would be quite different if the cross-sections of the states were competitive. A new state could be then detected either by a drop or an increase in the increment of the total cross-section [112].

In this section we evaluate the contributions to the study of ionization that have been made by the use of monoenergetic electrons. Therefore the experiments mentioned will be mostly chosen from those using either monoenergetic electrons or a technique somehow improving the effective electron energy spread. The reader is also referred to the chapter by McDowell [96] and the book by McDaniel [97].

## A. ATOMS

### 1. *Hydrogen*

The theoretical formulation of the problem of the collision of an electron with an atomic system, consisting of more than two bodies cannot be specified exactly. Even assuming that the wave functions are known, the scattering problem is still too difficult to be solved exactly, for the quantum mechanical treatment leaves an infinite set of coupled equations. The validity of the approximations used in the solution of these equations is known but qualitatively. In addition, comparison of a theoretically predicted cross-section with an experiment usually involves target systems for which the wave functions are not known.

Study of collisions on atomic hydrogen however does give the possibility of comparing theory and experiment and of observing the validity of the scattering approximations used in theory. These considerations give to the study of collisions on atomic hydrogen a great fundamental importance.

A first attempt to measure the absolute ionization cross-section of atomic hydrogen was made in 1955 by Boyd, Fite and Green at University College London. Successful results were published by Fite and Brackmann [95] using the following technique (see Fig. 9-5-1). Similar but more or less improved versions of this apparatus were later used by other workers [99–101]. The technique used was as follows:

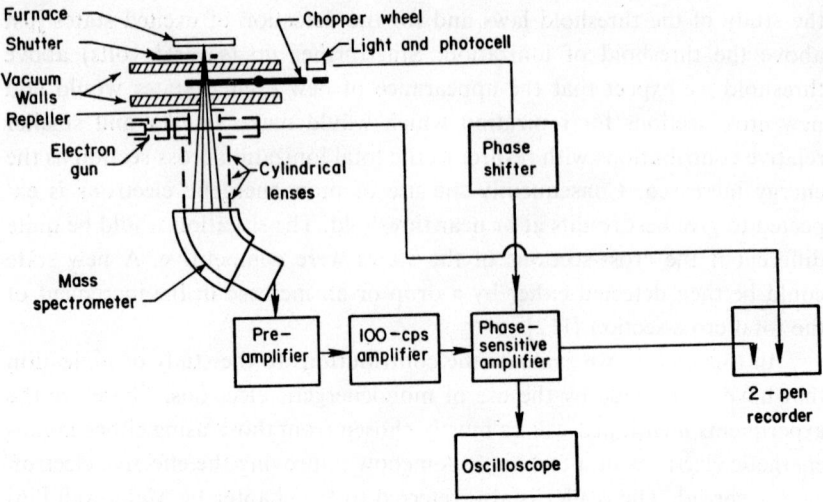

Fig. 9-5-1. Schematic diagram of ionization experiment made by Fite and Brackman [95] to measure ionization cross-section for atomic hydrogen.

Diatomic hydrogen molecules were dissociated on tungsten inside a hot furnace. The temperature was evaluated using thermocouples and optical pyrometers. The differentially pumped mixture of molecular and atomic hydrogen forming a beam was modulated 100 times per second by a chopper wheel. Care were taken that the background pressure did not fluctuate with the modulated beam.

This neutral chopped beam then crossed a non-modulated electron beam and the ions produced were mass analysed and detected.

The phase and the frequency of the amplified ion signal were electronically compared with a signal coming from the chopper wheel (through a modulated light, photocell and phase shifter) and the coherent signal was recorded simultaneously with the electron current.

It was observed that when the mass spectrometer was tuned to the proton peak and the neutral beam was adjusted to contain more than about 50% H atoms, atomic ions arising through dissociative ionization of $H_2$ necessitated corrections of less than 2% of the signal. Consequently relative cross-sections could be easily determined.

To measure the absolute cross-section Fite and Brackmann first evaluated the ratio of the atomic to the molecular cross-section at a given electron energy and then used the existing knowledge of the absolute molecular ionization cross-section to obtain the atomic cross-section. Thermodynamic considerations also led them to find the same absolute cross-section from their data. Fig. 9-5-2, taken from Omidvar [102] (compare Fite and Brackmann [95]) gives results near threshold (Fig. 9-5-2a curve FB) and in the range from 0–250 V (Fig. 9-5-2b curve I). The latter may be compared with the first Born approximation (Fig. 9-5-2b curve B). Complete agreement between the first Born approximation is found only above about 250 eV. At this energy Boyd [98] and Boksenberg [99] give results in fair agreement although the experimental cross-sections measured are somewhat lower. At lower energy, near maximum cross-section, the first Born approximation

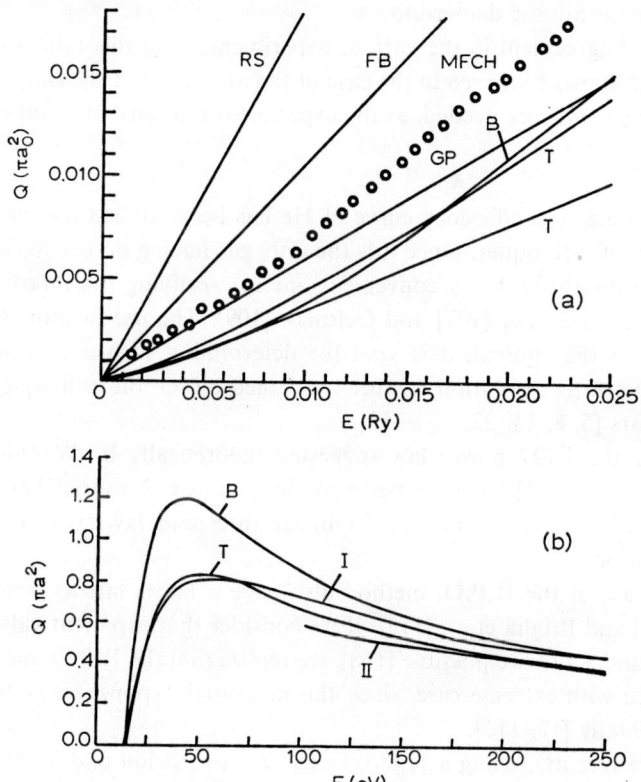

Fig. 9-5-2. Ionization of atomic hydrogen as obtained by different workers:
(a) ionization at thresholds $T_1$ and $T_2$ are the plot of Omidvar's functions [102]. GP and RS are calculations of Geltman [109] and Peterkop [168] and Rudge and Seaton [104]. FB and MFCH are the measurements of Fite and Brackmann [95] and McGowan, Fineman, Clarke and Hanson [145];
(b) ionization: Omidvar's method [102] and the Born approximation B are compared with experiment I [95] and experiment II [100].

gives a cross-section which is too large, but Omidvar [102] using an improvement of Vainshtein's recent calculation [103] predicts a cross-section well in agreement with experiment data in this region.

Figure 9-5-2a shows the cross-section near threshold. Experimental results of Fite and Brackmann [95], Boksenberg [99] and the theoretical results of Rudge and Seaton [104], Peterkop [105] and Geltman [109] are in favor of a linear threshold. More recent results by McGowan [101] et al. using monoenergetic electrons from an electron selector are in favor of the 1.13 power law as predicted by Wannier [107] and in agreement with Omidvar's [108] recent calculations. McGowan [101] also finds a similar behaviour for atomic deuterium.

The disagreement in the various experimental and theoretical slopes in Fig. 9-5-2a shows that even in the case of the ionization of the simplest atom confirming results are needed, as the experiment is a fairly difficult one.

## 2. *Helium*

The ionization efficiency curve of He has been studied for many years by all sort of techniques. Since it is the only gas having no excited state just above the threshold it is a convenient one for verifying the threshold laws predicted by Wannier [107] and Geltman [109]. The use of monoenergetic electrons or the equivalent is vital for determining such a threshold law. Improved energy resolution has been obtained by several techniques during recent years [6, 8, 14, 22, 110–112].

After the 1.127 power law suggested theoretically by Wannier [107], Hickam et al. [113] found experimentally a linear threshold law for He. Then in 1956 Geltman deduced a linear threshold law from theoretical considerations.

The use of the R.P.D. method also gave a linear law as observed by Fox [114] and Briglia et al. [115]. If we consider that exponent values other than 1.0 and 1.127 are possible [168], we realize that the R.P.D. method has to be used with extreme care, since the measured exponent can be varied experimentally [12, 116].

More recently, using a 127° velocity selector, Brion and Thomas [117] have found that Wannier's 1.127 power law best fits the experimental curve for $He^+$. Very recently Marchand, Paquet and Marmet [112] using an elaborate instrument and putting their data in a computer which was programmed to find the power law for the best fit, found consistently on all individual experimental curves a power $1.155 \pm 0.01$ law for helium. This value was found by the computer on all sets of data taken weeks apart for a range located between threshold and the first volt above threshold. The computer also confirmed, using another program, that the residual electron energy spread did not shift the calculated exponent outside the statistical error. This was done by calculating the new power law after the energy spread had been artificially increased by the computer by an amount equal to the experimental value. It was also shown that the difference between the power law calculated and all the observed points was only due to the statistical error in the data so that the calculated power law is as good as data permits. In Fig. 9-5-3a

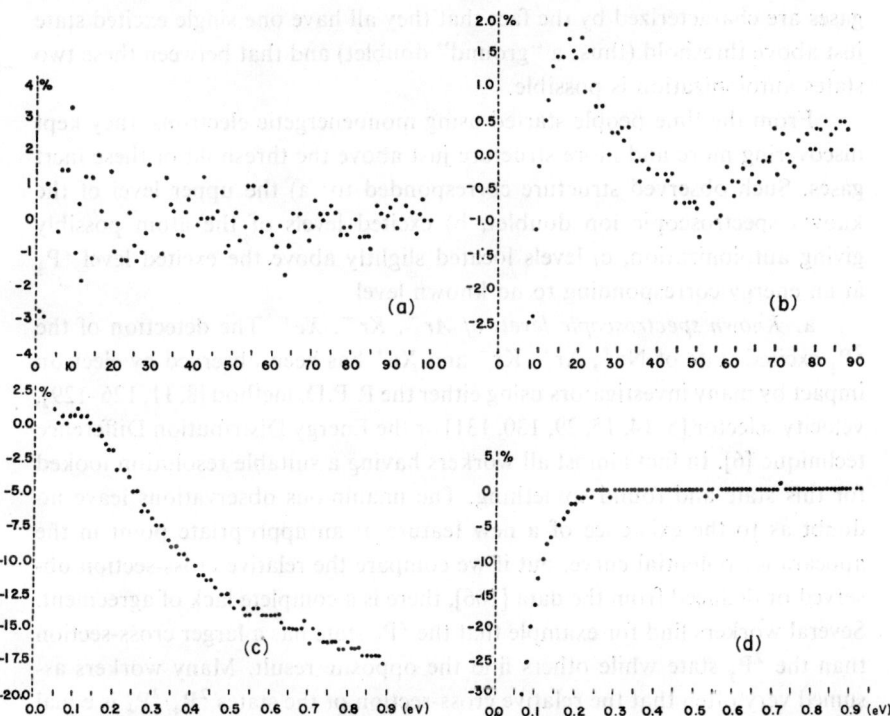

Fig. 9-5-3. (a) $(U-V)/U$ for the first volt above threshold of $He^+$ where $U$=experimental point, $V$=theoretical function of the form $A(C-V_0)^B$ giving the best fit. Exponent $1.155 \pm 0.01$. Mean deviation of all points $\pm 1\%$. (Abscissa in channel numbers; 100 channel numbers $= 0.89$ eV.)
(b) $(U-V)/U$ for the first volt of Ar when all the points were used to obtain the best fit. Note wide ($\pm 3\%$) point scatter at beginning of curve. Poor fit. (Abscissa as in (a)).
(c) $(U-V)/U$ for the first volt of argon where the points used for the best fit are from 0.05 eV to 0.15 eV above threshold.
(d) $(U-V)/U$ for the first volt of argon when the points used for the best fit are from 0.23 eV to 1.0 V. (c) and (d) demonstrate that the curve for $Ar^+$ is composed of two exponential segments with pertinent exponents, cross-sections and threshold energies.

are shown the differences between the calculated exponential law and the experimental points. The corresponding result for $Ar^+$ seen in Fig. 9-5-3b shows that a simple power law does *not* give a good fit in this case. Although the reproducibility is excellent we feel that unpredictable systematic errors could modify slightly the absolute value.

3. *Ne, Ar, Kr, Xe*

Since the structure of $Ne^+$, $Ar^+$, $Kr^+$ and $Xe^+$ near threshold have many similar characteristics they will be considered simultaneously. These gases are characterized by the fact that they all have one single excited state just above threshold (thus: a "ground" doublet) and that between these two states autoionization is possible.

From the time people started using monoenergetic electrons, they kept discovering more and more structure just above the threshold of these inert gases. Such observed structure corresponded to: a) the upper level of the known spectroscopic ion doublet, b) excited levels of the atom possibly giving autoionization, c) levels located slightly above the excited level $^2P_{\frac{1}{2}}$ at an energy corresponding to no known level.

a. *Known spectroscopic levels of $Ar^+$, $Kr^+$, $Xe^+$.* The detection of the $^2P_{\frac{1}{2}}$ excited state of $Ne^+$, $Ar^+$, $Kr^+$ and $Xe^+$ has been observed by electron impact by many investigators using either the R.P.D. method [8, 11, 126–129], velocity selector [5, 14, 15, 29, 130, 131] or the Energy Distribution Difference technique [6]. In fact almost all workers having a suitable resolution looked for this state and found something. The unanimous observations leave no doubt as to the existence of a new feature at an appropriate point in the appearance potential curve, but if we compare the relative cross-section observed or deduced from the data [106], there is a complete lack of agreement. Several workers find for example that the $^2P_{\frac{3}{2}}$ state has a larger cross-section than the $^2P_{\frac{1}{2}}$ state while others find the opposite result. Many workers assumed very often that the relative cross-section of the states $^2P_{\frac{3}{2}}/^2P_{\frac{1}{2}}$ is equal to two. Very recently Amme and Haugsjaa [137] measured this ratio and their result is different from the value assumed and used by many workers. It has been observed that the same apparatus can give different relative cross-sections on different runs the same day. To be reliable a structure must not only be observed but the cross-section must be reproducible (unless a precise reason explains what causes the non-reproducibility of the result). When unexplained changes occur in the measured cross-section they must have another origin. Many studies [45, 118] of surfaces done under mass spectrometric conditions show that unexpected reactions occur on surfaces.

These surface reactions are not usually under quantitative control. The great importance of surfaces is known to many investigators and different geometries have been used for the ion chambers as well as a great variety of techniques for plating, cleaning, or outgassing the source. All these precautions reveal the influence of the experimental surface conditions on the ionization curve.

Very recently Marchand, Paquet and Marmet [121] designed an apparatus "without" surfaces and found in $Ar^+$ at least above the $^2P_{\frac{1}{2}}$ state, no other structure than the one corresponding to the $^2P_{\frac{3}{2}}$ and $^2P_{\frac{1}{2}}$ states.

The apparatus used was as follows:

A differentially pumped non-modulated ribbon-shaped molecular beam was crossed by a monoenergetic electron beam coming from a 127° selector. The current-stabilized electron beam coming from the selector crossed the molecular beam at 90°. The electrons were then collected at the bottom of a deep well so that the interaction of its bombarded surface with the ionization region was negligible. It can be shown that the momentum of the ionized molecules (or atoms) suffices to drift the ions outside the ionizing region directly into a wide-aperture electron lens which preceeds the large aperture of a high-efficiency quadrupole mass filter [119]. They are then counted by a sensitive ion counting detector [120].

It was verified that the ions collected came essentially from the molecular beam rather than from the background gas pressure. This was done by closing the slit where the gas normally enters and letting the gas enter the system through another aperture so that the pressure measured in the ionizing region was the same. The ion current dropped by a factor of from 10 to 30 when the gas was thus at the same pressure in the source but introduced by an aperture other than the slit of the molecular beam.

In this way the ionization curve was formed from 90 to 97% of ions coming from the direction given by the molecular beam, not having touched a surface, and from 3 to 10% of ions from the diffused gas some of which may have touched a surface. Reactions with walls were thus minimized. Other experiments also showed that the electron well reduced to a negligible amount the influence caused by the usual wall trap. The results obtained with this apparatus were analysed by a computer and are perfectly reproducible month after month. In the case of $Ar^+$ and $He^+$ comparative data are shown in Fig. 9-5-3 (b and a).

These data show that when we examine the ionization efficiency curve, structure is undistinguishable on the direct curve and is also invisible on its first derivative. Comparative results between $He^+$ and $Ar^+$ curves show (Fig. 9-5-3a and b) that a simple power law is an excellent function for representing the first volt above threshold of the ionization efficiency curve of $He^+$, but it is very poor in the case of $Ar^+$.

The features seen in Fig. 9-5-3 are reproduced all the time. We see in the He curve that the energy spread does not produce a large deviation of measured points from a pure exponential law even at a few hundredths of a volt above threshold. The statistical error in the curve for $He^+$ is larger than the one for $Ar^+$ simply because the number of ions counted was smaller. As mentioned, the exponent for $He^+$ was $1.155 \pm 0.01$.

In the search for the best fit in the first part of the argon curve (Fig. 9-5-3c) we used only 10 points spaced by 0.01 V located in the central part of the 0.177 eV gap between the doublet states in order to minimize the influence of the electron energy spread at the doublet extremities. Above the $^2P_{\frac{1}{2}}$ state the first five points were omitted in the calculation of the best fit for the same reason. Curve 9-5-3c shows that for the $^2P_{\frac{1}{2}}$ state the rate of increase of the cross-section is smaller. This observation is completely unexpected but it has been obtained from all the data obtained with this instrument, when $(U-V)/U$ is plotted as a function of electron energy where:

$U$ = experimental point,
$V$ = point obtained from the threshold law giving the best fit (minimum in the sum of the quadratic errors) to the measured curve. It is found that this last method is a more sensitive one for finding structure than taking the derivative of the curve. We feel however that it can be applied almost exclusively to ions having a very simple structure.

This experiment cannot reject the possibility of autoionization between the two states of argon because the proximity of the doublet levels (0.177 eV) is only a few times the electron energy resolution, but no evident sign of autoionization was detected.

Finally the excellent $(U-V)/U$ fit (Fig. 9-5-3d) above the $^2P_{\frac{1}{2}}$ state shows with high sensitivity the absence of any other structure.

b. *Autoionization of $Ar^+$, $Kr^+$, $Xe^+$.* There are excited states of Ar, Kr and Xe having an energy between the $^2P_{\frac{3}{2}}$ and $^2P_{\frac{1}{2}}$ doublet states of their respective ions. Autoionization is then possible through the reaction $Xe + e \rightarrow Xe^* \rightarrow Xe^+ + 2e^-$ but the relative ionization probability with respect to direct ionization is unknown. Morrison [121] published data showing those states leading to $Xe^+$ and $Kr^+$. Unfortunately the signal-to-noise ratio is very small and not much can be said about the relative cross-sections for these autoionizing states. The energy resolution available is now good enough to show clear structure between the doublets if the relative cross-sections are large enough [131]. It seems that these data have not been easily reproduced in other laboratories although Williams [134] has recently verified them and measured charge-exchange cross-sections. In a more recent paper Morrison [138] notes that there may be pronounced effects due to configuration interaction.

c. *Unexplained states of $Ar^+$, $Kr^+$, $Xe^+$.* Several states observed experimentally by electron impact were unknown to spectroscopy and could not be predicted theoretically. The observed levels are numerous [6, 15, 29, 126, 127, 129–131] and have been observed in one case in the photon absorption spectrum [122]. Recent experiments [45] show that it is possible that some of these levels can be explained in terms of 1) surface reactions, 2) different lives of autoionizing states [145], 3) ultraviolet radiation coming from the bombardment of electrons and interacting with the gas. We have already noted that the experiments give non-consistent relative ionization cross-sections for the doublet states of $Ar^+$, $Kr^+$ and $Xe^+$. There is, we believe, a serious possibility that one or other of the processes mentioned is responsible for the measured structure observed by so many workers. Such processes would produce more structure in the ionization efficiency curves than that accruing from free gas atoms.

4. *Double ionization of inert gases*

Double ionization of He was first measured by Bleakney and Smith [132] and much later interpreted by Geltman [109] as being consistent with a threshold law varying as the square of excess energy above threshold. More recent work by Fox [123] agrees with this result. Others, for example Krauss, Reese and Dibeler also obtained a good fit using a square law [124]. It is of fundamental interest to find how close to power two the best fit is on the experimental curves since it is the only doubly-charged particle without any possible excited states.

Clarke [5] was in fact the first to observe the square threshold law (for $Xe^{++}$). The threshold law of doubly-charged ions has been studied for various inert gases by several workers. Morrison and Nicholson [133] and Dorman [135] observed straight segments on the first derivative of the ionization efficiency curve of $Ne^{++}$, $Kr^{++}$, $Ar^{++}$ and $Xe^{++}$. This is in agreement with the square law. These results neglect the contribution of the autoionizing levels [121] seen by the same authors. Furthermore they lead to the conclusion that the presence of several excited states of the doubly charged

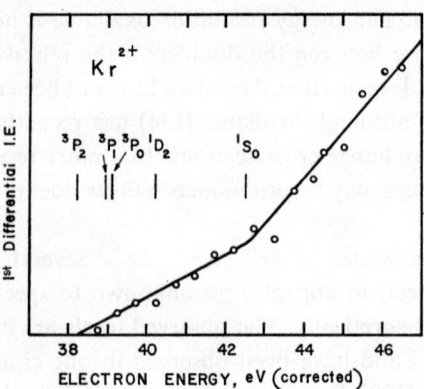

Fig. 9-5-4. First differential of the I.E. curve for $Kr^{2+}$ between 38 eV and 46 eV for the electron energy as obtained by Dorman and Morrison [135].

ion just above the threshold does not change the power law resulting from the sum of these states. Several individual excited states are indicated in the $Ar^{++}$, $Kr^{++}$ and $Xe^{++}$ as seen, for example, in Fig. 9-5-4 but no evident reason explains why they do not modify the threshold law for the ensemble.

## 5. *Higher degree of ionization of inert gases*

Very few experiments have been performed on atoms ionized 3 or more times. A good example is presented by Dorman et al. [139]. It shows that the theoretical prediction on multiply-ionized atoms is valid for Xe and the power observed corresponds to the degree of ionization at least up to $Xe^{6+}$.

Finally the data show no evident contribution from autoionization. The understanding of the behaviour of those reactions is far from good, since Hickam, Fox and Kjeldaas [136] have claimed that their experiments on $Ne^{++}$, $Ar^{++}$, $Kr^{++}$ and $Xe^{++}$ required a linear threshold law.
The deviation from the theoretically predicted power is such that it is at least the closest integer. This is surprising, as autoionization is possible. Many such states are also surely excited, and yet the threshold law is verified as far as 60 V above ionization for $Xe^{++}$.

Monoenergetic electrons were not necessary for studying multiply-ionized atoms in the past because they were studied at several volts above threshold where the cross-section is much larger. The theory will be verified only if ionization is studied in detail very close to threshold. In this case monoenergetic electrons are required.

## B. Diatomic Molecules

### 1. *Hydrogen*

The simplest molecular ion has been widely studied near threshold. Before the use of monoenergetic electrons a simple linear threshold law was observed by Morrison [140] in 1954. In 1960 Marmet and Kerwin [141] using the newly-developed electron selector observed structure in the ionization efficiency curve just above the threshold of $H_2$. The positions of the observed "breaks" were located at energies very close to those predicted by the theory for vibrational levels. In addition the relative cross-sections ("Franck-Condon factors") for the different vibrational states whose theoretical values were unknown to the authors at the time, were determined from the data for the first four vibrational levels. Values obtained from these data gave the third state ($v=2$) as being the most probable, in agreement with theory, and the relative cross-sections given are all in agreement with more recently calculated factors.

Those data were compared [142] with the Briglia and Rapp results [143] using the R.P.D. method and those of McGowan and Fineman [144] using a molecular beam. As the energy scale is very different in these three sets of data, and the number of points obtained per volt is also largely different, comparison is difficult. If, as discussed by Kieffer and Dunn [142], the results are not in agreement we deduce at least that the observed ionization efficiency curve must be influenced by a change in the geometry of the ionizing source, resulting in changing surface reactions or by the contribution from autoionizing states or stray U.V. radiation. The same sort of influence was mentioned earlier in the chapter in the study of inert gases. The real problem is not only getting monoenergetic electrons but of finding a source free of such influences.

One of the most recent works on the subject is by McGowan et al. [158, 144]. Their experiment is described in detail, including the efforts made to reduce secondary reactions by making use of a molecular beam, an electron selector and a mass spectrometer. Results are shown in Fig. 9-5-5.

They find that the tail of the ionization efficiency curve for $H_2^+$ is longer than that for $H^+$. They first believed that the threshold was slightly below the spectroscopic threshold but later [145] the electron energy scale was shifted to conform with a non-linear ionization threshold law for atomic hydrogen. This led them to believe that the structure produced was autoionization. Although there is some resemblance between the photoionization curve taken by Dibeler et al. [167] and the derivative of the electron impact

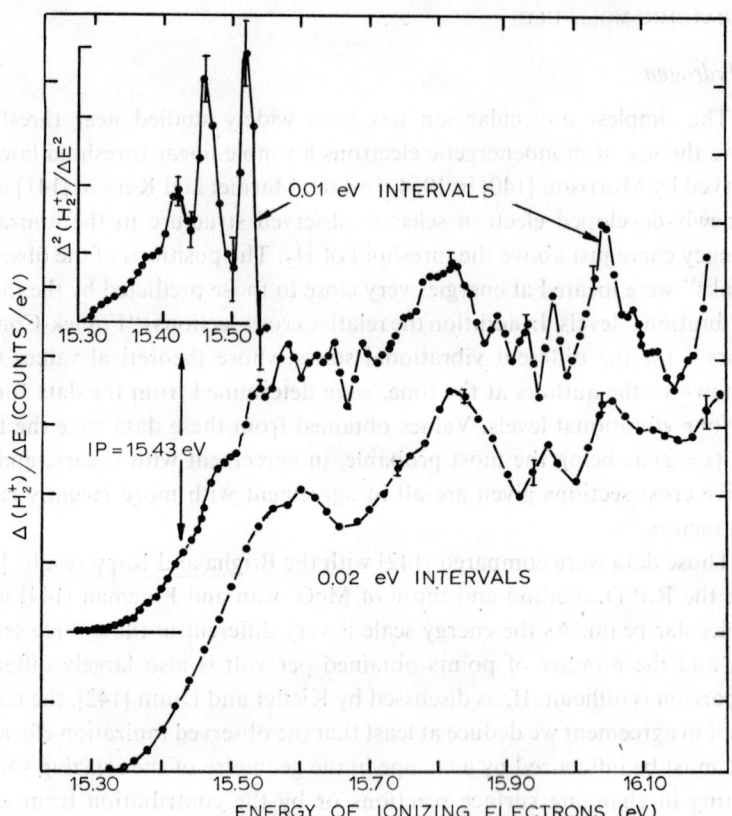

Fig. 9-5-5. Derivative of the ionization efficiency curve for $H_2$ from McGowan and Fineman [101].

data, it is not clear why the curve taken with 0.01 eV intervals presents some dissimilarities with the one given using 0.02 eV intervals. Recently a new experiment [165] has been done to verify the Franck-Condon principle by studying the double ionization of $H_2$. The corresponding theory has been done by McCullogh [166].

## 2. Nitrogen

Among the first ionization efficiency curves obtained using monoenergetic electrons are the ones published by Clarke [5] in 1954. Apart from the determination of a new reliable dissociation energy for $N_2$, a new feature was observed as a wide hump at one volt above threshold attributed to the state $A\ ^2\Pi_u$ identified earlier by Meinel [146] and Douglas [147] at 16.692 eV.

The same structure has been seen by so many workers [6, 148–151] in all sort of experimental conditions that it is probably real. The unresolved hump may include the optically allowed states observed in the U.V. absorption work of Huffman, Tanaka and Larrabee [152]. These ions would appear from autoionizing states.

Fineman et al. [153] took the derivative of the ionization efficiency curve they obtained on $N_2$. They find several peaks below the threshold of the $N_2^+$ (A $^2\Pi$) state which they interpret as being autoionization from the reaction $N_2(X'\Sigma_g^+; v''=0, J'') + e \rightarrow N_2^* + e$ followed by: $N_2^* \rightarrow N_2^+ (X'\Sigma_g^+; v', J') + e$. The possibility of getting direct ionization is not considered in their paper because the relative cross-sections for direct ionization to different vibrational levels using the Franck-Condon principle as calculated by Nicholls [154] do not agree with the obtained data. It is also suggested that the major portion of the area under the fairly broad peak at the beginning of the curve maybe due primarily to direct vibrational excitation but having an initial tail due to direct rotational excitation. Except for the identification of the big "hump" the agreement between the results of different workers is not very good.

### 3. Oxygen

The ionization efficiency curve for the oxygen molecular ion has been studied by several workers [149, 150, 155–159].

Franck-Condon probabilities for transitions to vibronic states have been calculated by Wacks and Krauss [160] using the Born-Oppenheimer approximation. One recent appearance potential curve using a 127° selector was obtained recently by Brion [150]. In Fig. 9-5-6 we see the $^2\Pi_g$, $^4\Pi_u$, $^2\Pi_u$, $^4\Sigma_g^-$ and the $^2\Sigma_g^-$ states. Good signal to noise ratio is very difficult to obtain and the exact contribution of each new excited state is difficult to evaluate. According to Wack's calculations the second vibrational ($v=1$) level of the ground state of the ion has a larger cross-section than the first ($v=0$) one. Consequently they could be seen in the ionization curve if the cross-section for autoionization is not too large.

### C. Triatomic Molecules

Several triatomic molecular ions have been studied, such as $CO_2$ [161, 162] and $N_2O$ [162, 163].

Four ionization curve breaks were observed in $CO_2^+$. Three were identified as being from the $^2\Pi_{\frac{3}{2}g}$, $^2\Pi_{\frac{3}{2}u}$, and $^2\Sigma_g^+$ states. In $N_2O^+$ the $^2\Pi$, $^2\Sigma_u^+$ and $^2\Sigma_g^+$ states were mentioned. There is no doubt that the interpretation of

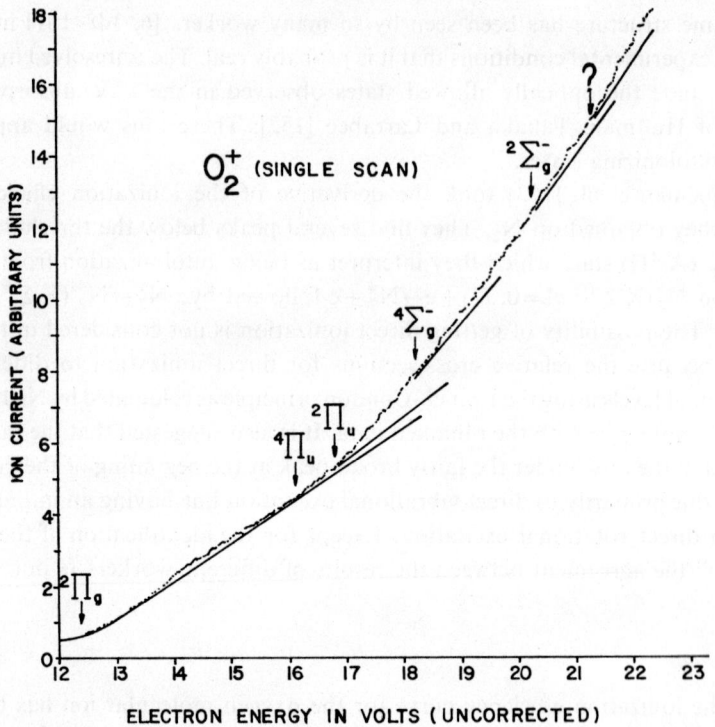

Fig. 9-5-6. Ionization efficiency curve of $O_2^+$ [150].

those states is extremely difficult and would be advanced if simple molecular ions were more completely resolved first.

### D. The negative ion $SF_6^-$

The negative ion formation of $SF_6^-$ from $SF_6$ occupies a special place in the technique of monoenergetic electrons. The shape of this resonant negative peak carefully studied in 1964 by Hickman and Fox [164] using the R.P.D. method shows that it appears at zero electron volts and that the profile of the peak may be used as a measure of the electron energy distribution.

## § 9–6. Conclusion

In résumé it may be stated that the last ten years has seen a remarkable series of developments in the instrumentation associated with electron spectroscopy, and that the arsenal of techniques now at our disposal should result in considerable advances in atomic physics over the next decade. The results obtained thus far with the improved instrumentation have been important, but relatively few species have been subjected to the detailed scrutiny now possible by electron bombardment, and much of the older work remains to be redone. In particular the divorcing of surface effects from phenomena in the pure gas phase must be carefully ensured.

The rich years for low energy electron collision physics should now be beginning.

### References

1. L. M. BRANSCOMB, Phys. Today **21** (1968) 36.
2. P. LENARD, Ann. de Physik **12** (1903) 714.
3. J. FRANCK and G. HERTZ, Verhandl. Deut. Phys. Ges. **16** (1914) 10.
4. H. D. SMYTH, Rev. Mod. Phys. **3** (1930) 347.
5. E. W. CLARKE, Can. J. Phys. **32** (1954) 764.
6. R. E. WINTERS, J. H. COLLINS and W. L. COURCHESNE, J. Chem. Phys. **45** (1966) 1931.
7. R. E. FOX, W. M. HICKAM, T. KJELDAAS and D. J. GROVE, Phys. Rev. **84** (1951) 859.
8. See, for example,
   R. E. FOX, W. M. HICKAM, D. J. GROVE and T. KJELDAAS, Rev. Sci. Inst. **26** (1955) 1101;
   R. E. FOX, W. M. HICKAM and T. KJELDAAS, Phys. Rev. **89** (1953) 555;
   R. E. FOX, J. Chem. Phys. **33** (1960) 200;
   R. E. FOX and R. K. CURRAN, J. Chem. Phys. **34** (1961) 1590, 1595.
9. See, for example,
   H. H. BRONGERSMA and L. J. OOSTERHOFF, Chem. Phys. Letters **1** (1967) 169;
   G. J. SCHULZ and R. E. FOX, Phys. Rev. **106** (1957) 1179;
   D. C. FROST and C. A. MCDOWELL, J. Chem. Phys. **29** (1958) 503;
   P. L. RANDOLPH and R. GEBALLE, U.S. Office Ordnance Res. Tech. Rept. **6** (1958);
   G. G. CLOUTIER and H. I. SCHIFF, J. Chem. Phys. **31** (1959) 793;
   J. F. BURNS, J. Chem. Phys. **23** (1955) 1347.
10. See, for example,
    L. G. CHRISTOPHOROU, R. N. COMPTON, G. S. HURST and P. W. REINHARDT, J. Chem. Phys. **43** (1965) 4273;
    S. TSUDA, C. E. MELTON and W. H. HAMILL, J. Chem. Phys. **41** (1964) 689.
11. D. C. FROST and C. A. MCDOWELL, Proc. Roy. Soc. **A232** (1955) 227.
12. P. MARMET, Can. J. Phys. **42** (1964) 2102.
13. A. L. HUGHES and V. ROJANSKY, Phys. Rev. **34** (1929) 284.
14. P. MARMET and L. KERWIN, Can. J. Phys. **38** (1960) 787.
15. See, for example,
    L. KERWIN, P. MARMET and E. M. CLARKE, Advanc. Mass Spectrometry **2** (1962) 522 (Pergamon Press, N.Y.);
    L. KERWIN and P. MARMET, J. App. Phys. **31** (1960) 2071;
    P. MARMET and J. D. MORRISON, J. Chem. Phys. **35** (1961) 746;
    J. W. MCGOWAN, P. MARMET and L. KERWIN, Proc. 3rd. Intern. Conf. on Physics of Electronic and Atomic Collisions (1963), ed. M. R. C. McDowell (North-Holland Publ. Co., Amsterdam, 1964) p. 854.
16. G. J. SCHULZ, Phys. Rev. **125** (1962) 229.
17. See, for example,
    J. D. CARETTE and L. KERWIN, Can. J. Phys. **42** (1964) 2022;

J. W. McGowan, E. M. Clarke and E. K. Curley, Phys. Rev. Letters **15** (1965) 917; **17** (1966) 66E;
J. W. McGowan, Phys. Rev. **156** (1967) 165;
J. P. Doering, J. Chem. Phys. **45** (1966) 1065.
18. M. Cotte, Compt. Rend. **265** (1967) 1388.
19. Y. Delage, P. Allard and J. D. Carette, Can. J. Phys. **47** (1969).
20. J. A. Simpson, Rev. Sci. Instr. **35** (1964) 1698.
21. C. E. Kuyatt and J. A. Simpson, Rev. Sci. Instr. **38** (1967) 103.
22. C. E. Kuyatt, J. A. Simpson and S. R. Mielczarek, Phys. Rev. **138** (1965) A385.
23. See, for example,
J. A. Simpson, G. E. Chamberlain and S. R. Mielczarek, Phys. Rev. **139** (1965) A1039;
M. G. Menendez and H. K. Holt, J. Chem. Phys. **45** (1966) 2743.
24. E. M. Purcell, Phys. Rev. **54** (1938) 818.
25. K. R. Spangenberg and M. L. Field, Elec. Commun. **21** (1943) 194.
26. D. A. Hutchison, J. Chem. Phys. **24** (1956) 628.
27. G. A. Harrower, Rev. Sci. Instr. **26** (1955) 850.
28. G. D. Yarnold and H. C. Bolton, J. Sci. Instr. **26** (1949) 38.
29. See, for example,
D. A. Hutchison, Symposium on Mass Spectrometry (Pergamon Press, Oxford, 1961);
S. N. Foner and B. H. Nall, Phys. Rev. **122** (1961) 512.
30. E. Blauth, Z. Physik **147** (1957) 228.
31. V. V. Zashkvara, M. I. Korsunskii and O. S. Kosmacker, Soviet Phys. - Tech. Phys. (transl. of Zh. Tekhn. Fiz.) **11** (1966) 96.
32. H. Z. Sar-el, Rev. Sci. Instr. **38** (1967) 1210.
33. J. A. Simpson, Rev. Sci. Instr. **32** (1961) 1283.
34. H. Hafner, J. A. Simpson and C. E. Kuyatt, Rev. Sci. Instr. **39** (1968) 33.
35. P. Marmet, Rev. Sci. Instr. **39** (1968) 1932.
36. C. Ramsauer, Ann. Phys. **64** (1921) 513; **66** (1921) 546.
37. E. O. Lawrence, Phys. Rev. **28** (1926) 947.
38. R. B. Brode, Rev. Mod. Phys. **5** (1933) 257.
39. W. B. Nottingham, Phys. Rev. **55** (1939) 203.
40. See, for example,
J. B. Farmer, Mass Spectrometry, ed. C. A. McDowell (McGraw-Hill Book Co., New York, 1963) ch. 2;
W. C. Wiley and I. H. McLaren, Rev. Sci. Instr. **26** (1955) 1150.
41. L. M. Chanin, A. V. Phelps and M. A. Biondi, Phys. Rev. **128** (1962) 219.
42. G. C. Baldwin and S. I. Friedman, Rev. Sci. Instr. **38** (1967) 519.
43. P. Marmet and J. D. Morrison, J. Chem. Phys. **36** (1962) 1238.
44. N. Papp and L. Kerwin, Can. J. Phys. **47** (1969).
45. Y. Petit-Clerc and J. D. Carette, Vacuum **18** (1968) 7.
46. J. D. Carette and L. Kerwin, Rev. Sci. Instr. **36** (1965) 537.
47. R. W. Ditchburn and U. Opik, Atomic and Molecular Processes, ed. D. R. Bates (Academic Press, New York, 1962).
48. S. Bashkin, Nucl. Instr. and Methods **28** (1964) 88.
49. D. Andrick and H. Ehrhardt, Z. Physik **192** (1966) 99.
50. A. Jacob and J. D. Carette, Annales de l'ACFAS **34** (1968) 156.
51. J. P. Doering and A. J. Williams, J. Chem. Phys. **47** (1967) 4180.
52. C. Ramsauer and R. Kollath, Ann. Physik **3** (1929) 536. (1965) 1010.
53. R. Kollath, Rev. Mod. Phys. **5** (1933) 258.

54. D. GOLDEN and M. W. BANDEL, Phys. Rev. **138** (1965) A14; Phys. Rev. Letters **14**
55. G. J. SCHULZ, Phys. Rev. Letters **10** (1963) 104.
56. J. AROL SIMPSON and U. FANO, Phys. Rev. Letters **11** (1963) 158.
57. G. J. SCHULZ, Phys. Rev. **136** (1964) A650.
58. U. FANO and J. W. COOPER, Phys. Rev. **137** (1965) A1364; **138** (1965) A400.
59. R. W. LABAHN and J. CALLAWAY, Phys. Rev. **135** (1964) A1539.
60. G. BREIT and E. P. WIGNER, Phys. Rev. **49** (1936) 519.
61. N. F. MOTT and H. S. W. MASSEY, The Theory of Atomic Collisions (3rd ed., Oxford University Press, London, 1965).
62. M. BARAT, J. BEAUDON and S. BLIMAN, Phys. Radium **28** (1967) 363; L. MOISEIWITSCH and S. J. SMITH, Rev. Mod. Phys. **40** (1968) 238.
63. A. JACOB and J. D. CARETTE, unpublished.
64. R. J. FLEMING, Proc. Phys. Soc. **81** (1963) 974.
65. E. B. WAGNER, F. J. DAVIS and G. S. HURST, J. Chem. Phys. **47** (1967) 3138.
66. J. WM. MCGOWAN, E. M. CLARKE and E. K. CURLEY, Phys. Rev. Letters **15** (1965) 917; **17** (1966) 66.
67. J. WM. MCGOWAN, Phys. Rev. **156** (1967) 165.
68. C. E. KUYATT, J. A. SIMPSON and J. MIELCZAREK, J. Chem. Phys. **44** (1966) 437.
69. J. WM. MCGOWAN, Phys. Rev. Letters **17** (1966) 1207.
70. G. J. SCHULZ, Phys. Rev. Letters **13** (1964) 583.
71. H. KLEINPOPPEN and V. RAIBLE, Physics Letters **18** (1965) 24.
72. M. J. W. BONESS and J. B. HASTED, 4th Intern. Conf. on the Physics of Electronic and Atomic Collisions, Quebec 1965 (Science Book-crafters, Hasting-on-Hudson N.Y., 10706) p. 130.
73. G. J. SCHULZ, Phys. Rev. **135** (1964) A988; **112** (1958) 150.
74. C. E. KUYATT, S. R. MIELCZAREK and J. A. SIMPSON, Phys. Rev. Letters **12** (1964) 293.
75. H. G. M. HEIDEMAN, C. E. KUYATT and G. E. CHAMBERLAIN, J. Chem. Phys. **44** (1966) 355.
76. J. A. SIMPSON and S. R. MIELCZAREK, J. Chem. Phys. **39** (1963) 1606; G. E. CHAMBERLAIN, J. A. SIMPSON, S. R. MIELCZAREK, J. Chem. Phys. **47** (1968) 4266; J. A. SIMPSON, J. A. MIELCZAREK and J. COOPER, J. Opt. Soc. Am. **54** (1964) 269; J. A. SIMPSON, M. G. MENENDEZ and S. R. MIELCZAREK, Phys. Rev. **150** (1966) 76.
77. R. H. MCFARLAND, Phys. Rev. **136** (1964) A1240.
78. A. KUPERMAN and C. F. RAFF, Disc. Faraday Soc. **35** (1963) 30; C. R. BOWMAN and W. O. MILLER, J. Chem. Phys. **42** (1965) 681.
79. J. D. CARETTE and L. KERWIN, Annales de l'ACFAS **32** (1966) 127.
80. A. SKERBELE and E. N. LASSETTRE, Abstract of 5th Intern. Conf. on the Physics of Electronic and Atomic Collisions (Nauka, Leningrad, 1967).
81. H. G. M. HEIDEMAN, C. E. KUYATT and G. E. CHAMBERLAIN, J. Chem. Phys. **44** (1966) 440.
82. H. H. BRONGERSMA and L. J. OOSTERHOFF, Chem. Phys. Letters **1** (1967) 169.
83. G. J. SCHULZ, Phys. Rev. **128** (1962) 174.
84. G. J. SCHULZ, J. Chem. Phys. **33** (1960) 1661; **34** (1961) 1778.
85. E. N. LASSETTRE, M. E. KRASNOW and S. M. SILVERMAN, J. Chem. Phys. **40** (1964) 1242; E. N. LASSETTRE, M. F. GLASER, V. D. MEYER and A. SKERBELE, J. Chem. Phys. **42** (1965) 3429; S. M. SILVERMAN and E. N. LASSETTRE, J. Chem. Phys. **40** (1964) 1265.
86. A. M. SKERBELE and E. N. LASSETTRE, J. Chem. Phys. **40** (1964) 1271; **42** (1965) 395.
87. E. N. LASSETTRE, A. SKERBELE and V. D. MEYER, J. Chem. Phys. **45** (1966) 3214.
88. V. D. MEYER, A. SKERBELE and E. N. LASSETTRE, J. Chem. Phys. **43** (1965) 805.
89. V. D. MEYER and E. N. LASSETTRE, J. Chem. Phys. **42** (1965) 3436.

90. E. N. LASSETTRE and J. C. SHILOFF, J. Chem. Phys. **43** (1965) 560.
91. F. C. Y. INN, K. WATANABE and M. ZELIKOFF, J. Chem. Phys. **22** (1954) 863; **21** (1953) 1643.
92. G. E. CHAMBERLAIN, Phys. Rev. Letters **14** (1965) 581;
    G. E. CHAMBERLAIN and H. G. M. HEIDEMAN, Phys. Rev. Letters **15** (1965) 337.
93. M. G. MENENDEZ and H. K. HOLT, J. Chem. Phys. **45** (1966) 2743.
94. E. EHRHARDT and F. LINDER, Phys. Rev. Letters **21** (1968) 419.
95. W. L. FITE and R. T. BRACKMANN, Phys. Rev. **112** (1958) 1141.
96. C. A. MCDOWELL, in: Mass Spectrometry (McGraw-Hill Book Co., N.Y., 1963) Ch. 12.
97. E. W. MCDANIEL, Collision Phenomena in Ionized Gases (John Wiley and Sons, 1964).
98. R. L. F. BOYD and A. BOKSENBERG, Proc. 4th Intern. Conf. on Ionization Phenomena in Gases, Uppsala, 1959 (North-Holland Publ. Co., Amsterdam, 1960) Vol. 1, p. 529.
99. A. BOKSENBERG, Electron Collision Processes in Dissociated Molecular Gases, Thesis University of London, June (1961).
100. E. W. ROTHE, L. L. MARINO, R. H. NEYNABER and S. M. TRUJILLO, Phys. Rev. **125** (1962) 582.
101. J. Wm. MCGOWAN and M. A. FINEMAN, 4th Intern. Conf. on the Physics of Electronic and Atomic Collisions, Quebec, Canada, 1965 (Science Bookcrafters Inc., New York 10706); Phys. Rev. **167** (1968) 43.
102. K. OMIDVAR, 5th Intern. Conf. on the Physics of Electronic and Atomic Collisions, July (1967).
103. L. A. VAINSHTEIN, L. PRESNYKOV and I. SOBELMAN, Zh. Eksperim. i Teor. Fiz. **45** (1963) 2015 (Engl. transl.: Soviet Phys. JETP **18** (1964) 1383).
104. M. R. H. RUDGE and M. J. SEATON, Proc. Roy. Soc. (London) **A213** (1965) 262; Proc. Phys. Soc. (London) **83** (1964) 680;
    M. R. H. RUDGE and S. B. SCHWARTZ, Proc. Phys. Soc. (London) **88** (1966) 579.
105. R. K. PETERKOP, Izv. Akad. Nauk. Latv. S.S.R. Riga **9** (1960) 79; **27** (1963) 1012; Proc. Phys. Soc. (London) **A77** (9161) 1220; Zh. Eksperim. i. Teor. Fiz **41** (1962) 1938; **43** (1962) 616; Optika i Spectroskopiya **13** (1962) 163.
106. S. N. FONER and B. H. NALL, Phys. Rev. **122** (1961) 512.
107. G. H. WANNIER, Phys. Rev. **90** (1953) 817.
108. K. OMIDVAR, Phys. Rev. Letters **18** (1967) 153.
109. S. GELTMAN, Phys. Rev. **102** (1956) 171.
110. D. A. HUTCHISON, Advan. Mass Spectrometry **2** (1963) 527.
111. T. M. SUGDEN and W. C. PRICE, Trans. Faraday Soc. **44** (1948) 116.
112. P. MARCHAND, C. PAQUET and P. MARMET, Phys. Rev. (1969).
113. W. M. HICKAM, R. E. FOX and T. KJELDAAS, Phys. Rev. **96** (1954) 63.
114. R. E. FOX, J. Chem. Phys. **35** (1961) 1379.
115. D. D. BRIGLIA and D. RAPP, Lockheed report No. 6-75-65-2, Palo Alto, Calif.
116. D. C. FROST and C. A. MCDOWELL, Proc. Roy. Soc. (London) **A232** (1955) 230.
117. C. E. BRION and G. E. THOMAS, 5th Intern. Conf. on the Physics of Electronic and Atomic Collisions, Leningrad, July (1967).
118. P. MARMET and J. D. MORRISON, J. Chem. Phys. **36** (1962) 1238.
119. P. MARCHAND and P. MARMET, Can. J. Phys. **42** (1964) 1914.
120. P. MARCHAND, C. PAQUET and P. MARMET, Rev. Sci. Instr. **37** (1966) 1702.
121. J. D. MORRISON, J. Chem. Phys. **22** (1954) 1219.
122. F. S. COMES and W. LESMANN, Z. Naturforsch. 16a, **1326** (1961) 127.
123. R. E. FOX, Advances Mass Spectrometry (Pergamon Press, London, 1959) pp. 397–412.
124. M. KRAUSS, R. M. REESE and V. H. DIBELER, J. Res. Natl. Bur. Std. **63A** (1959) 201.
125. J. D. MORRISON and A. J. C. NICHOLSON, J. Chem. Phys. **31** (1959) 1320.

126. C. E. MELTON and W. H. HAMILL, Notre Dame Rept. (1963).
127. Y. KANIKO, J. Phys. Soc. Japan **16** (1961) 1587.
128. R. E. FOX, Westinghouse Res. Lab. Rept. **60**, 94439-4-R2.
129. C. E. MELTON and W. H. HAMILL, J. Chem. Phys. **41** (1964) 546.
130. C. E. BRION, D. C. FROST and C. A. MCDOWELL, Advan. Mass Spectrometry, Proc. Conf. Montreal, Canada, June 1964, J. Chem. Phys. **44** (1966) 1034.
131. M. HUSSAIN and L. KERWIN, Can. J. Phys. **44** (1966) 57.
132. W. BLEAKNEY and L. G. SMITH, Phys. Rev. **49** (1936) 402.
133. J. D. MORRISON and A. J. C. NICHOLSON, J. Chem. Phys. **31** (1959) 1320.
134. J. F. WILLIAMS, Can. J. Phys. **46** (1968) 2339.
135. R. H. DORMAN and J. D. MORRISON, J. Chem. Phys. **34** (1961) 578.
136. W. M. HICKAM, R. E. FOX and T. KJELDAAS, Phys. Rev. **96** (1964) 63.
137. R. C. AMME and P. O. HAUGSJAA, Phys. Rev. **165** (1968) 63.
138. J. D. MORRISON, J. Chem. Phys. **40** (1964) 2488.
139. F. H. DORMAN, J. D. MORRISON and A. J. C. NICHOLSON, J. Chem. Phys. **31** (1959) 1335.
140. J. D. MORRISON, J. Chem. Phys. **22** (1954) 1219.
141. P. MARMET and L. KERWIN, Can. J. Phys. **38** (1960) 972.
142. L. J. KIEFFER and G. H. DUNN, Rev. Mod. Phys. **38** (1966) 31.
143. D. D. BRIGLIA and D. RAPP, Phys. Rev. Letters **14** (1965) 245.
144. J. W. MCGOWAN and M. A. FINEMAN, Phys. Rev. Letters **15** (1965) 179.
145. J. W. MCGOWAN, Report No. G.A.-8034, Gulf General Atomic Inc., P.O. Box 608, San Diego, Calif. 92112.
146. A. B. MEINEL, Astrophys. J. **114** (1951) 431.
147. A. E. DOUGLAS, Astrophys. J. **117** (1953) 380.
148. Y. KANEKO, J. Phys. Soc. Japan **16** (1961) 1587.
149. J. D. MORRISON, Bull. Soc. Chim. Belges **73** (1964) 99.
150. C. E. BRION and C. E. THOMAS, Intern. Report Dept. of Chemistry, U.B.C., Vancouver, Canada.
151. P. MARMET, Doctoral Thesis Laval University (1960).
152. R. E. HUFFMAN, Y. TANAKA and J. C. LARRABEE, Disc. Faraday Soc. **37** (1964) 159.
153. M. A. FINEMAN, E. M. CLARKE, H. P. HANSON and J. W. MCGOWAN, 4th Intern. Conf. on the Physics of Electronic and Atomic Collisions (Sciences Bookcrafters Inc., Hastings on Hudson, N.Y. 10706).
154. R. W. NICHOLLS, J. Res. Natl. Bur. Std. **65A** (1961) 451.
155. D. C. FROST, Ph. D. Thesis, University of Liverpool (1958)
156. C. E. MELTON and W. H. HAMILL, J. Chem. Phys. **41** (1964) 546.
157. C. E. BRION, D. C. FROST and C. A. MCDOWELL, J. Chem. Phys. **44** (1966) 1034.
158. J. W. MCGOWAN, E. M. CLARKE, H. P. HANSON and R. F. STEBBING, Phys. Rev. Letters **13** (1964) 620.
159. D. C. FROST and C. A. MCDOWELL, J. Amer. Chem. Soc. **80** (1958) 6183.
160. M. E. WACKS and M. KRAUSS, Letters to the Editor, page 1902, received July 7th, 1961.
161. J. COLLIN, J. Chem. Phys. **29** (1960) 424.
162. J. D. CARETTE, Can. J. Phys. **45** (1967) 2931.
163. R. M. CURREN and R. E. FOX, J. Chem. Phys. **34** (1960) 1590.
164. W. M. HICKMAN and R. E. FOX, J. Chem. Phys. **25** (1964) 642.
165. K. E. MCCULLOGH and H. N. ROSENSTOCK, J. Chem. Phys. **48** (1968) 2084.
166. K. E. MCCULLOGH, J. Chem. Phys. **48** (1968) 2090.
167. V. DIBELER, R. REESE and M. KRAUSS, J. Chem. Phys. **42** (1965) 2045.
168. R. K. PETERKOP, Izv. Akad. Nauk Latv. SSR **9** (1960) 79; Izv. AN SSSR **24** (1966) 947;
    A. TEMKIN, Phys. Rev. Letters **16** (1966) 835.

# AUTHOR INDEX

ABERTH, W., 320, *334*
ABRINES, R., 284, *331*, 346, 376, *396*
AFROSIMOV, V. V., 401, 403, 407, 413, 425, 426, 446, 453, *460–2*
AITKEN, K. L., 273, *334*
AKERIB, R., 132, 134, *166*
ALBRITTON, D. L. ,23, 28, *46*
ALEKSAKHIN, I. S., 77, *97*
ALLARD, P., 536, *578*
ALLEN, J. S., 403, *461*
ALLIS, W. P., 477, *522*
ALLISON, D. S. C., 218, 222, *247*
AMME, R. C., 568, *581*
AMUSIA, Ya. M., 404, 446, *460–1*
ANDERSON, C. D., 171, *245*, 326, *333*
ANDRICK, D., 544, 548, 551, 553, *578*
ARMSTEAD, R. L., 219, 227, *247*
ARTHURS, A. M., 15, *45*
ASKIN, J., 369, *397*

BACON, F. M., 292, 300, 326, *331–4*
BAILEY, C. A.,383, *397*
BAKER, F. A., 323, *332*
BALDWIN, G. C., 541, 552, *578*
BANDEL, M. W., 547, *578*
BARAT, M., 548, *579*
BARDSLEY, J. N., 471, 473, *522*
BARKER, M. I., 245, *248*
BARNARD, G. A., 327, *333*
BARNES, W. S., 23, 34, *46*
BARNETT, C. F., 129, *166*, 330, *334*
BARTKY, C. D., 49, *96*, 345, 385, *396*
BARTON, R. S., 327, *333*
BASHKIN, S., 543, *578*
BASSEL, R. H., 118, *166*
BATES, D. R., 49, 56, 57, 82, *96*, 119, 124, 138, *166, 167*, 288, *332*, 346, 359, *396*, 468, 471, 474, 503, *522*
BAUER, E., 49, *96*, 345, 385, *396*
BAYFIELD, J. E., 120, *166*
BEATY, E. C., 19, *45*
BELL, J. S., 64, *97*
BELY, O., 93, *97*, 288, 297, 311, *331, 332, 334*

BELYAEV, V. A., 318, *332*
BENNETT, W., 193, 233, *246*
BERKNER, K. H., 129, *166*
BERRY, H. W., 441, *462*
BETHE, H. A., 55, 67, *96*, 338, 344, 363, 369, *396*
BEYNON, J. H., 494, *523*
BIERMANN, L., 82, *97*
BINGHAM, F. W., 403, 441, *461*
BIONDI, M. A., 23, *46*, 465, 466, 467, 468, 480, 484, 487, 489, 496, 502, 504, 512, 516, 518, *521–3*, 541, *578*
BJORKEN, J. D., 173. *246*
BLACKETT, P. M. S., 171, *245*, 339, *395*
BLANKENBECLER, R., 196, 239, *246*
BLAUTH, E., 441, *462*, 538, *578*
BLIMAN, S., 548, *579*
BLOCH, F., 446, *462*
BLEAKNEY, W., 388, *397*
BOECKNER, C., 54, *96*
BOERBOOM, A. J. H., 287, *331*, 358, *397*
BOHME, D. K., 23, *46*
BOHR, N., 344, *395*
BOKSENBERG, A., 563, *580*
BOLAND, B. C., 325, *333*
BOLTON, H. C., 267, *331*, 537, *578*
BONESS, M. J. W., 553, *579*
BONSEN, T. F. M., 345, 365, 367, 369, 388, *396*
BOROWITZ, S., 132, 134, *166*
BOWMAN, C. R., 556, *579*
BOYD, A. H., 49, 61–62, 82–84, *96*, 287, *332*
BOYD, R. L. F., 563, *580*
BRACKMANN, R. T., 127, 152, *166–7*, 266, 284, 291, *331*, 386, *397*, 563, *580*
BRANSCOMB, L. M., 256, 262, 285, 306, 326, *331*, 528, *577*
BRANSDEN, B. H., 114, 128–129, 138, 158, *166*, 200, 226, 227, 236, 240, 245, *246–8*
BREIT, G., 548, *579*
BREZHNEV, B. G., 319, *332*
BRIGLIA, D. D., 566, 573, *580–1*
BRINK, G. O., 49, 72, *96*, 394, *398*

# AUTHOR INDEX

BRINKMAN, H. C., 118, *166*
BRION, C. E., 566, 568, 571, 575, *580–1*
BRODE, R. B., 540, *578*
BRONGERSMA, H. H., 530, 556, *577–9*
BROUILLARD, F., 320, *332*
BROWN, S. C., 456, 466, 468, 479, 484, 494, 503, 512, 518, *521–2*
BROWNE, J. C., 19, *45*
BROWNING, R., 271, 325, *333*
BRUBAKER, W. M., 495, *523*
BUCHSBAUM, S. J., 478, *522*
BURGESS, A., 283, 297, *331*, 345, 395, *396–7*
BURKE, P. G., 101, *165*, 192, 211, 219, *240–7*, 284, 287, 297, *331–2, 334*, 390, *397*
BURNS, J. F., 530, *577*
BYRON Jr., F. W., 295, *332*

CALLAWAY, J., 228, 230–231, *247*, 548, *579*
CAPLINGER, E., 76, *96*
CARBONE, R. J., 401, 413, 440, *460–1*
CARETTE, J. D., 532, 536, 543, 544, 550, 556, 575, *577–80*
CARLETON, N. P., 292, 304, *332*
CARTER, V. L., 49, 50, 52, 57, 58–62, 69, *96*
CASTELLEJO, L., 199, *246*
CATLOW, G., 49, 78, 87, *96*, 345, 385, *396*
CERMAK, V., 2, 37, *45*
CHADWICK, J., 338, 339, *395*
CHAMBERLIAN, G. E., 392, *397*, 536, 556, 561, *578–80*
CHAMPION, F. C., 339, *396*
CHANDRASEKAR, S., 344, *396*
CHANG, E. S., 49, 55, 59–60, 63–69, *96*
CHANIN, L. M., 23, *46*, 541, *578*
CHAPMAN, S., 9, *46*
CHEN, J. C. Y., 216, *247*, 473, *522*
CHEREPKOV, N. A., 446, *462*
CHESHIRE, I. M., 118, 126–130, 138, 158, *166*, 236, *247*
CHEW, G. F., 107, 110, 112, 133, *166*
CHRISTOPHOROU, L. G., 530, *577*
CLARKE, E. W., 528, 531, 532, 544, 553, 575, *577–8, 581*
CLOUTIER, G. G., 530, *577*
CODY, W. J., 200, 220, 226, *246*
COHEN, M. H., 235, *247*
COLEMAN, J. P., 104, 111, 114, 118, 120, 122–123, 126, 128, 134–135, 146, 157, 160, 163, *166–7*, 290, *334*
COLLIN, J. D., 575, *581*
COLLINS, C. B., 488, 511, *523*

COLLINS, J. H., 529, *577*
COMES, F. S., 570, *580*
COMPTON, D. M. J., 271, *333*
COMPTON, R. N., 530, *577*
CONNALLY, R. E., 403, *461*
CONNER, T. R., 467, 489, 516, *522*
CONWAY, D. C., *46*
COOK, C. J., 321, *334*, 346, 359, *396*
COOPER, J., 49, 55, 61, 71–72, *96*, 548, 550, *579*
CORBEN, H. C., 177, *246*
COTTE, M., 536, *578*
COURCHESNE, W. L., 529, *577*
COWLING, T. G., *45*
CRAGGS, J. D., 479, *522*
CROMPTON, R. W., 4, 28, *46*
CROTHERS, D. S. F., 139, 145, 149, 151–153, 163, *167*
CURLEY, E. K., 532, 553, *578–9*
CURRAN, R. K., 530, *577*
CURREN, R. M., 575, *581*
CUTHBERT, J., 323, *333*

DALGARNO, A., 7, 14, 15, 19, 21, 37, *45*, 124, *166*, 200, *246* 274, *522*
DALY, N. R., 273, 299, 308, 330, *332*
DANCE, D. F., 262, 287, 291, 293, 306, 312, *331–2*
D'ANGELO, N., 466, 504, 509, *522*
DAVIS, F. J., 552, *579*
DE BENEDETTI, S., 177, *246*
DE HEER, F. J., 369, 386, *397*
DEICHSEL, H., 392, *396*
DELAGE, Y., 536, *578*
DEMKOV, Yu. N., 202, *244*
DEUTSCH, M., 177, 180, *246*
DIBELER, V. H., 571, 573, *580–2*
DIETZ, L. A., 330, *334*
DIRAC, P. A. M., 171, 173, *245*
DITCHBURN, R. W., 49, *96*, 543, *578*
DODD, L. R., 157, *167*
DOERING, J. P., 532, *578*
DOLDER, K. T., 77, 93, 97, 256, 261, 266, 267, 272, 281–284, 287, 290, 293, 325, *331–4*
DORMAN, R. H., 571, *581*
DOUGLAS, A. E., 574, *581*
DRACHMAN, R. J., 227–229, 231, 235, 242, *247–8*
DRAWIN, H. W., 385, *397*
DRELL, S. D., 173, *246*
DRISKO, R. M., 130, *166*

DUCKWORTH, H. E., 327, *333*
DUFF, B. G., 245, *248*
DUGAN, J. V., 345, *396*
DUNN, G. H., 74, 77, 92, *97*, 252, 257, 263, 267, 285, 312, 326, *332*, *334*, 573, *581*
DUSHMAN, S., 327, *333*
DUTTON, J., 22, *45*
DUVENECK, F. B., 340, *395*
DUXLER, W. M., 228–230, *247*

ECONOMIDES, D. G., 93, *97*
EDELSON, D., 23, *46*
EDWARDS, A. K., 445, *462*
EHRHARDT, H., 123, *166*, 390, *397*, 544, 548, 551, 553, 562, *578*, *580*
ELFORD, M. T., 4, 28, *45*, 275, 326, *334*
ELLIS, C. D., 338, *395*
EL-SHERBINI, Th. M., 391, *396*
EMEL'YANOV, A. M., 324, *333*
ENGLANDER-GOLDEN, P., 285, *331*, 388, *397*
ENSKOG, G., 9, *46*
ERASTOV, E. M., 319, *333*
ERDELYI, I. A., 161, *167*
EVANS, R. D., 459, *462*
EVERHART, E., 401, 403, 407, 411, 413, 417, 420, 437, 440, 445, 451, 456, *460–1*
EWALD, H., 269, *331*
EYRING, H., 57, *96*

FALCK, W. P., 184, 193, 195, 223, 234, *246*
FANO, U., 55, *96*, 300, *334*, 367, 369, 371, *397*, 404, 443, 448, 450, *460*, 548, 550, *579*
FARKAS, Gy., 489, *523*
FARMER, J. B., 541, *578*
FARREN, J., 323, *333*
FEDORENKO, N. V., 401, 403, 407, 413, 426, 431, 453, *460–1*
FEENLY, R. K., 275, *334*
FELDMAN, P., 270, *331*
FELS, M. F., 242, *248*
FERGUSON, E. E., 2, 37, *45*
FERRELL, R. A., 178, *246*
FESER, S., 299, 324, *332*
FESHBACH, H., 186, 190, 198, *246*
FIELD, M. L., 536, *578*
FINEMAN, M. A., 563, 565, 573, 575, *580–1*
FIRSOV, O. B., 440, *462*
FITE, W. L., 127, 128, 152, *166–7*, 266, 284, 291, *331*, *334*, 386, 397, 563, *580*
FLEMING, R. J., 552, *579*
FLETCHER, W. H. W., 325, *333*
FOCK, V., 59, *96*

FONER, S. N., 538, 543, 568, *578*, *580*
FOWLER, R. H., 466, *522*
FOX, J. N., 520, *523*
FOX, R. E., 529, 530, 566, 568, 571, 572, 575, 576, *577–81*
FRANCK, J., 528, *577*
FRASER, P. A., 184, 195, 219, 223, 234, 242–245, *246–8*
FREEMAN, N. J., 330, *333*
FRIEDMAN, L., 3, 37, *45*
FRIEDMAN, S. I., 541, 552, *578*
FRISCHE, C. A., 401, *460*
FROMHOLD, L., 480, 484, 487, 503, 516, *522–3*
FROST, A. A., 178, *246*
FROST, D. C., 530, 566, 568, 571, 575, *578*, *580–1*
FULS, E. N., 401, 440, *460–1*

GABRIEL, A. H., 325, *334*
GAILITIS, M., 210, 215, 239, *247*
GAILY, T., 124, *167*
GALLAHER, D. F., 119, 120, 123–124, *166*
GARCIA, J. D., 345, 360, 384, 388, *396*
GASCOIGNE, J., 28, *46*
GATLAND, I. R., 30, *46*
GEBALLE, R., 124, *167*
GELTMAN, S., 214, 247, 566, 571, *580*
GERBER, R. A., 23, *46*, 488, 500, 504, 511, *523*
GERJUOY, E., 118, *166*, 345, 360, 384, 388, *396*
GERTLER, F. H., 178, *246*
GILBODY, H. B., 121, 126, 138, *166*, 271, 326, *333*, 386, *397*
GIOVANELLI, R. G., 466, 479, *522*
GLASER, M. F., 560, *579*
GOLDBERGER, M. L., 104, 112, *165–6*
GOLDEN, D. E., 23, *46*, 547, *579*
GORDEEV, Yu.S., 403, 407, 413, 426, 431, 446, *460–2*
GOULD, L., 479, *522*
GRAY, E. P., 416, 480, 503, 511, 518, *522*
GREEN, L. G., 223, *247*
GREEN, T. A., 154, *164*
GREGOIRE, D., 381, *397*
GREIDER, K. R., 157, *167*
GRIEM, H. R., 325, *334*
GROVE, D. J., 529, 530, *577*
GRYZINSKI, M., 338, 344, 358, 385, *395–7*
Gorokhov, L. N., 324, *333*
GUIDINI, J., 318, *332*
GUTHRIES, A., 327, *333*

HAFNER, H., 539, *578*
HAHN, Y., 199, 212, 213, 215, 220, 222, 227, 228, 237, 239, *246–7*
HAMILL, W. H., 530, 568, 571, *577*, *581*
HANGSJAA, P. O., 568, *580*
HANSEN, W. W., 340, *395*
HANSON, H. P., 575, *581*
HARMER, D. S., 23, *46*
HARRISON, M. F. A., 252, 260, 261, 262, 263, 273, 291, 293, 206, 311, 314, 370, 328, *330–334*
HARROWER, G. A., 267, *331*, 537, *578*
HASSÉ, H. R., 8, *45*
HASTED, J. B., 23, *46*, 323, *332–3*, 402, *461*, 553, *579*
HEADRICK, L. B., 251, *331*
HEARN, A. G., 474, *522*
HEICHE, G., 15, *45*
HEIDEMAN, H. G. M., 390, 391, *397*, 553, 556, 561, *579–80*
HEIL, H., 49, 72, *96*, 394, *398*
HENRY, R. J. W., 15, *45*
HERMAN, L., 466, 504, *522*
HERTZ, G., 528, *577*
HEYMANN, F. F., 245, *248*
HICKAM, W. M., 529, 530, 566, 572, 576, *577*, *580–1*
HICKS, B., 341, *395*
HICKS, W. T., 51, *96*
HINNOV, E., 509, *523*
HIRSCHBERG, J. G., 509, *523*
HOBSON, R. M., 520, *523*
HOLSTEIN, T., 14, 17, *45*
HOLT, H. K., 536, 561, *578*, *580*
HOLT, R. B., 466, 479, 504, 514, 519, *522–3*
HONIG, R. E., 53, *96*
HOOPER, J. W., 252, 256, 267, 272, 275, 284, 292, 300, 303, 326, *330–4*
HORNBECK, J. A., 54, *96*
HOSELITZ, K., 23, *45*
HOUSTON, S. K., 222, 223, *247*
HOWLAND, B., 479, 514, *522*
HUDSON, R. D., 49, 50, 52, 57, 58–62, 69, *96*
HUFFMAN, R. E., 571, *581*
HUGHES, A. L., 341, *395*, 531, *577*
HUGHES, V. W., 193, 233, *246*
HULTHÉN, L., 202, *246*
HUMMER, D. G., 127, *166*, 297, *332*, 386, *394*
HUND, F., 447, *462*
HURST, G. S., 530, 552, *579*
HURT, W. B., 488, 511, *523*

HUSSAIN, M., 568, 571, *581*
HUTCHINSON, D. A., 537, 538, 563, *578*, *580*
HUXLEY, L. H. G., 4, *45*
HYLLERAAS, E., 178, *246*

INN, F. C. Y., 560, *580*
INOKUTI, M., 178, *246*, 285, 311, *332–4*, 365, 369, 370, 389, *397*
IONOV, N. I., 326, *333*
IRELAND, J. V., 386, *397*

JACKSON, J. D., 124, 130, *166*
JACOB, A., 544, 550, *578–9*
JAHODA, F. C., 325, *333*
JAMINKER, D. M., 401, 440, *460–1*
JAUCH, J. M., 173, 176, *246*
JAUNCEY, G. E.M., 341, *395*
JOACHAIN, C. J., 306, *332*
JOHNSON, R. A., 466, 504, *522*
JONES, G., 184, 193, 195, 223, 234, *246*
JONES, P. R., 401, 440, *460–1*
JONES, T. J. C., 325, *333*
JORGENSEN, T., 404, 441, 443, *461*
JORTNER, J., 235, *247*
JUNDI, Z., 200, 227, 240, *246–7*
JUTSUM, P. J., 49, *96*

KANEKO, Y., 23, *46*, 568, 571, 575, *581*
KAPLAN, S. N., 129, *166*
KASNER, W. H., 467, 496, 518, *522–3*
KAUPILLA, W. E., 299, *334*
KEIFFER, L. J., 74, 77, 92, *97*, *332*, 573, *581*
KELLER, G. E., 23, *46*
KELLY, T. M., 195, 245, *246*, *248*
KERR, D. E., 466, 480, 503, 504, 511, 518, *522*
KERWIN, L., 532, 543, 544, 556, 568, 571, 573, *577–81*
KESSEL, Q. C., 403, 411, 419, 420, 427, 445, 451, *460–1*
KESTNER, N. R., 235, *247*
KHODEYEV, Yu.S., 324, *333*
KIHARA, T., 12, *45*
KIM, Y-K., 285, 311, *332*, *334*, 370, *397*
KIMBALL, G. E., 57, *96*
KINGSTON, A. E., 345, *396*, 474, *522*
KINNEY, J. D., 49, 72–76, 84–86, *96*, 286, *332*
KIRKPATRICK, P., 341, *395*
KIRZHNITS, D. A., 446, *462*
KISTEMAKER, J., 285, 327, *331*, *333*, 358, 369, *397*

KJELDAS, T., 529, 566, 572, *578*, *580–1*
KLEIMAN, C. J., 199, 220, 227, *246–7*
KLEINPOPPEN, H., 553, *579*
KLEMPERER, O., 257, 327, *331*
KNYSTANTAS, E. J., 403, 437, 451, *460*
KOHL, W. H., 330, *333*
KOHN, W., 202, *246*
KOLLATH, R., 401, *460*, 546, *578*
KORCHEVOI, Yu. P., 76, 97, 394, *398*
KORSUNSKII, M. I., 538, *578*
KOSMACKER, O. S., 538, *548*
KOYAMA, K., 403, *461*
KRAIDY, M., 223, 234, 236, 242, 245, *247–8*
KRAMERS, H. A., 118, *166*
KRASNOW, M. E., 560, *579*
KRAUSS, M., 571, 573, 575, *580–1*
KUNKEL, W. B., 474, *522*
KUNZE, H. J., 325, *334*
KUPERMAN, A., 556, *579*
KUPRIYANOV, S. E., 270–272, *331*
KUYATT, C. E., 327, *333*, 539, 553, 556, *578–80*
KYLE, H. L., 146, 153, *167*

LABAHN, R. W., 228, 230, 231, 245, 548, *579*
LAMB, W. E., 292, *332*
LAMKIN, J. C., 200, *246*
LANDAU, L. D., 216, *246*
LANGEVIN, P., 6, 9, 12, *45*
LANGHUS, L., 392, *397*
LANGLEY, R. A., 129, *166*
LARRABEE, J. C., 571, *581*
LASSETTRE, E. N., 556, 560, *579–80*
LATYPUV, Z. Z., 269, 278, *331*
LAWRENCE, E. O., 540, *578*
LAWSON, J., 200, 220, 226, 233, *246–7*
LEE, A. R., 292, 304, *332*
LEFFEL, C. S., 466, 511, *522*
LENARD, P., 528, *577*
LENNON, J. J., 479, *522*
LEVY, G., 271, 326, *333*
LEWIS, J. T., 322, *333*
LEWIS, M. N., 223, *247*
LICHTEN, W., 299, *332*, 404, 448, 450, *461*
LIFSHITZ, E. M., 216, *246*
LINDER, F., 392, *397*, 562, *580*
LINEBERGER, W. C., 252, 256, 267, 272, 284, 326, *330–1*
LIPSKY, L., 299, *332*
LIVINGSTON, M. S., 369, *397*

LLEWELLYN-JONES, F., 22, *45*
LOEB, L. B., 6, 23, 37, *45*
LOMBARD, F. J., 330, *334*
LORENTS, D. C., 321, *334*
LOVELL, S. E., 119, 121, *166*
LOWE, J. P., 178, *246*
LOZOVIK, Yu. E., 446, *462*
LUBECK, K., 82, *97*
LUCAS, C., 324, *333*
LUND, R. E., 501, *523*
LYNN, N., 200, *246*

MADSON, J. M., 23, *46*
MAGNUS, W., 161, *167*
MAMAEV, Yu-A., 403, 453, *461*
MANN, M. M., 341, *395*
MANSON, S., 49, 55, 61, 71–72, *96*
MANUS, C., 318, *332*
MAPLETON, R. A., 112, 129, 131, 134, *166*, 346, 359, *396*
MARCHAND, P., 535, 563, 566, 569, *580*
MARDER, S., 193, 233, *246*
MARGENAU, H., 486, *522*
MARINO, L. L., 563, *580*
MARMET, P., 530, 536, 539, 540, 543, 544, 563, 566, 573, 575, *577–81*
MARR, G. V., 49, 53, 69, *96*
MARTIN, D. W., 23, 28, 30, 37, *46*
MARTIN, F., 330, *334*
MARTIN, S. O., 77, *97*, 283, 286, 291, 326, *331*
MASON, E. A., 12, 17, 21, *45*
MASSEY, H. S. W., 111, 161, 166, *167*, 171, 182, 185, 192, 196, 199, 200, 216, 220, 226, 234, 238, 243, *245–7*, 296, *332*, 346, 353, 396, 405, *460*, 468, *522*, 548, *555*, *579*
MAY, R. M., 125, *166*
MEGILL, L. R., 23, *46*
MEHR, F. J., 484, 487, 517, *522–3*
MEINEL, A. B., 574, *581*
MELI, J. A., 452, *462*
MELTON, C. E., 530, 568, 571, *577*, *581*
MENENDEZ, M. G., 556, *579*
MESSIAH, A., 160, *166*, 192, *246*, 346, *396*
MERZBACHER, E., 182, 185, *246*
MEYER, V. D., 560, *579*
MIELCZAREK, S. R., 536, 553, 556, *578–80*
MILLER, T. M., 23, 28, 30, 37, *46*
MILLER, W. F., 344, 367, 368, 369
MITTLEMAN, M. H., 213, 215, 242, *247*, 248, 473, *522*

MITTLESTADT, V. R., 466, 484, 518, *522*
MOHLER, F. L., 54, *96*
MOHR, C. B. O., 161, *167*, 171, 236, 243, 245
MOISEIWITSCH, B. L., 202, 218, 222, 235, 247, 548, *579*
MØLLER, C., 340, *395*
MOLNAR, J. P., 54, *96*
MOLYNEAUX, L., 266, *331*
MOORES, D. L., 385, *334*
MORGAN, G. H., 401, *460*
MORINO, L. L., 76, *96*
MORRISON, J. D., 532, 543, 544, 568, 569, 571, 573, 575, *577–81*
MORSE, P. M., 190, 198, *246*
MORUZZI, J. L., 23, *46*
MOSELEY, J. T., 23, 28, 30, 37, *46*
MOSES, H. A., 443, *462*
MOSHARRAFA, M. A., 494, *523*
MOTT, N. F., 111, *166*, 182, 185, 238, *246* 338, 346, 351, *396*, 405, *460*, 548, 555, *579*
MOUSSA, A. H. A., 102, 196, 199, 201, 216–217, 235, 236, *246–7*
MOUSTAFA, H. R., 369, 386, *397*
MULDER, M. M., 223, *247*
MULLIKEN, R. S., 447, *462*, 509, *523*
MUNSON, R. J., 23, *45*
MYERSCOUGH, V. P., 49, 80, *96*
MCAFEE, K. B., 23, 38, *46*
MCCARROLL, R., 118, 119, 149, 151–153, 158, *166–7*
MCCAUGHEY, M. P., 403, 413, 431, 437, 445, 451, *461*
MCCLURE, B. T., 466, 479, 504, 574, *522*
MCCRACKEN, G. M., 327, *333*
MCCULLOGH, K. E., 574, *581*
MCDANIEL, E. W., 2, 8, 15, 18, 21, 23, 28, 34, 37, *45*, 252, 256, 267, 272, 284, 326, *330–1*, 563, *580*
MCDOWELL, C. A., 530, 563, 566, 568, 571, 575, *577–81*
MCDOWELL, M. R. C., 14, *45*, 49, 55, 59, 63, 78, 87, 93, *96*, *97*, 104, 112, 114, 118–120, 122, 128, 134, 146, 153, 158, 160, 163, *166–7*, 285, 290, 309, 311, *332–4*, 345, 385, *396*
MCECHRAN, R. P., 219, *244*
MCELROY, M. B., 119–121, *166*
MCFARLAND, R. A., 49, 52, 72, 84, *96*, 286, *332*, 345, 396, 556, *579*

MCGOWAN, J. W., 532, 544, 553, 563, 565, 573, 575, *577–81*
MCGUIRE, E. J., 55, 63, *96*
MCINTYRE, H. A. J., 218, 222, *247*
MCKNIGHT, L. G., 38, *46*
MCLACHLAN, H. W., 497, *523*
MCLAREN, I. H., 541, *578*
MCVICAR, D. D., 287, *332*
MCWHIRTER, R. W. P., 325, *333*, 474, *522*

NAKAI, M. Y., 129, *166*
NAKSBANDI, M. M., 23, *46*
NALL, B. H., 538, 543, 568, *578*, *580*
NEFF, S. H., 283, 326, *331*
NESMEYANOV, A. N., 51, *96*
NEYNABER, R. H., 320, *332*, 563, *580*
NEWTON, R. G., 104, *165*, 238, *247*
NICHELSON, A. I. C., 571, *581*
NIELSEN, K. O., 280, 314, *331*
NIJBOER, B. R. A., 372, *397*
NILES, F. E., 488, 504, *523*
NORCROSS, R., 54, *97*
NORDSIECK, A., 161–165, *167*
NOTTINGHAM, W. B., 540, *578*
NOVICK, R., 270, *331*
NYGAARD, K. J., 49, 76–77, 88, *96*

OBERHETTINGER, F., 161, *167*
OCCHIALINE, G. P. S., 171, *245*
OCHKUR, V. I., 345, 385, *396*
O'MALLEY, T. F., 205, 213, 215, 220, 239, *247*
OMIDVAR, K., 49, 57, 58, 78, 84, 86, *96*, 144, 148, *167*, 287, *334*, 565–6, *580*
ONG, P. P., 23, *46*
OOSTERHOFF, L. J., 530, 556, *577*, *579*
OPIK, U., 543, *578*
OPPENHEIMER, J. R., 118, *166*
OPYKHTIN, U., 146, *167*
ORE, A., 175, 192, *246*
ORMONDE, S., 101, *165*, 299, *332*, 390, *397*
ORTH, P. H. R., 184, 195, 223, 234, *246*
OSKAM, H. J., 23, *46*, 466, 477, 480, 484, 488, 494, 501, 505, 511, 513, 518, *521–3*
OSMON, P. E., 195, 245, *246*, *248*
OTT, W. R., 299, *334*
OXLEY, C. L., 123, *166*

PACE, M. O., 303, *334*
PACK, J. L., 23, *45*
PANOV, M. N., 403, 407, 413, 426, 431, 453, *460–2*

PAPP, N., 543, *578*
PAQUET, C., 535, 563, 566, 569, *580*
PAUL, D. A. L., 195, *246*
PAUL, W., 495, *523*
PEACH, G., 49, 61–63, 78, 80–82, 93, *95*, 288, *332*
PEART, B., 77, 93, *97*, 267, 272, 281, 283, 286, 287, 291, 326, *331*, *334*
PEEK, J. M., 316, *332*
PERCIVAL, I. C., 199, *246*, 284, *331*, 345, 376, 395, *398*
PEREL, V. I., 12, *45*
PERSSON, K. B., 478, *522*
PETERKOP, R., 111, *166*, 192, *246*, 566, *580–1*
PETERSON, J. R., 321, *334*
PETIT-CLERC, Y., 543, *578*
PETRASHEN, M. J., 56, *96*
PETRUNKIN, A. M., 345, 385, *396*
PHELPS, A. V., 23, *46*, 488, 494, *522–3*, 541, *578*
PIERCE, J. R., 757, *331*
PLATZMAN, R. L., 285, *332*, 365, 370, *397*
POWELL, A. L. T., 325, *333*
POWELL, J. L., 175, *246*
POWELL, R. E., 299, 324, 330, *332*
PRADHAN, T., 112, 118, 131, *166*
PRAHALLADARAO, B. S., 323, *333*
PRASAD, K., 345, *396*
PRASAD, S. S., 49, 82, *96*, 287, *332*, 345, *396*
PRESENT, R. D., 7, *45*
PRESNYAKOV, L., 101, 138ff, 146, 153, 165, *167*, 566, *580*
PROZONSKI, A. M., 76, *97*, 394, *398*
PU, R. T., 228, 230, *247*
PURCELL, E. M., 536, *578*
PYLE, R. V., 129, *166*

RAFF, C. F., 556, *579*
RAHMAN, A., 372, *397*
RAIBLE, V., 553, *579*
RAMSAUER, C., 401, *460*, 540, 546, *578*
RANDOLPH, P. L., 530, *577*
RAPP, D., 285, *331*, 388, *397*, 566, 573, *580–1*
RAU, A. R. P., 367, *397*
REDFIELD, A., 519, *523*
REDHEAD, R. A., 299, 324, *332*
REEH, H., 200, *246*
REES, W. D., 22, *45*
REESE, R. M., 571, 573, *580–1*

REICHERT, E., 392, *398*
REINHARD, H. P., 495, *523*
REINHARDT, P. W., 530, *577*
REYNOLDS, H. K., 129, *166*
RICE, S. A., 235, *247*
RICHARDSON, J. M., 479, 514, *522*
RITLAND, R. O., 341, *395*
RIVIERE, J. C., 327, *333*
ROBBEN, F., 474, *522*
ROBERTSON, W. W., 488, 504, *523*
ROBINSON, B. B., 345, *396*
ROELLIG, L. O., 195, 245, *246*, *248*
ROGERS, W. A., 467, 489, 496, 504, *522–3*
ROHRLICH, F., 173, *246*
ROJANSKY, V., 531, *577*
ROMAN, P., 104, *165*
ROSE, D. J., 465, *521*
ROSEBURY, F., 330, *333*
ROSENBERG, L., 205, 217, *247*
ROSENSTOCK, H. N., 574, *581*
ROSS, P. A., 341, *395*
ROTHE, E. W., 319, *332*, 563, *580*
ROUSE, A. G., 401, *460*
RUARK, A. E., 177, *246*
RUDD, M. E., 381, *397*, 404, 441, 443, 445, *460–1*
RUDGE, M., 284, *331*, *334*, 566, *580*
RUNDEL, R. D., 262, 273, 306, 312, 323, 329, *332–4*
RUSSEK, A., 403, 427, 452, 452, *461–2*
RUTHERFORD, E., 338, *395*
RUTHERFORD, J. A., 271, *333*
RYDING, G., 121, 126, 138, *166*

SALIN, A., 152, 157–8, *167*
SALPETER, E. E., 55, *97*
SAMSON, J. A. R., 49, 50, 54, *96*
SANDLER, S. I., 17, 21, *45*
SAPOROSCHENKO, M., 23, *46*
SAR-EL, H. Z., 538, *578*
SAUTER, G. F., 23, *46*, 488, 500, 504, 511, *523*
SAUTTER, C. A., 383, *397*
SCHAMP, H. W., 12, *45*
SCHEY, H., 219, *247*
SCHIFF, H., 124, 130, *166*
SCHIFF, H. I., 530, *577*
SCHRAM, B. L., 285, *331*, 358, 369, 389, *397*
SCHULZ, G. J., 547, 551, 553, 556, *577–9*
SCHUTTEN, J., 386, *397*
SCHWARTZ, C., 218, *247*

SCHWARTZ, S. B., 93, *97*, 311, *334*
SCHWINGER, J., 202, *246*
SCOTT, B., 49, 72, *96*, 394, *398*
SEATON, M. J., 55, 60, 61, *96*, 197, 244, 287, 303, *332*, 566, *580*
SEWELL, K. G., 59, 60, *96*
SEXTON, M. C., 479, *522*
SHARMA, R. R., 178, *246*
SHEFTEL, S. I., 446, *462*
SHELDON, J. W., 345, *396*
SHEPNIK, O. B., 49, 72, 76–77, *96*
SHILOFF, J. C., 560, *580*
SHIMON, L. L., 303, *332*
SILVERMAN, S. M., 560, *579*
SINDA, T., 318, *332*
SINNOTT, G., 23, *46*
SIMPSON, J. A., 327, *333*, 536, 538, 539, 553, 556, *578–80*
SIPLER, D. P., 23, 38, *46*
SKERBELE, A., 556, 560, *579*
SKINNER, M., 292, *332*
SKULLERUD, H. R., 22, *45*
SLATER, J. C., 476, *522*
SLOAN, I. H., 93, *97*
SMIT, C. H. R., 391, *397*
SMIT, J. A., 391, *397*
SMITH, A. C. H., 76, *96*, 128, *166*, 262, 287, 270, 292, 312, *331–2*
SMITH, F. J., 346, 359, *396*
SMITH, K., 192, 200, 208, 220, 226, 235, *246–7*, 284, *332*
SMITH, P. T., 72, *97*, 384, *397*
SMITH, S., 252, *331*, 548, *579*
SMYTH, H. D., 528, *577*
SNODGRASS, H. B., 178, *246*
SNUGGS, R. M., 23, 30, 37, *46*
SOBLEMAN, I., 101, 138ff., 144, 146, 148, 149, 151, 153, *165*, *167*, 566, *580*
SPAGENBERG, K. R., 257, *331*, 536, *578*
SPRUCH, L., 178, 199, 202, 205, 213, 215, 217, 220–222, 227, 239 *246–7*
SQUIRES, E. J., 64, *97*
STABLER, R. C., 345, 355, 385, *396*, 473, *522*
STARR, M. A., 341, *395*
STEBBINGS, R. F., 123, 127, 128, *166*, *167*, 386, *397*
STEINWEDEL, H., 495, *523*
STEWART, A. L., 49, 59, 60, *96*
STONE, G., 401, 413, 440, *460–1*
STONE, P., 54
SUGAR, R., 196, 239, *246*

SULLIVAN, E., 49, 78, 84–86, *96*

TAIT, J. H., 69–71, *96*
TALBOT, L., 474, *522*
TANAKA, Y., 571, *581*
TATE, J. T., 72, *97*, 384, *397*
TAYLOR, A. J., 101, *165*, 210, *247*, 284, 297, *331*
TEMKIN, A., 200 *246*, 566, *581*
TEUSCH, W. B., 193, *246*
THOMAS, G. E., 566, 575, *580*
THOMAS, L. H., 340, 342, 345, 359, *395*
THOMAS, M. T., 404, 452, *461*
THOMPSON, N. J., 257, *331*
THOMPSON, D. G., 234, *247*
THOMSON, J. J., 339, *395*
THONEMANN, P. C., 261, 272, *331*
TISONE, G. C., 256, 262, 285, 306, 326, *331*
TOBUREN, L. H., 129, *166*
TRELEASE, S., 124–128, *166*
TRICOMI, F. G., 163, *167*
TRIPATHY, D. N., 118, *166*
TRUJILLO, S. M., 319, *332*, 563, *580*
TSUDA, S., 530, *577*
TULLEY, J. A., 297, *332*
TUNITSKII, N. N., 269, 272, *331*
TUNSTED, J., 49, 53, *96*
TURNBULL, A. H., 327, *333*
TURNER, B. R., 271, *333*
TUUL, J., 495, *523*

VAINSHTEIN, L., 101, 138ff, 146–149, 151, 153, *165*, *167*, 566, *580*
VALENTINE, N. A., 284, *331*, 345, 376, *396*
VAN DER WIEL, M. J., 369, 389, *397*
VAN ZYL, B., 252, 257, 263, 267, 285, 312, 326, *331*
VARGA, P., 489, *523*
VARNEY, R. N., 23, *46*
VEATCH, G. E., 511, 515, *523*
VEIT, J. J., 245, *248*
VELDRE, V., 111, *166*, 192, *246*
VOLZ, D. J., 404, 441, 443, *461*
VON ARDENNE, M., 257, 269, 314, 325, *331*
VON ZAHN, U., 495, *523*
VOSHALL, R. E., 23, *46*
VRIENS, L., 345, 350, 354, 365, 369, 381, 384, 386, 389, *396*, *397*
VROSHEVICH, V., 403, 453, *461*

WACKS, M. E., 575, *581*

WAGNER, E. B., 552, *579*
WALKER, J. C. G., 346, 359, *397*
WALTER, J., 57, *96*
WALTON, D. S., 281, 287, *334*
WANNIER, G. H., 10, 11, 35, *45*, 566, *580*
WARDLE, C., 220, 235, *247*
WAREING, J. B., 256, 267, 282, 284, *331*
WATANABE, K., 560, *580*
WATEL, G., 318, *332*
WATSON, K. M., 104, *165*
WEBSTER, D. L., 340, *395*
WELKER, J., 345, 360, 384, 388, *396*
WELLER, C. S., 467, *522*
WELSH, L. M., 129, *166*
WENTZEL, G., 161, *167*
WEST, S. S., 341, *395*
WHEALTON, J. H., 31, *45*
WHEELER, J. A., 178, *246*
WHITAKER, W., 101, *165*, 299, *332*, 390, *397*
WIELCAUS, W. F., 245, *248*
WIGNER, E. P., 548, *579*
WILETS, L., 119–121, 123–124, *166*
WILEY, W. C., 541, *578*

WILLIAMS, A., 14, *45*
WILLIAMS, E. J., 340, 344, *395*
WILLIAMS, E. M., 22, *45*
WILLIAMS, J. F., 571, *581*
WILLIAMSON, J. H., 309, *332*
WILLIAMSON, R. E., 344, *396*
WINTERS, R. E., 529, *577*
WITTING, H. L., 76, *97*
WITTKOWER, A. B., 121, 126, 138, *166*
WOLL, J. W., 223, *247*
WOO, S. B., 31, *45*
WU, C. S., 193, 233, *246*

YARNOLD, G. D., 267, *331*, 537, *578*
YOUNG, R. A., 123, 124, *166*

ZANDBERG, E. Ya., 326, *333*
ZAPESOCHNYI, I. P., 49, 72, 76–77, *96*, *97*, 303, *332*
ZARE, R. N., 312, *332*
ZASHKVARA, V. V., 538, *578*
ZELIKOFF, M., 560, *580*
ZIEMBA, F. P., 401, 440, *460–1*

# SUBJECT INDEX

Alkalis, review of studies of ionization, 49
Analysis of space-time behavior of ion swarm, 29–37
Annihilation of electron-position pair, 172–176
Antiparticle, 171
Arrival time spectra for ions in drift tubes, examples, 26, 35
Atomic form factor, 338, 339

Beam experiments, energy limitation, 18
Beam-foil technique, 543
Binary-encounter collision theory – basic assumptions, 337
Binding energy of molecular ions, 23
Blanc's law, 15–17
Born approximation (first), 79, 338
Brueckner-Goldstone perturbation theory, 64

Chapman-Enskog theory, 9–10
Charge analysis of coincident ions, 458–459
Classical scattering theory, 376–382
Coincidence apparatus, 407–410
Coincidence experiments, argon–argon, 413–430
– – electron–ion, 456
– – krypton–krypton, 436–438
– – neon–argon, 438–440
– – neon–neon, 430–436
– – oxygen–argon, 440
Coincidence measurements of total scattering cross sections, 453–455
Coincidence resolving time, 459–460
Collective oscillations of electrons in atoms, 446–447
Colliding beams experiments, beam analysis, 267–269
– – – beam modulation, 261–267
– – – calculation of cross sections, 252–254
– – – effects of space charge, 257–260
– – – excited states in primary beams, 269–271

– – – experimental arrangements, 252, 272, 279, 281, 293, 308, 320, 321
– – – form factors, 254–257
– – – separation of signals from background, 260–267
Collisional-radiative recombination, 473–476
Conductivity (electrical) of gas, 477
Correlation effects in photoionization, 63–72
Coulomb functions, 160–161
Crossed beam experiments, *see* colliding beams experiments
Cross section for ionization of atom by electron impact, equation for, 79

Detachment of electrons from $H^-$ by electron impact, 306–311
– – – – negative ions by electron impact, 306–311
– – – – $O^-$ by electron impact, 277
Detector for photons, 294
Diffusion coefficient, 9
– – longitudinal, 21
– – measurement techniques, 20–22, 40–42
– – transverse, 21
Diffusion cross section, 9–10
Dissociation of $H_2^+$ ions by electron impact, 311–317
– – – – – $N_2^+$ impact, 318
Dissociative, recombination, 468–473
Drift tube measurements, general considerations, 17–19
Drift tubes, description, 23–29
Drift velocity, definition, 5
– – measurement techniques 19–20, 38–40

$E/N$, significance of, 4
$E/p$, significance of, 4
Einstein relation, 9, 21
Electron density, microwave average of, 477
Electron guns, 326, 327

Electron-ion recombination, mechanisms, 468–476
Electron spectroscope, 544–546
Electron velvet, 532
Electrostatic selector, 127°, 530–536
– – cylindrical, 538–539
– – monokinetron, 539–540
– – parallel-plate, 537–538
– – spherical, 536–537
Energies of electrons ejected in heavy particle collisions, 441, 443–445
Energy level crossings of molecular wave functions, 447–451
Energy of ion at high $E/N$, 11
Excitation of $Ba^+$ by electron impact, 301–304
– – $He^+$ by electron impact, 276, 291–300
– – $N_2^+$ by electron impact, 304–306
Excitation of positive ions by electron impact, 291–306
Excited states in mobility measurements, 19
Extended impulse approximation, 154–160

Field energy of ions, 4
Formal theory of scattering, 104–107
Frequency shift in microwave cavity, 477

Generalized oscillator strength, 363–372
Green's function in elastic scattering of positrons, 188–189

High-field region, definition, 5
Hornbeck-Molnar process, 54

Impulse approximation, definition, 107
– – electron capture, 115–131
– – excitation, 131–135
– – formal derivation, 112–115
– – ionization, 135–136
– – post form, 114–115
– – prior form, 115
Impulse hypothesis, 107–109
Inelastic energy loss, 404–405
Inelastic scattering of electrons, 555–562
Interaction potential between ion and molecule, 10
Ionization by electron impact, theory, 78–93
Ionization cross sections for alkali atoms by electron impact, 74, 76, 78, 83–88
– – – – $Mg^+$ by electron impact, 78
Ionization of alkali atoms by electron impact, experimental studies, 72–78
– – $Ba^+$ by electron impact, 273, 275, 289
– – H by electron impact, 382, 563–565
– – $H_2$ by electron impact, 573, 574
– – $H_2$ by proton impact, 383
– – He by electron impact, 561–563, 566–568
– – $He^+$ by electron impact, 273–274, 284
– – He by proton impact, 384
– – $Li^+$ by electron impact, 273, 274, 285, 286
– – $Mg^+$ by electron impact, 273, 275, 280, 287
– – $Mg_2^+$ by electron impact, 273, 275, 290
– – $N_2$ by electron impact, 569–570, 574, 575
– – Ne, Ar, Xe by electron impact, 568–570
– – noble gas atoms by electron impact, multiple, 571–572
– – $O_2$ by electron impact, 575
– – positive ions by electron impact, 272–291
– – triatomic molecules by electron impact, 575–576
Ion-molecule reactions, effects in drift tube measurements, 19–20
Ion sources, 27, 325, 326

Kihara theory of mobility, 12

Langevin theories of mobility, 6–9
Light spectrometer techniques for studying recombination, 487–494
Low-field region, definition, 5

Magnetic electron momentum selectors, 540, 541
Mason and Schamp's mobility calculation, 12–13
Mass spectrometer techniques for studying recombination, 494–503
Merging-beam techniques, 3, 318, 319
Microwave cavity method of measuring recombination coefficients, 476–487
Minimum principles in scattering theory, 210–216
Mobility, applications, 5–6
– definition, 5
– temperature variation, 10, 13
Mobility theory, classical, 6–13
Mobility, theory for molecular gases, 15

## SUBJECT INDEX

Mobility theory, quantum mechanical 13–15
Models for ionization in heavy ion collisions, 445–452
Monokinetron, 539, 540
Motion of slow ions through gases, 3–6
Mutual neutralization of $H^+$ and $H^-$, 321

Negative ion formation in $SF_6$, 576
Nitrogen ions in nitrogen gas, drift tube studies, 37–44
Nordsieck's integration technique, 161–165
Notation for atomic collision, 402

Particle detectors, 327–330
Peaking approximation, 110–112
Perel theory of mobility, 13
Photoabsorption cross section, 50
– – – for Na and Li, 52
Photoionization cross section, equation for, 55
– – – of K, 54
– – – of Li, 53, 58, 60, 61, 70
– – – for Na, 60, 62, 72
Photoionization, experimental studies, 50–54
– theoretical studies, 54–63
Polarization attraction, 7
Polarization limit of mobility, 8
Perveance, 258
Polarized orbital method, 225–235
Positronium, 177–180
– formation, 235–243
– scattering, 243–245
Positrons, elastic scattering, 184–235
Promotion of molecular orbitals, 447

Quadrupole mass filter, 495–503
Quality of microwave cavity, 476
Quenching of positronium, 181–182

Reaction frequency, 30
Reaction rate coefficients, measurement techniques, 22–23, 42–44
Recombination coefficient, definition, 465
– cross section, 465
– involving argon ions, 518–520
– – helium ions, 503–512

– – krypton and xenon ions, 520, 521
– – neon ions, 512–518
Reduced mobility, definition, 6
Reduced pressure, $p_0$, 4
Resonances in electron scattering, 390–393, 546–555
Retarding potential difference (RPD) method, 529–530
Rutherford cross section, 338

Scaling laws, 284, 385–390
Selection rules for electron–positron pair, 176–177
Sequential mass spectrometry, 323–324
Space charge effects in drift tubes, 18, 21–22
Space charge spreading of charged particle beam, 257–260
Statistical models for ionization, 451, 452
Stopping power, 369–370
Sum rules, 370–372
Surface ionization detector efficiency, 75
– – ion sources, 270, 326
Surface potentials, 543
Symmetry effects between ion and molecule, 11

Thermionic ion sources, 270, 278, 326
Thomson ionization cross section, 339–340
Time-of-flight electron selector, 541, 542
Townsend unit (Td), 4
Transition matrix element, 79, 105–106
Transport equation for ions in drift tube, high $E/p_0$, 35,
– – – – – – – low $E/p_0$, 31
Tyndall grid, 27

Vacuum techniques, 23–27, 327
Vainshtein, Presnyakov and Sobelman approximation, 138–154
Vapour pressure of Li and $Li_2$, 51
Variational method, applications, 216–225
Variational methods in scattering theory, 202–216
Velocity distribution of atomic electrons, 362
Virtual positronium formation, 200–202

Wall recombination, 465
Wannier theory of mobility, 10–12

QC
173
M224
v.1

JUN 3 1977